CONTENTS

1 HOW TO USE THIS BOOK

1·1 Structure ■ ■ ■ ■ ■ ■ ■ ■ ■ ■ ■ ■ ■ ■ ■ ■

Why do you need a text book? Perhaps you will read *A-level Biology* from cover to cover, but it is more likely that you will read material set by your teacher, or dip into the book for information about a chosen topic, or to find the answer to a particular question.

A-level Biology has been written with this selective use in mind and is intended as a versatile reference book rather than as a course book to be followed in a predetermined way. The overall aim is to simplify the student's task in acquiring the knowledge and understanding required by new A-level Biology syllabuses. This introductory Unit outlines the structure of the book and explains how you can use it to good effect.

Sections

The book is divided into five sections as follows.

Section A: Cell Biology and Biochemistry This section describes the chemical structures of the important types of molecules found in living organisms and gives an account of cell structure as revealed by the electron microscope. Units 2–7 should be studied early in your A-level course. Units 8 and 9, which together describe protein synthesis, are in a logical sequence but could also serve as a starting point for the study of genetics. Units 10 and 11, which outline the biochemical pathways of respiration and photosynthesis, build on the concepts developed in the preceding units of the section, but would often be studied later in the course when you are more familiar with the structure and physiology of animals and plants.

Section B: Physiology This is the largest section of the book, consisting of 24 units which relate structure to function for fundamental physiological processes. The examples chosen to illustrate these processes refer mainly to mammals or to flowering plants. This restricted range of examples is deliberate. Understanding the concepts and processes involved in, say, the transport of respiratory gases is difficult enough without at once introducing the complications associated with circulatory systems in invertebrate groups like annelids or arthropods, or with transport systems in plants. Such comparisons are important for a balanced understanding, but you will find them easier to make when you know the details of one system well.

Section C: Genetics and Evolution Evolution is the most important linking theme in the study of Biology. This section develops evolutionary ideas through an investigation of the genetic principles which affect inheritance. A thorough knowledge of genetics will help you to understand how variation is produced, and to explain the effects of natural selection. Units 40 and 41 discuss how new species arise and how existing species are kept distinct. It is important to appreciate that these two processes are both logically and functionally separate. Unit 42 concludes the section with a review of the evidence for evolution.

Section D: Origin and Diversity of Life Knowing something about the structure and life styles of different organisms is interesting and worthwhile. The 5-Kingdom classification scheme used here emphasizes the evolutionary relationships of the various groups. The problems associated with increasing body size, the advantages of coelomate organization, and the adaptations needed for successful colonization of terrestrial habitats, are all discussed in detail.

Section E: Ecology Ecology is concerned with the interactions between individual organisms and their surroundings, and with the interactions between different populations and species of organisms. This section outlines the nature of ecosystems and gives examples of the techniques used in ecological investigations.

Units

The five sections of *A-level Biology* are subdivided into Units. These are short and most can be read in an hour or less.

The **objectives** at the beginning of each Unit set you a definite task. Always read the objectives first. Keep them in mind and refer to them frequently to check your understanding of the text. The objectives have been carefully chosen to emphasize what is important. Use them not just to check your progress, but also as a guide for note-taking.

Having a book organized in this way gives you considerable flexibility in the order in which individual topics are covered. Many teachers favour a comparative approach for topics such as nutrition, gas exchange, or transport. This can be easily achieved by selecting the appropriate Units from different sections. On the other hand, some syllabuses restrict study of these functions to mammals and flowering plants. Similarly, syllabuses vary in the emphasis given to different groups of organisms.

Index–glossary

The index–glossary is a special feature of *A-level Biology*. It is intended firstly as a reference entry point to the text but also gives simple definitions of basic terms. In addition, you will find explanations of chemical and physical terms.

The index–glossary will help you as a source for writing essays. For example, under the index–glossary entry for 'skeleton' you are directed to the main parts of the text where the structure and functions of the human skeleton are described. You will also find references to the skeletons of arthropods, earthworms, and to support in plants. Given an essay title such as 'Discuss the advantages and disadvantages of endoskeletons and exoskeletons', you will be able to access the source material quickly and easily.

1·2 Syllabus and examinations ■ ■ ■ ■ ■ ■ ■ ■ ■ ■ ■

The GCE Examination Boards of England, Wales and Northern Ireland have agreed a **common core** for A-level Biology. *A-level Biology* covers this core content, amounting to 60% of the total syllabus, in full. There is significant variation between Boards in the content which forms the remainder of the syllabus. Within the constraints of the space available, every effort has been made to cover this additional material. The book will also prove useful to students of A-level Human Biology, Social Biology, and for AS-level and Higher Grade courses.

Find out from your teacher the name of your syllabus and Examination Board. Obtain a copy of the syllabus direct from the Examination Board and, at the same time, purchase copies of the previous two or three years' examination papers. (You will find the appropriate address under 'Examination Boards' in the index–glossary). It is very much in your own interest to obtain a syllabus and past papers. The syllabus tells you precisely what you need to know to prepare yourself for a particular examination. Familiarity with the format of examinations used by your Board gives you confidence and allows you to do your best. Although past papers are not expensive, your school or college may lack the resources to provide them for every student.

The questions provided at the end of each section of the book are intended to test your knowledge and understanding of the material covered. All Examination Boards require basic mathematical skills including drawing and interpreting bar charts and graphs. The questions have been selected to test a range of skills. Statistical terms and statistical tests are included in the index–glossary.

1·3 Course work ■ ■ ■ ■ ■ ■ ■ ■ ■ ■ ■ ■ ■

Weekly programme

A-level courses are much more demanding than GCSE courses. If you are to be successful, you must take responsibility for your own progress from the beginning. Consistent work is essential. Devise a weekly study programme and stick to it. Each week should include:

1 Checking that your class notes are accurate and complete.

2 Reading the parallel Unit(s) in *A-level Biology*.

3 Testing your knowledge using short answer, interpretation or essay questions from past examination papers.

4 Checking that your record of practical work is accurate and complete.

Try to develop your interest in Biology by reading articles from newspapers and from magazines such as *New Scientist*, or *Scientific American*, available in most schools and colleges. The benefits of this approach will be confirmed by any teacher.

Practical work

Practical work is an integral part of every A-level Biology course. Detailed instructions and 'recipes' for practical work have not been included in *A-level Biology* partly from pressure of space, but also because the practical work you are able to do depends to such an extent on the resources and equipment available in your school or college. Find out what your Examination Board expects and take advice from your teacher in developing your practical skills and learning techniques.

Revision

Provided you have kept pace with the work during your course, revision will be straightforward.

1 Plan a revision timetable leading up to the examinations.

2 Divide the total syllabus into manageable topics.

3 Identify the main headings and sub-headings within each topic.

4 Using your class notes and the objectives from *A-level Biology* write summary notes under these headings.

5 Practise questions from past papers, making sure that you are familiar with the format of examinations set by your board.

However organized your revision period, your understanding and enjoyment of biology, and so your examination result, will depend on working consistently through the course.

Section A

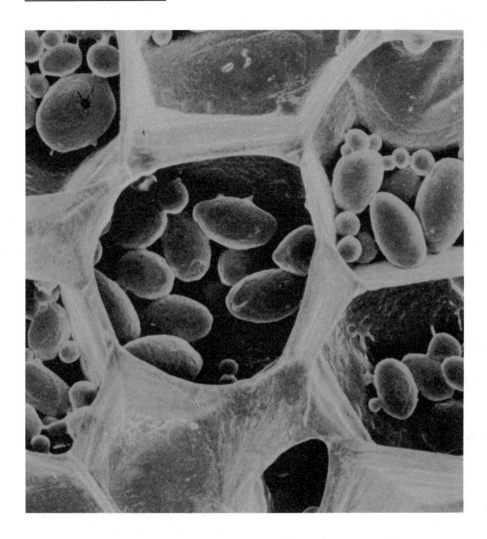

CELL BIOLOGY
AND BIOCHEMISTRY

2 CHEMICAL COMPOSITION OF LIVING ORGANISMS

Objectives

After studying this Unit you should be able to:–

- Define a bioelement and list in order of abundance the major bioelements found in all living organisms

- Describe the structure of water molecules and comment on their interactions in liquid water and ice

- Explain why water is a good solvent

- Write definitions of an acid, a base, pH, and pH buffers (check your definitions with the index-glossary at the back of the book)

- List the biologically important physical properties of water

- Explain why the element carbon is of fundamental importance to living systems

- Name the four major classes of important macromolecules and the basic units from which they are constructed

2·1 Bioelements ■ ■ ■ ■ ■ ■ ■ ■ ■ ■ ■ ■ ■ ■ ■

Hundreds of chemical reactions are involved in maintaining life activities of even the simplest organisms. In view of this, it is something of a surprise to find that, of the 92 naturally occurring chemical elements, only 16 are commonly used in forming the chemical compounds from which living organisms are made. These 16 elements and a few others which occur in particular organisms are called **bioelements**.

In the human body, only six bioelements account for 99% of the total mass. This is shown in Figure 2·1. The remaining common bioelements are also listed.

Although oxygen, carbon, hydrogen and nitrogen are among the commonest elements in the environment, they are present in living things in quite different proportions. For example, carbon forms 0.03% of the Earth's crust, but represents almost 20% of living organisms. On the other hand, some chemical elements such as silicon (which forms 27.7% of the Earth's crust) are extremely common but scarcely occur in living organisms.

The fact that the same 16 chemical elements occur in all organisms, and the fact that their proportions differ from those in the non-living world, implies that the bioelements are not in any way an arbitrary collection – each has special properties which make it particularly appropriate as a basis for life.

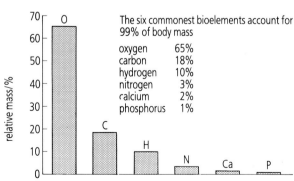

The six commonest bioelements account for 99% of body mass

oxygen	65%
carbon	18%
hydrogen	10%
nitrogen	3%
calcium	2%
phosphorus	1%

Other bioelements (about 1%) – potassium (0.35%), sulphur (0.25%), chlorine (0.15%), sodium (0.15%), magnesium (0.05%), iron (0.004%), copper (trace), manganese (trace), zinc (trace), iodine (trace).

Fig 2·1 Percentage composition by mass of a human being

5

2·2 Water molecules, water and ice ■ ■ ■ ■ ■ ■ ■ ■ ■ ■

Most of the oxygen and hydrogen present in living organisms is accounted for by water. Water forms about 75% of most cells, but its contribution to the total mass of an organism depends on whether skeletal materials such as bone are present. In organisms without hard tissues, the overall proportion is often close to 75% by mass, but in man it is about 60%.

Water is a unique chemical substance with a range of important properties, most of which derive from the way in which the oxygen and hydrogen atoms are linked. Each hydrogen atom shares a pair of electrons with the oxygen atom, forming a linkage called a **covalent bond**. As illustrated in Figure 2·2, the three atoms comprising a water molecule are not in a straight line. Instead, the two hydrogens form bonds which are at an angle shown to be 104.5° by X-ray crystallography. (You will find this technique explained in the index-glossary at the back of the book.)

The nucleus of the oxygen atom has a strong positive charge which tends to pull electrons away from the smaller hydrogen atoms so that they are not shared equally in the covalent bonds. As a result, the water molecule develops partial positive charges near each hydrogen and corresponding negative charges close to the oxygen atom. Such molecules are called **polar molecules**. The fact

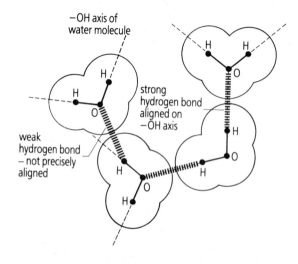

Fig 2·3 Hydrogen bonding between water molecules

that water has polar molecules is the key to many of its unusual properties.

Because they are polar, the molecules attract each other. This explains why water is a liquid at ordinary temperature and pressure. Other compounds made with hydrogen and elements from Group VI of the Periodic Table, like hydrogen sulphide (H_2S), are gases because their molecules are not polar and do not attract to anything like the same extent. The electrostatic attraction between water molecules gives rise to loose linkages called **hydrogen bonds**. These bonds are strongest when they are lined up along the axes of the O–H bonds in adjacent water molecules (see Fig. 2·3). Electrostatic attraction still occurs when the water molecules are not lined up, but the hydrogen bond formed is weaker.

In ice, all of the bonds are in their strongest form so that the molecules are arranged in a precise **lattice** structure. In liquid water, up to 80% of the molecules are still hydrogen-bonded in the same way, with detached water molecules filling the spaces between the larger associations. The random arrangement of the detached molecules in liquid water allows them to be more closely packed than in the structured ice lattice. Consequently, ice is less dense than water, so that it floats as water freezes. Water is most dense at 4°C: while the surface temperature may change, water in the depths of the ocean remains constantly at this temperature.

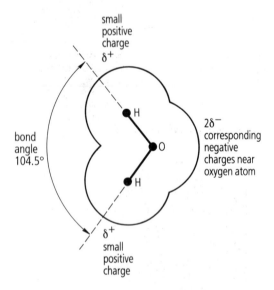

Fig 2·2 Arrangement of atoms in a water molecule

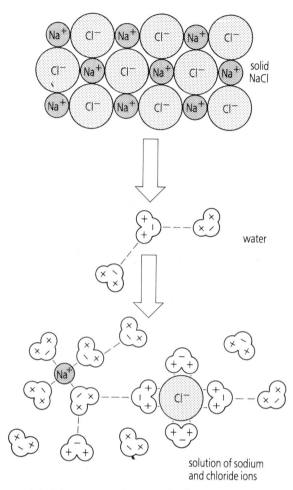

Fig 2·4 Solvent action of water molecules

2·4). Each ion goes into solution carrying with it a shell of water molecules, and is therefore called a **hydrated ion**.

Water readily dissolves a large number of non-ionic organic substances, such as sugars, because these contain polar side groups. Water molecules form electrostatic links with these groups and thereby dissolve the substance. In the same way, water molecules can gather round really large molecules, like proteins, forming a special kind of solution, called a **colloid**. As shown in Figure 2·5, a colloid is relatively fluid when the solute particles present are dispersed throughout the liquid. This arrangement is called a **sol**. Alternatively, the particles can become attached to each other, forming a loose network which restricts the movement of the solute molecules. In this case, the colloid is viscous and jelly-like, and is called a **gel**. Some colloids have the ability to change reversibly from the sol state to the gel state.

Water is an excellent solvent. The main reason for this is that water molecules can form electrostatic links with molecules of the dissolving substance, or **solute**. Common salt, sodium chloride, is virtually insoluble in non-polar liquids such as chloroform but it dissolves easily in water. The polar water molecules cluster round the sodium and chloride ions in the salt lattice and form linkages with them, thereby overcoming the electrostatic attraction which holds the crystal together (Fig.

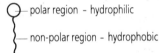

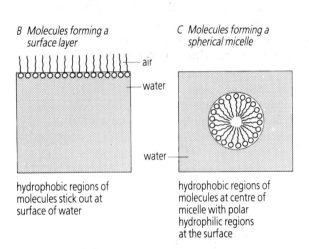

A Single molecule

○— polar region - hydrophilic

—— non-polar region - hydrophobic

B Molecules forming a surface layer

— air
— water

— water

hydrophobic regions of molecules stick out at surface of water

C Molecules forming a spherical micelle

hydrophobic regions of molecules at centre of micelle with polar hydrophilic regions at the surface

Fig 2·6 Detergent molecules and water

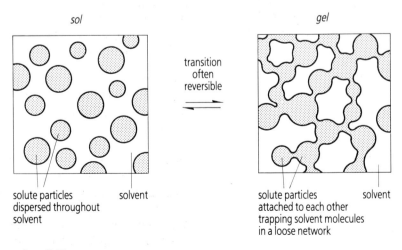

sol

solute particles dispersed throughout solvent

solvent

transition often reversible

gel

solute particles attached to each other trapping solvent molecules in a loose network

solvent

Fig 2·5 The two forms of a colloid

The protoplasm of most cells is colloidal in nature. The possibility of interchange from sol to gel may be important in movements of the cell, as in *Amoeba* and white blood cells. Blood clotting also involves a change from sol to gel (Unit 18).

Certain organic molecules including detergents and phospholipids (a major component of cell membranes) contain both a polar region and a non-polar region. The polar region is readily soluble in water and is therefore said to be **hydrophilic** ('water-liking'). On the other hand, the non-polar region is insoluble and is said to be **hydrophobic** ('water-hating'). Molecules of this type can form thin layers on the surface of water, or can be dispersed as minute spherical structures called **micelles** (see Fig. 2·6). In each micelle, the non-polar regions are grouped together in the centre, while the polar groups are arranged on the outside, where they are attracted to the water molecules. Details of cell membrane structure are given in Unit 6.

2·4 Acids, bases, and pH ■ ■ ■ ■ ■ ■ ■ ■ ■ ■ ■ ■ ■ ■

In a given volume of water, a small but definite proportion of water molecules split, or **dissociate** to form ions. This occurs when a hydrogen atom from one water molecule is transferred to the oxygen atom of a second. The reaction yields hydroxyl ions (OH^-) and hydroxonium ions (H_3O^+). Each hydroxonium ion can be thought of as a hydrogen ion (H^+) and a water molecule combined as a single unit. Hydrogen ions in water are always present in this combined form and do not exist separately. Nevertheless, to simplify the description of many similar reactions, it is convenient to write the equation for the ionization of water as:

$$H_2O \rightleftharpoons H^+ + OH^-$$

Pure water contains an exactly equal number of H^+ and OH^- ions, in addition to many undissociated water molecules. However, substances dissolved in water often affect the balance of H^+ and OH^- ions. For example, an acid substance increases the number of H^+ ions in solution, while a base, or alkali, combines with H^+ ions and, in effect, removes them from the solution. The **pH** of the solution is formally defined as:

$$pH = - \log_{10}[H^+]$$

where $[H^+]$ is the concentration of H^+ ions in moles per litre. (Chemical units and other terms with which you may be unfamiliar are defined in the index-glossary at the back of the book.) Practical values of pH range from 0–14. At 25°C pure water has a pH=7, called neutral pH, acid solutions have a pH less than 7, while an alkaline solution has a pH greater than 7. Strong acids, like hydrochloric acid, HCl, dissociate almost completely when dissolved in water, while in a weak acid, like carbonic acid, H_2CO_3, many molecules remain in combined form.

Certain solutions, called **pH buffers**, resist changes in pH, so that the pH of the solution tends to remain constant. For example, hydrogencarbonate ions can combine with H^+ ions:

$$\underset{\text{hydrogen-}}{HCO_3^-} + H^+ \rightleftharpoons \underset{\text{carbonic acid}}{H_2CO_3}$$
hydrogen-
carbonate

or with OH^- ions:

$$\underset{\text{hydrogen-}}{HCO_3^-} + OH^- \rightleftharpoons H_2O + \underset{\text{carbonate}}{CO_3^{2-}}$$
hydrogen-
carbonate

In this way, H^+ and OH^- ions added to a solution containing hydrogencarbonate ions are absorbed and do not affect the overall pH level. Other pH buffers in living systems include phosphate ions, $H_2PO_4^-$, amino acids and proteins. The buffering action of amino acids is explained in Unit 4.

2·5 Water in living organisms ■ ■ ■ ■ ■ ■ ■ ■ ■ ■ ■ ■ ■

The main importance of water to living systems is as a solvent. By dissolving many of the body chemicals, even those with large molecules such as proteins, water provides a medium in which chemical reactions can occur. Its role in metabolism is therefore vital.

Water can also take part directly as a **reactant** in chemical processes. One important type of reaction is called **hydrolysis**, where large molecules are broken down into simpler ones by the addition of water. The complementary process is called **condensation**, where larger molecules are built up from smaller ones by removing water. Almost all the large molecules produced in metabolism are built up by condensation reactions.

The physical properties of water, like its chemical properties, derive mainly from the polar nature of water molecules. Some of these properties and the reasons for their importance are set out in Table 2·1.

Table 2·1 *Physical properties of water.*

PROPERTY	EXAMPLE OF BIOLOGICAL IMPORTANCE
density	When displaced, water gives good support for aquatic organisms.
surface tension	Strong surface film allows small organisms to be attached below or supported above.
capillarity	Because they are polar, water molecules are attracted to many surfaces. Therefore water can enter very narrow spaces, such as those between cells, even against the pull of gravity. This is called capillary attraction, or **capillarity**, and may be involved in transporting water in the stem of a plant.
compressibility	Water is incompressible. This is important in transport systems and as a means of support in organisms with a 'hydrostatic' skeleton.
specific heat	The high specific heat of water means that organisms gain and lose heat slowly. This is helpful in temperature control.
heat of vapourization	High heat of vapourization gives efficient cooling of the body as sweat evaporates.
electrical conductivity	Pure water has a low conductivity but dissolved ions make cytoplasm quite a good conductor. This is important for the functioning of many cells, such as nerve cells.

2·6 Carbon compounds ■ ■ ■ ■ ■ ■ ■ ■ ■ ■ ■ ■ ■

Carbon accounts for about 18% by mass of the human body. It forms hundreds of different compounds, each with its own particular part to play in the metabolism and life of the organism. These compounds are so important that living systems are often said to be **carbon-based**.

With a few exceptions, including carbon dioxide and carbonates, compounds which contain the element carbon are called **organic** compounds. Over two million such compounds are known, although many of them do not occur in organisms. It is clearly important to consider the properties of carbon which enable it to form so many compounds.

Carbon has the atomic number 6 and each atom of carbon has two electrons in its first 'shell', and four electrons in its second, outermost, 'shell'. Thus, each atom has four electrons which can be shared, allowing four covalent bonds to be formed with other atoms. The other atoms include hydrogen, oxygen and nitrogen atoms and, most significantly, additional carbon atoms.

Table 2·2 *Bond energies*

BOND	BOND ENERGY IN KILOJOULES PER MOLE ($kJ\,mol^{-1}$)
C–C (single bond)	345
C=C (double bond)	610
C–H	413
C–N	304
C–O	358
H–H	435
N–N	163
O–O	146

The strength of covalent bonds varies depending upon the atoms which are involved. In general, small atoms form stronger bonds. As indicated in Table 2·2, those involving carbon are especially strong. The significance of these high bond energies is that carbon forms extremely stable compounds. In particular, the high bond energies for carbon-carbon bonds mean that carbon atoms can form stable chain and ring structures. Sometimes, the chains and rings may contain other atoms such as nitrogen and oxygen. The strength of C−N and C−O bonds is sufficient to ensure the stability of the molecules produced.

It is no coincidence that together hydrogen, oxygen, nitrogen and carbon form more than 95% by mass of living material. These elements are the four smallest atoms in the Periodic Table which can form covalent bonds, both between atoms of the same element (as in C−C), or between atoms of different elements (as in C−H). Hydrogen can form only one covalent bond, oxygen forms two, nitrogen three, and carbon four such bonds.

The organic compounds in cells are a mixture of numerous small molecules and very large molecules called **macromolecules**. There are just four classes of important macromolecules, as shown in Table 2·3. Each of these classes is described separately in its own Unit (carbohydrates – Unit 3, proteins – Unit 4, lipids – Unit 6, nucleic acids – Unit 8). Steroids and pigment groups are related to and can conveniently be considered along with lipids.

Many of the smaller molecules present in cells, such as glucose and amino acids, are the subunits of macromolecules. On the other hand, some small molecules are important in their own right. **Adenosine triphosphate (ATP)** acts as a store of chemical energy which can be directly supplied when needed for the cell's activities. Other small molecules can act as **carriers**, transferring electrons, ions or small molecules from one place to another in the cell. Yet others are **enzyme cofactors** which bind to enzymes and thereby ensure that they will function correctly. (This action is explained in Unit 5.) Plants are able to make their own cofactors from simple raw materials, but animals often need to obtain them as '**vitamins**' in their diets.

Table 2·3 Macromolecules found in cells.

CLASS	ELEMENTS PRESENT	BASIC UNITS	MACROMOLECULES
carbohydrates	C, H, O	monosaccharides	polysaccharides
proteins	always – C, H, O, N sometimes – S, P	amino acids	polypeptides – proteins
lipids	always – C, H, O sometimes – N, P	glycerol, fatty acids	fats, oils, waxes (steroids, pigment groups)
nucleic acids	C, H, O, N, P	⌐pentose sugar, ├ phosphate group, └organic bases nucleotides	RNA (ribonucleic acid) DNA (deoxyribonucleic acid)

2·7 Inorganic substances ■ ■ ■ ■ ■ ■ ■ ■ ■ ■ ■ ■

The simplest substances in cells are the various inorganic ions which are obtained from the environment. Ions which carry a positive charge are called **cations**; those required by living organisms include calcium (Ca^{2+}), potassium (K^+), sodium (Na^+), magnesium (Mg^{2+}), and iron (Fe^{2+} or Fe^{3+}). The trace elements copper, manganese and zinc also form cations. Ions which carry a negative charge are called **anions**. Phosphorus is present in the form of phosphate ions ($H_2PO_4^-$),

sulphur as sulphate ions (SO_4^{2-}), chlorine as chloride ions (Cl^-), and iodine as iodide ions (I^-). Particular organisms may require minute traces of additional elements such as molybdenum and cobalt.

These ions have a wide variety of functions as will be discussed in other Units. The inorganic elements important in the human diet are listed in Table 12·3.

3 CARBOHYDRATES

Objectives

After studying this Unit you should be able to:–

- Briefly describe the functions of carbohydrates in living organisms

- Explain the terms monosaccharide, disaccharide, and polysaccharide

- Distinguish between aldose and ketose sugars

- Draw diagrams to show the three alternative molecular structures of glucose and distinguish between its α- and β-ring forms

- Describe glycoside linkage as an example of a condensation reaction

- Outline the structures of maltose, lactose, and sucrose and distinguish between α1,4 ; β1,4; and α1,2 glycoside linkages

- Explain how the properties and uses of starch, glycogen, and cellulose are related to their molecular structure

- List the common chemical tests for different carbohydrates

3·1 Introduction ■ ■ ■ ■ ■ ■ ■ ■ ■ ■ ■ ■ ■ ■ ■ ■

This Unit considers the chemical structure and function of the sugars and other carbohydrates which are essential components of living systems.

All carbohydrates contain the chemical elements **carbon, hydrogen,** and **oxygen**. The atoms of these elements are usually combined to form molecules according to the **general formula** $C_x(H_2O)_y$, where x and y can vary (e.g. $C_6H_{12}O_6$, $C_{12}H_{22}O_{11}$). Thus, whatever the number of carbon atoms, hydrogen and oxygen are present in the proportion 2:1, as in water.

The most important function of carbohydrates is the storage and supply of energy. During photosynthesis, light energy is converted into stored chemical energy in the form of carbohydrates. These are used in the plant as fuel for respiration, releasing energy to drive the other chemical reactions of metabolism. As a result, the plant can manufacture the amino acids, proteins, and other substances needed for growth. Complex carbohydrates in plants include **starch**, which provides a means of long-term energy storage, and

cellulose, which is the main structural material of plant cell walls.

When plant material is eaten, the stored chemical energy it contains is transferred to animals. This process of transfer can continue from one animal to another, so that, whether or not they eat plants directly, all animals ultimately depend on plants for their carbohydrate substances. Like plants, animals use simple sugars, such as **glucose** as the main fuel for respiration while complex carbohydrates, such as **glycogen**, allow for energy storage. Carbohydrates are less important as structural materials in animals but are sometimes used in modified form, as in the exoskeletons of insects.

To summarise these functions, it can be said that simple carbohydrates act as an **energy source**, while complex carbohydrates act as an **energy store**, or as **structural materials**. In addition, some carbohydrates form an integral part of the structure of nucleic acids (described in Unit 8).

3·2 Structure of monosaccharide sugars ■ ■ ■ ■ ■ ■ ■ ■

Carbohydrate molecules vary greatly in size but are not difficult to classify. There are three main groups, namely **monosaccharides, disaccharides,** and **polysaccharides**. The monosaccharides are 'simple sugars' which have between 3 and 10 carbon atoms per molecule. Disaccharides are described as 'compound sugars' and each molecule consists of two monosaccharide units joined together. In a polysaccharide molecule, many monosaccharide units are joined to form a chain.

The monosaccharides are normally classified according to the number of carbon atoms they contain. The most important types in living organisms are 3-carbon sugars, called **trioses,** 5-carbon sugars, or **pentoses,** and 6-carbon sugars, called **hexoses**.

Within these groups, the various atoms present can be bonded together in different ways, often giving different chemical structures even when the number of carbon, hydrogen, and oxygen atoms remains the same. These alternative structures are called **structural isomers**. One of the most important variations is illustrated by the structures of **glucose** and **fructose**, as shown in Figure 3·1. Both of these sugars contain 6 carbon atoms and have the overall formula $C_6H_{12}O_6$. In addition, each molecule contains a chemically reactive double-bonded oxygen atom (=O) which gives it its characteristic chemical properties. In the molecule of glucose, this atom is located at the end of the chain of carbon atoms, where it forms part of a H−C=O, or **aldehyde** group. However, in fructose, the double-bonded oxygen is attached to the second carbon atom in the chain, forming a C=O, or **ketone** group.

These isomeric alternatives give rise to two different series of monosaccharide sugars, each with slightly different chemical properties. The **aldose** series includes sugars such as glyceraldehyde, ribose, glucose, and galactose, which have the aldehyde group, while the **ketose** series includes dihydroxyacetone, fructose, and other sugars which contain the ketone group.

The aldehyde and ketone groups both act as strong chemical **reducing agents**, meaning that they tend to donate electrons to the atoms or molecules of other substances. As a result, monosaccharides and most disaccharides have reducing properties 'nd are called **reducing sugars**. Such sugars are identified in tissue extracts by heating with a standard test reagent called **Benedict's solution**; the solution contains Cu^{2+} ions which are reduced to Cu^+ ions, giving a brick-red precipitate of copper (I) oxide when reducing sugars are present.

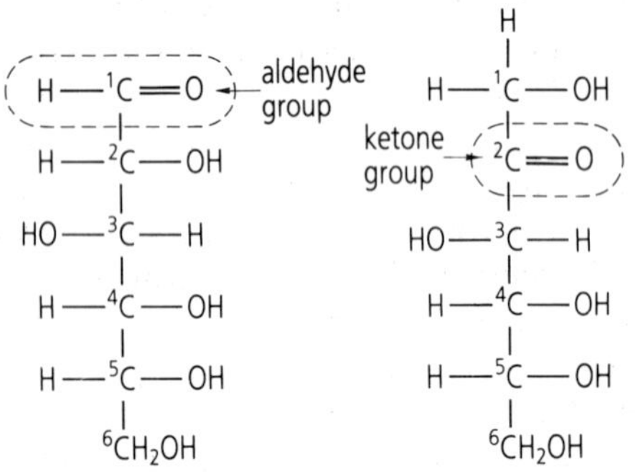

Fig 3·1 Chemical structure of glucose and fructose

3·3 Ring structures of monosaccharide sugars ■ ■ ■ ■ ■ ■

Most pentose and hexose sugar molecules can exist in the form of rings in addition to their straight chain form. This is a very important feature of carbohydrate chemistry because the ring structures produced are the subunits needed for building up complex carbohydrates.

Powdered dry glucose exists mainly in straight chain form. However, when glucose molecules are dissolved in water, two different ring structures are formed, as shown in Figure 3·2. These ring structures are more stable in solution, so that, at equilibrium, almost all of the molecules are present as rings, with the straight chain form being a relatively short-lived intermediate.

By convention, the carbon atoms in straight chain glucose are numbered from the end nearest the aldehyde group. This numbering convention is carried through in representing the two ring structures; carbon atom number 1 is drawn at the

right hand end of the molecule, and carbon atoms 2 and 3 are pictured as being nearer than carbon atoms 5 and 6, which thus lie at the back of the ring.

Drawn in this way, the structures of **α-glucose** and **β-glucose** differ only in the position of the −OH and −H groups attached to carbon atom number 1. In α-glucose, the −OH is pictured as being 'below' the ring, while in β-glucose, it is 'above' the ring. This difference may seem slight but, as you will see later in this Unit, the two ring structures behave quite differently as subunits for complex carbohydrates.

Two other monosaccharides which form rings are **fructose** and **galactose**. Fructose forms a five-sided ring, while galactose forms a six-sided ring. This differs from glucose by having a different arrangement of groups fixed 'above' and 'below' the ring.

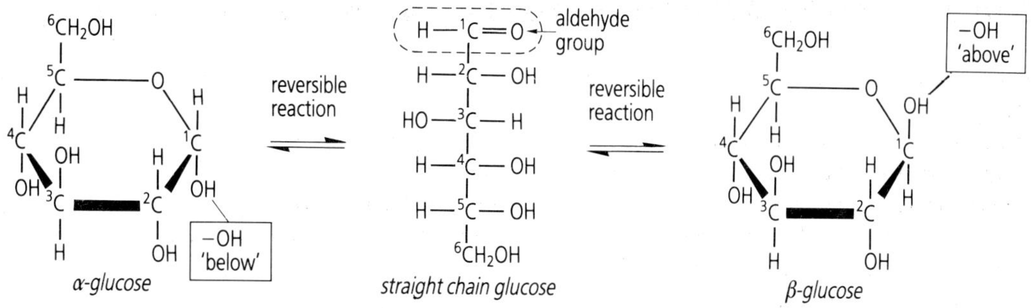

Fig 3·2 Three alternative structures of glucose

3·4 Disaccharides ■ ■ ■ ■ ■ ■ ■ ■ ■ ■ ■ ■ ■ ■ ■

Disaccharides are sugars which consist of two monosaccharide units joined to form a single molecule. They often occur as intermediates either in the breaking down or building up of polysaccharides. For example, **maltose** is found in the human alimentary canal as the first product of starch digestion and is further broken down to glucose before being absorbed into the body and used for respiration. Other disaccharides must also be broken down.

Maltose consists of two molecules of α-glucose joined by a **glycoside linkage**. In living organisms, linkages of this type are formed in several distinct steps each controlled by its own enzyme but, as outlined in Figure 3·3, the net result is to remove a molecule of water. Reactions of this type are called **condensation** reactions and are important in building up many complex biological molecules, including disaccharides, polysaccharides, nucleic acids, and proteins. The complementary process, whereby complex molecules can be split into their component parts, is called **hydrolysis** and requires the addition of water.

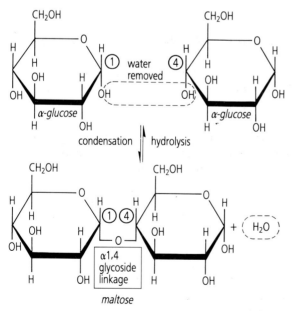

Fig 3·3 Formation of maltose

In drawing the molecular structures of maltose and other disaccharides the structural formulae of the monosaccharide subunits are simplified by omitting the carbon atoms in the main ring: these are now 'understood' to be present where bonds from the various different side groups join the ring. In representing polysaccharides, the formulae are further simplified by omitting all but the most important side groups. The numbering system for carbon atoms developed for monosaccharides is still used so that the linkages between the subunits can be referred to in a precise way.

Significantly, not all glycoside linkages are identical; there are several different types depending upon the particular ring structures of the monosaccharides involved. In maltose, carbon atom 1 of the first glucose is linked through its α-position (i.e. from 'below' the ring) to carbon atom 4 of the second glucose unit. Thus, maltose is said to contain an α1,4 glycoside linkage. To visualise this structure, the glucose molecules can be thought of as coins which are linked together in the 'heads up' position.

Figure 3·4 illustrates the structure of two other important disaccharides. **Lactose**, found in the milk of mammals, consists of galactose and glucose molecules joined by β1,4 glycoside linkage. This linkage changes the orientation of the second subunit so that it is 'upside down' compared with the second glucose unit in maltose. Using the coins analogy, the β1,4 linkage joins molecules which are alternately in the 'heads' and 'tails' positions. (Note that the projecting −CH₂OH groups of lactose lie respectively 'above' and 'below' the plane of the molecule.)

Sucrose consists of glucose and fructose joined by an α1,2 glycoside linkage. This particular linkage affects the chemical properties of the monosaccharide subunits so that their reducing properties are lost. Sucrose is the only common **non-reducing sugar** and must be hydrolysed into its monosaccharides before it will give a positive result with Benedict's test. In many plants, sucrose is used for transporting food reserves, often from the leaves to other parts of the plant. It is extracted from sugar cane and sugar beet and is important to humans both as a sweetener and as a raw material in the production of ethanol.

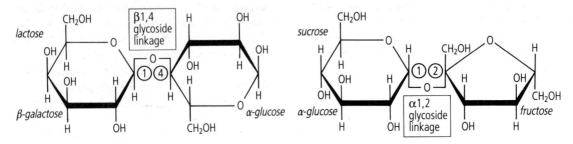

Fig 3·4 Disaccharides and glycoside linkage

3·5 Polysaccharides ■ ■ ■ ■ ■ ■ ■ ■ ■ ■ ■ ■ ■ ■ ■

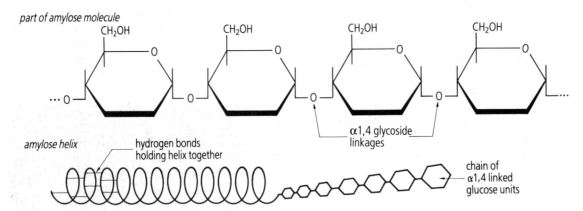

Fig 3·5 Structure of amylose

Polysaccharides are complex carbohydrates with large molecules consisting of chains of monosaccharide units joined together. They do not taste sweet and, unlike sugars, are either insoluble in water or form colloidal solutions only. As a result, they can accumulate without affecting normal metabolism and are widely used as storage and structural materials. The most important polysaccharides are **starch**, **glycogen**, and **cellulose**.

Starch is the main food storage material of plants. It is not a pure substance but is instead a mixture consisting of the unbranched chain molecules of a polysaccharide called **amylose** and the branched chain molecules of a second polysaccharide, called **amylopectin**.

Part of the molecular structure of amylose is shown in Figure 3·5. Such molecules are called **polymers**, meaning that they consist of a chain of repeating units. The repeating units in this case are all individual glucose molecules. These are joined by α1,4 linkages so that all of the glucose units are in the 'heads up' position. This configuration allows the amylose molecule, containing a total of 250–300 glucose units, to adopt a stable coiled shape, called a **helix**. This structure has six glucose units per turn and is held together by **hydrogen bonds** between adjacent turns. By chance, the space in the middle of the helix has just the correct dimensions to accommodate iodine molecules which interact with the glucose units in the helix, causing a slight change in their position. This leads in turn to the formation of a characteristic blue-coloured complex and provides the basis for the procedure of testing with iodine solution for the presence of starch.

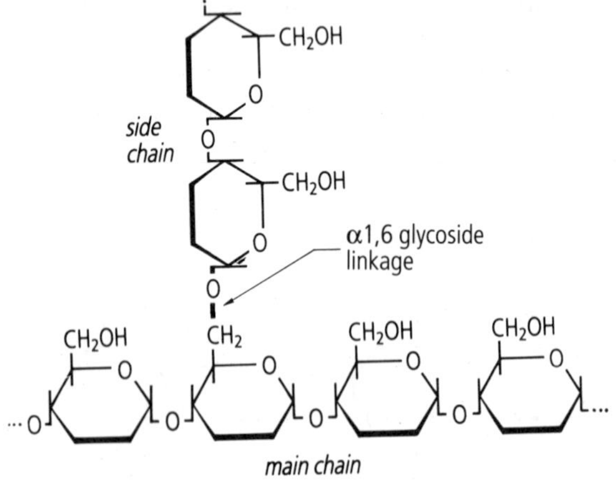

Amylopectin differs from amylose only by having a branching structure. As shown in Figure 3·6, each branch starts with a α1,6 glycoside linkage. These occur at intervals of about once every 24–30 glucose units, greatly increasing the

Fig 3·6 Branching in amylopectin

number of terminals where additional glucose units can be added, or where enzyme breakdown can begin. Amylopectin, which forms up to 80% of starch, can be rapidly built up or degraded (by **amylase** enzymes) to keep supplies of simple sugars in the plant at the correct level.

Starch is ideal as a storage material because it cannot diffuse out of the cell and has little or no osmotic effect. It is stored in the form of large aggregations of molecules called **starch grains**. These occur particularly in chloroplasts or in specialized structures such as seeds and potato tubers (see Fig. 3·7).

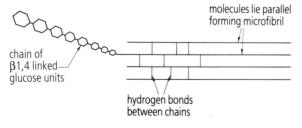

Fig 3·7 Scanning electronmicrograph (SEM) showing starch grains in potato tuber cells

Glycogen is the storage polysaccharide most common in animals. It has a molecular structure very similar to that of amylopectin but branches more often, about once every 8–12 glucose units. It is found in liver and muscle cells.

Cellulose is the most abundant of all organic compounds in living organisms because it is the main structural material in plant cell walls. Like amylose and amylopectin, cellulose is a polymer of glucose units, but, as illustrated in Figure 3·8,

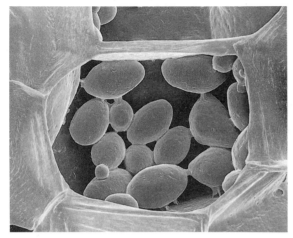

Fig 3·8 Structure of cellulose

these are attached to each other by β1,4 linkages, making the glucose molecules alternately 'heads up' and 'tails up'. This comparatively small change completely transforms the properties of the resulting polysaccharide. Individual molecules of cellulose, containing from 300–15,000 glucose units, cannot be coiled to form a helix but are instead flat ribbons. This structure encourages hydrogen-bonding between adjacent parallel molecules so that long threads called **microfibrils** are formed. These are completely insoluble and ideal as structural components. As illustrated in Figure 3·9, the cellulose microfibrils in the cell wall of a green plant are arranged in overlapping layers. This gives a particularly tough and rigid structure. Sometimes cell walls are reinforced with a second polysaccharide material called **lignin**. This is deposited in the spaces between the cellulose microfibrils.

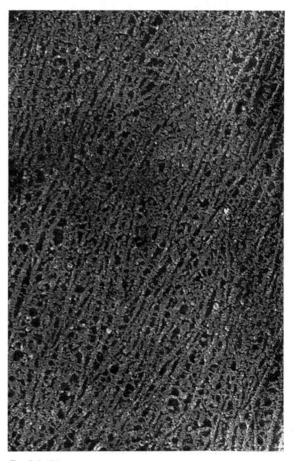

Fig 3·9 Electronmicrograph showing cellulose microfibrils in the cell wall of a green plant

The structural material of insect exoskeletons is a carbohydrate derivative called **chitin**. Like cellulose, this is a polymer of β1,4 linked subunits but the subunits themselves differ from glucose by having one of their −OH groups replaced by a more complex $-NH.CO.CH_3$ group. This difference means that there is even more hydrogen-bonding between chains than in cellulose, making chitin extremely tough and resilient.

15

3·6 Chemical tests for carbohydrates ■ ■ ■ ■ ■ ■ ■ ■ ■

The common chemical tests used to distinguish
different carbohydrates are listed in Table 3·1.

Table 3·1 Chemical tests for carbohydrates

CARBOHYDRATE	TEST	CONDITIONS	RESULT
reducing sugars glucose fructose maltose lactose	Benedict's solution	Grind tissue sample in a small quantity of water, filter. To 2 cm^3 of the liquid extract in a test tube add 2 cm^3 of Benedict's solution, heat to 95°C for 2 minutes.	colour change or brick-red precipitate
	Fehling's solutions I and II	To 2 cm^3 of a liquid extract add 1 cm^3 of Fehling's I and 1 cm^3 of Fehling's II, heat to 95°C for 2 minutes.	colour change or brick-red precipitate
non-reducing sugars sucrose	Benedict's solution	Provided an initial test for reducing sugar gives no colour change, heat 2 cm^3 of fresh extract with 1 cm^3 of dilute hydrochloric acid to hydrolyse sucrose into its monosaccharides, neutralise with sodium hydroxide, test for reducing sugars as above.	colour change or brick-red precipitate
starch	Iodine in KI solution	add solution to crushed tissue sample at room temperature	blue-black colour
glycogen	Iodine in KI solution	add solution to crushed tissue sample at room temperature	red-violet colour
cellulose	Schultz' solution	mount tissue section in solution for light microscopy	violet colour
lignin	Acidified phloroglucin solution	mount tissue section in solution for light microscopy	bright red colour

4 STRUCTURE OF PROTEINS

Objectives

After studying this Unit you should be able to:–

- Briefly describe the different types of proteins and list some of their functions

- Write the general formula for an amino acid

- Comment on the classification and importance of the $-R$ group components of amino acids

- Describe the role of amino acids as 'buffers'

- Write an equation to show how a peptide bond is formed and define the terms dipeptide, tripeptide, and polypeptide

- Distinguish between the primary, secondary, tertiary, and quaternary structures of protein molecules

- Describe the α-helix and the β-pleated sheet as examples of secondary structure in fibrous proteins

- Describe the structures of myoglobin and haemoglobin as examples of tertiary and quaternary structure in globular proteins

- List the different types of chemical bonds involved in maintaining protein structure

- Explain how denaturation can occur and distinguish between denaturation and renaturation

- List the common chemical tests for proteins

4·1 Types of proteins and their functions ■ ■ ■ ■ ■ ■ ■ ■

Proteins make up about half the carbon compounds in living organisms. Although they share many basic structural features, their overall organization is extremely flexible so that individual proteins can have very different and highly specialized functions.

All proteins are polymers made up of subunits called **amino acids** joined together in long chains which may be folded or twisted in various ways. Two main groups of proteins can be distinguished according to how the chains are arranged. In **fibrous proteins** the chains are extended so that the molecules are long and thread-like. Partly as a result of this, fibrous proteins tend to be insoluble and comparatively resistant to changes in

temperature and pH. These properties make the fibrous proteins ideal as **structural materials**; important examples are collagen and elastin (major proteins in skin and connective tissue), and keratin (in hair, horns, nails, and wool).

The second major group of proteins are the **globular proteins** in which the chains are folded in a more complex way to produce compact rounded molecules. These dissolve easily to form colloidal solutions and are often susceptible to small changes in temperature and pH. Globular proteins are the functional proteins of metabolism. In other words, they carry out the many specialized tasks which enable the organism as a whole to function properly. Many of the globular proteins are **enzymes**: these act as specific catalysts, altering the rates of most of the chemical reactions of the body. More than 2000 enzymes have been isolated in pure form. They are particulary important because, as will be discussed in Unit 5, they provide the normal means whereby the genetic instructions coded in the nucleus of a cell are carried out.

Globular proteins also act as hormones, antibodies, blood transport pigments, light-sensitive pigments, and blood-clotting agents. Some of these proteins, and many enzymes as well, cannot function unless joined to a non-protein component, called a **prosthetic group**. Such proteins are called **conjugated proteins** and include haemoglobin, the pigment in red blood cells. Haemoglobin molecules are able to transport oxygen because they contain prosthetic groups

called **haem groups** as an integral part of their structure. Each haem group contains an iron atom which forms a loose association with a molecule of oxygen, enabling the oxygen to be carried to the tissues. (The transport of oxygen is described in detail in Unit 18.)

Summarizing the main roles of proteins it can be said that the fibrous proteins usually act as **structural materials**, while the globular and conjugated proteins act as the essential **functional components** of metabolism. Occasionally, proteins can also have a role as **storage materials**, as in seeds and eggs. The remainder of this Unit considers how the structure of protein molecules is adapted to carry out these different functions. It is useful to begin with a study of the amino acid subunits from which proteins are made.

4·2 Amino acids ■ ■ ■ ■ ■ ■ ■ ■ ■ ■ ■ ■ ■ ■ ■ ■

Amino acids are the subunits of all proteins and each has the general formula given in Figure 4·1. This shows a central carbon atom, called the α-**carbon**, attached in turn to −H, −NH₂ (**amino**), and −COOH (**carboxyl**) groups, and also to a variable group referred to as an −**R group**. The hydrogen, amino, and carboxyl groups are fixed parts of the molecule present in all amino acids. All amino acids (and therefore all proteins) must contain the chemical elements **carbon**, **hydrogen**, **oxygen**, and **nitrogen**. The −R group has a different structure in each of the 20 biologically important amino acids and determines their individual chemical properties. Two amino acids contain **sulphur** as part of their −R groups.

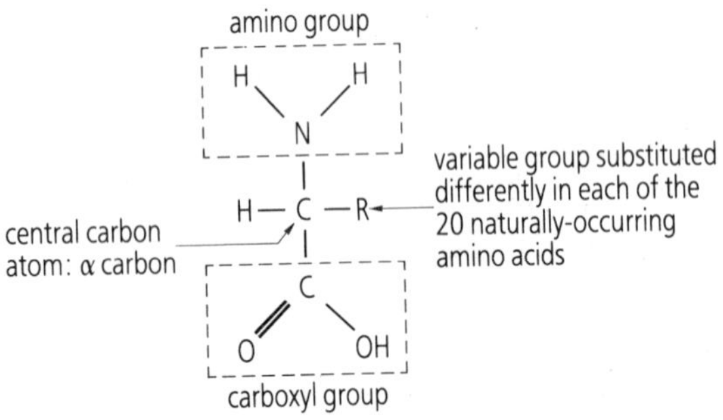

Fig 4·1 General formula of an amino acid

Table 4·1 Amino acids classified according to their −R groups

AMINO ACIDS WITH NON-POLAR −R GROUPS	AMINO ACIDS WITH POLAR −R GROUPS	AMINO ACIDS WITH ACIDIC −R GROUPS (negatively charged)	AMINO ACIDS WITH BASIC −R GROUPS (positively charged)
example: alanine	*example:* serine	*example:* aspartic acid	*example:* lysine
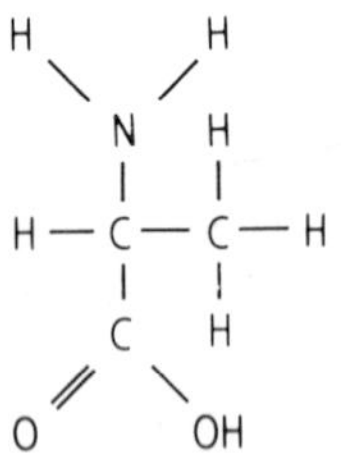	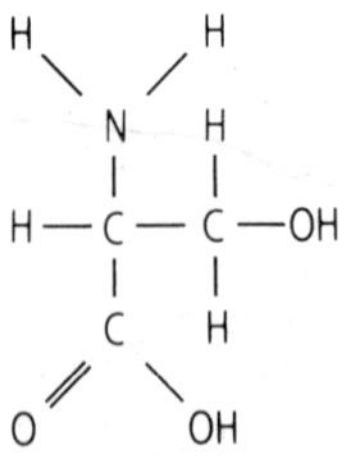	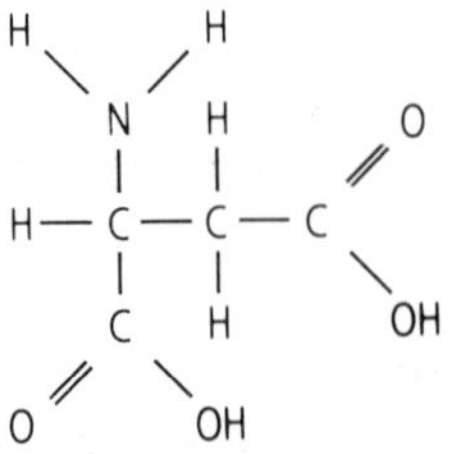	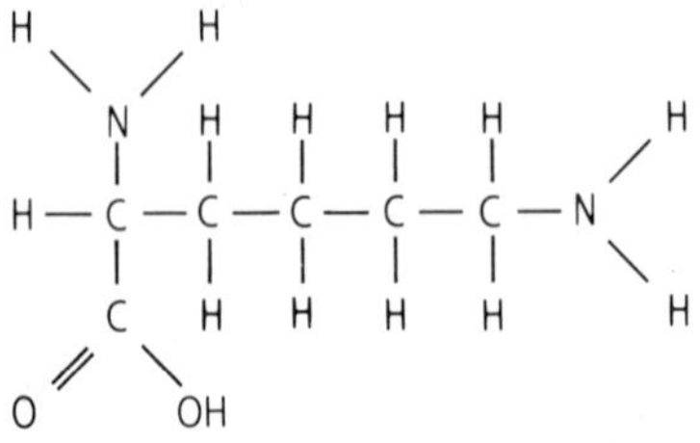
other amino acids in the group	*other amino acids in the group*	*other amino acids in the group*	*other amino acids in the group*
valine leucine isoleucine proline phenylalanine tryptophan methionine	glycine threonine cysteine tyrosine asparagine glutamine	glutamic acid	arginine histidine

−R group classification

Four different types of amino acid are recognized according to the −R groups they possess (see Table 4·1). Simple **non-polar** hydrocarbon groups occur in the −R positions of alanine, valine, and several other amino acids. A large proportion of these amino acids in a protein makes its molecules insoluble and rather unreactive. They occur mainly in structural proteins such as collagen. On the other hand, amino acids with **basic** and **acidic** −R groups form negatively and positively charged ions and are strongly hydrophilic ('water-loving'). As a result, the proteins which contain them are fairly soluble. In globular proteins, these charged −R groups are also important in forming bonds between different parts of the molecule, helping to

hold it in its proper shape. Amino acids with **polar** −R groups, that is, groups which develop partial charges but do not lose or gain electrons to form ions, also increase protein solubility and allow hydrogen-bonding between chains.

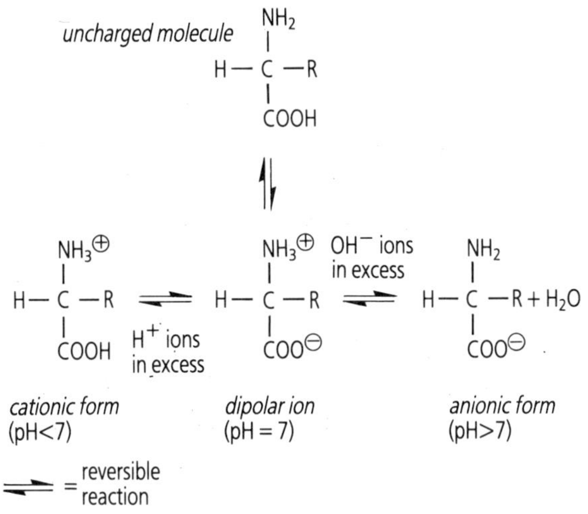

Amino and carboxyl groups

While the −R groups of different amino acids vary, the amino and carboxyl groups are common to all: they are important for two reasons. Firstly, as shown in Figure 4·2, they tend to dissociate when dissolved in water so that many amino acid molecules become **dipolar ions**, each containing oppositely charged COO^- and NH_3^+ groups. These ions give a solution of amino acids a **'buffering'** effect, that is, the solution opposes changes in pH and tends to remain at or close to neutral pH. This happens because the charged groups of the amino acid are formed reversibly and can reassociate when conditions are altered, mopping up the H^+ or OH^- ions, whichever is in excess. Amino acids, proteins, phosphate ions, and hydrogencarbonate ions play an important role as buffers in tissue fluid and in the cytoplasm of most cells, maintaining the pH within the narrow limits needed for normal metabolism and correct enzyme function.

The second function of the carboxyl and amino groups is to form linkages called **peptide bonds** which join the amino acids together to make a chain. As illustrated in Figure 4·3, the overall chemical process is a condensation reaction which leads to the formation of a linkage called a **peptide bond**. Note that, in living systems, the

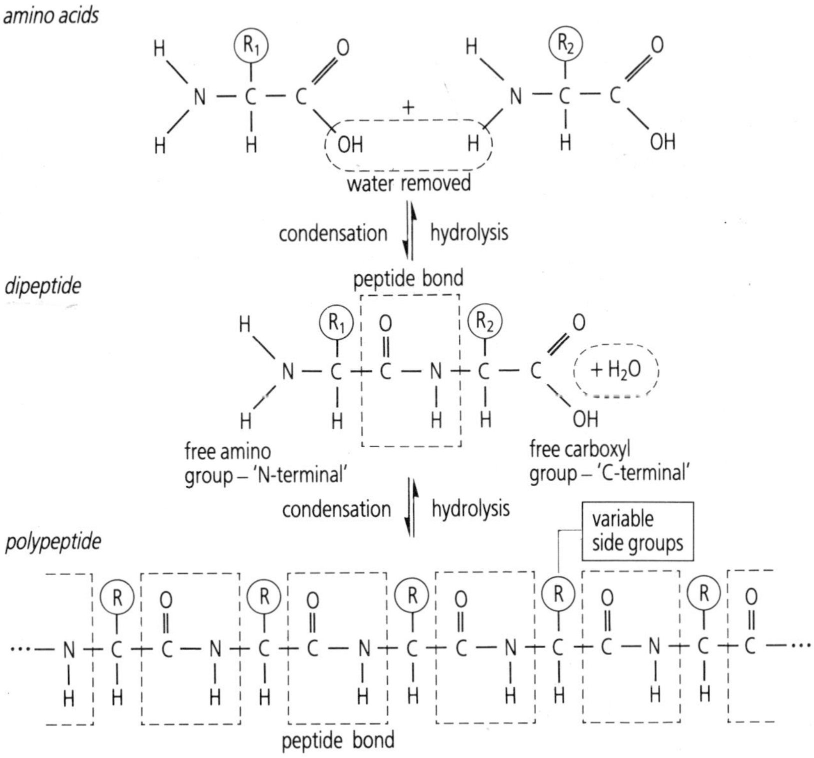

Fig 4·2 Ionic forms of amino acid molecules

Fig 4·3 Formation of dipeptides and polypeptides

reaction which forms the bond occurs indirectly in several distinct steps each controlled by its own enzyme. Two amino acids joined together form a molecule called a **dipeptide**, while three form a **tripeptide** and so on. Long chains, which may contain 500 or more amino acids in any order, are called **polypeptides**. There is virtually unlimited potential for variation in the sequence of amino acids and many thousands of different structures are known. To form a protein, these polypeptides must be folded and twisted in an appropriate way, as described in the next section.

4·3 Organization of protein molecules ■ ■ ■ ■ ■ ■ ■ ■ ■

The **primary structure** of a protein is defined as the sequence of amino acids in its molecule. This sequence is vital because, through predictable interactions between different parts of the protein chain, it determines the three-dimensional shape of the protein molecule on which its properties depend. This shape is of obvious importance in enzymes where the configuration determines whether the molecules on which the enzyme acts (that is, its substrates) will be able to fit in, and therefore whether or not the enzyme will work.

The **secondary structure** of a protein refers to the regular arrangement of its polypeptide chains. While the primary structure of a protein is determined by chemical analysis, this secondary structure is usually studied by the technique of **X-ray crystallography**, that is, by passing X-rays through a purified crystal of the protein. When this is done, the X-rays are scattered by the crystal and form a characteristic pattern which can be recorded on a photographic plate. Using appropriate mathematical techniques, the structure of the protein can be inferred from the pattern it produces. The first proteins to be successfully investigated in this way were fibrous proteins such as keratins (found in hair, nails, and wool),

collagen (found in skin), and fibroin (found in silk). These proteins are easily crystallized and have a comparatively simple and regular three-dimensional structure.

The most common type of secondary structure, first discovered in keratin, is illustrated in Figure 4·4. This shows the polypeptide chain loosely coiled in a regular spiral shape called an α-**helix**. The twisting of the chain necessary to form the helix involves the bonds on either side of the α-carbon atoms of the amino acids rather than the peptide linkages themselves, which tend to be quite rigid. Within the helix, the C=O and N−H groups from the peptide bond regions are held near to each other, so that they form numerous hydrogen bonds. These hold the α-helix together and make it a stable structure. In many fibrous proteins, the α-helices are themselves coiled together in a ropelike arrangement, giving greater overall strength.

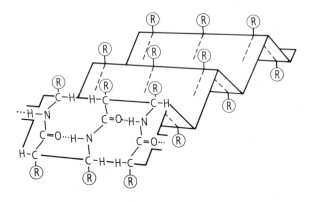

Fig 4·5 Secondary structure – the β-pleated sheet

Figure 4·5 shows another type of secondary structure in which the polypeptide chains are more extended and lie parallel with hydrogen-bonding between chains. This structure is called a **β-pleated sheet** and occurs, for example, in fibroin, the main protein component of the silk threads used by caterpillars to spin their cocoons.

Both these secondary structures are formed by hydrogen-bonding between the peptide bond regions of the polypeptide chain. The variable −R groups of the protein are *not* involved in forming its secondary structure. There are many hydrogen bonds in the α-helix but their individual bond energy is quite low so that the α-helix can be

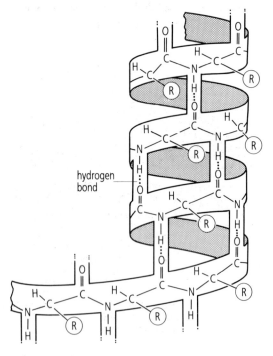

hydrogen bond

Fig 4·4 Secondary structure – the α-helix

extended like a spring simply by pulling its ends. This property provides an explanation at the molecular level for the elasticity of natural fibres such as wool. On the other hand, in fibroin, the polypeptide chains are already extended: the β-pleated sheet allows a silk thread to be folded into any position but it breaks easily if stretched.

The structure of globular proteins which form crystals can also be worked out by X-ray analysis although this is more complex than in the case of fibrous proteins. **Myoglobin**, the red oxygen-carrying pigment found in vertebrate muscle, has a structure which is typical. As shown in Figure 4·6, each of its molecules contains several lengths of α-helix, but these are folded together rather like links in a string of sausages to give a compact overall shape. The folding process gives the protein what is known as its **tertiary structure**.

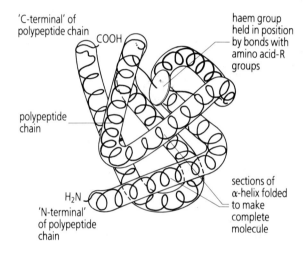

Fig 4·6 Tertiary structure – the myoglobin molecule

This depends particularly on the properties of the different −R groups in the polypeptide chain.

The −R groups can contribute to tertiary structure in several ways. For example, the −R groups of certain amino acids such as **proline** prevent the formation of the α-helix so that their position in the polypeptide chain determines where folding will occur. Similarly, some of the −R groups are strongly hydrophilic ('water-loving') and will usually be located on the outside of the molecule, while other hydrophobic ('water-hating') groups are located on the inside, protected from exposure to water. In other cases, as shown in Figure 4·7, there is actual bonding between −R groups. The amino acid **cysteine** is particularly important in this respect because its −R group contains an −SH, or thiol group. The −SH groups can be linked to form a strong covalent −S−S− connection, called a **disulphide bridge**, between chains, as in molecules of the protein hormone insulin. Other weaker connections include **hydrogen bonds** and **electrovalent bonds**, which occur between oppositely charged −R groups. In the structure of myoglobin there

are no disulphide bridges so that the whole of the molecule is held in its correct configuration by these other forces.

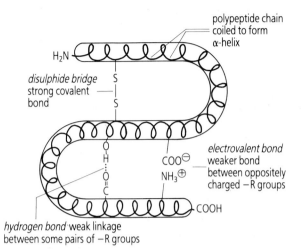

Fig 4·7 Chemical bonds involved in maintaining tertiary structure in a protein

Some proteins show what is called **quaternary structure**, that is, they consist of two or more different polypeptide chains fitted together to make a larger protein assembly. The structure of **haemoglobin** shown in Figure 4·8 illustrates this. The molecule consists of two pairs of subunits called, **α-chains** and **β-chains** respectively. Each of these chains is similar to an individual molecule of myoglobin and contains a haem group which binds reversibly with oxygen. However, the amino acid sequences of the α- and β-chains differ in detail so that the two types have slightly different overall shapes which allow them to be fitted tightly together, forming a stable structure. This is the result of many weak −R group interactions including the formation of hydrogen bonds and does not involve disulphide bridges or any other covalent links between the subunits.

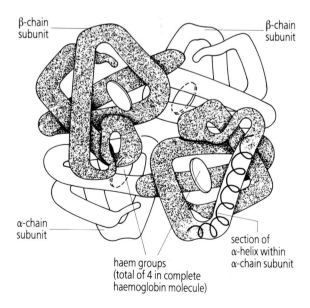

Fig 4·8 Quaternary structure – the haemoglobin molecule

4·4　Denaturation and renaturation ■ ■ ■ ■ ■ ■ ■ ■ ■ ■

Most proteins lose their biological activity and are said to become **denatured** in unfavourable conditions of temperature and pH. Denaturation can occur at temperatures of 50–70°C. It does not normally affect the covalently bonded backbone of the protein or any of its disulphide linkages, but the weaker hydrogen bonds and electrovalent bonds are broken so that the polypeptide chain unwinds, forming loose random coils. The complex shape of the protein is lost and it no longer functions correctly.

When the white of an egg is heated, the molecules of the albumen protein it contains are denatured and their polypeptide chains become entangled, making the change which takes place a permanent one. However, in many cases, denaturation is a reversible process and the properties of the protein can be restored simply by returning it to normal conditions. This process, called **renaturation**, involves the spontaneous recoiling of unfolded protein molecules into their correct configuration. It follows from this that the sequence of amino acids in a protein is, by itself, sufficient to determine the overall shape of the entire protein molecule. Thus, by specifying these sequences, the genetic material in the cell nucleus is able to direct the synthesis of the many enzymes and other proteins on which living processes depend. This is a very important concept which will be referred to again in the later Units of this Section.

4·5　Chemical tests for proteins ■ ■ ■ ■ ■ ■ ■ ■ ■ ■ ■

There are only two common chemical tests for the presence of protein. These are listed in Table 4·2.

Table 4·2 Chemical tests for proteins

TEST	CONDITIONS	RESULT
Millon's reagent (solution of mercuric nitrate and nitrous acid)	Add 1 cm^3 of reagent to 2 cm^3 of tissue extract in a test tube, heat to 95°C for 2 minutes	pink or red precipitate
Biuret test	Add 2 cm^3 of potassium hydroxide solution to 2 cm^3 of tissue extract in a test tube, add 1–2 drops of copper sulphate solution and mix.	violet colour

5 ENZYMES AND METABOLISM ☐ ☐

Objectives

After studying this Unit you should be able to:–

- State the two fundamental roles of enzymes in living systems

- Describe the basic mechanisms of enzyme action and distinguish between the 'lock and key hypothesis' and the concept of 'induced fit'

- Define activation energy and explain why reduced activation energy is important for the catalytic action of an enzyme

- Sketch the graph of the relationship between temperature and reaction rate for a typical enzyme and explain its shape

- Similarly, sketch and comment on the relationships between pH and reaction rate

- Explain how changes in substrate concentration and enzyme concentration affect the rate of enzyme-controlled reactions

- Distinguish between competitive inhibition and non-competitive inhibition and give examples of enzyme inhibitors of each type

- Outline the role of enzyme cofactors

- Discuss some of the ways in which control and coordination of enzyme activity is achieved

- Explain the naming and classification of enzymes and list the six major enzyme types

5·1 Mechanisms of enzyme action ■ ■ ■ ■ ■

At any moment, hundreds of chemical reactions are in progress in a typical living cell. These reactions provide energy and help to maintain supplies of the many different substances needed for growth and repair. The remarkable fact that all these reactions contribute to an organized overall **metabolism** is due almost entirely to the production of special substances called **enzymes**. The purpose of this Unit is to outline the nature and properties of these enzymes and to explain their role in regulating metabolism.

All enzymes are globular proteins and within the cell they have two fundamental roles:

1 They act as highly specific **catalysts**, greatly speeding up chemical reactions which would otherwise be hopelessly slow.

2 They provide a mechanism whereby individual chemical reactions can be controlled, the available quantity of an enzyme determining the rate of the corresponding reaction.

You can see from Figure 5·1 that the specificity of an enzyme's action arises because each enzyme has a definite three-dimensional shape which is complementary to that of its reacting molecules, or **substrates**. The first step in any reaction catalysed by an enzyme is the formation of a specific association between the molecules called an **enzyme-substrate complex**. This is made possible by the fact that the configuration of the enzyme matches the shape of the substrates over a relatively large area called the **active centre** of the enzyme. When enzyme and substrate molecules

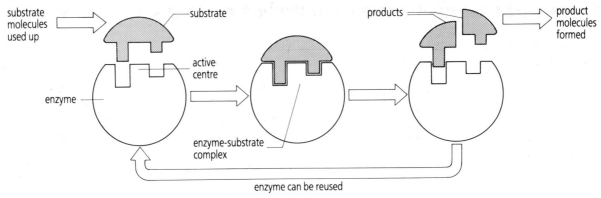

Fig 5·1 Outline mechanism of enzyme action

collide in the correct orientation, the substrates become attached and are held temporarily in position at the active centre. The enzyme and substrate molecules now interact in such a way that a chemical reaction involving the substrates takes place and the appropriate **products** are formed. Immediately afterwards, the products leave the active centre and the enzyme is again available completely unchanged and ready to combine with new substrate molecules.

This sequence of events can be repeated very rapidly. The action of the enzyme **catalase** provides a dramatic example. Catalase is found in the body tissues of many animals and plants and breaks down molecules of hydrogen peroxide (H_2O_2), releasing oxygen gas and water (see Fig. 5·2). Each molecule of the enzyme can combine with up to 100 000 molecules of hydrogen peroxide per second, speeding up the rate of this particular reaction by about 10 000 times. Like many enzymes, catalase will work with only a single substrate but some enzymes are less specific. **Chymotrypsin**, for example, splits peptide linkages by hydrolysis and is effective between many different pairs of amino acids in a polypeptide chain.

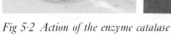

Fig 5·2 Action of the enzyme catalase

The mechanism of enzyme action described above is known as the **lock and key hypothesis**, but this concept is helpful only up to a point. The 'lock' and 'key' in this case are not static structures but interact with each other so that the shapes of both are changed. A better concept is that of '**induced fit**', meaning that the configuration of

the enzyme becomes altered as the substrate binds at the active centre. At the same time, the binding process stretches or compresses one or more of the chemical bonds in the substrate molecule so that a particular chemical reaction becomes much more likely. When the products are released, the enzyme returns to its normal shape.

The induced fit action of an enzyme can be explained further in terms of its effect on the **activation energy** of the reaction it controls. In any chemical reaction, the molecules involved must acquire a certain amount of energy from their surroundings before chemical changes are possible. This energy is called activation energy. For reactions which occur spontaneously at normal temperatures sufficient thermal energy is always available but, in many other cases, the activation energy represents a 'barrier' which slows down or prevents the reaction (refer to Fig. 5·3). This

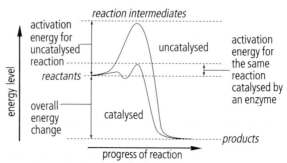

Note that the overall energy change is the same for the catalysed and uncatalysed reactions.

Fig 5·3 Activation energy and the action of catalysts

barrier can be overcome by increasing the temperature and pressure, increasing the kinetic energy of the reacting molecules and making collisions more likely. The reaction is also speeded up if a suitable enzyme or other catalyst is available. Note that the curve for the enzyme catalysed reaction must have at least two peaks. The reaction is made easier because of a change in mechanism. The first peak represents the formation of the enzyme–substrate complex, while the second larger peak corresponds to the reaction which generates the products. The reaction is made easier by alterations in the shape of the substrate molecules.

5·2 Factors which affect enzyme-controlled reactions ▪ ▪ ▪ ▪

Enzyme-controlled reactions have a number of important characteristics related to the mechanisms of enzyme action described above, and to the fact that all enzymes are proteins. The effects of temperature, pH, substrate and enzyme concentration and those of inhibitors and cofactors are of particular biological interest and can be summarised as follows:

Temperature

The relationship between temperature and reaction rate for a typical enzyme is illustrated in Figure 5·4. The initial increase in reaction rate with rising temperature is the expected result for any chemical reaction because the increased heat energy raises the kinetic energy of the reacting molecules, leading to more frequent collisions. In this case, more enzyme-substrate complexes are formed per unit time, with a corresponding increase in catalysis and in the number of product molecules.

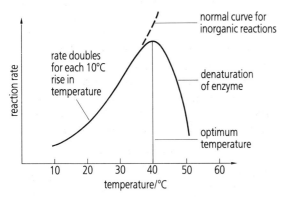

Fig 5·4 *Effect of temperature on the rate of an enzyme-controlled reaction*

With many reactions involving an inorganic catalyst, this increase would continue indefinitely, but in an enzyme-controlled reaction, an **optimum temperature** is soon reached corresponding to the maximum reaction rate. Above the optimum temperature, the reaction rate falls off sharply. This is because enzyme molecules, like all proteins, owe their shape to weak attractive forces like hydrogen bonds: these bonds become unstable at higher temperatures, causing the enzyme to become denatured. The active centre loses its correct configuration and fails to fit the substrate so that the enzyme no longer functions as a catalyst.

Most enzymes have optimum temperatures between 40–50°C but there are some specialised types which function best either above or below this range. For example certain kinds of bacteria survive in hot springs above 85°C and require particularly stable enzymes (see Unit 45). At the other extreme, the ice fish of Antarctic waters has enzymes which function efficiently at −2°C.

pH

Most enzymes have an **optimum pH** close to pH 7, which is the normal intracellular pH. This is illustrated in Figure 5·5. Enzymes which work extracellularly can have very different pH requirements. A notable example is the enzyme **pepsin** which is found in the stomach and works best in highly acidic conditions in the range pH 1–2.

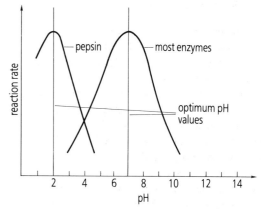

Fig 5·5 *Effect of pH on the rate of an enzyme-controlled reaction*

pH changes away from the optimum can affect enzyme activity adversely in either of two ways. In cases where the binding or catalytic sites within the active centre take the form of charged ions, some values of pH are inhibitory because they cause these ions to reassociate. The uncharged groups formed as a result can no longer interact with the substrate and so catalytic activity is lost. A second possibility is that the enzyme molecule can change shape and become denatured; this is more likely to happen at extreme values of pH which tend to weaken the forces holding the enzyme molecule together.

Substrate concentration and enzyme concentration

The reaction rate for most enzyme-controlled reactions varies with the concentration of the available substrate in the way shown in Figure 5·6. You can see from either graph *A* or graph *B* that increasing the concentration of substrate can give a corresponding increase in reaction rate, but only when the substrate concentration remains comparatively small. When larger substrate concentrations are used, the reaction rate becomes less dependent upon the concentration of substrate and tends towards a fixed maximum determined by the amount of enzyme present. These observations can be interpreted in a simple way. At low substrate concentrations many of the available enzyme molecules will have active centres which are unoccupied and the restricted supply of

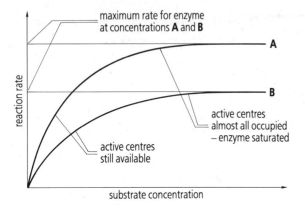

Fig 5·6 Effect of substrate concentration and enzyme concentration on the rate of an enzyme-controlled reaction

substrate molecules largely determines the reaction rate. At large substrate concentrations almost all of the active centres will be occupied so that the number of enzyme molecules becomes the limiting factor. When this number is increased, there is a proportionate increase in the maximum rate (this effect can be seen by comparing curves *A* and *B* in Fig. 5·6).

In the context of cell metabolism, these relationships are potentially important as a means of controlling the rate of different reactions – for some reactions substrate concentration will normally be the more important factor, but for others enzyme availability will be critical.

Enzyme inhibitors

Inhibition occurs when the action of an enzyme is slowed down by another substance. There are two types of inhibitor:

1 Competitive inhibitors. These substances are chemically closely related to the normal substrate and have a shape which is very similar. When both substrate and inhibitor molecules are present, they 'compete' for the active centre so that the catalytic action of the enzyme is slowed down.

One example of competitive inhibition is provided by the enzyme **succinic dehydrogenase** which is the catalyst for one of the many steps in the process of energy release in cells. As illustrated in Figure 5·7, the normal substrate of the enzyme, **succinic acid**, is converted to fumaric acid by removing two hydrogen atoms. The enzyme normally releases these products and can then be reused. **Malonic acid** acts as a competitive inhibitor for this enzyme by temporarily occupying the active centre in the same way. The difference is that malonic acid cannot be broken down. In this particular reaction, the enzyme-inhibitor complex is more stable than the enzyme-substrate complex. This means that the two substances do not compete on equal terms – small amounts of inhibitor greatly reduce reaction rates because the inhibitor molecules bind to the enzyme for longer. Competitive inhibitors have no permanent effect on the enzyme molecules and the inhibition they cause can be overcome either by increasing the concentration of the substrate or by reducing the concentration of the inhibitor substance.

2 Non-competitive inhibitors. These substances are distinguished from competitive inhibitors because they do not combine with the active centre of the enzyme and are unaffected by substrate concentration. The commonest types include **heavy metal ions** such as mercury (Hg^{2+}) and silver (Ag^+) ions. These combine with the enzyme molecule, producing a change which indirectly affects the shape of the active centre, so that the enzyme is unable to interact correctly with its substrate (see Fig. 5·8). This effect, where a chemical reaction involving one region of a protein molecule alters the shape and properties of a second region, is known as an **allosteric** effect. Many poisons such as arsenic and cyanide act in this way, but it would be a mistake to think that all non-competitive inhibitors must be poisonous. On the contrary, binding of inhibitors at 'allosteric sites' represents one of the most important ways of regulating enzyme activity. Some enzyme

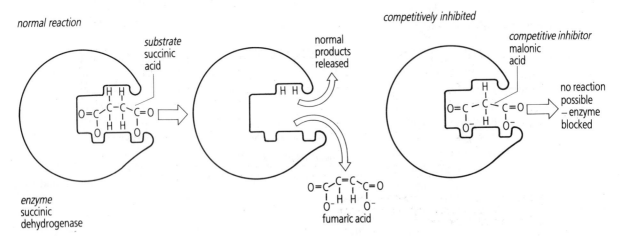

Fig 5·7 Competitive inhibition of an enzyme

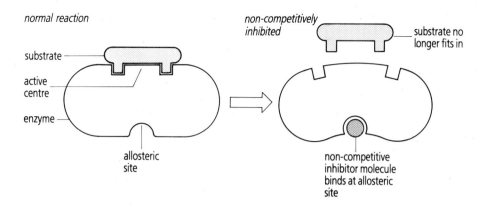

Fig 5·8 Non-competitive inhibition of an enzyme

molecules possess an allosteric site which must be occupied by an **activator** substance to give the enzyme its correct overall shape. Such activators are equally important in regulating enzyme action.

Enzyme cofactors

Many enzymes will not work correctly unless a smaller non-protein substance, called a **cofactor**, is present within the active centre. As shown in Figure 5·9, the cofactor usually acts as a 'bridge' between the enzyme and its substrate; often it contributes directly to the chemical reactions which bring about catalysis. Sometimes the cofactor provides a source of chemical energy, helping to drive reactions which would otherwise be difficult or impossible.

Some enzymes need **metal ions** as cofactors – examples include Mg^{2+}, Fe^{2+}, and the ions of trace elements such as Zn^{2+} or Cu^{2+}. Alternatively, the cofactor can be a small organic molecule. These substances are called **coenzymes** and are often closely related to **vitamins**, which represent the essential raw materials from which coenzymes are made. Only small quantities of vitamins are needed because, like enzymes, coenzyme substances can be reused again and again. Sometimes, the coenzyme is covalently bonded to the enzyme in a more or less permanent way so that it forms the **prosthetic group** of a conjugated protein. However, about 80% of enzymes are pure proteins consisting of polypeptide chains; any coenzymes required are bound loosely and possibly only briefly while the enzyme carries out its catalytic function.

All the factors discussed above help to explain what individual enzymes are and how they work within cells. Even in simple cells hundreds of different enzymes are required and all of these must be regulated and coordinated, one with another, so that they act only where and when they are needed. How this occurs is a question of great interest because the capacity for internal control is the key characteristic of living systems. Some of the mechanisms involved are outlined in the next section.

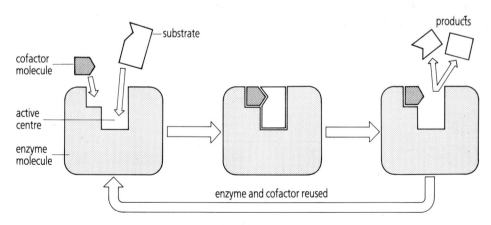

Substrate molecules will not fit correctly at the active centre and there will be no catalytic action unless the cofactor molecule is also present.

Fig 5·9 Action of an enzyme cofactor

5·3 Regulation of enzyme activity ■ ■ ■ ■ ■ ■ ■ ■ ■ ■

Some enzymes are potentially damaging if they become active in the wrong place, giving rise to a need for **containment**. For example, **pepsin** is a powerful protein-digesting enzyme and is quite capable of disrupting a cell's internal structure. Accordingly, the stomach cells from which pepsin is obtained produce an inactive form, called **pepsinogen**. This substance is released from the cells and only becomes active when exposed to strong acid conditions. The cells lining the stomach are protected from acid and enzymes alike by a thick coating of mucus and so digestion of the proteins in food can be safely carried out. Many cells produce similar enzymes which are not exported but remain within the cell. Cell damage in this case is prevented by enclosing the enzymes inside small membrane-bound structures called **lysosomes** (discussed fully in Unit 7).

A different principle is seen in cases where a series of chemical reactions must take place in a particular order to complete a metabolic pathway such as respiration or photosynthesis. The successive enzymes controlling these reactions are normally present together in a precise **spatial arrangement** such that substrate molecules can be literally 'handed on' from one enzyme to another. In this way, the products from one step in the pathway are transferred to the enzyme catalysing the next step. Most enzymes do not float about in a kind of cytoplasmic 'soup' but are attached to membrane systems inside the cell in specific and orderly arrangements. Mitochondria and chloroplasts (considered in Units 10 and 11) provide clear examples of this.

Precise localization increases efficiency but does not provide moment to moment, or '**dynamic**', control of enzyme action and for this further mechanisms are needed. Enzyme pathways can be regulated in a more flexible way according to the availability of other substances in the cell: two possibilities are illustrated in Figure 5·10. When the final product of a sequence of reactions starts to accumulate, the enzyme pathway leading to its manufacture can be shut down by a process called **feedback inhibition**. The final product acts as a non-competitive inhibitor for one of the enzymes at the beginning of the pathway, that is, the activity of the enzyme is prevented because the product molecule binds at an allosteric site. The second possible means of control arises when accumulation of a potential substrate causes a particular reaction pathway to be opened up. This is called **precursor activation** and also involves an allosteric site, however, the binding molecule activates rather than inhibits the enzyme involved. Mechanisms of this kind are important because they help the cell to make the best possible use of whatever substances are available.

The final level of enzyme control is provided by the genetic information stored in the nucleus of the cell. While the action of individual enzymes and pathways can be modified by other factors, the genes present in the nucleus ultimately control which enzymes will be synthesized and therefore determine the limits of the cell's metabolism. Further details of this kind of control are given in Unit 9, but you should now begin to realise how sophisticated and intricate the cell's system of metabolism must be.

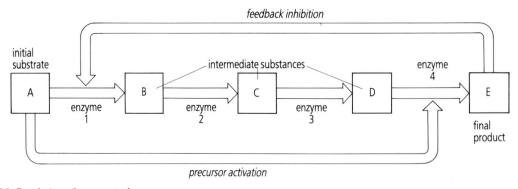

Fig 5·10 Regulation of enzyme pathways

5·4 Naming and classifying enzymes ■ ■ ■ ■ ■ ■ ■ ■ ■ ■

Individual enzymes usually have names which end in **-ase** – common examples include **maltase** and **sucrase**, which act on the disaccharides maltose and sucrose, and **amylase**, which acts on starch. These names are to some extent arbitrary and, as more and more enzymes have been discovered, a

systematic way of naming and classifying enzymes has become necessary. The six major groups of enzymes listed in Table 5·1 are now recognized by international agreement and all enzymes are now named within this scheme.

Table 5·1 International classification of enzymes

CLASS OF ENZYME	TYPES OF REACTION CATALYSED
1. Oxidoreductases	These enzymes catalyse oxidation-reduction reactions, that is, they transfer hydrogen atoms or electrons from their substrates to acceptor molecules.
2. Transferases	Small groups of atoms are transferred from one substrate to another.
3. Hydrolases	These enzymes split chemical bonds by hydrolysis. Different types attack peptide bonds, glycoside linkages and ester linkages.
4. Lyases	These enzymes add new groups to a substrate by breaking a double bond. Alternatively, they may catalyse the formation of double bonds.
5. Isomerases	These catalyse internal rearrangement of the atoms in a substrate, that is, they convert between one isomer and another.
6. Ligases	New chemical bonds are formed by these enzymes; energy from ATP is needed to make the new bonds. Ligases help in the synthesis of carbohydrates, proteins and other macromolecules.

6 LIPIDS AND CELL MEMBRANES

Objectives

After studying this Unit you should be able to:—

- Draw the structural formula for glycerol and distinguish between saturated and unsaturated fatty acids

- Explain how fatty acids and glycerol are joined together to make molecules of fats and oils

- Outline some of the functions of fats, oils, and waxes in living systems

- List the common chemical tests for identification of fats and oils

- Describe the structure of a phospholipid molecule and explain the terms hydrophilic and hydrophobic

- Explain the structure of a lipid bilayer

- List the functions of some important steroid substances

- Draw and describe the Davson-Danielli model of plasma membrane structure and explain why this is now rejected

- Draw and describe the fluid mosaic model of plasma membrane structure

- Discuss diffusion, active transport, and cytosis as methods of transferring substances across the plasma membrane

6·1 Introduction ■ ■ ■ ■ ■ ■ ■ ■ ■ ■ ■ ■ ■ ■ ■ ■

Compared to carbohydrates, **lipids** are a rather mixed collection of substances and have a range of different functions in living systems. Nevertheless, they form a group which can be considered together because their molecular structures all contain a high proportion of CH_2 groups. This means that lipids have a low solubility in water but a high solubility in non-polar solvents such as ethanol and chloroform. Thus, when cells are broken up finely in a non-polar solvent, the lipid substances present are extracted together. The most important types include **fats** and **oils**, **waxes**, **phospholipids**, and **steroids** such as **cholesterol**.

The purpose of this Unit is to describe the structures and functions of these different types of lipids and to explain how two of them, namely phospholipids and cholesterol, contribute to the formation of cell membranes. Some of the more important roles of the cell's outer membrane, called the **plasma membrane**, will also be discussed.

6·2   Fats, oils, and waxes ▪ ▪ ▪ ▪ ▪ ▪ ▪ ▪ ▪ ▪ ▪ ▪ ▪

The molecules of fats, oils, and waxes contain the chemical elements **carbon**, **hydrogen**, and **oxygen**. Precisely the same elements occur in carbohydrates but the relative amounts in fats and oils are quite different: in particular, far less oxygen is present. This greatly reduces the possible number of polar $-OH$ groups in molecules of fat and helps to explain their low solubility in water.

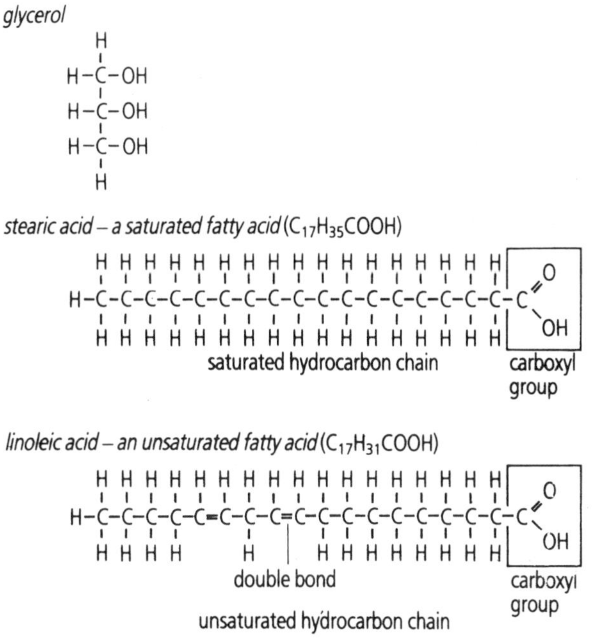

glycerol

stearic acid – a saturated fatty acid $(C_{17}H_{35}COOH)$

linoleic acid – an unsaturated fatty acid $(C_{17}H_{31}COOH)$

Fig 6·1 *Chemical subunits of fats and oils*

Fats and oils are built up from two basic types of subunits, namely **fatty acids** and **glycerol**. The structures of these subunits are illustrated in Figure 6·1. All fatty acid molecules consist of a $-COOH$, or carboxyl, group attached to one end of a hydrocarbon chain. This contains between 4 and 24 carbon atoms and is normally unbranched, although it may be either **saturated**, or **unsaturated**. Saturated chains contain repeating CH_2 groups joined by single bonds, as in the molecule of **stearic acid**, while unsaturated chains contain double bonds in the form of $-CH=CH-$ groups, as in **linoleic acid**.

Structures called **mono-**, **di-**, and **triglycerides** are formed when a single glycerol molecule is joined to one, two, or three of the fatty acid subunits. As indicated in Figure 6·2, the triglyceride structure typical of fats and oils is formed when condensation reactions take place between the $-OH$ (alcohol) groups of the glycerol molecule and the $-COOH$ groups of the three fatty acid molecules. (In living systems, several enzyme-controlled steps are required.) The new linkages formed are consistent with the general formula for an **ester**, usually written as R.COOR′, where R and R′ can be replaced by any organic

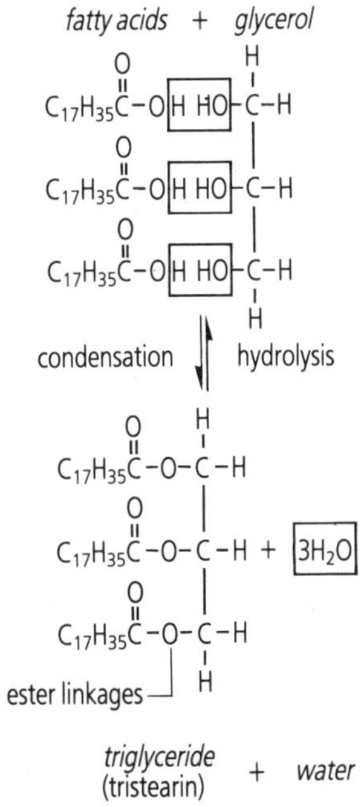

fatty acids + glycerol

Fig 6·2 *Formation of a triglyceride*

group, and they are therefore called **ester linkages**. Although the three fatty acids are often identical a mixture of types is possible so that many different triglyceride structures can be produced.

Fats are defined as triglycerides which are solids at room temperature; they often contain only saturated fatty acids. **Oils** are liquids at room temperature and usually contain unsaturated fatty acids. **Waxes** also contain ester linkages, but each molecule contains only one fatty acid subunit and this is linked with the molecule of a long chain alcohol instead of with glycerol.

Fats and oils function efficiently as **storage materials**, partly because of their limited solubility in water, and partly because of the numerous $C-C$ and $C-H$ bonds which characterise their structure. These bonds represent a large reservoir of stored chemical energy which can be released and used by the cell when required. In this context, it is interesting to note that fats and oils provide about 38 kJ per gram, while carbohydrates, which may also be used for storage, provide only 17 kJ per gram. In many organisms, fat is deposited when other energy-rich substances are in surplus and it can then serve additional functions such as **thermal insulation**, and **mechanical protection** of delicate organs. Fats, oils, and especially waxes are also used to 'waterproof' the external surfaces of both plants and animals, usually in order to reduce water loss.

Two common chemical tests are available for the identification of fats and oils – these are listed in Table 6·1.

Table 6·1 Chemical tests for fats and oils

TEST	CONDITIONS	RESULT
emulsion test	grind tissue sample in ethanol to dissolve any fats and oils. Filter off and add 2 cm³ of extract to 2 cm³ of water in a test tube	milky-white emulsion formed

TEST	CONDITIONS	RESULT
Sudan III	mount tissue section in solution for microscopic examination or add to tissue sample and rinse	lipid-containing structures are stained red

6·3 Phospholipids ■ ■ ■ ■ ■ ■ ■ ■ ■ ■ ■ ■ ■ ■ ■ ■

Phospholipids are the most important group of lipids because, along with proteins, they are the essential components of all cell membranes. A typical phospholipid has the structure illustrated in Figure 6·3. This shows two molecules of fatty acid linked to a molecule of glycerol in the same way as in fats and oils. The third position on the glycerol molecule is occupied by a **phosphate group**, which links the glycerol to an additional molecule, usually a complex alcohol.

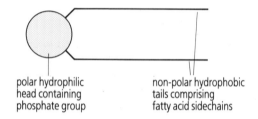

polar hydrophilic head containing phosphate group

non-polar hydrophobic tails comprising fatty acid sidechains

Fig 6·4 Simplified diagram of a phospholipid molecule

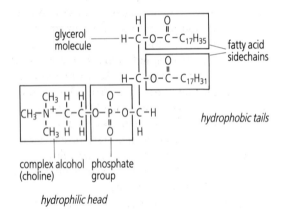

Fig 6·3 Structure of a phospholipid molecule (phosphatidylcholine)

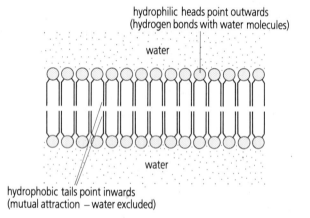

hydrophilic heads point outwards (hydrogen bonds with water molecules)

water

water

hydrophobic tails point inwards (mutual attraction – water excluded)

Fig 6·5 Structure of a phospholipid bilayer

In the molecule of a fat, the non-polar hydrocarbon chains of the fatty acid molecules attract each other and, as a result, they all lie in the same direction. However, either of the C—C bonds in the glycerol molecule can be twisted, so that, in the molecule of a phospholipid, the strongly polar phosphate group projects in the opposite direction from the two fatty acid side chains (see Fig. 6·3). This gives the phospholipid molecule a different shape and different properties to a molecule of fat. One end of the molecule, containing the phosphate group, is **hydrophilic**. In other words, it is polar and readily soluble in water. The other end, containing the fatty acid side chains, is **hydrophobic**, that is, non-polar and insoluble. Phospholipid molecules are represented more simply as shown in Figure 6·4.

When phospholipids are mixed with water, they can form **surface layers** and **micelles**, as in the case of detergent molecules (discussed briefly in Unit 2). A far more important arrangement is the **bilayer** structure illustrated in Figure 6·5. This comprises two 'back-to-back' layers of phospholipid molecules. The hydrophilic heads point outwards, forming hydrogen bonds with the surrounding water molecules, while the mutually attracting hydrophobic tails point inwards. Individual molecules can move about from side to side quite freely within their own layers, so that the configuration is a 'fluid' rather than a static one. Nevertheless, the bilayered arrangement is stable and not easily disrupted. Although their detailed composition varies, bilayers provide the structural basis for all cell membranes.

6·4 Steroids ■ ■ ■ ■ ■ ■ ■ ■ ■ ■ ■ ■ ■ ■ ■ ■

Unlike other groups of lipids, steroids have a molecular structure which contains rings of atoms. They include **cholesterol**, which is a second important lipid component of many cell membranes. Cholesterol is quite abundant in the body, but other steroids are only present in minute amounts. These substances are not used as structural components but may act as **hormones** or **vitamins**. Some important examples are listed in Table 6·2.

Table 6·2 Some important steroids

STEROID	FUNCTION
cholesterol	major component of cell membranes, raw material for many other steroids
bile acids	in the form of bile salts, they help in emulsifying fats during digestion (Unit 13)
corticosteroids	hormones manufactured in the cortex region of the adrenal gland – they are involved in stress responses (Unit 25)

Steroid substances are rare in plants

STEROID	FUNCTION
oestrogens, progesterone	reproductive hormones in female mammals (Units 29, 30)
testosterone	reproductive hormone in male mammals (Unit 29)
calciferol (Vitamin D_2)	promotes calcium and phosphate absorption from the small intestine (Unit 12)
ecdysone	hormone causing moulting (ecdysis) in insects (Unit 49)

6·5 Structure of cell membranes ■ ■ ■ ■ ■ ■ ■ ■ ■ ■ ■

A knowledge of the molecular structure and properties of phospholipids gives some clues to the likely nature of cell membranes. The cell's outer membrane, or plasma membrane, is particularly important and, taking additional experimental evidence into account, several possible structures have been suggested. The purpose of this section is to evaluate two of these, called the **Davson-Danielli model**, and the **fluid mosaic model**.

Davson-Danielli model

Red blood cells swell up and burst when placed in water and their plasma membranes, called 'ghosts', can then be easily separated from the other parts of the cell. Chemical analysis of the membrane fraction obtained shows that these membranes consist of about 40% lipid, 0–10% carbohydrate, and 50–60% protein. (The carbohydrate present forms the prosthetic groups of conjugated proteins and is not of direct structural importance.) Calculations based on such analyses suggested that the quantity of lipid present was just enough to allow for a bilayer structure over the whole of the cell surface. This finding led to the development of the **Davson-Danielli** model of plasma membrane structure, as illustrated in Figure 6·6. This shows the membrane as a continuous lipid bilayer with the protein as a surface layer on both sides, holding the lipid molecules in a more or less rigid arrangement. To account for its permeability properties, the membrane was presumed to contain protein-lined channels, or '**pores**'.

This simple model received some support from **electron microscope** studies of membrane structure. As outlined in the index-glossary, this instrument allows extreme magnifications, up to 250,000 times or more. With appropriate staining techniques, the plasma membrane consistently appears as two dense lines separated by a space.

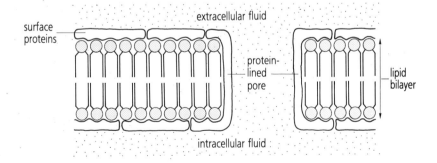

Fig 6·6 Davson-Danielli model of plasma membrane structure

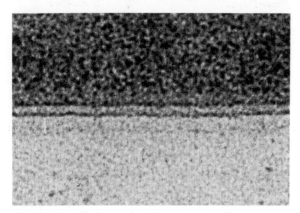

Fig 6·7 Electronmicrograph of the plasma membrane

This is illustrated by the electron microscope photograph (**electronmicrograph**) given in Figure 6·7. The dimensions of the space (which represents the hydrophobic tails of the phospholipid molecules) and the width of the dense lines (which represent the hydrophilic heads and the protein layers) are both consistent with the Davson-Danielli model.

Fluid mosaic model

More recently, other evidence has accumulated which has given a new insight into membrane properties and led to the development of an improved model of plasma membrane structure. This questions the idea of the membrane as a rigid structure and is a much more useful model, providing possible explanations for a whole range of membrane properties. Two important aspects of this evidence will be briefly discussed. The first concerns the 'fluidity' of membrane structure.

The membranes of different species of organisms carry **species-characteristic proteins** as an integral part of their structure. Under certain conditions, mouse and human cells (which have distinctly different proteins) can be made to fuse together to form a single large cell. In these circumstances, the Davson-Danielli model predicts that the species-characteristic proteins should remain at opposite ends of the new cell. However, as illustrated in Figure 6·8, the membrane proteins

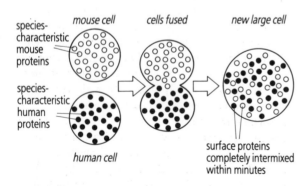

Fig 6·8 Fusing mouse and human cells

are not locked rigidly in position but can move freely. As a result, mouse and human proteins are completely intermixed in minutes.

A second line of evidence comes from a new technique of preparing material for the electron microscope, called the **freeze-etch technique**. Small pieces of tissue are frozen rapidly in liquid nitrogen (−180°C) and split in two under vacuum. An extremely thin film of carbon and platinum is then deposited onto the exposed face of the tissue. Finally, the tissue is digested away by enzymes to leave a carbon-platinum 'replica' which exactly mirrors the structure of the fractured surface. The overall process is similar in principle to taking a plaster cast of a person's footprints. Electron microscope study of membrane replicas can give results like that shown in Figure 6·9. The fracture plane in this case passes right through the plasma membrane, separating the two halves of the lipid bilayer and exposing the membrane proteins as 'lumps' within the membrane structure.

Fig 6·9 Electronmicrograph of freeze-etched replica of plasma membrane

From these observations, a new model of plasma membrane structure, called the **fluid mosaic model**, has been proposed. This is illustrated in Figure 6·10. The lipid bilayer is still seen as forming the structural framework of the membrane, but it is recognised that individual phospholipid molecules can move quite freely provided they remain correctly orientated within their own half of the bilayer. Cholesterol, which is sometimes present, restricts the movement of the phospholipids to some extent and therefore tends to stabilize the membrane structure. The membrane proteins include both globular and fibrous types. Some of them, called **intrinsic proteins**, pass straight through the membrane and have both an intracellular and an extracellular portion. Others, called **extrinsic proteins**, are fixed in one half of the bilayer, or attached to its surface. Most of the proteins can move laterally but they remain held in the membrane by mutual attractive forces. These arise between hydrophobic amino acid −R groups projecting from the proteins and the hydrophobic tails of the lipid molecules.

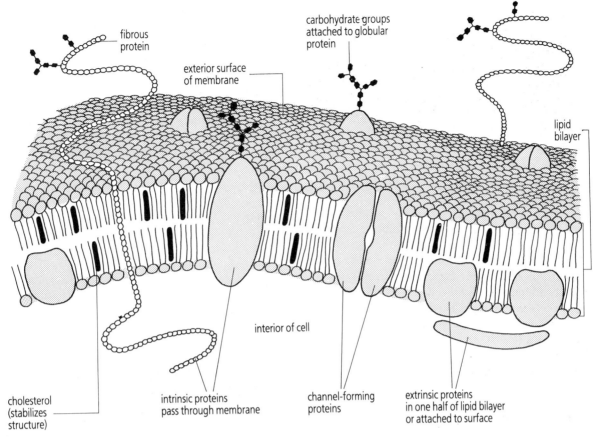

Fig 6·10 Fluid mosaic model of plasma membrane structure

labels:
- fibrous protein
- exterior surface of membrane
- carbohydrate groups attached to globular protein
- lipid bilayer
- interior of cell
- cholesterol (stabilizes structure)
- intrinsic proteins pass through membrane
- channel-forming proteins
- extrinsic proteins in one half of lipid bilayer or attached to surface

6·6 Functions of the plasma membrane ■ ■ ■ ■ ■ ■ ■ ■ ■

The main function of a cell's plasma membrane is to act as a **selective barrier** between the inside of the cell and the external fluid. In other words, the membrane regulates the transfer of substances into and out of the cell. This process of transfer is necessary to keep the cell supplied with raw materials and to remove waste products; it can involve many different mechanisms varying from simple **diffusion** to sophisticated methods of **active transport**.

Diffusion is the normal method whereby cells gain and lose small molecules such as dissolved oxygen, carbon dioxide, and water molecules. Small ions, like sodium, potassium, and chloride ions, can also cross the membrane. Diffusion relies entirely on differences in concentration and the substances involved always move down a 'concentration gradient', that is, from high to low concentration. This net movement occurs simply because atoms and molecules are in a state of continuous random motion, and leads to an equilibrium, usually when equal concentrations are established. Diffusion can occur directly through the lipid bilayer, but, more importantly, it appears that some of the proteins present in the membrane act as 'carriers' or 'channels' through which movement takes place. Some channels are permanently open but others can be opened and closed, as in the case of those which regulate the movement of ions across nerve cell membranes (Unit 22). The rate of diffusion is directly related to surface area and, in a small structure such as a cell, there is more than enough surface membrane

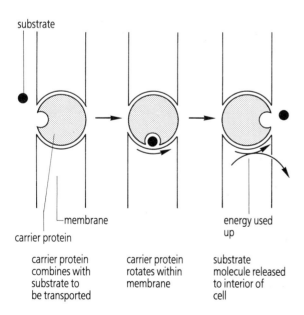

labels:
- substrate
- membrane
- carrier protein
- energy used up
- carrier protein combines with substrate to be transported
- carrier protein rotates within membrane
- substrate molecule released to interior of cell

Fig 6·11 Possible mechanism for active transport

in relation to volume for diffusion to provide an efficient exchange of materials. Diffusion is always a **passive** process, that is, it does not require energy expenditure by the cell.

Cells can gain or lose water by **osmosis**. This process is defined as the net transfer of water molecules across a differentially permeable membrane. The relationship between diffusion and osmosis is explained in detail in Unit 33.

Active transport always requires membrane proteins, but, in this case, energy is used and substances are often transported against their concentration gradients. A possible mechanism is illustrated in Figure 6·11. This shows a carrier protein which rotates within the membrane and thereby transfers its 'substrate' from the outside to the inside of the cell. The same result could be achieved by a change in shape of the membrane protein without actual rotation. Most cells use energy to pump ions against their concentration gradients and to take up organic molecules like glucose and amino acids. Active transport is also involved in many specialised activities such as absorption of digested food substances, excretion, and transmission of nerve impulses.

A third major type of transport across membranes is called **cytosis** and may take the

form either of **endocytosis** (inward transport), or **exocytosis** (outward transport). As shown in Figure 6·12, endocytosis occurs when the plasma membrane becomes 'invaginated', forming a pocket which encloses fluid or solid material. This becomes an intracellular **vesicle** when it is closed off and detached from the plasma membrane. Two types of endocytosis are recognised: when large particles such as bacteria are engulfed by white blood cells, the process is called **phagocytosis**, but when tiny fluid-filled vesicles are formed, it is called **pinocytosis**. Exocytosis is the reverse process and takes place when vesicles fuse with the plasma membrane and expel their contents to the outside. Some of the cells lining the alimentary canal release digestive enzymes in this way.

In addition to their plasma membranes, all cells possess internal membranes which may serve a variety of additional functions. The most important of these are the subdivision of the cell into different 'compartments' and the provision of surfaces for enzyme attachment. The next Unit discusses these internal membranes in the context of a detailed study of cell structure and it will become clear that they are every bit as vital as the plasma membrane itself.

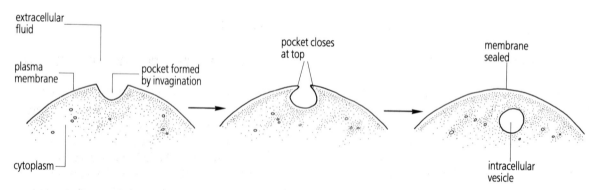

Fig 6·12 Uptake by endocytosis

7 CELLS AND CELL STRUCTURE

Objectives

After studying this Unit you should be able to:–

● List the three basic structures common to all types of cell

● Compare the structures of animal and plant cells as observed with the light microscope

● Draw and label a diagram to show the generalised structure of a eukaryotic cell as observed with the electron microscope

● Describe the intracellular membrane systems of the cell including the rough ER, Golgi body, smooth ER, and the nuclear envelope; comment on the functions of these regions in the living cell

● Outline the structure and function of each of the following cell organelles: lysosomes, nucleus, nucleolus, mitochondria, chloroplasts, microfilaments, microtubules, centrioles, cilia, and flagella

● Summarize your knowledge of cell ultrastructure in the form of a table

7·1 Introduction ■ ■ ■ ■ ■ ■ ■ ■ ■ ■ ■ ■ ■ ■ ■ ■ ■

All living organisms are composed of cells. Many simple organisms, including bacteria and protozoa like *Amoeba*, are single-celled, or unicellular. More complex organisms are multicellular and each individual may consist of anything from a few tens of cells to many millions all working together as an organised whole. The human body, for example, contains at least 10^{12} cells. These cells and virtually all other cells share three basic structural features:

1 All cells are surrounded by a **plasma membrane** which acts as a barrier between the inside and outside of the cell and helps to regulate its internal composition.

2 All cells have a **nucleus** or **nuclear material** which contains the cell's genetic information and ultimately controls all cellular activities.

3 All cells contain a fluid or jelly-like medium, called **cytoplasm**, where the chemical reactions of metabolism take place and where the enzymes, proteins, and other substances needed by the cell, can be manufactured.

These structures and a few additional ones can be easily seen with the light microscope, as shown in Figure 7·1. You will probably be familiar with the structure of cells at this level. Unfortunately, the **resolving power** of the light microscope (that is, its ability to distinguish detail) is rather limited, giving a maximum useful magnification of about 1500 times. Although the light microscope is very useful for studying the ways in which cells are arranged to make tissues and organs, the structures visible within individual cells give little indication of how the cellular machinery is organised to carry out its various tasks. A far more useful instrument in this context is the **electron microscope** (described in the index-glossary). This has some drawbacks – for example, only a handful of staining techniques can be used, and, as electron beams can only travel effectively in a vacuum, only dead tissue can be examined. Nevertheless, with useful magnifications in excess of 250 000 times, a comprehensive and convincing picture of detailed cell structure, called **ultrastructure** has been built up. The remainder of this Unit presents a composite of this picture for the case of a 'typical' **eukaryotic** cell. Such cells have a distinct nucleus surrounded by a pair of membranes which form the **nuclear envelope**. (As explained in Unit 45, **prokaryotic** cells, which include bacteria and blue-green algae, have nuclear material, but do not have a true nucleus bounded by membranes.)

ANIMAL CELLS	PLANT CELLS

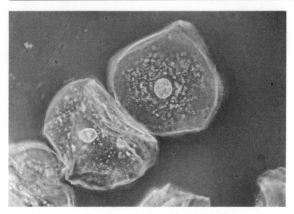

human cheek cells (×1500)

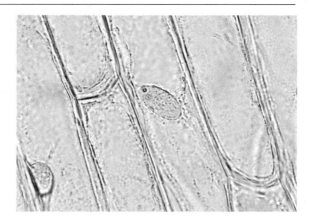

onion membrane cells (×70)

general characteristics of animal cells	*general characteristics of plant cells*
1 possess cell membrane, nucleus and cytoplasm	1 possess cell membrane, nucleus and cytoplasm
2 heterotrophic nutrition	2 autotrophic nutrition
3 smaller (20 μm diameter)	3 larger (50 μm diameter)
4 irregular shape	4 regular shape
5 often able to move about (motile)	5 rarely motile
6 no chloroplasts	6 chloroplasts usually present
7 no large sap vacuole	7 large sap vacuole in centre of cell
8 food storage in glycogen granules	8 food storage in starch grains
	9 cellulose cell wall

Fig 7·1 Structure of animal and plant cells as seen with the light microscope

7·2 Ultrastructure of eukaryotic cells ■ ■ ■ ■ ■ ■ ■ ■ ■

The important structures of a eukaryotic cell are shown diagrammatically in Figure 7·2; electronmicrographs of plant and animal cells are provided for comparison (Figs 7·3 and 7·4).

The most striking feature of cells seen for the first time at the resolution of the electron microscope is their high degree of structural organisation. The cytoplasm is very far from the featureless 'soup' originally envisaged. Broadly speaking, the structures present can be divided into two groups. The first group includes **membrane systems**, comprising the **nuclear envelope, endoplasmic reticulum, Golgi body**, and various structures derived from them. The remainder of the cytoplasm, not enclosed by membranes, is called the **cytoplasmic matrix** and contains a number of **organelles**, among which **ribosomes, mitochondria**, and **chloroplasts** are particularly important. These groups of structures will now be considered.

Membrane systems

All the membranes within the cell share the same basic structure as the plasma membrane and, in the evolutionary sense, appear to have been derived from it. This close similarity gave rise to the idea that all these membranes (sometimes collectively called **'unit' membranes**) were in fact directly joined to each other, making all the spaces inside them continuous with the extracellular environment. This view is not now generally accepted, but the different parts of the membrane system are nevertheless known to be structurally and functionally related. For the purposes of description, it is best to consider the different parts separately but you should not forget that they contribute to an organized overall whole.

1 Endoplasmic reticulum. The main intracellular component of the cell's membrane system is the **endoplasmic reticulum (ER)**. This consists of a series of flattened cavities, called **cisternae**, which are lined with membranes and often interconnected with one another. The ER is most highly developed in secretory cells and this provides a clue to one of its main functions, namely that of manufacturing proteins for export from the cell. Associated with this function is the fact that most of the endoplasmic reticulum, called **rough ER**, is encrusted with small particles called **ribosomes**. These particles are bound to the outer surfaces of the membranes of the ER and are therefore exposed to the cytoplasmic matrix. They are responsible for making proteins. Each ribosome

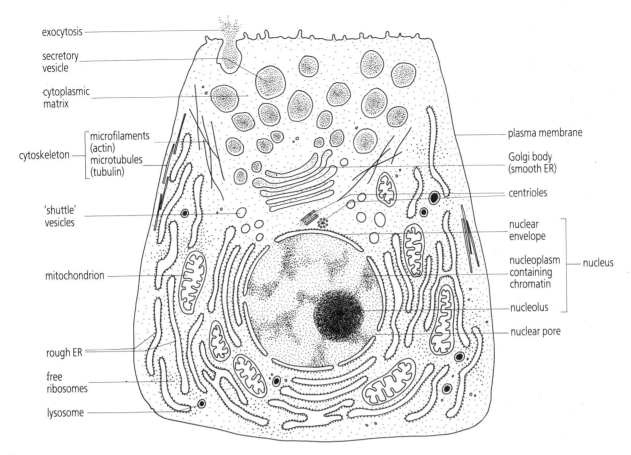

Fig 7·2 Ultrastructure of a eukaryotic cell

Labels (Fig 7·2):
exocytosis
secretory vesicle
cytoplasmic matrix
cytoskeleton — microfilaments (actin), microtubules (tubulin)
'shuttle' vesicles
mitochondrion
rough ER
free ribosomes
lysosome
plasma membrane
Golgi body (smooth ER)
centrioles
nucleus — nuclear envelope, nucleoplasm containing chromatin, nucleolus, nuclear pore

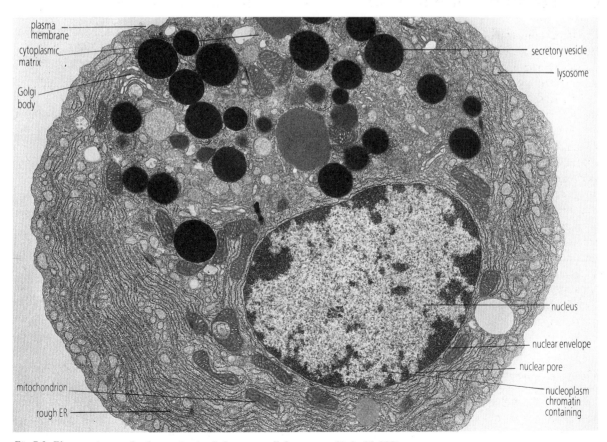

Fig 7·3 Electronmicrograph of an animal cell (secretory cell from stomach) (×10 000)

Labels (Fig 7·3):
plasma membrane
cytoplasmic matrix
Golgi body
mitochondrion
rough ER
secretory vesicle
lysosome
nucleus
nuclear envelope
nuclear pore
nucleoplasm containing chromatin

39

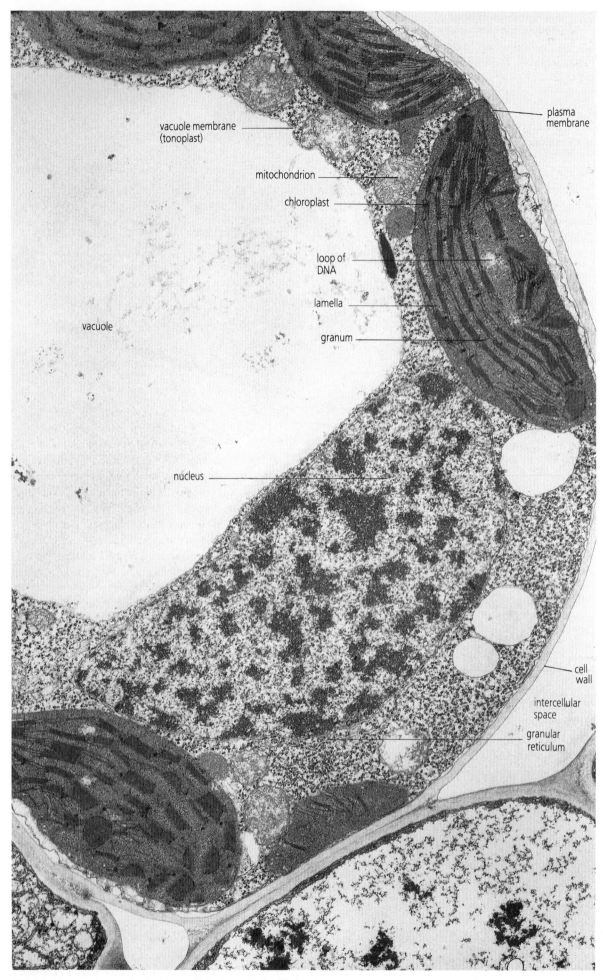

plasma membrane

vacuole membrane (tonoplast)

mitochondrion

chloroplast

loop of DNA

lamella

granum

vacuole

nucleus

cell wall

intercellular space

granular reticulum

Fig 7·4 Electronmicrograph of a plant cell (leaf mesophyll cell) (×14 800)

makes just one molecule of protein at a time, and, as this is progressively assembled from amino acids, it passes *through* the membrane and into the cisternal space. In this way, the proteins manufactured by these ribosomes accumulate within the cavities of the rough ER.

2 Golgi body. If secretory cells are supplied with radioactive amino acids, the proteins inside the rough ER become radioactively 'labelled' and their progress through the rest of the cell can then be followed. Within minutes of their synthesis, proteins for export are found in the second major division of the cell's intracellular membrane system, namely the **Golgi body**. This comprises a stack of flattened disc-shaped cisternae lined with **smooth ER**, that is, membranes which lack the attached ribosomes of rough ER. Almost all the proteins present in cell secretions are **glycoproteins**, and the main function of the Golgi body is to add a carbohydrate prosthetic group to protein components manufactured by the rough ER, thereby forming complete molecules of glycoprotein.

As illustrated in Figure 7·5, **'shuttle' vesicles** containing the protein component are budded off from the rough ER and fuse with the cisternae at

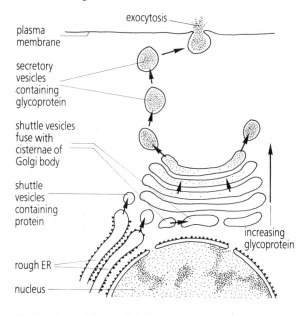

Fig 7·5 Rough ER and Golgi body – export of glycoprotein

the bottom of the Golgi body. Protein molecules are transferred between the individual cisternae, possibly by means of small vesicles, and are progressively converted into glycoprotein as they move up through the stack. At the top of the stack, **secretory vesicles** are budded off and pass through the cytoplasmic matrix, finally fusing with the plasma membrane and discharging their contents to the exterior of the cell (exocytosis).

3 Lysosomes. The Golgi body is involved in the manufacture of additional substances which include **polysaccharides**, and some hormones, notably **insulin** and **gastrin**. It is also the source of small membrane-bound structures called

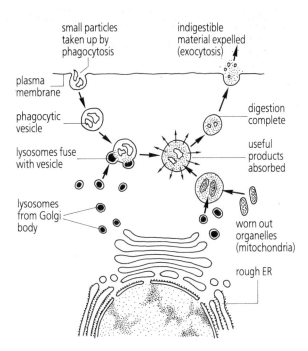

Fig 7·6 Action of lysosomes – intracellular digestion

lysosomes. These arise in exactly the same way as secretory vesicles but are usually retained in the cell cytoplasm. Lysosomes contain a variety of powerful enzymes and carry out **intracellular digestion**. As outlined in Figure 7·6, they break down food materials taken in by phagocytosis, and attack worn out cell organelles. The useful substances which result from digestion, such as amino acids and monosaccharides, are absorbed into the cytoplasm, while any solid material leaves the cell when the remnants of the lysosomal vesicle fuse with the plasma membrane.

Lysosomes are also involved in the rapid breakdown of cellular materials after cell death. This process is called **autolysis** and is advantageous in as much as it clears the way for new growth. It occurs, for example, during insect metamorphosis, or when a tadpole reabsorbs its tail. Lysosome malfunction is thought to be a contributing factor in certain types of cancer.

4 Smooth ER. The Golgi body is a specialised region of the smooth ER with a definite structure. Most cells contain additional membrane regions extending from the rough ER but lacking its attached ribosomes. These membranes are also classified as smooth ER and are concerned with the synthesis of various lipids including steroids. Such regions are extensively developed in cells like those of the adrenal cortex, which produce steroid hormones.

5 Nuclear envelope. The nuclear envelope of eukaryotic cells provides a barrier which separates the cytoplasmic matrix from the fluid interior, or **nucleoplasm**, of the nucleus. Although the membranes of the nuclear envelope are penetrated by **nuclear pores** (see Fig. 7·7), these appear to be plugged with proteins, suggesting that the constant traffic of substances between the nucleus and cytoplasm is actively regulated.

The cavity between the inner and outer membranes of the envelope is continuous with the cisternae of the rough ER and the two regions have overlapping functions. The cytoplasmic surface of the nuclear envelope is often encrusted

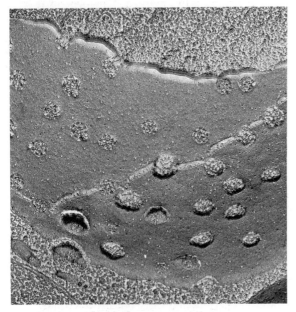

Fig 7·7 Freeze-etched replica showing nuclear pores

with ribosomes and its internal cavity can be shown to contain the same proteins as the endoplasmic reticulum.

The many processes described above are only possible because the membranes of the cell are in a continual state of activity and change. There is a constant need for new membranes to be produced and for old ones to be broken down. The main site of synthesis is the rough ER, and materials from here are incorporated into other membranes including the Golgi body and the plasma membrane itself. Material from these other membranes is broken down and reused in a cyclical way, so that the overall structure and division of the cell into its different compartments is maintained.

Organelles

The term **organelle** is used to describe any self-contained structure found inside living cells. Most organelles are enclosed by their own membranes and these share the same basic features as the plasma membrane. Examples of organelles include the **nucleus, mitochondria, chloroplasts,** and membrane-derived structures such as **lysosomes** and **secretory vesicles.**

1 Nucleus. The importance of the nucleus in directing the activities of the cell can be illustrated by a simple experiment in which the unicellular organism *Amoeba* is cut into two pieces, one with,

and one without, the nucleus (Fig. 7·8). Both parts of the organism round up and their plasma membranes seal together to enclose the cytoplasm. The part with the nucleus grows and develops normally and eventually reproduces by **binary fission**, that is, it divides to give two genetically identical 'daughter' cells. The part without the nucleus is able to move and may take in food, but it does not continue to produce the enzymes needed for digestion and cannot grow or reproduce: it dies when its stored food reserves have been used up.

Experiments of this kind show that the nucleus has two main functions – it controls the chemical activities of the cell, and it carries the hereditary information transmitted to daughter cells when the cell divides. The nuclear structures ultimately responsible for these functions are the **chromosomes**, each of which comprises a series of hereditary 'factors', or **genes**. Chemically the genes consist of a substance called **deoxyribose nucleic acid** or **deoxyribonucleic acid (DNA)**. This is associated with specialised proteins called **histones** and, together, the DNA and the histones form a material called **chromatin**. The chromatin is characteristically acidic in nature, so that, when tissue sections are prepared for light microscopy, the cell nuclei can be selectively stained using basic stains such as **methylene blue**.

2 Nucleolus. Within the nucleus, one or more dense spherical nucleoli are usually present. These structures are composed of a second type of nucleic acid, called **ribose nucleic acid** or **ribonucleic acid (RNA)**, along with some DNA and proteins. Their main function is the **manufacture of ribosomes**. Each nucleolus produces the subunits needed to make ribosomes and these are then passed to the cytoplasm for final assembly.

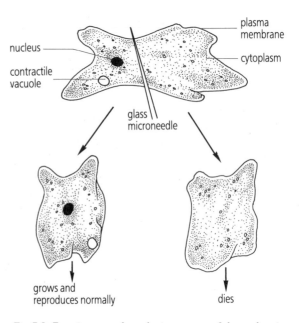

Fig 7·8 Experiment to show the importance of the nucleus in the single-celled organism Amoeba

42

The nucleoli do not have their own membranes and they break down and disappear when the cell divides, only being reformed when the daughter cells produced by division have separated.

3 Mitochondria. Mitochondria are small round or sausage-shaped organelles 2–5 μm long. They are found in all eukaryotic cells and are of vital importance because, through the chemical process of **respiration**, they provide the energy which powers the cell's activities. More specifically, the mitochondria make **adenosine triphosphate (ATP)**. This substance provides energy in a useful form and it is the universal energy currency of all living organisms. Using an enzyme called **ATP-ase**, cells break down ATP to provide energy wherever and whenever it is needed.

The structure of an individual mitochondrion is illustrated in Figure 7·9. As you can see, the whole of this structure is surrounded by a pair of membranes. The outer membrane is a smooth surface layer enclosing the mitochondrion, while the inner one is folded to form numerous plate-like extensions reaching into the interior. These are called **cristae** (singular **crista**), and they provide the mitochondrion with a large internal surface where some of the enzymes involved in respiration are located. High power electronmicrographs reveal that the membranes of each crista are densely studded with **stalked particles** made of protein. As will be discussed in Unit 10, these particles are believed to contain the

enzymes for making ATP in a precise arrangement which helps them to function correctly.

The number of mitochondria present in each cell varies from 50–1000. Particularly active cells, such as liver and muscle cells, have the largest numbers and, as in the case of heart muscle, the interior of the individual mitochondria contains more cristae (and therefore more enzymes) than in less active tissues. Inside the cell, mitochondria are located near to the sites of maximum energy utilization, for example, in liver cells, the mitochondria are interspersed with the rough ER, where they supply the energy needed for protein synthesis.

4 Chloroplasts. Chloroplasts are large cell organelles 4–10 μm in diameter. They contain the green pigment **chlorophyll** and are found inside the cells of green plants. The function of chloroplasts is to trap light energy for **photosynthesis** – through this process, they help the plant to make carbohydrate food molecules from simple raw materials like carbon dioxide and water.

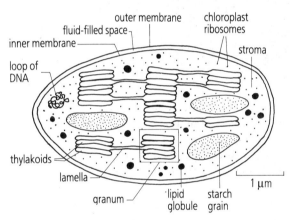

Fig 7·10 Structure of a chloroplast

The basic structure of a chloroplast is illustrated in Figure 7·10. Like mitochondria, chloroplasts are surrounded by paired membranes, but, in this case, both membranes are smooth. Additional membranes inside the chloroplast give a well organised internal structure consisting of **grana** and **lamellae**. Each **granum** is made up of a series of disc-shaped cavities, called **thylakoids**, stacked together like a pile of coins. The lamellae form an interconnecting network between the grana. **Starch grains** and other granules are found in the fluid interior, or **stroma**, of the chloroplast.

Mitochondria and chloroplasts share a number of features which have fascinating implications for the evolution of eukaryotic organisms. For example, both types of organelle contain their own DNA and ribosomes and can direct some of their own protein synthesis. Furthermore, they can divide and reproduce independently of the rest of the cell. As explained in Unit 45, accumulated evidence of this sort strongly suggests that mitochondria and chloroplasts are the intracellular

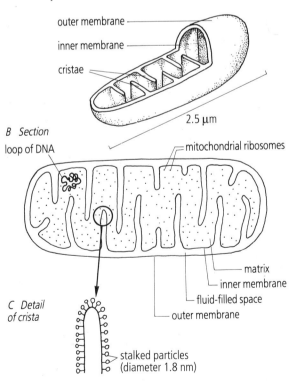

Fig 7·9 Structure of a mitochondrion

descendants of prokaryotic cells which were incorporated into primitive eukaryotic cells, presumably by phagocytosis. This hypothesis provides a simple explanation for the paired membranes of these organelles: the outer membrane originates from the eukaryotic cell, while the inner one corresponds to the plasma membrane of the ingested microorganism.

5 Microfilaments and microtubules.

Microfilaments are long thin threads 4–6 nm in diameter composed of a protein called **actin**. Their occurrence is associated with cell movements and they are involved in cell division and endocytosis as well as in the migratory movements of whole cells such as white blood cells.

Microtubules are long hollow tubes typically about 25 nm in diameter. They are composed of **tubulin** protein and can be rapidly built up or broken down when needed. In contrast to microfilaments, microtubules appear to be rigid and act as supporting structures within the cell, forming a kind of intracellular skeleton, or **cytoskeleton**. Microtubules are sometimes involved in transporting substances from place to place inside the cell and can also be important as subunits for other cell organelles.

6 Centrioles, cilia, and flagella. These all have a structure involving microtubules.

Animal cells contain a pair of **centrioles** which usually lie at right angles to each other adjacent to the nuclear envelope. As shown in Figure 7·11, each centriole consists of nine groups of microtubules in a circular arrangement. Only the first microtubule in each group forms a complete tube, while the remaining two are incomplete U-shaped additions. During cell division (described in Unit 36), the centrioles divide and migrate to opposite poles of the cell where they act as a focus for the formation of the **spindle**. The fibres of the spindle play an essential part in division, pulling different groups of chromosomes apart to form the nuclei of the daughter cells.

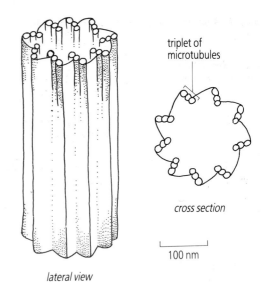

triplet of microtubules

cross section

100 nm

lateral view

Fig 7·11 Structure of centrioles

Centrioles do not occur in most plants, although cell division depends on spindle formation in a similar way.

Cilia and **flagella** are hair-like extensions of the cell which have the same basic structure as centrioles. Both cilia and flagella are involved in cell locomotion and transport. Cilia tend to be shorter than flagella and act together in functional groups, while each flagellum is larger and usually acts individually. Cilia are found in large numbers lining tubes and ducts in the bodies of many animals, where their beating action helps to maintain a flow of fluid. The detailed mechanisms underlying movement of cilia and flagella will be described elsewhere (Unit 46).

The structures described above do not represent an exhaustive list of those present: most cells contain a further assortment of granules, vacuoles, and minor organelles. Table 7·1 gives an overall summary of the structures found inside cells and includes some of these additional components.

Table 7·1 Summary table of cell ultrastructure

STRUCTURE	DESCRIPTION	FUNCTIONS
MEMBRANE SYSTEMS		
plasma membrane	'unit' membrane with fluid mosaic structure 10 nm thick	selective barrier, retains cell contents
rough ER	intracellular membrane system encrusted with ribosomes	synthesis of proteins for export, intracellular transport
smooth ER	intracellular membrane system without ribosomes	synthesis of lipids including steroids
Golgi body	specialised smooth ER forming a stack of disc-shaped cavities (cisternae)	synthesis of glycoproteins, polysaccharides and hormones, production of lysosomes
nuclear envelope	paired membranes surrounding nucleus	regulates exchange betweeen nucleus and cytoplasm, some protein synthesis
ORGANELLES		
ribosomes	small particles with complex structure 25 nm diameter	synthesis of proteins
lysosomes	spherical vesicles containing enzymes 500 nm diameter	intracellular digestion
nucleus	major cell organelle containing chromatin (DNA and histones) 20 μm diameter	regulates cell activities, carries hereditary information
nucleolus	specialised region of nucleus, not surrounded by membranes 2 μm diameter	synthesis of RNA and ribosomes
mitochondrion	cell organelle with inner and outer membranes carrying enzymes for respiration 2–5 μm diameter	production of ATP
chloroplast	cell organelle containing chlorophyll and enzymes for photosynthesis 4–10 μm diameter	production of carbohydrates from simple raw materials (CO_2 and H_2O)
microfilaments	solid protein threads (actin) 5 nm diameter	cell movements
microtubules	protein tubes (tubulin) 25 nm diameter	intracellular support 'cytoskeleton'
centrioles	rod-like structures containing microtubules 200 nm diameter	cell division in animal cells
cilia and flagella (Unit 46)	fine hairs projecting from cell surface 9+2 arrangement of microtubules	cell locomotion, transport of extracellular materials
cellulose cell wall (Unit 32)	layered structure of cellulose microfibrils found in plants 5 μm thick	support, protection
vacuoles	variable	storage of food and waste materials
granules	variable	storage: starch grains in plants, glycogen in animals

8 NUCLEOTIDES AND NUCLEIC ACIDS □

Objectives

After studying this Unit you should be able to:–

- Draw and label a diagram to illustrate the general structure of a nucleotide, and comment on the importance of each of its three parts

- Describe the structure of ATP and explain its role as an energy source within the cell

- Describe the structure of NAD and explain the function of hydrogen carriers as coenzymes in the cell

- Describe how chains of nucleotides are built up using sugar-phosphate linkages

- Explain the significance of base-pairing and list the base pairs formed in DNA and RNA

- Describe the structure of DNA and explain how replication takes place

- Describe a nucleosome and discuss the structural relationships between DNA and chromosomes

- Outline the structures of mRNA, tRNA, and rRNA

8·1 Introduction ■ ■ ■ ■ ■ ■ ■ ■ ■ ■ ■ ■ ■ ■ ■ ■

Carbohydrates, proteins, and lipids form the essential structure and working parts of most cells, but do not ultimately control their operation. This function is carried out by a fourth main group of biological molecules, called **nucleic acids**. As their name suggests, these substances were first discovered and identified in the nucleus. It is now known that they occur in the cytoplasm as well, but the name remains an appropriate one, not least because of the central role played by nucleic acids in regulating cell activity.

This Unit describes the chemical nature of the subunits, called **nucleotides**, from which nucleic acids are built up and outlines the importance of certain closely related substances, including ATP. Its main purpose, however, is to explain the structures of **DNA** (**deoxyribose nucleic acid**, or **deoxyribonucleic acid**), and **RNA** (**ribose nucleic acid**, or **ribonucleic acid**). This basic knowledge is a necessary preliminary for an understanding of protein synthesis, which is the single most important activity carried out by the cell, and is responsible for producing all the enzymes needed to control metabolism.

8·2 Nucleotides ■ ■ ■ ■ ■ ■ ■ ■ ■ ■ ■ ■ ■ ■ ■ ■

Nucleotides are the subunits for all nucleic acid substances. As you can see from Figure 8·1*A*, each nucleotide consists of three smaller parts linked together. The first of these parts is a **phosphate group**, represented by $-\text{P}$ (Fig. 8·1*B*). This is an important feature because the phosphate group is chemically reactive and allows new groups to be added by condensation reactions. More complex structures can be built up, and the nucleotides can be linked together to make a chain.

The second component of each nucleotide is a molecule of a **5-carbon sugar** (pentose sugar). In

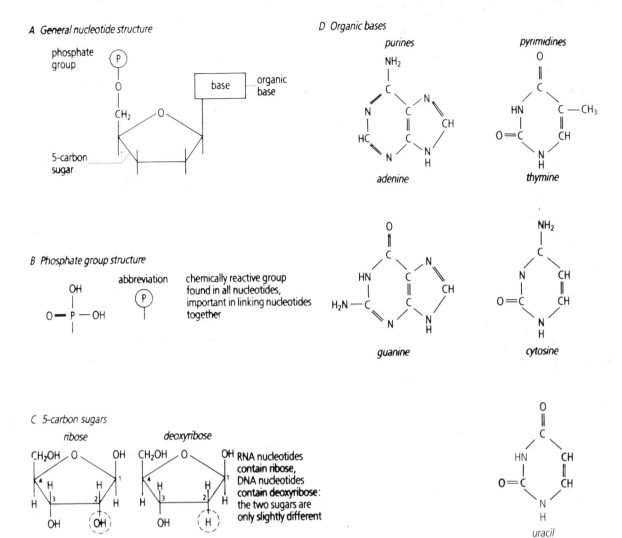

A *General nucleotide structure*

phosphate group (P)

base — organic base

CH₂ — O

5-carbon sugar

B *Phosphate group structure*

OH

O = P — OH

abbreviation (P)

chemically reactive group found in all nucleotides, important in linking nucleotides together

C *5-carbon sugars*

ribose   deoxyribose

RNA nucleotides contain ribose, DNA nucleotides contain deoxyribose: the two sugars are only slightly different

D *Organic bases*

purines   pyrimidines

adenine   thymine

guanine   cytosine

uracil

Fig 8·1 *Nucleotide building blocks*

RNA molecules, the sugar component is always a molecule of **ribose**, while in DNA, it is a molecule of the closely related sugar **deoxyribose** (Fig. 8·1*C*). These sugars are essential because, like the phosphate group, they are involved in linking different nucleotides together.

The third part of each nucleotide, called an **organic base**, gives each nucleotide its particular character. The organic base is linked to the sugar ring via the carbon atom in position 1 and corresponds roughly to the variable −R group of an amino acid. However, while there are 20 common −R groups giving 20 different amino acids, there are only five different organic bases, namely **adenine (A)**, **thymine (T)**, **guanine (G)**, **cytosine (C)**, and **uracil (U)**. Four of these, adenine, thymine, guanine, and cytosine occur in DNA, while a slightly different group of four, comprising adenine, guanine, cytosine, and uracil, occur in RNA.

The organic bases belong to two different chemical families, as outlined in Figure 8·1*D*. The **purine** bases, adenine and guanine, have a double-ringed structure, while the **pyrimidine** bases, including thymine, cytosine, and uracil, have a single-ringed structure. Both types of bases have rings containing nitrogen atoms as well as carbon atoms and this makes them quite different from the compounds previously discussed, although they can be synthesized starting from amino acids.

The combination of a 5-carbon sugar and an organic base with no attached phosphate group is sometimes called a **nucleoside**. The compound formed from ribose and adenine is called **adenosine** and the derived nucleotide is **adenosine monophosphate (AMP)**. Cells continually produce nucleotides and these form a 'pool' from which nucleotides can be used up as required for manufacturing DNA, RNA, or a variety of other substances.

8·3 Simple nucleotide substances – ATP and NAD ■ ■ ■ ■ ■

Adenosine triphosphate

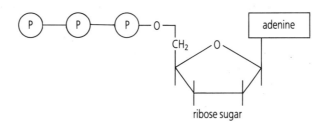

Fig 8·2 Adenosine triphosphate

Nucleotide substances have many vital functions in the cell in addition to their role as subunits for nucleic acids. **Adenosine triphosphate (ATP)** provides an important example.

As illustrated in Figure 8·2, ATP consists of a molecule of adenosine linked to a chain of three phosphate groups. ATP is an effective **phosphorylating agent**. Under suitable conditions, the terminal phosphate group can be transferred to a molecule of a different substance. This increases the reactivity of the second substance and allows useful work to be done. The overall reaction involves the hydrolysis of ATP, as summarized in the following equation:

$$\text{ATP} + \text{H}_2\text{O} \xrightarrow{\text{hydrolysis}} \text{ADP} + \textcircled{P} + \text{energy}$$

| adenosine triphosphate | | adenosine diphosphate | inorganic phosphate | (31 kJ mol⁻¹) |

Note that energy is available as a result of the *transfer* of the phosphate group. The transfer of the second phosphate also yields 31 kJ per mole. On the other hand, hydrolysis of the covalent bond linking the first phosphate to the ribose sugar provides a maximum of 12 kJ per mole.

Within the cell, the breakdown of ATP is **coupled** to other reactions which require energy. As a result, much of the energy released is immediately transferred into new chemical bonds, with only the minimum amount being allowed to escape as heat. Virtually all the complex biological compounds found in cells are built up using energy obtained from ATP.

In order to renew their supplies of ATP, cells carry out the chemical process of **respiration** (described in detail in Unit 10). In the laboratory, energy can be released from a carbohydrate such as glucose simply by burning it in air or oxygen. However, in the living cell, the breakdown of glucose is a stepwise process during which the glucose molecule is progressively dismantled. This gradual breakdown requires the presence of

nucleotide substances called **hydrogen carriers**. These include **NAD** (nicotinamide adenine dinucleotide), **NADP** (Nicotinamide adenine dinucleotide phosphate), and **FAD** (flavin adenine dinucleotide). The structure of NAD is illustrated in Figure 8·3 and comprises two nucleotides linked via their phosphate groups. The nicotinamide part of the structure can combine reversibly with atoms of hydrogen, allowing the hydrogen to be transferred from one compound to another within the cell. Because of their ability to combine with hydrogen, NAD, NADP, FAD, and related substances function as **coenzymes** for enzymes which are involved in removing hydrogen atoms from their substrates.

The dietary factors called **vitamins** are often needed as starting materials for making coenzymes. For example, **niacin** (vitamin B) is needed for making NAD and NADP.

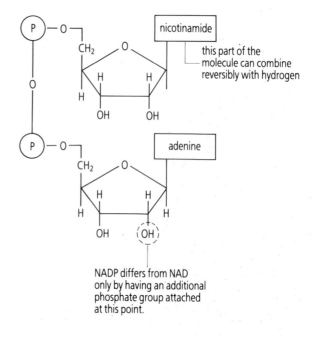

Fig 8·3 Structure of NAD

8·4 Building up nucleic acids ■ ■ ■ ■ ■ ■ ■ ■ ■ ■ ■ ■ ■

Two types of linkage between individual nucleotides are involved in building up molecules of nucleic acid. In the first type of linkage, a connection is formed between the sugar group of one nucleotide and the phosphate group of the next. As illustrated in Figure 8·4, this results in a covalently bonded chain held together by a **sugar-phosphate backbone**. The structures of DNA and RNA both involve chains joined in this way. The sugar-phosphate link can be used to join the nucleotide subunits in any desired order or sequence. This is of crucial importance in determining the shape and properties of the molecule produced.

The second type of linkage between nucleotides is called **base-pairing**. This is illustrated in Figure 8·5 and takes the form of hydrogen-bonding between pairs of organic bases projecting from the sugar-phosphate backbones of parallel strands of nucleic acid. Base-pairing stabilizes the molecule and helps to give it its overall three dimensional configuration.

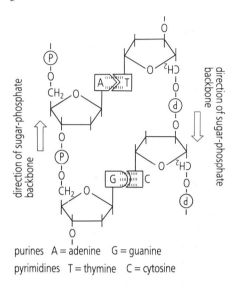

purines   A = adenine   G = guanine
pyrimidines   T = thymine   C = cytosine

Fig 8·5 Base-pairing in nucleic acids

There are two important constraints on the way in which base-pairing can occur. The first of these is that the sugar-phosphate backbones of the two strands must run in opposite directions. This allows the projecting nucleotides to be fitted together. The second constraint is that base-pairing normally occurs only between particular pairs of bases. In DNA the possible pairs are **adenine-thymine (A–T)** and **guanine-cytosine (G–C)**, while in RNA the pairs are **adenine-uracil (A–U)** and **guanine-cytosine (G–C)**. In molecules of DNA, adenine and thymine are linked by two hydrogen bonds, while guanine and cytosine are linked by three hydrogen bonds. Consequently, it is impossible for pairing to take place between A and G or between C and T.

As you can see from Figure 8·5, in each pair one base belongs to the double-ringed purine family, while the other belongs to the single-ringed pyrimidine family. It follows that the overall length of all the base pairs is the same so that the linked strands are held a constant distance apart. The individual hydrogen bonds are weak but, provided the sequences of bases on the two strands are **complementary** (i.e. provided the strands fit together correctly), extremely stable three-dimensional structures can result. The most important of these, called a **double helix**, is formed when the two strands twist together. Regions of double helix occur in both DNA and RNA.

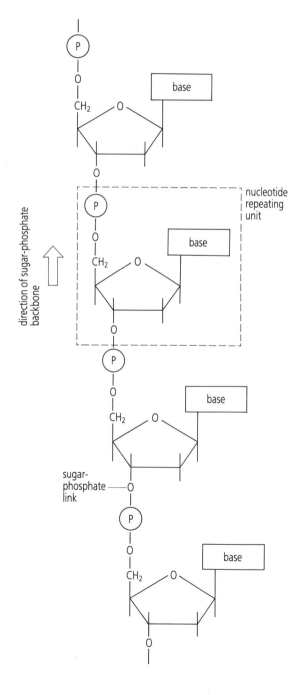

Fig 8·4 Backbone structure of nucleic acid chains

8·5 Structure of DNA ■ ■ ■ ■ ■ ■ ■ ■ ■ ■ ■ ■ ■ ■ ■

Molecules of DNA are built up from nucleotides containing the 5-carbon sugar **deoxyribose** and they are always **double-stranded**. As illustrated in Figure 8·6, the two strands are fully complementary so that the entire molecule, which may contain millions of nucleotides, is held together by base-pairing. At the same time, the molecule is twisted into the double helix configuration. The full elegance and compactness of this structure, called the **Watson-Crick model** of DNA, can be appreciated from Figure 8·7, which shows the positions of all the various atoms in a short length of DNA. This structure was first determined by analysing crystals of DNA using the technique of X-ray crystallography.

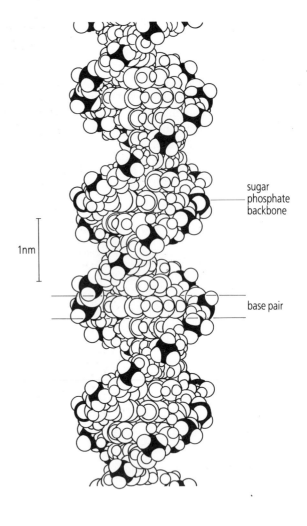

Fig 8·7 Space-filling molecular model of DNA

The significance of the model is that it suggests a mechanism whereby new copies of DNA molecules can be produced. Because the two strands are fully complementary to each other, each can act as a **template** (or 'mould') for making the other. With the aid of an enzyme, called **DNA polymerase**, DNA strands can copy themselves, or **replicate**, as outlined in Figure 8·8. The process illustrated is called **semi-conservative** replication because half of the original parent molecule is retained, or 'conserved', in each of the new strands produced. Replication is essential so that the cell can provide identical copies of all its genetic instructions to be passed on to daughter cells. During a normal cycle of cell growth, the DNA doubles so that, when the chromosomes separate at cell division, each daughter cell receives a genetically identical package. Further details of cell division are given in Unit 36.

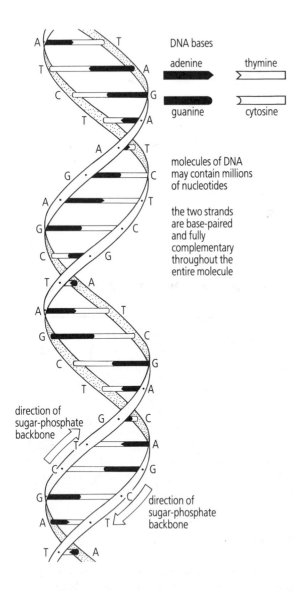

Fig 8·6 Double helix structure of DNA

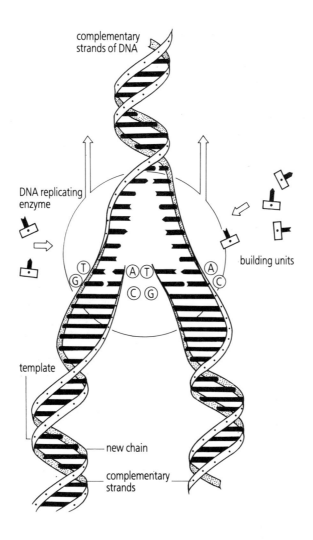

Fig 8·8 Replication of DNA

8·6 DNA and chromosomes ■ ■ ■ ■ ■ ■ ■ ■ ■ ■ ■ ■ ■ ■

The total length of the DNA molecules in a typical mammalian cell has been calculated to be about 2 metres, giving a staggering total of more than 10 *billion* km of DNA in the human body as a whole. Each chromosome in an individual cell is believed to contain just one single long molecule of DNA so that the 46 chromosomes of a human cell correspond to 46 individual molecules of DNA each about 4 cm long.

To prevent the interior of the nucleus from becoming a hopeless tangle, it is necessary for the DNA threads of the individual chromosomes to be coiled in some organized way. This is the function of the specialized **histone** proteins which are associated with DNA to form the **chromatin** of the nucleus. The DNA and the histones are now known to form particles called **nucleosomes**. These have the structure shown in Figure 8·9, and each consists of a length of DNA double helix containing about 200 base pairs wound around a 'bead' of histone protein.

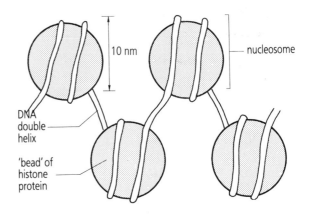

Fig 8·9 Structure of nucleosomes

In an actively growing cell, the chromosomes in the nucleus are mainly present as long chains of nucleosomes. However, when the cell is preparing to divide, the chromosomes coil up much more tightly and become visible by a process called

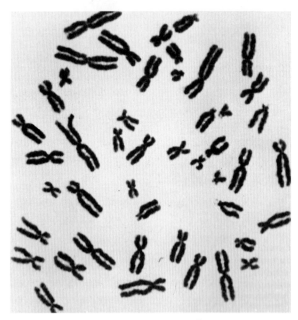

Fig 8·10 *Chromosomes from a human cell*

proteins, but the mechanisms involved are not yet known in detail. In fully condensed chromosomes (Fig. 8·10), the original DNA molecule is looped and coiled to form a structure which is only about 6 μm long and about 0.8 μm in diameter. In other words, the final chromosome is about 5000 times shorter than the DNA it contains. Eukaryotic cells carry out cell division with the chromosomes in this densely packed form.

The chromosomes transmit the hereditary or 'genetic' information to the daughter cells and are said to contain the **'genetic material'** of the cell. DNA fulfils all the fundamental requirements of the genetic material. It is, firstly, an extremely stable substance, so that the coded instructions it contains normally remain intact from one generation to the next. Secondly, DNA can replicate and is therefore able to provide new copies of the genetic instructions for each new cell. The third essential requirement for the genetic material is that it should be able to control the cell's activities. DNA achieves this control primarily by directing the synthesis of proteins and it is aided in this function by several types of RNA. The chemical structure of these different substances is outlined below.

condensation. This is partly controlled by interactions between the histone proteins of individual nucleosomes and partly by non-histone

8·7 Structure of RNA ■ ■ ■ ■ ■ ■ ■ ■ ■ ■ ■ ■ ■ ■ ■

Molecules of RNA are built up from nucleotides containing the 5-carbon sugar **ribose** and are always **single-stranded**, although the strand may be folded back on itself to give an apparently double-stranded structure. All cells contain three important types of RNA:

1 Messenger RNA, or mRNA, is produced in the nucleus from coded instructions in the DNA and then passes into the cytoplasm where it becomes associated with ribosomes and starts the process of protein synthesis. Molecules of mRNA are 75–3000 nucleotides long and are not folded in any special way (Fig. 8·11*A*).

2 Transfer RNA, or tRNA. These are small RNA molecules containing 75–90 nucleotides. All cells have at least 20 different kinds of tRNA molecules, all of which are very similar in shape. As illustrated in Figure 8·11*B*, each molecule is held together by a precise base-pairing arrangement between different parts of the RNA strand. When the molecule is 'flattened out', it has a clover-leaf shape, but in its normal form, each of the base-paired regions is twisted to form a double helix. As a result, the molecule as a whole adopts an approximately L-shaped configuration.

Transfer RNAs play an essential role in protein synthesis – each tRNA combines with a particular amino acid at one end and with a molecule of messenger RNA at the other. In this way, the tRNA molecules of the cell allow the coded

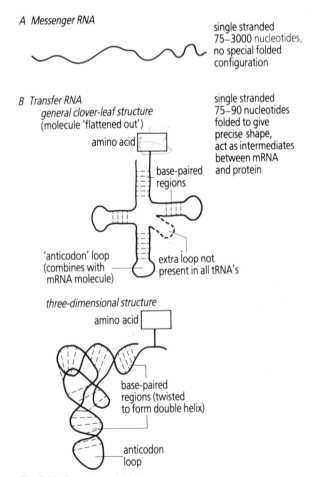

A Messenger RNA

single stranded 75–3000 nucleotides, no special folded configuration

B Transfer RNA
general clover-leaf structure (molecule 'flattened out')

amino acid

base-paired regions

'anticodon' loop (combines with mRNA molecule)

extra loop not present in all tRNA's

single stranded 75–90 nucleotides folded to give precise shape, act as intermediates between mRNA and protein

three-dimensional structure

amino acid

base-paired regions (twisted to form double helix)

anticodon loop

Fig 8·11 *Structure of RNA*

instructions carried on the DNA to be 'translated' into a sequence of amino acids forming a protein chain.

3 Ribosomal RNA, or **rRNA,** is made inside the nucleus within the nucleoli and is a major component of ribosomes. There are several different kinds of rRNA and each of these is essential in order for the ribosomes to function correctly. The precise configuration of rRNA molecules is unknown but, as outlined in Figure 8·11C, they are very large molecules containing thousands of nucleotides. Each molecule consists of a single strand with regions of base-paired double helix projecting on either side.

The importance of these different types of RNA will be discussed further in Unit 9, which considers the mechanisms of protein synthesis in greater detail.

C Ribosomal RNA

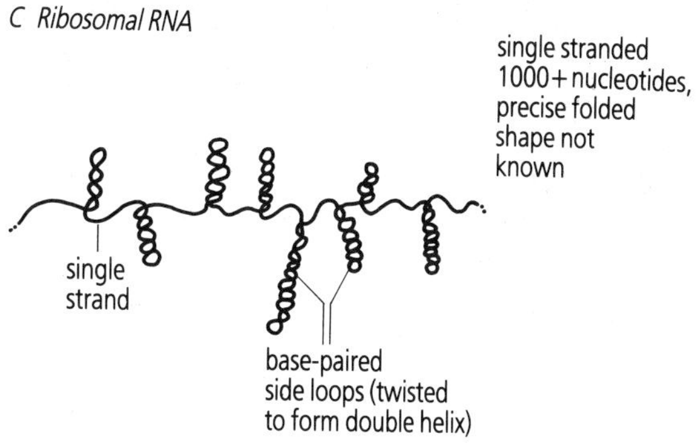

single stranded 1000+ nucleotides, precise folded shape not known

single strand

base-paired side loops (twisted to form double helix)

Fig 8·11 (cont.) Structure of RNA molecules

9 SYNTHESIS OF PROTEINS

Objectives

After studying this Unit you should be able to:–

- Define a gene and describe how genetic instructions are coded in the DNA of a single chromosome

- Describe the process of transcription leading to the formation of messenger RNA

- Explain the term translation and describe the role of ribosomes and transfer RNA in carrying out this process

- Comment on the significance of amino acid activation and polyribosomes

- Explain what is meant by the 'central dogma' of protein synthesis and describe the process of information flow in cells

- Describe the operon system of controlling gene expression in bacterial cells

9·1 DNA and the genetic code ■ ■ ■ ■ ■ ■ ■ ■ ■ ■ ■ ■

As explained in Unit 8, the chemical structure of DNA gives stability and allows DNA molecules to make new copies of themselves by replication. This Unit describes how DNA controls protein synthesis. To help you understand the processes involved, Figure 9·1 gives an outline summary showing the relationships between the two major stages, called **transcription** and **translation**. Transcription occurs in the nucleus and is the copying of coded instructions from DNA to RNA. It produces 'messenger' RNA molecules which pass to the cytoplasm. Translation takes place inside ribosomes where protein molecules are assembled by interpreting the coded message. Transfer RNA, ATP, and a supply of amino acids are needed.

A detailed description of protein synthesis is best begun by considering how molecules of DNA are able to carry information. This is possible using a remarkable molecular code language, or **genetic code**, in which the nucleotides in an individual strand of DNA represent a sequence of 'codewords' following one after another, rather in the manner of computer instructions recorded on magnetic tape.

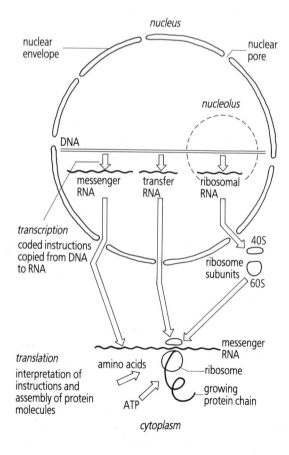

Fig 9·1 Summary of protein synthesis

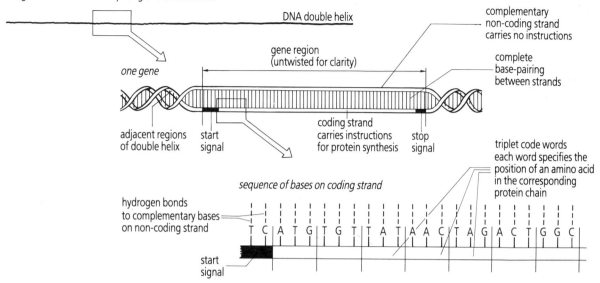

long molecule of DNA comprising one chromosome

DNA double helix

complementary
non-coding strand
carries no instructions

gene region
(untwisted for clarity)

complete
base-pairing
between strands

one gene

adjacent regions
of double helix

start
signal

coding strand
carries instructions
for protein synthesis

stop
signal

triplet code words
each word specifies the
position of an amino acid
in the corresponding
protein chain

sequence of bases on coding strand

hydrogen bonds
to complementary bases
on non-coding strand

T C A T G T G T T A T A A C T A G A C T G G C

start
signal

Fig 9·2 Information coding in DNA

The single long molecule of DNA in each chromosome comprises numerous shorter sections called **genes** each of which contains the instructions for making one protein. Protein molecules can contain any of 20 different amino acids and these are joined together in an exact sequence which determines the properties of the protein as a whole. Consequently, the coded instructions in each gene must specify the overall length of the protein chain and the precise position of amino acids within it. From Figure 9·2 you can see that, within a given gene, these instructions are coded onto only *one* of the two strands of the DNA molecule. The other strand is simply the complementary sequence of nucleotides needed to complete the molecule and stabilize its structure and to allow for replication and repair.

Each gene begins with a special sequence of nucleotides, called a **promoter region**, which signals a 'start', and terminates with a second sequence called a 'stop'. Between these 'start' and 'stop' signals there is a long chain which may contain as many as 3000 nucleotides, depending on the length of the corresponding protein chain. All of these nucleotides are arranged in a definite order so that each of the attached organic bases represents a 'letter' in a molecular code language which leads directly to the synthesis of proteins.

The English language uses 26 letters which may be linked together to make words of any length. Within an accepted system of rules, these words can be used to convey a great deal of information. The 'language' of protein synthesis, usually called the genetic code, is much more restricted. Firstly, there are only four 'letters', corresponding to the four different organic bases found in DNA, namely adenine (A), thymine (T), guanine (G), and cytosine (C). Secondly, all the 'words' are of constant length and contain just three letters, so that the code as a whole contains only **64 triplet words,** or **codons.** Despite this restricted vocabulary, the genetic code provides more than enough words to specify the positions of the 20 different amino acids in a protein chain and leaves some sequences over for use as 'start' and 'stop' signals.

9·2 Transcription ■ ■ ■ ■ ■ ■ ■ ■ ■ ■ ■ ■ ■ ■ ■

With the exception of small quantities of DNA contained in cell organelles, the cell's DNA is entirely confined to the nucleus. Nevertheless, protein synthesis occurs mainly in the cytoplasm. For this to be possible, there must clearly be some way of transferring coded information from the chromosomal DNA across the nuclear membranes and into the cytoplasm where the actual assembly of protein molecules takes place. This function is carried out by a special kind of RNA molecule, appropriately called **messenger RNA (mRNA).**

Messenger RNA is manufactured in the nucleus directly from the 'template' provided by the coding strand of the DNA molecule. This process is called **transcription** because it results in the coded message in the DNA being 'transcribed' (i.e. copied across) into the new molecule of mRNA. Figure 9·3, shows that transcription can occur only with the aid of an enzyme, called **DNA-dependent RNA polymerase,** which acts as a catalyst, linking RNA nucleotides together in a chain.

The first step in transcription occurs when the enzyme 'recognizes' the 'start' sequence in the

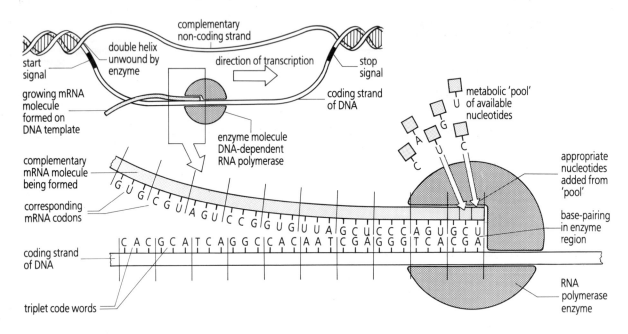

Fig 9·3 Transcription

DNA coding strand at the beginning of the gene and becomes attached to the DNA at this point. This seems to break the hydrogen bonds linking the two strands of the DNA, so that the double helix unwinds in the gene region. The DNA-dependent RNA polymerase molecule now travels along the gene and new nucleotides, complementary to those in the DNA strand are inserted into the growing mRNA. When the enzyme encounters the base thymine (T), adenine (A) is inserted, and, similarly, when cytosine (C) or guanine (G) occur in the DNA, G or C respectively are inserted into the mRNA molecule. Thymine does not occur in RNA and positions in the mRNA strand opposite adenine in the DNA template are occupied instead by uracil (U). These changes do not affect the accuracy of the transcription process – the final mRNA molecule is precisely matched to the DNA template and contains an exactly equivalent set of instructions.

At the end of transcription, the polymerase enzyme recognizes the 'stop' sequence and becomes detached from the gene region. This allows the DNA molecule to rewind into its normal double helix configuration. The completed mRNA molecule is released and can travel out through one of the pores in the membranes of the nuclear envelope. In this way, the instructions needed for protein synthesis are transferred into the cytoplasm.

9·3 Translation ■ ■ ■ ■ ■ ■ ■ ■ ■ ■ ■ ■ ■ ■ ■ ■

In the next stage of protein synthesis, the coded instructions in a molecule of mRNA must be 'translated' into a sequence of amino acids forming a protein chain. This process requires **transfer RNA (tRNA)** molecules of various kinds and can only take place when the mRNA molecule is attached to a **ribosome**.

Ribosomes are small cell organelles about 20 nm in diameter, each containing about 50% **ribosomal RNA (rRNA)** and about 50% protein. They consist of two particles of different sizes, called **40 S** and **60 S** subunits. (These names reflect the relative masses of the particles in Svedberg units (S) – see index-glossary.) Both kinds of subunit are assembled in the nucleolar regions of the nucleus and later transferred out into the cytoplasm.

The two subunits of the ribosome remain separate until a molecule of mRNA is available.

Then they lock on to a special sequence of nucleotides at one end of the mRNA, called a **ribosome binding site**. Translation can now begin.

Within the ribosome there are two **tRNA sites** where a triplet code word, or **codon**, from mRNA can become attached by base-pairing to a molecule of tRNA (refer to Fig. 9·4A). Each tRNA has a special triplet of nucleotides called an **anticodon** at one end for precisely this purpose. At its opposite end, the tRNA is linked to a specific amino acid, which is therefore brought into position at the beginning of the protein chain (Fig. 9·4B). When two amino acids are brought together in this way, they become linked through the action of an internal enzyme (Fig. 9·4C). At the same time, the first tRNA is disconnected from its amino acid and leaves the ribosome which moves one 'notch' along the mRNA strand to bring the

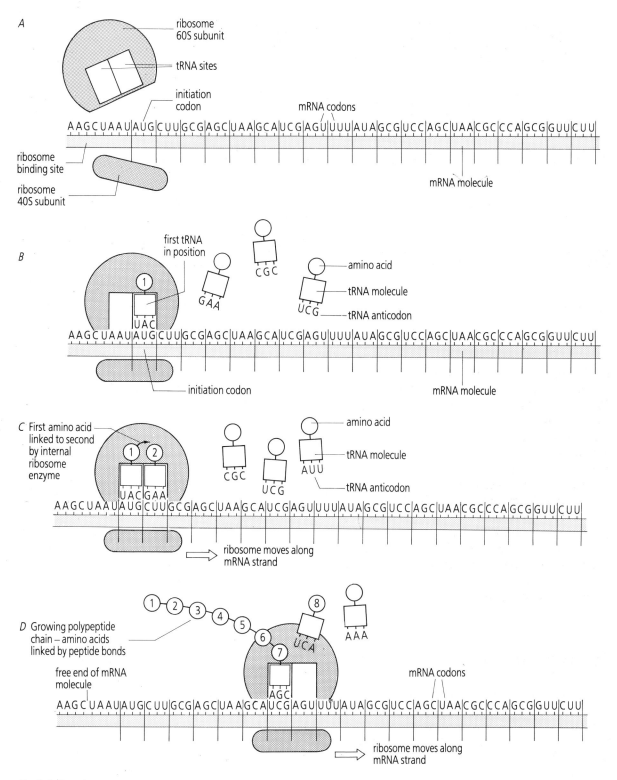

ribosome
60S subunit

tRNA sites

initiation
codon

mRNA codons

AAG|CUA|AU|AUG|CUU|GCG|AGC|UAA|GCA|UCG|AGU|UUU|AUA|GCG|UCC|AGC|UAA|CGC|CCA|GCG|GUU|CUU

ribosome
binding site

ribosome
40S subunit

mRNA molecule

B

first tRNA
in position

CGC

amino acid

tRNA molecule

GAA

UCG

tRNA anticodon

UAC

AAG|CUA|AU|AUG|CUU|GCG|AGC|UAA|GCA|UCG|AGU|UUU|AUA|GCG|UCC|AGC|UAA|CGC|CCA|GCG|GUU|CUU

initiation codon

mRNA molecule

C First amino acid
linked to second
by internal
ribosome
enzyme

amino acid

tRNA molecule

1 2

CGC

UCG

AUU

tRNA anticodon

UAC GAA

AAG|CUA|AU|AUG|CUU|GCG|AGC|UAA|GCA|UCG|AGU|UUU|AUA|GCG|UCC|AGC|UAA|CGC|CCA|GCG|GUU|CUU

ribosome moves along
mRNA strand

D Growing polypeptide
chain – amino acids
linked by peptide bonds

free end of mRNA
molecule

1 2 3 4 5 6 7 8

UCA

AAA

AGC

mRNA codons

AAG|CUA|AU|AUG|CUU|GCG|AGC|UAA|GCA|UCG|AGU|UUU|AUA|GCG|UCC|AGC|UAA|CGC|CCA|GCG|GUU|CUU

ribosome moves along
mRNA strand

Fig 9·4 Translation

next codon into position. As this process continues, the ribosome travels along the mRNA strand adding more amino acids to the growing protein chain (Fig. 9·4*D*).

Accurate interpretation of the mRNA message is only possible because, without exception, each tRNA molecule has a particular anticodon specific for only one amino acid. Consequently, a given mRNA codon is always translated in the same way. The 'meanings' of the 64 different mRNA codons appear to be the same for almost all living organisms. They are listed in Table 9·1. For reasons which are not fully understood, the first codon to be translated from an mRNA molecule is always the sequence AUG. This corresponds to the amino acid **methionine** and is called the **initiation codon**. Methionine therefore occurs at the beginning of every protein chain, although, in many cases, it is removed by enzyme action before the protein is folded into its final shape.

Table 9.1 The genetic code

AMINO ACID	mRNA codons used to specify the position of this amino acid		AMINO ACID	mRNA codons used to specify the position of this amino acid	
alanine	GCA, GCC, GCG, GCU	4	lysine	AAA, AAG	2
arginine	AGA, AGG, CGA, CGC, CGG, CGU	6	methionine	AUG (initiation codon)	1
asparagine	AAC, AAU	2	phenylalanine	UUC, UUU	2
aspartic acid	GAC, GAU	2	proline	CCA, CCC, CCG, CCU	4
cysteine	UGC, UGU	2	serine	AGC, AGU, UCA, UCC, UCG, UCU	6
glutamic acid	GAA, GAG	2	threonine	ACA, ACC, ACG, ACU	4
glutamine	CAA, CAG	2	tryptophan	UGG	1
glycine	GGA, GGC, GGG, GGU	4	tyrosine	UAC, UAU	2
histidine	CAC, CAU	2	valine	GUA, GUC, GUG, GUU	4
isoleucine	AUA, AUC, AUU	3	'stop' codons	UAA, UAG, UGA	3
leucine	CUA, CUC, CUG, CUU, UUA, UUG	6	total number of different mRNA codons –		64

9·4   Amino acid activation ■ ■ ■ ■ ■ ■ ■ ■ ■ ■ ■ ■

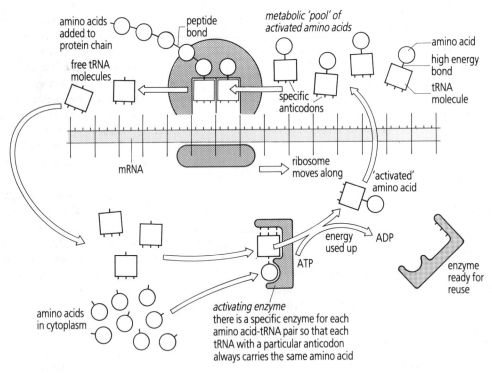

Fig 9·5 Amino acid activation

Protein synthesis directed by mRNA can continue in this way provided supplies of tRNA molecules carrying the appropriate amino acids remain available. The tRNA molecules needed are first synthesised in the nucleus and become linked to amino acids in the cytoplasm. This linking process, called **amino acid activation**, is shown in Figure 9·5. It uses up ATP and requires a specific enzyme for each of the possible tRNA-amino acid pairs. Energy obtained from ATP is transferred into the formation of a high energy bond between each tRNA and its corresponding amino acid. This bond energy is greater than that required for making a peptide bond, so that the formation of peptide bonds within the ribosome becomes a 'downhill' process. tRNA molecules released from the ribosomes are reused in exactly the same way, and, as a result, the cell's supplies of tRNA are economically recycled.

9·5 Polyribosomes ■ ■ ■ ■ ■ ■ ■ ■ ■ ■ ■ ■ ■ ■ ■

Within limits, messenger RNA molecules can be reused because they pass through the ribosomes completely unchanged. The free end of an mRNA molecule frequently becomes associated with a second ribosome before translation is complete and, depending on the length of the mRNA, this can lead to formation of small groups of ribosomes, called **polyribosomes**, or **polysomes**. As illustrated in Figure 9·6, protein chains in different stages of completion emerge from the ribosomes according to how far they have travelled along the mRNA strand. In this way, polyribosomes increase the rate of protein synthesis. In a eukaryotic cell, several hundreds or thousands of protein molecules can be manufactured from a single molecule of mRNA

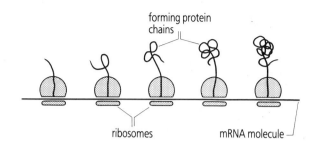

Fig 9·6 Polyribosome structure

before this is degraded. Nevertheless, an individual mRNA molecule can survive for only a few hours, making its lifetime quite short compared to that of other types of RNA.

9·6 Control of protein synthesis ■ ■ ■ ■ ■ ■ ■ ■ ■ ■ ■ ■

The process of protein synthesis described above depends on the transfer of information within the cell. This transfer or information 'flow' can be summarized in the form of a simple rule:

DNA ——→ RNA ——→ proteins

During the early development of molecular biology, this rule provided a useful generalization and became known as the '**central dogma**' of protein synthesis. More recently, it has become clear that information transfer is not exclusively a one-way process and that conditions inside the cell and in its surroundings can in turn influence the cell's DNA. A more modern statement of information flow would be:

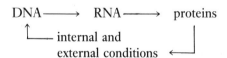

This pattern of flow gives a 'feedback' control of protein synthesis which enables the cell to adapt speedily and efficiently to changing circumstances.

The behaviour of *E. coli* bacterial cells illustrates this principle very well. *E. coli* cells can be grown quite satisfactorily in a 'minimal' culture medium containing only glucose and a few inorganic salts. When such a culture is transferred to a different medium containing the disaccharide sugar lactose, the cells soon begin to produce a new enzyme, called **β-galactosidase**, which breaks the β1,4 glycoside linkage between the subunits of lactose. Experiments of this kind show that β-galactosidase is only produced when it is needed, although the corresponding gene is always present within the bacterial chromosome. The gene for the enzyme is said to be '**induced**' by the presence of the appropriate substrate.

Figure 9·7 illustrates the control system which underlies this behaviour. It shows several genes which together form a functional unit called an **operon**. The first of these genes, called the **repressor gene**, produces molecules of mRNA which code for the synthesis of a **repressor protein**. In the absence of a suitable substrate (in this case lactose) the repressor protein binds to the

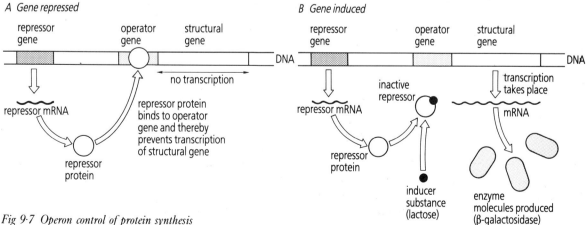

Fig 9·7 Operon control of protein synthesis

cell's DNA in the **operator gene** region. This blocks the 'start' sequence for the adjacent **structural gene**, so that, when the repressor protein is in position, transcription cannot occur.

When lactose becomes available, it combines with the repressor protein, forming an **inducer-repressor complex**. This alters the shape of the repressor protein which then becomes inactive and is no longer capable of attaching to the cell's DNA. Transcription of the structural gene now occurs in the normal way and β-galactosidase is produced. On the other hand, if the amount of lactose available starts to fall, or if the cells are transferred back to a minimal medium, the number of inducer-repressor complexes is reduced, and the structural gene is again repressed.

In eukaryotic organisms, the mechanisms controlling gene expression are more complex than in bacteria. While all the cells in an organism's body contain the same DNA, different groups of genes can be expressed in each. Thus, as development proceeds in different parts of the organism, some genes become permanently 'switched on', while other are permanently repressed. This process leads to cell specialization and to '**differentiation**' of tissues and organs. Although the mechanisms which underly development and differentiation are not yet understood, it is clear that DNA and the genetic information it contains are at the heart of these processes.

10 BIOCHEMISTRY OF RESPIRATION [   ]

Objectives

After studying this Unit you should be able to:–

- Define respiration and write the chemical equations which summarise aerobic respiration, lactic fermentation, and alcoholic fermentation

- Distinguish between the three different types of oxidation reaction and outline the role of hydrogen carriers

- Explain reaction coupling and write the chemical equations which describe the formation of ATP and NAD

- Give a simple account of glycolysis and explain how it is modified during anaerobic respiration

- Give a simple account of Krebs cycle as a source of ATP and NADH

- Discuss how the structure of the mitochondrion is related to the biochemical events of respiration and outline the chemiosmotic theory linking the hydrogen transport pathway with ATP production

- Draw a flow diagram to show the relationships between the three phases of aerobic respiration leading to the formation of ATP

10·1 Basic concepts ■ ■ ■ ■ ■ ■ ■ ■ ■ ■ ■ ■ ■ ■ ■

Respiration is the chemical process of energy release from organic compounds – in particular from glucose. Its essential purpose is to make ATP (adenosine triphosphate) from ADP and inorganic phosphate. As indicated in Unit 8, ATP is the basic energy currency of the cell used to transfer freely available energy from place to place, providing energy for protein synthesis and muscle contraction, or for pumping ions against a concentration gradient.

In most organisms, **aerobic respiration** occurs if oxygen is available, converting glucose to carbon dioxide and water. If oxygen is not available, or if it is in short supply, **anaerobic respiration** takes place, and glucose is only partly broken down, generally to **lactate**, the ionic form of lactic acid, or to **ethanol**. Some organisms such as yeast can survive for long periods without oxygen, but others can tolerate only a temporary shortage. A few organisms, like some of the bacterial species described in Unit 45, are poisoned by oxygen and can survive only in anaerobic conditions.

Before describing the biochemistry of these different types of respiration in detail, three important concepts must be learned, namely **energy-providing reactions, oxidation-reduction**, and **reaction coupling**.

Energy-providing reactions

Energy is more easily released from some chemical bonds than from others and can be transferred when these bonds are broken. A glucose molecule is 'energy-rich' because it contains many C–C, C–H, and C–OH bonds capable of supplying energy. As summarized below, these bonds are broken when glucose is used for aerobic respiration and are replaced by less easily broken C=O and O–H bonds in carbon dioxide and water:

aerobic respiration:

$$C_6H_{12}O_6 + 6O_2 \longrightarrow 6CO_2 + 6H_2O$$

glucose   oxygen   carbon   water
dioxide

$$\Delta G = -2870 \text{ kJ mol}^{-1}$$

The quantity of $\Delta G = -2870$ kJ mol^{-1} given in this equation is called the **free energy change** of the reaction, and is a measure of the energy available for transfer. (In this context, ΔG is determined at pH 7 with the reactants at appropriate biochemical concentrations.) ΔG is given a negative value because the substances present at the end of the reaction have in effect lost energy compared with those present at the beginning.

Energy-releasing reactions of this type are described as **exergonic**. They are always capable of proceeding spontaneously, although, as in the case of glucose, an initial quantity of energy, called **activation energy**, may be needed to set the reaction in motion. The opposite kind of reactions, for which ΔG is positive, are described as **endergonic** or energy-consuming and cannot proceed at all unless energy is supplied continuously from an external source.

The two forms of anaerobic respiration, known as **lactic fermentation** and **alcoholic fermentation**, are exergonic but release much less energy than aerobic respiration, as indicated by their equations:

lactic fermentation:

$$C_6H_{12}O_6 \longrightarrow 2C_3H_5O_3^- + 2H^+$$
glucose   lactate

$$\Delta G = -150 \text{ kJ mol}^{-1}$$

alcoholic fermentation:

$$C_6H_{12}O_6 \longrightarrow 2C_2H_5OH + 2CO_2$$
glucose   ethanol   carbon
                                            dioxide

$$\Delta G = -210 \text{ kJ mol}^{-1}$$

Oxidation-reduction

The chemical process of **oxidation** is defined as the addition of oxygen, the removal of hydrogen, or the loss of electrons. The complementary reaction, called **reduction**, is defined as the addition of hydrogen, the removal of oxygen, or the gain of electrons. The following examples illustrate the meanings of these definitions.

An oxidation reaction takes place when a substance combines directly with oxygen:

$$A + O_2 \longrightarrow AO_2$$

Equally, an oxidation reaction takes place when hydrogen is removed:

$$AH_2 + B \longrightarrow A + BH_2$$

In this case, A has been **oxidized** by transferring hydrogen to a second substance, called B, which thus acts as a **hydrogen carrier**. Oxidation reactions of this type are of crucial importance in respiration because they allow hydrogen atoms to be removed from glucose, at the same time releasing energy in a useful form. The most important hydrogen carrier in the living

cell is **nicotinamide adenine dinucleotide (NAD)**. NAD passes on hydrogen atoms to a system of carrier molecules located in the mitochondria. As hydrogen is transferred from one carrier to the next, energy is released and can be stored in the form of ATP. At the end of the chain, hydrogen atoms are combined with oxygen to make water.

The third form of oxidation involves removing an electron from a charged ion:

$$X^+ \longrightarrow X^{2+} + e^-$$
                        electron

In biological systems, these different types of oxidation reaction are always exergonic, or **energy-releasing**, whereas reduction reactions are normally endergonic, or **energy-consuming**. All three forms of oxidation play an essential part in the complete breakdown of glucose during respiration.

Coupled reactions

Burning glucose in oxygen is a simple but dramatic event which releases large amounts of energy all at once. Respiration, on the other hand, is a gradual process consisting of about 30 small steps, each of which is controlled by its own enzyme. Energy is released only a little at a time, and much of it is transferred into the formation of new chemical bonds.

In the living cell, this can be achieved because chemical reactions are frequently **coupled** together so that the energy obtained from an energy-providing or oxidation reaction can be used to drive a parallel energy-consuming or reduction reaction. Two particularly important types of coupled reactions occur during both aerobic and anaerobic respiration – the first channels energy into making ATP and the second leads to the formation of NADH, the reduced form of NAD. Through hydrogen transport, NADH is used to make more ATP so that, in the final analysis, *all* the useful energy released by respiration is stored in the form of ATP.

ATP formation requires an energy input of 31 kJ per mole, as follows:

$$\text{ADP} + \textcircled{P} \longrightarrow \text{ATP} + H_2O$$
adenosine   inorganic   adenosine   water
diphosphate   phosphate   triphosphate

$$\Delta G = +31 \text{ kJ per mole}$$

ATP molecules can therefore be made whenever this reaction can be successfully coupled to an energy-releasing step in the breakdown of glucose yielding 31 kJ or more.

NAD is usually present in aqueous solution in the form of a charged ion, NAD^+, and the reduced form NADH is made according to the equation:

$$NAD^+ + 2(H) \longrightarrow NADH + H^+$$

$$\Delta G = +219 \text{ kJ per mole}$$

NADH can be formed during glucose breakdown whenever a pair of hydrogen atoms can be removed with an energy yield for the reaction of 219 kJ or more.

The biochemical pathways of aerobic and anaerobic respiration which lead to energy release are almost identical except that anaerobic respiration stops before glucose breakdown is complete. Aerobic respiration is subdivided into three separate processes, namely **glycolysis**, **Krebs cycle**, and **hydrogen transport**. The remaining sections of this Unit describe these processes in turn and discuss how they are carried out within the mitochondria, the cell's essential respiratory organelles. As you study these sections, you will find it helpful to keep in mind that the most useful products of respiration are ATP and NADH.

10·2 Glycolysis ■ ■ ■ ■ ■ ■ ■ ■ ■ ■ ■ ■ ■ ■ ■ ■ ■

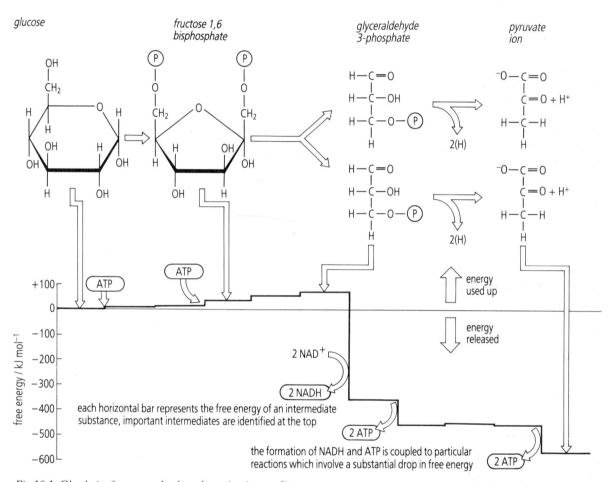

Fig 10·1 Glycolysis: free energy levels and reaction intermediates

The process of **glycolysis** (literally 'glucose-splitting') can be carried out by virtually all living organisms. As shown in Figure 10·1, the first few steps in glycolysis *require* energy in the form of ATP. This is used to increase the free energy of the glucose molecule, at the same time converting it into a substance called **fructose 1,6 bisphosphate**. Fructose 1,6 bisphosphate is a more reactive molecule than glucose and can be split by enzyme action into two molecules of the 3-carbon sugar **glyceraldehyde 3-phosphate (GALP)**. This substance is now oxidized by removing a pair of hydrogen atoms which are used to convert NAD^+ to NADH (refer to Fig. 10·1). Note that the oxidation is accompanied by a large change in free energy.

The remainder of glycolysis consists of progressively converting molecules of GALP into molecules of **pyruvate**, the ionized form of pyruvic acid. Two of the steps in this process release enough energy for ATP to be formed by a coupled reaction. Since this happens for each molecule of GALP, the two molecules of ATP initially used up are recovered and two additional molecules are produced. Glycolysis shows a net gain of two molecules of ATP and two molecules of NADH for each molecule of glucose broken down.

You will notice that, although glucose is partly oxidized during glycolysis, oxygen itself is not involved. Under aerobic conditions (Fig. 10·2A), the NADH produced by glycolysis is transferred to the hydrogen transport pathway. This frees NAD^+ so that the vital coupled reaction which removes

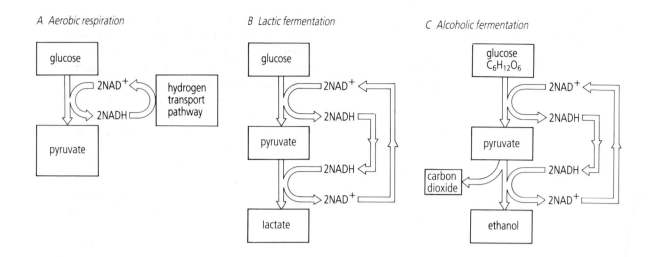

A Aerobic respiration

B Lactic fermentation

C Alcoholic fermentation

Fig 10·2 *Recycling NAD$^+$ during aerobic and anaerobic respiration*

hydrogen from GALP can continue. Under anaerobic conditions, that is, when oxygen is not available, the hydrogen transport pathway is blocked (for reasons to be explained shortly) and the cell's initial supply of NAD$^+$ is rapidly used up. When this happens, NADH normally transfers its hydrogen to form either **lactate** from lactic acid, or **ethanol** (Fig. 10·2B and 10·2C), depending upon which type of anaerobic respiration is involved. This frees NAD$^+$ but represents a substantial loss of energy. Anaerobic respiration is a wasteful process and traps only about 2% of the available energy stored in the

glucose molecule. Exclusively anaerobic cells use 19 times more glucose than aerobic cells to carry out the same work, and release large amount of lactate or ethanol as waste products.

Alcoholic fermentation by **yeast** is turned to commercial advantage in making wine and beer. However, like most organisms, yeast cells can extract far more energy from glucose if oxygen is present. The commercial user encourages ethanol production by excluding oxygen, but in natural conditions, yeast cells often respire aerobically. In this case, glucose breakdown is completed by way of Krebs cycle, as described in the next section.

10·3   Krebs cycle ■ ■ ■ ■ ■ ■ ■ ■ ■ ■ ■ ■ ■ ■

Krebs cycle (also called the **citric acid cycle** or the **tricaroboxylic acid cycle**) converts pyruvate obtained from glycolysis to molecules of carbon dioxide, which the organism releases as a waste product. In this process, water is used up and hydrogen atoms are made available for transfer to hydrogen carriers, as follows:

Krebs cycle:

$$C_3H_3O_3^- + H^+ + 3H_2O \longrightarrow 3CO_2 + 10(H)$$

| pyruvate | water | carbon dioxide | hydrogen for transfer |

The cycle is shown in more detail in Figure 10·3. In the first step, a molecule of carbon dioxide is removed from pyruvate and a pair of hydrogen atoms is transferred to make NADH. What remains is a 2-carbon fragment (CH$_3$CO), called an **acetyl group**, or **ethanoyl group**. This becomes joined to a vitamin related organic molecule, called **coenzyme A** or **CoA** for short, providing a vitally important substance called **acetyl CoA**. The role of CoA is shown in Figure

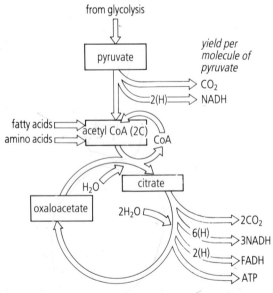

Fig 10·3 *Krebs cycle*

10·4. Within the active centre of the enzyme **citrate synthetase** CoA transfers the 2-carbon acetyl group to a 4-carbon molecule of

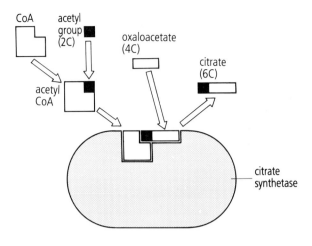

Fig 10·4 Acetyl CoA and citrate synthetase

oxaloacetate, to make a molecule of citrate. The remaining steps in Krebs cycle (refer again to Fig. 10·3) progressively break down the 6-carbon molecule of citrate until oxaloacetate is reformed, already to repeat the cycle with a new molecule of acetyl CoA.

Some steps in the cycle use up water, while others release carbon dioxide. ATP is made by a coupled reaction at one point and, elsewhere, several pairs of hydrogen atoms are transferred to carrier molecules. NAD$^+$ is normally used, but at one of the steps this is replaced by an alternative carrier, called **FAD (flavin adenine dinucleotide)**. Krebs cycle as a whole uses up 6 molecules of water and shows a net gain of 8 molecules of NADH, 2 molecules of FADH, and 2 molecules of ATP per molecule of glucose. Note that each molecule of glucose passing through glycolysis produces *two* molecules of pyruvate entering Krebs cycle.

Acetyl CoA serves as a link between many different pathways of metabolism. Depending on the circumstances, acetyl CoA is used as a subunit in building up larger chemical structures, providing a whole range of carbon compounds needed in the cell. Alternatively, when energy is in short supply, amino acids from proteins and fatty acids from lipids can be broken down to provide acetyl CoA for use in respiration.

10·4 Hydrogen transport and mitochondria ■ ■ ■ ■ ■ ■ ■

Chemical reactions can take place only when the reacting atoms or molecules collide with each other. The probability of such collisions varies in a predictable way with factors such as temperature and substrate concentration. Consequently, the rates of reaction for substances dissolved in water can be accurately calculated. Such calculations show conclusively that the numerous reactions of respiration could scarcely proceed at all if the 30 or so enzymes and all the intermediate substances required were simply dissolved in free solution.

The reactions of respiration only occur because many of the enzymes are either tightly bound together to make large 'multi-enzyme

complexes' or are fixed near to each other in the membranes of the mitochondria. In either case, intermediate substances in the breakdown of glucose are handed on directly from one enzyme to the next without ever being released into free solution. As a result, each reaction proceeds much more quickly than would otherwise be possible. The details of this arrangement are not yet fully understood but its main features can be described, as outlined in Figure 10·5. This shows the mitochondrion in a very diagrammatic form, and you should refer briefly to Figure 7·9 to identify the structures represented.

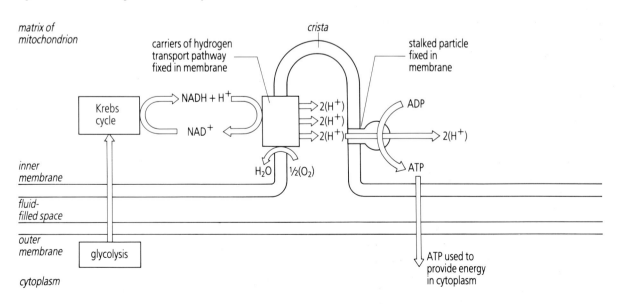

Fig 10·5 Respiration and the mitochondrion (chemiosmotic theory)

Glycolysis occurs by means of enzymes located in the cytoplasm, producing pyruvate which diffuses into the fluid interior, or **matrix**, of the mitochondrion. Here pyruvate is converted into acetyl CoA which enters the Krebs cycle. NADH produced in the matrix is taken up by the carrier molecules of the **hydrogen transport pathway** which lie firmly fixed in the inner membrane of the mitochondrion. The energy released is not used to make ATP directly but instead helps to transfer H^+ ions outwards from the matrix into the surrounding fluid-filled space. This creates a difference in pH and a strong electrical potential across the inner membrane. Both these forces drive H^+ ions back into the matrix by means of numerous stalked particles which are also fixed in the inner membrane. Each of the stalked particles makes one molecule of ATP for each pair of H^+ ions pushed through. This system for coupling the manufacture of ATP to hydrogen transport is known as the **chemiosmotic theory**. It is one of the most important ideas in modern biology and has given a fresh insight into the functioning of cells at the molecular level.

The hydrogen transport pathway is shown in more detail in Figure 10·6. It consists of a chain of carrier molecules. At the beginning of the chain, hydrogen atoms are transferred from NADH to an enzyme called **NADH dehydrogenase**. This releases NAD^+, which can be reused in Krebs cycle. It is the first step in a sequence of alternate reduction and oxidation reactions which are linked to the production of ATP.

NADH dehydrogenase passes on electrons, rather than complete hydrogen atoms, to a carrier substance called **ubiquinone**, leaving behind an equal number of H^+ ions which are pumped into the fluid-filled space between the membranes of the mitochondrion. Ubiquinone in turn transfers electrons to a group of important protein substances,

called **cytochromes**. Each of these has an iron-containing **haem group** as an integral part of its structure and, as electrons pass from one molecule to the next, the iron atoms are alternately reduced and oxidized according to the following equation:

$$Fe^{3+} + e^- \underset{\text{reduction}}{\overset{\text{oxidation}}{\rightleftharpoons}} Fe^{2+}$$

The various steps in the reduction-oxidation sequence release energy which is used to pump more H^+ ions across the membrane. A total of six H^+ ions are pumped across if the carrier chain starts with NADH. If FADH is the initial hydrogen donor, there is insufficient energy for the first of the pumping stages, so that only four H^+ ions are transferred.

At the end of the transport chain, the enzyme **cytochrome oxidase** takes up electrons, together with H^+ ions and combines them with oxygen to form water. This final step in the chain is the *only* reaction in the whole of respiration which directly involves oxygen. Nevertheless, if oxygen is not available to perform this final oxidation, the hydrogen transport pathway and Krebs cycle are blocked completely, leaving the organism with anaerobic glycolysis as its only method of releasing energy from organic materials.

The H^+ ions pumped out drive ATP production by way of the stalked particles in the inner membrane, which contain the enzyme **ATP synthetase**. For each pair of H^+ ions passing through, one molecule of ATP is produced. Therefore, three molecules of ATP are made if the carrier chain starts with NADH, but, if FADH is used, only two molecules of ATP are formed.

The relationship of hydrogen transport to the other phases of aerobic respiration is summarized

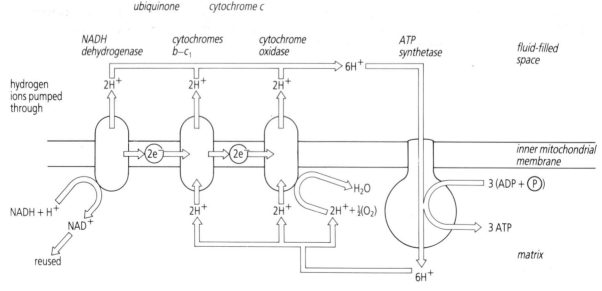

Fig 10·6 Hydrogen transport and ATP synthesis (chemiosmotic theory)

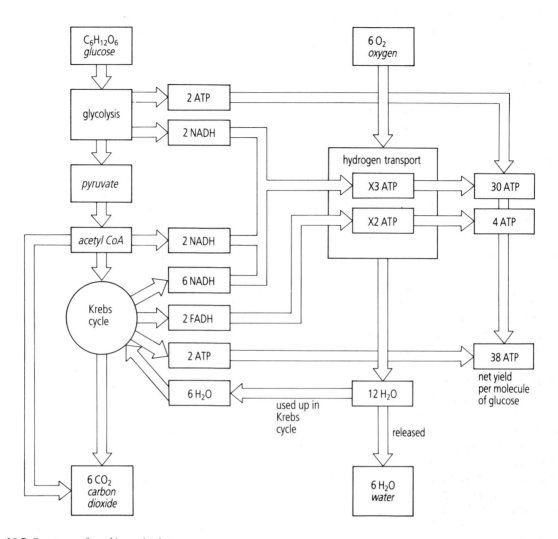

Fig 10·7 Summary of aerobic respiration

in Figure 10·7. You should study this diagram carefully to confirm that you understand the process of respiration as a whole. The most important steps are those which lead to the production of ATP. Altogether, 38 molecules of ATP are made for each molecule of glucose broken down. This provides 38 × 31 = 1178 kJ of usable energy per mole of glucose. The complete oxidation of glucose has an energy yield of 2870 kJ per mole so it is clear that the coupling reactions do not work with 100% efficiency. The 'missing' energy is released as heat and can help in maintaining a constant body temperature. Respiration has a 'fuel efficiency' of about 40% and compares favourably in this respect with the best oil-fired power stations. Internal combustion engines have fuel efficiencies of 20–25%.

11 BIOCHEMISTRY OF PHOTOSYNTHESIS

Objectives

After studying this Unit you should be able to:–

- Write the summary equation for photosynthesis

- Describe an experiment to measure the rate of photosynthesis in the water plant *Elodea* and draw a large labelled diagram of the apparatus used

- Sketch graphs of the rate of photosynthesis plotted against light intensity for two different temperatures and explain why the results obtained support the hypothesis that photosynthesis is carried out in two separate stages

- Outline the chemical changes which take place during the light reactions and dark reactions of photosynthesis and draw a flow diagram to summarize the relationships between these two processes

- Describe the structure of chloroplasts and outline their role in photosynthesis

- Explain the terms action spectrum and absorption spectrum

- Describe the light reactions of photosynthesis and explain their relationship to the chemiosmotic mechanism for making ATP

- Distinguish between non-cyclic and cyclic phosphorylation

- Draw and explain a flow diagram of the biochemical pathways involved in the dark reactions of photosynthesis

- Explain the term compensation point

- Explain the term photorespiration

- Briefly describe the biochemistry of C_4 photosynthesis and explain why this is important

11·1 Introduction ■ ■ ■ ■ ■ ■ ■ ■ ■ ■ ■ ■ ■ ■ ■ ■

Photosynthesis is the process by which green plants trap light energy from the sun and transform it into chemical energy stored in molecules of carbohydrate. This process is of great importance because, directly or indirectly, it gives rise to almost all the carbon compounds in living organisms.

The overall process of photosynthesis is represented in the **summary equation** opposite.

You will note that this equation is precisely the reverse of the summary equation for respiration

$$6CO_2 + 6H_2O \xrightarrow{\text{energy from sunlight}} C_6H_{12}O_6 + 6O_2$$

carbon dioxide   water   absorbed by chlorophyll   glucose   oxygen

and the two processes are in many respects complementary. It is useful to refer back to the biochemical pathways of respiration described in Unit 10 to help you understand the similar pathways involved in photosynthesis.

11·2 Factors limiting the rate of photosynthesis ■ ■ ■ ■ ■ ■

The summary equation for photosynthesis reveals nothing about the intermediate steps in the process. However, by means of simple experiments, it can be demonstrated that at least two separate sets of reactions must be involved.

Figure 11·1 shows the apparatus needed for measuring the rate of photosynthesis of the water plant *Elodea* (Canadian pond weed). A freshly cut strand of the plant is suspended upside down in a weak solution of sodium hydrogencarbonate ($NaHCO_3$), which acts as a source of carbon dioxide. Under these conditions, a healthy strand of *Elodea* produces bubbles of oxygen gas when brightly illuminated. The bubbles emerge from the cut end of the stem and are collected in the bulb at the base of the apparatus. From here, the oxygen can be drawn into the capillary tube by means of the syringe. The volume of gas collected in a fixed time interval gives a direct measurement of the rate of photosynthesis. The effect of different light intensities can be investigated by moving the light source closer to or farther away from the apparatus. (Note that the intensity of light reaching the *Elodea* strand is inversely proportional to the square of its distance from the source of illumination.) Similarly, the effect of temperature can be investigated by changing the temperature of the surrounding water.

The results obtained from two series of experiments carried out at different temperatures are summarised in Figure 11·2. From the Figure you can see that, at low levels of illumination, the rate of photosynthesis can be increased by increasing the light intensity. This corresponds to the initial sloping part of each of the two curves. At higher levels of illumination, a plateau level is reached representing the maximum rate of photosynthesis which can be achieved under the particular conditions of the experiment. Once this maximum has been reached, further increases in light intensity have no effect.

At low levels of illumination, the two curves follow one another exactly. In this region the entire process of photosynthesis is held in check by the availability of light, so that light is said to be the **limiting factor** for photosynthesis. On the other hand, at higher levels of illumination, the difference in temperature becomes important, curve *B* (35°C) reaching a higher plateau than curve *A* (20°C). Thus, the rate of photosynthesis is no longer limited by light but depends on temperature which thereby becomes a limiting factor. These observations give good support to the hypothesis that photosynthesis occurs in two separate stages. The first stage is assumed to involve only **photochemical reactions** because the rates of such reactions depend directly on light intensity and are unaffected by temperature. The second stage involves ordinary non-photochemical

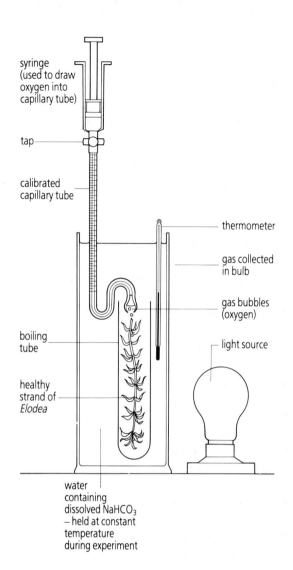

Fig 11·1 *Measuring the rate of photosynthesis in a water plant (*Elodea*)*

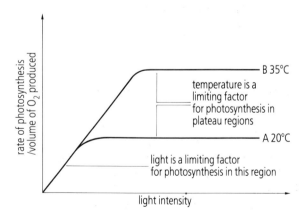

Fig 11·2 *Limiting factors for photosynthesis*

enzyme-controlled reactions which are strongly influenced by temperature.

Similar experiments carried out with different amounts of carbon dioxide available to the plant show that, at low light intensities, photosynthesis is unaffected by carbon dioxide concentration, while, the plateau levels reached at higher light intensities vary in the same way as with temperature. It is concluded that carbon dioxide is used in the second stage of photosynthesis, because, like temperature, carbon dioxide concentration acts as a limiting factor for the second stage.

11·3 Light and dark reactions of photosynthesis ■ ■ ■ ■ ■ ■

Further evidence that photosynthesis is a two stage process comes from experiments in which different **isotopes** of elements such as carbon and oxygen are used to 'label' the substances needed as raw materials for photosynthesis. For example, it was assumed initially that the oxygen produced by photosynthesis originated from carbon dioxide. However, experiments using $^{18}O_2$ (a 'heavy' isotope of oxygen in which the atoms have a relative atomic mass of 18 rather than 16, as in the common isotope of oxygen) show that this is not the case. Plants provided with $C^{18}O_2$, that is, carbon dioxide made with ^{18}O atoms, combine these atoms into carbohydrate, whereas plants provided with $H_2^{18}O$ release the ^{18}O atoms as oxygen gas.

It is now clear that *all* the oxygen released by photosynthesis comes from water, according to the equation:

$$H_2O \xrightarrow[\substack{\text{absorbed by} \\ \text{chlorophyll}}]{\substack{\text{energy from} \\ \text{sunlight}}} \underset{\substack{\text{electrons}}}{2H^+ + 2e^-} + \underset{\substack{\text{oxygen} \\ \text{gas}}}{\tfrac{1}{2}(O_2)}$$

This reaction is called **photolysis** (literally **'light-splitting'**) of water and it is the key chemical change in a series of reactions called the **light reactions**, or **photo-reactions** of *photosynthesis*. These reactions provide the energy to make ATP from ADP and inorganic phosphate and finally transfer hydrogen ions (H^+) and electrons (e^-) to **NADP (nicotinamide adenine dinucleotide phosphate)**, forming **NADPH**:

$$2H^+ + 2e^- + NADP^+ \longrightarrow NADPH + H^+$$

NADPH functions as a **hydrogen carrier** for photosynthesis in exactly the same way as NADH acts as a hydrogen carrier during respiration. (The chemical structures of NAD and NADP are given in Unit 8 – NADP differs from NAD only by having an additional phosphate group.)

The raw material carbon dioxide is used in a completely separate series of reactions called the **dark reactions**, or **synthesis reactions**, of *photosynthesis*. These reactions do not need light directly but use energy from ATP and NADPH to build up carbohydrates. The relationship between the light and dark reactions is summarized in

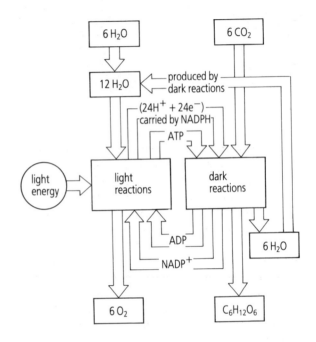

Fig 11·3 Summary of photosynthesis

Figure 11·3 and their separate equations are balanced to give the overall equation for photosynthesis as follows:

light reactions:
$$12H_2O \longrightarrow (24H^+ + 24e^-) + 6O_2$$

dark reactions:
$$6CO_2 + (24H^+ + 24e^-) \longrightarrow C_6H_{12}O_6 + 6H_2O$$

overall reaction:
$$6CO_2 + 12H_2O \longrightarrow C_6H_{12}O_6 + 6H_2O + 6O_2$$

In this equation water appears both as a raw material and as a product. This is correct in terms of the mechanism of photosynthesis. However, the water produced can be recycled within the plant so that the *net* requirement for water is only $6H_2O$; cancelling out the 'extra' water on each side gives the more familiar summary equation for photosynthesis, as already stated.

The remaining sections of this Unit describe the biochemistry of the light and dark reactions and discuss the relationship between chloroplasts and photosynthesis. Details of leaf structure and other photosynthetic adaptations of plants are described in Units 31, 32, and 33.

11·4 Chloroplasts and photosynthesis ■ ■ ■ ■ ■ ■ ■ ■ ■

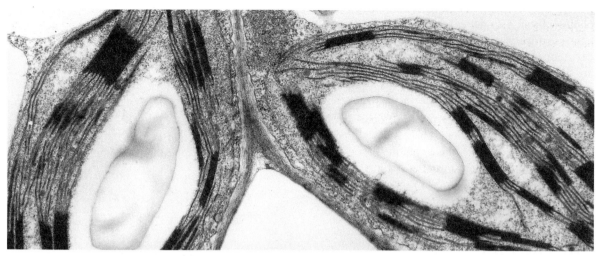

Fig 11·4 Electronmicrograph of chloroplasts

Chloroplasts are the cell organelles of green plants which carry out photosynthesis. The basic structure of chloroplasts has been outlined in Unit 7 and is shown in the electronmicrograph of Figure 11·4. Like mitochondria, chloroplasts are surrounded by a pair of membranes. However, within the fluid interior, or **stroma**, of the chloroplast there is an additional system of disc-shaped cavities called **thylakoids**. The thylakoids are stacked together like coins to make structures called **grana** (singular **granum**) and these in turn are interconnected by membranes called **lamellae** (refer to Fig. 7·10).

components needed for absorbing light and for carrying out the light reactions of photosynthesis, while the dark reactions are carried out in the stroma. Glucose made by the dark reactions is stored in starch grains which can be formed within the chloroplast itself.

Chloroplast structure is thought to be linked to the biochemistry of photosynthesis as shown in Figure 11·5. The details of this arrangement are not yet fully understood but it is known that H^+ ions are transferred into the interior of the thylakoids, thereby creating a difference in pH and a strong electrical potential which drive the

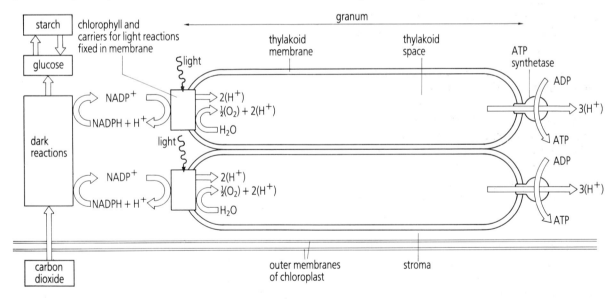

Fig 11·5 Photosynthesis and the chloroplast (chemiosmotic theory)

Freeze fracture electronmicrographs (obtained by the technique described in Unit 6) show that the membranes of the thylakoids possess many large particles distributed within the lipid bilayer and projecting from its inner and outer surfaces. The biochemical evidence available indicates that the thylakoid membranes contain all the

production of ATP. Thus, ATP formation depends on a **chemiosmotic** mechanism very similar to that used during respiration. For reasons which remain to be discovered, 3 H^+ ions are needed for each molecule of ATP formed instead of 2, as in the case of respiration.

11·5 Biochemistry of the light reactions ■ ■ ■ ■ ■ ■ ■ ■ ■

The **light reactions,** or **photo-reactions,** of photosynthesis trap the energy of sunlight using light-sensitive pigments including **chlorophylls, carotenoids** and **xanthophylls.** These substances are fixed in the thylakoid membranes, but can be extracted by grinding the leaves of a plant in a suitable solvent such as acetone. The solution obtained is then filtered free from leaf debris and concentrated by evaporation. The various pigments can then be separated by the technique of **chromatography,** as illustrated in Figure 11·6.

The molecule has a 'lollipop' shape; the flat pigment group (porphyrin) contains a magnesium atom at its centre and corresponds to the head of the lollipop; the long hydrocarbon side chain corresponds to the stick

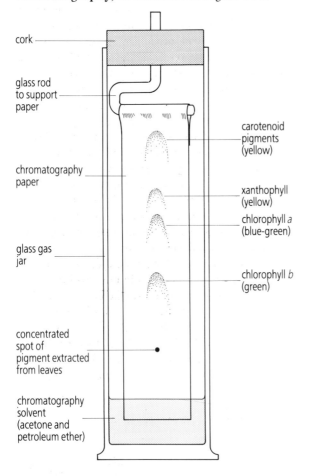

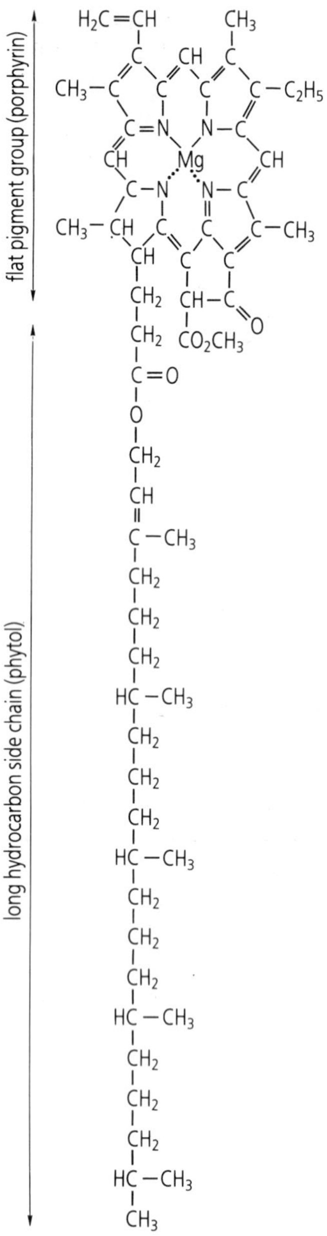

Fig 11·6 Chromatography of leaf pigments

Fig 11·7 Structure of chlorophyll a

The most important pigment is **chlorophyll a,** which has the flat 'lollipop' shape shown in Figure 11·7 and contains a magnesium (Mg) atom located at its centre. This structure enables electrons to wander freely over the surface of the chlorophyll molecule, rather than being held rigidly in place. As a result, when a 'packet', or **photon,** of light of the correct wavelength strikes the molecule, an electron can be temporarily displaced to a higher energy level. Stated in another way, the chlorophyll molecule is said to become '**excited**' by **absorbing** light energy.

The rate of photosynthesis varies according to the different colours of light supplied. This can be demonstrated experimentally by measuring the amount of oxygen produced by a plant exposed to

a range of different light wavelengths. Presented graphically, the results of this experiment show the **action spectrum** for photosynthesis (see Fig. 11·8). Measuring the amount of light of different wavelengths which passes through a solution of chlorophyll molecules extracted from a leaf shows that blue (short wavelength) and red (long wavelength) light are absorbed, while green light is reflected. This information is presented graphically as an **absorption spectrum.** The absorption spectrum closely follows the action spectrum for photosynthesis, providing good evidence that chlorophyll is important.

In the thylakoid membranes, chlorophyll molecules are grouped together into clusters which contain several hundred molecules. Each cluster,

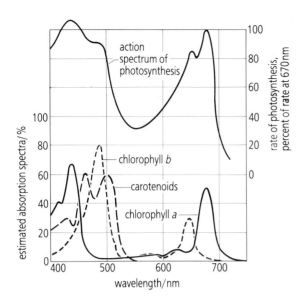

Fig 11·8 The action spectrum of photosynthesis compared to the absorption spectrum for the pigment molecules present

called an **antenna complex**, traps light over the whole of its surface and is arranged in a special way so that, no matter where light energy is received, it is channelled to just one molecule, called the **reaction centre chlorophyll**. Light absorbed by an antenna complex is converted into a useful form when high energy electrons from the 'excited' reaction centre chlorophyll are exchanged for low energy electrons at the beginning of a chain of carrier molecules, also located in the thylakoid membranes (refer to Fig. 11·9).

You can see from Figure 11·9 that, following the absorption of light energy, two distinct pathways of electron movement are possible. Both

cause H+ ions to be pumped into the thylakoid spaces. In **non-cyclic electron flow**, light energy is used for photolysis of water, providing electrons which can be transferred to chlorophyll and raised in energy. This reaction is made possible by a large protein complex known, for historical reasons, as **photosystem II**. From photosystem II, electrons pass from carrier to carrier along a chain which includes a second light sensitive complex, called **photosystem I**. At the end of the chain the enzyme **NADP reductase** transfers electrons to $NADP^+$ to make NADPH.

Plastoquinone, **plastocyanin**, and **ferredoxin** are small carrier molecules which take up electrons and move about at random within the membrane until they collide with other carriers able to receive electrons. These movements provide the basis for a second pattern of electron flow, called **cyclic electron flow**. For this to occur, electrons from photosystem I must pass from ferredoxin to plastoquinone before returning to photosystem I.

Both cyclic and non-cyclic electron flow release energy which is used by a group of cytochrome proteins, called the **cytochrome b_6-f complex**, to pump H^+ ions across the membrane. The H^+ gradient created drives ATP production by the enzyme **ATP synthetase**. Because ATP is made by adding a phosphate group to ADP, the reaction is called phosphorylation. **Cyclic phosphorylation**, resulting from cyclic electron flow, produces ATP only, while **non-cyclic phosphorylation**, from non-cyclic electron flow, produces ATP and NADPH. The dark reactions of photosynthesis described in the next section require far more ATP than NADPH so that both types of electron flow must occur if these substances are to be produced in the appropriate amounts.

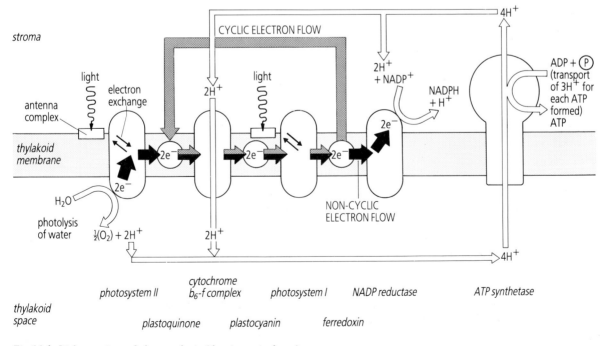

Fig 11·9 Light reactions of photosynthesis (chemiosmotic theory)

11·6 The dark reactions ■ ■ ■ ■ ■ ■ ■ ■ ■ ■ ■ ■ ■ ■

The **dark reactions**, or **synthesis** reactions, of photosynthesis build up molecules of carbon dioxide to form glucose. This process transfers hydrogen atoms from NADPH and requires energy in the form of ATP.

The biochemical pathway leading to glucose is shown in outline in Figure 11·10 and was first worked out by allowing plants to photosynthesise using carbon dioxide 'labelled' with the radioactive isotope [14]C. A suspension of single-celled algae was exposed to a brief flash of light and then killed by running the suspension into hot alcohol. This instantly stopped chemical reactions so that any radioactive substances found to be present had to be part of the photosynthesis pathway. By varying the duration of the light flashes the correct sequence of events could be determined – substances which occurred near the beginning of the pathway were formed even during the shortest flashes, while other substances, which occurred later, were only formed after exposure to longer flashes of light.

The first step in the reaction sequence consists of linking carbon dioxide to a molecule of the 5-carbon sugar **rubulose bisphosphate (RuBP)**, using an important enzyme called **ribulose bisphosphate carboxylase**. As a result of its action, a highly unstable 6-carbon structure is formed which immediately splits into two molecules of a 3-carbon glycerate 3-phosphate (GP). Molecules of GP are, in turn, promptly converted into **glyceraldehyde-3-phosphate (GALP)**. The cell's supplies of GALP form a 'pool' which may be utilized in various ways. Glucose can be built up from GALP by a pathway which exactly reverses the pathway of glucose breakdown by glycolysis in respiration. Alternatively, GALP can be chanelled into forming fats or proteins, or may be broken down to pyruvate and used for respiration in the

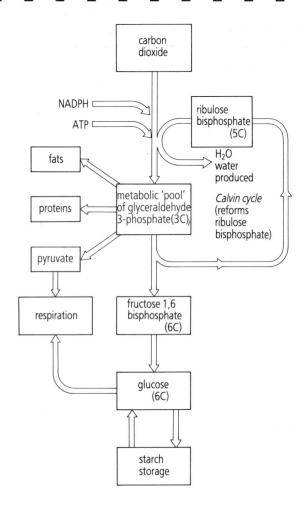

Fig 11·10 Dark reactions of photosynthesis

mitochondria. Most importantly, however much of the cell's GALP is converted once again to ribulose bisphosphate so that more carbon dioxide can be taken up. This completes a cycle of events known as the **Calvin cycle**. The cell's supplies of glucose may be used for respiration or stored as starch, depending on current needs.

11·7 Oxygen and carbon dioxide balance ■ ■ ■ ■ ■ ■ ■ ■ ■

Compensation point

Aerobic respiration occurs in the mitochondria of plant cells at all times, with the result that O_2 is used up and CO_2 is continually being produced. In bright sunlight, chloroplasts use up CO_2 and release O_2 at a far greater rate. Consequently, the plant as a whole takes in CO_2 and gives out O_2. In darkness this situation is reversed.

If the light intensity is gradually increased, a particular level can be found where there is no net uptake or release of O_2 or CO_2 so that respiration and photosynthesis are exactly in balance. This is called the **compensation point**. For many plants, it is equivalent in intensity to about 2% of full sunlight. The exact point varies from species to

species and is affected by environmental temperature and CO_2 concentration.

Net manufacture of glucose and increase in dry mass are only possible when light intensity is above the compensation point.

Photorespiration

Oxygen and carbon dioxide balance also depends on a remarkable process called **photorespiration**.

As already explained, ribulose bisphosphate (RuBP) acts as the carbon dioxide acceptor in the first stage of the dark reactions of photosynthesis. RuBP and CO_2 become the substrates for the enzyme ribulose bisphosphate carboxylase (RuBP-carboxylase). The products of the reaction are converted by a further reaction to give GALP.

Unfortunately, RuBP-carboxylase also acts as the catalyst for another quite different reaction for which *oxygen* and RuBP are the substrates. In this case, one of the products is a 2-carbon molecule, **phosphoglycolate**, which cannot be converted to GALP.

The phosphoglycolate formed is useless to the plant and must be removed. It is oxidized to CO_2 by a series of reactions which do not yield energy in a useful form. Photorespiration uses up O_2 and releases CO_2 and is a wasteful process for the plant.

The relative levels of photorespiration and normal CO_2 fixation by RuBP-carboxylase depend on several factors. Oxygen is a competitive inhibitor for the enzyme. In other words, it reduces the effectiveness of the enzyme by 'competing' with its normal substrate to occupy the active centre. (The action of these inhibitors is explained in Unit 5.) Consequently, photorespiration increases when O_2 concentration is high, or when CO_2 concentration is low. Both processes are light-dependent and take place more rapidly in bright light because of increased availability of ribulose bisphosphate.

It is estimated that photorespiration reduces the potential yield of photosynthesis by as much as 30–40%. Any mechanism which limits this loss is likely to be of tremendous benefit to the plant.

11·8 C_3 photosynthesis and C_4 photosynthesis ■ ■ ■ ■ ■ ■

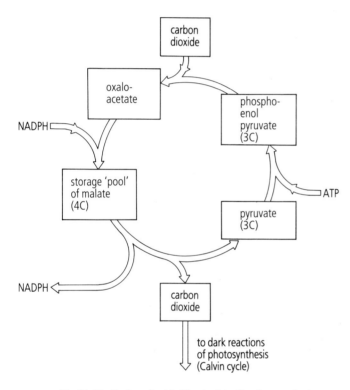

Fig 11·11 *Carbon dioxide 'fixation' in C_4 photosynthesis*

Plants which combine CO_2 from the air directly with RuBP produce a 3-carbon compound (namely GALP) as the first stable intermediate in photosynthesis. For most plants from temperate regions, this is the only method of using CO_2. These plants are therefore described as C_3 **plants**.

An interesting alternative mechanism for fixing CO_2 occurs in some plants, called C_4 **plants**, which are adapted for life in hotter or tropical regions. In these plants, which include maize and sugar cane, the stomata on the leaves must be kept closed for much of the time in order to conserve water. The light intensity is high but CO_2 concentrations inside the plant fall to very low levels. These circumstances would favour photorespiration at the expense of normal CO_2 fixation except for an additional pathway for photosynthesis which begins when molecules of CO_2 are combined with molecules of a 3-carbon substance derived from pyruvate. This substance, called **phosphoenol pyruvate**, combines with CO_2 to form molecules of the 4-carbon compound **oxaloacetate** instead of GALP as the first stable intermediate. As outlined in Figure 11·11, oxaloacetate is converted into a second substance, called **malate**, which acts as a temporary store for CO_2.

In many C_4 plants, malate is transferred from one cell to another and becomes concentrated in particular tissues where the other reactions of photosynthesis can be carried out with maximum efficiency. For example, in a maize leaf, malate is transferred from mesophyll cells to **bundle sheath cells** surrounding the vascular bundles, where O_2 concentration is kept low.

In desert plants, such as cacti, the stomata are opened at night and closed during the day. CO_2 is fixed as malate at night and is used during the day when the light-dependent reactions of photosynthesis are carried out.

C_4 photosynthesis operates efficiently at very low CO_2 concentrations and at high light intensities. In appropriate conditions, C_4 plants make more glucose per unit leaf area and grow much more quickly than comparable C_3 plants.

Section A Questions ■ ■ ■ ■ ■ ■ ■ ■ ■ ■ ■ ■

Short-answers and interpretation

1 **a** Define a bioelement. [1]
b List the six most common bioelements in order of their abundance in the human body. [6]
c What are hydrogen bonds? [2]
d Why is water a good solvent? [2]
e What is a colloid? [2]
f What is meant by pH? [1]
g On the pH scale, what values would represent
  i a strong acid
  ii a neutral solution
  iii a strong alkali [1]

2 **a** What is the general formula for molecules of carbohydrate? [1]
b Give the general formula for a hexose sugar. [1]
c Hexose sugars are known as monosaccharides and form the basic units of more complex carbohydrates.
  i What is the name of the reaction in which two hexose sugars combine to form a disaccharide? [1]
  ii What is the name given to the type of bond which joins them together? [1]
  iii What is produced along with the disaccharide in the reaction? [1]
d Name three biologically important disaccharides and state where they can be found. [3]
e Starch and cellulose are high molecular weight polysaccharides. Which hexose sugar forms their basic unit? [1]
f What is the essential structural difference between starch and cellulose? [1]
g Starch is an important storage compound. In what structure is it characteristically located in plant cells? [1]
h Name the storage polysaccharide found in mammals. [1]
  i Name the organ where it occurs in large amounts. [1]
  ii Where is it deposited in the cell? [1]
Welsh Joint Education Committee

3

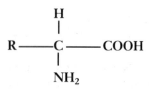

a What general type of molecule is shown in the diagram above? [1]
b What is the simplest form of **R**? [1]

c Which part of the structure gives acidic properties to the molecule? [1]
d Which part of the structure gives basic properties to the molecule? [1]
e It may be said that because molecules of this type can show polymerization they are very important biologically.
  i What is meant by polymerization? [1]
  ii If molecules of this type polymerize, what will be formed? [1]
f With the aid of a diagram, illustrate the product when two of the units shown have joined. [2]
g What general type of biochemical reaction is this? [1]
h What name is given to the type of bond formed between the two units? [1]
University of Oxford Delegacy of Local Examinations

4 The table shows the rate of activity of an enzyme at different temperatures.

temperature/°C	rate/mg of product per minute
0	1.8
5	2.4
10	3.7
15	4.9
20	7.4
25	9.3
30	13.4
35	17.2
40	19.0
45	19.0
50	8.1
55	1.7
60	0

a Draw a graph of these results. [4]
b i State the optimum temperature for this enzyme. [1]
  ii Explain the rate of activity at 17°C. [1]
  iii Explain the results at temperatures above 45°C. [2]
c Name two factors other than temperature which would affect the rate of enzyme activity. [2]
d If the enzyme used in the experiment was amylase, name
  i the substrate, [1]
  ii the products. [1]
Joint Matriculation Board

5 **a** Make a labelled diagram to represent the way in which molecules are arranged in the cell membrane. [5]
b Name two sites inside a cell where you would expect to find such a membrane. [2]
c Give three functions of the external cell membrane. [3]
Associated Examining Board

tube number	contents of tube (cm³)					appearance of tube contents after ten minutes at room temperature
	catechol	apple extract	buffer (pH 7)	dilute acid	dilute alkali	
1	2	–	5	–	–	colourless
2	2	2	3	–	–	dark brown
3	2	2	–	3	–	colourless
4	2	2	–	–	3	light brown
5	–	2	5	–	–	light brown
6	2	2*	3	–	–	colourless

*extract had been previously boiled

6 The browning which occurs when many types of vegetable and fruit are peeled is caused by enzymes called phenol oxidases. These catalyse the relatively slow conversion of naturally occurring phenolic compounds into dark brown melanins.

phenols → quinones → melanins
(colourless) (yellow) (dark brown)

The results in the table above were obtained from an investigation into the browning of apple.

  a Use the results of this investigation to
    i suggest two ways in which apples, once peeled, can be prevented from turning brown. [2]
    ii state what the apple extract contains. [2]
  b From the information given, what type of substance do you think catechol is, and what purpose does it serve in this investigation? [2]
  c Describe simple experiments which you might carry out to show that
    i the contents of tube 1 were effectively buffered [2]
    ii oxygen is necessary for browning reactions to occur. [2]
      Joint Matriculation Board

7 a Name the four nitrogenous bases found in DNA. [4]
  b What is meant by a nucleotide? [2]
  c Describe the structure of a molecule of DNA. [4]
  d Give two ways in which the structure of DNA differs from that of messenger RNA. [2]
  e By means of a simple flow diagram, indicate the sequence in which the following are involved in protein synthesis: polypeptide, DNA, transfer RNA, amino acids, messenger RNA. [3]
  University of London School Examinations Board

8 The diagram opposite shows part of an electronmicrograph of a cell.
  a Identify each of the structures labelled A–H [4]

b i Give three features of the cell which indicate that it is an active secretory cell. [3]
ii Give one reason which suggests that this is an animal cell. [2]
    Joint Matriculation Board

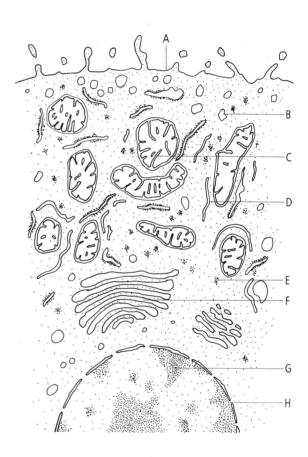

9 a Define the term enzyme. [3]
  b Describe the role of the active site in enzyme activity. [3]
  c Explain each of the following characteristics of enzymes in relation to their structure.
    i specificity, [3]
    ii denaturation by heat, [3]
    iii competitive inhibition. [3]
  University of London School Examinations Board

10 a Read through the following passage and then write down the most appropriate word or words to fill in the blanks.

The DNA molecule consists of a large number of nucleotides; each nucleotide consists of a nitrogenous base joined to a (i) sugar and a (ii). The DNA consists of two strands running (iii) to each other and coiled in a double helix. The strands are held together by (iv) bonds and there are (v) nucleotides for each turn of the helix.

In RNA, the base thymine is replaced by (vi) and the sugar present is (vii). In the cell, there are three types of RNA present called (viii), (ix) and ribosomal RNA. The type of RNA called (x) is synthesized in the nucleus of the cell. [10]

b i Biochemical analysis of a sample of DNA extracted from a cell showed that 33% of the nitrogenous bases were cytosine. What percentage of the bases in the DNA would be thymine? Give reasons for your answer. [2]

ii Given the following triplet sequence in DNA:

AGT/ACC/ATG/TAA/CAT/CAA/ATA

What sequence would result if the guanine bases were deleted? [1]

What is the likely effect of these changes? [3]

iii Starting from the original sequence, what sequence would result after substitution of guanine by cytosine bases? [1]

iv Write down the transcribed mRNA sequence for this modified sequence of DNA. [2]

v State one effect of the substitution given in iii [1]

Oxford and Cambridge Schools Examination Board

11 The diagram below shows an outline of cellular respiration.

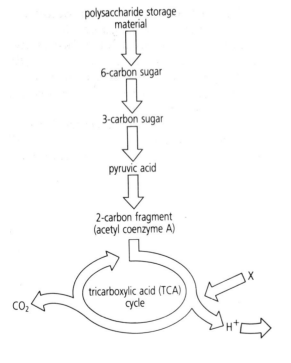

polysaccharide storage material

6-carbon sugar

3-carbon sugar

pyruvic acid

2-carbon fragment (acetyl coenzyme A)

tricarboxylic acid (TCA) cycle

CO₂

X

H⁺

a Name a polysaccharide commonly stored in
i green plants, [1]
ii mammals. [1]

b i Name the process by which the 6-carbon sugar is converted to pyruvic acid. [1]
ii Where in the cell does this process occur? [1]
iii Why is ATP used in this process? [2]

c Name the compound formed from pyruvic acid in muscle cells under conditions of oxygen debt. [1]

d Name the type of enzyme involved at stage X. [1]

e What happens finally to the hydrogen ions released from the tricarboxylic acid cycle? [2]

f Make a labelled drawing of a mitochondrion, and on your drawing indicate where ATP synthesis occurs. [4]

University of London School Examinations Board

12 Plant species A and B grow naturally in different habitats. In an experiment the exchange of carbon dioxide between the atmosphere and species A and B was determined over a range of light intensities from darkness to the equivalent of mean noon sunlight. A constant temperature was maintained throughout the experiment. The amount of carbon dioxide absorbed or released was determined by measuring the carbon dioxide concentration in a stream of air before and after it had passed over the plants. The data obtained are given below.

Light intensity/ % mean noon sunlight	Net CO₂ absorption/ arbitrary units	
	Species A	Species B
0	−0.1	−0.8
10	+3.0	+0.5
20	+5.3	+3.5
30	+6.5	+7.0
40	+6.5	+9.3
50	+6.7	+11.5
60	+6.8	+13.2
70	+7.0	+15.0
80	+6.5	+17.0
90	+6.8	+18.0
100	+6.7	+19.0

a Plot these data on a single set of axes [5]

b Discuss the extent to which species A and species B might be able to grow in the same habitat. [4]

c i What is meant by the term 'compensation point'? [1]
ii Clearly indicate on your graph the compensation point for species B. [1]

d i What is meant by the term 'limiting factor'? [1]
ii From your knowledge of photosynthetic pathways, explain precisely how three named factors can be limiting in photosynthesis. [6]

e Distinguish between C₃ and C₄ plants. [2]

Associated Examining Board

13 **a** Draw a labelled diagram of a chloroplast as seen with the electron microscope. Show clearly, on the diagram, where the chlorophyll is found within the chloroplast. [7]
  b i Name the metallic element that occurs in chlorophyll *a* and *b*. [1]
   ii Name two other pigments that occur in association with chlorophyll *a* and *b* in higher plants. [2]
  c Give two advantages of chlorophyll occurring in chloroplasts rather than being diffuse in the cytoplasm. [2]
  d Explain briefly how you would demonstrate the presence of the photosynthetic pigments in plant material. [4]
  University of London School Examinations Board

Essays

1 'Water is essential for life'. Explain in what ways this statement is true for plants and animals. [20]
  University of London School Examinations Board

2 **a** Give an account of the chemical nature and variety of carbohydrates. [10]
  b Outline the role of carbohydrates in the life of a plant. [8]
  Cambridge University Local Examinations Syndicate

3 **a** Outline the structure of a protein molecule. [4]
  b Describe four different functions of proteins. [8]
  c Explain how the generalized structure of a protein molecule can be varied to suit different functions. [6]
  Cambridge University Local Examinations Syndicate

4 Discuss the roles of proteins in the human body. [25]
  Associated Examining Board

5 **a** What is an enzyme? [3]
  b Describe the effects on enzyme action of
   i pH,
   ii temperature,
   iii substrate concentration. [12]
  c Explain the significance of enzymes in metabolic pathways. [5]
  University of London School Examinations Board

6 **a** Using suitable examples, give an account of the properties of enzymes and describe the main types of reactions carried out by them. [13]
  b Discuss fully an hypothesis explaining how enzyme reactions are brought about. [7]
  Welsh Joint Education Committee

7 **a** Discuss the structure and distribution of the membranes of cells. [6]
  b How do the different chemical components of membranes affect the properties of membranes? [6]
  c Describe mechanisms by which extracellular material may enter a cell. [8]
  Joint Matriculation Board

8 Give an illustrated account of the structure and function of the following cell organelles:
   i rough endoplasmic reticulum, [7]
   ii cilia, [6]
   iii chloroplasts. [7]
  Welsh Joint Education Committee

9 **a** With reference to two named types of animal cell, draw annotated diagrams of the various organelles which would be present. [20]
  b What clues from the overall appearance of each cell give an indication of its function? [5]
  Cambridge University Local Examinations Syndicate

10 **a** Describe briefly the structure of the DNA molecule [5]
  b i Explain clearly how RNA differs in molecular structure from DNA.
   ii State where RNA may be found in a cell. [6]
  c Outline how RNA is involved in protein synthesis. [9]
  University of London School Examinations Board

11 **a** Define the term anaerobic respiration. [3]
  b Outline the principal stages of aerobic respiration in cells. [10]
  c Discuss the roles of adenosine triphosphate (ATP) in cells. [5]
  Cambridge University Local Examinations Syndicate

12 **a** The uptake of carbon dioxide by an actively-photosynthesizing flowering plant can be influenced by a number of factors. Discuss how each of these exerts its influence. [10]
  b i Explain how a carbon atom taken in by such a plant may later appear in a molecule of ribulose bisphosphate.
   ii Indicate briefly three other possible fates of this carbon atom. [10]
  Joint Matriculation Board

Section B

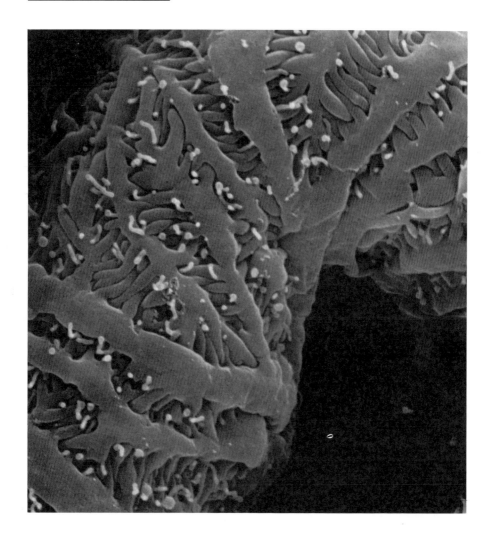

PHYSIOLOGY –
THE FUNCTIONING
OF ORGANISMS

12 NUTRITION AND DIET

Objectives

After studying this Unit you should be able to:–

- Draw a diagram classifying the possible types of nutrition and give definitions of each main type

- List the main substances present in a balanced diet

- Discuss carbohydrates and lipids as substances which provide energy in the diet

- Describe how the energy content of foodstuffs can be measured

- Discuss proteins and other substances needed in the diet as material for growth and repair

- List the main vitamins in the human diet and outline their importance

- List the main inorganic elements in the human diet and give a function for each

12·1 Modes of nutrition ■ ■ ■ ■ ■ ■ ■ ■ ■ ■ ■ ■ ■ ■ ■

All living organisms must obtain a variety of chemical substances from their surroundings. These substances are essential because, in order to maintain itself in a working condition, every organism must carry out a continual cycle of breakdown and repair at the molecular level. Waste products are lost from the organism and new materials are constantly needed for growth and repair, and to provide energy. The process of obtaining and using these substances is called **nutrition**; the main types of nutrition are listed in Figure 12·1 and can be described as follows:

autotrophic nutrition is **photosynthesis** carried out by green plants. The energy-rich substances produced by green plants represent a reservoir of stored chemical energy which can be exploited in various ways by other organisms.

A different form of autotrophic nutrition is shown by **chemosynthetic** bacteria which obtain energy by the oxidation of simple inorganic substances in the environment. For example, as outlined in Unit 62, **nitrifying bacteria** play a vital role in the nitrogen cycle by oxidizing ammonium ions (NH_4^+) to form nitrites (NO_2^-) and nitrates (NO_3^-).

Autotrophic nutrition

Autotrophic (or 'self-feeding') organisms obtain energy directly from their surroundings and use simple inorganic substances, such as carbon dioxide, water, and nitrate or ammonium salts, to build up the complex biological substances needed to stay alive. The most important form of

Heterotrophic nutrition

Heterotrophic (literally 'other-feeding') organisms obtain energy-rich substances by breaking down carbon compounds obtained from the bodies of other organisms. This breakdown, called **digestion**, also provides building blocks, such as

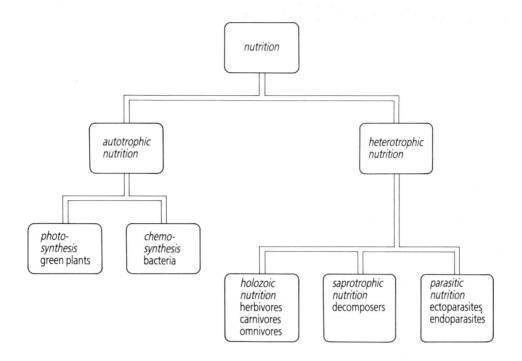

Fig 12·1 Classification of nutrition

amino acids and monosaccharides, which the heterotrophic organism can use to make its own proteins and carbohydrates. There are several different types of heterotrophic nutrition, each representing a different strategy for obtaining food.

1 Holozoic nutrition. The term 'holozoic' means literally 'feeding like an animal'. Thus, holozoic organisms feed on solid organic material which is taken into the body and digested internally to give smaller chemical fragments suitable for absorption.

Most familiar animals are holozoic. Rabbits, cattle, and other animals which survive mainly by eating plant material are called **herbivores**. Animals which eat meat, often after catching and killing their prey, are called **carnivores**, and animals like man, which have a mixed diet, containing some animal and some plant material, are called **omnivores**.

Holozoic feeders are among the most advanced of all living creatures. Characteristically, they have well developed sense organs and nervous systems for detecting and recognising food and can move actively from place to place to keep themselves supplied. The search for food is so important that many of the advanced features of mammals to be discussed in the later Units of this Section can be traced directly to their method of nutrition.

2 Saprotrophic nutrition. Saprotrophic organisms feed exclusively on dead or decaying organic material. They include many fungi and bacteria. These organisms release digestive enzymes to break down organic material in their surroundings and survive by taking up the simpler soluble substances produced.

Within natural ecosystems, saprotrophic organisms are important because they act as **decomposers** and thereby allow essential nutrients to be recycled. Without the activity of the decomposers, such substances would be permanently trapped in the corpses of dead animals and plants which, in a short time, would be littered everywhere.

3 Parasitic nutrition. A parasite is defined as an organism which obtains food materials from the living body of another organism, called the **host**.

The relationships between parasites and their hosts vary considerably. A mosquito taking a blood meal from a man is acting as an external parasite, or **ectoparasite**, the association is fleetingly temporary and no real damage is inflicted. At the other end of the scale, the malarial parasite, which the same mosquito might transmit, is an internal parasite, or **endoparasite**, which lives permanently in the host and may inflict considerable damage, leading even to the death of the host.

All heterotrophic organisms are ultimately dependent on autotrophic organisms as a source of pre-formed organic materials, while the autotrophic organisms are dependent on the decomposers for a continuing supply of the simple substances needed as raw materials. Interactions of this kind form the essential subject matter of ecology and are discussed in some detail in Section E of this book.

The remainder of this Unit and the following three Units consider specifically the main aspects of nutrition in humans and other mammals. This study begins by considering the substances typically required in the human diet.

12·2 The balanced human diet ■ ■ ■ ■ ■ ■ ■ ■ ■ ■ ■

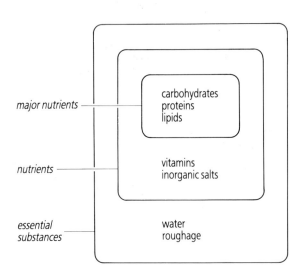

Fig 12·2 Components of a balanced diet

A **balanced diet** is one which provides all the materials needed for normal growth and development and for maintaining the body in a healthy condition. Within this definition, there are many diets any of which is equally satisfactory from a nutritional point of view. The individual foodstuffs which comprise a balanced diet can vary widely but the groups of substances shown in Figure 12·2 are always included.

The **nutrient** substances participate directly in the chemical reactions of the body and can contribute to growth. Carbohydrates, proteins, and lipids are classed as **major nutrients** because they are taken in the largest amounts and because each of these substances can be broken down to provide energy. The major nutrients are also those which require to be digested by enzyme action before being absorbed into the body.

Water is not usually described as a nutrient because, although it can participate in chemical reactions, it is present mainly as a solvent and is not a raw material for growth. **Roughage** or **dietary fibre** is the name given to indigestible material, mainly cellulose, which passes through the digestive system unaltered. It is important because it gives 'bulk' to the food mass inside the intestine, allowing it to be efficiently moved along by muscular contractions. Although it has no nutritional value, roughage is an essential component of a healthy diet.

All the different nutrient substances in a person's diet fulfil at least one of three main functions. They can be used for respiration, that is, as a **source of energy**, they are used as **building materials** and they can act as **control factors**. These control factors include substances such as vitamins and many inorganic salts which are needed only in small amounts. They determine the overall pattern of growth without themselves being used as building materials. For example, vitamin D is needed for healthy bone growth but does not form a significant part of bone structure. The requirements of a balanced diet are described in terms of these three main functions.

12·3 Providing energy from the diet ■ ■ ■ ■ ■ ■ ■ ■ ■ ■

Carbohydrates and lipids are the materials most often used for respiration and energy production. Carbohydrate foods are cheap to obtain and, in poorer countries, they provide as much as 80% of the total energy in people's diets.

As discussed in Unit 3, carbohydrates include monosaccharide sugars, such as **glucose** and **fructose**, disaccharide sugars (**maltose, lactose** and **sucrose**), and polysaccharides (**starch, glycogen** and **cellulose**).

Among these substances, polysaccharides occur in the diet in the largest amounts. Starch is the main carbohydrate storage material in plants and forms up to 50% by mass of foods such as bread and roughly 20% by mass of potatoes. Cellulose is the structural material of plant cell walls. It cannot be broken down effectively in the human digestive system but it is used as an energy source in the diet of herbivorous animals such as sheep and cattle. Sucrose is the common form of sugar, used extensively in cooking and as a sweetener. Within the body, all these different carbohydrates are converted into glucose, the main fuel for respiration.

The lipids most often present in the diet include **fats** and **oils**. These can be used for respiration by conversion to glucose or acetyl-CoA. The chemical structures of lipids are described in Unit 6. Foods such as butter, margarine, cheese, and milk are well known to contain fat, but eggs, fish, meats, and bread also contain significant amounts. Oils are mainly of vegetable origin and are used for cooking. In developed countries, consumption of fats and oils has steadily increased and now provides as much as 40% of the daily energy requirement. Saturated fats increase the amount of cholesterol in the bloodstream and are thought to be a contributing factor in the occurrence of gallstones and heart attacks.

The energy content of different foodstuffs can be estimated using a **calorimeter**. Such an instrument works by collecting the heat produced when a weighed food sample is burned, usually in an oxygen atmosphere. One simple design of

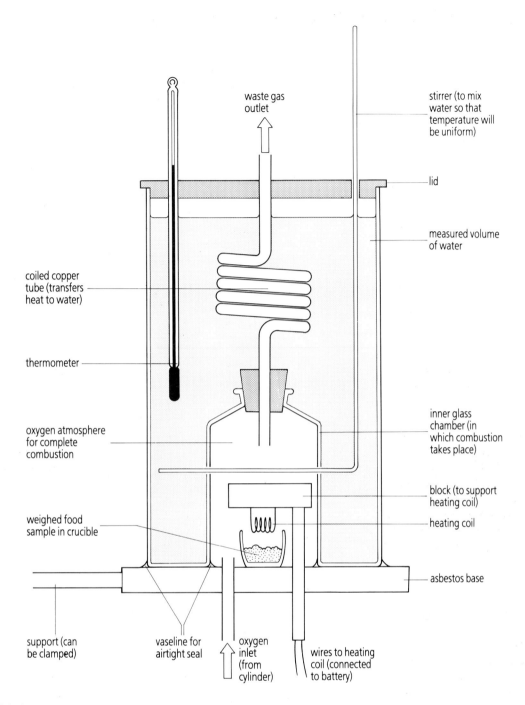

waste gas outlet

stirrer (to mix water so that temperature will be uniform)

lid

measured volume of water

coiled copper tube (transfers heat to water)

thermometer

oxygen atmosphere for complete combustion

inner glass chamber (in which combustion takes place)

block (to support heating coil)

heating coil

weighed food sample in crucible

asbestos base

support (can be clamped)

vaseline for airtight seal

oxygen inlet (from cylinder)

wires to heating coil (connected to battery)

Fig 12·3 Simple calorimeter

calorimeter is illustrated in Figure 12·3. The food sample is placed in the crucible and a gentle flow of oxygen through the apparatus is established. Once the temperature is constant, the food is ignited by means of the electric coil. While combustion is taking place, the water in the surrounding vessel is stirred to ensure even mixing and the final temperature reached is noted. From this result and the known volume of the water, the amount of heat produced can be calculated. Until recently, the energy content of food substances was measured in units called **calories**, where one calorie is defined as the quantity of heat energy needed to raise the temperature of 1 cm³ of water by 1°C. The calorie has since been replaced by the **joule** as a unit of energy and energy contents are

now usually given in units of kilojoules per gram (kJ g⁻¹). Converting to the new units is a straightforward process of multiplication: 1 calorie = 4.18 joules.

Estimated in this way, carbohydrates have an average content of 17 kJ g⁻¹ while fats provide 38 kJ g⁻¹. An adult male requires around 12 000 kJ per day and could therefore satisfy his energy needs by eating 700 g of carbohydrate or 300 g of fat. Proteins in the diet also have an energy value measured at 17 kJ g⁻¹, but the amino acids obtained from protein breakdown are not used for respiration unless surplus to the body's other requirements. Proteins are far more important in terms of the second main function of substances in the diet, as described below.

12·4 Materials for growth and repair ■ ■ ■ ■ ■ ■ ■ ■ ■ ■ ■ ■

Proteins form about half of the dry mass of the human body and much of the remaining dry mass is accounted for by the bony skeleton, consisting mainly of calcium phosphate. Even when not growing actively, the body's tissues are continually being renewed. Depending on the rate of this 'turnover', it is to be expected that the body will require significant amounts of **protein, calcium**, and **phosphate** in the diet, and these substances are in fact the most important materials needed for growth and repair.

Many thousands of different types of protein molecules exist, each suited to its own particular function in the life of an organism. The amino acid contents of these proteins vary widely so that the mixture of amino acids present in a person's diet may be quite different from that needed by his body. In a similar way, a person's amino acid requirements may vary from time to time, depending on which proteins are being replaced and on whether growth is taking place. Fortunately, the diet does not always need to provide precisely the correct mixture of amino acids because body tissues, and in particular the liver, have the ability to convert some amino acids into others. The amino acids which the body cannot manufacture, and which must therefore be present in the diet, are called **essential amino acids**, while the remaining amino acids are described as **non-essential** (see Table 12·1).

The only protein foods known to provide nearly the correct mixture of amino acids for human growth are eggs and human breast milk. All other protein foods are deficient in one or other of the essential amino acids. In general, foods from animal sources approximate more closely to human requirements than foods from plants. The nutritional value of wheat, for example, is reduced

Table 12·1 Amino acids in the human diet

ESSENTIAL AMINO ACIDS	NON-ESSENTIAL AMINO ACIDS
leucine	alanine
isoleucine	asparagine
lysine	aspartic acid
methionine	cysteine
phenylalanine	glutamic acid
threonine	glutamine
tryptophan	glycine
valine	proline
	serine
	tyrosine

ESSENTIAL FOR CHILDREN ONLY
arginine
histidine

because it is seriously deficient in the amino acid lysine. Taking account of these problems, the recommended daily intake of protein for an adult is around 70 g. Protein is more often deficient in the diet than carbohydrates and, in tropical countries, a lack of protein can give rise to diseases such as **kwashiorkor**. This occurs when infants are first weaned from breast milk onto low-protein cereal foods. It can easily be corrected by adding suitable protein-rich foods to the diet.

The 'turnover' of materials in bone is slower than the turnover of proteins so that calcium and phosphate are required in the diet in much smaller amounts. A daily calcium intake of 500 mg is recommended for adults. Milk and cheese are the best sources. Phosphate is present in virtually all foodstuffs and is rarely, if ever, deficient in the diet. As well as being used to make bones, phosphate is essential for making ATP and the nucleic acids, DNA and RNA.

12·5 Control factors in the diet ■ ■ ■ ■ ■ ■ ■ ■ ■ ■ ■ ■

The control factors in the diet include **vitamins** and **inorganic salts**. Vitamins are relatively complex carbon compounds which cannot be manufactured in the body but which are essential for normal metabolism. Table 12·2 summarizes their main functions in the human diet. The different functions of inorganic salts are listed in Table 12·3.

From an analysis of all the different food substances described in this Unit, it is a relatively easy matter to construct a balanced diet capable of supporting a healthy life. Comparing this with what people actually eat, nutrition becomes an even more fascinating study, for it is clear that people

do not automatically choose the best diet available. Taste, smell and colour all influence food choice and certain excellent foods may or may not be eaten by reason of religious belief or simple prejudice. In developed countries, advertising and social pressures condition eating habits. For example, 70% of the bread sold in the U.K. is made with white flour although, even with additives, white bread is nutritionally inferior to brown and wholemeal breads. Taking the numerous and often contradictory factors which affect food choice into account, it is a remarkable fact that most people do still manage to eat moderately good, if not ideal, diets.

Table 12·2 Vitamins in the human diet

VITAMIN			CHEMICAL NAME	FUNCTION IN BODY	DEFICIENCY DISEASE
WATER SOLUBLE	B Complex	B_1	thiamine	part of coenzyme thiamine pyrophosphate for respiration	BERI-BERI: nerve and heart disorders
		B_2	riboflavin	part of coenzyme FAD needed for respiration	ARIBOFLAVINOSIS: skin and eye disorders
		B_{12}	cyanocobalamin	coenzyme needed for making red blood cells	PERNICIOUS ANAEMIA: bone, blood and nerve changes
		B_3	nicotinic acid ('niacin')	part of coenzymes NAD, NADP used in respiration	PELLAGRA: skin, gut and nerve disorders
	C		ascorbic acid	not precisely known	SCURVY: degeneration of skin teeth and blood vessels
FAT SOLUBLE	A		retinol	not fully known but part of visual pigment rhodopsin	XEROPTHALMIA: 'dry eyes'
	D		cholecalciferol	stimulates calcium absorption by small intestine, needed for proper bone growth	RICKETS: bone deformity
	E		tocopherol	not precisely known	INFERTILITY: in rats (not shown for humans)
	K		phylloquinone	involved in blood clotting	possible HAEMORRHAGE

Table 12·3 Inorganic elements in the human diet

ELEMENT	COMMON ION	FUNCTIONS IN THE HUMAN BODY
calcium	Ca^{2+}	Bone contains large amounts of calcium phosphate. Calcium ions are needed for stability of cell membranes, as cofactors for some enzymes and are involved in muscle contraction, and blood clotting.
phosphorus	$H_2PO_4^-$	Bones, component of many organic molecules, DNA, RNA, ATP – phosphate group very reactive.
potassium sodium chlorine	K^+ Na^+ Cl^-	These ions are important in determining the balance of electrical charges in body fluids. This balance affects many processes including the production of nerve impulses.
sulphur	usually combined in organic molecules	Disulphide bridges are essential for protein structure.
magnesium	Mg^{2+}	Enzyme cofactor, involved in transmission of nerve impulses.
iron	Fe^{2+}, Fe^{3+}	Component of haemoglobin and cytochrome molecules.
iodine	I^-	Component of hormone thyroxin
copper manganese zinc	Cu^{2+} Mn^{2+} Zn^{2+}	Trace elements: usually enzyme cofactors, for example, Cu^{2+} is a cofactor for cytochrome oxidase.

13 HUMAN DIGESTIVE SYSTEM ⬚

Objectives

After studying this Unit you should be able to:–

- Draw and label a diagram showing the important structures of the human digestive system

- State the three main functions of the digestive system

- Give an account of the structure of human teeth

- Describe the actions of the teeth, tongue, and saliva on food materials in the mouth

- Give an account of digestion in the human stomach

- Draw and label a diagram showing the arrangement of tissues in the wall of the small intestine as seen with the light microscope

- Outline the processes involved in the digestion and absorption of food materials in the small intestine

- Comment on the structure and functions of the large intestine

- Distinguish between nervous and hormonal control of the alimentary canal

- Draw up a summary table listing all the important digestive processes taking place in the different regions of the human alimentary canal

13·1 Introduction ■ ■ ■ ■ ■ ■ ■ ■ ■ ■ ■ ■ ■ ■ ■ ■ ■

The human digestive system is illustrated in Figure 13·1. It consists of a long tube, called the **alimentary canal**, together with its associated organs such as the **liver** and **pancreas**. The alimentary canal runs continuously from the mouth to the anus and, in an adult person, it is 4·5 m long. Different parts of the canal are specialised for different functions but its walls share a similar structure throughout.

The functions of the digestive system are the **mechanical treatment** of food, followed by its **chemical digestion** and, finally, **absorption**. Mechanical treatment of food breaks it down into smaller pieces but does not alter its chemical structure. Chemical digestion occurs when large insoluble food molecules are acted on by digestive enzymes to form small soluble molecules capable of being absorbed, or, in other words, taken up across the walls of the alimentary canal into the

body. This division of functions roughly corresponds to the subdivision of the alimentary canal itself. Thus, the mouth, and in particular the teeth, are specialized for the mechanical treatment of food, while the stomach and the first part of the small intestine release enzymes for chemical digestion. Absorption takes place farther down the alimentary canal, particularly in the **ileum**.

This Unit considers these processes in turn and, at the same time, describes the structures involved in carrying them out. In following through the different stages in food breakdown, you should remember that the essential function of digestion is to produce molecular fragments small enough to be absorbed. In the case of the major nutrients, digestion is completed when all the carbohydrates have been broken down to **monosaccharides**, all the proteins to **amino acids**, and all the fats and oils to **fatty acids** and **monoglycerides** (glycerol

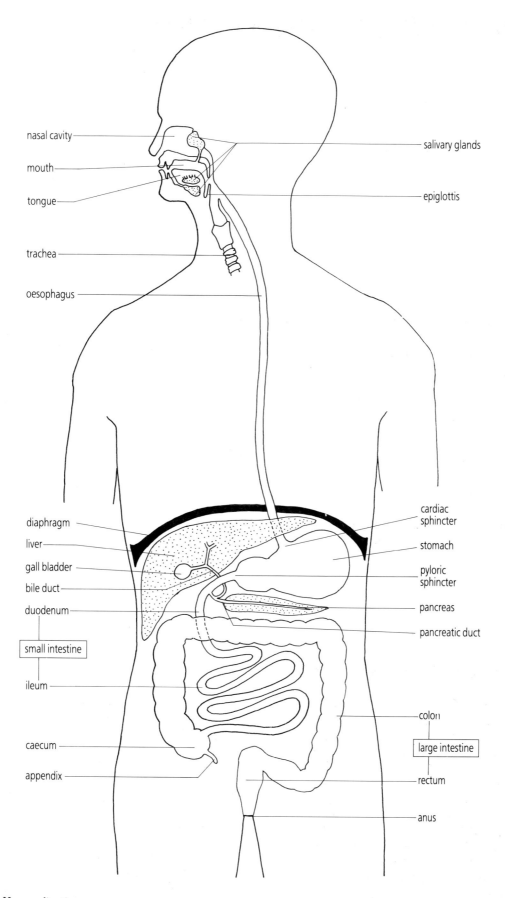

nasal cavity
mouth
tongue
trachea
oesophagus

salivary glands
epiglottis

diaphragm
liver
gall bladder
bile duct
duodenum
small intestine
ileum
caecum
appendix

cardiac sphincter
stomach
pyloric sphincter
pancreas
pancreatic duct
colon
large intestine
rectum
anus

Fig 13·1 Human digestive system

combined with one molecule of fatty acid). Larger molecules, representing intermediate stages in the digestion of major nutrients, *cannot* be absorbed. Vitamins and inorganic ions are usually absorbed directly, without chemical modification, and simple mechanical treatment of the food is enough to make them available for absorption.

Digestion in the mouth involves the **teeth**, the **tongue**, and the production of **saliva**.

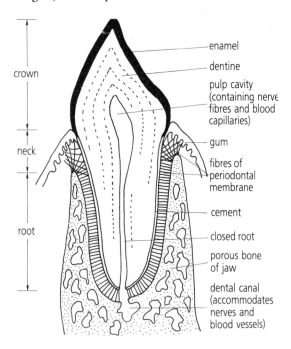

Fig 13·2 Cross section of human incisor tooth

Teeth

The human dentition consists of different types of teeth; chisel-shaped **incisors** and **canines** at the front of the mouth, giving way to flattened **premolar** and **molar** teeth, used for grinding, at the back. This arrangement is described as a **heterodont** dentition. Only a very few mammals, such as the dolphin, have the more primitive **homodont** dentition, consisting entirely of simple peg-like teeth. Homodont dentition is appropriate for dealing with slippery prey but, because the peg-like teeth are not capable of cutting or grinding, the food must be swallowed whole.

The structure of an individual human incisor tooth is illustrated in Figure 13·2 Like all other mammalian teeth, it consists of a number of distinct layers:

1 Enamel. This forms the outermost layer of the tooth and is the hardest substance produced by any living organism. It consists mainly of long thin crystals of calcium phosphate laid down side by side in a precise arrangement. The special cells, called **ameloblasts**, which deposit the enamel lie outside the tooth during its development so that they die when the tooth erupts (that is, when it emerges from the gum). Consequently, damaged enamel cannot be replaced; it is very resistant to fracture and abrasion, but can be attacked by acids in the mouth. Sugary foods encourage the growth of acid-producing bacteria and can lead to tooth decay, or **dental caries**, unless the teeth are properly cleaned.

2 Dentine. This material resembles bone and forms the bulk of the inside of the tooth. It is produced by cells called **odontoblasts** which remain alive even when the tooth is fully developed. Dentine retains a limited capacity for repair and can protect the tooth if the enamel becomes cracked or damaged.

3 Pulp cavity. The pulp forms the living centre of the tooth and consists of connective tissue containing **nerve fibres** and **blood capillaries**. In human teeth, the opening to the pulp cavity narrows considerably and the teeth are said to have **closed roots**.

4 Cement. Cement is a specialised form of bone which surrounds the roots of teeth and helps to hold them in position. Between the cement and the jaw bone itself lies the **periodontal membrane**. This membrane contains short fibres made of the tough fibrous protein called **collagen**. The fibres are embedded in cement at one end and are attached directly to bone at the other. In this way, the tooth is held securely but is able to move slightly in its socket. Such movements absorb sudden stress and make the tooth less likely to break during feeding.

Human teeth, like those of other mammals, are replaced during life. The 20 temporary or 'deciduous' teeth called **milk teeth** erupt during the first two years of life, and replacement begins from the age of six years. The final arrangement, comprising 32 **permanent teeth**, is shown in Figure 13·3. The premolar teeth have single roots and replace milk teeth in the corresponding position. They have two small pointed projections called **cusps** on their grinding surfaces. Molar teeth have either two or three roots and appear only in the permanent dentition; all of them have four cusps. (The last molars to appear are sometimes called 'wisdom teeth'). The chisel-shaped incisors and the pointed canines each have only one cusp.

The lower jaw in man is moved by four main groups of muscles. The **temporalis** muscles run under the cheek bone (zygomatic arch) on each side of the head and are attached in the region of the temples (see Fig. 13·3A). These muscles normally contract together to produce a powerful upward movement of the lower jaw, or **mandible**. The **masseter** muscles are enclosed in the flesh of the cheeks and assist in closing the mouth, bringing the upper and lower teeth into contact. The side to side movements needed for grinding, or **mastication**, of food are produced mainly by alternate contractions of the **medial pterygoid** muscles. These are attached at one end to the *inside* surfaces of the mandible, and at the other to the bony floor of the skull. Their position and action can be visualized looking up into the skull from behind (Fig. 13·3D). The mouth is opened by

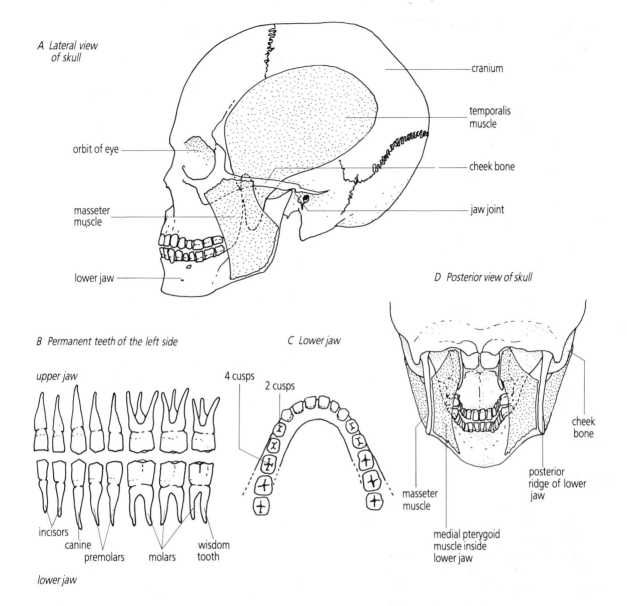

A *Lateral view of skull*

cranium
temporalis muscle
cheek bone
orbit of eye
jaw joint
masseter muscle
lower jaw

D *Posterior view of skull*

B *Permanent teeth of the left side*

C *Lower jaw*

upper jaw

4 cusps
2 cusps

cheek bone

incisors
canine
premolars
molars
wisdom tooth

lower jaw

masseter muscle

posterior ridge of lower jaw

medial pterygoid muscle inside lower jaw

Fig 13·3 Human skull and dentition

the **lateral pterygoid** muscles which pull forwards on the mandible at the gliding joint formed where it articulates with the skull.

Tongue

The tongue is a mass of tissue attached to the floor of the pharynx at the back of the mouth. It contains numerous internal muscles and its main function in digestion is to move the food around in the mouth so that it can be acted on by the teeth. **Taste buds** located in the tongue play a part in food selection and stimulate the production of saliva by reflex action.

Saliva

Saliva is a watery liquid released from the salivary glands. Flow is continuous at a rate of about 0·05 cm³ per minute, but increases when food is present. Water in the saliva helps to dissolve soluble materials in the food and **mucus** lubricates the food mass, enabling it to be swallowed. Saliva is nearly neutral (pH 7) and contains an enzyme, **salivary amylase** (sometimes called **ptyalin**), which acts on starch in the food, breaking it down into maltose. Food remains in the mouth only briefly, but, when the food mass is swallowed, it is shaped into a ball, or **bolus**. Salivary amylase continues to work inside such a bolus even when it is transferred to the acid environment of the stomach (pH 2). Eventually, when the acid penetrates to the centre of the bolus, the action of the enzyme is inhibited but by this time significant amounts of starch have been digested.

Like many other body secretions, saliva contains the enzyme **lysozyme**, which attacks bacterial cell walls. Its action helps to keep the tongue and teeth clean and free from infection.

Swallowing is a complex reflex activity which occurs automatically when a bolus of food reaches the back of the mouth. The food mass is propelled down the **oesophagus** by a wave of muscular contraction, called **peristalsis**, and enters the stomach when the muscles of the **cardiac sphincter** relax.

The human stomach is a hollow J-shaped bag closed at the top by the cardiac sphincter and at the lower end by the **pyloric sphincter**. Food taken in at a meal can be stored in the stomach for several hours. During this time, the food mass is churned by rhythmic waves of contraction which pass along the stomach about once every 20 seconds. At the same time, the food is acted on by fluid secretions, collectively called **gastric juice**, produced from glands in the stomach wall.

The structure of a small part of the stomach wall is illustrated in Figure 13·4. The surface epithelium is infolded to give numerous tubular glands called **gastric pits**. (There are as many as 10 000 gastric pits per square centimetre.) Inside these pits, the **oxyntic cells** manufacture hydrochloric acid (HCl) and the **chief cells** manufacture **pepsinogen**. These substances are released in the gastric juice together with copious quantities of mucus from the **neck cells** surrounding the openings of the gastric pits. The presence of food in the stomach stimulates **endocrine cells** in the stomach wall to release a hormone called **gastrin**. This substance circulates in the bloodstream and stimulates further release of gastric juice.

The acid environment of the stomach denatures proteins, that is, it causes individual protein molecules to unwind, thereby making it easier for food to be broken into smaller pieces by the churning action of the stomach. At the same time, HCl converts the harmless pepsinogen into a powerful enzyme called **pepsin** by removing a molecular fragment masking the active centre. Pepsin acts on the proteins in the food, producing short peptide chains, usually containing from 4 to 12 amino acids. The peptide bonds broken are mainly in the middle of the protein chain and pepsin is therefore described as an **endopeptidase** enzyme. **Exopeptidase** enzymes, some of which occur in other parts of the alimentary canal, work only at the extreme ends of the protein chain. A thick layer of mucus protects the lining of the stomach from attack by pepsin.

The combined action of acid and enzymes kills bacteria in the food and produces a semi-fluid mixture called **chyme**. The materials in chyme have been only partly digested and, as a result, there is little or no absorption of organic nutrients from the stomach. Chyme enters the small intestine via the pyloric sphincter at a rate which is controlled by nerves and hormones.

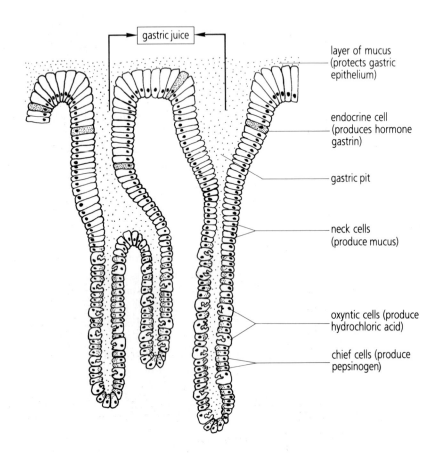

Fig 13·4 Diagram to show the arrangement of cells in gastric pits

13.4 Digestion in the small intestine ■ ■ ■ ■ ■ ■ ■ ■ ■ ■

The human small intestine is a coiled tube 2·75 m long and 4 cm in diameter. It is the most important region of the alimentary canal for chemical digestion and absorption.

The first 20 cm of the small intestine, called the **duodenum**, receives digestive secretions from the **liver** and from the **pancreas**, as well as producing some enzymes of its own. These secretions neutralise the stomach acid and rapidly break down partly digested food molecules into fragments suitable for absorption. In the human alimentary canal, the duodenum merges without abrupt change into the second, much longer section of the small intestine, called the **ileum**.

The structure of the ileum, illustrated in Figure 13·5, is adapted particularly for absorption. The surface epithelium of the ileum is folded to form numerous finger-like processes called **villi** and, in addition, the membranes of the epithelial cells are folded to form **microvilli**. This folding increases the surface area in contact with food materials by as much as 600 times compared with a simple tube of the same dimensions. The **lacteal** in the centre of each villus is a narrow tubular sac joined at its lower end to one of the many **lymphatic vessels** found in the walls of the small intestine. When the muscles in the villi contract, fluid is squeezed out of the lacteals and drains into the lymphatic system. The **blood capillary network** serving the villi drains to the **hepatic portal vein** which transports blood directly to the liver. Between the villi are tubular glands called **crypts of Lieberkuhn**. These secrete **intestinal juice** consisting of mucus and enzymes believed to come from the **Paneth cells** lining the base of the crypts.

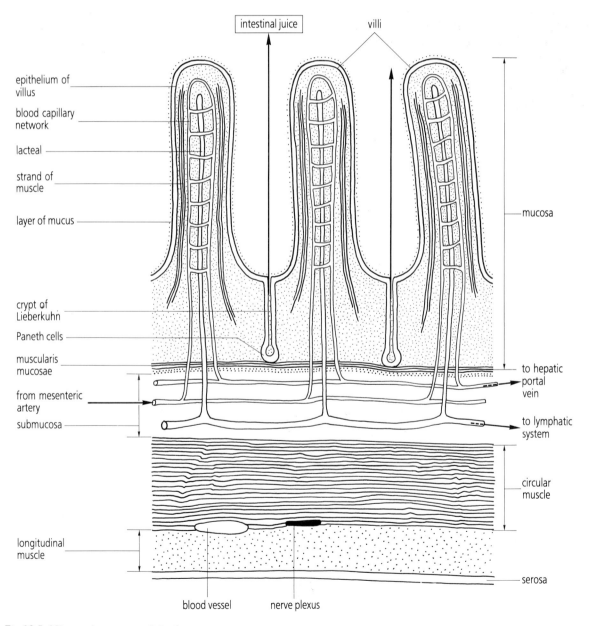

Fig 13·5 Microscopic structure of the ileum

These structures together with a thin layer of muscle (**muscularis mucosa**) form a distinct surface layer, or **mucosa**, on the inside of the ileum. Beneath this layer is the **submucosa** containing blood and lymph vessels and nerve fibres. The submucosa is followed by a thick layer of **circular muscle** and a thinner layer of **longitudinal muscle** which together provide for peristalsis. A final layer of connective tissue called the **serosa** forms the outermost covering of the ileum. This layered structure is found in all parts of the alimentary canal. In life, the serosa is attached to sheets of connective tissue called **mesenteries**, which hold the different parts of the canal in position within the body cavity.

A considerable number of different digestive processes take place in the small intestine, as summarized in Table 13·1. It is best to consider these processes in terms of the digestion and absorption of the major nutrients and vitamins, as follows:

Table 13.1 *Summary of digestion in the human alimentary canal*

SECRETION	SOURCE	SITE OF ACTION	ACTION ON CARBOHYDRATES	ACTION ON PROTEINS	ACTION ON FATS AND OILS	OTHER DIGESTIVE FUNCTIONS
SALIVA	salivary glands	mouth				
salivary amylase			starch ► maltose			
lysozyme						attacks bacterial cell walls
mucus						lubrication of food mass
GASTRIC JUICE	stomach wall	stomach				
hydrochloric acid				denatures protein pepsinogen ► pepsin		
pepsin				protein ► peptides		
mucus						protection of stomach lining
intrinsic factor						absorption of vitamin B_{12}
BILE	liver	small intestine				
bile salts					emulsification, formation of micelles	
sodium hydrogen-carbonate						neutralisation of stomach acid
PANCREATIC JUICE	pancreas	small intestine				
pancreatic amylase			starch ► maltose			
trypsin				protein ► peptides		
chymotrypsin				protein ► peptides		
carboxypeptidase				peptides ► amino acids		
lipase					fats ► fatty acids, monoglycerides	
nuclease						DNA, RNA ► nucleotides
sodium hydrogen-carbonate						neutralization of stomach acid

SECRETION	SOURCE	SITE OF ACTION	ACTION ON CARBOHYDRATES	ACTION ON PROTEINS	ACTION ON FATS AND OILS	OTHER DIGESTIVE FUNCTIONS
INTESTINAL JUICE	wall of small intestine	small intestine				
disaccharidase enzymes						
maltase			maltose ▶ glucose + glucose			
sucrase			sucrose ▶ glucose + fructose			
lactase			lactose ▶ glucose + galactose			
enterokinase				trypsinogen ▶ trypsin		
aminopeptidase				peptides ▶ amino acids		
lipase					fats ▶ fatty acids, monoglycerides	
nucleotidase						breakdown of nucleotides
mucus						protection, lubrication of food mass

Carbohydrates

Carbohydrates are digested first in the mouth by the action of salivary amylase, and subsequently by **pancreatic amylase** present in the pancreatic juice. Polysaccharides are converted into short chains of glucose units and disaccharides, mainly maltose. Lactose and sucrose may be present in the diet. The membranes of the epithelial cells lining the small intestine contain **disaccharidase** enzymes as an integral part of their structure. The monosaccharides formed when these enzymes act combine immediately with specific carrier proteins and are actively transported into the cytoplasm of the epithelial cells. The absorbed monosaccharides enter the blood capillary network of the villi and are transported to the liver.

Proteins

The breakdown of proteins begun by pepsin in the stomach continues in the duodenum with the action of **trypsin** and **chymotrypsin**. Like pepsin, these powerful enzymes are first released in a harmless form. **Trypsinogen** is converted into its active form, trypsin, by another enzyme called **enterokinase** (molecules of enterokinase are embedded in the membranes of the cells lining the small intestine). Trypsin and chymotrypsin are endopeptidase enzymes and act to produce short protein fragments such as dipeptides and tripeptides. The final stage in protein digestion involves **exopeptidase** enzymes; **carboxypeptidase**, present in the pancreatic juice, splits off amino acids one at a time from the carboxyl or C-terminal end of a protein chain, while **aminopeptidase**, located in the membranes of the epithelial cells, splits off amino acids from the amino or N-terminal end. The free amino acids produced are actively transported into the villi and enter the blood capillaries for transport to the liver.

Fats and oils

The triglyceride molecules of fats and oils are insoluble in water and tend to become separated from other components of the food, forming large droplets. No significant digestion of fats takes place until the duodenum; the processes which follow are outlined in Figure 13·6. **Bile salts**, released from the liver in the bile, are steroid substances manufactured from cholesterol. One end of the bile salt molecule is hydrophobic ('water-hating') and lipid soluble, while the other end contains a negatively charged ionic group and is strongly hydrophilic ('water-loving'). The bile salt molecules therefore dissolve in the fat droplets with the charged groups pointing outwards, forming a surface layer. Small droplets formed by mechanical agitation do not join together again because the external charged surfaces repel each other. In this way, bile salts act to produce small lipid droplets (1 μm in diameter) suspended in water; this process is called **emulsification**.

Emulsification enormously increases the surface area available for enzyme attack and **pancreatic lipase** now breaks down exposed triglycerides to produce fatty acid and monoglyceride molecules.

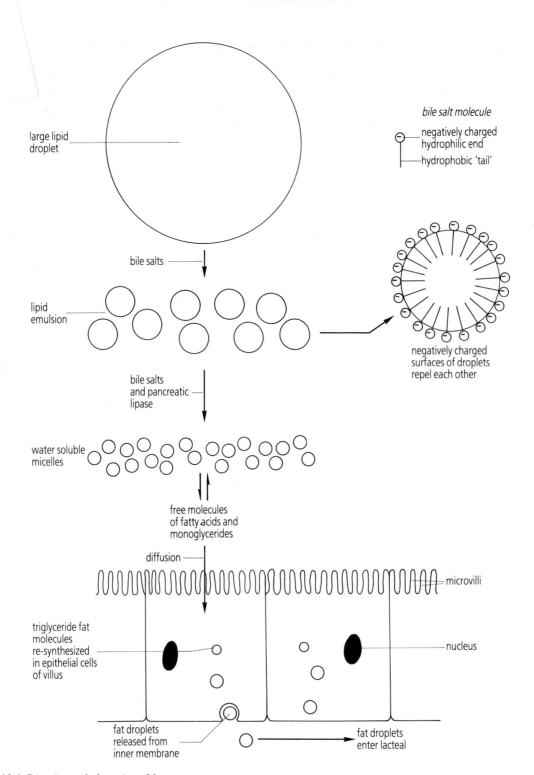

Fig 13·6 Digestion and absorption of fats

Bile salts form these molecules into minute particles (3 to 10 nm in diameter) called **micelles** and, from these, fatty acid and monoglyceride molecules diffuse into the epithelial cells lining the intestine. Inside the epithelial cells, triglyceride fat molecules are re-synthesized and minute fat droplets are released into the interior where they enter the lacteals. After transport through the lymphatic system, fat droplets enter the bloodstream near the subclavian veins and are distributed to all parts of the body. Surplus fats are stored by **adipose cells**, which form a layer under the skin.

Vitamins

Most vitamins are actively absorbed by the epithelial cells without chemical modification. Vitamin B$_{12}$ (cyanocobalamin) provides an interesting exception. This vitamin can only be absorbed in combination with a special protein called **intrinsic factor**. Intrinsic factor is released along with mucus by the neck cells in the stomach lining and binds to the epithelial cells in the ileum, stimulating membrane endocytosis (described in Unit 6). In this way, vitamin B$_{12}$ molecules cross the membrane and are absorbed into the body.

13·5 The large intestine ■ ■ ■ ■ ■ ■ ■ ■ ■ ■ ■ ■ ■

The human large intestine is 1·2 m long and 6 cm in diameter. It is subdivided into two sections, called the **colon** and the **rectum**. The main function of the colon is to absorb water from the undigested material inside, concentrating it into a semi-solid mass called **faeces**. Water absorption is achieved by actively transporting sodium ions out of the colon. This reduces the concentration of dissolved substances remaining, with the result that water is absorbed into the body by osmosis.

The faeces are stored temporarily in the rectum before being passed out of the body via the anus. The removal of faeces is called **defecation**.

13·6 Control of the alimentary canal ■ ■ ■ ■ ■ ■ ■ ■ ■ ■

All the different digestive processes described in this Unit must be coordinated together so that the individual cells of the body can obtain a steady supply of the raw materials they need. There are two separate control systems for this purpose. **Nervous control** determines hunger levels, and thereby regulates food intake, and is also involved in reflex activities such as salivation, swallowing, secretion from digestive glands, and peristalsis. **Hormonal control** acts mainly to modify the long term activities of the alimentary canal, dampening down or speeding up nervous activity, and acting directly on secretory cells to control their rates of production. Some aspects of these control mechanisms are discussed in Unit 23, and in Unit 15, which deals with the liver and pancreas. The next Unit considers how the basic structure of the mammalian digestive system is altered to cope with the very different specialized diets of carnivores and herbivores.

14 CARNIVORES AND HERBIVORES

Objectives

After studying this Unit you should be able to:–

- Describe how the teeth and skull of a mammalian carnivore (dog) are adapted for eating meat

- Describe how the teeth and skull of a mammalian herbivore (cow) are adapted for eating grass

- Draw diagrams to illustrate the structure of a high-crowned molar tooth

- Comment on the different roles of temporalis, masseter, and medial pterygoid muscles in carnivores and herbivores

- Give an account of the chemical digestion of cellulose in herbivores

- Discuss the specializations of the alimentary canal found in carnivorous, ruminant, and non-ruminant animals

14·1 Introduction ■ ■ ■ ■ ■ ■ ■ ■ ■ ■ ■ ■ ■ ■ ■

In Unit 13, the structure and function of the human digestive system was described. The human system is rather unspecialized and copes moderately well with a wide range of different food materials. Carnivores and herbivores have much more specialized digestive systems which deal with very restricted diets. **Carnivores** eat meat: this material is mainly protein and, since its chemical composition is similar to that of the animal's own body, it is needed only in small amounts. It is easy to digest but difficult to obtain, at least in as much as prey organisms evade capture. **Herbivores** eat grass or other vegetation. This is easy to obtain but is very difficult to digest and is therefore eaten in large amounts. These differences in diet have a dramatic effect on the structure of the organisms involved, especially in relation to the teeth and skull.

14·2 Teeth, skull, and digestion in a carnivore ■ ■ ■ ■ ■ ■

Figure 14·1 shows the teeth and skull of a dog, well adapted for a carnivorous mode of life. Domestic dogs do not catch and kill prey in the same way as their wolf-like ancestors but retain many typical carnivore features. At the front of the mouth, the **incisor** teeth are chisel-shaped and meet with a nibbling action which helps in removing scraps of meat from the bone. The prominent curved **canine** teeth are pointed and used to pierce the prey, holding it firmly. Carnivores often kill their prey by vigorous shaking at the neck. The jaws of the dog close with a powerful scissor-like action, the **premolars** and the large **carnassial** teeth being superbly adapted for bone-splitting and cutting the meat into smaller pieces. At the back of the mouth, the **molar** teeth are small and little used. The teeth of a carnivore merely slice the food into pieces small enough to swallow and there is little or no mastication. Although the teeth are exposed to large forces during feeding, the food is not abrasive, so that they do not wear excessively. Like human teeth, they have closed roots.

In carnivores the **temporalis** muscles are

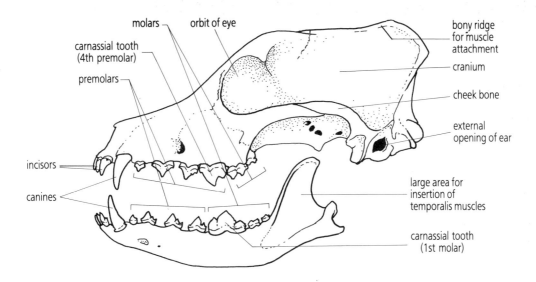

molars — orbit of eye

carnassial tooth
(4th premolar)

premolars

incisors

canines

bony ridge
for muscle
attachment

cranium

cheek bone

external
opening of ear

large area for
insertion of
temporalis muscles

carnassial tooth
(1st molar)

Fig 14·1 Teeth and skull of a carnivore (dog)

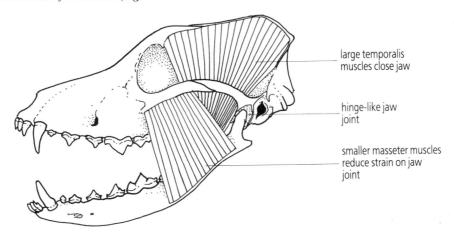

large temporalis
muscles close jaw

hinge-like jaw
joint

smaller masseter muscles
reduce strain on jaw
joint

Fig 14·2 Jaw muscles in a carnivore (dog)

enlarged (see Fig. 14·2) and run all the way from the lower jaw to the top of the skull where there is a pronounced ridge for muscle attachment. The shape of the lower jaw gives the temporalis muscles good leverage so that the jaw closes with considerable force. The **masseter** and **medial pterygoid** muscles are less well developed but also contract when the jaws close, pulling the jaw forward and thereby reducing the strain at the jaw

joint. The joint itself is hinge-like and unlikely to become dislocated during feeding.

The remainder of the alimentary canal in the dog is quite similar to that in humans but somewhat shorter. Carnivores chasing a mobile prey would be at a disadvantage weighted down with food and therefore tend to eat occasional large meals followed by periods of inactivity, during which digestion takes place.

14·3 Teeth, skull, and digestion in a herbivore ■ ■ ■ ■ ■ ■

Herbivores experience different problems from carnivores in collecting and digesting food. The teeth and skull of a cow (Fig. 14·3) illustrate the characteristic herbivore pattern. At the front of the mouth, the **incisors** and the incisor-like **canines** of the lower jaw bite against a horny pad on the surface of the upper jaw and, assisted by the tongue, mouthfuls of grass are ripped off. Further back in the mouth, separated from the front teeth by a space called the **diastema**, lies a battery of **premolar** and **molar** grinding teeth. These have a

very specialized structure to cope with the difficult diet.

As illustrated in Figure 14·4, herbivore teeth are high-crowned and have open roots providing a rich blood supply, so that they continue to grow throughout life. During development, the outer surface of the tooth is covered with **cement** which wears away on the grinding surface. Continued abrasion exposes **ridges of enamel** separated by slightly softer regions of **dentine**. The ridges of the adjacent teeth run from the front to the back

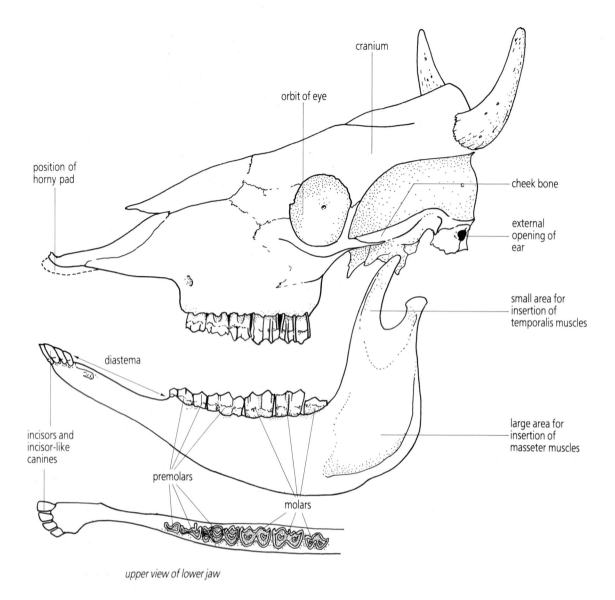

Fig 14·3 *Teeth and skull of a herbivore (cow)*

of the mouth so that sideways movements of the jaws produce an efficient grinding action. The alternate arrangement of harder and softer regions gives a pattern of **uneven wear** which maintains a good grinding surface throughout the life of the animal.

The main muscles closing the jaws are the **medial pterygoid** muscles and the **masseter** muscles located in the cheeks (Fig. 14·5). Both groups of muscles are inserted over a large area of the lower jaw and have good leverage compared with the smaller **temporalis** muscles. As explained in Unit 13, the medial pterygoid muscles are attached to the inside surfaces of the lower jaw and are the major muscles involved in producing side to side motion. The jaw joint is very loose. Viewed from the front, the lower jaw moves in a circular path, either clockwise or anti-clockwise, so that the horizontal surfaces of the teeth grind against each other. This grinding action reduces vegetation to cellular fragments, releasing the cell contents and breaking up cellulose cell walls into small pieces. The diastema allows food materials to

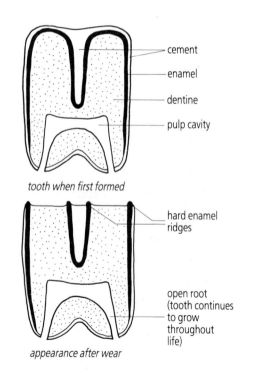

Fig 14·4 *Formation of a high-crowned molar tooth*

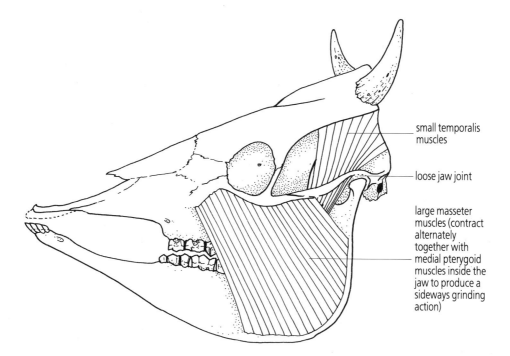

Fig 14·5 Jaw muscles in a herbivore (cow)

be transferred by the tongue into the spaces under the cheeks, ready to be acted on by the teeth. Mastication is the most important role of the teeth in herbivores for, without it, adequate chemical digestion of the food is impossible.

The chemical digestion of cellulose in the alimentary canal of herbivores is a remarkable achievement. As described in Unit 3, cellulose is a complex carbohydrate consisting of glucose units linked by 1β – 4 glycoside linkages. These linkages are quite different in shape compared with the 1α – 4 linkages in starch and glycogen and cannot be broken down by the same enzymes. Thus, although cows and other herbivores produce amylase enzymes, these are quite useless as a means of digesting cellulose. The **cellulase** enzymes needed are produced not by the cow but by the microorganisms, mainly bacteria, living in large numbers in specialized regions of the alimentary canal. Inside the alimentary canal, oxygen is in short supply, so that the bacteria use cellulose as the raw material for **anaerobic respiration**, releasing several different substances, particularly **fatty acids**, as waste products. The fatty acids are absorbed into the bloodstream of the herbivore and are converted to form other organic compounds or used directly for **aerobic respiration** within the tissues of the animal, where oxygen is now available. An interesting consequence of the use of fatty acids for respiration is that cows often survive with very low levels of glucose circulating in the bloodstream.

In rabbits and horses, the stomach remains unspecialised, but the **colon** and **caecum** are enormously enlarged to accommodate microorganisms. This arrangement is not very efficient because, within the time available the food is only partly digested, and because absorption from the large intestine is limited. To improve the yield of useful substances, rabbits often eat their own faecal pellets so that food materials pass through the alimentary canal for a second time.

The best adapted grazing or browsing herbivores are the **ruminant** mammals such as sheep and cows. These animals have a four-chambered 'stomach', as shown in Figure 14·6. The first three compartments, the **rumen**, **reticulum**, and **omasum**, develop from the oesophagus, while the fourth, called the **abomasum** corresponds to the true stomach. The rumen is a large sac with a capacity of about 150 litres in the cow. During the first phase of feeding, grass is collected and churned together with saliva in the rumen. When the rumen is full, the animal stops feeding and mouthfuls of grass are regurgitated for fine grinding in the mouth. This process, called 'chewing the cud', or **rumination**, speeds up the digestion of cellulose and helps to provide good conditions for microbial growth. The rumen contents may contain as many as 10^{11} bacteria and 10^6 protozoa per cm^3.

Partly digested food from the rumen passes through the reticulum into the omasum where excess fluid is squeezed out. The semi-solid mass entering the abomasum is acted on by hydrochloric acid and pepsin, as in the case of food entering the human stomach. The acid and enzymes kill and digest many of the bacteria, and this provides the largest part of the animal's protein needs. As outlined in Figure 14·7, protein in the diet is not used directly, but is first taken up by microorganisms and either incorporated into new bacterial protein or 'deaminated' to release

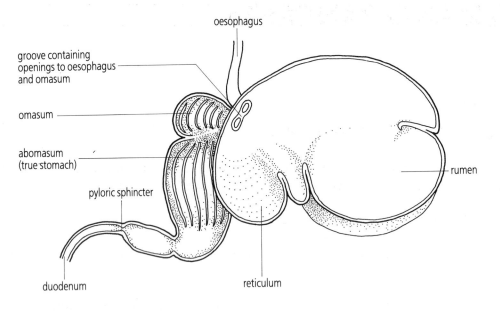

Fig 14·6 The 'stomach' of a cow opened to show its four compartments

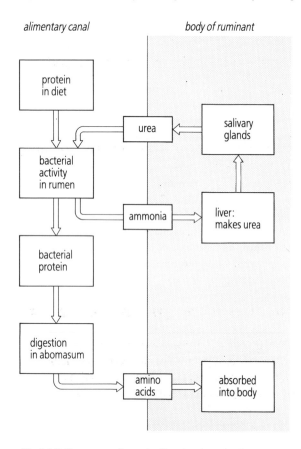

Fig 14·7 Summary of protein digestion in a ruminant

ammonia. The ammonia is recycled through the body of the ruminant and returns to the alimentary canal in the form of urea present in the saliva. The urea is then used to make new proteins which are ultimately digested and absorbed into the body. In this way, herbivores make the best possible use of nitrogenous substances and very little dietary

protein is wasted. The remaining sections of the alimentary canal in ruminants are long but not specially modified for bacterial action.

The association between a herbivore and its microorganisms benefits both partners and is an example of **mutualism**, as discussed in Unit 57. As well as helping in carbohydrate and protein digestion, bacteria in the alimentary canal produce some of the vitamin B substances, which are then absorbed and used by their hosts. Other organisms benefit from gut bacteria in a similar way. For example, mutualistic bacteria in the human alimentary canal produce vitamin K, and in the colon there is even a limited breakdown of cellulose. Thus, herbivores are not unique organisms, but merely specialized for one particular form of digestion. In the same way, carnivore specializations are best thought of as modifications to a basic plan which is common to most mammals and other vertebrates.

In young ruminants, the abomasum produces a special enzyme called **rennin** which clots milk. This is important because it allows the milk to be held in the stomach for longer and prepares the way for chemical digestion by other enzymes. (In human infants, milk is clotted by hydrochloric acid alone and the enzyme rennin does not occur.)

The next Unit describes the structure and functions of the liver and pancreas. These organs play an important part in digestion in all vertebrates and also help to regulate the detailed composition of the blood. In this way, the activities of the liver and pancreas affect all parts of the body. In mammals, these regulatory functions are particularly important.

15 LIVER AND PANCREAS

Objectives

After studying this Unit you should be able to:–

- Describe the location of the liver in the human body and draw a diagram to show how it is linked to the circulatory system and alimentary canal

- Give a clear account of the arrangement of cells within a liver lobule

- Describe the flow of blood through the liver sinusoids and comment briefly on the formation of bile

- Explain how the production and release of bile is controlled by secretin and cholecystokinin (CCK)

- Define homeostasis

- List the regulatory functions of the liver and discuss the main factors which affect blood glucose concentration

- List and explain the excretory functions of the liver

- Draw a diagram to illustrate the arrangement of cells in the pancreas and comment on the production of pancreatic juice and pancreatic hormones

- Briefly describe the disease diabetes mellitus

15·1 Introduction ■ ■ ■ ■ ■ ■ ■ ■ ■ ■ ■ ■ ■ ■ ■ ■ ■

The digestion of food is not achieved by the alimentary canal alone but depends on the activity of other organs and associated glands. The two most important organs associated with the gut are the liver and the pancreas. These structures have major digestive functions, but their importance is not limited to digestion for they carry out many other vital tasks needed to keep the body as a whole functioning correctly.

15·2 Structure of the liver ■ ■ ■ ■ ■ ■ ■ ■ ■ ■ ■ ■ ■

The liver is the largest organ in the body of all vertebrates. The human liver has a mass of about 1·5 kg and is located under the diaphragm at the top of the abdominal cavity. The relationship of the liver to the alimentary canal and blood vessels is illustrated in Figure 15·1. Blood flows into the liver from the **hepatic artery** at a rate of approximately 21 litres per hour, and from the **hepatic portal vein** at approximately 66 litres per hour. Almost all the water entering the liver in the bloodstream remains in the blood and passes out via the **hepatic vein**, which leads directly to the inferior vena cava and the heart. However, many different dissolved substances can be removed from or added to the blood so that its composition leaving the liver is altered. Some substances leave the liver in the **bile**, but the amount of fluid involved is relatively small (only about 0.03 litres

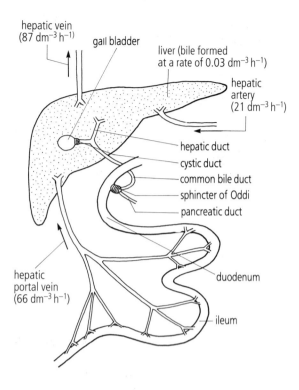

hepatic vein
(87 dm⁻³ h⁻¹)

gall bladder

liver (bile formed
at a rate of 0.03 dm⁻³ h⁻¹)

hepatic
artery
(21 dm⁻³ h⁻¹)

hepatic duct

cystic duct

common bile duct

sphincter of Oddi

pancreatic duct

duodenum

hepatic
portal vein
(66 dm⁻³ h⁻¹)

ileum

Fig 15·1 Blood supply to the liver

per hour). Bile passes down the **hepatic duct** and may be temporarily stored in the **gall bladder** before being discharged via the **common bile duct** and the **sphincter of Oddi** into the duodenum.

The external shape of the liver is of little importance, but inside the liver the various blood vessels branch repeatedly and the liver cells, called **hepatocytes**, are arranged in a precise way forming more or less distinct structures called **lobules** held in a connective tissue matrix (see Fig. 15·2, 15·3 and 15·4). The **interlobular veins** are branches of the hepatic portal vein and the **central vein** of each lobule drains to the hepatic vein. Radiating from the central vein like the spokes of a wheel is a network of tiny blood spaces called **sinusoids**. As illustrated in Figure 15·4, blood enters the sinusoids from the hepatic portal vein *and* from the hepatic artery and slowly percolates through to the central vein.

The cells lining the sinusoids are in direct contact with the liquid part of the blood, called blood plasma, and their surface area is increased by numerous **microvilli**. Between these cells, fitted into grooves in the membranes of adjacent cells, there are tiny channels called **bile canaliculi** into which the bile is secreted. There are no direct connections between the canaliculi and the sinusoids so that any substances extracted from the blood must pass through the cytoplasm of the liver cells before entering the bile. The canaliculi form an interconnected network which drains to the **bile ducts** located in the interlobular connective tissue. Bile is produced continuously and is eventually collected and passed out for storage in the gall bladder.

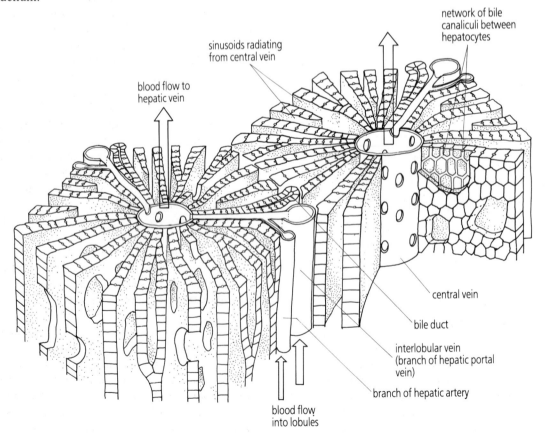

network of bile
canaliculi between
hepatocytes

sinusoids radiating
from central vein

blood flow to
hepatic vein

central vein

bile duct

interlobular vein
(branch of hepatic portal
vein)

branch of hepatic artery

blood flow
into lobules

Fig 15·2 Diagram to illustrate the structure of liver lobules

104

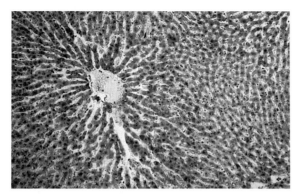

Fig 15·3 Microscopic structure of liver lobules (×1000)

The sinusoids also contain numerous **Kupffer cells**. These are vigorously phagocytic, engulfing worn out red blood cells and any bacteria present in the blood coming from the small intestine.

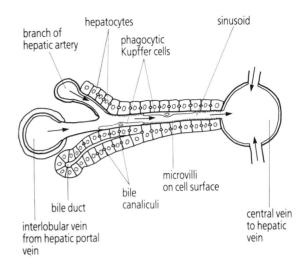

Fig 15·4 Arrangement of cells in part of a liver lobule

15·3 Functions of the liver ■ ■ ■ ■ ■ ■ ■ ■ ■ ■ ■ ■ ■

The functions of the liver can be grouped under three main headings, as follows:

Digestive functions

The digestive functions of the liver are carried out by bile. As described in Unit 13, **bile salts** play an essential role in the digestion of fats and, in addition, **sodium hydrogencarbonate** (NaHCO$_3$) helps to neutralize hydrochloric acid entering the duodenum from the stomach. The production and release of these substances is controlled by two main hormones, namely **secretin** and **cholecystokinin (CCK)**. Secretin is released into the bloodstream from endocrine cells in the small intestine when acid is present and CCK is released in response to amino acids and fatty acids. Both hormones speed up the secretion of hydrogencarbonate ions into the bile, while CCK stimulates contraction of the gall bladder and causes the sphincter of Oddi, joining the common bile duct to the duodenum, to relax. 95% of the bile salts released into the small intestine are reabsorbed into the body and can be reused. The liver synthesizes new bile salts to replace the 5% lost from the body in the faeces.

Regulatory functions

To keep them alive, all the cells of an organism's body must be bathed in a fluid medium. In vertebrates, this fluid, called **tissue fluid**, is derived from the blood and its composition is very like that of the blood plasma. The composition of plasma and of tissue fluid is kept constant so that the individual cells of the body are maintained in ideal conditions for growth and development. The maintenance of constant conditions, often described as the maintenance of a constant **internal environment**, is called **homeostasis**. The liver is very important for homeostasis because it directly determines the concentrations of many different substances in the blood plasma. These activities, collectively called the **regulatory functions** of the liver, are described below:

1 Regulation of glucose. The body obtains glucose by the digestion of disaccharides and polysaccharides in the diet. In addition to glucose, the diet provides a mixture of other monosaccharides but these are converted to glucose in the liver. The concentration of glucose in the blood is not absolutely constant but varies between about 0.6 and 1.4 g dm^{-3}, depending on the conditions. After a meal, the level rises, while, after fasting, it becomes lower. Glucose is the *only* fuel taken up by the specialised cells of the brain and nervous system and the minimum level of 0.6 g dm^{-3} is just enough to keep these essential regions properly supplied.

Glucose levels are kept within the normal limits in several ways (see Fig. 15·5). After a meal, the liver receives more glucose from the hepatic portal vein than is required in the general circulation. Under these circumstances, glucose is converted to the polysaccharide storage material called **glycogen**. The amount of glycogen formed is controlled by a hormone called **insulin**. Insulin is released from the pancreas and acts mainly by promoting the uptake of glucose by the membranes of liver and muscle cells. Up to 60 g of glucose can be stored as glycogen in the liver, and up to 150 g in the muscles. Above this limit, excess glucose is converted into molecules of **fat**. Droplets of fat leave the liver via the hepatic vein and are taken up for storage by **adipose cells**,

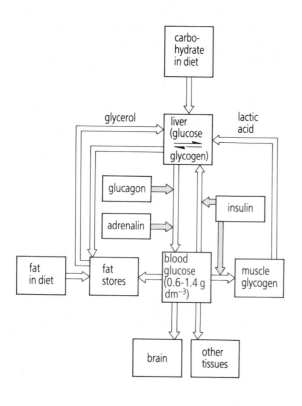

Fig 15·5 Regulation of blood glucose

mostly under the skin. Storing surplus energy in this way is a great advantage for a mobile animal because fat contains more than twice as much energy per gram as carbohydrates or proteins.

Except in the period following a meal, the liver releases more glucose than it receives, breaking down its reserves of glycogen and also manufacturing new glucose from other organic compounds. Glycogen breakdown, called **glycogenolysis**, is controlled by two hormones, **glucagon** from the pancreas and **adrenalin** from the adrenal glands. Both hormones are released in response to a fall in blood glucose, but adrenalin release can also be speeded up via the nervous system so that it boosts the availability of glucose at times of sudden stress. The liver's supply of glycogen is limited and must be supplemented by the manufacture of new glucose after only a few hours. The new glucose is made mostly from lactic acid, released from the muscles, and from glycerol produced from the breakdown of fats. During prolonged fasting, amino acids are also used. The manufacture of new glucose, called **gluconeogenesis**, is promoted by glucagon and up to 180 g per day can be produced. The biochemical pathway used reverses the process of glycolysis by which glucose is broken down in respiration, as described in Unit 10.

2 Regulation of amino acids. The cells of the body use amino acids for protein synthesis and

as raw materials for building up other organic compounds. The liver processes amino acids received from the digestive system and ensures that appropriate amounts of all the different amino acids circulate in the bloodstream. Some amino acids can be converted into others, which may be in short supply, and the liver can also make new amino acids using glucose and fatty acids. These interconversions are possible because the biochemical pathways used to break down these different substances all lead to the important intermediate substance acetyl CoA (see Unit 10). Many of the pathways are reversible so that acetyl CoA is often used as a subunit in building up more complex compounds. Depending on the circumstances, acetyl CoA obtained from glucose breakdown can be used to make amino acids or fatty acids. Alternatively, acetyl CoA obtained from fatty acids or amino acids can be broken down to release energy.

Surplus amino acids, not immediately required by the body, are **deaminated**, that is, the NH_2 or amino group is removed to form ammonia. This process leaves behind a selection of other molecular fragments, called **keto acids**, which are converted into acetyl CoA and used directly for respiration or converted into glucose or fat, depending on current needs. The body cannot store surplus amino acids or proteins in the same way as surplus carbohydrates or fats.

3 Regulation of plasma proteins. Blood plasma contains about 70 g dm^{-3} of proteins in colloidal solution. In addition to the blood-clotting protein **fibrinogen**, these proteins include **albumin** and many different **globulin** proteins covering a wide range of functions. Collectively, the plasma proteins act as pH 'buffers' and, as discussed in Unit 17, they help to regulate the distribution of liquid between the blood and tissue fluid. Almost all plasma proteins are manufactured and later broken down by the liver.

4 Regulation of lipids. Lipids are needed as building materials for cell membranes and for making some hormones. The liver produces fatty acids and cholesterol and interconverts existing lipid substances absorbed from the diet. Surplus lipids are stored as fat.

5 Storage of vitamins and inorganic elements. Liver cells take up and store vitamin B_{12} (cyanocobalamin) and the fat soluble vitamins A, D, E, and K. These substances are released whenever required. Iron is the main inorganic element stored but copper, zinc and trace elements, including cobalt and molybdenum, are all stored in significant quantities by the liver cells.

6 Production of heat. The liver is a major heat source for the body. Heat produced by all the chemical reactions going on in the liver is distributed to the rest of the body in the bloodstream. In mammals and birds, this heat helps to maintain a constant body temperature,

thereby providing near optimum conditions for enzyme activity.

7 Storage of blood. The liver acts as a blood reservoir storing up to 1 litre of blood. The amount of blood retained depends on the diameter of the various blood vessels leading to and from the liver and this in turn is controlled by the nervous system and by hormones. During exercise, the liver provides extra blood to increase oxygen transport to the muscles.

Excretory functions

Excretion is the removal from the body of the waste products of metabolism and of substances surplus to the body's requirements. The liver participates in excretion in several important ways:

1 Production of urea. As already stated, surplus amino acids cannot be stored in the body and are deaminated. The ammonia resulting from this process is extremely poisonous and it is immediately combined with carbon dioxide to form urea. This reaction occurs by means of a pathway called the **ornithine cycle**, and is summarized by the following equation:

$$CO_2 + 2NH_3 \longrightarrow CO(NH_2)_2 + H_2O$$

$$\text{ammonia} \qquad\qquad \text{urea}$$

Urea can only be manufactured in the liver and, once formed, it passes into the general circulation and is later removed by the kidneys, finally leaving the body dissolved in the urine.

2 Breakdown of red blood cells. Red blood cells are continually produced in the bone marrow and have a life span of about 120 days. Old red blood cells are destroyed by groups of phagocytic cells together known as the 'reticulo-endothelial system'. These cells occur in many different parts of the body and are represented in the liver by the Kupffer cells. All reticulo-endothelial cells break down haemoglobin obtained from the red cells to form a yellow pigment called **bilirubin**. This is released into the bloodstream and is responsible for the yellow colour of blood plasma. **Iron** is sent for storage in the liver, while the protein residue of the haemoglobin molecule is broken down into amino acids. The hepatocyte cells of the liver extract bilirubin from the blood and transfer it to the bile, forming the major part of the **bile pigments**. The bile pigments are released into the small intestine along with the rest of the bile but have no digestive function. Bilirubin is converted by bacterial action into **stercobilinogen**, a brown pigment which gives faeces their characteristic colour.

The symptoms of **jaundice**, a yellow colour of the skin and eyes, occur when bilirubin accumulates in the blood. This can be caused by an over-rapid breakdown of red cells, by liver disease or by a blockage of the bile ducts. Such blockages result when cholesterol secreted in the bile is deposited to form solid particles, or **gallstones**, which become lodged in the bile ducts.

3 Detoxication. The liver disposes of spent hormones, drugs and many poisons by detoxication. This process usually consists of attaching, or **conjugating**, the unwanted substance to another organic group (such as **glycine** or **glucuronic acid**). This group acts as a molecular 'label' enabling the substance to be recognised by the kidneys as waste material and excreted. In other cases, the liver breaks down poisons directly into harmless substances which can be used in metabolism.

All these different functions make the liver a very versatile organ and easily the most important single structure involved in maintaining homeostasis. It is interesting that much of this versatility comes from the activities of the individual liver cells, while the structure of the organ as a whole is comparatively uncomplicated.

15·4 Pancreas ■ ■ ■ ■ ■ ■ ■ ■ ■ ■ ■ ■ ■ ■ ■ ■ ■ ■

The pancreas produces digestive enzymes and hormones and its structure is illustrated in Figures 15·6 and 15·7.

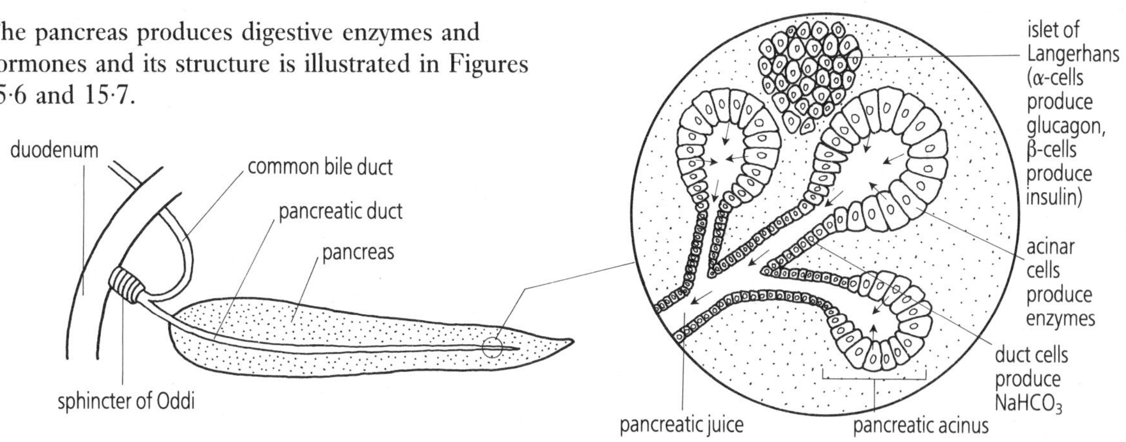

Fig 15·6 Diagram to illustrate the structure of the pancreas

The enzyme-producing cells of the pancreas are grouped together at the ends of minute ducts where they form structures called **acini**, rather like bunches of grapes surrounding a hollow core. In the pancreas of mammals, the acini are grouped together to form irregular **lobules**, separated from each other by connective tissue. All the various ducts arising from the acini drain eventually to the main **pancreatic duct** carrying a mixture of enzymes and other substances, called **pancreatic juice**, towards the duodenum. Like the release of bile, secretion of pancreatic juice is stimulated by nervous activity, and by the hormones **secretin** and **CCK**. The acinar cells produce the various enzymes required, while the cells lining the interconnecting ducts produce sodium hydrogencarbonate. In man, 0·5 to 0·8 litres of pancreatic juice is released daily.

Between the pancreatic lobules lie small patches of tissue called **islets of Langerhans**. Within the islets two main groups of cells, called **α-cells** and **β-cells**, can be identified. The α-cells produce the hormone **glucagon** and the β-cells produce **insulin**. These hormones are both involved in regulating blood glucose, as described earlier in this Unit. They have opposite actions and their 'push-pull' effects help to hold blood glucose levels steady. The disease **diabetes mellitus** occurs if too little insulin is produced. The initial symptoms of diabetes are **hyperglycaemia**, that is, a raised blood glucose level, and a loss of glucose in the urine. If the condition is untreated, abnormal quantities of salts and water are lost, leading to dehydration and eventually to coma and death. A daily injection of insulin obtained from cows or sheep, or synthesized by gentically-engineered bacteria (Unit 45), can substitute for human insulin and, given this treatment, most diabetics can lead fairly normal lives.

This description of the liver and pancreas concludes the main account of the alimentary canal and its related structures given in this book. Some of the food gathering techniques used by invertebrates are described in Section D which discusses the range of living organisms and their different specializations. Unit 16 describes gas exchange while the following three Units describe the main features of the circulatory system. Like the digestive system, these systems contribute in a major way to maintaining homeostasis.

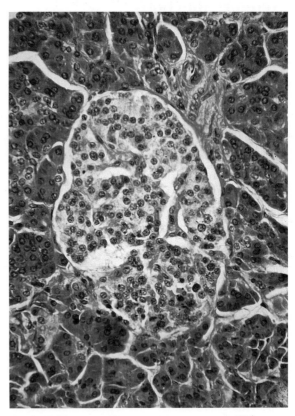

Fig 15·7 Microscopic structure of the pancreas (×750)

16 RESPIRATION AND GAS EXCHANGE

Objectives

After studying this Unit you should be able to:–

- Define internal respiration and external respiration

- List and explain the important properties of gas exchange surfaces

- Describe in detail the structure of the gas exchange system of a bony fish and explain how ventilation takes place

- Distinguish between parallel flow and counterflow, and explain the importance of counterflow in gas exchange systems

- Describe the main structural features of the human gas exchange system and outline the mechanisms involved in its ventilation

- Draw and label a diagram to illustrate the microscopic structure of human lungs

- List and discuss the main mechanisms involved in regulating gas exchange

- Draw a diagram of a respirometer and explain how this apparatus is used to measure the respiratory rate of small invertebrates

- Explain the terms respiratory quotient (RQ) and basal metabolic rate (BMR)

16·1 Introduction ■ ■ ■ ■ ■ ■ ■ ■ ■ ■ ■ ■ ■ ■ ■ ■ ■

The term 'respiration' is applied to two separate biological processes. Firstly, it refers to the chemical process of energy release from organic compounds such as glucose. This type of respiration is called **internal respiration** and may take place aerobically, if oxygen is available, or anaerobically, usually if oxygen is absent or in short supply. For aerobic respiration to continue in the cells of an organism, a supply of oxygen must be maintained, and the waste product carbon dioxide must be removed. **External respiration, or breathing,** refers to the process of exchanging these gases between the organism and its surroundings. The various structures and mechanisms involved form the subject matter of this Unit.

The external gills of this amphibian provide a large surface for gas exchange.

16·2 Gas exchange surfaces ■ ■ ■ ■ ■ ■ ■ ■ ■ ■ ■ ■ ■

The structures used to carry out external respiration make up the **respiratory system**, or **gas exchange system**, of the organism. The most important component of this system is the **gas exchange surface**, which acts as a limiting boundary between the external environment and the living interior.

It is important to appreciate that dissolved gases can cross the gas exchange surface only by diffusion. Unicellular and small multicellular organisms are able to carry out gas exchange directly across the body surface because they have a large surface area relative to their volume. In addition, dissolved gases can diffuse rapidly to and from all parts of the interior because the distances involved are small. In contrast, larger multicellular organisms must develop additional structures uniquely specialised for gas exchange. These structures share several characteristics, each of which helps to make the process of diffusion more efficient:

1 Large surface area. The organism's ability to exchange gases with the environment depends directly on the surface area available for diffusion.

2 Moist permeable surface. A film of moisture is necessary at the surface to dissolve the gases and allow them to pass through. For most terrestrial organisms, plants as well as animals, this presents a problem, for the gas exchange surface inevitably becomes a region of potential water loss.

3 Ventilation. As oxygen diffuses inwards it tends to become depleted immediately next to the gas exchange surface. The supply of oxygen from the external environment can be much more efficiently maintained if air, or water containing dissolved oxygen, is constantly replaced. A flow of this kind is called **ventilation** and it creates the maximum possible concentration difference, or **diffusion gradient**, directly across the gas exchange surface.

The structures used for ventilation vary from organism to organism, but muscular movements of some kind are normally involved. The flow itself can be **continuous** or **tidal**: continuous flow occurs in aquatic organisms as, for example, when a current of water flows over the gills of a fish. Tidal flow is typical of air-breathing organisms and consists of intermittent flow into and out of gas exchange structures such as lungs.

4 Good capillary blood supply. In many organisms, blood flowing through capillaries brings carbon dioxide to the gas exchange organs and transports dissolved oxygen rapidly away. As in the case of ventilation, this helps to keep the diffusion gradients for both gases steep and promotes an efficient interchange. Furthermore, the capillaries run very close to the gas exchange surface, keeping the diffusion distances as short as possible.

Insects provide an important exception to this general rule. The gas exchange system of insects (described in Unit 49) is called the **tracheal system** and consists of finely divided tubes which reach all parts of the interior. Thus, the body tissues exchange gases directly with the environment.

5 Respiratory pigments. Respiratory pigments combine reversibly with oxygen and greatly increase the oxygen-carrying capacity of the blood. Significantly, this removes oxygen from free solution in the blood plasma so that the gradient for inward diffusion of oxygen becomes even steeper.

The commonest respiratory pigment is **haemoglobin** which is found in most vertebrates and many invertebrates. Similar iron-containing pigments called **haemerythrin** and **chlorocruorin** are found in some annelids, while copper-containing **haemocyanin** is found in some molluscs and a few arthropods.

In the following sections of this Unit, the sophisticated gas exchange systems of bony fish and of mammals are discussed in detail. As you study these systems, you should gain an understanding of the important differences and similarities between organisms specialized for gas exchange in two very different environments – water and air. Water is a dense medium in which diffusion is slow. It provides support but it contains only small amounts of dissolved oxygen, usually less than 1 cm^3 per 100 cm^3 of water. On the other hand, air contains 21 cm^3 oxygen per 100 cm^3 and diffusion takes place very rapidly (in the case of oxygen, about 300 000 times more quickly than in water).

16·3 Gas exchange in bony fish ■ ■ ■ ■ ■ ■ ■ ■ ■ ■ ■ ■ ■

Bony fish Class Osteichthyes are among the most active and successful aquatic organisms; more than 20 000 different species have been described. They inhabit the oceans in vast numbers, where they represent a major source of human food, and are also abundant in almost every type of fresh water habitat. To some extent this success is due to a sophisticated and efficient gas exchange system which extracts as much as 80% of the available oxygen from the water passing through.

Structure

The structure of the gas exchange system of the herring, a typical bony fish, is illustrated in Figure 16·1. The gas exchange organs are the **gills**. There are four gills on each side, and these are separated from each other by cavities called **gill slits** (see Fig. 16·1D). Each gill is supported by a curved and jointed rod of bone called a **gill bar**, or **visceral arch**. Where the gill slits open into the **pharynx**, the gill bars give rise to a series of stiff elongated projections, the **gill rakers**, which act as a screen, protecting the delicate gills from the entry of food and other particles which might damage them. The gill bars also give rise to bony rods called **gill rays** which provide support for the flattened **gill filaments**. Externally, the gills are protected by a bony cover, the **operculum**. A flap of tissue round the posterior edge of the operculum acts as a valve which can be used to seal the opening and prevent the backflow of water into the **opercular cavity**.

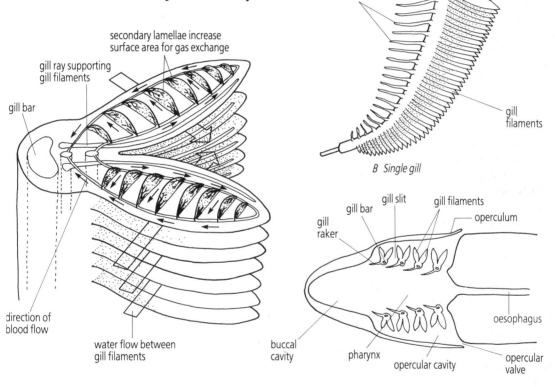

A Head with operculum removed

B Single gill

C Detail of gill filaments

D Horizontal section through gill region

Fig 16·1 Gas exchange system of a bony fish (herring)

Ventilation

Water will flow over the gills whenever the water pressure in the buccal and pharyngeal cavities is greater than the water pressure in the opercular cavity (see Fig. 16·1D). In bony fish this pressure difference can be achieved either by the action of a positive pressure pump, the **buccal pressure pump**, in front of the gills, or by a negative pressure pump, the **opercular suction pump**, acting after the gills.

These two pumps operate at different times during the ventilation cycle, as shown in Figure 16·2. During **inspiration**, the volume of the pharynx is increased mainly by muscular contractions which lower its floor. This process reduces the water pressure inside the cavity so that water enters via the mouth. However, at exactly the same time, the operculum on each side bulges *outwards*, closing the opercular valves and greatly reducing the water pressure in the opercular cavities. These movements produce a suction force

111

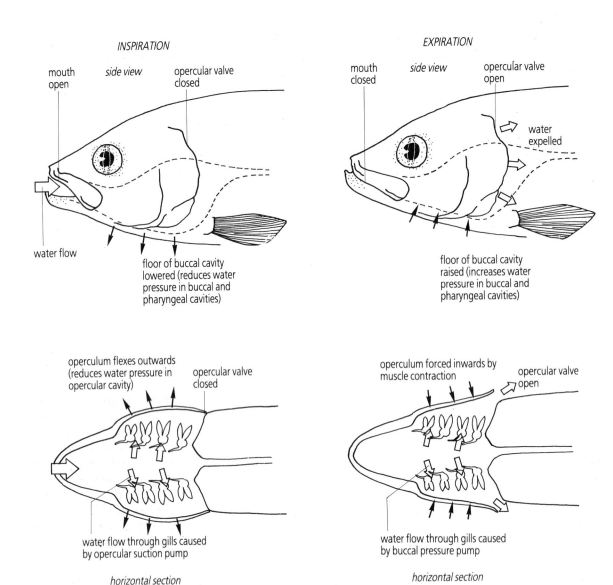

INSPIRATION

mouth open   *side view*   opercular valve closed

water flow

floor of buccal cavity lowered (reduces water pressure in buccal and pharyngeal cavities)

EXPIRATION

mouth closed   *side view*   opercular valve open

water expelled

floor of buccal cavity raised (increases water pressure in buccal and pharyngeal cavities)

operculum flexes outwards (reduces water pressure in opercular cavity)   opercular valve closed

water flow through gills caused by opercular suction pump

horizontal section

operculum forced inwards by muscle contraction   opercular valve open

water flow through gills caused by buccal pressure pump

horizontal section

Fig 16·2 Ventilation of the gills in a bony fish

(opercular suction pump) so that water continues to flow through the gills in spite of the reduced pressure in the pharyngeal cavity. During **expiration**, the mouth is closed and the floor of the pharynx is rapidly raised to create a strong positive pressure (buccal pressure pump) which drives water through the gill slits. Simultaneously, each operculum is forced inwards by muscular contractions which also open the opercular valves and cause water to be expelled. The net effect of all this is to increase the efficiency of the gills by maintaining an almost continuous flow of water.

Gas exchange

The detailed structure of the gills is illustrated in Figure 16·1C. This shows two sets of flattened gill filaments, or **primary lamellae**, projecting alternately at different angles in a V-shaped arrangement. Water passes through between the gill filaments and over the numerous **secondary lamellae**, where gas exchange takes place. The total surface area of the secondary lamellae varies

from one species of fish to another, being the greatest in the largest and most active species. Each lamella is well supplied with blood capillaries arising from the **afferent branchial artery** within the gill. Oxygen diffuses into the blood stream of the fish and combines with haemoglobin to form a loose compound called **oxyhaemoglobin**. Oxygenated blood, containing oxyhaemoglobin, is then collected in the **efferent branchial artery** and eventually passes to a main blood vessel, the **dorsal aorta**, for distribution to the body tissues. Deoxygenated blood returns to the heart and then passes again to the gills. This pattern of blood flow is characteristic of a **single circulation**, as discussed further in Unit 17.

A most important observation is that blood flow in the capillaries of the secondary lamellae is always opposite in direction to the flow of water passing over the surface. This situation, called **counterflow**, promotes a more efficient exchange of gases than the alternative arrangement, called **parallel flow**, where the blood and water flow in

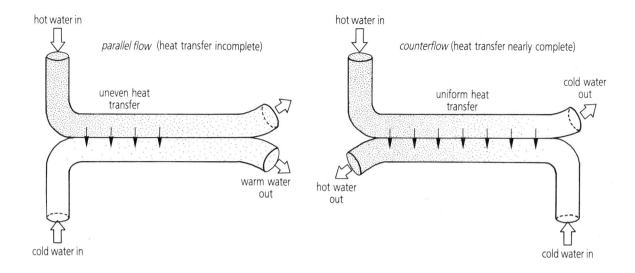

Fig 16·3 Parallel flow and counterflow

the same direction. You can easily understand why counterflow is important by considering the model systems shown in Figure 16·3. In both systems, heat transfer from the water flowing in one pipe to the water in the other depends on the magnitude of the temperature difference between them. In the case of parallel flow, heat transfer is rapid at first. However, farther along the pipes, the hot water has cooled down while the cold water is warmer. As a result, heat transfer becomes progressively slower and finally stops when the temperature of the water in the two pipes is the same. Note that, at this point, only a fraction of the available heat

energy has been transferred across from one pipe to the other. In the counterflow system, the cold water runs alongside progressively hotter water, while the hot water encounters progressively colder water. In this way, a useful temperature difference is maintained all the way along the pipes and almost all of the available heat energy can be transferred. Precisely the same principle applies to maintaining a favourable diffusion gradient for the uptake of oxygen by the gills of a fish. Counterflow ensures that as much oxygen as possible is transferred from the water and diffuses inwards to become trapped in the blood capillaries.

16·4 The human gas exchange system ■ ■ ■ ■ ■ ■ ■ ■ ■ ■

Structure

The gas exchange organs in humans and other mammals are the **lungs**. The lungs develop as outgrowths from the pharyngeal region of the alimentary canal and, together with the heart, they fill almost all the available space inside the thorax (Fig. 16·4).

The conducting part of the respiratory system, through which the air passes, begins with the nasal cavity where the air is warmed and moistened, and continues across the pharynx to a branching system of tubes inside the lungs. The largest of these tubes, the **trachea**, or windpipe, is typically about 2 cm in diameter and 10 cm long. The trachea is supported at intervals by **C-shaped cartilages** embedded in its walls. Each cartilage extends around the trachea, its ends being separated by a short gap located at the back. This structure makes the trachea extremely flexible and allows it to be stretched or compressed during movements of the lungs. In addition, because the cartilages are incomplete, it is possible to increase the diameter of the airway by coughing, making it easier to free an obstruction.

At the top of the trachea, where it joins the pharynx, is the **larynx**. This structure is visible externally as the 'Adam's apple' and has two important functions: 1 It acts as a valve which is closed during swallowing, sealing the trachea against the entry of food. 2 In humans, a complex control system has developed so that the **vocal folds** within the larynx can be used in the production of speech. The **epiglottis**, a flap of tissue supported by cartilage, projects forwards from the larynx and assists in directing food towards the oesophagus. However, it is not essential for this purpose.

At its base, the trachea divides into a **left** and **right principal bronchus**, each of which is further subdivided into smaller tubes also called bronchi and supported by cartilage in a similar way to the trachea. This branching pattern is not random but is consistent from one person to another and divides each lung into a number of distinct **lobes**. (Note that the left and right lungs differ in shape so as to accommodate the heart.) Within each lobe, the tubes of the bronchial tree continue to subdivide and eventually give rise to

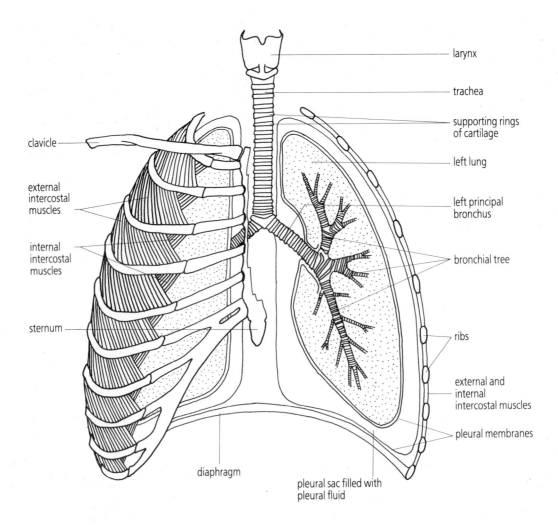

Fig 16·4 Structure of the human gas exchange system

narrow unsupported tubes less than 1 mm in diameter. These are the **bronchioles** which subdivide yet again and finally terminate in minute air-filled sacs called **alveoli**. The alveoli are arranged in **lobules**, like small bunches of grapes, and are well supplied with blood capillaries so that gas exchange can take place. Although each alveolus is only 100–300 μm in diameter, the two lungs of a human contain more than 700 million of them, providing a total surface area recently estimated at about 140 m². To accommodate this vast surface area, the lungs have a highly organized internal structure supported by connective tissue. As a result they are compact organs, relatively firm to the touch, but highly elastic to allow expansion when filled with air.

Ventilation

Air enters the lungs whenever the air pressure inside the alveoli is reduced below atmospheric pressure – this is called **inspiration**. The reverse process, **expiration**, occurs when the pressure of the alveolar air exceeds atmospheric pressure.

The lungs themselves are not muscular so that these changes in pressure must be achieved

indirectly. This is possible only because of the **pleural sac** surrounding each lung (Fig. 16·4). Each sac consists of two **pleural membranes**: the outer pleural membrane is firmly attached to the inside of the thorax and tends to be pulled away from the lungs when the volume of the thorax increases, so that the pressure of the **pleural fluid** drops below atmospheric pressure. This pressure difference is transmitted to the lungs by the inner pleural membrane and the lungs are therefore pulled outwards, nearly filling the enlarged thoracic cavity. The importance of the pleural sac in keeping the lungs distended becomes clear if air at atmospheric pressure is allowed to enter the intrapleural space, for example, as the result of a stab wound. If this happens, the highly elastic lung immediately collapses and is useless for gas exchange.

It should now be clear that, under normal circumstances, changes in the volume of the thoracic cavity are transmitted to the lungs by means of the two pleural sacs. Ventilation of the alveoli in humans thus depends on methods of increasing and decreasing the volume of the chest cavity. There are two such methods, as follows:
1 Contraction of the diaphragm. The diaphragm is a dome-shaped sheet of muscle and

connective tissue separating the thoracic and abdominal cavities. When the muscle fibres contract, the diaphragm becomes flatter and pushes downwards against the contents of the abdomen. This increases the volume of the thoracic cavity and causes the lungs to fill with air. When the muscle fibres relax, the diaphragm is forced to return to its domed shape partly because of the upward pressure of the abdominal organs and partly by elastic forces associated with the lungs. Thus, expiration is a passive process, at least during quiet breathing, when the diaphragm is often the only muscle involved.

2 Contraction of the intercostal muscles.

There are two sets of intercostal muscles located between the ribs as shown in Figure 16·4. They are used mainly during deep breathing. When the **external intercostal muscles** contract, the entire rib cage is lifted upwards and outwards increasing the volume of the thoracic cavity, so that these muscles are responsible for inspiration. Expiration is partly the result of passive forces but can be assisted by contraction of the **internal intercostal muscles**, which act to lower the ribs.

The changes in lung volume resulting from these ventilation movements are investigated using a **spirometer** (Fig. 16·5). This consists of an oxygen-filled chamber floating over a water bath. The lid of the chamber is hinged at one side so that, when a person using the apparatus breathes in and out, the movements of the lid correspond exactly to changes in the volume of air held in his lungs. Carbon dioxide released is removed by passing expired air through soda lime before it is returned to the main chamber. The recordings obtained show that with each normal breath approximately 500 cm³ of air, called the **tidal**

volume, enters the respiratory system. However, only about 350 cm³ of this is useful for gas exchange, the remaining 150 cm³ being trapped in **dead space**, that is, inside the various air passages leading to the alveoli. During deep breathing an additional 2·5 litres of air, called the **inspiratory reserve volume**, can be inhaled, while an extra 1·5 litres, the **expiratory reserve volume**, can be expelled. This gives a total 'vital capacity' of 4·5 litres, corresponding to the maximum volume of air which can be exchanged in a single breath. Provided the pleural sacs are intact, the lungs cannot be completely emptied so that there is always a **residual volume** amounting to about 1 litre. Even after death, 100 cm³ of air, called **minimal air**, remains inside the lungs and cannot be expelled.

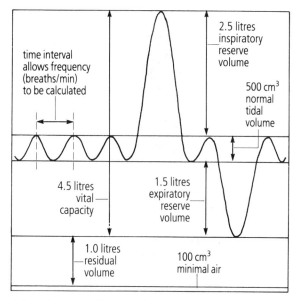

B Recording obtained interpreted to show lung volumes

Fig 16·5 Using a spirometer to estimate lung volumes

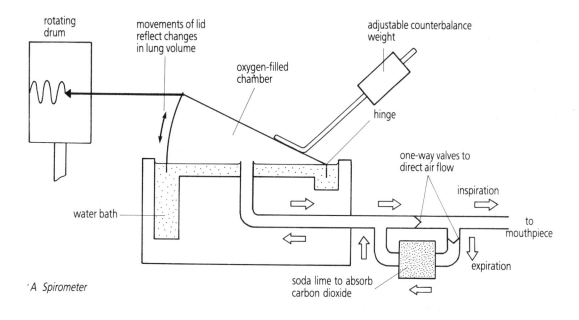

A Spirometer

Microscopic structure and gas exchange

Figure 16·6 shows the microscopic structure of a terminal bronchiole together with its associated alveoli. The bronchiole, like the other air passages of the lungs, is lined with **ciliated cells** and **goblet cells** producing mucus. The beating of the cilia maintains a steady current of mucus so that dust particles and bacteria become trapped and are carried upwards to the pharynx. Here the mucus is swallowed and any bacteria are killed by acid and enzymes in the stomach. Smoke from a single cigarette prevents the cilia from beating for several hours, making smokers particularly liable to chest infection. The surfaces of the alveoli are kept clean by scavenging phagocytic cells called **macrophages**.

During ventilation, the air in the alveoli is only partly mixed with fresh incoming air. However, because oxygen diffuses very rapidly from place to place in air, it continues to reach the alveoli and its concentration next to the gas exchange surface remains approximately constant (see Table 16·1). Oxygen dissolves in the moist lining of the alveolus and diffuses across the epithelium directly into the blood capillaries. The total surface area of the capillaries in each adult lung is estimated to be 60 m^2 and the diffusion distance is only 2–3 μm,

less than half the diameter of a red blood cell. Within the capillaries, oxygen is taken up by the haemoglobin inside the red cells and transported rapidly away. Interestingly, the blood pressure in lung capillaries is kept very low. This is essential to prevent the blood plasma from leaking outwards and accumulating in the alveoli. The thin layer of moisture which is present contains numerous molecules of **pulmonary surfactant**, a phospholipid-protein substance which prevents collapse of the alveoli by greatly reducing the surface tension.

Oxygenated blood collected from the capillaries enters the **pulmonary veins** and passes to the heart before being distributed at high pressure around the body. Deoxygenated blood returning from the tissues passes to the heart for a second time and is then pumped to the lungs via the **pulmonary arteries**. This pattern of blood flow is characteristic of a **double circulation**, as found in mammals and birds.

Table 16.1 Composition of atmospheric and alveolar air

GAS	ATMOSPHERE	ALVEOLI
nitrogen	78%	75%
oxygen	21%	13%
carbon dioxide	0.04%	5%
water vapour	0.8%	6%

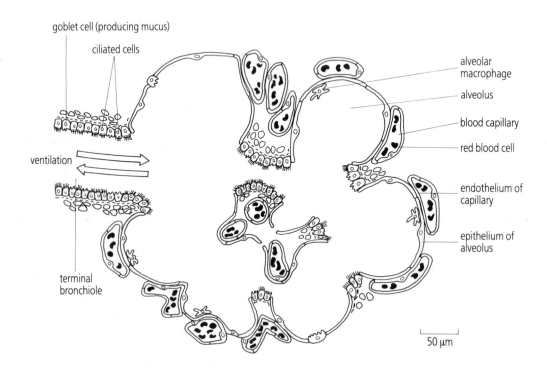

Fig 16·6 Microscopic structure of human lungs showing terminal bronchiole and alveoli (from electronmicrographs)

Control of respiratory gases

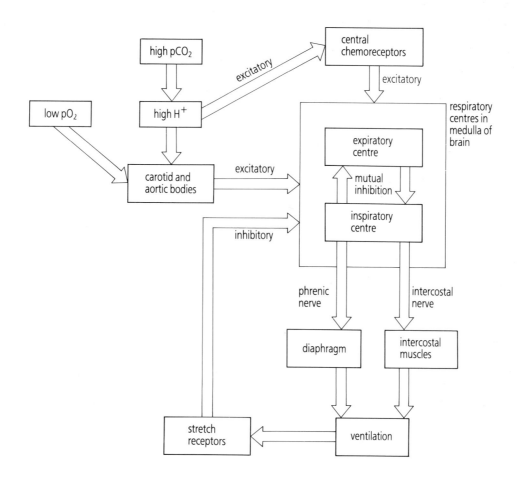

Fig 16·7 Flow diagram showing control system for ventilation of the lungs in humans

The essential purpose of the respiratory system is to provide oxygen for the process of aerobic respiration taking place in the tissue cells, and, at the same time, to remove carbon dioxide. The tissue cells are remote from the gas exchange structures so it becomes important to ask how gas exchange is controlled to meet the varying needs of the tissues. For example, during exercise, the muscles use more oxygen and produce more carbon dioxide. This leads, appropriately, to an increase in both tidal volume and frequency of breathing, but how is the change in oxygen demand signalled to the respiratory system? You can understand some of the processes involved by considering the flow diagram of Figure 16·7.

The most vital part of this control system is the **inspiratory centre** located in the **medulla** of the brain. The inspiratory centre consists of nerve cells which generate periodic bursts of nerve impulses. These impulses pass down the **phrenic** and **intercostal nerves**, causing the contractions of the diaphragm and external intercostal muscles which bring about inspiration. Expiration, as you will recall, can occur passively, but in some cases it is positively assisted by nerve impulses from the

expiratory centre which cause contraction of the internal intercostal muscles. The inspiratory and expiratory nerve cells inhibit each other so that they cannot be active simultaneously; together the two centres are responsible for the basic rhythm of breathing movements. If the phrenic and intercostal nerves are cut, breathing stops immediately.

The remaining elements of the control system act to modify the basic pattern of nerve impulses produced by these centres. In this way, the respiratory system is able to respond appropriately to stimuli received by each of the following sense organs:

1 Stretch receptors. The walls of the various air passages in the lungs contain numerous stretch receptors which become active during inspiration. Signals from these receptors help to terminate the bursts of impulses from the inspiratory centre and thus prevent over-expansion of the lungs.

2 Carotid and aortic bodies. These are located as shown in Figure 16·8. The carotid bodies are more important, but both types of structure contain specialised **chemoreceptor** cells. These cells are capable of responding to just two different stimuli. Firstly, they are sensitive to pO$_2$, the partial pressure of oxygen dissolved in the blood plasma. Low levels of pO$_2$ stimulate the receptor cells and lead via the medulla to increased ventilation of the lungs.

The second, but more important, stimulus for the chemoreceptor cells is pCO$_2$, the partial pressure of carbon dioxide. Carbon dioxide dissolves in the blood plasma to produce carbonic acid according to the following equation:

$$CO_2 + H_2O \rightleftharpoons H_2CO_3 \rightleftharpoons HCO_3^- + H^+$$

From this it follows that whenever pCO$_2$ is changed, H$^+$, the hydrogen ion concentration, changes in a precisely corresponding way and it is in fact to H$^+$ that the carotid chemoreceptors respond. Even slightly raised levels of H$^+$ are very effective in stimulating the receptor cells and, like low pO$_2$, are followed by an increase in ventilation. It is important to note that, because the receptors are so sensitive to H$^+$, the concentration of carbon dioxide in the blood becomes the main factor controlling gas exchange.

3 Central chemoreceptors. These are located in the medulla itself and respond to H$^+$ ions in the same way as the carotid bodies. They act directly on the respiratory centres to increase ventilation whenever the level of H$^+$ starts to rise.

Although the system described above is complicated, it is only a small part of the body's mechanism for regulating the supply of oxygen and the removal of carbon dioxide from its tissues. A whole series of parallel changes take place in the circulatory system as, for example, when the heart rate and blood pressure increase during exercise.

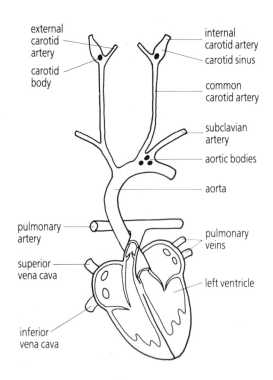

Fig 16·8 *Location of carotid and aortic bodies*

Equally important are the subtle interactions between oxygen, haemoglobin, and carbon dioxide, which affect the transport of respiratory gases. Some of these processes will be discussed in subsequent Units. The complexity of the total system is not altogether unexpected in view of the vital importance of the chemical reactions of respiration in supplying energy for all other life activities.

16·5 Experimental study of respiration ▪ ▪ ▪ ▪ ▪ ▪ ▪ ▪ ▪ ▪

The **respiratory rate** of an organism is usually equivalent to the rate of aerobic respiration taking place within its tissues. The chemical equation for this process is as follows:

$$C_6H_{12}O_6 + 6O_2 \longrightarrow 6CO_2 + 6H_2O$$
$$\Delta G = -2870 \text{ kJ per mole}$$

From the equation you can see that respiratory rates can be estimated in several ways, for example, by measuring the rate of oxygen consumption, the rate of carbon dioxide production, or the rate of energy production. In practice, oxygen uptake is normally measured, using an instrument called a **respirometer**. For purposes of comparison, the respiratory rates of different tissues are expressed in terms of the amount of oxygen used up in a given time by 1 gram of the tissue held at a constant temperature.

One possible design of respirometer is illustrated in Figure 16·9. The small invertebrates, or other respiring tissues to be investigated, are placed in a perforated gauze basket inside tube *A*. Beneath the basket, a rolled filter paper soaked in potassium hydroxide solution acts to remove carbon dioxide from the surrounding air. This means that any carbon dioxide which is produced by respiration is immediately absorbed so that it does not affect the volume of air remaining. Therefore, as you can confirm by referring to the equation for aerobic respiration, any changes in volume which do take place must be due to the uptake of oxygen. These changes are measured by means of the manometer and the calibrated scale. To avoid errors, temperature fluctuations during the experiment are reduced to a minimum by immersing the apparatus in a constant temperature water bath. Minor variations should be cancelled out by tube *B*, which acts as a control. The syringe is used to level the manometer fluid after each measurement.

In humans samples of expired air are easily collected and can be compared to atmospheric air

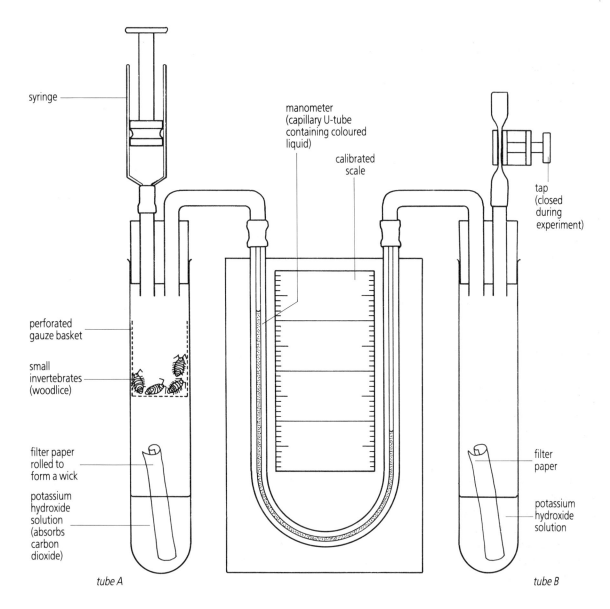

Labels on the figure:

syringe

manometer (capillary U-tube containing coloured liquid)

calibrated scale

tap (closed during experiment)

perforated gauze basket

small invertebrates (woodlice)

filter paper rolled to form a wick

filter paper

potassium hydroxide solution (absorbs carbon dioxide)

potassium hydroxide solution

tube A

tube B

Fig 16·9 Respirometer

by **gas analysis**. A known volume of air is exposed firstly to potassium hydroxide (which absorbs carbon dioxide) and, secondly, to **alkaline pyrogallol** (which absorbs oxygen). The changes in volume which take place are noted and allow the approximate composition of the gas sample to be calculated. From results of this kind, the person's **respiratory quotient (RQ)** can be determined, where

$$RQ = \frac{\text{volume of } CO_2 \text{ produced}}{\text{volume of } O_2 \text{ used up}}$$

The values of RQ to be expected vary depending on which substances are broken down by respiration. Glucose gives a theoretical value RQ = 1.0, while for fats RQ is approximately 0.7, and for proteins approximately 0.8. Under normal conditions, the human RQ is in the range 0.8–0.9, indicating that some fats and proteins, as well as carbohydrates, are used for respiration. Values of RQ greater than 1.0 are obtained when anaerobic respiration is in progress.

An alternative method of investigating the rate of respiration in a person's tissues involves measuring the amount of energy which is released in a given time. All of this energy eventually leaves the body as heat, and can be collected using a special form of **calorimeter** large enough to accommodate people (see Fig. 16·10). Water flowing through the apparatus is warmed in proportion to heat lost from the body, so that the rate of heat production, called the person's **metabolic rate**, can be calculated. This rate varies widely depending on activity – it is lowest during sleeping or resting, but increases sharply during exercise.

The minimum rate of energy release needed to maintain essential body processes is called the **basal metabolic rate (BMR)** and is usually expressed in kilojoules of heat released per square metre of body surface per hour ($kJ\ m^{-2}h^{-1}$). Basal metabolic rate is measured under standard conditions with the person completely at rest 12–18 hours after a meal. For an adult male,

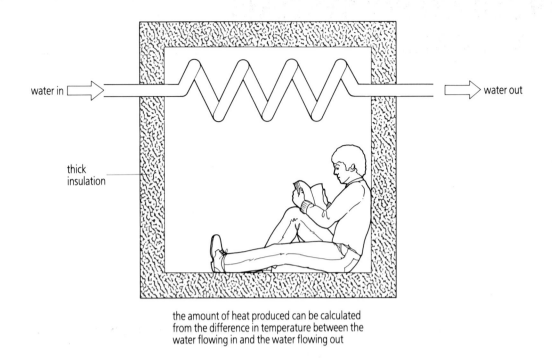

water in → ← water out

thick insulation

the amount of heat produced can be calculated from the difference in temperature between the water flowing in and the water flowing out

Fig 16·10 Human calorimeter

BMR is typically about 165 kJ m^{-2}h^{-1}, while for a female, lower values of about 150 kJ m^{-2}h^{-1} are normal. Young children have much higher BMR's, but, in adults, a high BMR is usually associated with fever or disease. BMR is strongly affected by thyroid hormone, so that an increase in heat production is one of the first symptoms of **hyperthyroidism**, an excess secretion of this hormone. Conversely, BMR drops below normal when insufficient thyroid hormone is produced – this is called **hypothyroidism**. BMR also depends on external temperature and can be increased in cold conditions, or reduced in hot conditions. This is one of the body's mechanisms for maintaining a constant body temperature and is discussed further in Unit 21.

17 CIRCULATORY SYSTEMS IN ANIMALS

Objectives

After studying this Unit you should be able to:–

- State the basic characteristics of circulatory systems

- Distinguish between open and closed circulatory systems

- Describe the pattern of blood flow in the single circulation of a fish

- Describe the pattern of blood flow in the double circulation of birds and mammals

- Draw and label a diagram to show the important structural features of the human heart

- Explain how blood flow through the heart is regulated by valves

- Outline the structure and properties of cardiac muscle

- Explain how the heart beat is coordinated by the pacemaker and conducting systems of the heart

- Briefly describe the human electrocardiogram (ECG)

- Relate the structure of arteries, arterioles, capillaries, venules, and veins to the different functions these blood vessels carry out

- Discuss the formation of tissue fluid and explain how water is returned to the capillaries at the venous end of the circulation

- Draw a flow diagram to show the important mechanisms involved in regulating the arterial pressure of the blood

17·1 Introduction – the need for a circulatory system ■ ■ ■ ■

A large multicellular organism needs a circulatory system for three main reasons:

1 The surface area to volume ratio of the organism is small so that diffusion of materials across the body surface cannot keep pace with demand.

2 A large part of the external body surface must often be kept impermeable to conserve water, particularly in terrestrial organisms.

3 Internal distances are too great for diffusion.

These problems are overcome firstly by developing specialized organs for functions such as gas exchange, digestion, and excretion, and secondly by linking these organs by means of a circulatory system. This moves substances rapidly from one place to another so that each of the specialized organs can function efficiently.

17·2 Types of circulatory system ■ ■ ■ ■ ■ ■ ■ ■ ■ ■ ■

All circulatory systems share several basic characteristics:

1 Circulating fluid. The circulating fluid, called **blood**, is used to transport dissolved oxygen, food, and waste products. In many organisms, there are special pigments, such as haemoglobin, to increase the oxygen-carrying capacity of the blood.

2 Pumping device. The pumping device, or **heart**, creates a pressure difference which forces the blood to flow.

3 Blood vessels. The blood must be at least partly contained in tubes in order to carry it towards the tissues and then back to the heart.

4 Valves. Valves are needed to ensure that the blood flows in the correct direction, and to prevent backflow. They form an essential part of the heart structure and are also found in blood vessels, usually where the blood is flowing slowly under low pressure.

Within the circulatory system these components are organized to promote an efficient exchange of materials between the blood and the tissue cells. This exchange can be achieved in either of two basic ways, as illustrated in Figure 17·1.

Some organisms, including insects and molluscs, have an **open circulatory system** (Fig. 17·1A). The blood is pumped at relatively low pressure from the heart into the main body cavity, called the **haemocoel**. The circulating fluid bathes the cells directly and only slowly percolates

through the tissues, returning to the heart by a system of collecting vessels. This makes open systems rather inefficient so that they are only suitable for small organisms. Significantly, the circulatory system is only incidental to gas exchange in insects, for the body cells exchange gases directly with the environment via the tracheal system (described in Unit 49). This makes gas exchange much more rapid so that some insects can lead more active lives than would otherwise be possible.

Almost all larger organisms, including all vertebrates, have **closed circulatory systems** where the blood circulates within a continuous network of tubes. This allows high pressures and a rapid rate of flow to be maintained. The tissue cells are not in direct contact with the blood itself, but are bathed in **tissue fluid**; this is derived from the blood by a filtering process across the walls of permeable blood vessels called **capillaries**. In vertebrates, most of the tissue fluid returns to capillaries at lower pressure, but some of it is collected into a separate system of drainage channels, called **lymph vessels**. These return fluid to the general circulation at a pressure which is only slightly below that of tissue fluid (see Fig. 17·1B). Circulatory systems constructed on this plan operate very efficiently and have been an important factor in the evolution of large active vertebrate organisms.

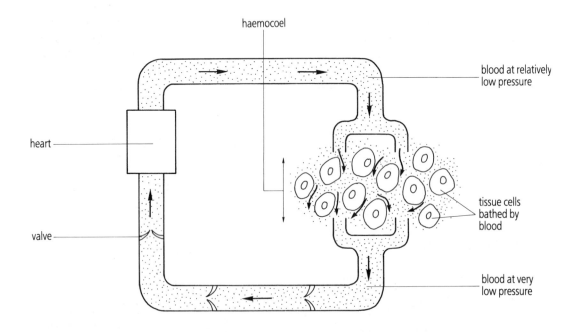

Fig 17·1A Open circulatory system

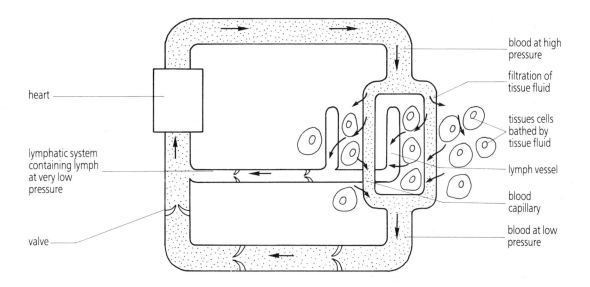

Fig 17·1B Closed circulatory system

Among the vertebrates, the circulatory system varies in structure according to the different needs of the organisms involved. Fish gain support from their environment and normally have a low body temperature, matching that of the surrounding water. Both of these factors greatly reduce their energy needs, so that, although fish are active organisms, their oxygen demands are very much less than in the case of mammals. Consequently, fish can survive with a simple form of closed circulation, called a **single circulation**, as illustrated in Figure 17·2. Blood leaves the heart at high pressure and passes to the gills via the **afferent branchial arteries**. Oxygenated blood is collected into the **efferent branchial arteries** which unite to form a single large blood vessel, the

dorsal aorta, running down the back of the animal. Branches from the dorsal aorta supply the body organs directly and deoxygenated blood is then collected at very low pressure into large blood spaces, called **sinuses**, in which the blood moves very slowly. The large size of the sinuses is necessary to provide a sufficient reservoir of blood from which the heart can be refilled. Note that, in this pattern of blood flow, the blood passes through the heart only once before reaching the body tissues.

More advanced vertebrates have a **double circulation** in which oxygenated blood from the gas exchange organs returns to the heart for a second time before being distributed to the body tissues. This increases the pressure of the blood

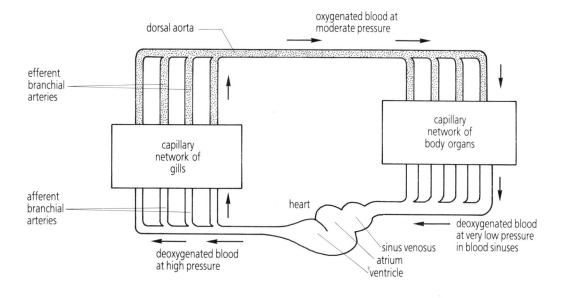

Fig 17·2 Single circulation as in fish

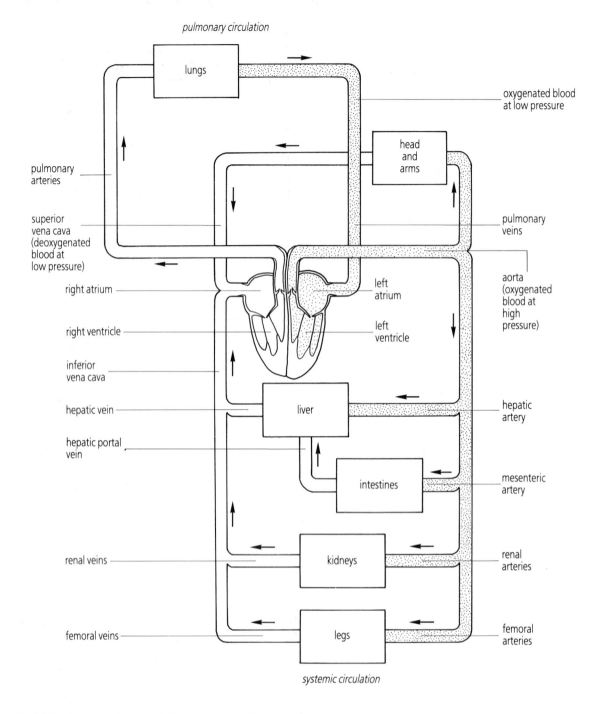

Fig 17·3 Double circulation as in birds and mammals

and the rate of flow, thereby speeding up delivery of oxygen to the tissues. Mammals and birds have a fully developed double circulation, as outlined in Figure 17·3. The heart consists of four chambers, comprising a thin-walled **atrium** and a thick-walled **ventricle** on each side. Deoxygenated blood collected from the body tissues passes into the right atrium. From here, it enters the right ventricle and is pumped to the lungs via the **pulmonary arteries**. In the capillary network of the lungs, carbon dioxide is released and oxygen is taken up. Blood, now loaded with oxygen, passes back to the heart via the **pulmonary veins**; this completes the **pulmonary circulation**. In the second part of the double circulation, called the **systemic circulation**, blood passes at high pressure from the left ventricle to the **aorta** for distribution to the body tissues. It returns eventually to the right atrium via the **superior vena cava** (from the head and arms), and the **inferior vena cava** (from the rest of the body, including the legs).

The remainder of this Unit describes the human double circulation in detail. From studying this, you should gain an understanding of the structures and mechanisms involved in an advanced mammalian system, as well as an appreciation of its important homeostatic role.

17·3 Structure and action of the heart ■ ■ ■ ■ ■ ■ ■ ■ ■ ■

Structure

The human heart is located inside the chest within a membrane-lined cavity, the **pericardium**. It is about 12 cm long and roughly cone-shaped, with the apex of the cone pointing obliquely down to the left. The apex lies nearest to the surface between the 5th and 6th ribs on the left side, 8–10 cm from the midline, where the beating of the heart can be easily felt. The heart has a mass of 300 g in an adult male, and about 250 g in an adult female.

The internal structure of the heart is illustrated in Figure 17·4. This shows the heart viewed from the front so that the right side of the heart, corresponding to the person's right hand side, is correctly drawn on the left of the diagram. As you can see, the two thin-walled atria at the top and the larger ventricles at the bottom are linked by a **fibrous skeleton** of tough connective tissue. This forms a thin dividing layer between the upper and lower chambers and is perforated by four valves, comprising one **atrioventricular** valve and one **semi-lunar** valve on each side. The left atrioventricular, or **bicuspid** valve, and the right

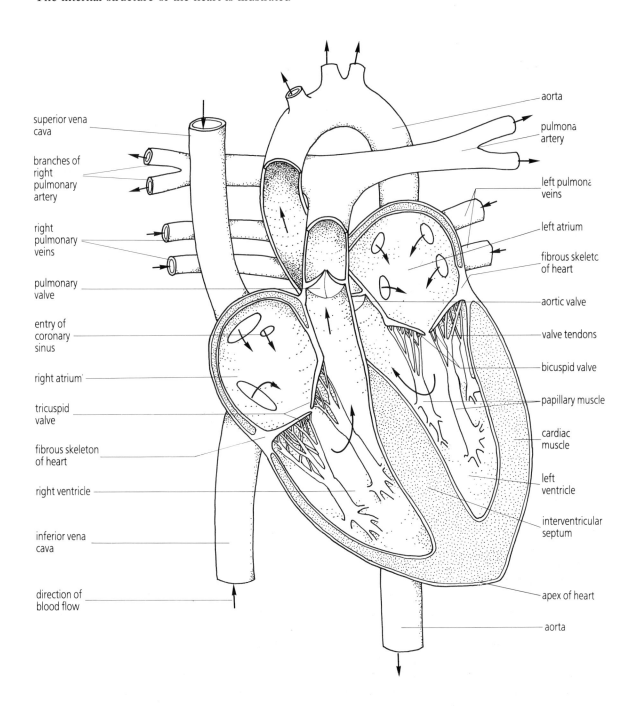

Fig 17·4 Structure of the human heart

atrioventricular, or **tricuspid** valve both open when the ventricles relax. When the ventricles contract, these valves close with a parachute-like action, forcing blood to leave the heart via the two semi-lunar valves (called respectively the **pulmonary** and **aortic** valves). The **valve tendons** (chordae tendinae) and **papillary muscles** on each side prevent the atrioventricular valves from 'blowing through' under pressure. This action is possible because the papillary muscles contract at the same time as the ventricles, effectively shortening the strings of the parachute as contraction continues. On the right side, blood flows out to the lungs, while on the left, blood flows into the aorta and out to the systemic circulation.

The operation of the semi-lunar valves is explained in Figure 17·5. Each valve consists of three flexible pocket-shaped flaps. When the blood flow is upwards, the flaps are pushed loosely to the side, but, as soon as the pressure in the ventricles drops, blood starts to flow down and the flaps fill up and bulge inwards to close the valve. The free edge of each flap is slightly thickened and has a small lump, or nodule, at its centre to ensure a perfect fit with its neighbours.

Interestingly, the two sides of the heart are not exactly alike. The reason for this is that the pressure needed to keep blood flowing in the pulmonary circulation (typically 4 kPa) is much less than that needed in the systemic circulation (typically 16 kPa). **Note**: 1mm Hg = 133.3 Pa. Thus, although the quantity of blood pumped in each case is the same, the walls of the left ventricle are thicker and more muscular than those of the right ventricle. This difference affects the shape of the ventricle cavities so that the right ventricle is twisted over the left, making the heart less symmetrical than suggested by Figure 17·4. You may discover this from dissection The same lack of symmetry applies to the aorta and the other main arteries, which have much thicker walls than the pulmonary arteries. In life, the pressure in the pulmonary capillaries must remain low to prevent accumulation of fluid in the alveoli of the lungs.

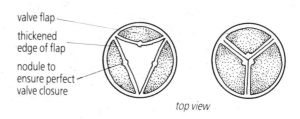

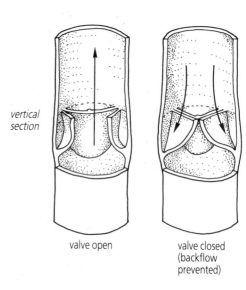

Fig 17·5 *Operation of semi-lunar valves*

Cardiac muscle

Cardiac muscle tissue is uniquely specialised for its role in a pumping organ and forms approximately 50% of the heart by mass. As illustrated in Figure 17·6, the individual muscle cells, or **myocytes**, are branched and joined together by a series of **intercalated discs** to form a dense interconnected network. As described in more detail in Unit 26, this structure allows electrical excitation to pass rapidly from cell to cell, so that the linked cells contract together almost simultaneously.

Once activated, an individual muscle cell reacts completely to produce its maximum contraction.

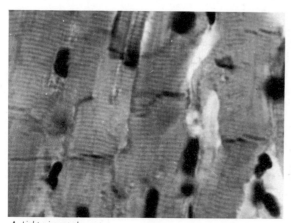

A Light micrograph

Fig 17·6 *Microscopic structure of cardiac muscle tissue*

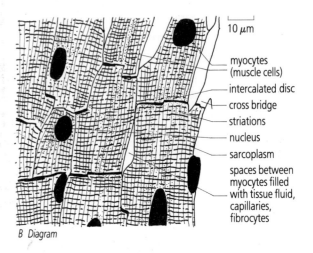

B Diagram

This is called the **all-or-none** effect. Furthermore, the cells show an unusually long **refractory period** during which they are completely unable to contract for a second time. This is very important because it ensures that each contraction, or **systole** of the heart will be separated from the next by a resting period, or **diastole**. During diastole the cardiac muscle relaxes, allowing the atria and ventricles to increase in volume passively under the pressure of inflowing blood, becoming filled with blood before the next contraction.

The high oxygen demands of the cardiac muscle are met by numerous blood capillaries. In the living heart, these receive blood from a pair of **coronary arteries** which arise from the aorta, immediately above the aortic valve, and spread over the surface, carrying oxygenated blood to all parts. Deoxygenated blood is collected by a system of **cardiac veins** into a single large **coronary sinus**, which empties into the right atrium.

Coordination of heart beat

The human heart beats about 70 times per minute, pumping 70 cm^3 of blood from each of its two ventricles each time it beats. Thus, in a single day, it beats 100 000 times and pumps 7000 litres of blood around the circulation. The sequence of events which takes place during each beat is called the **cardiac cycle** and is controlled by the **pacemaker** and **conducting systems** of the heart, as illustrated in Figure 17·7.

The **sinoatrial node (SAN)** is a small patch of specialised cardiac muscle tissue located in the wall of the right atrium near the entrance of the superior vena cava. Normally, it acts as the pacemaker for the rest of the heart, determining how often it will beat. The cells of the SAN are **spontaneously active**, or **myogenic**, so that they contract rhythmically in the absence of stimulation. The SAN continues to function, and the heart beats normally, even after its nerve supply has been cut. The concept of spontaneous activity is quite simple: the SAN is little more than a 'clock' which sets the rhythm of the rest of the heart by its ticking.

Excitation originating in the SAN spreads out smoothly across the atria, producing a uniform contraction, called **atrial systole**, which helps to fill the ventricles with blood. The contraction is *prevented* from spreading directly to the ventricles by the thin layer of connective tissue which forms the fibrous skeleton of the heart, as described earlier. Consequently, excitation must pass to the ventricles by way of the **atrioventricular node (AVN)**, a second group of specialized cells located near the base of the atria. From the AVN conducting fibres in the **bundle of His** and the left and right **bundle branches** carry signals downwards to the base of the heart. Here the signals are transmitted to a network of **Purkyne fibres** within the ventricle walls. Thus the main contraction, called **ventricular systole**, starts at the bottom of the heart and spreads upwards, squeezing blood out from the ventricles into the aorta and pulmonary artery.

This method of coordinating the heart beat has two enormous advantages. It ensures that atrial and ventricular systole each involve the combined activity of large numbers of muscle cells for maximum pumping effect, and, secondly, it delays ventricular systole for approximately 0.1 s, allowing time for the ventricles to fill completely before contraction takes place.

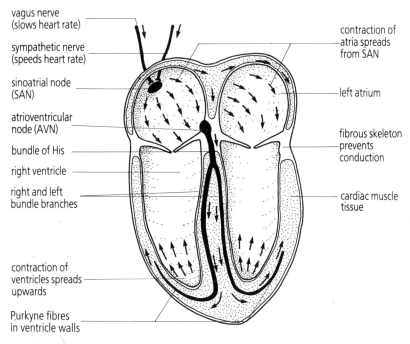

Fig 17·7 Pacemaker and conducting systems of the heart

Electrocardiogram (ECG)

The electrical activity of a person's heart can be investigated by placing metal electrodes on the external surface of his body, usually on the legs, arms, and chest. The minute electrical signals obtained are amplified and displayed on the screen of an oscilloscope, or may be permanently recorded on moving paper. This gives a graph of voltage variations against time, called the person's **electrocardiogram**, or **ECG**. The normal ECG pattern is shown in Figure 17·8. Each part of the ECG is given a letter according to an international code. The **P-wave** represents the electrical activity and contraction of the atria as excitation spreads outwards from the SAN, while the **QRS complex** corresponds to excitation of the ventricles. The final **T-wave** signals recovery of the ventricles at the end of contraction.

The ECG provides a useful method of monitoring changes in heart rate and activity and is often used in conjunction with exercise machines.

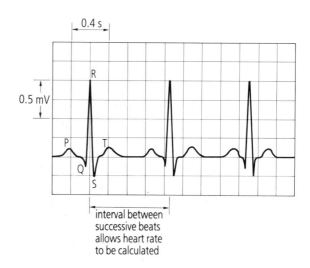

Fig 17·8 Electrocardiogram (ECG)

It is invaluable in the diagnosis of heart disease, for many common abnormalities produce characteristic alterations in the ECG pattern.

17·4 Blood vessels and blood flow ■ ■ ■ ■ ■ ■ ■ ■ ■ ■ ■

Both the pressure of the blood and the velocity of its flow vary greatly. The highest pressures occur in the systemic circulation following contraction of the left ventricle. Like all fluids, blood flows from high pressure to low pressure regions. Accordingly, the hydrostatic pressure in blood vessels decreases progressively and is least in the great veins and atria of the heart. The velocity of flow changes round the circulation: it is fastest in the arteries, slows down markedly in the capillaries and speeds up again towards the heart. This may seem surprising, but is simply the result of the large total cross-sectional area of the capillaries compared to that of the veins.

The different structure of the blood vessels, illustrated in Figure 17·9, reflects their position in the circulation, as follows:

1 Arteries. Arteries carry blood away from the heart and have thick walls to resist the pressure of blood inside. The innermost layer of the wall, or **tunica intima**, comprises a single layer of cells, the **endothelium**, supported by connective tissue. The endothelium is common to all parts of the circulatory system, including the heart, and provides a smooth lining which offers the least possible frictional resistance to blood flow. The **tunica media** of the arteries nearest the heart consists of up to 40 **elastic lamellae** which stretch outwards with the pressure of the heart beat. In arteries farther away from the heart, where the pressure is less, much of the elastic tissue is replaced by **muscle** cells. The outermost layer of the artery, called the **tunica adventitia**, forms a thin fibrous covering.

Energy is stored in the walls of the arteries as they stretch outwards during ventricular systole. As the ventricles relax, the aortic valve closes and the elastic fibres in the artery walls shorten, releasing the stored energy and thereby maintaining the arterial pressure at a high level. This phenomenon is called **elastic recoil** and it helps to convert the discontinuous pumping of the heart into a continuous pressure which keeps the blood flowing evenly.

2 Arterioles. These are small blood vessels, less than 100 μm in diameter, which arise by repeated branching from the arteries. The walls of arterioles contain numerous muscle cells, with little or no elastic tissue. The diameter of the individual arterioles can be precisely controlled by the action of nerves and hormones and, as you will discover shortly, this plays a major part in regulating blood flow to the body organs. The nerves reaching the arterioles originate from the sympathetic nervous system which forms one part of the autonomic nervous system, responsible for the automatic control of body functions.

3 Capillaries. The capillaries are the smallest and most numerous blood vessels, having an average diameter of 8 μm and a wall thickness of only 0.2 μm. The human body contains about 100 000 km of capillaries which penetrate almost every tissue, so that no cell is more than a short distance from a capillary.

The capillaries are the most important part of the circulation for they are the site of exchange with the tissue cells. This exchange takes place in

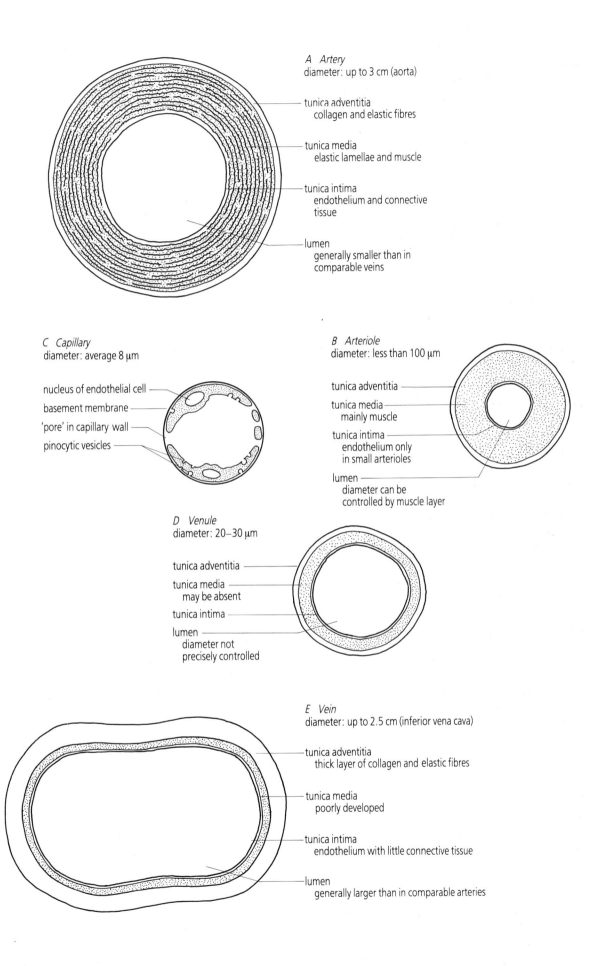

A Artery
diameter: up to 3 cm (aorta)

tunica adventitia
    collagen and elastic fibres

tunica media
    elastic lamellae and muscle

tunica intima
    endothelium and connective
    tissue

lumen
    generally smaller than in
    comparable veins

C Capillary
diameter: average 8 μm

nucleus of endothelial cell

basement membrane

'pore' in capillary wall

pinocytic vesicles

B Arteriole
diameter: less than 100 μm

tunica adventitia

tunica media
    mainly muscle

tunica intima
    endothelium only
    in small arterioles

lumen
    diameter can be
    controlled by muscle layer

D Venule
diameter: 20–30 μm

tunica adventitia

tunica media
    may be absent

tunica intima

lumen
    diameter not
    precisely controlled

E Vein
diameter: up to 2.5 cm (inferior vena cava)

tunica adventitia
    thick layer of collagen and elastic fibres

tunica media
    poorly developed

tunica intima
    endothelium with little connective tissue

lumen
    generally larger than in comparable arteries

Fig 17·9 Structure of blood vessels

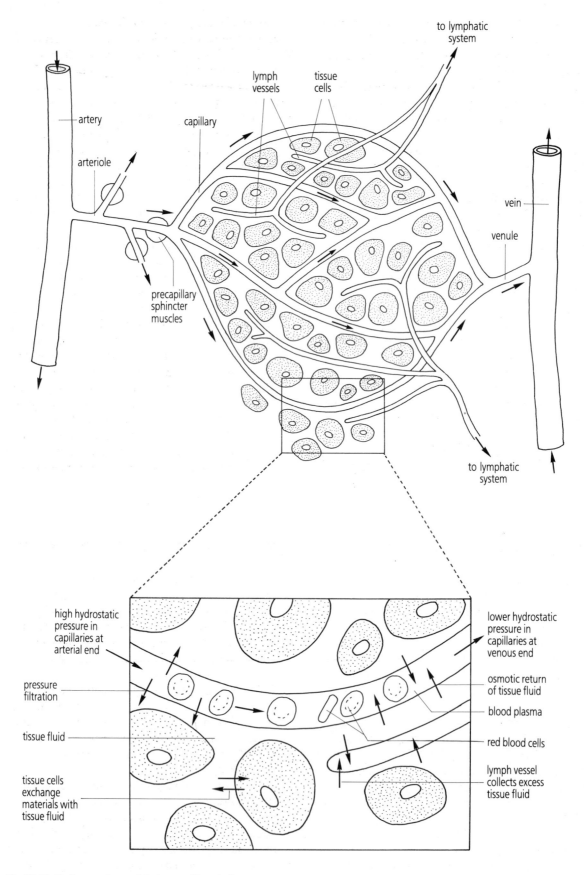

Fig 17·10 Exchange of materials in a capillary bed

capillary beds, as outlined in Figure 17·10. Tissue fluid, which bathes the tissue cells, is formed by **pressure filtration** across the capillary walls. These act as a selectively permeable barrier, allowing water, dissolved gases, ions, and soluble food molecules to pass, but restricting the movement of larger molecules such as proteins. In addition, certain substances appear to be actively transported by the movement of **pinocytic vesicles** directly through the endothelial cells.

The hydrostatic pressure in capillaries is much greater than in the tissues so it is easy to understand how tissue fluid is formed. What is less clear is how fluid is transferred back into capillaries at the venous end of the capillary bed. The hydrostatic pressure here still exceeds tissue fluid pressure so why is fluid transferred in the reverse direction? The answer is that the blood plasma has a much higher concentration of dissolved molecules than tissue fluid, so that the water is drawn into the capillaries by osmosis. The additional molecules are mainly plasma proteins, especially **albumins**. At the arterial end of the capillary bed, the hydrostatic pressure of the blood is sufficient to overcome the osmotic effect, but, at the venous end, the osmotic effect is more important, leaving a net tendency for water to enter the blood. Substances such as CO_2, and other waste products of metabolism, leave the tissues and enter the capillaries by diffusion. A small proportion of tissue fluid drains to lymph vessels, as already mentioned.

The quantity of blood flowing through the capillary bed is regulated at local level by various stimuli. Reduced O_2 and increased CO_2 result in dilation of the arterioles and **precapillary sphincter muscles**, leading to an improved blood flow. In this way, increased metabolic demands by the tissue cells are immediately met by an additional supply of oxygen. Similarly, accumulating waste products are rapidly removed.

An additional role of the blood capillaries is to allow the migration of certain types of white blood cells, or leucocytes, into the tissues. This process is part of the body's defence system against disease, and is discussed further in Unit 19.

4 Venules. These are small blood vessels, $20-30$ µm in diameter, formed where the capillaries join together. The first venules lack the tunica media layer and are still important sites of leucocyte migration and exchange with the tissues. Larger venules have a thin tunica media containing muscle fibres.

5 Veins. Veins are large blood vessels which carry blood towards the heart. Unlike other blood vessels, they have valves to prevent backflow. The walls of veins, like those of arteries, have three layers, an innermost endothelium, a middle layer, the tunica media, and an external fibrous covering, the tunica adventitia (refer to Fig. 17·9E). The tunica media in veins contains muscle and other fibres but is thin and weak compared to the tunica media of arteries. The tunica adventitia in veins is often the thickest of the three layers and consists mainly of collagen fibres, with some elastic and muscle fibres. The space, or **lumen**, in the centre of a vein tends to be larger in diameter than in the equivalent arteries, reflecting a slower rate of flow.

17·5 Regulation of the circulation ■ ■ ■ ■ ■ ■ ■ ■ ■ ■ ■

The most important role of the circulatory system is to distribute dissolved oxygen to all the body organs. Some organs can tolerate a temporary shortage of this vital substance, but others must have a continual supply. The brain is irreversibly damaged if its blood flow is interrupted for more than 3–4 minutes. What happens, therefore, when the oxygen demand of one part of the body suddenly increases? For example, when the arterioles of the muscles dilate during exercise, large quantities of blood are diverted from the general circulation. Inevitably, this tends to reduce the pressure in the arterial system, leaving less blood available for the other body organs. This kind of problem is overcome by a control system which regulates arterial pressure within narrow limits, as outlined in Figure 17·11.

The two main factors which determine arterial pressure are **cardiac output** and **arteriolar resistance**. Cardiac output is defined as the volume of blood pumped per unit time and depends both on heart rate (number of beats per minute) and stroke volume (volume of blood pumped with each beat). During exercise, cardiac output increases automatically, as the result of several processes:

1 Any fall in arterial pressure is detected by special pressure receptors, or **baroreceptors**, located in the walls of the aorta and carotid arteries. These send impulses to the medulla of the brain, suppressing activity in the vagus nerve and speeding up the heart rate by way of the sympathetic nerve.

2 At the same time, the sympathetic nervous system acts on the medulla of the adrenal glands, stimulating the release of the hormone **adrenalin** which circulates in the bloodstream and acts directly on the heart to cause an increase in cardiac output.

3 Sympathetic nerve activity, contraction of the skeletal muscles and dilation of the muscle arterioles (which allows blood to flow through more freely) all increase the pressure of blood in the venous system leading to the heart. As a result, the ventricles fill more completely during diastole so that the volume of blood pumped per beat increases. The increased pumping action is possible because the heart muscle responds to increased stretching at rest by more powerful contraction when stimulated. This relationship is known as **Starling's Law**, and is an interesting and important effect for it allows transplanted

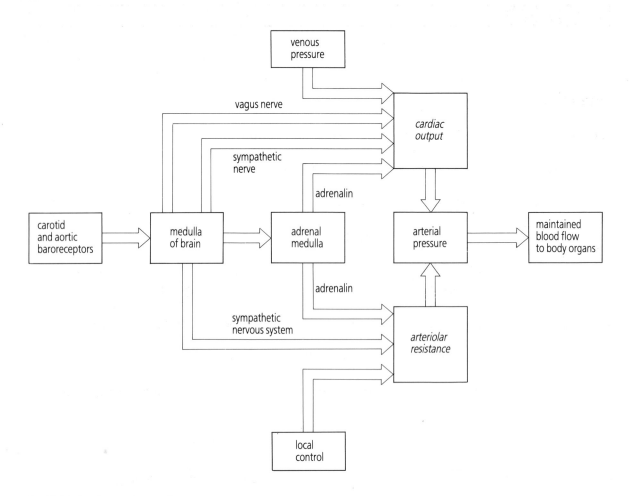

Fig 17·11 Regulation of arterial pressure

hearts, which lack a nerve supply altogether, to respond appropriately to different conditions, keeping the patient's tissues properly supplied with oxygen.

The second factor determining arterial pressure is arteriolar resistance, that is, the opposition to blood flow caused by narrowing of arterioles. During exercise, most arterioles, except those leading to the brain, muscles, and skin, are constricted. This happens partly by sympathetic nerve activity and partly by hormone action, adrenalin being especially important.

18 BLOOD

Objectives

After studying this Unit you should be able to:—

- List the main components of blood and outline their importance

- List and describe the functions of blood plasma

- Describe the development of red blood cells

- Discuss fully the role of haemoglobin in the transport of oxygen and explain the effects of pH and temperature on the oxygen dissociation curve for haemoglobin

- Discuss the mechanisms involved in the transport of carbon dioxide in the blood

- Outline the processes involved in the destruction of red cells

- Comment on the role of other oxygen-carrying pigments including myoblobin and fetal haemoglobin

- List the different steps which contribute to the prevention of blood loss

- Draw and explain a flow diagram to summarize the process of blood clotting

18·1 Introduction ■ ■ ■ ■ ■ ■ ■ ■ ■ ■ ■ ■ ■ ■ ■

The fluid flowing in the circulatory system is called **blood**. In mammals, the functions of blood can be divided into two groups: **transport functions** and **homeostatic functions**. The transport functions are concerned mainly with the supply of food materials and oxygen, and the removal of waste products from the body cells. The homeostatic functions include the regulation of tissue fluid, regulation of pH, distribution of heat, and defence against disease. This Unit describes all the main components and functions of blood except those concerned with defence. The defence functions are considered along with the body's other methods of protection in Unit 19.

18·2 Components of blood ■ ■ ■ ■ ■ ■ ■ ■ ■ ■ ■ ■ ■

The components of human blood are normally kept well mixed by constant motion within the circulatory system. However, if a sample of blood (treated to prevent clotting) is placed in a test tube and allowed to stand, it separates into two fractions, called **formed elements** and **plasma**. The formed elements settle to the bottom of the tube, forming a red layer which accounts for about 45% of the total blood volume. As summarised in Table 18·1, the formed elements consist almost entirely of red cells, or **erythrocytes**, with various different kinds of white cells, or **leucocytes**, and tiny cell fragments, called **platelets**. Together, the leucocytes and platelets form only about 1% of the blood volume.

The leucocytes are divided into two main groups called **granulocytes** and **agranulocytes**. Granulocytes have an irregular lobed nucleus and cytoplasm containing prominent granules. As shown in Table 18·1, the granulocytes are further subdivided into **neutrophils**, **eosinophils**, and **basophils**. These types of cell are so named because their granules react in different ways when a dried blood film is treated with **Leishman's stain**, a mixture of an acid dye (eosin) and a basic dye (methylene blue). The granules of the neutrophils remain unstained, while those of the eosinophils stain red, and those of the basophils stain blue. The neutrophil granules are now known to be lysosomes and they contain powerful

Table 18.1 Components of blood

FORMED ELEMENTS (45% of blood volume)

ERYTHROCYTES: 4 000 000–6 000 000 per mm³ of blood

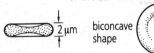

origin: bone marrow
functions: transport of O_2, CO_2, pH buffering
life-span: 90–120 days

LEUCOCYTES:

1. Granulocytes
   (i) *neutrophil:* 3000–6000 per mm³ of blood

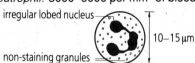

irregular lobed nucleus
non-staining granules
10–15 µm

origin: bone marrow
functions: phagocytosis of microorganisms
life-span: 12 hrs–3 days

   (ii) *eosinophil:* 100–400 per mm³ of blood

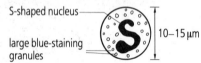

nucleus with two lobes
large red-staining granules
10–15 µm

origin: bone marrow
functions: not precisely known
life-span: 3–5 days

   (iii) *basophil:* 25–200 per mm³ of blood

S-shaped nucleus
large blue-staining granules
10–15 µm

origin: bone marrow
functions: develop into mast cells in connective tissue
life-span: 9–18 months

2. Agranulocytes
   (i) *lymphocyte:* 1500–2700 per mm³ of blood

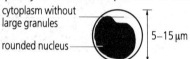

cytoplasm without large granules
rounded nucleus
5–15 µm

origin: lymph system
functions: phagocytosis, antibody production
life-span: 100–300 days

   (ii) *monocyte:* 100–700 per mm³ of blood

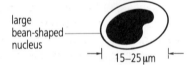

large bean-shaped nucleus
15–25 µm

origin: bone marrow
functions: develop into tissue macrophages
life-span: 100–300 days

PLATELETS: 250 000–500 000 per mm³ of blood

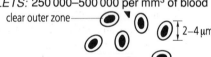

clear outer zone
blue-staining inner zone
2–4 µm

origin: bone marrow
functions: involved in haemostasis and blood clotting
life-span: 8–14 days

PLASMA (55% of blood volume)

COMPOSITION	FUNCTIONS
water (90% of plasma)	solvent, transport of heat
solutes (1–2% of plasma) Na^+, K^+, Ca^{2+}, Mg^{2+}, Cl^-, HCO_3^-, $H_2PO_4^-$, glucose, urea, fats, amino acids, vitamins	maintenance of osmotic pressure, pH buffering, nutrients, wastes
plasma proteins (6–8 per cent of plasma) albumins, globulins, fibrinogen	maintenance of osmotic pressure, pH buffering, defence, transport, hormones, blood clotting

enzymes which help to digest microorganisms following phagocytosis. Agranulocytes have a smooth rounded or bean-shaped nucleus, and cytoplasm with no obvious granules. They include **lymphocytes** and **monocytes**. All the different types of white blood cells are involved in defence of the body against disease, and are discussed further in Unit 19.

18·3 Functions of blood plasma ■ ■ ■ ■ ■ ■ ■ ■ ■ ■

Transport of materials in solution

The composition of plasma is outlined in Table 18·1. Water forms 90% of the plasma and dissolves the many other substances listed so that they are carried around in the general circulation. Substances including glucose, amino acids, and vitamins, form a reservoir of raw materials which can be tapped by the tissue cells according to their needs, while waste products, like urea, circulate until being removed by the excretory organs.

Some substances such as iron and hormones are transported by **carrier proteins**, which belong to the **globulin** group. Iron ions combine with a particular protein called **transferrin**. This effectively prevents the iron from combining with other components of the plasma, and delivers it in a useful form to the liver, where it can be stored, and to the bone marrow cells, where it can be used for making haemoglobin. Similarly, hormones are transported to all parts of the body and are taken up by their appropriate target organs.

Plasma is not the main vehicle for the transport of oxygen, this function being carried out largely by the red cells. However, oxygen must be released from the red cells and dissolved in the plasma before it can become available in the tissue fluid, dissolved oxygen reaching the tissue cells by diffusion. Carbon dioxide accumulates in the tissue fluid and diffuses in the opposite direction; most of it is carried in the plasma in the form of hydrogencarbonate ions.

In addition to its role as a solvent for small molecules, the water in plasma allows large protein molecules to form a colloidal solution (described in Unit 2). It also provides a good medium for the dispersal and transport of the erythrocytes and other formed elements of the blood. Furthermore, water has a high specific heat capacity and transports heat effectively from metabolically active regions, such as the liver and muscles, to other parts of the body.

Regulation of tissue fluid

The concentration of dissolved substances in the blood plasma is much higher than that in the tissues. Consequently, as outlined in Unit 17, water returns from the tissue fluid to the capillaries at the venous end of the circulation by osmosis. This process depends on the difference in concentration of the blood plasma and the tissue fluid. Accordingly, the overall concentration of the blood plasma must be regulated within narrow limits to maintain a proper balance of fluids between the tissues and the circulation. The most important solutes in this connection are the **plasma proteins** and the numerous different ions present in the plasma. Most types of plasma

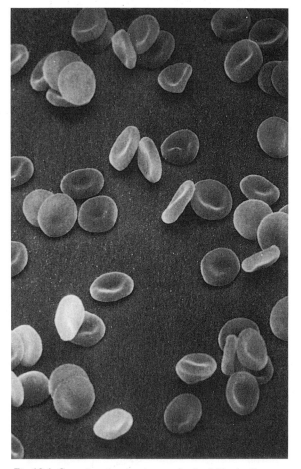

Fig 18·1 Scanning electronmicrograph of red blood cells

protein are manufactured by the liver, which is thus able to regulate their concentration. The **albumins** are the most plentiful of the proteins and these are particularly important in raising the osmotic pressure of the plasma above that of the tissue fluid. They do not normally cross the capillary walls, but remain inside, helping to ensure the return of water by osmosis. If the concentration of plasma proteins falls, which sometimes happens as a result of liver or kidney malfunction, or as a result of malnutrition, this osmotic effect is reduced and the tissues swell with accumulated fluid, a condition known as **oedema**.

Regulation of pH

The concentration of H^+ ions affects many different chemical reactions and plays a major role in determining the shape and activity of protein molecules, including enzymes. Consequently, pH is one of the most important factors involved in maintaining homeostasis. The normal pH of human plasma and tissue fluid is 7.4 and, in a healthy person, the pH of the plasma varies by no more than about 0.04 units from this value.

H^+ ions are formed in the body mainly by aerobic respiration. This process generates large amounts of carbon dioxide which dissolve in the

135

body fluids, forming carbonic acid, which dissociates in turn to yield H^+ ions:

$$CO_2 + H_2O \rightleftharpoons H_2CO_3 \rightleftharpoons H^+ + HCO_3^-$$

carbonic acid

Within the plasma, **sodium hydrogencarbonate, protein**, and **phosphate** ($H_2PO_4^-$) act as **pH buffers** (see Unit 2),

temporarily taking up any excess H^+ ions, and, at the same time, keeping the pH constant. H^+ ions are removed from the body in two ways. Some of them are used up when CO_2 is excreted from the lungs since this process reverses the chemical changes which lead to H^+ production. Alternatively, H^+ ions can be excreted directly by the kidneys, as described in Unit 20.

18·4 Red blood cells ■ ■ ■ ■ ■ ■ ■ ■ ■ ■ ■ ■ ■ ■ ■

Development

In the human embryo, red blood cells are formed initially in the yolk sac, liver, and spleen. Later in development, this function is carried out by the bone marrow, and, from birth onwards, this becomes the only site of production. In young children, the red bone marrow is distributed throughout the skeleton, but, in adults, it is confined to the bones of the vertebral column, sternum, and ribs, and to the ends of the long bones in the arms and legs. The development of red cells, called **erythropoiesis**, takes from 5 to 9 days and begins from undifferentiated cells, called **stem cells**. These multiply and then start to produce the large amounts of RNA needed for synthesis of haemoglobin. As soon as the developing cell is filled with haemoglobin, RNA production stops and the nucleus of the cell is extruded. When the mature erythrocyte enters the circulation, it is a biconcave disc 8·5 μm in diameter and has no nucleus (see Fig. 18·1). Each erythrocyte contains about 280 million molecules of haemoglobin. The membrane surrounding the cell contains a wide variety of proteins, including several enzyme systems. It also displays on its surface the glycolipid blood group substances, or **ABO antigens**, which are discussed in Unit 19.

The rate of production of red cells is controlled by a hormone, **erythropoietin**, released from the kidneys. A decrease in the amount of oxygen reaching the kidney cells results in the release of more erythropoietin, thereby stimulating the production of new cells. Normally, the number produced is equal to the number of red cells destroyed, but, if a person is exposed to new conditions, as in the case of oxygen shortage at high altitudes, the number of red cells slowly increases to meet the new demands. The development of erythrocytes also depends on the availability of a wide range of raw materials, any shortage leading to a failure in normal production, and to symptoms of **anaemia**, one cause of which is a lack of iron in the diet. Another form of anaemia, called **pernicious anaemia**, results from lack of vitamin B_{12}, which is used as a coenzyme in the synthesis of nucleic acids.

Transport of oxygen

More than 98% of the oxygen transported in human blood is carried by haemoglobin. The molecules of this remarkable protein each consist of four subunits, two α-chains, and two β-chains, fitted together to give a compact overall structure. Furthermore, as described in Unit 4, each of the subunits contains a special pigment group, or **haem group**, with an iron atom located at its centre. A single iron atom binds one molecule of oxygen, so that each molecule of haemoglobin can combine reversibly with up to four molecules of oxygen. Four different kinds of **oxyhaemoglobin** can be formed, as follows:

$$\begin{array}{ccc} \text{Hb} & + O_2 \rightleftharpoons & HbO_2 \\ \text{haemoglobin} & & \text{oxyhaemoglobin 1} \end{array}$$

$$\begin{array}{ccc} HbO_2 & + O_2 \rightleftharpoons & HbO_4 \\ \text{oxyhaemoglobin 1} & & \text{oxyhaemoglobin 2} \end{array}$$

$$\begin{array}{ccc} HbO_4 & + O_2 \rightleftharpoons & HbO_6 \\ \text{oxyhaemoglobin 2} & & \text{oxyhaemoglobin 3} \end{array}$$

$$\begin{array}{ccc} HbO_6 & + O_2 \rightleftharpoons & HbO_8 \\ \text{oxyhaemoglobin 3} & & \text{oxyhaemoglobin 4} \end{array}$$

The individual steps in this reaction sequence are not all equally easy to carry out. Oxygen is taken up and released least readily by haemoglobin itself. The different types of oxyhaemoglobin combine with oxygen progressively more easily and bind it less tightly. Thus oxyhaemoglobin 4 dissociates more easily than oxyhaemoglobin 3 and so on. This phenomenon gives a characteristic sigmoid, or S-shape, to the **oxygen dissociation curve for haemoglobin**, as illustrated in Figure 18·2. This shows the percentage saturation of a solution of haemoglobin, that is, the percentage of available binding sites which are occupied, plotted against the concentration of oxygen. This is given by pO_2, the partial pressure of oxygen dissolved in the surrounding fluid. At very low levels of pO_2, below 2 kPa, the curve rises relatively slowly. Above this, the curve rises steeply, reflecting an easy uptake and release of oxygen. In the plateau region saturation approaches 100%.

The air in the lungs has a pO_2 of about 13 kPa. This is much higher than the pO_2 in blood

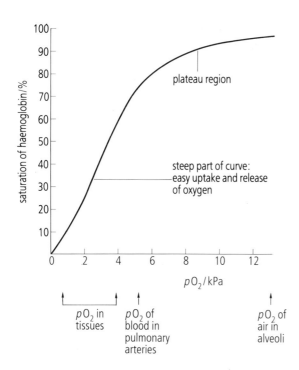

Fig 18·2 *Oxygen dissociation curve for haemoglobin (at pH 7.4 and 38°C)*

entering the lungs from the pulmonary arteries, typically only about 5 kPa. Therefore, dissolved oxygen diffuses rapidly across the alveoli into the blood plasma and then into the interior of the red cells, where it combines with haemoglobin. The process of diffusion is assisted by the intimate contact between the walls of the alveoli and the blood capillaries (described in Unit 16), and also by the biconcave shape of the red blood cells. This shape gives the red cells a particularly large surface area to volume ratio and promotes a rapid exchange of substances across the cell membrane.

Interestingly, the biconcave shape also protects the red cells against rupture, or **haemolysis**, caused by variations in the osmotic concentration of the blood plasma. Cell membranes tear very easily if stretched, so that a spherical cell placed in a weak solution bursts instantly. However, because of its shape, a red blood cell can take up water and swell, equalising internal and external concentrations of solutes without unduly stretching its membrane. Therefore, red cells can tolerate an extended range of external concentrations, although they do burst if placed directly in water.

On leaving the lungs, the blood is almost completely saturated with oxygen, and carries the equivalent of 200 cm^3 of gaseous oxygen per litre. Only about 3 cm^3 of this is dissolved in the plasma, the remainder being carried as oxyhaemoglobin within the red cells. The normal pO_2 in the lungs is well to the right on the plateau of the oxygen dissociation curve (refer to Fig. 18·2). Thus, if the alveolar pO_2 falls, as often happens during exercise, the amount of oxygen carried by the blood is only slightly reduced.

Oxygen supply to the tissues can easily be maintained by increasing the rate of blood flow through the lungs.

The pO_2 in the tissues is very variable in the range 1–4 kPa, depending on the oxygen demand. As oxygen is used up, the pO_2 falls causing oxyhaemoglobin in the blood capillaries to dissociate. This releases additional oxygen which can be used by the tissue cells. Note that lower levels of pO_2 coincide with the steepest parts of the dissociation curve, so that slight reductions in pO_2 cause the release of large amounts of oxygen. Furthermore, the shape of the curve is altered by pH, which falls in the presence of accumulating CO_2, and by changes in temperature, in such a way that rapidly respiring tissues receive the maximum amount of oxygen from the blood passing through. These relationships are illustrated in Figure 18·3. The effect of CO_2 is called the **Bohr effect** after its discoverer. The mechanism responsible for this effect is explained in the next section.

A *Effect of pH* (temperature = 38°C)

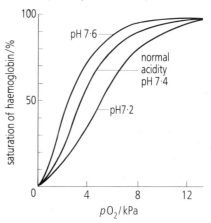

The pH in the body tissues is affected by CO_2 – accumulating CO_2 causes a fall in pH shifting the dissociation curve to the right and thereby causing more oxygen to be released. This effect of CO_2 is called the *Bohr effect*.

B *Effect of temperature* (pH = 7·4)

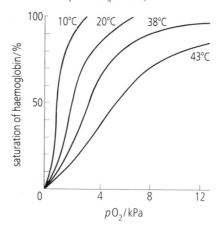

Fig 18·3 *Effect of pH and temperature on the oxygen dissociation curve for haemoglobin*

Transport of CO$_2$

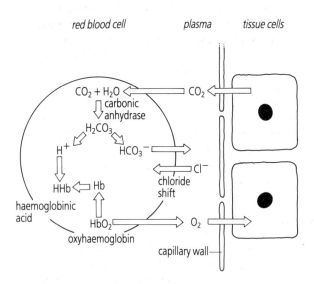

Fig 18·4 Transport of carbon dioxide in the blood

The transport of CO$_2$ is closely linked to the transport of oxygen. The process is outlined in Figure 18·4. CO$_2$ produced in the tissue cells dissolves in the tissue fluid and diffuses across the capillaries into the blood plasma, and finally into the individual red cells. As already explained, CO$_2$ reacts with water to form carbonic acid in the blood plasma. Inside the red cells, this process is speeded up enormously by the enzyme **carbonic anhydrase**. As blood flows through the tissue capillaries, the enzyme acts to produce large amounts of carbonic acid which immediately dissociates to form H$^+$ ions and hydrogencarbonate ions (HCO$_3^-$). Most of the hydrogencarbonate ions diffuse back into the plasma. This outward flow is balanced by an inward movement of Cl$^-$ ions, called the **chloride shift**. Note that equal numbers of ions move in each direction, so that the overall distribution of electrical charges remains the same. The H$^+$ ions produced inside the red cells combine with

haemoglobin to form a very weak acid, called **haemoglobinic acid**. This reaction uses up some of the haemoglobin inside the red cells, causing further dissociation of oxyhaemoglobin. In this way, an increase in the number of H$^+$ ions brings about an increase in the amount of oxygen released. This provides an explanation for the Bohr effect linking an accumulation of CO$_2$ to an improved supply of oxygen to the tissues.

Destruction of red cells

Human red blood cells have a life span of 90–120 days. They cannot divide, and, because they have no nucleus and no rough ER, they are unable to make new enzymes or structural proteins. As a result, they deteriorate progressively. Old red blood cells are recognised and eventually destroyed by large phagocytic cells, called **macrophages**. The macrophages develop from monocytes in the bloodstream and are found throughout the body, being concentrated in the liver, spleen, and bone marrow. Collectively, they form the **reticulo-endothelial system** of the body.

When a red cell is engulfed, its haemoglobin is broken down by lysosomal enzymes, as outlined in Figure 18·5. The protein, or globin, parts of the molecule are broken down into individual amino acids which can be reused. Iron atoms are removed from each of the four haem groups and pass into the blood plasma where they combine with transferrin and circulate until reaching the liver. Here, iron is selectively taken up and stored in combination with a second protein carrier substance, called **ferritin**. Iron from ferritin can be mobilised as required and transferred to the bone marrow for use in the manufacture of new red blood cells. The residue of the haem group is released from the macrophages as the yellow-red pigment **bilirubin**. Bilirubin is responsible for the yellowish colour of blood plasma. It circulates in the blood until it is removed by the liver cells, as outlined in Unit 15. Eventually, bilirubin leaves the body in the form of bile pigments.

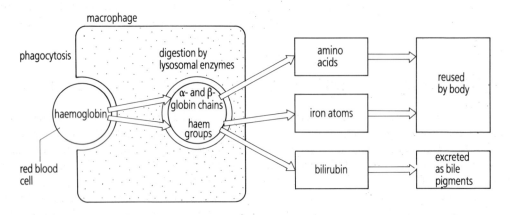

Fig 18·5 Destruction of red blood cells by tissue macrophages

18·5 Other oxygen-carrying pigments ■ ■ ■ ■ ■ ■ ■ ■ ■ ■

In the human body, haemoglobin occurs only in the bloodstream and in the bone marrow, where red cells develop. A second oxygen-carrying pigment, called **myoglobin**, is found in large quantities in the body muscles. As described in Unit 4, each molecule of myoglobin consists of a haem group attached to a globin protein chain, the complete molecule comprising only one such unit, rather than four, as in the case of haemoglobin. However, there is an important difference in the oxygen dissociation curve for a solution of myoglobin molecules, as illustrated in Figure 18·6.

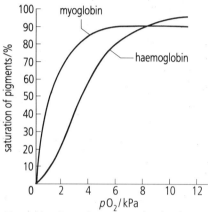

Myoglobin releases its oxygen at low levels of pO_2 and acts as a store of oxygen within the muscles.

Fig 18·6 Oxygen dissociation curves for haemoglobin and myoglobin

As you can see, oxygen is bound to myoglobin much more tightly and is released only at very low pO_2 levels. Consequently, myoglobin is most useful during strenuous activity, when insufficient oxygen reaches the muscles from the blood. Under these conditions, pO_2 in the muscles falls towards zero and oxygen is given up by myoglobin, thereby allowing aerobic respiration to continue for as long as possible. Myoglobin makes a significant contribution to sustained muscular effort. When its reserves of oxygen are exhausted, the muscle begins to respire anaerobically, and lactic acid is produced.

Another situation which calls for a different oxygen-carrying pigment occurs during pregnancy, when oxygen must be transferred across the placenta to the developing fetus. To accomplish this, **fetal haemoglobin** has a molecular structure incorporating different globin proteins, called **γ-chains**. A molecule of fetal haemoglobin comprises two α-chains and two γ-chains instead of two α-chains and two β-chains, as in adult haemoglobin. This alteration shifts the oxygen dissociation curve for fetal haemoglobin towards the left, like that of myoglobin. Thus fetal haemoglobin binds oxygen more tightly than adult haemoglobin, so that oxygen can be transferred easily from mother to fetus. After birth, large numbers of red cells are destroyed and gradually replaced with new cells containing adult haemoglobin. This sometimes leads to symptoms of jaundice, caused by an accumulation of bilirubin in the blood.

Abnormal haemoglobins include **haemoglobin S** which causes the inherited disease **sickle cell anaemia**. The β-chains of haemoglobin S have only a single alteration in the normal sequence of 146 amino acids, the glutamic acid in position 6 of the normal β-chain being replaced by valine. This change corresponds to a single base substitution in the DNA which codes for the β-chains. Nevertheless, haemoglobin S becomes insoluble at low levels of pO_2 and crystallizes out, greatly altering the normal shape of the red cells. The deformed cells block the body capillaries, causing impaired blood flow and tissue damage. Sickle cell anaemia is a relatively common condition in areas affected by malaria for reasons explained in Unit 39. **α-thalassemia** and **β-thalassemia** are similar inherited diseases, which are caused by partial or complete failure to produce normal α-chains and β-chains respectively.

A whole range of other oxygen-carrying pigments occur in other vertebrates and invertebrates, and even in some plants. Subtle changes in molecular structure ensure that the normal forms of these pigments carry out their own particular roles with optimum efficiency.

18·6 Prevention of blood loss ■ ■ ■ ■ ■ ■ ■ ■ ■ ■ ■ ■

The prevention of blood loss, called **haemostasis**, is essential for the correct functioning of the circulatory system. Even when there is no external injury, the capillaries and other small blood vessels rupture frequently under pressure, allowing blood to leak into the tissues. The mechanisms of haemostasis block these small leaks and help to keep the blood properly confined within the circulation. Without haemostasis, a mammal would

die within a few days from internal haemorrhage. This fact is exploited by the rat poison **warfarin**, which prevents haemostasis so that the rats bleed to death from the combined effect of numerous tiny internal injuries.

The events of haemostasis are listed below. The first three steps interact to seal small blood vessels in seconds. More permanent repair is effected afterwards by blood clotting.

1 Vessel constriction. When any muscular blood vessel is damaged, there is an immediate constriction which restricts blood loss, or, in small arterioles, closes the vessel completely. The initial constriction is enhanced at a later stage by chemicals released from the blood platelets.

2 Sticking of endothelia. The surface properties of the capillary endothelial cells are altered by injury so that cells forced together by constriction stick tightly to each other, keeping small vessels sealed, even when the initial constriction starts to subside.

3 Formation of a platelet plug. Platelets are small non-cellular fragments 2–4 μm in diameter consisting of a clear outer zone surrounding a granular interior. They do not adhere to the smooth undamaged endothelium lining the blood vessels, but, if the endothelium is damaged, exposing the collagen fibres of the underlying connective tissue, the platelets stick tightly to form a semi-solid mass which plugs the vessel. At the same time, the chemical contents of the platelet granules are released, causing more platelets to become attached to the 'plug' and promoting further vasoconstriction. The platelet plug forms quickly, but is mechanically weak and is replaced by a more robust blood clot, as described in the next section.

4 Blood clotting. The most important steps in blood clotting are summarised in Figure 18·7. A clot is formed when the soluble plasma protein **fibrinogen** is converted to insoluble threads of **fibrin**. These threads form a mesh which traps the formed elements of the blood and changes it from a free-flowing liquid into a solid gel. The formation of fibrin is catalysed by the enzyme **thrombin**: this is not normally present in the blood, but appears after vessel damage. As you can see from the diagram, thrombin production does not occur directly but follows a chain, or 'cascade', of linked reactions. These involve plasma protein **clotting factors**, most of which are manufactured by the liver. At each step, the inactive form of an enzyme is converted to an active form which catalyses the next step in the chain. Some of the enzymes require Ca^{2+} ions as cofactors. Others require phospholipids which are released from the platelets as they accumulate to form a plug. Vitamin K does not function as a cofactor but is needed in the synthesis of **prothrombin** and certain other clotting factors.

Within the body, clotting is normally confined to the immediate vicinity of vessel damage. This localisation is possible because the blood plasma and tissues contain additional enzymes which inactivate the various clotting factors almost as soon as they are formed. Contrary to popular belief, exposure to air has nothing whatever to do with blood clotting.

5 Clot retraction. Once formed, blood clots are reduced in size by a process called retraction. This action depends on the platelets, which send out cytoplasmic projections, or pseudopodia, which adhere to the fibrin threads. The cytoplasm of the platelets contains contractile protein strands which then contract, pulling the threads closer together to form a stronger denser clot. Fibrin is eventually broken down by the enzyme **plasmin**. This enzyme is produced from its inactive form, **plasminogen**, when clotting first takes place, and acts very slowly within the clot, progressively dissolving the fibrin and promoting an efficient final repair of the damaged vessel.

Diseases associated with haemostasis include various forms of **haemophilia** in which the blood fails to clot. The condition occurs in the absence of any one of the many essential protein clotting factors normally present in blood plasma. The commonest type is caused by a sex-linked recessive gene, as explained in Unit 38.

Excessive clotting gives rise to **thrombosis**, or blockage of vessels, and is particularly serious when it affects the supply of blood to the heart or brain.

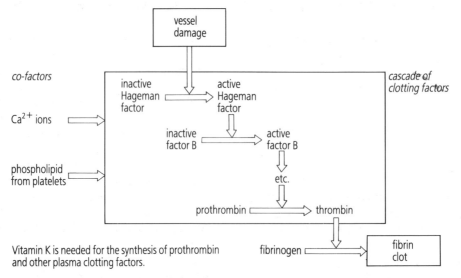

Fig 18·7 Simplified summary of blood clotting

19 DEFENCE OF THE BODY AGAINST DISEASE

Objectives

After studying this Unit you should be able to:–

- List the main environmental and internal causes of disease and give examples of each

- Describe the first lines of defence of the body against disease and summarize these defences in the form of a diagram

- Describe the structure of the lymphatic system and draw and label a diagram of a lymph node

- List the effector cells of the immune system and explain the difference in development between T lymphocytes (T cells) and B lymphocytes (B cells)

- Describe the non-specific immune responses of the body including the role of inflammation, phagocytosis, pyrogen, and interferon

- Describe the structure of antibody molecules and define an antigen

- Explain the role of T lymphocytes and B lymphocytes in the specific immune response

- Explain how immunity can be produced by vaccination, and distinguish between active and passive immunity

- Describe the ABO and Rhesus blood grouping systems

19·1 Introduction – causes of disease ■ ■ ■ ■ ■ ■ ■ ■ ■

Normally, the body is maintained in a healthy state by all the various processes which contribute to homeostasis. Disease occurs whenever homeostasis is upset or abnormal. Such failures in homeostasis arise in several different ways which can be listed as follows:

Environmental causes of disease

1 Absence of an essential chemical or condition. Autotrophic organisms obtain energy and simple raw materials from the environment, but heterotrophic organisms must obtain ready-made food materials from a variety of different sources. Disease can result if any of these necessary substances are absent or in short supply.

Such diseases are called **deficiency diseases**. They include various forms of malnutrition and protein deficiency diseases as well as vitamin deficiency diseases (see Table 12.2·).

2 Presence of a damaging chemical or condition. All organisms become poisoned or damaged in an unfavourable environment. In developed countries, the human population is continually exposed to damage from food additives, lead in petrol, toxic industrial wastes, and to increasing levels of radiation. In addition there are self-imposed hazards such as cigarette smoking, drugs, and excessive alcohol intake. Occupational diseases include the serious lung condition **pneumoconiosis**, found among coal miners, **asbestosis**, and some types of **leukaemia**, as well as many minor ailments.

3 Parasites. Organisms which cause disease are called **pathogens**. They may be divided into **microorganisms**, or 'microbes', including bacteria, viruses, protozoa, and fungi, and **macroorganisms**, such as the parasitic flatworms. It is estimated that at least 2 million people die annually from malaria, and this is just one of a whole range of human diseases caused by parasites. You will find details of the life history of the malarial parasite in Unit 46, and descriptions of other parasites in Units 45, 46, 47, 48, and 52.

Internal causes of disease

1 Genetic abnormality. Inherited, or genetic, diseases are usually the result of gene mutations which affect the production of enzymes or other proteins. For example, the disease **galactosaemia** occurs when the body is unable to make the enzyme **galactose-1-phosphate uridyl transferase**. This enzyme forms part of the normal pathway for converting galactose into glucose for use in respiration. In its absence, the intermediate substance galactose-1-phosphate accumulates in the blood, causing liver damage and mental retardation. In the U.K., newborn babies are routinely screened for the disease by pricking the baby's heel and collecting a drop of blood, which is then tested for the presence of the enzyme. Completely excluding lactose (from milk and milk products) and galactose from the diet allows affected infants to develop normally.

Other genetic abnormalities include **phenylketonuria (PKU)**, which affects the metabolism of the amino acid phenylalanine, **sickle cell anaemia**, and the sex-linked conditions **haemophilia** and **colour blindness**. Chromosome abnormalities give rise to conditions such as **Down's syndrome, Turner's syndrome**, and **Klinefelter's syndrome**; these are discussed more fully in Unit 38.

2 Somatic mutations. These are alterations in the DNA of individual body cells. Such mutations are not inherited but occur continually from exposure to radiation and toxic chemicals, or as a result of chance errors in DNA replication before cell division. While the body possesses sophisticated mechanisms for the repair of damaged DNA, some somatic mutations cause permanent alterations in the DNA, resulting in abnormal cell activity. An important example is the uncontrolled cell division of some types of cancer cells which leads to the development of tumours.

3 Autoimmune diseases. The cells of the immune system normally attack only foreign material, but sometimes they malfunction and attack the body's own tissues, causing autoimmune disease. An increasing number of human diseases are thought to be caused in this way, including certain types of **anaemia**, **diabetes mellitus**, **multiple sclerosis**, **hyperthyroidism**, and some types of muscle and kidney disease.

4 Wear and tear diseases. The body has a remarkable capacity for self-repair, but many parts slowly deteriorate with age. The circulatory system in particular is liable to serious damage. **Atherosclerosis** is the progressive thickening of artery walls with abnormal muscle cells and fatty deposits. It is a primary cause of heart attacks and cerebrovascular stroke. A different condition follows from weaknesses in the vessel walls – these allow swellings, or **aneurysms**, to develop which can rupture with potentially fatal results. In the case of the digestive system, muscle weakness often leads to the formation of small pouches in the gut wall, called **diverticula**, which are prone to infection and interfere with the proper digestion and absorption of food. These conditions and many others can be considered as 'wear and tear' diseases which are inevitable in the course of everyday life. Their progress can sometimes be slowed by exercise and proper attention to diet, but they cannot be halted altogether.

It must be emphasised that many diseases do not have a single cause but arise from a combination of factors. Autoimmune disease, for example, can be triggered off by virus infection, while dietary excess of cholesterol or genetic factors may hasten the development of atherosclerosis. The remainder of this Unit discusses all the body's defence mechanisms against disease. These mechanisms consist of a series of external barriers, which prevent the entry of microorganisms, and a sophisticated internal **immune system** which disposes of 'worn-out' or damaged body cells and helps to protect against cancer, as well as against infection by parasites of all kinds.

19·2 First lines of defence ■ ■ ■ ■ ■ ■ ■ ■ ■ ■ ■ ■

Bacteria and viruses are among the most important human parasites. They are transmitted from one host to another via the digestive, respiratory, and reproductive systems, or sometimes, directly across the skin. Each of these regions has a number of important protective mechanisms to prevent or reduce the entry of microorganisms. Together, these mechanisms constitute the body's first lines of defence, as summarized in Figure 19·1.

The digestive system is protected from parasites transmitted in food and water by the enzyme **lysozyme** which is present in saliva. Lysozyme

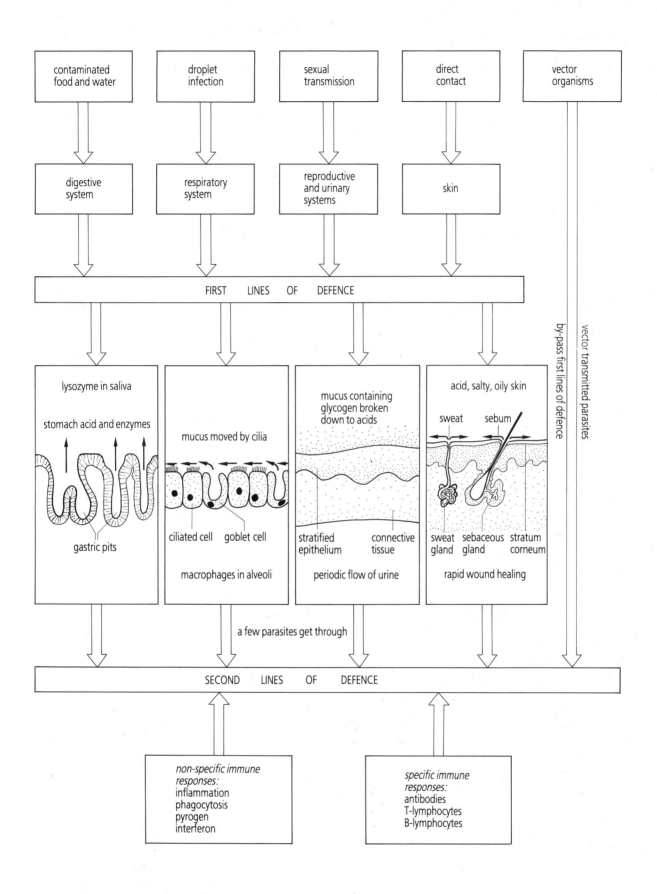

Fig 19·1 Summary of body defences against disease

attacks the polysaccharide carbohydrates of bacterial cell walls and kills the cells by causing them to split open. At the same time, bacteria and other potential pathogens become trapped in mucus and mixed in with food materials which are then swallowed and enter the stomach. Only the most resistant pathogens can survive the powerful proteolytic enzymes and highly acid conditions of the stomach.

Cholera, typhoid, and food-poisoning provide examples of bacterial diseases transmitted in food and water. Also transmitted in this way is amoebic dysentery, caused by the protozoan *Entamoeba histolytica*. Macroorganisms which affect the digestive system include the human tapeworm, *Taenia solium*, described in Unit 47. Most digestive infections can be treated successfully, using drugs.

The respiratory system is rather vulnerable to attack. Many different bacteria and viruses can tolerate drying out and are small enough and light enough to be carried around by air currents. Characteristically, they are transmitted by **droplet infection** from one person to another – coughs and sneezes spread diseases! To counteract these organisms, the trachea, bronchi, and bronchioles have a **ciliated epithelium** containing numerous **goblet cells** which produce mucus. Dirt particles and germs become trapped in the sticky mucus and, through the beating action of the cilia, are moved along at a rate of 1–3 cm per hour. When it reaches the back of the pharynx, the mucus is swallowed, exposing the trapped organisms to the acid and enzymes of the stomach. During an infection, the quantity of mucus increases and its upward flow is accelerated by coughing and sneezing. Pathogens which penetrate this line of defence may reach the lining of the alveoli, but this is protected by phagocytic **macrophage** cells.

Tuberculosis, diphtheria, pneumonia, and whooping cough are examples of bacterial diseases transmitted by droplet infection, while viral diseases include influenza and the common cold.

The male and female urinary and reproductive systems are protected by mucous membranes and by the periodic flow of urine, which flushes out bacteria. Bacterial growth is inhibited by the acid pH and high solute concentration of urine. In females, the moist lining of the vagina contains **glycogen** which is broken down to lactic acid by harmless bacteria which occur naturally in this region. This reduces pH and so prevents the growth of other, potentially pathogenic, types of bacteria.

Sexually-transmitted, or venereal diseases include **gonorrhoea** and **syphilis**, which are caused by bacteria, and **genital herpes** which is viral in origin. Antibiotics like penicillin are an effective treatment for gonorrhoea and syphilis, although there may be serious complications, including sterility, if the diseases are not treated early. At present there is no cure for genital herpes.

Acquired immune deficiency syndrome, or **AIDS**, is caused by a virus first isolated and described by Luc Montagnier at the Pasteur Institute in Paris in May 1983, and now known as **human immunodeficiency virus (HIV)**. The virus attacks one type of T cells, essential for normal functioning of the immune system, and leaves victims without effective defences against infection. Full-blown AIDS is invariably fatal. HIV was first identified among homosexual men in Europe and the U.S.A. but appears to have originated in Africa. Anal intercourse between homosexual men is just one way in which the virus can be spread. In Africa, heterosexual intercourse is the main method of transmission and it is clear that HIV can be passed to women in semen and to men in vaginal secretions. Reuse of infected hypodermic needles and transfusions with infected blood have helped to spread the disease further in the population. It is estimated that at least a million Africans will die from AIDS in the next few years. In all countries, urgent action is needed to prevent the spread of AIDS until a cure is found.

The skin is the most effective barrier against the entry of microorganisms mainly because of its tough outer layer of closely packed dead cells, called the **stratum corneum**. As explained in Unit 21, these cells are filled with the fibrous protein keratin and stick together in flakes, called **squames**, which are continually shed from the surface. The stratum corneum is kept supple and waterproof by **sebum** released from the **sebaceous glands**. This is an oily mixture containing unsaturated fatty acids which favour the growth of certain kinds of beneficial bacteria. The bacteria release acids as by-products of their metabolism and give skin a characteristically acid pH between 3 and 5. Sweat contains lysozyme, the anti-bacterial enzyme which is also found in saliva. Intact skin is virtually impenetrable but damaged skin is vulnerable and must be repaired. This is accomplished by blood clotting mechanisms and rapid healing of wounds.

The skin is less effective against larger organisms like mosquitoes and ticks which use their mouthparts to penetrate the outer layers and obtain a blood meal. Often these blood-sucking arthropods act as **vectors** for other parasites, as in the case of the *Anopheles* mosquito which transmits malaria. Parasites transmitted in this way have an enormous advantage because they completely bypass the body's first lines of defence and gain entry directly to the blood stream where they may multiply rapidly before the body's second lines of defence are fully mobilised. Vector-borne diseases are among the most damaging human diseases and among the most difficult to control.

The following sections describe the mechanisms which provide the second lines of defence, namely the **immune system** of the body.

19·3 The immune system ■ ■ ■ ■ ■ ■ ■ ■ ■ ■ ■ ■ ■

Lymphatic system

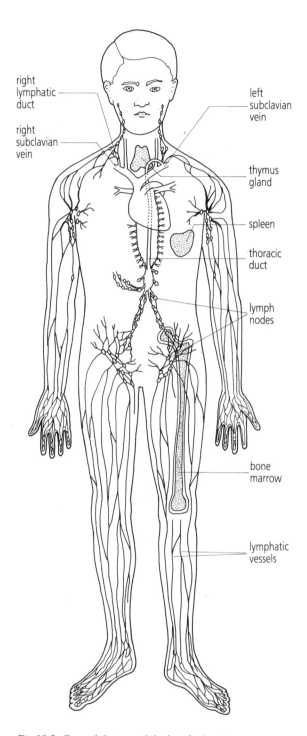

Fig 19·2 General features of the lymphatic system

The general features of the lymphatic system are shown in Figure 19·2. The main function of the system is to provide a series of drainage channels for excess tissue fluid not drained by the blood capillaries (Fig. 17·10). Transport in the lymphatic system is one-way and starts from minute blind-ended tubes including **lacteals** in the digestive system, and **lymph capillaries** in other tissues. These small tubes unite to form larger and larger

vessels called **lymphatics**. The flow of lymph is slow and occurs mainly when the vessels are squeezed by contractions of nearby body muscles; as in veins, valves are needed to prevent backflow. Eventually, the lymphatics from all parts of the body join together to form two large ducts, the **right lymphatic duct** and the **thoracic duct**, which drain respectively into the left and right **subclavian veins**.

Lymph nodes

These are oval or bean-shaped structures which occur at intervals along the lymphatics, often at the junction of a number of **afferent lymphatic vessels** (Fig. 19·3). Inside each node there is a distinct **cortex** layer containing **lymph nodules**, and a central **medulla**. At the centre of each nodule there is a paler region, the **germinal centre**, where new **lymphocyte** cells are produced. In the medulla, the lymphocytes form strands called **medullary cords**. Lymph entering the node percolates through a network of fibres in the **sinuses** between the nodules. These filter out bacteria and other foreign particles. Lymph is then collected into one or more **efferent lymphatic vessels**.

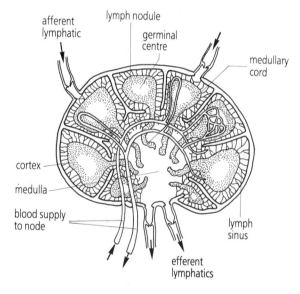

Fig 19·3 Structure of a lymph node

Effector cells

The effector cells of the immune system include **granulocytes**, that is, neutrophils, eosinophils, and basophils, and **agranulocytes**, comprising lymphocytes and monocytes (refer to Table 18·1). All of these cells are involved in the defence of the body against disease, but they have different roles. Granulocytes and monocytes are involved in **non-specific immune responses**. This means that they react to general stimuli such as the release of

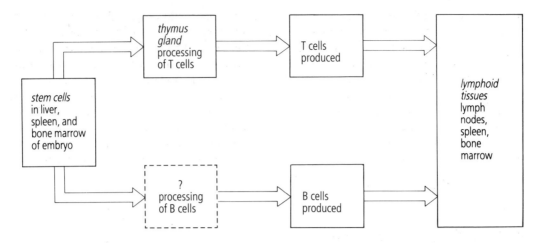

Fig 19·4 Origin of T cells and B cells

chemicals from damaged tissue, rather than to individual types of disease organism. Lymphocytes, on the other hand, mediate **specific immune responses**, including the production of **antibodies**.

Two different forms of lymphocytes may be distinguished according to their development from **stem cells** located in the liver, spleen, and bone marrow of the embryo. As outlined in the flow diagram of Figure 19·4, the stem cells multiply to produce two groups of lymphocytes which pass into the blood stream. One group, called **T lymphocytes**, or **T cells**, travel to the **thymus gland** where they proliferate before passing to other lymphoid tissues. The second group, called **B lymphocytes**, or **B cells**, travel to the same sites by different and as yet unexplained routes. The important point is that there are two separate populations of lymphocytes each of which makes a different contribution to the immune response.

19·4 Non-specific immune responses ■ ■ ■ ■ ■ ■ ■ ■ ■ ■

When any microbe penetrates the body's first lines of defence, it immediately triggers a series of changes leading to **inflammation** at the point of entry. One of the most important changes is the release of **histamine** from connective tissue **mast cells** and other tissue cells. Histamine causes local vasodilation and an increase in the permeability of blood vessel walls which allows proteins to leak out into the tissue fluid. The damaged tissues swell and, as the inflammation develops, the effects of histamine are enhanced by other chemical mediators, including **kinins** and **complement** proteins, formed in the blood plasma. These chemicals attract neutrophils by a process called **chemotaxis**. The neutrophils first stick to the walls of damaged blood vessels and then penetrate into the tissues by squeezing between the endothelial cells lining the vessel walls. In the tissues, they attack both bacteria and damaged tissue cells, engulfing them in cytoplasmic vacuoles (Fig. 19·5). Inside the vacuoles, bacteria are normally killed and digested by lysosomal enzymes. Dead bacteria, tissue cells, and neutrophils may accumulate to form **pus** at the site of injury.

Complete tissue repair normally follows phagocytosis, but there may be **abscess** or **granuloma** formation. An abscess is a bag of pus surrounded by fibrous tissue; sometimes, it will not heal unless drained artificially. A granuloma is a similar structure but contains live microorganisms.

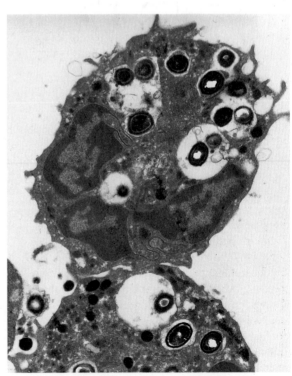

Fig 19·5 Electronmicrograph showing phagocytosis of bacteria by a neutrophil

Tuberculosis bacteria, for example, are capable of surviving inside neutrophils but can be safely tolerated in the body provided they are walled off inside the granuloma.

The responses so far described occur only within damaged tissue but there are other non-specific responses. Virtually all infections trigger a rise in body temperature leading to **fever**. This response is caused by **pyrogen**, a protein substance released from neutrophils during inflammation. Pyrogen acts on the hypothalamus of the brain, temporarily resetting the body 'thermostat'. Pyrogen also stimulates the liver to take up iron from the blood plasma and thereby slows the reproduction of bacteria, most of which need iron for cell division. Viral infections cause several different types of body cells to produce **interferon**. This substance is released into the circulation and then acts on other cells, making them immune to viral attack. Interferon stimulates cells to produce antiviral proteins which interfere with viral DNA and RNA replication, and is now the subject of intensive study because its use as a drug offers exciting possibilities in the treatment of virus diseases and certain types of cancer.

19·5 Specific immune responses ■ ■ ■ ■ ■ ■ ■ ■ ■ ■

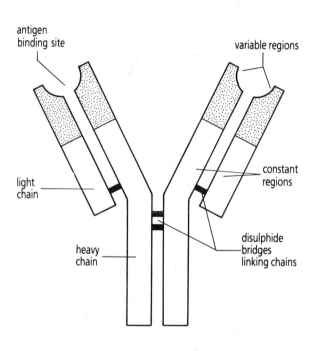

Fig 19·6 Structure of an antibody molecule

Specific immune responses depend on the production of **antibodies** by lymphocyte cells. Each antibody is a Y-shaped protein with the structure illustrated in Figure 19·6. The molecule consists of four polypeptide chains, comprising two **heavy chains** and two **light chains**, linked by disulphide bridges. Part of the structure of the chains is constant, but the terminal portions, near the top of the 'Y', are variable and form **binding sites** capable of reacting with other chemicals, called **antigens**. This happens in much the same way as an enzyme combines with its substrate.

Typically, antigens are 'foreign' substances like proteins or polysaccharide carbohydrates exposed on the surface of invading microorganisms. The antigen stimulates production of its own antibody, and the specific antigen-antibody binding which follows leads eventually to the destruction of the microbe. Rather surprisingly, single lymphocytes produce just one kind of antibody. Thus, to protect the body against all the possible kinds of diseases, there must be thousands of different lymphocytes, each capable of recognising and responding to a unique antigen. Precisely how all these different cells are formed remains a mystery.

When a T lymphocyte is stimulated by the correct antigen, it enlarges and then divides repeatedly to produce a series of genetically identical cells, called a **clone**. The clone cells form antibodies, but these remain fixed on the surfaces of the cells and are *not* released. Instead, the T cells, now called **killer T cells**, migrate to the site of injury where they attack microorganisms directly, releasing chemicals which kill the pathogens. The same chemicals also attract additional lymphocytes and phagocytic macrophage cells (Fig. 19·7).

When B cells are stimulated, they enlarge and reproduce to form clones in the same way as T cells, but the clone cells differentiate further into **plasma cells** which remain within the lymphoid tissue. They have a well developed rough ER (Fig. 19·8) and are specialised to produce antibodies, which are secreted at a rate of up to 2000 molecules per second from each cell. This rate of production is maintained for several days, after which the cell dies. The antibodies pass into the blood stream and can have a number of different effects ranging from destruction, or **lysis**, of the microbe, to **opsonization**, in which antibodies attached to the surface of a microorganism make it more susceptible to phagocytosis. **Memory cells** from each clone remain in the lymphoid tissue so that a second exposure to the same antigen causes a much more rapid and vigorous response. Occasionally, the response to repeated exposure is so dramatic as to produce an **allergic reaction**, as in the case of hay fever.

The presence of circulating antibodies and of memory cells in the lymph nodes gives lasting protection, or **immunity**, against disease organisms. The same immunity is now commonly

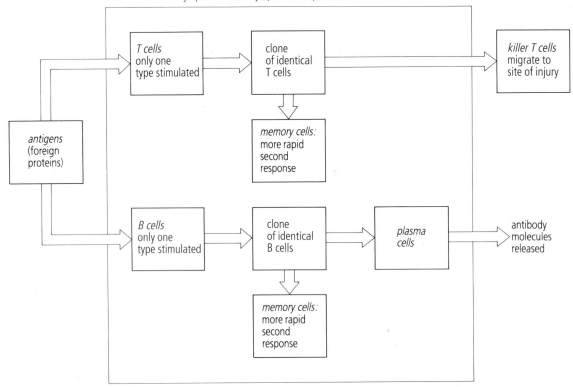

Fig 19·7 Stimulation of T cells and B cells

produced by **vaccination**, in which dead or weakened microbes, or purified antigens isolated from them, are injected into the body. This stimulates the appropriate lymphocytes just as effectively as the normal live pathogens. People who are vaccinated cannot normally catch the particular disease and, equally important, they cannot transmit it to others, so that mass vaccination campaigns can eliminate it altogether. Smallpox has been completely wiped out in this way and worldwide efforts are being made to eliminate polio. Many serious diseases, such as tuberculosis and diphtheria, have been effectively controlled in human populations.

Vaccination produces what is called **active immunity**, meaning that the body produces its own antibodies against a particular antigen. An alternative form of immunity is **passive immunity**, where ready-made antibodies enter the blood stream from a different individual. This occurs naturally across the placenta from mother to foetus and is an important means of protecting the infant in the first few months of life. Passive immunity is also established artificially by injecting specific antibodies in **serum** prepared from the blood of humans, horses, or cows. Serum antibodies are used to treat people exposed to serious diseases, like botulism, infectious hepatitis, and rabies, and give instant protection. Passive immunity is relatively short-lived because the antibodies slowly break down in the blood stream and cannot be replaced from within the body.

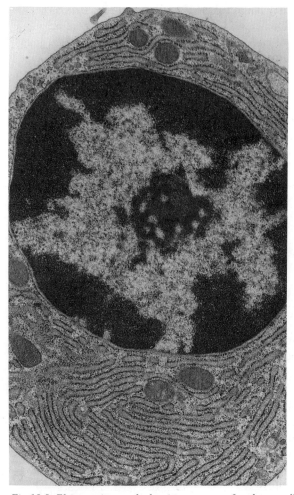

Fig 19·8 Electronmicrograph showing structure of a plasma cell

19·6 Blood groups and transplantation ■ ■ ■ ■ ■ ■ ■ ■ ■

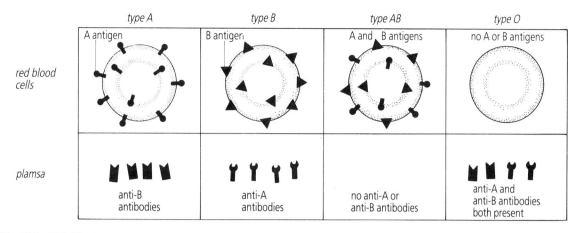

Fig 19·9 ABO blood groups

Blood grouping is based on the occurrence of a number of substances fixed to the membranes of red blood cells. These substances may act as antigens in the event of blood transfusion from one individual to another. The most important examples are the glycolipid **ABO antigens**. As outlined in Figure 19·9, a person whose red blood cells carry only **A antigens** is said to have blood type A, individuals with only **B antigens** have blood type B, while those with both the A and B antigens have blood type AB. If the red blood cells carry neither A or B antigens, the person has blood type O. In each case, the blood plasma contains antibodies against the blood type antigens which are *not* present. For example, type A blood contains **anti-B** antibodies, and type B contains **anti-A**. Type AB has neither of the antibodies, and type O contains both anti-A and anti-B. Surprisingly, the production of anti-A and anti-B antibodies is genetically determined and does not depend upon exposure to antigens. The reason for their presence is not known.

The consequences of this arrangement for blood transfusion are summarised in Table 19·1. If

Table 19·1 Blood transfusions and ABO blood group

patient's blood group	antibodies present in patient's blood serum	blood groups which may be used for transfusion
A	anti-B	A, O
B	anti-A	B, O
AB	–	A, B, AB, O
O	anti-A, anti-B	O

blood of an inappropriate type is given, for example, if type B blood is transfused to a type A person, there is an immediate reaction in which the antibodies in the recipient's blood attack red cells from the donor. This causes the red cells to **agglutinate**, that is, to stick together in clumps, resulting in blocked blood vessels, kidney damage,

or even death. The parallel reaction, in which donor antibodies attack the recipient's red cells, also occurs but the number of antibody molecules transferred is so small that this effect can be neglected.

In practice blood of the matching type is always used for transfusion and must be matched, not only for the ABO group, but also for **Rh**, or **rhesus**, type. The rhesus 'factor', first discovered in the blood of rhesus monkeys, is another antigen which may be present in the membranes of red cells. An individual possessing the antigen is said to be **rhesus positive** (Rh+), while a person who lacks the antigen is **rhesus negative** (Rh−). In the U.K., 85% of the population are Rh+. Exposure to Rh+ blood causes a Rh− person to produce **anti-Rh** antibodies. This can occur during the pregnancy of a Rh− mother with a Rh+ fetus. If Rh antigens from the foetus cross the placenta, the mother will make antibodies capable of destroying the fetal red cells. Serious effects are possible, particularly in a second or subsequent pregnancy, for the anti-Rh antibodies remain in the mother's blood stream. Fortunately, red cells from the fetus do not normally cross the placenta, so that maternal antibodies are not usually formed.

Successful organ transplantation from one individual to another is much more difficult to accomplish than successful blood transfusion. This is because the transplanted organs, like the heart or kidneys, contain many more antigens than either the blood plasma or the surfaces of the red cells, so that they are rejected by the body's immune system. In transplant surgery this problem has been partly overcome by careful **tissue typing**, which matches the donor and recipient antigens as exactly as possible, and by the use of **immunosuppressive drugs**. These techniques are most effective for kidney transplants. For reasons explained in Unit 24, the interior of the eyes shows no immune response. Consequently, lens and corneal transplants are successful even between unmatched individuals.

20 EXCRETION AND OSMOREGULATION □

Objectives

After studying this Unit you should be able to:—

- Define and explain the terms homeostasis, dynamic equilibrium, excretion, and osmoregulation

- List the main substances excreted by living organisms and state how they are produced

- Draw and label diagrams to illustrate the structure and internal anatomy of the mammalian kidney and of a kidney tubule (nephron)

- Describe in detail how ultrafiltration takes place across the basement membrane of Bowman's capsule

- Give an account of reabsorption and explain the role of the countercurrent multiplier mechanism in the reabsorption of water

- Explain the roles of osmoreceptors, baroreceptors, antidiuretic hormone (ADH), and aldosterone in regulating water and salt balance

- Describe the special adaptations shown by fish, desert mammals, and flowering plants in relation to osmoregulation

20·1 Homeostasis ■ ■ ■ ■ ■ ■ ■ ■ ■ ■ ■ ■ ■ ■ ■ ■

The term **homeostasis** is used to describe the maintenance by the body of a constant **internal environment**. This concept has particular meaning for multicellular organisms because only these organisms possess an internal fluid medium which is kept separate and distinct from the external environment. This fluid, called **extracellular fluid** or **tissue fluid**, acts as the 'internal environment' on which the individual cells of the body depend for all their requirements.

Homeostasis places multicellular organisms at a considerable advantage because it provides their body cells with relatively constant conditions. Consequently, they can tolerate a much wider range of external conditions than single-celled organisms, which remain at the mercy of even small changes in their environment. Effective homeostasis involves the regulation of the temperature, osmotic concentration, and detailed composition of tissue fluid. As the surroundings and the activity of the organism change, so continual adjustments must be made. Homeostatic mechanisms are therefore **dynamic** systems which

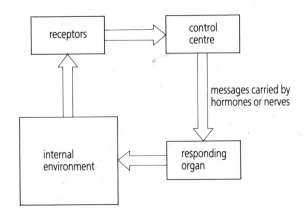

Fig 20·1 Components of a dynamic homeostatic control system

act to produce a state of balance, described as a **dynamic equilibrium**. Typically, the balancing process involves three functional parts, as shown in Figure 20·1. As you can see, **receptors** constantly monitor the internal environment and pass information to a **control centre**: this initiates the correct response by sending out nervous or hormonal messages. The effect of these signals upon the **responding organ** and consequently upon the tissue fluid is detected once again by the receptors, providing 'feedback' control so that the correcting response ceases when optimum conditions are restored.

One important homeostatic mechanism has already been described in Unit 15, that is, the maintenance of a constant blood sugar level. The subject of this Unit is the balance of water and dissolved solutes in the body; two major processes are involved in maintaining this balance, namely **excretion** and **osmoregulation**. Excretion is the removal from the body of the waste products of metabolism and of substances surplus to the body's requirements. Osmoregulation is the process which controls the relative quantities of water and dissolved solutes in the tissue fluids so that their osmotic concentration remains at the appropriate level. The mechanisms and structures which carry out these processes are closely related, as will be seen in the following sections.

20·2 Excretory substances ■ ■ ■ ■ ■ ■ ■ ■ ■ ■ ■ ■ ■ ■

New substances are continually acquired by body cells by uptake from the extracellular fluid, or as products of metabolism. Similarly, excess quantities of some chemicals may be present in the diet. Some of these substances are useful but others, called **excretory** substances, are useless or even damaging and must be removed. The main excretory substances of a mammal, and their origins and destinations are listed below.

1 Carbon dioxide. Carbon dioxide is formed in all body cells as a waste product of aerobic respiration. As explained in Unit 18, it diffuses into the tissue fluid and enters the blood, where a series of reversible reactions enable it to be transported around the body, eventually reaching the lungs. Most of the body's carbon dioxide is excreted at the lung surface into the alveolar air spaces, but a small proportion follows an alternative course and is excreted by the kidneys in the form of hydrogencarbonate ions (HCO_3^-).

2 Nitrogenous waste from protein metabolism. Surplus amino acids in the diet cannot be stored by the body but are broken down by the liver through the process called **deamination** (described more fully in Unit 15). In this process, the nitrogen-containing —NH_2 amino group of each amino acid molecule is removed and **ammonia** is formed. This substance is highly toxic and must be removed immediately.

Small organisms such as *Amoeba* or *Hydra* can excrete ammonia by simple diffusion across their body surface, but most larger animals must convert it into less harmful substances, such as **urea**, which are allowed to circulate in the body before being excreted. In mammals, urea is formed in the liver by a series of reactions known as the **ornithine cycle**. It enters the blood stream and is finally excreted by the kidneys; a small quantity of urea is also present in sweat. Other organisms, notably many insects, some reptiles, and all birds excrete **uric acid**. This substance is solid and insoluble so that very little water is needed to eliminate it from the body.

3 Water and inorganic ions. Water enters the body by drinking and eating and is also a product of aerobic respiration. Inorganic ions are present in the diet and are absorbed from the intestines. The relative amounts of water and inorganic ions retained in the body are controlled by the kidneys, which regulate the osmotic concentration of the blood, at the same time excreting the waste and surplus materials. The skin also loses water and inorganic ions by producing sweat, but this is secondary to the function of temperature control (described in Unit 21) and is not a method of osmoregulation.

4 Bilirubin. Bilirubin is the yellow pigment which results from the breakdown of haemoglobin. As outlined in Unit 15, it is extracted from the blood by the hepatocyte cells of the liver and excreted in the bile.

5 Products of detoxication. Hormones, antibiotics like penicillin, other drugs, and many poisons are removed from the general circulation and made harmless, or 'detoxified', by the liver. Some of the substances produced are excreted by the kidneys.

20·3 The mammalian kidney ■ ■ ■ ■ ■ ■ ■ ■ ■ ■ ■

The two kidneys of a mammal form part of its **urinary system**, as shown in Figure 20·2, and are located on either side of the vertebral column. Human kidneys are bean shaped organs about 12 cm long and 7 cm wide. They are covered by a layer of fat and loosely joined to the dorsal body wall by connective tissue.

The internal structure of a mammalian kidney

is illustrated in Figure 20·3. If a fresh kidney is cut vertically, the different regions can be clearly seen. The outer **cortex** has a dark red colour while, inside this layer, the **medulla** is paler and the collecting funnel, or **pelvis**, of the **ureter** is white. In life, the pelvis forms a collecting region for urine which drains from thousands of **collecting ducts** located in the kidney **pyramids**. A human kidney contains at least a million **kidney tubules**, or **nephrons**, each of which is connected to a collecting duct. Urine passes from the ureter to the **bladder**, and finally to the outside by means of the **urethra**. The whole system of interconnected tubes leading from the kidney tubules to the exterior is called the **urinary tract**.

The nephrons are the functional units of the kidney. As illustrated diagramatically in Figure 20·4, the first section of each nephron is located in the cortex of the kidney and consists of a cup-shaped structure called **Bowman's capsule**, enclosing a dense network of capillaries, called the **glomerulus**. The glomerulus receives blood at high pressure from a branch of the renal artery and fluid from the bloodstream is forced across into the Bowman's capsule by a process called **pressure filtration**, or **ultrafiltration**. The build up of pressure necessary for this filtration of fluid is facilitated in two ways. Firstly, the renal arteries

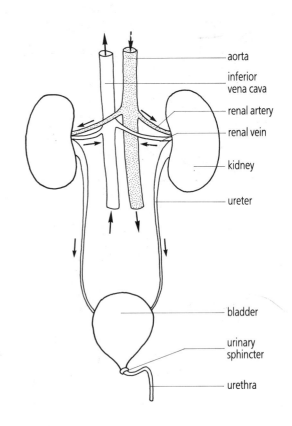

Fig 20·2 General anatomy of the human urinary system

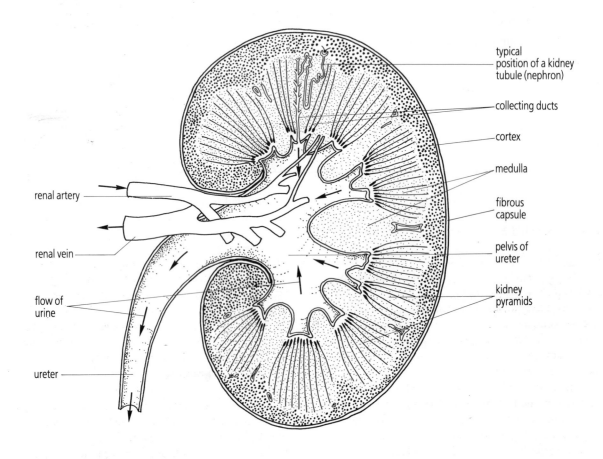

Fig 20·3 Vertical section of a human kidney

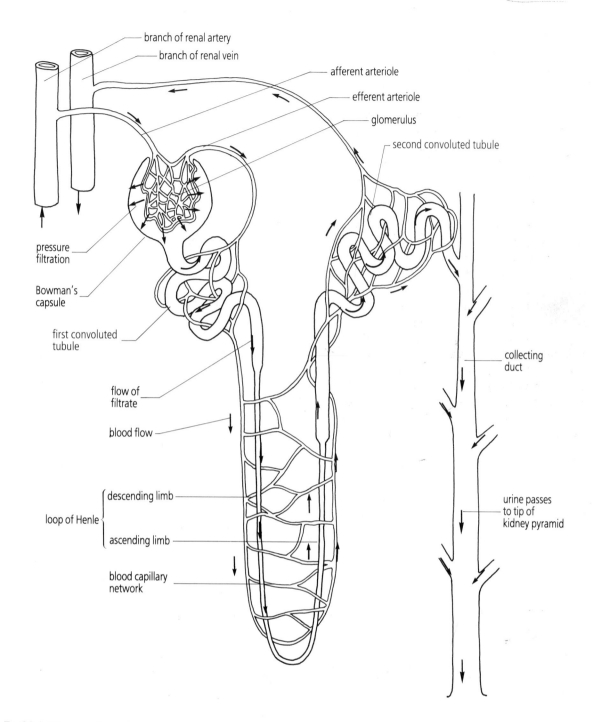

Fig 20·4 *Diagram of a nephron*

Labels in figure:
- branch of renal artery
- branch of renal vein
- afferent arteriole
- efferent arteriole
- glomerulus
- second convoluted tubule
- pressure filtration
- Bowman's capsule
- first convoluted tubule
- collecting duct
- flow of filtrate
- blood flow
- loop of Henle
  - descending limb
  - ascending limb
- urine passes to tip of kidney pyramid
- blood capillary network

branch directly from the aorta so that blood enters the kidneys at relatively high pressure. Secondly, the efferent arteriole leaving each glomerulus can be constricted so that it becomes narrower than the afferent vessel, and therefore represents a 'bottle neck' for blood flow. This causes the maximum pressure to develop inside, rather than after, the glomerulus.

The transfer of fluid into the kidney tubules is assisted by the rather unusual cellular arrangement at the junction of the epithelial cells of Bowman's capsule and the endothelial lining cells of the blood capillaries. The detailed structure of this junction is illustrated in Figure 20·5. As you can see, the endothelial cells of the capillaries are perforated by tiny **pores** which allow the blood to come into direct contact with the underlying **basement membrane**. This membrane forms the only barrier between the blood and the cavity of the nephron, for the epithelial cells, or **podocytes**, of Bowman's capsule are structured so as to allow the free passage of filtrate between their foot-like processes. The basement membrane acts as a selective barrier allowing water and soluble substances with a relative molecular mass of up to 68 000 to cross. Therefore, all the components of blood plasma except protein molecules pass through into the nephron, forming a fluid called **glomerular filtrate** (see Table 20·1).

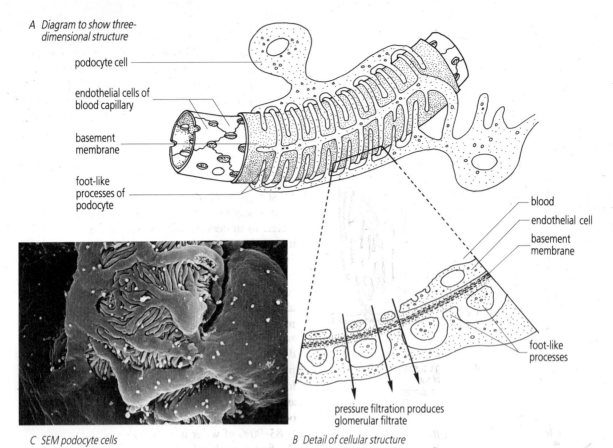

A Diagram to show three-dimensional structure

podocyte cell

endothelial cells of blood capillary

basement membrane

foot-like processes of podocyte

blood

endothelial cell

basement membrane

foot-like processes

pressure filtration produces glomerular filtrate

C SEM podocyte cells

B Detail of cellular structure

Fig 20·5 Structure of the blood/nephron barrier

Table 20·1 Quantities of water and solutes in blood plasma compared with quantities present in glomerular filtrate and urine during a 24 hour period

CHEMICAL	QUANTITY IN 180 LITRES OF BLOOD PLASMA	QUANTITY IN GLOMERULAR FILTRATE PER DAY	QUANTITY REABSORBED PER DAY	QUANTITY IN URINE PER DAY
water (litres)	180	180	178–179	1–2
solutes (grams)				
proteins	7000–9000	10–12	10–20	0
sodium (Na⁺)	540	540	537	3
chloride (Cl⁻)	630	630	625	5
hydrogencarbonate (HCO₃⁻)	300	300	299.7	0.3
glucose	180	180	180	0
urea	53	53	28	25
potassium (K⁺)	28	28	24	4
uric acid	8.5	8.5	7.7	0.8
creatinine	1.4	1.4	0	0

20·4 Reabsorption by the kidney tubules ■ ■ ■ ■ ■ ■ ■ ■ ■

The two kidneys of an adult human filter about 60 litres of blood per hour and produce glomerular filtrate at a rate of approximately 7·5 litres per hour. Within the same period, only about 0·07 litres of urine is formed, a volume which corresponds to less than 1% of the glomerular filtrate. Thus, at least 99% of the water present in the filtrate is **reabsorbed** from the kidney tubules back into the bloodstream. A comparison of the composition of glomerular filtrate with that of blood plasma and urine (Table 20·1) shows that many solutes are likewise reabsorbed. On the other hand, some substances are concentrated by their passage through the kidney tubules so that they are removed from the body when urination occurs.

Reabsorption of solutes

Glucose present in the glomerular filtrate is actively reabsorbed in the next section of the nephron, called the **first convoluted tubule**. Active reabsorption in this region is facilitated by

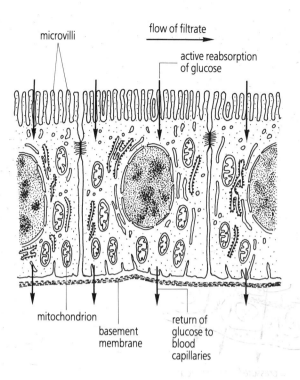

microvilli

flow of filtrate

active reabsorption
of glucose

mitochondrion

basement
membrane

return of
glucose to
blood
capillaries

Fig 20·6 Ultrastructure of tubular cells

the structure of the epithelial lining cells of the tubule (see Fig. 20·6). These cells have surface membranes with **microvilli**, which increase the surface area in contact with the tubule contents, and contain numerous **mitochondria** capable of supplying energy. The reabsorption process involves specific carrier molecules located in the membranes of the epithelial cells, and glucose is transported to the opposite sides of the cells, where it is taken up by the blood capillaries surrounding the tubule. Glucose is reabsorbed completely in normal circumstances and only appears in the urine when its concentration in the blood plasma exceeds the kidneys' capacity for reabsorption, as in the disease **diabetes mellitus** (described in Unit 15).

Sodium ions are also actively reabsorbed from the first convoluted tubule. In this case, reabsorption is less complete, but still accounts for up to 90% of the Na^+ ions present in the glomerular filtrate. As Na^+ ions are pumped out, they carry with them an equal number of negatively charged anions, chiefly Cl^- ions. At the same time, large quantities of water are reabsorbed following the concentration gradient established by the active pumping of ions. Other ions actively reabsorbed in the first convoluted tubule include potassium (K^+), phosphate ($H_2PO_4^-$), and hydrogencarbonate (HCO_3^-) ions. Most of the remaining sodium ions are reabsorbed in other regions of the tubule, so that, of the 540 g per day typically present in glomerular filtrate, only about 3 g per day is excreted.

The majority of the urea present in the

glomerular filtrate remains within the tubule but some is lost as it diffuses passively outwards. Some substances, including a number of toxins and drugs such as penicillin pass across the epithelial cells in the opposite direction, that is, they are transported from blood capillaries *into* the tubule for eventual excretion.

Reabsorption of water

Most regions of the kidney tubule are concerned (in one way or another) with the reabsorption of water. In understanding the mechanisms involved, it is vital to remember that water moves *passively* from regions where the solute concentration is low to regions where it is higher, that is, the water tends to follow the actively transported ions. It is also important to note that the rate of water movement depends on the permeability of the tubule. The first convoluted tubule and the descending limb of the **loop of Henle** are freely permeable to water, but the ascending limb of the loop of Henle, the **second convoluted tubule**, and the collecting duct, all have a restricted permeability which is subject to hormonal control.

85–90% of water is reabsorbed in the region of the first convoluted tubule. This is sometimes called **obligatory reabsorption** and occurs for two reasons. Firstly, liquid and small solutes are lost from the blood passing through the glomerulus while plasma protein molecules are retained. This makes the blood relatively more concentrated so that it can absorb water by osmosis once the capillary network surrounding the tubule is reached. Secondly, the active pumping of sodium ions across the first tubule increases the osmotic concentration of the surrounding tissues, again causing water to be reabsorbed by osmosis.

Much of the remaining water is reabsorbed across the second convoluted tubule, or from the collecting ducts as a result of a remarkable mechanism which will now be explained.

If the contents of the tubule are analysed at different points, it is found that the contents of the first convoluted tubule have the same overall concentration as the surrounding tissue fluid. On the other hand, the contents of the second convoluted tubule are *less* concentrated while the urine is *more* concentrated than tissue fluid. It can be concluded that (1) the filtrate is diluted in the loop of Henle, and (2) that water is extracted from the urine as it passes through the collecting duct. The explanation of these facts requires an understanding of the **countercurrent multiplier mechanism** of the nephron which is described below. The system depends upon an active pumping mechanism which forces Na^+ and Cl^- ions out of the ascending limb of the loop of Henle. The effect of this is illustrated by Figure 20·7 and is best understood if considered in stages.

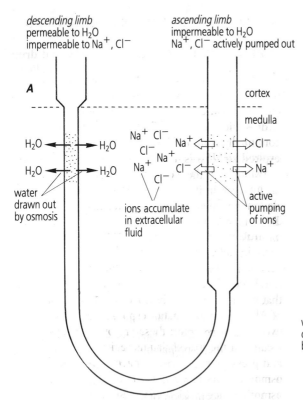

descending limb
permeable to H_2O
impermeable to Na^+, Cl^-

ascending limb
impermeable to H_2O
Na^+, Cl^- actively pumped out

A

cortex

medulla

H_2O ← → H_2O

H_2O ← → H_2O

Na^+ Cl^- Na^+ ← ⇨Cl^-
Cl^-
Na^+
Na^+ Cl^- ← ⇨Na^+
Cl^-

water drawn out by osmosis

ions accumulate in extracellular fluid

active pumping of ions

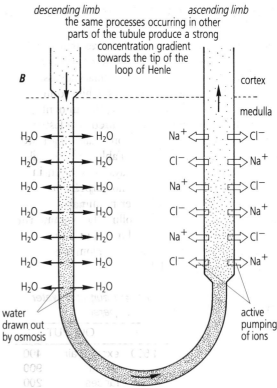

descending limb   *ascending limb*
the same processes occurring in other parts of the tubule produce a strong concentration gradient towards the tip of the loop of Henle

B

cortex

medulla

H_2O ← → H_2O Na^+ ← ⇨Cl^-

H_2O ← → H_2O Cl^- ← ⇨Na^+

H_2O ← → H_2O Na^+ ← ⇨Cl^-

H_2O ← → H_2O Cl^- ← ⇨Na^+

H_2O ← → H_2O Na^+ ← ⇨Cl^-

H_2O ← → H_2O Cl^- ← ⇨Na^+

H_2O ← → H_2O

water drawn out by osmosis

active pumping of ions

Fig 20·7 Countercurrent multiplier mechanism

1 Imagine that the fluid in the loop of Henle is stationary. As Na^+ and Cl^- ions are pumped out of the ascending limb they accumulate in the extracellular fluid. The tendency is for water to follow the ions out of the tubule, but this is prevented in the ascending limb by its low permeability to water. The descending limb, however, is freely permeable to water but impermeable to Na^+ and Cl^-. Therefore, water is drawn out by osmosis causing the fluid inside the descending limb to become more concentrated.
2 Now consider what happens when new glomerular filtrate enters the tubule and starts to flow around the loop of Henle. Inevitably, as fluid flows down the descending limb water continues to be withdrawn so that the contents of the tubule become progressively more concentrated. On the other hand, in the ascending limb, the active pumping of ions causes the fluid inside to become less concentrated. The net result of this is to produce the maximum concentration at the extreme tip of the loop of Henle both inside the tubule, and outside in the extracellular fluid.

These events are of great importance because, as shown in Figure 20·8, the combined action of numerous loops of Henle produces a strong concentration gradient inside the kidney, with the highest concentration in the medulla near the tips of the kidney pyramids. These high concentrations cause large quantities of water to be reabsorbed by osmosis from the collecting ducts, thereby concentrating the urine as it passes through.

The ability to produce a concentrated urine allows the organism to conserve water and is a

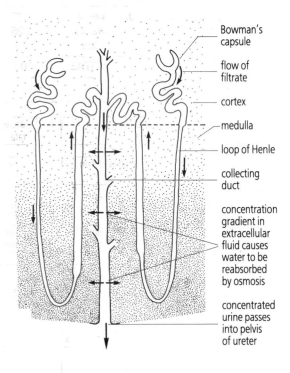

Bowman's capsule

flow of filtrate

cortex

medulla

loop of Henle

collecting duct

concentration gradient in extracellular fluid causes water to be reabsorbed by osmosis

concentrated urine passes into pelvis of ureter

Fig 20·8 Reabsorption of water from collecting ducts

useful adaptation for terrestrial life. Among the vertebrates, only birds and mammals have loops of Henle as part of their kidney structure and are able to concentrate their urine in this way. In mammals, the quantity of water extracted from the collecting ducts can be precisely controlled, thereby providing the organism with an important method of osmoregulation as outlined in the following section.

20·5 Osmoregulation in mammals ■ ■ ■ ■ ■ ■ ■ ■ ■ ■ ■

Osmoregulation in a mammal ensures that the total volume of its blood plasma and the concentration of dissolved substances in the plasma and tissue fluids all remain constant. This is achieved in two basic ways, namely, by controlling the amount of water, and the amount of salt gained and lost by the body. An analysis of a person's daily input and output of water and salts (see Table 20·2) reveals that the production of urine plays a large part in water and salt losses, and it is no surprise to discover that humans and other mammals osmoregulate mainly by controlling the volume and concentration of their urine. The important mechanisms involved in osmoregulation are summarised in Figure 20·9.

Table 20·2 Typical daily input and output of water and salts for an adult person

	INPUT		OUTPUT	
Water cm³ per day	diet	2500	expired air	400
	oxidation of hydrogen in food	500	sweat	900
			faeces	200
			urine	1500
	TOTAL	3000	TOTAL	3000
Salts g per day	food	10.5	sweat	0.25
			faeces	0.25
			urine	10.00
	TOTAL	10.5	TOTAL	10.5

Regulation of water

The two main factors which affect the amount of water lost in the urine are the osmotic concentration of the blood plasma, and the hydrostatic pressure of the blood. Water is lost continually from the body by sweating and in expired air. If it is not replaced, there is a drop in the volume of the blood plasma, accompanied by an increase in its osmotic concentration, and a decrease in blood pressure (refer to Fig. 20·9). Both these changes lead to an increased secretion of **antidiuretic hormone (ADH)** from the posterior lobe of the pituitary gland. This hormone acts by increasing the permeability to water of the distal convoluted tubules and collecting ducts, so that more water is reabsorbed. Increased secretion of ADH is an automatic response which involves two sorts of receptor: these are **osmoreceptors** located in the hypothalamus region of the brain, and pressure receptors, or **baroreceptors**. The osmoreceptors respond to an increase in the osmotic concentration of the blood plasma and stimulate the production of ADH via their connections with the pituitary gland. At the same time, the osmoreceptors are thought to act on control centres in the hypothalamus, causing increased feelings of thirst.

Baroreceptors are found in many regions of the circulatory system and play an essential part in the

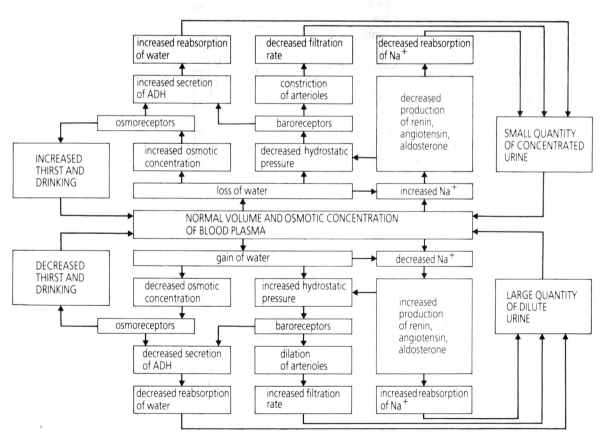

Fig 20·9 Summary of mechanisms involved in osmoregulation in a mammal

157

regulation of blood pressure (Unit 17). Their activity increases in response to increased blood pressure, and decreases if blood pressure falls. Reduced activity leads to a general constriction of arterioles. In the case of renal arterioles, this results in diminished blood flow, so that less glomerular filtrate is formed. One group of baroreceptors embedded in the wall of the left atrium of the heart produces impulses which *inhibit* the production of ADH. When blood pressure falls, this inhibiting effect is lifted, and more ADH is produced. On the other hand, a raised blood pressure results from an increase in the total volume of blood plasma and often indicates that the plasma has been diluted. By inhibiting ADH secretion, the atrial baroreceptors cause more water to be lost in the urine, reducing plasma volume and returning its concentration to the proper level.

The disease **diabetes insipidus** arises from a failure to produce ADH, usually through the hypothalamus being damaged. In this condition, large quantities of very dilute urine are produced and the body's water and salt balance must be artificially maintained by regulating input, rather than output.

Regulation of salt

The salt concentration in plasma and tissue fluids, although indirectly affected by ADH, is under the direct control of the hormone **aldosterone**, produced by the **adrenal cortex**. As indicated in Figure 20·9, aldosterone is secreted when levels of salt in the blood fall, and stimulates Na^+ reabsorption in the distal convoluted tubules and collecting ducts.

The chain of events which leads to secretion of aldosterone begins with the decrease in plasma volume which is the inevitable result of salt shortage. In the kidney, special cells lining the arterioles react to such a decrease by releasing an enzyme called **renin** into the blood stream. Renin catalyses the conversion of a blood protein **angiotensinogen** to **angiotensin**, and it is the presence of this substance in the blood which causes aldosterone to be released from the adrenal cortex. Angiotensin also affects the diameter of arterioles and influences the brain, causing sensations of thirst.

When the body suffers a prolonged shortage of salt, the osmoreceptors fail to stimulate the pituitary, so less ADH is produced and a greater quantity of dilute urine is formed. In this way, the body maintains the osmotic pressure of its fluids at the expense of their volume. In sweating, both water and salt are lost, but often only water is replaced. The consequence may be muscular cramp, caused by a drop in the level of Na^+ ions in tissue fluids below that which is necessary for muscle action.

Excessive intake of either water or salt causes an increased urine output in order to eliminate the surplus. Although the human body can tolerate relatively large increases in water, it does not have the same facility to excrete extra salt. This is because salt must be removed in solution and, even at its most concentrated, human urine can carry only about 5 g salt per litre of water. This explains why the stranded sailor cannot survive by drinking sea water. Sea water contains about 10 g salt per litre, so, for every litre of water obtained, it would require 2 litres of urine to eliminate the unwanted salt.

20·6 Special adaptations for osmoregulation ■ ■ ■ ■ ■ ■ ■

Fish

The maintenance of a constant osmotic concentration in tissue fluids presents a problem for any aquatic organism which needs to exchange materials with the surrounding water. For example, gas exchange in fish requires a large surface area across which diffusion can occur. However, from an osmoregulatory point of view, the gills represent an unprotected area where water may be gained or lost according to the osmotic gradient between internal and external fluids.

Fresh water fish gain water by osmosis from their surroundings and therefore require to eliminate excess water. This is achieved by a high rate of filtration through huge numbers of glomeruli, resulting in large quantities of extremely

dilute urine. The urine produced is mostly water because salts are very efficiently reabsorbed by the kidney tubules. Small salt losses are made good by an active uptake mechanism at the gills. The kidneys are not important in the excretion of nitrogenous waste because this is mostly eliminated as ammonia by diffusion across the gill surface.

Marine fish normally have tissue fluids which are *less* concentrated (hypotonic) compared with the external environment and therefore tend to lose water by osmosis. This situation can be described as '**physiological drought**' and requires efficient mechanisms for conserving water. These mechanisms are entirely different in the two major groups of marine fish as described below.

1 Chondrichthyes. Cartilaginous fish, such as sharks and dogfish, are able to tolerate concentrations of urea in the blood 150 times greater than those found in the blood of other vertebrates. This creates an artificially high blood osmotic concentration and may reduce the osmotic gradient across the gills to zero. Apart from the unusual degree of urea reabsorption, their kidney function is similar to that of mammals.

2 Osteichthyes. Bony fish, like cod and herring, replace water lost through the gills by drinking sea water and, consequently, require mechanisms to cope with an excessive salt intake. By far the most important of these mechanisms is the active excretion of excess salt at the gills, a reversal of the salt pump possessed by fresh water bony fish. The kidneys of marine bony fish have a very small fluid output and usually lack glomeruli. Waste products are selectively extracted from the blood by the tubule cells and ultrafiltration does not occur. One further consequence of physiological drought is the inability to excrete nitrogenous waste as ammonia which, being toxic, needs to be diluted by large quantities of water for expulsion. To avoid this ammonia is converted into the non-toxic substance **trimethylamine oxide** which is removed by the kidneys. Migratory bony fish such as eels and salmon can change physiologically from salt water to fresh water and back again. This ability is poorly understood but implies the possession of two alternative sets of genes controlling osmoregulation.

Desert mammals

Desert mammals show a remarkable ability to control water loss in dry environments. The camel, for example, has a number of typical and a few rarer physiological modifications for desert life. It can eat dry food and drink water equivalent to 33% of its body weight at one session, absorbing the water only slowly from its alimentary canal to avoid dilution of the blood. It should be noted that the chemical process of respiration produces significant amounts of water. In the camel, this 'respiratory water' supplements dietary intake to such an extent that in winter, when fresh green plants are available, the animal has no need to drink, while in summer it only drinks infrequently. The camel's hump is a store of fat which can be metabolized to provide water in times of shortage.

A further important modification, common to many desert mammals, concerns the loops of Henle and the collecting ducts of the kidney tubules. These are much longer than in other mammals, enabling a very concentrated urine to be formed. Although primarily an adaptation to conserve water, this mechanism also increases the animals ability to excrete salt so that the camel can drink salty water (up to 75% sea water). As in other ruminants (Unit 14), the need for

nitrogenous excretion is reduced by the secretion of urea into the alimentary canal. Bacteria in the rumen convert the urea to protein which is then utilized as food.

Flowering plants

In common with all land organisms, terrestrial flowering plants must conserve water in order to survive. Water is lost through the leaves during gas exchange and a continual flow of water must be maintained to meet the plant's needs for transport and structural support.

In general, plants can control the amount of water lost by opening and closing their stomata according to external conditions. In cases where an

Fig 20·10A The sunflower is a mesophyte

excess of water accumulates in a plant, special water-expelling glands called **hydathodes** (described in Unit 32) may occur. During periods of drought, the plant may cease activity or die, leaving resistant seeds which germinate when favourable conditions return. Physiological drought occurs whenever the ground is frozen, so many trees and shrubs living in temperate climates survive the winter by shedding their leaves, thereby reducing water loss to a minimum.

Plants may be classified into three groups according to their modifications for water conservation.

Fig 20·10B The cactus is a xerophyte

1 Mesophytes. These are plants which live in areas where there is a continual supply of water. They may lose their leaves in winter or die back in drought, but generally the surface area of their leaves is not restricted with a view to water conservation (Fig. 20·10*A*).

2 Xerophytes. These are plants which show modifications for water conservation and are specialized for life in hot dry climates or areas where the ground is frozen for considerable periods. The most extreme examples of xerophytes

are cacti in which leaves are reduced to form non-photosynthetic spines while the stem is swollen by water storage tissue (Fig 20·10*B*). The roots of xerophytes may be very deep to reach underground water sources or alternatively may spread out laterally to catch every drop of available rainfall. Their external surfaces have a thick, waxy cuticle and a reduced number of stomata. Some xerophytes show physiological as well as structural adaptations. Cacti, for example, have a specialised type of photosynthesis (described in Unit 11) which allows them to open their stomata only during the night.

Fig 20·10C The mangrove is a halophyte.

3 Halophytes. These are plants such as the many different kinds of mangrove trees (Fig 20·10*C*), which are modified to withstand the physiological drought imposed by living in salty areas around the margins of the sea. In addition to some of the xerophytic features mentioned above, they usually possess physiological adaptations to tolerate high salt levels and absorb water against an unfavourable osmotic gradient.

Plants, like animals, are capable of homeostatic control. This ability enables them to withstand a greater degree of fluctuation in environmental conditions and hence to occupy a wider ecological zone. It is a feature common to all advanced organisms that the greater the degree of homeostasis, the more control they have over their environment, and the greater their range of habitats. The next Unit describes another kind of homeostasis, that is, the ability to control body temperature.

21 SKIN AND TEMPERATURE CONTROL

Objectives

After studying this Unit you should be able to:–

- Distinguish between ectotherms and endotherms, and explain the advantages and disadvantages of their differing adaptations

- Explain what is meant by thermogenesis

- Comment on radiation, conduction, convection, and evaporation as methods of heat exchange with the environment

- Describe the functions of an organism's thermal shell and thermal core

- Explain the importance of surface area to volume ratio in relation to temperature control

- List the main functions of human skin

- Draw and label a diagram to show the detailed structure of human skin

- Describe the structure and function of skin components, including epidermis, dermis, sense organs, hair follicles, sebaceous glands, and sweat glands

- Draw a flow diagram to summarise the mechanisms of temperature regulation in humans, and explain how these mechanisms operate

- Give specific examples of the adaptations used by organisms which live in hot and cold environments

21·1 Ectotherms and endotherms ■ ■ ■ ■ ■ ■ ■ ■ ■ ■ ■

Natural air temperatures vary from about −60°C or lower at the poles, to about 55°C in the Sahara desert. Temperatures in the surface waters of the oceans vary from −2°C in the Arctic and Antarctic, to about 35°C in shallow regions of the Persian Gulf. Some living organisms are adapted for life in extreme cold, while others are adapted to tolerate hot conditions. The lower limit consistent with sustained activity is about −2°C. The upper limit for most types of organisms is about 40°C and is partly determined by the thermal stability of enzymes, and possibly also by the melting of body lipids. As explained in Unit 5, most enzymes are completely denatured at 50°C, although a few are more stable, as in the case of bacteria which thrive at 85°C in the water from hot springs.

Organisms use two different strategies for surviving these extremes of temperature, and for coping with the more modest variations typical of temperate habitats. One group, called **ectotherms**, or **poikilotherms**, derive their body heat mainly from their surroundings and are physiologically adapted to tolerate a wide range of body temperatures. Single-celled organisms, plants, fungi, invertebrates, and many vertebrates, including fish, amphibians, and reptiles, are all ectotherms. The temperature inside a single-celled organism is normally equal to that of the environment, but many larger ectotherms are able to maintain their body temperatures a few degrees higher than the surrounding air by means of their behaviour. This is called **behavioural thermoregulation**. A lizard, for example, heats up early in the day by exposing its body sideways to the sun, absorbing heat energy from the sun's rays. Later in the day, it avoids extremes of temperature by facing the sun head on, or by seeking shade. In this way, the lizard can keep its body close to the optimum for activity during much of the day. However, when night falls, its body loses heat, and the lizard becomes torpid, that is, sluggish and slow-moving.

Endotherms, or **homoiotherms**, produce most of their body heat internally and possess physiological mechanisms for maintaining a constant body temperature. They are said to show **physiological thermoregulation**. Only birds and mammals belong to this group. The main advantage of endothermy is that it provides nearly optimum conditions for enzyme-controlled reactions. Consequently, the metabolism of the organism can become increasingly sophisticated and intricate. It is no coincidence that endotherms have developed the most effective homeostatic mechanisms and achieved the greatest degree of homeostatic control over their body functions. Muscles and nerves, for example, work best at 35–40°C. Coordination and sensory processing are speeded up at these temperatures, allowing endotherms to become highly efficient predators, or fleet-footed herbivores well able to evade capture. On the other hand, endotherms require a far greater energy input than ectotherms and are forced to spend much of their time in search of food. Typically, at least 80% of the energy derived from food is used for maintaining body temperature. In habitats where the availability of food is limited, as in deserts, ectotherms compete very successfully with endotherms and they should not be thought of as primitive or ill-adapted organisms.

21·2 Gain and loss of heat ■ ■ ■ ■ ■ ■ ■ ■ ■ ■ ■ ■

To keep body temperature within survival limits, organisms must achieve an overall balance between heat gain and heat loss. Heat is generated in body tissues as a result of respiration and other exergonic reactions. This is called **thermogenesis**, and is the main source of heat for endotherms. The rate of heat production is called the metabolic rate of the organism (see Unit 16). Heat can be lost or gained through the surface layers of the body in several ways:

1 Radiation. Thermal radiation is the process whereby energy in the form of electromagnetic radiation of infra-red wavelengths is emitted by a heated surface, and travels off in all directions. The amount of energy emitted or absorbed depends upon the temperature of the surface and other characteristics like colour and texture. A dark-coloured surface gains and loses radiant heat quickly, while a light-coloured surface absorbs and emits heat slowly. Air absorbs very little thermal radiation and is a minor obstacle to heat transfer between objects.

Ectotherms absorb radiated solar energy either directly by basking behaviour, or indirectly from rocks and other objects. Some lizards can change their skin colour by means of pigmented cells, or **chromatophores**. These are extended to produce a dark-coloured surface early in the morning, and contracted to give a lighter colour later in the day. Endotherms typically lose heat by radiation. For example, in normal conditions, about 60% of a person's overall heat loss occurs in this way.

2 Conduction. Conduction of heat energy occurs between objects which are in direct contact. Air is a poor thermal conductor, but water is an excellent thermal conductor, so that transfer by conduction is the most important method of heat exchange for aquatic organisms. Many terrestrial organisms increase their heat loss in extreme conditions by wallowing in water. Conduction can be an important method of heat exchange for terrestrial ectotherms through contact with the soil, but conductive losses from terrestrial endotherms are usually small.

3 Convection. Convection occurs when a current of air or water flows over the surface of the body. Air warmed by the body of a terrestrial organism rises and is constantly replaced by cooler air. Fur, feathers, and clothing trap a layer of still air next to the body so that, in endotherms, convection is often a minor method of heat exchange.

4 Evaporation. It is impossible for an organism to gain heat by evaporation, but it is an important method of heat loss in most terrestrial organisms. The change of state from liquid to vapour requires an input of energy, called **latent heat of vaporisation**. When water evaporates from the body surface, the necessary energy is absorbed from the body tissues. The latent heat varies with the temperature of the skin, but at 30°C, 2·45 kJ is taken up for each 1 cm³ of water evaporated, producing a significant cooling effect.

Both ectotherms and endotherms tend to restrict water loss to a minimum, but both can increase heat loss by evaporation, for example, by panting in extreme conditions. A person's evaporative heat loss normally accounts for 35–40% of the total, but this can increase dramatically in hot conditions.

The body of an organism can be thought of as consisting of two parts, a central **thermal core**, in which heat is generated and stored, and an outer **thermal shell**, consisting of the skin and superficial tissues. The degree of heat loss or gain through the shell depends upon its nature and thickness and also upon the organism's surface area to volume ratio. Small organisms have a greater surface area relative to their volume than large organisms, and therefore exchange heat much more rapidly with their surroundings.

In endotherms, temperature regulation involves changes in the characteristics of the thermal shell which increase or decrease heat loss. The most important organ of temperature control in mammals is the skin. The structure of human skin is described in the next section, and must be understood before the physiological mechanisms of temperature regulation in humans and other mammals can be usefully discussed.

21·3 Human skin ■ ■ ■ ■ ■ ■ ■ ■ ■ ■ ■ ■ ■ ■ ■ ■ ■ ■

The skin of an adult person has a surface area of about 2 m^2 and varies in thickness from 0·5 mm in most body regions to 3 mm on the soles of the feet. It is tough and flexible and provides mechanical protection for the underlying tissues. In addition, it is a major sensory surface, manufactures vitamin D, screens the body from harmful ultraviolet radiation, and prevents the entry of bacteria and other microorganisms (this defence function is described in Unit 19). It plays an essential part in temperature regulation.

Two distinct regions, the **epidermis**, and the **dermis**, are easily recognised in a vertical section of human skin (see Fig. 21·1).

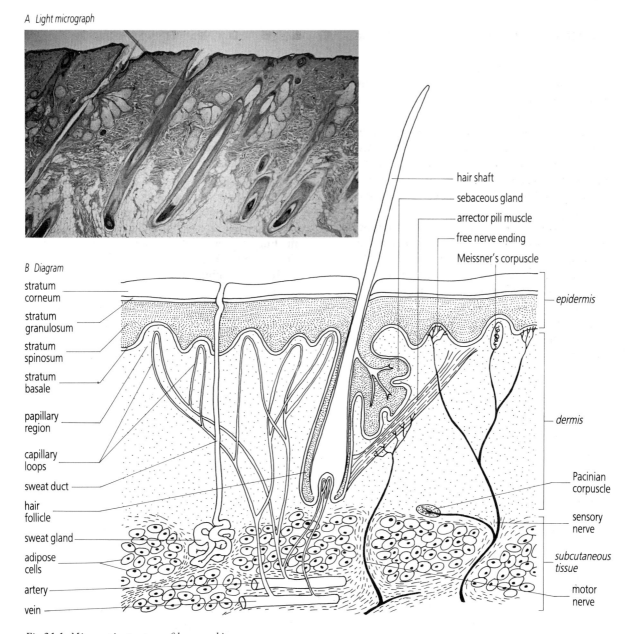

A Light micrograph

B Diagram

stratum corneum
stratum granulosum
stratum spinosum
stratum basale
papillary region
capillary loops
sweat duct
hair follicle
sweat gland
adipose cells
artery
vein

hair shaft
sebaceous gland
arrector pili muscle
free nerve ending
Meissner's corpuscle
epidermis
dermis
Pacinian corpuscle
sensory nerve
subcutaneous tissue
motor nerve

Fig 21·1 Microscopic structure of human skin

Epidermis

The outer epidermis is firmly cemented to the underlying connective tissue dermis, and itself consists of several cell layers. The much folded **stratum basale** consists of a single layer of cells which remain capable of cell division. Division results in new cells which are pushed upwards and gradually pass through the remaining layers before being shed at the skin surface. Together, the stratum basale and the **stratum spinosum**, which contains 8–10 layers of cells, form the germinative layer of the skin. This is followed by a third cell layer, the **stratum granulosum**, which consists of several rows of flattened cells containing granules of **keratohyalin**. The nuclei of these cells start to disintegrate, as the keratohyalin is converted first to **eleidin** in the **stratum lucidum**, and finally into the waterproofing protein **keratin**, which fills the cells of the outermost **stratum corneum**. The flat dead cells of the stratum corneum form flakes, or **squames**, which are continually shed. A large proportion of household dust consists of human squames. It is estimated that human epidermal cells take approximately 27 days to move from the stratum basale to the surface, so that, depending on its thickness, the entire epidermis can be replaced in this time.

The colour of the skin is due to cells called **melanocytes**, which are located under or between the dividing cells of the stratum basale. The black pigment **melanin**, produced by these cells, is taken up by cells in the remainder of the epidermis, helping to screen out ultraviolet radiation. The number of melanocytes in people of different races is approximately the same, but they are much more active in black and coloured-skinned people. The amount of melanin produced is increased by melanocyte stimulating hormone (MSH), but the importance of this is not fully understood.

Dermis

The inner dermis consists of a connective tissue matrix containing collagen and elastic fibres. Its upper region, called the **papillary region**, gives rise to finger-like projections, the **dermal papillae**, which extend upwards into the epidermis. The papillae contain capillary loops and touch sensitive nerve endings. Within the remainder of the dermis lie numerous structures including sense organs, nerves, blood vessels, lymph vessels, hair follicles, and glands. Some of these are described below:

1 Sense organs. The sensory structures present include free nerve endings sensitive to touch, pain, and temperature, Meissner's corpuscles, sensitive to touch, and Pacinian corpuscles sensitive to pressure changes (see Unit 24). The distribution and density of sense organs varies. They are most numerous in the lips and finger tips, and sparse in regions like the forearm or shoulder.

2 Hair follicles. The hair inside each tubular follicle is formed by cell division from the hair papilla at its base. Hairs grow at a rate of approximately 0·3 mm/day. Old hairs are shed periodically and are normally replaced by new hairs which develop in the same follicles. Baldness is due to the progressive loss of active follicles. The follicles lie at an angle to the skin surface, but are attached to strands of muscle, called **arrector pili** muscles. These can be contracted to hold the hair shafts at right angles to the skin, producing small bumps, or 'goose pimples'. Body hair is quite useless for temperature control in humans, but, in animals with fur, the raised hairs trap a thicker layer of air which acts as an insulator to reduce heat loss.

3 Sebaceous glands. Sebaceous glands are usually found inside hair follicles and secrete an oily fluid called **sebum**. This helps to keep the hair and skin surface waterproof and flexible, and may have some function in preventing the growth of harmful bacteria.

4 Sweat glands. Two categories of sweat glands are distinguished: **apocrine** glands, and **eccrine** glands. They have quite different functions. Apocrine glands are found in large numbers in the armpit and groin regions of human skin, and around the nipples. They secrete a whitish liquid containing chemicals called **pheromones**.

In many mammals, pheromones are used as chemical signals for marking out territory, or for attracting the opposite sex, and it may be that they have a similar, if subconscious, role in humans. The unpleasant smell, body odour, arises when apocrine secretion is decomposed by bacteria. Ironically, it is sometimes combatted by perfumes which contain pheromone secretion from the apocrine glands of other mammals. Modified apocrine glands in the external ear produce ear wax, or **cerumen**.

Eccrine glands produce the watery liquid known as sweat. These glands are very common in human skin, but are absent from the skin of most hairy mammals, except from friction surfaces, such as the hands and feet of apes and monkeys.

Beneath the dermis is a layer of subcutaneous tissue containing large numbers of **adipose cells**. These cells store fat and form an important insulating layer, preventing heat loss in humans, as well as a protective cushion against knocks.

21·4 Temperature regulation ■ ■ ■ ■ ■ ■ ■ ■ ■ ■ ■

Normal adults maintain core body temperature in the range 36–37.5°C. There is some variation between individuals, for example, children typically have higher body temperatures than adults. A daily variation of 1.5°C can be observed in any normal person (Fig. 21·2). Core temperature can be kept within these limits over a wide range of external conditions by means of precise regulatory mechanisms. It is particularly important to prevent overheating. Most people would suffer convulsions at 41°C, while body temperatures above 43°C are fatal.

The important mechanisms of temperature regulation in humans are summarised in Figure 21·3. At all times a balance must be maintained between heat gain and heat loss. This is achieved by increasing heat production in the cold, and by regulating the rate of heat loss from the body.

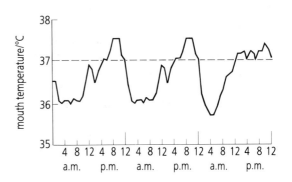

Fig 21·2 Daily variation of human body temperature

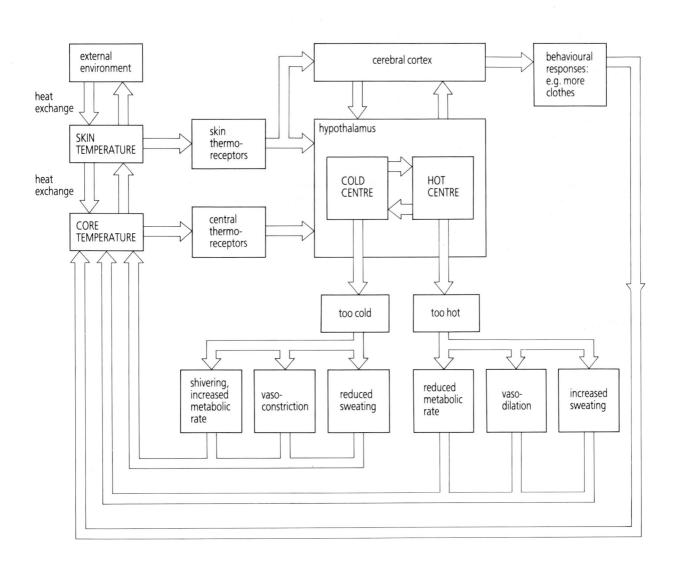

Fig 21·3 Summary of human temperature regulation

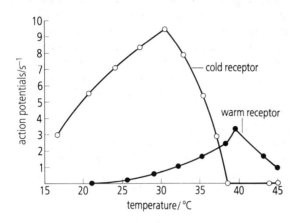

Fig 21·4 Responses of cold and warm receptors

Such adjustments are possible because body temperature is constantly monitored by heat sensitive sense organs, called **thermoreceptors**. There are two types of thermoreceptors in the skin (Fig. 21·4). Cold receptors show a maximum response at about 30°C, while warm receptors have their maximum response at about 40°C. Signals from these receptors pass to the hypothalamus and cerebral cortex regions of the brain, giving rise to conscious sensations of cold and warmth (refer to

Fig. 21·3). The temperature of the body core is monitored by central thermoreceptors located in the hypothalamus. These respond to changes in the temperature of the blood passing through. The hypothalamus is believed to contain separate cold and warm control centres; each centre inhibits the activity of the other so that they cannot be active at the same time. Depending on the signals arriving from the sense organs, either centre may initiate a whole series of appropriate responses, as explained below:

1 Thermogenesis. The chemical reactions of metabolism generate heat, especially in active organs like the liver. The amount of heat given out in this way cannot be reduced below a definite minimum level, known as the **basal metabolic rate** (see Unit 16). On the other hand, heat production is increased by muscle contraction during exercise. Brief repeated contractions, called **shivering**, can be very effective in helping to prevent a drop in body temperature in cold conditions.

Metabolic rate is also affected by hormones, notably adrenalin, and thyroxin. These may play some part in long term adjustment, or acclimatisation to new conditions, but are of minor importance in short term responses.

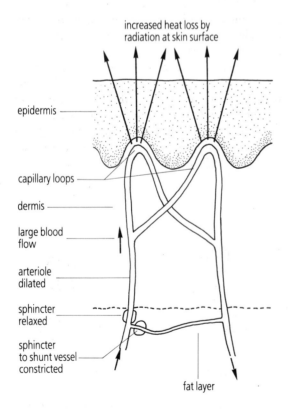

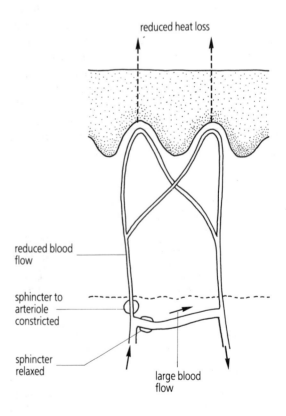

Fig 21·5 Vasodilation and vasoconstriction

2 Vasodilation and vasoconstriction.

Heat is transferred from the thermal core of the body to the surface layers mainly by the blood. In hot conditions, the arterioles leading to the capillary loops of the skin are dilated, allowing a large volume of blood to flow through. This response is called **vasodilation** (Fig. 21·5*A*). Heat transferred from the blood to the epidermis is lost at the surface mainly by radiation. In cold conditions, the opposite response, called **vasoconstriction**, takes place (Fig. 21·5*B*). As you can see, the arterioles leading to the capillary loops are constricted by sphincter muscles, causing blood to be diverted through **shunt vessels** located in the subcutaneous layer. In this way, very little heat is lost. In some parts of the body, blood flow to the skin can be reduced by as much as 99%, as in the case of fingers exposed to the cold.

3 Sweating.

Evaporative heat loss from the skin occurs continually by diffusion of water through the skin. This loss may amount to 600–800 cm³ per day, and is called **insensible perspiration**. In hot conditions, evaporative heat loss is increased by secretion of sweat from eccrine glands. Sweat is a very dilute solution containing salt, urea, and lactic acid. In extreme conditions, as much as 12 litres per day of sweat is produced, and may evaporate from the body surface. By altering the amount of sweat, heat loss can be controlled over an extended range of external temperature. When the external temperature exceeds core temperature, evaporation becomes the only method of heat loss. In dry conditions, a person can tolerate temperatures of up to 120°C without any increase in core temperature. On the other hand, if the air is humid, evaporation is reduced or prevented and body temperature rises quickly, even when external temperature in the range 40–55°C.

The physiological mechanisms of temperature control provide good examples of homeostasis. Like those concerned with osmoregulation, they act to produce a dynamic equilibrium in the face of constantly changing conditions. Look again at Figure 21·3 and identify the receptors, control centre, and effectors involved. The effectors modify the core temperature of the body, resulting in feedback, or 'closed loop', control, as explained in Unit 20.

21·5 Adaptations for life in hot climates ■ ■ ■ ■ ■ ■ ■ ■

The surface temperature of sand in the Sahara desert can change from 0°C at midnight to 55°C at noon. Very few organisms can tolerate such a wide range and most desert animals survive by restricting their activities to times when conditions are less extreme. Kangaroo rats, for example, are largely nocturnal and spend most of the day in deep burrows where the temperature stays approximately constant. Many larger mammals, such as camels, have thick non-wettable fur. This is pale coloured to reflect heat, insulates the body from excessive heat, and allows water to evaporate from the skin surface, so that cooling occurs. The same result can sometimes be achieved by increasing surface area. This is why elephants have large ears, and why similar species of hares adapted to different climates have differently-sized ears (Fig. 21·6). In both cases, the ears are important sites of heat exchange. There are many examples of similar changes in body size and body colour with climate.

Ectotherms avoid temperature extremes in the same way. Some reptiles and insects have become very well adapted for desert life, where their lower metabolic rate enables them to survive for long intervals without food. Desert plants, like cacti, are poorly adapted for temperature control, but highly adapted to reduce water loss (see Unit 20).

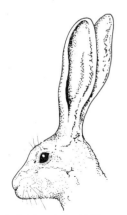

A *Antelope jack rabbit* (Lepus alleni) *from Arizona.*

B *Black-tailed jack rabbit* (L. californicus) *from Oregon.*

C *Snowshoe hare* (L. americanus) *from British Columbia.*

D *Arctic hare* (L. arcticus) *from tundra region.*

Fig 21·6 Variation in ear size in hares from different climates

21·6 Adaptations for life in cold climates ■ ■ ■ ■ ■ ■ ■ ■

Mammals which live in very cold environments are usually well insulated by a thick layer of subcutaneous fat. Their ability to conserve heat is often increased by body shapes and sizes which result in smaller surface area to volume ratios. Whales, seals, and polar bears, for example, have rounded bodies consisting of a thick thermal shell enclosing a substantial core. Animals which are not adapted in this way cannot survive the coldest seasons unless they migrate to warmer places, or hibernate.

Hibernation drastically reduces an organism's energy needs, and enables it to survive when food is not available. It occurs in several groups of mammals including insectivores, like the hedgehog, rodents, bats, and bears. In **deep hibernators**, body temperature falls to near freezing, and the organism sleeps for prolonged periods, surviving on stored glycogen and fat. **Light hibernators** have higher core temperatures and awake more frequently, often feeding on food previously collected and stored in the nest or burrow.

Changes in day length and climate trigger hormonal changes which slow down metabolism and lead to hibernation. The dormouse is stimulated to begin hibernation by temperatures below 16°C. As the mouse falls asleep, its core temperature fluctuates from 30–35°C, but, as the external temperature drops below zero, the metabolic rate falls rapidly. Shivering and vasoconstriction are inhibited and body temperature declines to about 4°C. At this level, homeostatic control is regained, so that at no time does the core temperature fall below 0°C.

Benefits similar to those of hibernation are sometimes obtained by cooling human tissues. For example, kidneys and hearts used for transplantation are invariably cooled, with the result that the organs remain viable for much longer. Body cells are usually destroyed by freezing because their structure is disrupted by the formation of ice crystals. In a few cases, including the storage of sperm cells, this problem can be overcome by the addition of glycerol.

22 NEURONS AND SYNAPSES

Objectives

After studying this Unit you should be able to:–

- State the importance of neurons and synapses in simple terms

- List the basic structural features of neurons

- Draw and label a diagram of a vertebrate motor neuron and describe the structure and function of each of its main parts

- Describe the role of Schwann cells in forming the myelin sheath and comment on the functions of other glial cells

- Describe the electrical properties of neurons and list in sequence the phases of the nerve impulse

- Discuss propagation of the nerve impulse in non-myelinated and myelinated axons

- Describe external and internal ion concentrations in a neuron and relate permeability differences of the neuron membrane to the resting potential and nerve impulse

- Describe the role of the Na^+-K^+ exchange pump

- Draw and label a diagram of a typical synaptic terminal and explain how synaptic transmission can occur.

22·1 Introduction ■ ■ ■ ■ ■ ■ ■ ■ ■ ■ ■ ■ ■ ■ ■

This Unit is the first of several dealing with the coordinating systems of the body, that is, with the **nervous** and **endocrine** systems. The description of these fascinating and important systems follows a sequence beginning with the nervous system, and in particular, with the individual cells from which it is made.

At its simplest, the correct operation of the nervous system in any animal depends on the transfer of electrical signals called **nerve impulses** to and from all parts of the body. This transfer involves the individual message-carrying cells, which are termed **neurons**, and small structures called **synapses**, which allow nerve impulses to be transmitted from one neuron to another. Any nervous pathway, whether simple or complex, is made up of a number of neurons connected together by synapses. The purpose of this Unit is to give a basic understanding of these different component parts.

22·2 Basic structure of neurons ■ ■ ■ ■ ■ ■ ■ ■ ■ ■ ■ ■

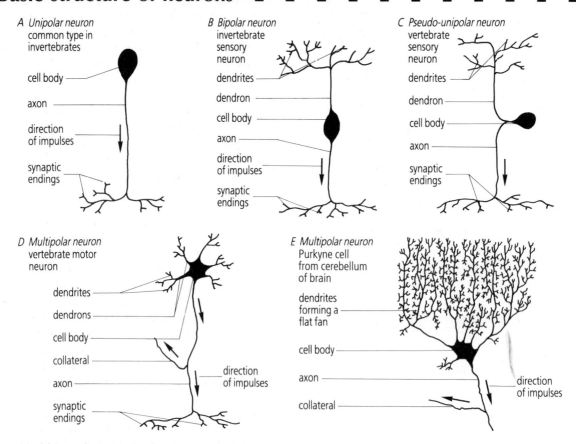

Fig 22·1 Range of form in neurons

The most striking feature of neuron structure is that, while their branches are only a few micrometres (μm) in diameter, some of them are extremely long cells. Just how long depends on their precise role in the nervous system. The **motor neurons** of vertebrates, also called **effector neurons**, conduct nerve impulses from the spinal cord to the various muscles of the body, thereby controlling muscle activity. It follows that, when the points to be connected are far apart, the motor neurons must be very long cells. Many large animals have motor neurons a metre or more in length; the motor neurons which supply the muscles in a giraffe's leg provide an extreme example. The **sensory neurons**, which carry impulses inwards from a sense organ to the spinal cord, are also frequently long cells. On the other hand, the **interneurons**, which link from one place to another within the nervous system, may be quite short.

The size and shape of neurons of each type vary considerably. Some indication of the possible range is given in Figure 22·1. In spite of the variation, some common features of neuron structure can be identified; these are all shown and labelled in Figure 22·1A. Each neuron has a prominent swelling; this is called the **cell body** and it contains the nucleus of the cell. Arising from the cell body are a number of large branches, loosely termed **nerve fibres**. Almost all neurons have a single **axon**, which is normally the longest of these fibres. The axon is extremely important because it is the only branch of the neuron which carries nerve impulses away from the cell body. Each of these impulses is conducted by the axon from its place of origin to the correct destination. Sometimes, axons divide, forming a number of **collaterals**, so that impulses from one neuron can reach several different destinations. The ends of the axon and its collaterals subdivide to form numerous smaller branches, each of which terminates in a minute swelling. These swellings are the **synaptic endings** and they connect the neuron to further neurons in the pathway, forming synapses so that impulses can be transmitted across from one neuron to the next.

A neuron in which the axon is the only large branch from the cell body is said to be a **unipolar neuron**. As Figure 22·1 shows, other neurons may have additional large branches, called **dendrons**. Thus, in the **bipolar neuron** shown in Figure 22·1B, there is one axon and one dendron, while in a **multipolar neuron**, such as the vertebrate motor neuron shown in Figure 22·1D, there is one axon and several dendrons. Generally, each dendron subdivides to form many smaller branches, called **dendrites**. These small branches are contacted by synaptic endings coming from other neurons, and they therefore form the receptive regions of the neuron.

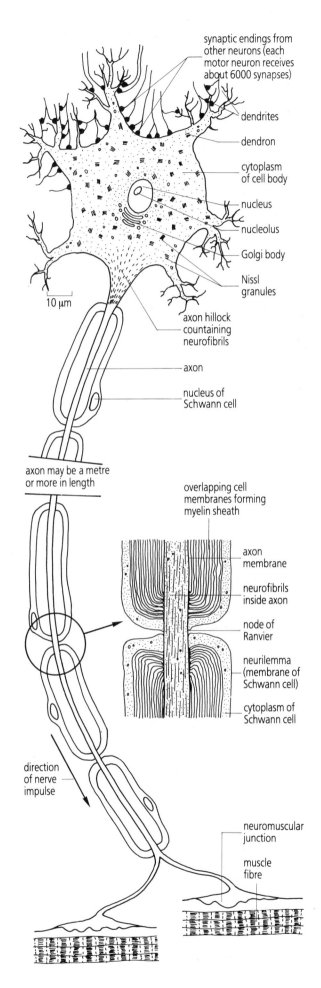

synaptic endings from other neurons (each motor neuron receives about 6000 synapses)

dendrites

dendron

cytoplasm of cell body

nucleus

nucleolus

Golgi body

Nissl granules

10 µm

axon hillock countaining neurofibrils

axon

nucleus of Schwann cell

axon may be a metre or more in length

overlapping cell membranes forming myelin sheath

axon membrane

neurofibrils inside axon

node of Ranvier

neurilemma (membrane of Schwann cell)

cytoplasm of Schwann cell

direction of nerve impulse

neuromuscular junction

muscle fibre

Fig 22·2 Structure of a vertebrate motor neuron

The role of the dendrites and other structures can be explained in more detail by using, as an example, the vertebrate motor neuron shown in Figure 22·2. The cell body of the motor neuron is located in the spinal cord and, the long thin axon travels out, eventually reaching the appropriate muscle.

As can be seen, the dendrites and the surface of the cell body of the motor neuron both receive messages from numerous synaptic endings from other neurons. In fact, there are usually about 6000 synapses, each of which can affect the activity of the motor neuron. Single synapses have only a small effect, but the signals from all the different synapses added together determine whether or not the motor neuron will **fire**, in other words, whether it will produce its own nerve impulses.

The addition, or **integration** of synaptic input received by the dendrites is focussed on a specialised region of the cell body called the **axon hillock**. This region appears to act as a trigger zone for the production of nerve impulses. When the combined input from the synapses reaches a certain level, called the **threshold** level, electrical impulses start in the axon hillock and then travel rapidly down the axon, at speeds of up to 100 m s^{-1}, reaching the muscle after only a short delay.

Many vertebrate neurons, including motor neurons, conduct these electrical impulses much more quickly than most invertebrate neurons. This can be attributed to the presence of the **myelin sheath** surrounding the axon. This sheath is formed by specialized **Schwann cells**. As indicated in Figure 22·3, during development the Schwann cell rotates around the axon, forming a 'swiss roll' structure consisting of overlapping layers of cell membranes. The lipid-protein mixture of these membranes is called **myelin**, and its functional significance is as an efficient electrical insulator. The myelin sheath is not continuous, but is interrupted at intervals of about 1 mm by **nodes of Ranvier** where the axon is uncovered. The sheath between the two nodes is formed by a single Schwann cell; the nodes themselves aid in the rapid transfer of impulses.

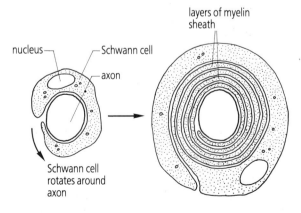

layers of myelin sheath

nucleus

Schwann cell

axon

Schwann cell rotates around axon

Fig 22·3 Development of the myelin sheath

The impulses which travel down the axon of the motor neuron have as their destination the synapses formed between the motor neuron and the muscle it supplies. The effect of the transmission at these synapses, called 'end plates', or **neuromuscular junctions**, is to cause the muscle to contract. Through its neuromuscular junctions, the motor neuron finally carries out its function of regulating muscle activity.

In order to maintain the impulse conducting functions described above, neurons require a number of specialised internal structures. Such structures contribute to the metabolism of the neuron, allowing the single nucleus in the cell body to exert metabolic control throughout the dendrites and the long axon. The special structures present include **Nissl granules** in the cell body, and **neurofibrils**, which occur throughout the cytoplasm. Each Nissl granule consists of membranes of rough endoplasmic reticulum which enable the neuron to be particularly active in protein synthesis, manufacturing within the cell body all the proteins needed elsewhere in the cell. The transport of these proteins, and other substances produced in the cell body, seems to be carried out by the neurofibrils. Both protein (from the Nissl granules), and glycolipids (from the Golgi body) are packaged into membrane bound vesicles before transport takes place.

While the neurons are the most important functional components of the nervous system, they are not the only cells present. In addition to blood vessels and connective tissue, there is always a mass of **glial** cells, collectively known as the **neuroglia**. While some of these glial cells, such as the Schwann cells, are known to have special functions, the majority seem to be 'packing' cells, giving mechanical, and possibly metabolic, support to the delicate network of neurons carrying the nerve impulses.

22·3 Electrical properties of neurons ■ ■ ■ ■ ■ ■ ■ ■ ■ ■

So far the nerve impulse has been rather vaguely described as an 'electrical message'. The first step towards a deeper understanding is to consider the electrical properties of neurons. These electrical properties have been studied using tiny probes called **microelectrodes**.

The commonest type of microelectrode is a minute glass pipette filled with the conducting solution of a salt such as potassium chloride. Electrical signals detected by the microelectrode are usually amplified and then passed to an **oscilloscope**, which displays the signals in a visible form.

The tip diameter of a glass microelectrode is often less than 1 μm and, with appropriate techniques, it can be placed inside the cell body or axon of any suitably sized neuron. The cell body of a vertebrate motor neuron is about 50–70 μm in diameter and therefore large enough to be penetrated successfully. However, much larger neurons are found in some invertebrate animals, particularly molluscs. The squid, for example, has a number of **giant neurons** with axons 1000 μm (=1 mm) in diameter and several cm in length. These giant axons have contributed significantly to present understanding of the nerve impulse.

Figure 22·4 shows a simple experiment which can be carried out using the squid giant axon. (The axon would normally be dissected out and placed in a bath of fluid for the purpose.) The

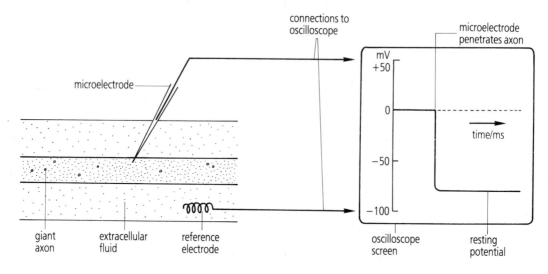

Fig 22·4 Electrical recording from the squid giant axon

microelectrode, mounted on a micromanipulator for fine adjustment, is manoeuvred until it penetrates the axon. When this happens, there is an immediate change in the oscilloscope display, showing that the microelectrode has entered a negatively charged region. In fact, the whole of the inside of the axon is negatively charged so that there is a constant electrical potential difference across the axon membrane. As a result, a steady negative potential, usually about -70 mV, is recorded. This steady potential is called the **resting potential**, and the mechanisms which underlie it are crucial for impulse production. Before considering these mechanisms further, it is helpful to know a little more about the electrical nature of the impulse itself.

The axon can be made to produce impulses if a second microelectrode is inserted and then used to deliver electrical pulses to the axon. Small negative and positive pulses, called **sub-threshold stimuli**, do not produce an impulse, but as the size of the positive pulses is increased, a definite **threshold** value is reached. Above this threshold size, each stimulus pulse causes a rapid **all-or-nothing** response in the axon, giving rise to the electrical event called a nerve impulse. The impulse travels rapidly along from its point of origin as a wave of electrical activity. Significantly, its size and shape remain constant so that the 'electrical message' reaching the far end of the axon is unaltered.

Figure 22·5 shows the electrical events which can be recorded at any point in the axon as the impulse travels past. The impulse begins with a phase of **depolarisation** during which the potential difference across the axon membrane is reduced. The potential across the membrane is then reversed, so that there is a definite 'overshoot', when the inside of the axon becomes positively charged. This is followed by a phase of **repolarisation**, which leads in turn to an 'undershoot', during which the potential recorded is more negative than the normal resting potential. The undershoot usually dies away gradually until the resting potential is restored.

During the early part of the undershoot, the axon is completely unable to produce a second nerve impulse, even if strongly stimulated. This is known as the **absolute refractory period**. During the latter part of the undershoot, called the **relative refractory period**, the axon can produce an impulse, but only if a stronger than normal stimulus is applied. These refractory periods are important because they affect the frequency at which nerve impulses can be carried. The normal maximum frequency of impulses is around 100 s^{-1} but can be as much as 1000 s^{-1} in some specialized neurons.

Impulses from many different animals correspond remarkably closely to this general pattern. The duration of the impulse is the commonest variable, with impulses recorded from most vertebrate neurons being significantly shorter than those from most invertebrate neurons.

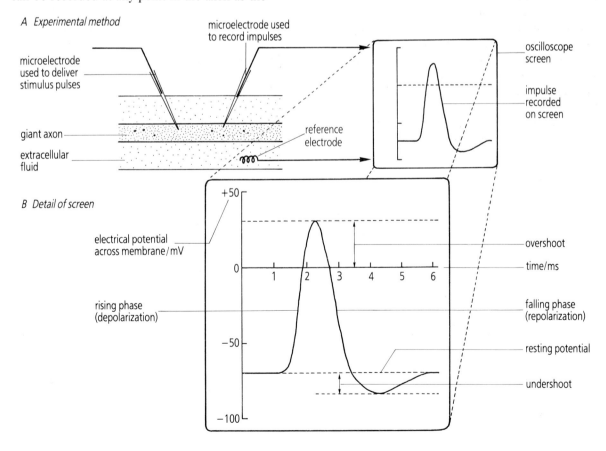

Fig 22·5 Recording impulses from the squid giant axon

22·4 Propagation of the nerve impulse ■ ■ ■ ■ ■ ■ ■ ■ ■ ■

The term **propagation** refers to the process whereby the nerve impulse travels along the axon of a neuron, without decreasing in size. In a non-myelinated axon, such as the squid giant axon, the impulse travels smoothly along. This process is illustrated in Figure 22·6.

The part of the axon involved in the impulse at any one time can be considered in three sections. There is first an **active zone** where the nerve impulse is at its peak. At this point the axon is positively charged inside. As a result, small positive electric currents (carried by ions in the cytoplasm of the axon) flow into the adjacent negatively charged regions of the axon and out across the axon membrane. Ahead of the peak impulse, this positive current flow acts in the same way as a stimulus pulse, and depolarises the next part of the axon, called the **depolarised zone**, to such an extent that it reaches and passes its threshold. When this happens, the depolarised zone becomes active and produces its own impulse, so that, as the process continues, the impulse passes along the

axon. Behind the impulse, in the **refractory zone**, the axon is temporarily unable to become active, and, consequently, any current flowing from the active zone has no effect. This is why the impulse normally travels in only one direction, namely from the cell body down the axon. If a resting axon is artificially stimulated somewhere along its length using a microelectrode, the nerve impulse travels off in both directions at once because there is no refractory zone to prevent it.

The impulse in a typical non-myelinated axon normally travels at a rate of about 1 m s^{-1} or less. This speed of propagation, also called the **conduction velocity**, can be increased by increasing the diameter of the axon. Where speed is particularly important, as in many escape responses, giant axons have developed. Thus, the giant axons of the squid are involved when it uses its siphon for a jet-propelled escape. The nervous systems of many other invertebrates also contain giant axons.

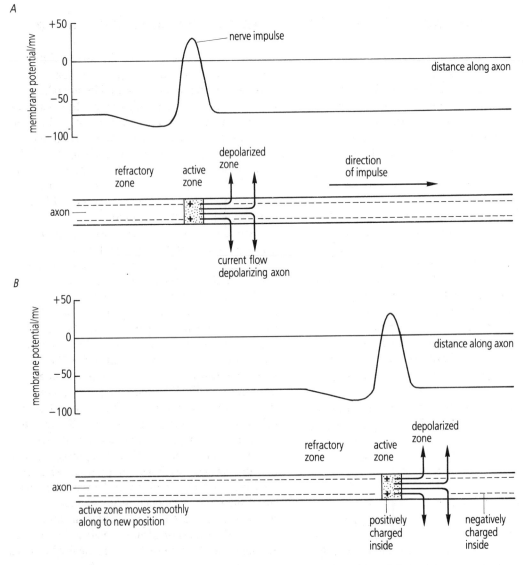

Fig 22·6 *Propagation of the nerve impulse in the squid giant axon: continuous conduction*

174

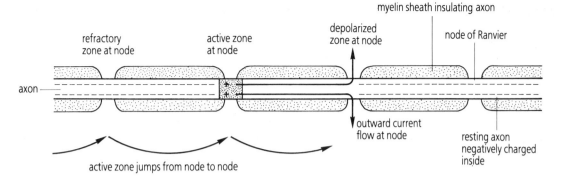

Labels on figure: refractory zone at node; active zone at node; depolarized zone at node; myelin sheath insulating axon; node of Ranvier; axon; outward current flow at node; resting axon negatively charged inside; active zone jumps from node to node

Fig 22·7 Propagation of the nerve impulse in a myelinated axon: saltatory conduction

In vertebrates, conduction velocity is greatly increased by the presence of insulating material round the axons of many neurons. This is the function of the myelin sheath. As shown in Figure 22·7, electric current can only leave the axon at a node of Ranvier where the axon is uncovered. The region between nodes does not produce an impulse, but allows ordinary electrical conduction as in a wire. Thus, the distance between nodes is crossed almost instantaneously, and the impulse 'jumps' from node to node as it travels along. This process is referred to as **saltatory conduction** and allows conduction velocities of up to 100 m s⁻¹ more. Myelination is effective even for small axons and, as a result, vertebrate nervous systems can contain many more cells than invertebrate systems. The combined advantages of efficiency and economy achieved with myelination have allowed vertebrate nervous systems to become extremely sophisticated, culminating in the intricate machinery of the human brain.

22·5 Ionic basis for the resting potential and nerve impulse ■ ■

The resting potential and nerve impulse both depend on the distribution of particular ions across the neuron cell membrane. The most important of these ions are potassium ions (K^+), and sodium ions (Na^+). Table 22·1 lists the internal and external concentrations of these ions for the squid giant axon. As a result of the radical differences in concentration, there is a strong tendency for potassium ions to diffuse out of the axon, and for sodium ions to diffuse in.

The membrane of a resting neuron is almost impermeable to sodium ions but quite permeable to potassium ions. Consequently, K^+ ions tend to diffuse out, leaving the inside of the axon negatively charged. This diffusion process reaches an equilibrium corresponding to the resting potential. At this potential, normally about -70 mV, the effect of the high internal concentration pushing K^+ ions out becomes balanced by the electrical attraction between the positive potassium ions and the negatively charged interior.

Table 22·1 Ion concentrations in a squid giant axon

ion	internal concentration in cytoplasm mmol dm⁻³	concentration in extracellular fluid mmol dm⁻³
K^+	400	10
Na^+	50	460

When an axon carries a nerve impulse, its membrane suddenly becomes permeable to sodium ions. These rapidly diffuse into the axon, not only abolishing the resting potential, but also causing the 'overshoot', during which the inside of the axon becomes positively charged. For complex reasons, the membrane permeability to sodium ions can be maintained only briefly. As it starts to decline, more potassium ions flow out of the axon, driving it once more towards the resting potential. This is the repolarisation phase; during it and the 'undershoot' period which follows, the potassium permeability of the membrane is greater than at other times. When it returns to its normal level, the resting potential is restored.

After a nerve impulse has passed, a neuron has *gained* a number of Na^+ ions and *lost* an approximately equal number of K^+ ions. For a single impulse these movements do not significantly affect the internal and external concentrations of ions. However, after a series of impulses, the concentrations will be affected. To maintain the correct concentrations, the neuron membrane has an ion pump called the **Na^+-K^+ exchange pump** (more loosely referred to as the '**sodium pump**'). The pump requires energy in the form of ATP and pumps Na^+ ions out of the cell, while at the same time pumping K^+ ions in. Generally speaking, the same number of ions are pumped in either direction. It is important to note that the Na^+-K^+ exchange pump does *not* directly contribute to the generation of the nerve impulse. So, if the pump is stopped using a metabolic poison such as ouabain, the squid giant axon can conduct thousands of impulses before the loss of K^+ ions and the gain of Na^+ ions finally becomes too great.

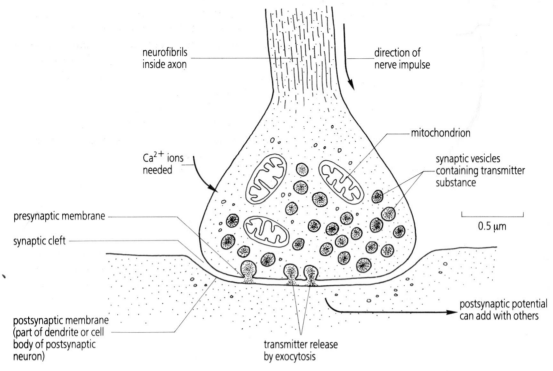

neurofibrils inside axon

direction of nerve impulse

mitochondrion

Ca^{2+} ions needed

synaptic vesicles containing transmitter substance

0.5 μm

presynaptic membrane

synaptic cleft

postsynaptic potential can add with others

postsynaptic membrane (part of dendrite or cell body of postsynaptic neuron)

transmitter release by exocytosis

Fig 22·8　Structure of a synapse

The points where the axon terminals of one neuron contact other neurons or muscle cells are called **synapses**. A typical synapse has the structure illustrated in Figure 22·8. The vast majority of synapses are one-way connections which rely for communication on chemical **transmitter substances**. The transmitter is contained in small **vesicles** in the presynaptic ending and, when a nerve impulse arrives at the ending, some of the vesicles fuse with the presynaptic membrane and release their contents into the small gap called the **synaptic cleft**. Each vesicle contains approximately the same amount of transmitter, this amount being termed a 'packet', or '**quantum**'. The release of a quantum of transmitter substance requires calcium ions (Ca^{2+}), but their precise role is not known.

One of the best known transmitter substances is **acetylcholine (ACh)**. When this is released from a synaptic ending, it rapidly diffuses across the synaptic cleft, and binds with receptors in the membrane of the postsynaptic neuron. As a result, there is a change in membrane permeability and sodium ions flow into the postsynaptic neuron, depolarising the neuron membrane and giving rise to an electrical event called a **postsynaptic potential**. This is smaller and longer lasting compared with the nerve impulse. Furthermore, postsynaptic potentials are not all-or-nothing events, but are instead graded in size, depending on the quantity of transmitter released.

The concentration of ACh acting on the

postsynaptic membrane does not remain high for long, because the synaptic cleft contains powerful **cholinesterase enzymes** which rapidly inactivate ACh, allowing the normal resting potential across the postsynaptic membrane to be restored, and 'clearing' the synapse so that the next impulse can be transmitted. There is a rapid return of ACh in inactivated form from the synaptic cleft back to the presynaptic ending, where it is reactivated and can be reused.

The effect of ACh release at a synapse is to cause depolarisation of the postsynaptic membrane, making impulse production in the postsynaptic neuron more likely. Synapses where ACh is the transmitter are therefore described as **excitatory** and give rise to **excitatory postsynaptic potentials** or **EPSPs**. Not all synapses are excitatory; **inhibitory** synapses producing **inhibitory postsynaptic potentials**, or **IPSPs**, also occur but require a different chemical transmitter. Most neurons receive synaptic endings which are a mixture of excitatory and inhibitory types.

Identifying the chemical transmitter at a particular synapse can be extremely difficult. A range of substances in addition to ACh are thought to be transmitters. Some of these substances and their presumed effects are listed in Table 22·2.

Many features of synaptic action are still not clearly understood. For example, the details of the mechanism whereby the arrival of a nerve impulse triggers transmitter release are still a mystery, as is

Table 22·2 *Chemical transmitter substances*

SUBSTANCE	PRESUMED EFFECT
cholinergic transmitters	
acetylcholine (ACh)	excitatory
adrenergic transmitters	
adrenalin (=epinephrine)	excitatory or inhibitory
noradrenalin (=norepinephrine)	excitatory or inhibitory
dopamine	excitatory or inhibitory
serotonin (=5-hydroxytryptamine)	excitatory or inhibitory
amino acids	
gamma-aminobutyric acid	inhibitory
glycine	inhibitory
glutamic acid	inhibitory
neuropeptides	
substance P	excitatory
encephalin	inhibitory

the role of synapses in learning and memory. Nevertheless, the basic reason for having chemically operated synapses instead of direct electrical contacts between neurons is quite clear. This is simply that chemical transmitters in minute quantities are much more effective in altering membrane potential than direct electrical stimulation by contact between cells would be. Thus, synaptic endings can be extremely small without becoming ineffective. This means that an individual neuron can receive synapses from many other neurons and, through its axon terminals, can pass on information to many more. The possibility for interconnection giving different pathways is almost limitless. This is one of the main reasons for the success of the nervous system in terms of information processing and flexibility. Some of the ways in which individual neurons can be arranged into pathways will be discussed in the next Unit, which deals with the overall organization of the nervous system.

23 ORGANIZATION OF THE NERVOUS SYSTEM

Objectives

After studying this Unit you should be able to:–

- Describe a simple nerve net

- Briefly outline the main stages in the evolution of complex nervous systems, including the formation of ganglia, centralization, and cephalization

- Briefly describe the embryological development of the vertebrate nervous system

- Draw and label a diagram of the human spinal cord as seen in cross section and explain the importance of the structures present

- Describe the nerve pathways involved in stretch reflexes and withdrawal reflexes and discuss the functions of their component cells

- Distinguish between spinal reflexes and cranial reflexes

- Draw and label diagrams to show the external and internal features of the human brain

- Name each of the main brain regions and state their functions

- Describe the structure of the autonomic nervous system and explain the roles of its two subdivisions, the sympathetic and parasympathetic nervous systems

- Comment on the structures and mechanisms used to protect the nervous systems of mammals and give an account of the blood-brain barrier

23·1 Evolution of the nervous system ■ ■ ■ ■ ■ ■ ■ ■ ■ ■

In Unit 22 the properties of individual nerve cells, or neurons, were discussed. This Unit considers how neurons are linked together to form the working **nervous systems** of different animals.

The simplest nervous systems consist of randomly connected neurons forming an array called a **nerve net**. Such networks are characteristic of Phylum Cnidaria, or coelenterates, the invertebrate group which includes *Hydra*, *Obelia*, sea anemones, and jellyfish. When the body surface of any of these organisms is touched, nerve impulses are transmitted in all directions, causing a contraction of the musculo-epithelial cells in the body wall, so that the animal moves away from the stimulus. In some coelenterates, additional '**through-conduction pathways**' have developed.

These are needed to coordinate complex movements, such as the swimming of jellyfish.

A major step in evolutionary development from simple nerve nets is the grouping of neuron cell bodies into structures called **ganglia** (singular **ganglion**). This shortens the communication distances between individual neurons and allows them to form more interconnections. In addition it provides the basis for two important trends in the evolution of the nervous system:

1 Centralization. This is the gathering together of ganglia to form a definite **central nervous system (CNS)** containing almost all the neuron cell bodies. The CNS communicates with the sense organs and with the rest of the body by means of the **peripheral nervous system**. This consists of bundles of fibres called **nerves**. Within the nerves, some fibres carry sensory information inwards towards the CNS (**afferent** fibres), while others carry motor information outwards to the muscles and glands (**efferent** fibres).

2 Cephalization. This is the tendency for more and more structures to be located at the front of the organism, forming a distinct head region. For the nervous system this means that increasing numbers of ganglia are concentrated together to form the **brain**. Afferent impulses reach the brain from sense organs such as the eyes and ears, and from various chemically-stimulated sense organs, which are also located in the head.

23·2 Basic organization of vertebrate nervous systems ■ ■ ■ ■

The basic structure of the vertebrate nervous system is best understood from its embryological development. As you can see from Figure 23·1, the nervous system develops initially from a fold in the outer cell layer, or **ectoderm**, of the embryo. This fold becomes a deep groove which is then closed off at the top to form a hollow **neural tube** running along the back of the animal. Thin rods of additional nervous tissue called **neural crest** material lie next to the neural tube on either side. As development proceeds, nerve fibres grow out from the neural tube and also from neuron cell bodies which originate in the neural crest. Eventually, these fibres extend to all parts of the body. The anterior end of the neural tube becomes the **brain**, with its corresponding **cranial nerves**, while the remainder of the tube develops into the **spinal cord**, with corresponding **spinal nerves**. This arrangement is illustrated in Figure 23·2

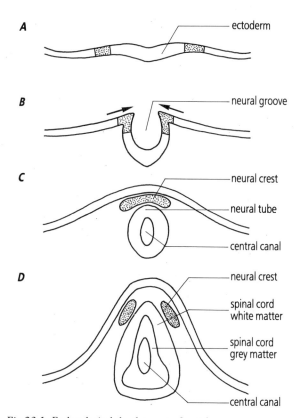

Fig 23·1 Embryological development of vertebrate nervous systems

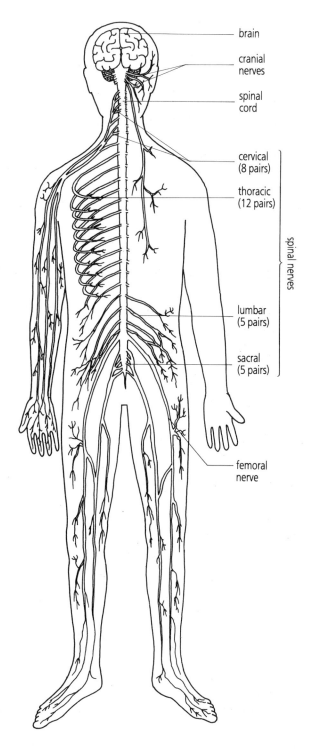

Fig 23·2 Mammalian nervous system

179

In an adult person, the spinal cord is a glistening white structure about 45 cm long and up to 2 cm in diameter. As shown in Figure 23·3, it passes through the **neural canal** present in each vertebra and is surrounded by bone for most of its length. The spinal nerves carrying nerve impulses to and from the cord emerge through small spaces between adjacent vertebrae. This pattern represents the remains of a segmented organisation which is clear in the embryo but becomes less visible as development proceeds.

Each spinal nerve is derived from the spinal cord by two separate connections, called the **dorsal root** and **ventral root**, which join together. The spinal nerves are **mixed** nerves, that is, each contains nerve fibres from both sensory and motor neurons. The sensory fibres enter the spinal cord exclusively via the dorsal root. This is because the sensory fibres arise in early development as outgrowths from the neural crest material. In the adult, their cell bodies are grouped separately from the CNS forming a swelling in each dorsal root called the **dorsal root ganglion**.

All the motor fibres leave the spinal cord via the ventral root. This has no swelling because the motor neurons develop from cells located in the neural tube of the embryo. Thus, in the adult spinal cord, the cell bodies of the motor neurons lie in a central butterfly-shaped region called the **grey matter**.

A cross section of the human spinal cord is shown in Figure 23·4. Within the grey matter, the motor neurons are restricted to the **ventral horn** and **lateral horn** regions while numerous interneurons are present in the **pars intermedia** and **dorsal horn** regions. Sensory fibres enter the dorsal horn. The dendrites and cell bodies of all these neurons lack myelin sheaths and are exposed for synaptic contact. Consequently, virtually all the synaptic connections in the spinal cord occur in the grey matter.

Surrounding the grey matter is a region of **white matter** containing numerous long nerve fibres. Many of the fibres are insulated by myelin sheaths and these are responsible for the whitish appearance of the region. **Ascending** fibres carry sensory impulses towards the brain, while **descending** fibres synapse with motor neurons to produce appropriate motor responses. As indicated in Figure 23·4, the ascending and descending fibres are organised into **tracts**, each tract having a definite origin and destination. This highly ordered structure contributes to an efficient routing of information which keeps individual communication distances as short and fast as possible.

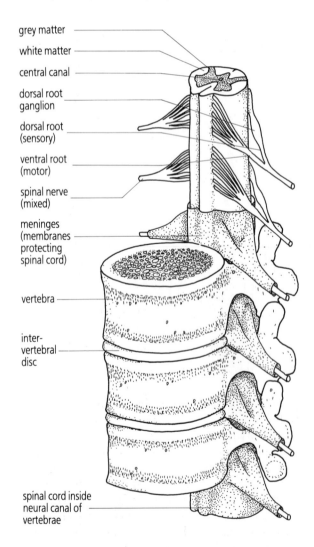

grey matter
white matter
central canal
dorsal root ganglion
dorsal root (sensory)
ventral root (motor)
spinal nerve (mixed)
meninges (membranes protecting spinal cord)
vertebra
inter-vertebral disc
spinal cord inside neural canal of vertebrae

Fig 23·3 Spinal cord and vertebral column

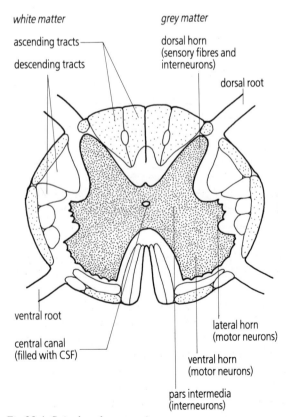

white matter
ascending tracts
descending tracts
grey matter
dorsal horn (sensory fibres and interneurons)
dorsal root
ventral root
central canal (filled with CSF)
lateral horn (motor neurons)
ventral horn (motor neurons)
pars intermedia (interneurons)

Fig 23·4 Spinal cord cross section

23·4　Spinal reflexes ■ ■ ■ ■ ■ ■ ■ ■ ■ ■ ■ ■ ■ ■ ■

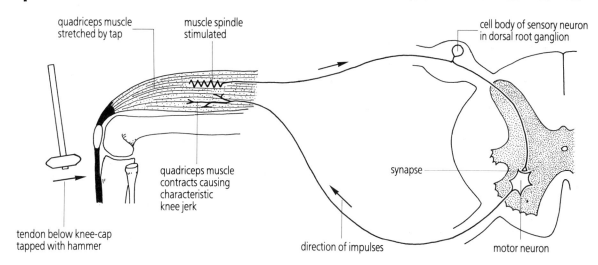

Fig 23·5 *Monosynaptic reflex: knee jerk reflex*

In the CNS of vertebrates, simple 'decisions' are made at the level of the spinal cord, while more complex 'decisions' are referred to the brain. The simplest responses are rapid and automatic and are termed **reflexes**.

One of the most familiar examples of reflex action is the 'knee jerk' reflex, or **stretch reflex**, produced when the tendon below the knee-cap is tapped. As you can see from Figure 23·5, the sense organs for this response are **muscle spindles** embedded in the large **quadriceps** muscle of the thigh. These consist of sensory nerves attached to small bundles of modified muscle fibres enclosed by connective tissue. They respond to stretching of the muscle. Tapping the tendon stretches the quadriceps muscle and stimulates the muscle spindles, producing a burst of sensory impulses which travel towards the spinal cord. Here, the sensory neurons synapse directly with the appropriate motor neurons so that, after a short delay, the quadriceps muscle contracts to give the characteristic knee jerk. Since the pathway of the reflex involves only a single set of synapses in the spinal cord, stretch reflexes are said to be **monosynaptic**.

The importance of stretch reflexes is that body muscles automatically adjust to the load placed upon them. Thus, a person jumping to the ground can avoid collapsing in a heap because his stretch reflexes create the necessary tension in the leg muscles. Stretch reflexes also play an essential part in controlling body movements and in maintaining basic muscle tone.

A second type of spinal reflex, called a **withdrawal reflex**, is illustrated in Figure 23·6, which shows what happens when a person steps onto a pin. The pathway normally involves one or more interneurons so that there may be several synapses in the spinal cord. Withdrawal reflexes are therefore said to be **polysynaptic**. In general, when any part of the foot or hand is exposed to a

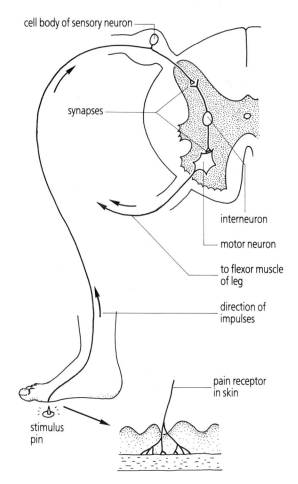

Fig 23·6 *Polysynaptic reflex: withdrawal reflex*

painful stimulus, the whole limb is rapidly withdrawn. Responses of this type have an obvious protective function.

Although the pathways of the reflexes described above can be represented as chains of individual cells, you should avoid thinking about reflexes in an over-simplified way. For example, impulses from a single muscle spindle arriving at a single synapse would scarcely affect an individual motor neuron, which receives up to 6000 synapses from various different sources. Similarly, each motor neuron connects to only a few muscle fibres so that, if only a single motor neuron was active, the contraction produced would be negligible. Reflex action is possible because there are large numbers of neurons in parallel and overlapping pathways. Thus, stretching a tendon stimulates many muscle spindles and generates impulses in many sensory neurons. Hundreds of motor neurons are active in producing the resulting muscle contraction.

Cranial reflexes include coughing, blinking and swallowing. These are very similar to spinal reflexes but have pathways which involve cranial nerves and the brain rather than spinal nerves and the spinal cord.

23·5 Structure and function of the brain ▪ ▪ ▪ ▪ ▪ ▪ ▪ ▪ ▪ ▪

The brain, like the spinal cord, is a hollow structure. During embryological development, the anterior end of the neural tube swells to form the **primary brain vesicle** (see Fig. 23·7). This structure then develops into three distinct regions, the **forebrain**, **midbrain**, and **hindbrain**. The enlarged cavities of these regions are called **ventricles**. They contain **cerebrospinal fluid (CSF)** and are continuous with the central canal of the spinal cord.

In lower vertebrates, such as fish, the three regions remain recognisable in the adult and each receives input from a principal sense organ. The forebrain is concerned with the sense of smell, the midbrain with vision, and the hindbrain with hearing, balance, and taste. The degree of development of each region corresponds to the importance of that sense in the life of the animal. For example, the dogfish depends largely on smell for finding food and has a large forebrain, but the trout depends more on vision and has a relatively larger midbrain.

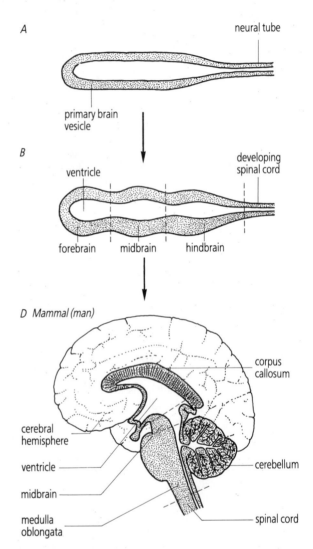

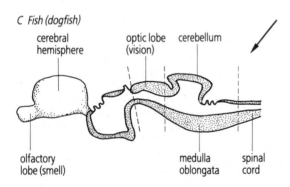

Fig 23·7 Development of the brain in different vertebrates

In the brains of mammals, this simple pattern is obscured because the forebrain is greatly enlarged to form the two **cerebral hemispheres**. These structures have a well developed surface layer of grey matter called the **cerebral cortex**. The cortex contains millions of neuron cell bodies which have migrated outwards from the centre of the neural tube. As indicated in Figure 23·8, the remaining cell bodies are located deep in the brain where they are grouped together in structures called **ganglia** or **nuclei**. Each nucleus acts as a relay station, gathering information from several sources and sending out impulses to some new destination. Between the cortex and the deep nuclei there are massive tracts of white matter carrying impulses from one point to another.

The structure of the human brain is illustrated in Figures 23·8 and 23·9 and the functions of the main regions are described below.

1 Cerebrum. This is the largest region of the brain in mammals and contains up to 90% of the neurons present in the nervous system as a whole. The cerebrum is subdivided to form two cerebral hemispheres linked by the **corpus callosum**, a large tract of white matter. As you can see, the

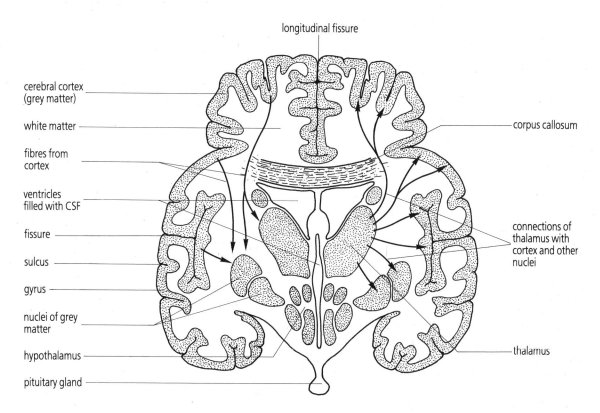

Fig 23·8 *Human brain: transverse section in the forebrain region*

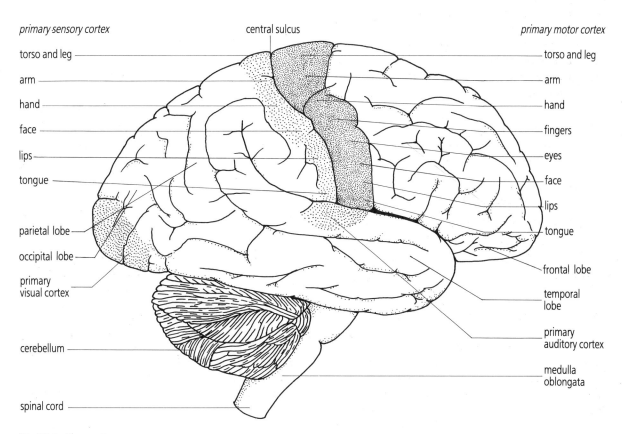

Fig 23·9 *Human brain: external features*

surface of the hemispheres is greatly folded, or **convoluted**. This enables the neuron cells bodies to lie near the surface of the brain where they can exchange materials with the cerebrospinal fluid.

Upfolds of the cortex are called **gyri** (singular **gyrus**), while shallow down folds are called **sulci** (singular **sulcus**). In several places deep **fissures** divide each hemisphere into distinct **lobes**.

Some areas of the cortex have fixed functions. The **primary motor cortex**, which lies immediately in front of the **central sulcus**, contains the cells controlling motor activity. These are arranged in an orderly way so that they form a 'map' of the body muscles on the surface of the cortex. Electrical stimulation at any point causes the corresponding body movement. For reasons which are not understood, motor instructions from the primary motor cortex are routed to the opposite side of the body via the spinal cord. Thus, the left hand side of the brain controls voluntary movements of the right hand side of the body and vice versa.

Immediately behind the central sulcus is the **primary sensory cortex** where incoming sensory signals from the opposite side of the body are 'mapped' in the same way. Electrical stimulation in these areas produces the sensation of touch or pain in the corresponding body region. Separate regions of the cortex receive sensory signals concerned with vision (**primary visual cortex**) and with hearing (**primary auditory cortex**).

The remainder of the cerebral cortex is called the **association cortex** and shows no clear cut links with distinct functions, although there can be localisation of a general sort. For example, electrical stimulation at any point in the **speech areas** of a person's brain does not produce recognisable sounds but speech may be prevented if these areas are damaged or surgically removed. Signals from many sources reach the association cortex and it seems that the ways in which these signals are processed give rise to many aspects of intelligent behaviour including conscious thought, learning, and memory.

The two cerebral hemispheres parallel each other in many of these functions but there are a number of intriguing differences. In right-handed people, the left hemisphere contains all the major speech areas and is said to be **language dominant**. This hemisphere also controls the dominant right hand. The right hemisphere is specialised for other functions including recognition of complex visual patterns. Thus, the two hemispheres tend to complement each other in their complex functions. (In a left-handed person, these functions are switched and the right hemisphere is language dominant and so on.)

2 Thalamus. The thalamus is the posterior part of the forebrain in higher vertebrates. It processes signals received from the sense organs and relays sensory information to the cortex. In conjunction with the **reticular formation**, an indistinct region of grey matter located in the brainstem, the thalamus helps to regulate the 'waking state' of the organism. When the reticular formation is active, the person is alert and wakeful. Conversely, if the reticular formation is suppressed, the person sleeps. The onset of sleep is thought to be caused by the chemical substance **serotonin** produced by **sleep centres**, also located in the brainstem.

3 Hypothalamus. The hypothalamus lies below the thalamus and forms the floor of the forebrain. It is one of the most important brain regions and has three main functions:
(i) The hypothalamus provides a link between the nervous system and the endocrine system. It produces a variety of hormones and other substances called **releasing factors**, which act on the pituitary gland, causing secretion of the appropriate pituitary hormones.
(ii) The hypothalamus is a major control centre for homeostasis. It is involved in regulating the osmotic balance of the body fluids (Unit 20), and in regulating body temperature (Unit 21). It contains the brain centres which control hunger and thirst.
(iii) Through its connections with other parts of the brain, the hypothalamus helps to control the **autonomic nervous system** (described later in this Unit). This regulates a wide range of involuntary activities including sweating, vasoconstriction, vasodilation, heart rate, dilation of the pupils, and shivering.

4 Midbrain. In lower vertebrates, the midbrain usually contains the **optic lobes**, which are the main brain area for visual processing. In mammals, this function is transferred to the cerebral hemispheres, so that the midbrain acts mainly as a relay centre for visual information.

5 Cerebellum. The cerebellum consists of a pair of swellings from the dorsal surface of the hindbrain. It is responsible for **balance** and **muscular coordination** in active movements as well as standing. In mammals, it has a well developed cortex layer containing numerous **Purkyne cells**, each of which forms as many as 100 000 synapses with other neurons. Input to the cerebellum comes from the parts of the inner ear concerned with balance, from muscle spindles in the body muscles, and from other parts of the brain. Voluntary movements are initiated in the cerebral hemispheres but the cerebellum produces the detailed instructions for the execution of each movement. It simultaneously coordinates activity in the other muscles, so that the body as a whole remains in balance. Output from the cerebellum passes to the motor neurons via the descending tracts of the spinal cord.

6 Medulla oblongata. The medulla oblongata, or **medulla**, forms a transition region between the brain and spinal cord. It contains the main nerve centres controlling breathing movements and blood pressure.

23·6 Autonomic nervous system ■ ■ ■ ■ ■ ■ ■ ■ ■ ■ ■

The autonomic nervous system controls many of the subconscious activities of the body, such as digestion and sweating. It consists almost entirely of efferent neurons, that is, neurons carrying impulses away from the CNS. As its name suggests, the autonomic nervous system coordinates many of its own activities, and is only partly under central control. Its effectors are the **visceral muscles** (also called **smooth, non-striated**, or **involuntary** muscles) and **glands**.

As indicated in Figure 23·10, the autonomic nervous system is divided into separate **sympathetic** and **parasympathetic** systems. These systems have opposite effects on the organs

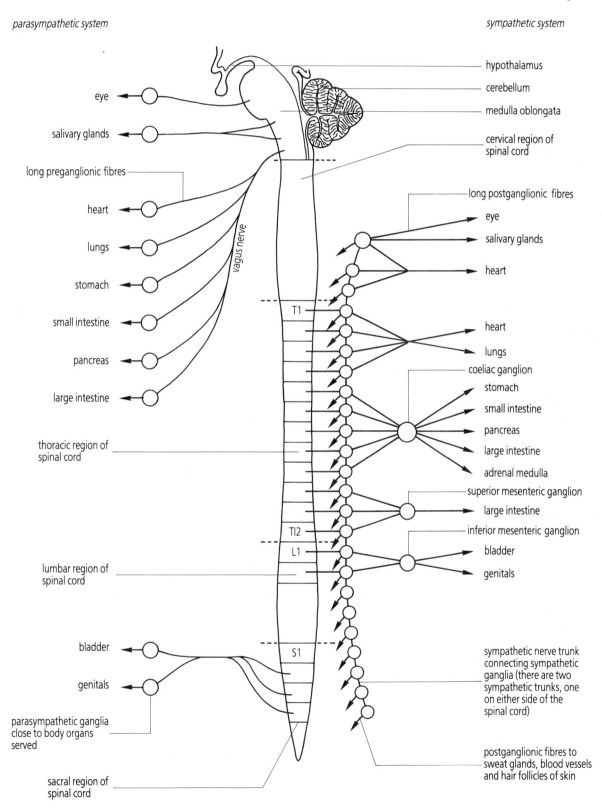

parasympathetic system

sympathetic system

eye

salivary glands

long preganglionic fibres

heart

lungs

stomach

small intestine

pancreas

large intestine

thoracic region of spinal cord

lumbar region of spinal cord

bladder

genitals

parasympathetic ganglia close to body organs served

sacral region of spinal cord

hypothalamus

cerebellum

medulla oblongata

cervical region of spinal cord

long postganglionic fibres

eye

salivary glands

heart

heart

lungs

coeliac ganglion

stomach

small intestine

pancreas

large intestine

adrenal medulla

superior mesenteric ganglion

large intestine

inferior mesenteric ganglion

bladder

genitals

sympathetic nerve trunk connecting sympathetic ganglia (there are two sympathetic trunks, one on either side of the spinal cord)

postganglionic fibres to sweat glands, blood vessels and hair follicles of skin

vagus nerve

T1

TI2

L1

S1

Fig 23·10 Autonomic nervous system

Table 23·1 Effects of sympathetic and parasympathetic nervous systems

	SYMPATHETIC	PARASYMPATHETIC
Action	prepares body for stress	reduces stress
Final transmitter substance	noradrenalin	acetylcholine
Effect on body organs		
Heart	accelerates heartbeat	slows heartbeat
Lungs	dilates bronchioles	constricts bronchioles
Intestines	decreased peristalsis	increased peristalsis
Blood vessels of alimentary canal	constriction	dilation
Salivary glands	reduced secretion	increased secretion
Adrenal medulla	secretion of adrenalin and noradrenalin	no effect
Blood vessels of skeletal muscles	dilation	no effect
Blood vessels of skin	constriction	no effect
Arrector pili muscles	contraction	no effect
Sweat glands	increased secretion	no effect

they serve (see Table 23·1). In general, the sympathetic nervous system prepares the body for stress, while the parasympathetic system reduces stress.

The main ganglia of the sympathetic nervous system are connected to each other in long chains, called **sympathetic nerve trunks**, which run parallel to the spinal cord on either side (only one of the sympathetic nerve trunks is shown in Fig. 23·10). The **preganglionic** fibres from the CNS to the sympathetic ganglia are short, while the **postganglionic** fibres connecting to the body organs are normally long. The **coeliac ganglion** receives input via the sympathetic ganglia on both sides of the spinal cord but is itself unpaired. **Acetylcholine** is the chemical transmitter used at

synapses within the sympathetic nervous system but **noradrenalin** is used at the terminal synapses with the effector cells. Because of this, the effect of sympathetic nerve activity is increased by the hormones **adrenalin** and **noradrenalin**, which circulate in the bloodstream.

The parasympathetic ganglia lie close to or inside the body organs they serve so that the preganglionic fibres are long and the postganglionic fibres very short. Acetylcholine is the chemical transmitter at all synapses, including those with effector cells. The system has central connections in the brainstem and at the base of the spinal cord. The **vagus nerve** is an important part of the system and connects to the heart, lungs, and other body organs.

23·7 Protection of the nervous system ■ ■ ■ ■ ■ ■ ■ ■ ■ ■

The CNS in mammals is protected in several ways. First, there are the bones of the skull and vertebral column. These provide a solid outer casing resistant to all but the most extreme mechanical forces. Within its bony casing, as illustrated in Figure 23·11, the brain is further protected by a series of three membranes called **meninges**. (Inflammation of these membranes gives rise to the symptoms of the disease **meningitis**.)

The outermost membrane, called the **dura mater**, forms a tough but loose connective tissue sheath enclosing the whole of the CNS. The dura mater is separated from the **arachnoid** membrane by a very narrow **subdural space**. The larger blood vessels supplying the brain are attached to the arachnoid and, below this, there is a larger **subarachnoid space** filled with cerebrospinal fluid. This acts as a shock-absorbing layer, distributing sharp forces over a wider area. The

third membrane is the **pia mater**. It contains numerous small blood vessels which supply the underlying nervous tissue.

Interestingly, the nervous system often requires metabolic protection as well as mechanical protection. For example, although the human brain accounts for more than 20% of the body's consumption of glucose and oxygen, it cannot store these substances. Furthermore, it is unable to tolerate high concentrations of other chemicals normally present in blood plasma. Accordingly, the passage of substances to the brain is regulated by the **blood–brain barrier**, as illustrated in Figure 23·12. This shows the cross section of a blood capillary passing through brain tissue. As you can see, the endothelial cells overlap so that the walls of the capillary lack the characteristic 'pores' found in other tissues. In addition brain capillaries are surrounded by processes ('feet') which arise from glial cells called **astrocytes**; up to 85% of the total

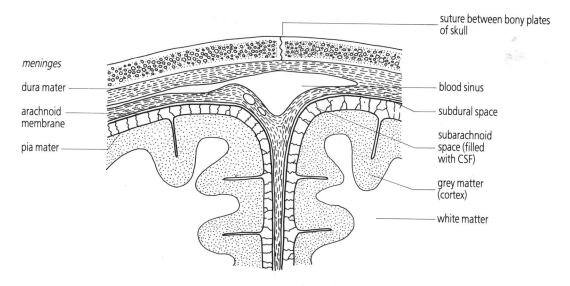

Fig 23·11 Protection of the nervous system: meninges of the brain

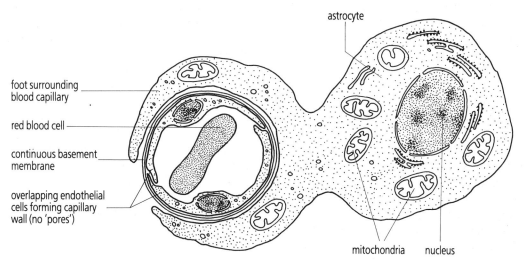

Fig 23·12 Anatomical basis of the blood-brain barrier

capillary surface may be covered in this way. The blood-brain barrier prevents or reduces the transfer of certain substances not involved in brain metabolism, including some proteins, some lipids, urea, and antibiotics. On the other hand, substances needed by the brain, such as glucose, oxygen, sodium ions, and potassium ions, pass through very rapidly.

24 SENSE ORGANS

Objectives

After studying this Unit you should be able to:–

- List the different types of receptor cells and describe the structure of a simple sense organ (Pacinian corpuscle)

- Draw and label a diagram to show the structure of the human eye, as seen in vertical section

- Describe the pathway of light in the human eye and explain the process of accommodation which focuses light from objects at different distances

- Describe the detailed structure of the human retina and comment on the functions of the different cell types present

- Explain how blue-, green-, and red-sensitive cone cells interact to produce colour vision

- Draw and label a diagram to show the important structures of the human ear and outline the functions of the outer ear, middle ear, and inner ear

- Give a detailed description of the structure of the cochlea and explain how it is thought to function in distinguishing sounds of different frequencies

- Describe the structure of the utricle, saccule, and semi-circular canals, and explain their roles in maintaining body equilibrium

24·1 Receptor cells ■ ■ ■ ■ ■ ■ ■ ■ ■ ■ ■ ■ ■ ■ ■ ■

To obtain information from its surroundings, the nervous system relies on many different kinds of **sensory cells**, or **receptors**. Each receptor responds to one particular type of environmental change, called a **stimulus**, and produces a corresponding pattern of nerve impulses which passes to the CNS.

The different types of receptor cells are named according to the stimulus for which they are specialised. **Photoreceptors** are stimulated by light, **chemoreceptors** by chemicals, and **thermoreceptors** by changes in temperature. **Mechanoreceptors** respond to physical stimuli such as touch, pressure, stretching, or vibration. Some of these receptors like the sensory cells of the eyes, ears, and the skin are located near the outside surface of the organism where they monitor changes in the external environment. Collectively, they are called **exteroceptors**. The

internal receptors, or **interoceptors**, respond to internal stimuli such as blood pressure and changes in the chemical composition of blood plasma. They play a vital role in the nervous control of homeostasis. **Proprioceptors** include muscle spindles and the parts of the inner ear concerned with balance; they provide information relating to body position and movement.

Figure 24·1 shows the structure of a **Pacinian corpuscle** which acts as a touch-pressure receptor in the skin. The sensitive part of the corpuscle is located at its core and is the modified terminal of a sensory neuron. Rapid changes in pressure are transmitted across the membranes of the capsule into the core, causing an increased permeability to sodium ions (Na^+). These flow inwards to give a depolarisation called the **receptor potential**, or **generator potential**. If the depolarisation exceeds threshold then nerve impulses are initiated. Since

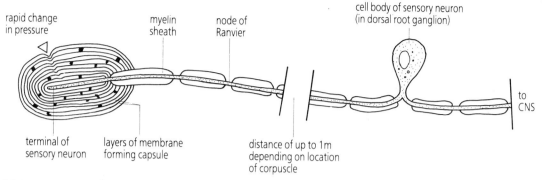

Fig 24·1 Pacinian corpuscle

the sensory neuron is stimulated directly, it is called a **primary receptor cell**. Slow changes in pressure or sustained pressures do not produce a response.

A Pacinian corpuscle together with its associated structures forms a simple **sense organ**. Complex sense organs contain many receptor cells grouped together. Sometimes, these are modified epithelial cells instead of sensory neurons and, in this case, they are called **secondary receptor cells**. The sensory hair cells of the ear are examples of secondary receptor cells. The remainder of this Unit discusses the structures of the human eye and ear in detail.

24·2 The human eye ■ ■ ■ ■ ■ ■ ■ ■ ■ ■ ■ ■ ■ ■

Structure

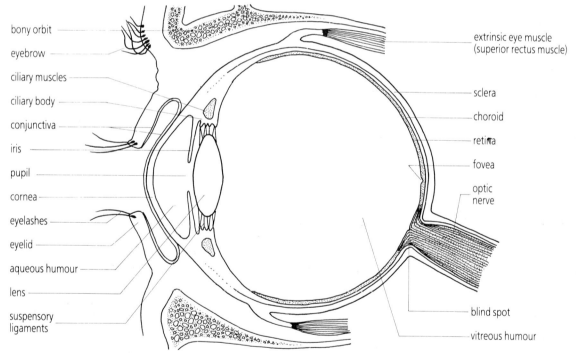

Fig 24·2 Structure of the human eye (vertical section)

An adult eyeball is about 2·5 cm in diameter and has the structure illustrated in Figure 24·2. Each eye is located in a bony socket, or **orbit**, formed in the skull. The eyeball is attached to the orbit and held in position by six **extrinsic eye muscles**, which allow the eye to be moved. The movements of the two eyes are coordinated for efficient binocular vision.

As you can see, the front surface of the eye is protected by the **eyelids** and **eyelashes**. The eyelids close periodically by the reflex action of 'blinking', and also in response to sudden bright light, air currents, or when the eyelashes are touched. The front surface of the eye, under the eyelids, is a thin transparent layer called the **conjunctiva**. This is kept moist by fluid secreted from the **lachrymal glands** (tear glands) which lie above and to the outside of each eye. The fluid contains the enzyme **lysozyme** which kills bacteria. After passing over the conjunctiva, it drains from the eyes into the nasal cavities through a system of **lachrymal canals**. Production of fluid increases when the eye is irritated by dirt particles or infection, or when crying.

The eyeball itself has a three-layered structure. The outermost layer, or **sclera**, is a tough connective tissue sheath which is whitish and opaque except towards the front where it merges with the transparent **cornea**. The middle layer, or **choroid**, contains numerous blood vessels and joins with the **ciliary body**. This structure, together with the **lens**, which is held in place by the **suspensory ligaments**, divides the interior of the eye into two compartments. The smaller front compartment is filled with **aqueous humour** secreted from the ciliary body. This liquid is similar to blood plasma except that it contains only a trace of protein and no antibodies. As a result, there is no immunological response in the interior of the eye, reducing the possibility of any interference with the passage of light. Aqueous humour supplies oxygen and nutrients to the cornea and lens, and removes waste products. It is under pressure and thus helps to maintain the shape of the eye. Eventually, it drains into the veins of the eyeball. Blockage of the drainage channels raises the internal pressure and gives rise to symptoms of the disease **glaucoma**.

The rear compartment of the eye is lined by a third layer, the **retina**. This is the light-sensitive layer of the eye and its detailed structure will be explained shortly. The cavity of the rear compartment is filled by jelly-like **vitreous humour** containing about 99% water, some salts, and **hyaluronic acid**, which forms the gel.

In nocturnal and deep sea vertebrates, the choroid layer is modified to form a reflective surface, the **tapetum**, just beneath the retina. This improves visual sensitivity by reflecting light back into the eye, a property which explains why cats' eyes shine in the beam of light from the headlamps of a car.

Pathway of light

Light is one form of **electromagnetic radiation** and, as such, it is described in terms of wavelengths and frequencies (see Fig. 24·3). The **wavelength** (measured in metres) is the distance between two successive peaks, while the **frequency** (measured in Hertz) is the number of peaks, or cycles, completed per second. The wavelength of visible light is between 400 and 700 nanometres (1 nm = 10^{-9} metres) and, within this range, different wavelengths produce different colour sensations.

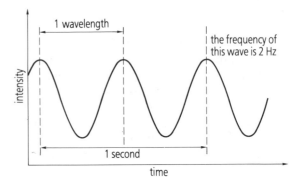

Fig 24·3 Electromagnetic radiation

Light waves normally travel in a straight line but they can be bent, or **refracted**, when passing from one medium to another, for example, from air to water. The amount of bending depends on the optical densities of the two media and also upon the angle at which light strikes the refractive surface. Different shapes of lenses produce different effects. A **convex** lens (Fig. 24·4*A*) causes light rays to **converge**, bringing them to a sharp focus. Fatter, more rounded lenses have more converging power than thin flat lenses. On the other hand, a **concave** lens (Fig. 24·4*B*) causes light rays to **diverge**, scattering light rays without bringing them to a focus.

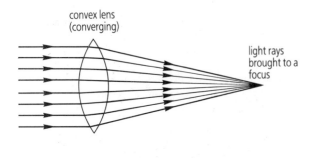

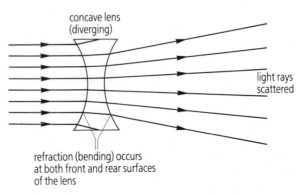

Fig 24·4 Action of convex (converging) and concave (diverging) lenses

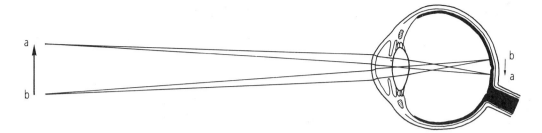

Fig 24·5 Pathway of light in the human eye

The optical system of the human eye focuses light from the surroundings to produce an inverted image on the retina, and in particular upon the **fovea** (Fig. 24·5). As you can see, refraction takes place at the front and rear surfaces of the cornea, and also at the front and rear of the lens. The cornea plays the greater part in focusing the light rays, but the shape of the lens can be changed to allow nearby or distant objects to be viewed.

The process of changing the focus of the eye is called **accommodation**, and takes place as follows. The ciliary body contains the **ciliary muscles** which form a ring. Because the contents of the eyeball are under pressure, the ciliary body is pushed outwards when the ciliary muscles are relaxed. As a result, the lens is pulled into a flat shape by the suspensory ligaments, causing light from distant objects to be focused on the retina. Nearby objects are focused when the ciliary muscles contract. Such contraction decreases the diameter of the muscle ring and pulls the ciliary body inwards, thereby reducing the tension on the suspensory ligaments. Consequently, the elastic lens is allowed to become more rounded and its converging power increases. In the normal eye, objects from about 10 cm to the far distance can be focused correctly.

Various visual defects arise when the elements of the optical system are not correctly matched to each other. Longsightedness, or **hypermetropia**, occurs when the eyeball is too *short* in relation to the converging power of the lens. Distant objects are focused correctly but, even in its maximum rounded shape, the lens is too weak to focus nearby objects. An additional converging lens in spectacles or contact lenses can correct this defect. Shortsightedness, or **myopia**, occurs when the eyeball is too *long* in relation to the lens. Nearby objects are focused correctly, but, even when the ciliary muscles are fully relaxed, the lens is not flat enough for distant objects to be seen correctly. This defect is corrected by a diverging lens.

In elderly people, the cells towards the centre of the lens die and the lens becomes less flexible so that accommodation over the whole range is difficult. This condition is called presbyopia and sometimes requires bifocal spectacles, comprising both converging and diverging lenses for each eye. **Cataracts** occur when the cells of the lens become

opaque; depending on the severity of this condition, the lens may have to be surgically removed.

The amount of light entering the eye is controlled by the **iris**, a pigmented ring located at the front of the lens with the **pupil** at its centre. The iris contains circular and radial muscle fibres which constrict or dilate the pupil according to conditions. A sudden increase in light intensity causes a rapid constriction (**pupillary reflex**), while in dim light, the pupil is dilated.

Retina

The detailed structure of the retina is outlined in Figure 24·6. It has a layered arrangement comprising several different cell types:

1 Photoreceptors. The light-sensitive cells of the retina are the **rods** and **cones**. Surprisingly, these are located farthest away from the surface of the retina so that light must travel *through* a surface network of blood vessels and all the other layers of the retina before being detected. This

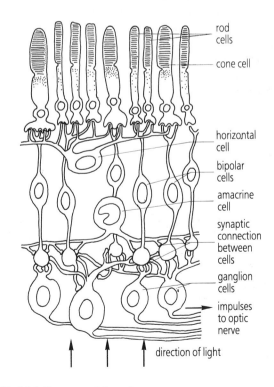

Fig 24·6 Structure of the retina

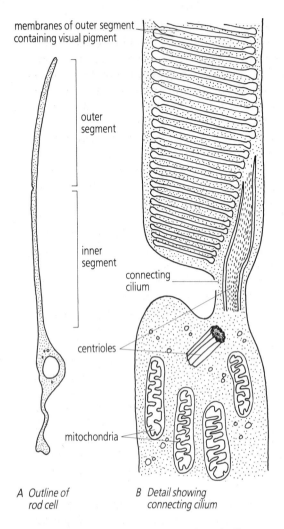

membranes of outer segment
containing visual pigment

outer
segment

inner
segment

connecting
cilium

centrioles

mitochondria

A Outline of
rod cell

B Detail showing
connecting cilium

Fig 24·7 Electronmicroscope structure of a rod cell

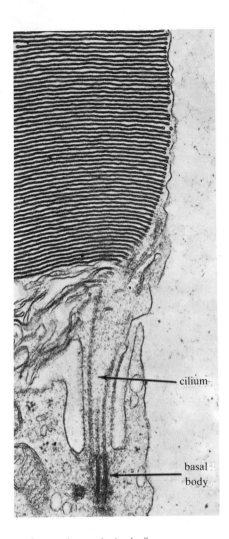

cilium

basal
body

C Electronmicrograph of rod cell

situation, which occurs in all vertebrates, is a
consequence of the way in which the eye is formed
as an outgrowth of the brain in the embryo.

Rods. The retina contains about 120 million
rods. They are sensitive to low levels of
illumination but not to colour and are used for
scotopic, or night vision. Rods are present in all
parts of the retina except the fovea.

The structure of an individual rod cell is shown
in Figure 24·7. The **inner segment** of the rod cell
is packed with mitochondria, while the **outer
segment** contains numerous membrane-lined discs
formed by invagination from the cell membrane.
Millions of molecules of the photopigment
rhodopsin are attached to the membranes, making
them light sensitive. Each molecule of rhodopsin
consists of a protein part, the **opsin**, connected to
a non-protein part, called the **chromophore**. In
rhodopsin, the chromophore is **retinal**, a derivative
of vitamin A. When light energy is absorbed by
this part of the molecule, it changes shape, making
the rhodopsin molecule unstable so that it splits
into separate opsin and chromophore parts. This
triggers a change in membrane permeability,
causing an electrical signal, the **generator
potential**. Signals generated in this way are

transmitted by synapses to **bipolar** and **horizontal**
cells (refer to Fig. 24·6).

Cones. The cones are responsible for colour
vision: they work best in bright light, giving
photopic, or daylight vision. The retina contains
about 5 million cones, 2000 of which are packed
together in the **fovea**. In this region, there are no
blood vessels and the surface layers of the retina
are particularly thin. Furthermore, the individual
cone cells are extremely small with a diameter of
as little as 1 μm, so that this area provides the
best possible definition for good discrimination of

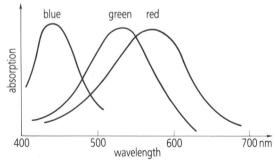

*Fig 24·8 Graphs to show the sensitivity of the three different
photopigments found in human cone cells*

192

detail. Objects from the centre of the visual field are focused on the fovea.

There are three types of cone cells containing photopigments sensitive to different wavelengths of light corresponding to the colours **blue** (445 nm), **green** (535 nm), and **red** (570 nm) – see Figure 24·8. The chromophore part of each photopigment is retinal as in the case of rhodopsin, but the opsin proteins are different so as to modify the overall response.

You should note that each of the cone types has a broad response so that their ranges overlap. This is very important because it allows the eye to discriminate other colours by comparing the activity of the different cone types. For example, the wavelength 550 nm stimulates green-sensitive and red-sensitive cones equally and is perceived as the colour yellow. Other wavelengths produce different colour sensations in a similar way. An interesting consequence of this system for detecting colours is that the sensation of *any* particular wavelength of light can be produced by stimulating the eye using an appropriate mixture of blue, green, and red light, as outlined in Figure 24·9. An equal mixture of green and red lights produces the sensation of yellow, while a mixture of blue and red produces magenta. Mixing blue, green, and red in equal proportions gives white.

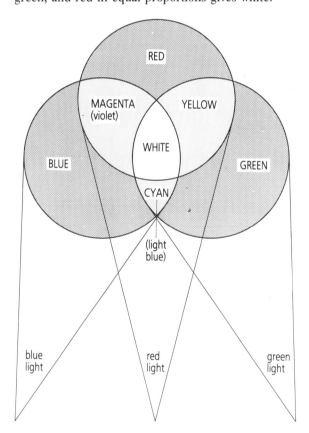

Fig 24·9 Colour mixing using red, blue, and green light

Other colours, such as orange, are produced by unequal mixtures of light.

Certain colours, including brown, and metallic colours like gold and silver, cannot be produced by simple mixtures of light. However, they can be represented in photographic transparencies and in television pictures which use only blue, green, and red. The details of the perception of these colours are not fully understood but must depend on the other cells of the retina which process the input received from the rods and cones.

The different kinds of **colour blindness** are explained by an absence or deficiency of one or other of the photopigments for the different cone types. For example, red-green colour blindness can be caused by a lack of red-sensitive cones (**protanopia**), or by a lack of green-sensitive cones (**deuteranopia**). For reasons to be explained in Unit 38, red-green colour blindness is much more common in males than in females.

2 Bipolar cells. The bipolar cells are neurons which gather information from the photoreceptors and transmit it to the next cell layer. Rod bipolar cells normally receive input from several rods, making it more likely that objects in dim light will be perceived. On the other hand, the cone bipolar cells are connected to just one or two cones so that good definition is preserved. This arrangement helps to explain why night vision is slightly 'fuzzy', while objects seen in daylight are sharply defined.

3 Horizontal and amacrine cells. These cells form lateral connections in the retina and help to coordinate the activity of the bipolar and ganglion cells.

4 Ganglion cells. These neurons have more complex connections than bipolar cells. Typically, they have circular **receptive fields**, that is, they receive information from a circular group of photoreceptors. Experimental studies using microelectrodes show that light stimuli round the edges of the receptive field can have different effects on the ganglion cell than stimuli directed at its centre. For example, some ganglion cells have an **'on-centre off-surround'** organization. In other words, stimuli round the edges of the receptive field reduce the rate of firing of the ganglion cell, while stimuli at the centre of the

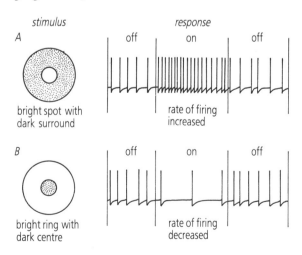

Fig 24·10 Receptive field of an 'on-centre off-surround' gangion cell

field increase activity (Fig 24·10). Other ganglion cells have the opposite **'off-centre on-surround'** organization. These types of cell respond to changes in the intensity of light without regard to colour, but there are ganglion cells which respond to colour. Most of these have receptive fields in which light of one colour causes increased activity, while light of another contrasting colour inhibits activity. Before any signals leave the retina, a significant amount of visual processing has taken place in the ganglion cells.

There are about 1 million ganglion cells and their axons travel over the surface of the retina towards the **blind spot** where they pass through, forming the **optic nerve** which carries information to the brain. You can confirm the existence of your own blind spot with the aid of Figure 24·11. With the left eye closed, hold the book at about 30 cms with the cross symbol directly ahead of the right eye. Keeping your right eye focused on the cross, move the page gradually towards you until the round symbol 'disappears'. When this happens, the image of the round symbol falls directly upon the blind spot. You may have to repeat the procedure to observe the effect.

Fig 24·11 Finding the blind spot

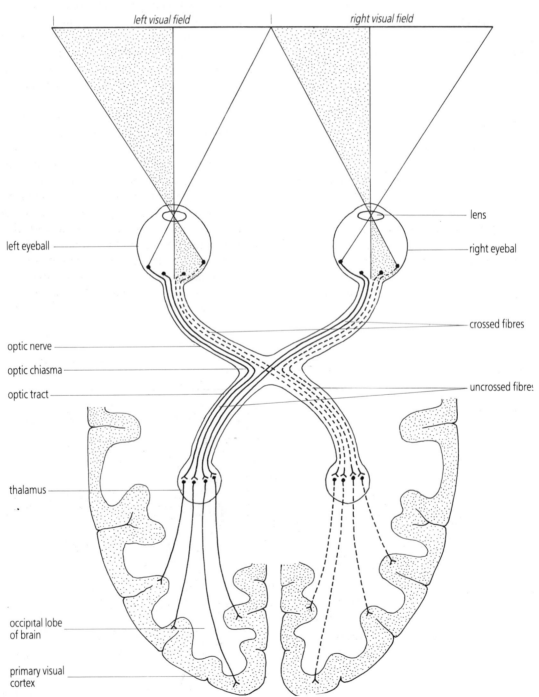

Fig 24·12 The visual pathway

The visual pathway to the brain is illustrated in Figure 24·12. At the **optic chiasma**, a proportion of the fibres from each optic nerve cross to the opposite side so that each half of the brain receives information from both eyes. The fibres in the **optic tract** synapse in the **thalamus**, and from here new fibres radiate to the **visual cortex** located on the posterior surface of the cerebral hemispheres. The right visual cortex receives signals from the right side of each eye, corresponding to the left half of each visual field, while the left visual cortex receives signals from the right half of each visual field.

At each level, further processing of visual information takes place. Most of the neurons in the thalamus have circular receptive fields like those of the ganglion cells, but their responses are more complex. In the visual cortex, individual cells respond to edges, lines, and borders, or to complex shapes. The interactions of these cells are responsible for all visual experience.

24·3 The human ear ■ ■ ■ ■ ■ ■ ■ ■ ■ ■ ■ ■ ■ ■ ■

The structure of the human ear is illustrated in Figure 24·13 and is best explained in three sections, namely, the **outer ear**, the **middle ear**, and the **inner ear**.

1 Outer ear. The outer ear consists of the **pinna** and the external ear tube, or **auditory meatus**, leading to the **tympanum** ('eardrum'). The pinna of the human ear plays only a minor role in channelling sound waves towards the tympanum, but, in many other mammals, the pinnae are relatively larger and form funnels which can be directed towards a sound source for maximum sensitivity. The opening of the meatus is protected by hairs which help to prevent the entry of dirt particles, while modified sebaceous glands in the walls of the tube produce ear wax, or **cerumen**. This traps dirt and bacteria so helping to keep the ear clean and free from infection. Sound waves, which are pressure waves transmitted in the air, cause the tympanum to vibrate.

2 Middle ear. This is an air-filled cavity connected to the pharynx via the **auditory tube (Eustachian tube)**. When swallowing, the entrance of the tube is opened, thereby equalising the pressures on either side of the tympanum, and allowing it to vibrate freely. The function of the middle ear is to transmit these vibrations to the inner ear by means of the **oval window**.

This transmission represents a mechanical problem for the fluid of the inner ear is much denser than air so that far more energy is required to make it vibrate. The difficulty is overcome by the three **ossicles** of the middle ear, called respectively the **malleus**, **incus**, and **stapes**, or, more commonly, the hammer, anvil, and stirrup bones. The ossicles form a lever system which decreases the amplitude of the vibrations. More importantly, the surface area of the stapes in contact with the oval window is much less than the surface area of the tympanum. As a result, about 30 times as much force per unit area is transmitted

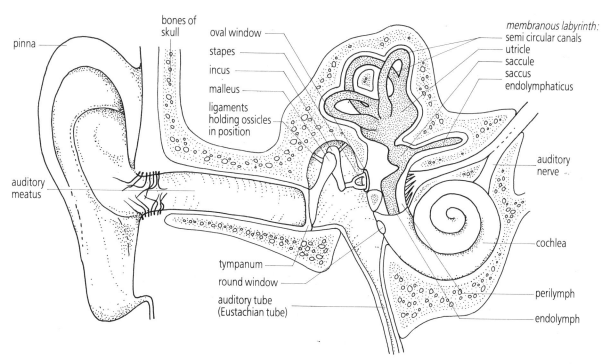

Fig 24·13 Structure of the human ear

to the inner ear compared with sound waves reaching the ear. Two tiny muscles are attached to the ossicles and act to limit their movements. The muscles contract automatically in response to sudden loud noises (**auditory reflex**), thereby protecting the inner ear from damage.

3 Inner ear. The receptors of the inner ear are located inside a system of interconnected tubes which form the **membranous labyrinth** (refer to Fig. 24·13). This consists of three **semi-circular canals** lying in different planes, two swellings, the **utricle** and **saccule**, and a coiled tube, the **cochlea**. All of these structures are filled with **endolymph** which is secreted by the walls of the labyrinth and circulates before being reabsorbed in the **saccus endolymphaticus**. The membranous labyrinth fits inside a larger cavity, the **bony labyrinth**, leaving a space which is filled with a different fluid, called **perilymph**. Perilymph is rich in Na^+ ions, while endolymph resembles intracellular fluid and is rich in K^+ ions. The membranous labyrinth is in contact with the middle ear at the **oval window**, and at the **round window**.

Hearing

The part of the inner ear concerned with hearing is the **cochlea** and its detailed structure is shown in Figure 24·14. You should examine this carefully before reading the remainder of this section.

Pressure waves received at the oval window are transmitted first to the perilymph within the **vestibular canal**, causing the **vestibular**

membrane and the endolymph inside the **cochlear duct** to vibrate in turn. Hearing depends mainly on the displacement of the **basilar membrane** which is stretched across the base of the cochlear duct. Near the oval window, the membrane is narrow and stiff but it becomes progressively wider and more elastic in the cochlear spirals. The result of this structure is that pressure waves caused by high frequency sounds produce maximum displacement near the oval window, while pressure waves corresponding to low frequency sounds produce maximum displacement near the top of the cochlea, where the membrane is wider.

These displacements stimulate the **sensory hair cells** which form part of the **organ of Corti**. Within this structure, the hair cells are fixed between the basilar membrane and a projecting shelf, the **tectorial membrane**, so that movements of the basilar membrane are detected by the hair cells and give rise to nerve impulses which travel towards the CNS via the **auditory nerve**. Complex sounds like music and human speech contain many different frequencies which are separated by the basilar membrane, producing signals which can be interpreted by the auditory cortex of the brain in a meaningful way.

Pressure waves transmitted to the vestibular canal pass to the tympanic canal either by way of the basilar membrane or by means of an interconnecting passage, the **helicotrema**, located at the apex of the cochlear spiral. Eventually, the vibrations are transmitted to the **round window** which bulges outwards into the middle ear cavity, so releasing the energy of the pressure wave.

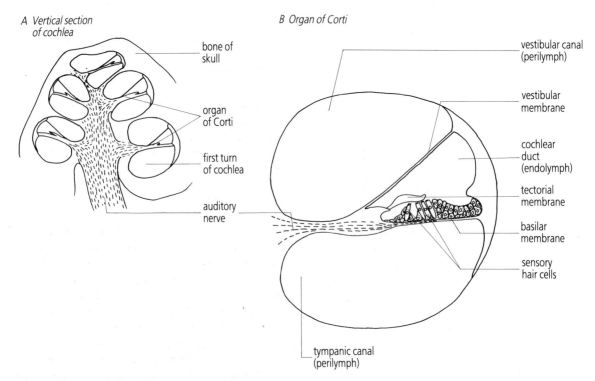

A Vertical section of cochlea

bone of skull

organ of Corti

first turn of cochlea

auditory nerve

B Organ of Corti

vestibular canal (perilymph)

vestibular membrane

cochlear duct (endolymph)

tectorial membrane

basilar membrane

sensory hair cells

tympanic canal (perilymph)

Fig 24·14 Structure of the cochlea

Equilibrium

The parts of the inner ear required for balance, or equilibrium, are the **utricle**, **saccule**, and the **semi-circular canals**.

The utricle and saccule contribute to **static equilibrium**, that is, they help in maintaining the position of the body relative to gravity. This is possible because each of these structures contains a special patch of sensory cells, called the **macula**. As shown in Figure 24·15, the hairs of the sensory cells are embedded in a jelly-like mass containing tiny particles of calcium carbonate, or **otoliths**. According to the position of the head, the jelly mass is pulled downwards by gravity and may bend the hairs, producing an appropriate pattern of nerve impulses transmitted to the brain.

The semi-circular canals are responsible for **dynamic equilibrium**, helping to preserve balance during body movements. The three canals on each side are located in different planes and contain fluid endolymph. When a rapid body movement takes place, the fluid tends to be left behind in its

A Semicircular canals

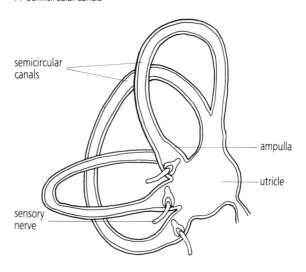

B Detail of ampulla

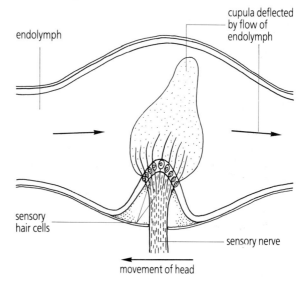

Fig 24·16 Structure and action of semi-circular canals

A Transverse section of utricle

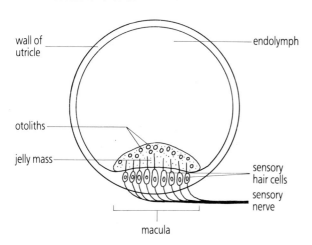

B Rotated through 90°

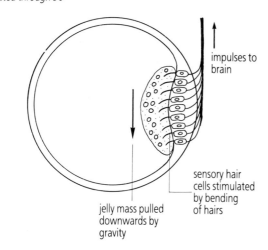

Fig 24·15 Structure and action of utricle and saccule

original position, and flows relative to the semicircular canals. This effect is due to the inertia of the fluid and can be easily demonstrated using a cup of tea or coffee – rotate the cup and the liquid inside remains stationary or moves to a much smaller extent. In a similar way, each semicircular canal responds to acceleration in its own plane.

Within the **ampulla** of each semi-circular canal there is a cone-shaped projection consisting of sensory hair cells and a jelly-like **cupula** (Fig. 24·16). Fluid movement deflects the cupula and stimulates the sensory cells. Impulses from the ampullae of all the semi-circular canals allow the brain to interpret body movement in three dimensions and are essential for coordinated muscle activity. They pass eventually to the cerebellum, where detailed instructions for muscle contraction are produced.

25 ENDOCRINE SYSTEM ⬚

Objectives

After studying this Unit you should be able to:–

- Define a hormone

- Distinguish between endocrine and exocrine glands

- List the principal endocrine glands of the human body and name the hormones they produce

- Describe the actions of adrenalin and noradrenalin in short term alarm responses to stress

- Outline the actions of ACTH, aldosterone, and glucocorticoid hormones in the longer term resistance responses to stress

- Name three processes normally associated with growth

- Draw and interpret human growth curves

- Comment on the functions of human growth hormone and thyroxin

- Briefly describe the 'second messenger' and 'gene activation' mechanisms of hormone action

25·1 Introduction ■ ■ ■ ■ ■ ■ ■ ■ ■ ■ ■ ■ ■ ■ ■ ■ ■

Hormones are 'messenger' substances produced in one part of the body and distributed to other parts. Chemically they may be **proteins, steroids,** or **amines** derived from amino acids. Each hormone regulates the activity of structures called **target organs,** which respond specifically, leaving other structures unaffected. Plants hormones are used to coordinate activities like growth, flowering, and dormancy: their action is described in Unit 34. In animals, the slow sustained responses produced by hormones complement much more rapid but short-lived responses controlled by the nervous system.

This Unit describes the hormone-producing system, or **endocrine system**, of the human body and discusses two important examples of endocrine control, namely, the stress reactions of the body, and the regulation of human growth. Appropriate references are given to help you find additional examples of hormone action described elsewhere in the book.

25·2 Human endocrine system ■ ■ ■ ■ ■ ■ ■ ■ ■ ■ ■ ■

As illustrated in Figure 25·1, the human endocrine system consists of widely separated **endocrine glands**. Each of these structures, also called **ductless glands,** manufactures one or more hormones and releases them directly into the bloodstream (Table 25·1). This mode of release distinguishes endocrine glands from **exocrine glands** like salivary glands, or the gastric pits of the stomach. Exocrine glands do not produce hormones and normally release their secretions by way of tubes, or ducts, leading into the internal body cavities, or, sometimes, to the exterior, as in the case of sweat ducts. The **pancreas** is unusual in this respect for it acts both as an endocrine gland and as an exocrine gland. The islets of Langerhans produce the hormones insulin and glucagon, while the remainder of the pancreas makes sodium hydrogencarbonate and enzymes which pass via the pancreatic duct into the duodenum.

The **pituitary gland**, nicknamed the 'master gland', is the most important endocrine organ because it coordinates the actions of many other endocrine glands and provides a vital link with the nervous system. It is a small structure about

1·3 cm in diameter located beneath the hypothalamus of the brain and connected by a short stalk, the **infundibulum**. As shown in Figure 25·2, the pituitary gland is divided into an anterior lobe, called the **adenohypophysis**, and a posterior lobe, the **neurohypophysis**. The adenohypophysis contains hormone-producing cells which respond to specific **releasing factors** from the hypothalamus. These substances are produced by specialised neurons and diffuse into the adenohypophysis where they trigger release of the appropriate hormone. As you can see from Table 25·1, five of the seven hormones produced are **trophic hormones**, meaning that they regulate the activity of other endocrine glands.

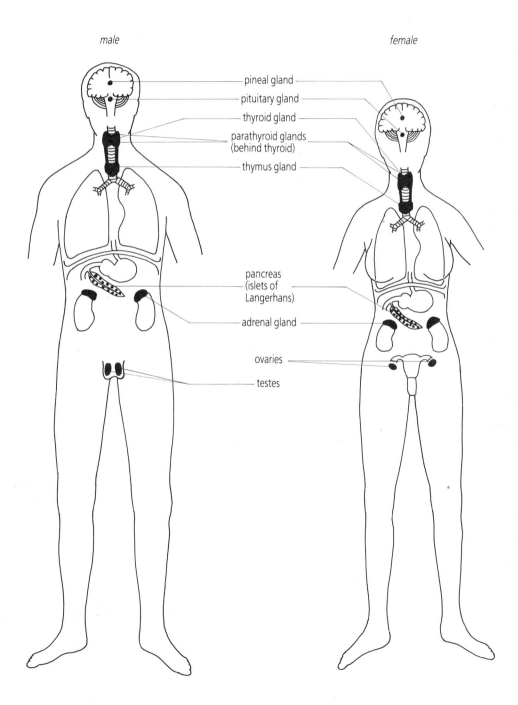

male *female*

- pineal gland
- pituitary gland
- thyroid gland
- parathyroid glands (behind thyroid)
- thymus gland
- pancreas (islets of Langerhans)
- adrenal gland
- ovaries
- testes

Fig 25·1 Human endocrine system

Table 25·1 Summary of human hormones

ENDOCRINE STRUCTURE	HORMONES PRODUCED	CHEMICAL NATURE	MODE OF ACTION	IMPORTANT ACTIONS	CROSS REFERENCES
PITUITARY GLAND 1 adenohypophysis	human growth hormone (HGH)	protein	cyclic AMP	stimulates protein synthesis and release of energy from fats	discussed in this Unit
	thyroid stimulating hormone (TSH)	glyco-protein	cyclic AMP	stimulates production and release of thyroid hormones	discussed in this Unit
	adrenocorticotrophic hormone (ACTH)	peptide	cyclic AMP	stimulates production and release of adrenal cortex hormones	discussed in this Unit
	follicle stimulating hormone (FSH)	glyco-protein	cyclic AMP	maturation of follicles in females, production of sperm in males	Unit 29
	luteinizing hormone (LH)	glyco-protein	cyclic AMP	triggers ovulation and development of corpus luteum	Unit 29
	prolactin (PR)	protein	–	stimulates milk production by mammary glands	Unit 30
	melanocyte stimulating hormone (MSH)	peptide	cyclic AMP	increases skin pigmentation	Unit 21
2 neurohypophysis	anti-diuretic hormone (ADH)	peptide	cyclic AMP	stimulates reabsorption of water by kidney tubules	Unit 20
	oxytocin	peptide	–	stimulates contraction of the uterus	Unit 30
PINEAL GLAND	melatonin	amine	–	possible inhibitory action on ovaries	–
THYROID GLAND	thyroxin	amino acid	cyclic AMP	increases metabolic rate, stimulates growth in infants	discussed in this Unit
	thyrocalcitonin	peptide	–	promotes calcium absorption by bones	Unit 27
PARATHYROID GLANDS	parathyroid hormone (PTH)	protein	cyclic AMP	promotes calcium absorption from intestine, stimulates calcium release from bones	Unit 27
THYMUS GLAND	thymosin	peptide	–	possible influence on B-lymphocytes	–
PANCREAS (islets of Langerhans)	insulin	protein	cyclic AMP	stimulates absorption of glucose into liver and muscle cells, formation of glycogen	Unit 15
	glucagon	peptide	cyclic AMP	increases blood glucose level, breakdown of glycogen	Unit 15
ADRENAL GLANDS 1 cortex	mineralocorticoids aldosterone	steroid	gene activation	stimulates reabsorption of sodium ions by kidney tubules, reduces reabsorption of potassium ions	Unit 20
	glucocorticoids hydrocortisone corticosterone cortisone	steroid steroid steroid	gene activation	reduce effects of stress responses	discussed in this Unit
2 medulla	adrenalin 80% noradrenalin 20%	amine amine	gene activation	increased heart and breathing rates, other 'fight and flight' responses	discussed in this Unit
OVARY	oestrogen	steroid	gene activation	development of female sexual characteristics, repair of uterus lining	Unit 29
	progesterone	steroid	gene activation	development of uterus ready for implantation	Unit 29
TESTIS	testosterone	steroid	gene activation	development of male sexual characteristics	Unit 29
PLACENTA	human chorionic gonadotrophin (HCG)	glyco-protein	–	maintenance of corpus luteum	Units 29, 30
	oestrogen progesterone	steroid steroid	gene activation	maintenance of pregnancy	Unit 29
DIGESTIVE SYSTEM	gastrin	peptide	cyclic AMP	stimulates production and release of pepsinogen and hydrochloric acid	Unit 13
	secretin	peptide	–	stimulates production of sodium hydrogencarbonate by the pancreas	Unit 15
	cholecystokinin (CCK)	peptide	–	stimulates release of bile	Unit 15
KIDNEY	erythropoietin	glyco-protein	–	promotes production of red blood cells (erythrocytes)	Unit 18
BODY TISSUES	prostaglandins	fatty acid	–	many local effects	–

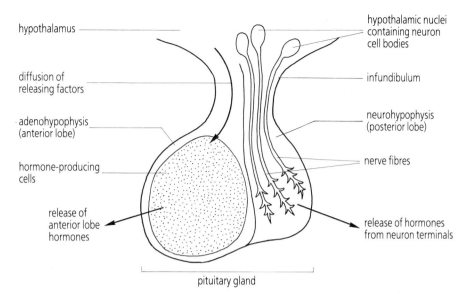

Fig 25·2 Structure of the pituitary gland

The neurohypophysis does not contain hormone-producing cells of its own, but receives nerve fibres from the hypothalmus. The terminals of these fibres are specialised to release two additional hormones, namely anti-diuretic hormone (ADH) and oxytocin.

25·3 Stress reactions ■ ■ ■ ■ ■ ■ ■ ■ ■ ■ ■ ■ ■ ■ ■

The stress reactions of the human body comprise a whole range of responses which prepare the body to resist any damaging, or potentially damaging, changes in the environment. An enormous variety of stimuli can act as **stressors** to trigger these responses, from injury to infection by microorganisms, or emotional factors like fear. The effectiveness of a particular stimulus varies from one person to another. Any change which is perceived as stressful produces signals which are routed to the hypothalamus where the stress reactions of the body are controlled. As outlined in Figure 25·3, these reactions include short term **alarm responses**, which prepare the body for immediate action, and longer term **resistance responses**, which promote an orderly return to normal activity.

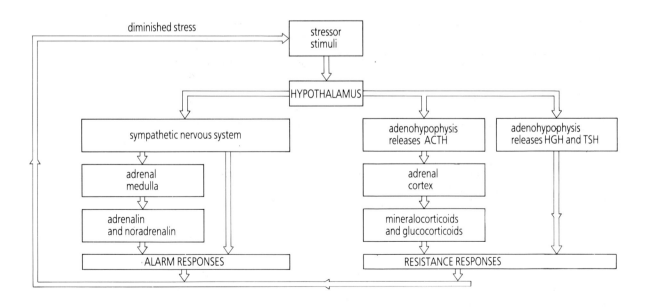

Fig 25·3 Summary of stress responses

Alarm responses

The hypothalamus causes a general increase in sympathetic nerve activity. This has a number of important effects, already described in Unit 23. They include increased heart and breathing rates, increased blood supply to the brain and body muscles, dilation of the bronchioles of the lungs, and increased sweating. Nerve impulses from the sympathetic nervous system also cause the release of the hormones **adrenalin** and **noradrenalin** from the **adrenal glands**. These structures are typically about 5 cm long and are located just above each kidney. They have a layered structure consisting of an outer **cortex** region surrounding an inner region, called the **medulla**. Both regions produce hormones – adrenalin and noradrenalin originate exclusively from the medulla, which normally releases a mixture containing about 80% adrenalin and 20% noradrenalin. These potent substances circulate in the blood, acting to enhance all the activities of the sympathetic nervous system and promoting the breakdown of glycogen in the liver and muscle cells. This liberates large amounts of glucose for increased respiration.

All of these effects, collectively called '**fight or flight**' responses, prepare the body for immediate action. They can be divided into two main categories, direct and indirect. The direct responses include increased heart and breathing rates, dilation of bronchioles, and glycogen breakdown. These provide the resources for rapid energy release by aerobic respiration and allow the body to perform far beyond its normal capacity. Indirect responses, like constriction of blood vessels leading to the skin and digestive organs, reduce the amount of energy used up for activities which will not contribute to diminishing stress.

Resistance responses

The longer term responses to stress are triggered by hormones from the anterior lobe (adenohypophysis) of the pituitary gland. Releasing factors from the hypothalmus stimulate the adenohypophysis to release ACTH (**adrenocorticotrophic hormone**) which acts on the cortex of the adrenal glands, causing increased secretion of **mineralocorticoid** and **glucocorticoid** hormones.

The mineralocorticoids help to counteract falling pH, or **acidosis**, of the body fluids, brought about by mobilisation of energy reserves and increased respiration, which tend to generate H^+ ions. The most important mineralocorticoid is **aldosterone** which acts on the kidney tubules, causing increased transfer of H^+ ions into the urine, so that they are eliminated from the body. (The mechanism of this action is described in Unit 20).

Glucocorticoids comprise **hydrocortisone**, **corticosterone**, and **cortisone**. These substances work together to increase the breakdown of body proteins into amino acids which can be used for making new enzymes, or may be converted into glucose for respiration. They promote constriction of peripheral blood vessels, helping to maintain high blood pressure and to limit blood loss from any injury. Finally, they reduce inflammation, thus preventing damage to healthy tissues. This anti-inflammatory action makes glucocorticoids and related compounds useful as drugs in the treatment of allergic diseases, like asthma, and of certain autoimmune diseases.

Resistance responses involve two further hormones released from the adenohypophysis. These are **human growth hormone (HGH)**, and **thyroid stimulating hormone (TSH)**, which acts on the **thyroid gland** causing the release of **thyroxin**. Together thyroxin and HGH increase supplies of glucose for respiration and produce a general increase in metabolic rate.

After a short time, resistance responses fade and normal conditions are restored. However, if stress-producing stimuli persist, the body may be subjected to an excessively high metabolic rate, high blood pressure, and a high heart rate for a prolonged period. This is particularly damaging for the circulatory system and, in developed countries, it is an important contributory factor increasing the risk of heart attacks and atherosclerosis.

25·4 Regulation of human growth ■ ■ ■ ■ ■ ■ ■ ■ ■

Growth normally involves three important processes, as follows:

1 Increase in dry mass. Food materials and simple substances obtained directly from the environment are used to build new tissues and organs.

2 Cell division. As the organism increases in size, the new cells needed arise by **mitosis**, a form of cell division which produces genetically identical daughter cells.

3 Cell specialization. When an organism increases in size, its structure becomes more complex as different tissues become specialized for different functions. This is achieved by specialization, or **differentiation**, of the individual tissue cells.

Animal and plant growth share these basic features, but there are important differences. Plants have specialized growth regions, called **meristems**, and often continue to grow throughout their lives: the factors which regulate their growth are described in Units 31 and 34. On the other hand, in animals, growth occurs in all parts of the body and stops when the individual organisms reach a fixed adult size.

The rate of growth is best determined by measuring the increase in dry mass of representative organisms. However, provided the water content of the organism remains fairly constant, increase in body mass, or increase in length, give an adequate indication of growth. The pattern of human growth is illustrated by the **growth curves** in Figure 25·4A. These show the changes in body mass of males and females (measured in kilograms) plotted against their ages (measured in years). The second set of curves, given in Figure 25·4B, show the corresponding **growth rates**, measured in kilograms per year (kg yr^{-1}). By studying these curves, you can confirm that the most rapid growth takes place in the first two years after birth. Thereafter growth slows until puberty is reached, when there is another burst of rapid growth. Final adult size is

general effect of promoting growth, but the type and effect of the growth response depends on the individual tissues and on their stage of development. For example, HGH causes elongation of bones in infants, but it cannot have this effect in adults because elongation is prevented by changes in the structure of the epiphyseal plates (details of bone growth are given in Unit 27).

Such differences in response help to explain the effects of a surplus or deficiency of the hormones at different ages. In children, too much HGH causes abnormal lengthening of the bones and leads to **gigantism**, while too little HGH causes the bones to remain short, resulting in **dwarfism**. The statures of fully developed pituitary giant, normal, and dwarf people are compared in Figure 25·5. In adults, excessive secretion of HGH causes symptoms of **acromegaly**, a thickening of the bones of the hands and feet and abnormal development of facial features.

Similar variations in response are observed with thyroxin. In children, too little thyroxin prevents

A Body mass

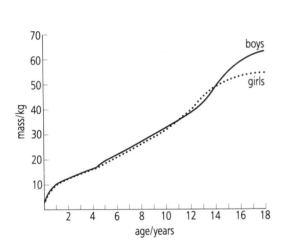

B Growth rate

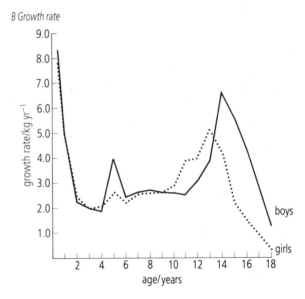

Fig 25·4 Human growth curves

normally achieved by age 20. On average, females are smaller from birth, reach puberty earlier and have a smaller adult size than males.

Many dietary factors influence the normal pattern of growth, particularly vitamins and inorganic salts. The most important hormones are **human growth hormone (HGH)** and **thyroxin**. HGH is released from the adenohypophysis and increases protein synthesis by stimulating the individual tissue cells to take up amino acids. It also stimulates the breakdown of fats and the synthesis of glucose for energy. Thyroxin increases the breakdown of carbohydrates and causes an increase in metabolic rate. These actions have the

normal development of bones and nervous tissue and may cause **cretinism**. In adults, the skeleton and nervous system are fully developed and are consequently unaffected by thyroxin. The symptoms of inadequate production, called **hypothyroidism**, are low metabolic rate, slow heart rate, high blood pressure and weight gain accompanied by tissue swelling, or **myxoedema**. Excess secretion, or **hyperthyroidism**, produces a high metabolic rate and symptoms which are opposite to those of hypothyroidism, including increased pulse, low blood pressure, and weight loss. Additional characteristic symptoms are bulging eyes and enlargement, or **goitre**, of the

Fig 25·5 Extremes of height: pituitary giants on left and twins exhibiting dwarfism on right (in this case not a result of pituitary deficiency).

thyroid gland, which may swell to several times its normal size. The condition is sometimes called **exopthalmic goitre** to distinguish it from **simple goitre**, a similar swelling caused by an insufficient supply of iodine in the diet.

25·5   How hormones act ■ ■ ■ ■ ■ ■ ■ ■ ■ ■ ■ ■ ■

Two important mechanisms of hormone action have been discovered.

1 Second messengers – cyclic AMP

As outlined in Figure 25.6*A*, the cell membranes of individual cells in the target organ contain protein **receptor substances** which are exposed on the cell surface and are able to combine with a specific hormone. The hormone acts as a **first messenger**. The **hormone-receptor complex** which is formed then triggers the activity of an enzyme, called **adenyl cyclase**, also located in the membrane. The enzyme converts ATP into a related substance, called **cyclic AMP (cAMP)**; this is not exploited as an energy source but is used instead as an **intracellular messenger**, or **second messenger**, within the target cell. It combines with specific enzymes to close down existing enzyme pathways, or open up new ones. For example, in liver and muscle cells, cyclic AMP triggered by adrenalin inhibits the enzyme needed for formation of glycogen and activates the enzymes needed for glycogen breakdown, thereby increasing the availability of glucose for respiration.

Many other hormones use cyclic AMP as a second messenger, as summarised in Table 25·1.

2 Gene activation

Other hormones, including many steroids, affect their target cells in a different way, as illustrated in Figure 25·6*B*. The hormone molecules cross the cell membrane and combine with specific **receptor protein** molecules within the cytoplasm of the cell. The **hormone-receptor complexes** formed enter the nucleus and act directly on the DNA of the target cell, triggering transcription of messenger RNA from a specific gene. The target cell makes a specific protein and consequently produces the response appropriate to the particular hormone. The details of gene activation by the hormone-receptor complex have not been determined, but one possible mechanism would be similar to the **operon** system described in Unit 9. You should refer to this Unit and follow through the process outlined in Figure 9·6, substituting a hormone molecule for the lactose substrate molecule illustrated.

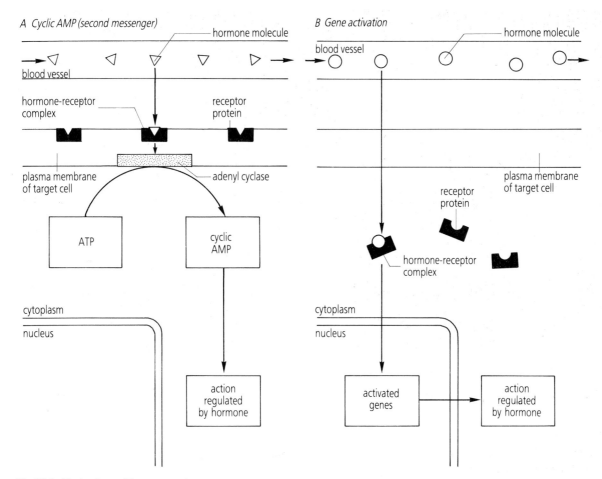

A   Cyclic AMP (second messenger)

hormone molecule

blood vessel

hormone-receptor complex

receptor protein

plasma membrane of target cell

adenyl cyclase

ATP

cyclic AMP

cytoplasm

nucleus

action regulated by hormone

B   Gene activation

hormone molecule

blood vessel

plasma membrane of target cell

receptor protein

hormone-receptor complex

cytoplasm

nucleus

activated genes

action regulated by hormone

Fig 25·6 Mechanisms of hormone action

26 STRUCTURE AND FUNCTION OF MUSCLE

Objectives

After studying this Unit you should be able to:–

- Draw and label diagrams to illustrate the external features and detailed internal structure of skeletal muscles

- Describe the ultrastructure of an individual muscle fibre and explain each of the following terms – sarcoplasm, myofibril, sarcolemma, T-tubule, sarcoplasmic reticulum, sarcomere, Z-disc, A-band, I-band, H-zone

- Give a detailed account of the sliding filament theory of muscle contraction and explain the changes in appearance of sarcomeres during contraction

- Describe the additional mechanisms needed to ensure that contraction will only occur following the arrival of nerve impulses at a neuromuscular junction

- Explain how motor units contribute to the behaviour of whole muscles

- Distinguish between twitch contraction and tetanic contraction

- Outline the structural differences between fast and slow muscle fibres

- Comment on the role of creatine phosphate

- List the factors which contribute to an oxygen debt

- Draw up a table to compare the specializations of skeletal, visceral, and cardiac muscle

26·1 Introduction ■ ■ ■ ■ ■ ■ ■ ■ ■ ■ ■ ■ ■ ■ ■ ■ ■ ■

Movement is a basic characteristic of living organisms and is particularly important for animals, many of which search actively for food and shelter, or for a mating partner. Plants and some single-celled organisms, like *Amoeba* and *Paramecium* (described in Unit 46), are able to move without muscles, but, in most multicellular animals, specialised **muscle tissue** has been developed. This forms about 40% of body mass in mammals and other vertebrates.

Three main types of muscles are distinguished in vertebrates. **Skeletal muscles** are attached to the bones of the skeleton and are adapted for rapid powerful contractions, resulting in body movement.

They are consciously controlled from the motor cortex of the brain. **Visceral muscles** occur in body organs like the intestines and are controlled by the autonomic nervous system. Their contractions are slower and more sustained, as in peristalsis. **Cardiac muscle**, found exclusively in the heart, is specialised for rhythmic contraction, producing an efficient pumping action. Of these three types, skeletal muscle is the most abundant, and, in the following sections, its structure and function will be described in detail. Although there are some structural differences, the basic mechanisms of contraction in visceral and cardiac muscle are very similar.

Structure

Individual skeletal muscles in the human body range in size from the tiny muscles of the middle ear, only a few millimetres in length, to the large muscles of the legs and arms, which may reach 40–50 cm. Even so, they share a number of important characteristics.

1 Each muscle has its own blood supply consisting of an **artery** carrying oxygenated blood, a **capillary network** deep within the muscle tissue, and a **vein** transporting deoxygenated blood back towards the heart.

2 Each muscle receives its own **motor nerve** (**effector neurone**) from the CNS.

3 The muscle exerts force only when it is stimulated to contract by the arrival of nerve impulses in the motor nerve.

4 The muscle is sheathed by connective tissue which extends at either end to form fibrous structures called **tendons**. These may be cord-like and circular in cross section, or broad flat bands. They serve to attach the muscle to a bone or sometimes to another muscle.

The structure of a typical skeletal muscle is illustrated in Figure 26·1 *A,B,C,* and *D.* Its most important components are the cylindrical **muscle fibres,** or **muscle cells,** which are 10–100 μm in diameter and up to 30 cm long. These are grouped together in bundles, called **fascicles,** and separated from one another by connective tissue containing blood capillaries and nerve fibres. During development, the individual muscle fibres are formed by the fusion of many shorter cells, each with its own nucleus. Consequently, they contain many nuclei, usually located in a thin surface layer of cytoplasm, called the **sarcoplasm.** This layer also contains many mitochondria. The interior of the fibre is packed with thread-like **myofibrils,** 1–2 μm in diameter, just visible with the light microscope. The myofibrils are the contractile elements of the muscle; they lie parallel and have prominent markings which give the fibre its characteristic **striped,** or **striated,** appearance. The outer membrane of the muscle fibre is called the **sarcolemma.**

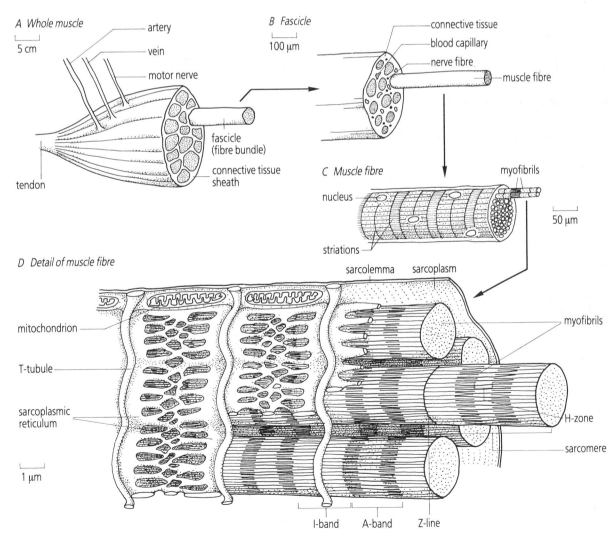

Fig 26·1 Structure of skeletal muscle

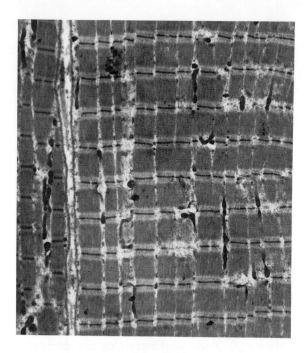

Fig 26·2 *Electronmicrograph of myofibrils*

Further details of the structure of fibres have been revealed by the electron microscope (Figs 26·1*D* and 26·2). As you can see, the sarcolemma is folded inwards at intervals to form long thin tubes, called **T-tubules**. These lie in close contact with well developed regions of **sarcoplasmic reticulum** which cover the myofibrils rather like a net stocking.

The markings of the myofibrils correspond to regular repeating units called **sarcomeres**, each about 2·5 μm long. The sarcomeres are joined end to end by thin plates known as **Z-lines**, or **Z-discs**. The dark stripes in the centre of adjacent sarcomeres have a pale central region called the **H-zone**, and are separated from the next dark stripe by lighter regions called **I-bands**. The significance of these regions is now fully understood and is explained in the next section.

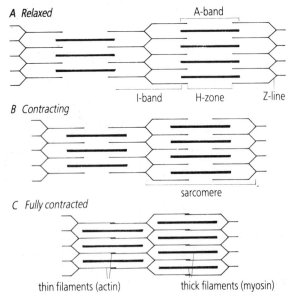

Fig 26·3 *Sliding filament theory of muscle contraction*

Mechanism of contraction

When a muscle fibre contracts, each of the sarcomeres in its myofibrils shortens by a corresponding amount. This is achieved by molecular interactions between the two major components of the sarcomere, called **thin filaments** and **thick filaments** (Fig. 26·3*A*). The thin filaments consist mainly of a protein called **actin** and are anchored at one end to the Z-discs. They are the only filaments in the I-band regions of the fibril and overlap with the thick filaments, which occupy the A-band. The thick filaments consist of a second protein called **myosin**. Within the H-zone, only thick filaments are present. Measurements taken from electronmicrographs of contracted muscles show that the width of the H-zone decreases, while the lengths of the filaments remain unaltered. It follows that, during contraction, the filaments slide past one another, as shown in Figure 26·3*B* and *C*. This proposed mechanism gives rise to the **sliding filament theory** of muscle contraction.

The forces which produce contraction depend on the formation of **cross-bridges** between actin and myosin molecules. Actin molecules are globular proteins which associate to form a twin-stranded helical structure which is the basic component of the thin filaments (Fig. 26·4). Each actin molecule has a specific **binding site** for myosin. Myosin molecules, which are larger, have a pair of globular **heads** attached to a long fibrous **tail** containing flexible 'hinge' regions (Fig. 26·5). Thick filaments are produced when the tails of many myosin molecules associate together, leaving their globular heads projecting outwards.

Cross-bridges between filaments are formed only when ATP is available (Fig. 26·6). ATP combines with the individual myosin heads and is then hydrolysed, the products of the reaction being temporarily retained by the myosin head. This hydrolysis causes a change in the shape of the head, so that it is now able to bind to the actin filament. As soon as this happens, there is a second change in shape which results in the release of ADP and inorganic phosphate. This change represents the power stroke of the system and it generates mechanical energy which is used to pull the filaments along. During contraction of a muscle fibre, this cycle of events is repeated many times as the heads of the myosin molecules from the thick filaments 'walk' along the actin molecules of the thin filaments.

Initiation of muscle contraction

ATP is not the only requirement for contraction in living muscle cells: **Ca²⁺** ions are also needed and in their absence contraction is blocked. This is explained by the presence of additional proteins in the thin filaments, namely **troponin** and

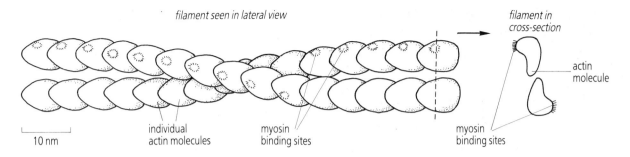

filament seen in lateral view

filament in cross-section

individual actin molecules

myosin binding sites

actin molecule

myosin binding sites

10 nm

Fig 26·4 Actin molecules and thin filaments

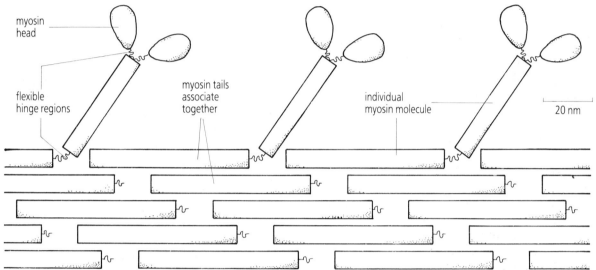

myosin head

flexible hinge regions

myosin tails associate together

individual myosin molecule

20 nm

The projecting parts of the myosin molecule are shown for the first row of molecules only. Thick filaments are densely covered by projecting myosin heads

Fig 26·5 Myosin molecules and thick filaments

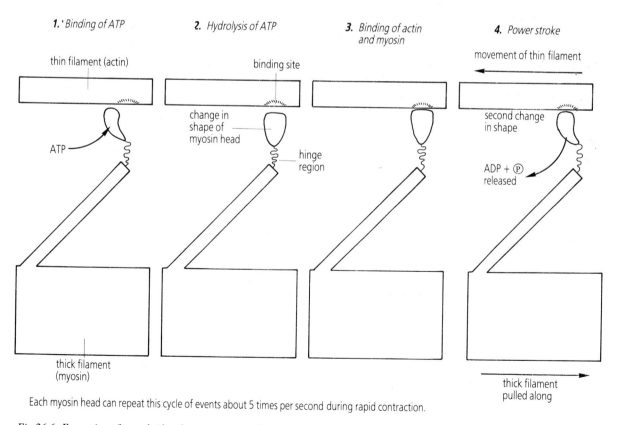

1. *Binding of ATP*

2. *Hydrolysis of ATP*

3. *Binding of actin and myosin*

4. *Power stroke*

thin filament (actin)

binding site

movement of thin filament

ATP

change in shape of myosin head

hinge region

second change in shape

ADP + $\textcircled{P}$ released

thick filament (myosin)

thick filament pulled along

Each myosin head can repeat this cycle of events about 5 times per second during rapid contraction.

Fig 26·6 Formation of cross bridges between actin and myosin

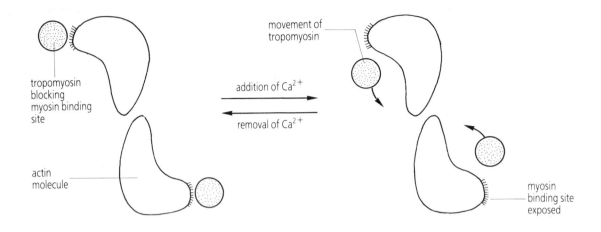

Fig 26·7 Calcium activation of contraction

tropomyosin. During inactive periods, the molecules of these proteins lie along the actin filament in such a way that all of the myosin binding sites are covered, making the formation of cross-bridges impossible. Ca^{2+} ions bind to the accessory proteins and twist them so as to expose the binding sites (Fig. 26·7). This remarkable mechanism provides the essential link between motor nerve impulses and muscle contraction.

Each skeletal muscle fibre is an electrically excitable cell receiving impulses across synapses called **neuromuscular junctions**. At these junctions, the transmitter substance **acetylcholine (ACh)** causes depolarisation of the muscle cell membrane (sarcolemma) and may trigger an all-or-none **action potential**. This spreads rapidly throughout the muscle cell and is conducted inwards by the T-tubules (refer to Fig. 26·1D), passing finally to the sarcoplasmic reticulum, where large numbers of Ca^{2+} ions are stored. The arrival of the action potential liberates these ions into the cytoplasm and thus triggers contraction of the myofibrils. The action potential is conducted so rapidly that all of the sarcomeres in the muscle fibre respond simultaneously. The contraction slowly subsides as Ca^{2+} ions are pumped back into the sarcoplasmic reticulum.

Physiology

The behaviour of the whole muscle depends on the combined effect of its individual muscle fibres, which are arranged into functional groups. Each muscle fibre contracts in an all-or-none fashion, as explained above, but the whole muscle can contract smoothly, with graded changes in tension. This is possible because the muscle receives impulses from many different motor neurons, each of which sends branches to its own group of individual

muscle fibres, collectively called a **motor unit**. The number of fibres varies from 150 or so per motor unit in a large muscle, to as few as 1 or 2 in human eye muscles. They are not confined to a single fascicle, but are distributed throughout the muscle and contract whenever the corresponding motor neuron is active. Thus, the tension in a particular muscle depends upon how many of its motor units are active. In muscles involved in maintaining posture, only a few motor units are used at any one time, producing a continuous low tension, but, during body movements, many motor units are activated. The tension in the muscle is monitored by sense organs called **muscle spindles** which send impulses via sensory nerves to the CNS where the output of the motor neurons is continually adjusted. Stretch reflexes, described in Unit 23, ensure that muscles adjust automatically to the load placed upon them.

The properties of skeletal muscles can be investigated by carefully removing a muscle from the body of an animal, such as a frog. The muscle is placed in a suitable fluid medium and held by its tendons so that changes in length or tension can be measured. The muscle is stimulated to contract by applying small electric shocks to its motor nerve, or directly to the muscle surface.

The normal response to a single stimulus is a rapid jerky contraction, called a **twitch contraction**. Figure 26·8 shows the relationship between this contraction and the action potential triggered by the stimulus. As you can see, the time course of the contraction is much slower than the action potential and is preceded by a brief latent period, corresponding to the time taken for the release of Ca^{2+} ions and the development of the first cross-bridges. The contraction reaches its peak after about 100 ms and then slowly subsides as the muscle relaxes. In skeletal muscle, repeated stimulation can cause a series of twitches which

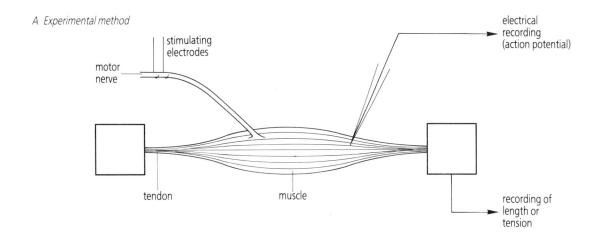

A Experimental method

motor nerve

stimulating electrodes

electrical recording (action potential)

tendon

muscle

recording of length or tension

B Twitch contraction

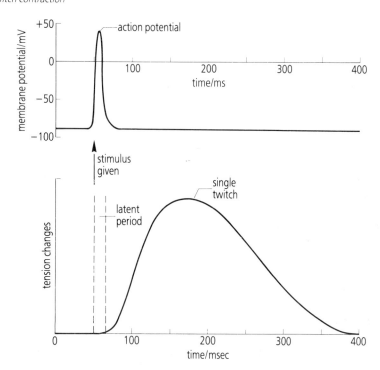

Fig 26·8 *Investigating the behaviour of whole muscle*

add together, or **summate**, to produce a more powerful sustained contraction, called a **tetanus**. This is illustrated in Figure 26·9. Normal contractions of skeletal muscles, which produce body movements, are generated in this way following rapid bursts of nerve impulses in their motor nerves. The ability of a muscle to produce a tetanus depends upon its **refractory period**, during which a second stimulus is ineffective. In skeletal muscle the refractory period is short but, in cardiac muscle, it is very much longer. This makes tetanic contractions impossible and is essential if the rhythmic beating of the heart is to be maintained.

The duration of a single twitch depends upon small variations in muscle structure. For example, in **fast**, or **white muscles**, like human eye muscles, there are fewer blood capillaries and the individual muscle fibres have a particularly well developed sarcoplasmic reticulum. This promotes rapid uptake and release of Ca^{2+} ions so that each contraction lasts only about 10 ms. On the other hand, in **slow**, or **red muscles**, like those of the arms and legs, individual twitch contractions last

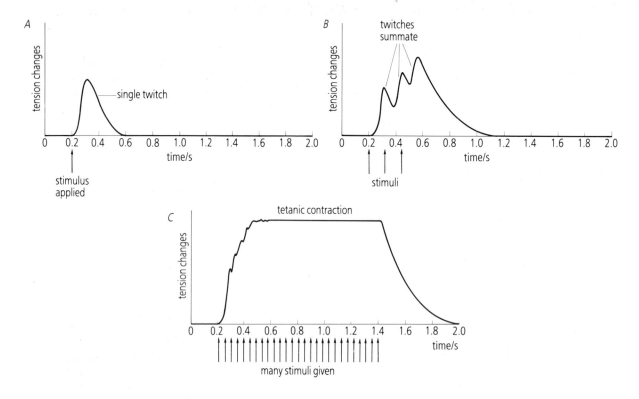

Fig 26·9 Summation and tetanic contraction

50 ms or more. These muscles have smaller fibres, more blood capillaries, more mitochondria, and large amounts of the red pigment **myoglobin** in their cytoplasm. As explained in Unit 18, the molecules of this substance bind reversibly with oxygen, which can be released at low partial pressures. Thus, myoglobin acts as a store of oxygen which can be used to maintain aerobic respiration during strenuous activity. The red appearance of slow muscles is due to their blood capillaries and to myoglobin.

Oxygen debt

Supplies of ATP needed for contraction are normally maintained by aerobic respiration and by a substance called **creatine phosphate**, which acts as a store of ATP. In resting muscle cells, ATP from respiration accumulates in the cytoplasm and is converted to creatine phosphate by the reaction:

$$ATP + creatine \rightleftharpoons ADP + creatine phosphate$$

As ATP in the cytoplasm is used up, the reaction is reversed liberating the stored ATP so that the period of contraction can be extended. During strenuous activity, the muscle's supplies of oxygen and creatine phosphate eventually become exhausted. In these circumstances, ATP is produced by anaerobic respiration and **lactate** is formed. Lactate leaves the muscles in the blood stream and is taken up by the liver.

To restore normal conditions, large amounts of oxygen are needed; this situation is known as an **oxygen debt**. During a race, an athlete's body may build up an oxygen debt of as much as 16 litres, which is then repaid by increased oxygen uptake over the next hour or more. The results of one experiment are illustrated in Figure 26·10. As you can see, oxygen uptake rises quickly after an initial delay, and continues at a high rate after the period of exercise, returning only slowly to the resting level. The extra oxygen is used partly to renew supplies of creatine phosphate in the body muscles and partly to convert lactate back to pyruvate, this is used for aerobic respiration, or may be converted into glucose and stored as glycogen in the liver. The loss of oxygen from myoglobin makes an additional contribution to oxygen debt.

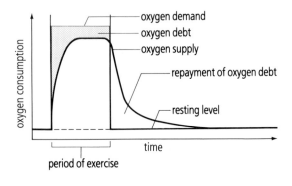

Fig 26·10 Oxygen debt

212

26·3 Specialization of visceral and cardiac muscle ■ ■ ■ ■ ■

A detailed comparison between skeletal, visceral, and cardiac muscle is given in Table 26·1. You should study this carefully and try to relate the specialisations of the different muscle types to their functions in the body. Visceral muscle should not be thought of as an inferior type because of its simpler structure. It is very well suited to maintaining steady tensions and uses far less energy during contraction than skeletal muscle. The role of the cardiac muscle in producing the heart beat is described in Unit 17.

Table 26·1 Comparison between skeletal, visceral and cardiac muscle

	SKELETAL	VISCERAL	CARDIAC
alternative names	striated, striped, voluntary	non-striated, unstriped involuntary, smooth	–
structure	fibres formed from many cells fused together, many nuclei in surface layer of sarcoplasm	fibres consist of individual cells with a central nucleus, fibres unbranched	individual cells with a central nucleus, cells branched and linked by intercalated discs
size	length 1 mm–30 cm diameter 10–100 μm	length 0.02–0.5 mm diameter 8–10 μm	length 0.06–0.08 mm diameter 10–15 μm
myofibrils	conspicuous	inconspicuous	conspicuous
physiology	contractions rapid and powerful but short-lived, brief refractory period allows tetanic contraction	contractions slow and sustained	rapid contractions spread through linked network, long refractory period prevents tetanic contraction
control	contraction triggered by motor neurons of CNS (neurogenic)	contractions triggered by impulses from the autonomic nervous system spread from cell to cell	spontaneous rythmic contractions (myogenic), rate regulated by autonomic nervous system
location	muscles attached to bones, sheets of muscle under skin, diaphragm	digestive, respiratory and urinogenital systems, blood vessels, ciliary muscles of eye, hair arrector muscles	heart only

27 SKELETON AND LOCOMOTION $\boxed{}$

Objectives

After studying this Unit you should be able to:–

- List the functions of skeletons

- Distinguish between hydrostatic skeletons, exoskeletons, and endoskeletons and give examples of each type

- Describe the structure of connective tissue

- Draw and label a diagram to illustrate the arrangement of cells in cartilage and identify two methods of cartilage growth

- Draw and label a diagram to illustrate the formation of bone by osteoblasts

- Explain the role of osteoclasts in remodelling of bone and describe the structure of compact bone

- Distinguish between intramembranous ossification and endochondral ossification

- Describe the development of long bones in the human skeleton

- List the main parts of the axial skeleton

- Describe the structure and specializations of cervical, thoracic, and lumbar vertebrae, and comment on the structure of the sacrum and coccyx

- List the main parts of the appendicular skeleton

- Distinguish between fibrous, cartilaginous, and synovial joints

- Draw and label a diagram of a synovial joint

- Outline the structure of gliding, hinge, pivot, ellipsoidal, saddle, and ball and socket joints and state where examples of each are located in the human skeleton

- Describe the action of antagonistic muscles in movements of the human arm

- State the meanings of flexion/extension, abduction/adduction, pronation/supination, and circumduction

- Explain the role of parathyroid hormone (PTH), vitamin D, and thyrocalcitonin in regulation of blood calcium

- Discuss methods of locomotion in fish, land vertebrates, and birds

27·1 Introduction ■ ■ ■ ■ ■ ■ ■ ■ ■ ■ ■ ■ ■ ■ ■ ■

The **skeleton** of an organism provides **support** and **protection**, and can usually be acted upon by mechanical forces, so as to produce **body movement**. Nearly all organisms have skeletons, although in a simple organism there may be no permanent skeletal materials like cartilage or bone. Three basic types are distinguished:

1 Hydrostatic skeletons. Liquid water has a high density and cannot be compressed. It forms 60–70% of most living organisms and provides support for all their internal organs, cells, and cell organelles. Consequently, any volume of fluid enclosed in the body of an organism acts as a hydrostatic skeleton.

In simple organisms, such skeletons often provide the only means of locomotion. For example, the contraction of actin filaments in one part of an *Amoeba* (described in Unit 46) exerts pressure on the fluid interior, causing the organism's cytoplasm to flow outwards as new pseudopodia develop. Similarly, in an earthworm (Unit 48), circular and radial muscles in the body wall act on the enclosed fluid in the body segments to produce changes in shape which result in body movement. Specialised examples of hydrostatic skeletons in vertebrates include the action of aqueous and vitreous humour in maintaining the shape of the eye, and the protective effects of cerebrospinal fluid, and amniotic fluid. In plants, hydrostatic forces resulting from **turgor pressure** (described in Unit 33) are vital for support and rapid movement.

2 Exoskeletons. These are hard external coverings like the thickened cuticles of arthropods and the shells of molluscs. The insect cuticle consists largely of **chitin**, an extremely tough polysaccharide material, which is made almost completely waterproof by a thin surface layer of wax (Unit 49). This combination provides a very versatile building material with almost ideal mechanical properties. Structures derived from the cuticle, including wings and mouthparts, allow insects to cope successfully with many different life styles and to exploit many different diets. On the other hand, exoskeletons work best for small organisms and become thicker, heavier, and less efficient as size increases. The largest present day insects are 10–15 cm long, but most species are much smaller.

3 Endoskeletons. Some protozoa, belonging to Phylum Actinopoda and Phylum Foraminifera, deposit internal skeletons of silica (SiO_2), strontium sulphate ($SrSO_4$), or calcium carbonate ($CaCO_3$), but there are few other examples of internal skeletons among the invertebrates. In contrast, many plants are supported by internal rods of vascular tissue (Unit 31). In trees, lignified xylem vessels form a solid core, allowing heights of 80 m to be reached, as in giant redwood (*Sequoia*) trees.

The largest animals are all vertebrates with an internal supporting framework made of **cartilage** or **bone**. Their organs and tissues are bound together and supported by **connective tissue**. The remainder of this Unit describes the structure and properties of these materials and explains how they are used in the human skeleton.

27·2 Skeletal materials ■ ■ ■ ■ ■ ■ ■ ■ ■ ■ ■ ■ ■ ■
Connective tissue

Figure 27·1 shows the appearance of one common type of connective tissue. As you can see, the individual cells are loosely scattered and are surrounded by a mixture of chemicals known as the **ground substance**, or **extracellular matrix**. There are several different kinds of cells – the most numerous are **fibroblasts** which secrete a mixture of protein molecules and polysaccharides. The most abundant proteins are **collagens**, while the polysaccharides belong to a group called **glycosaminoglycans**. Together, the molecules of these substances form a linked network which gives the matrix a gel-like consistency, but allows the diffusion of dissolved gases, nutrients, hormones, and other soluble chemicals. Collagen fibres and elastic fibres are embedded in the gel and improve its physical properties. The remaining cells include phagocytic **macrophages**, which attack invading microorganisms, and **mast cells**,

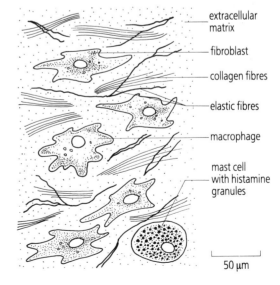

- extracellular matrix
- fibroblast
- collagen fibres
- elastic fibres
- macrophage
- mast cell with histamine granules

50 µm

Fig 27·1 Structure of connective tissue

which release histamine during imflammation (see Unit 19).

Connective tissue forms a large part of the skin, and is an integral part of many body organs. It holds together and supports organs like the lungs, and the liver, and forms a sheath surrounding the nervous system. Cartilage and bone, described in the next two sections, are specialised types of connective tissue.

Cartilage

Cartilage cells, or **chondrocytes**, produce a resilient matrix consisting of a dense network of collagen fibres and elastic fibres embedded in a tough gel. Each chondrocyte occupies its own space, or **lacuna**, in the matrix, and receives nutrients by diffusion from distant blood capillaries, which do not penetrate the matrix. The cartilage mass is normally surrounded and enclosed by a compact layer of connective tissue called the **perichondrium** (Fig. 27·2).

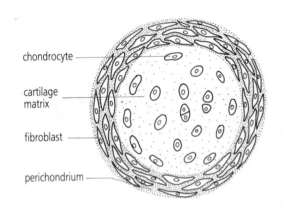

Fig 27·2 Structure of cartilage

Cartilage growth can occur in two ways. Cell division and the secretion of new matrix by chondrocytes within the existing cartilage mass lead to growth from within, while new chondrocytes develop from fibroblasts recruited from the perichondrium at the surface. The perichondrium acts as a jacket helping to determine the shape of the structure produced.

Three types of cartilage are distinguished. **Hyaline cartilage** is the most abundant type in the human body and has a characteristic bluish-white appearance. Its matrix is translucent when seen with the light microscope. The skeletons of all vertebrate embryos consist initially of hyaline cartilage which, in most cases, is gradually replaced by bone. In adults, hyaline cartilage is found where bones meet at movable joints, where it is called **articular cartilage**, at the ends of the ribs, and as supporting tissue in the nose, larynx, trachea, and bronchi.

Fibrocartilage has a matrix containing prominent bundles of collagen fibres and occurs in intervertebral discs. It combines strength and rigidity. As its name suggests, **elastic cartilage** contains a high proportion of elastic fibres. It is found in the external pinnae of the ears, and in the epiglottis.

Bone

The matrix produced by bone-forming cells, or **osteoblasts**, is called **osteoid** (Fig. 27·3). It contains collagen and **osteonectin** proteins, which are thought to act as sites for the growth of inorganic crystals. At first, the osteoid matrix is uncalcified but insoluble minerals are rapidly deposited from the blood. The commonest bone minerals are **hydroxyapatites** formed from calcium and phosphate ions. These have the general formula $Ca_{10}(PO_4)_6(OH)_2$ and form an extremely hard crystalline matrix. Other ions, including carbonate, magnesium, and fluoride, may also contribute to bone structure.

Once trapped in hard matrix, bone cells, now called **osteocytes**, are unable to divide. Each occupies a small space, or **lacuna**, as in the case of cartilage, but, unlike chondrocytes, osteocytes are linked to neighbouring cells by cytoplasmic threads which lie inside tiny channels known as **canaliculi**.

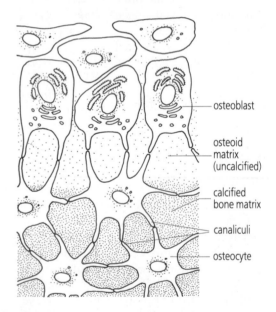

Fig 27·3 Formation of bone by osteoblasts

It might be thought that bone tissue would become permanent once formed. Nothing could be further from the truth. A second type of bone cells, called **osteoclasts**, continually erode the bone matrix. The osteoclasts are large multinucleate cells similar to macrophages. Like other macrophages they develop from monocytes (see Unit 18). Working in small groups, osteoclasts tunnel through bone, leaving cavities which are invaded by blood capillaries and new bone-forming osteoblasts. This process is illustrated in Figure 27·4. Within the tunnel, new bone matrix is laid down in concentric rings, forming an arrangement known as a **Haversian system**. The results of this activity are clearly seen in sections of compact bone (Fig. 27·5). Note that new tunnels have complete Haversian systems, while older systems may have been cut through and partly destroyed.

The factors which regulate the balance between bone destruction and rebuilding are not fully understood, but it does appear that osteoblasts and osteoclasts are able to respond to mechanical stresses, so that the shapes of bones are determined by the load placed upon them. Constant use leads to strengthening of bones, while disuse results in weakness. In this way, the bones of the skeleton are constantly **remodelled**. In an adult mammal, 5–10% of bone is replaced each year. It does not seem surprising that every person has a skeleton of the correct size and proportions to provide support, but you should appreciate that this can only be achieved by means of sophisticated internal control mechanisms. When conditions are altered dramatically, the existence of such mechanisms becomes more obvious. For example, exposed to weightless conditions, astronauts start to lose large amounts of calcium.

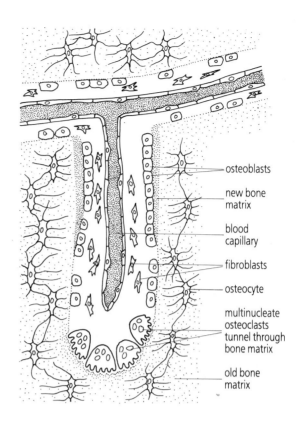

osteoblasts

new bone matrix

blood capillary

fibroblasts

osteocyte

multinucleate osteoclasts tunnel through bone matrix

old bone matrix

Fig 27·4 Remodelling of bone by osteoclasts

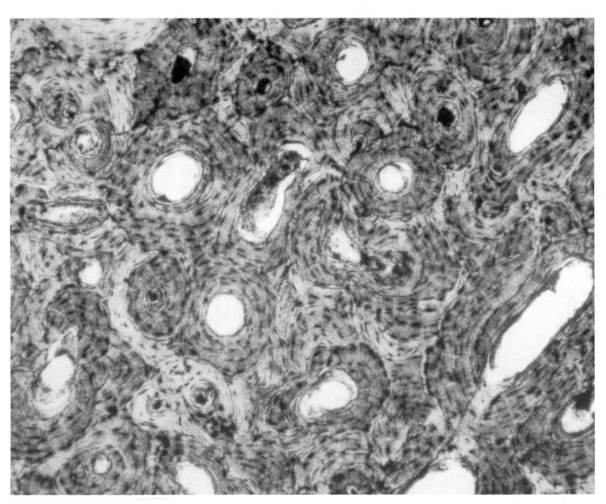

Fig 27·5 Microscopic structure of compact bone

Development

The bones of the human skeleton originate in two ways. The thin bony plates of the skull and parts of some other bones are formed directly by clusters of osteoblasts which appear inside fibrous membranes. The strands of bone formed by different clusters are called **trabeculae** (singular trabecula) and become linked to form a loose network described as **spongy bone**. The whole process is known as **intramembranous ossification**. As development continues, remodelling of skull plates converts some of the spongy bone to compact bone, and allows the plates to reach their adult size and shape.

The second method of bone formation involves the replacement of cartilage and is called **endochondral ossification** (see Fig. 27·6). As stated earlier, the skeleton of a vertebrate embryo consists mainly of hyaline cartilage. Each cartilage element is surrounded by a connective tissue jacket, the perichondrium, and is the same shape as the bone which will replace it (Fig. 27·6A). In this way, the cartilage acts as a 'model' for bone formation.

Ossification of bones like the long bones of the arms and legs begins when blood vessels penetrate the perichondrium midway along the shaft, or **diaphysis**, of the cartilage model (Fig. 27·6B). This stimulates some of the cells of the perichondrium to become osteoblasts which produce a collar of compact bone in the shaft region. The jacket of connective tissue covering the developing bone is now called the **periosteum**. A **primary ossification centre** appears inside the shaft and is progressively invaded by blood vessels and osteoclasts (Fig. 27·6C). The matrix of the cartilage tends to become calcified by deposition of calcium and phosphate, but is eroded by the osteoclasts, leaving spaces which eventually fuse to form the marrow cavity. Cartilage continues to grow at either end producing an increase in length. Most of it is later replaced by spongy bone.

In mammals, **secondary ossification centres** develop in the swollen ends, or **epiphyses**, of the cartilage models of long bones. These ossify more or less completely except for a thin layer of cartilage, called an **epiphyseal plate**, separating each epiphysis from the main shaft, and a thin surface layer of articular cartilage. This arrangement gives the bone strong rounded ends, but permits continued increase in length (Fig. 27·6D). Increase in diameter of the bone shaft is achieved by continual remodelling and deposition of new bone by the periosteum. As maturity

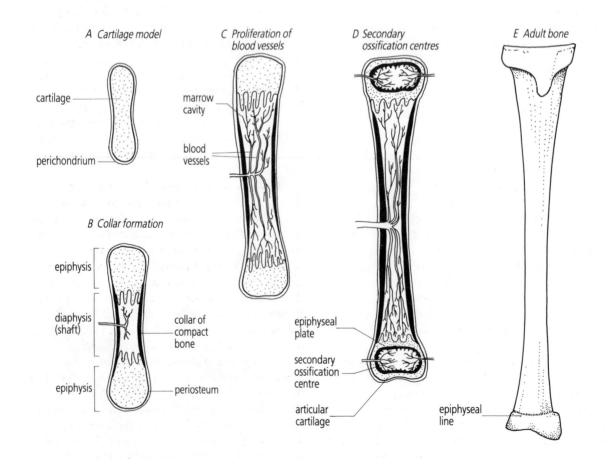

Fig 27·6 Endochondral ossification

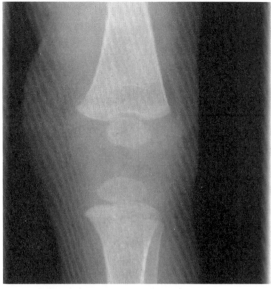

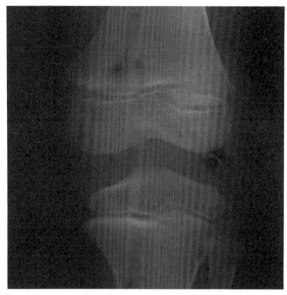

A Age 5 months

B Age 5 years 6 months

Fig 27·7 Development of long bones in the human leg

approaches, the thickness of the epiphyseal plates is reduced and, finally, the epiphyses and the bone shaft fuse completely, leaving a faint **epiphyseal line** (see Fig 27·6*E* and Fig. 27·7). Ossification of all bones in the human skeleton is normally complete by age 25.

Structure

The skeleton of an adult man is illustrated in Figure 27·8. It is convenient to describe its structure in two sections

1 Axial skeleton. The axial skeleton consists of the skull, the vertebral column, the sternum, and the ribs.

The skull comprises the **cranium** and the **face**. The fused bony plates of the cranium form a completely enclosed box, which supports and protects the brain. The face is supported by additional bones which protect the eyes and also provide sites of attachment for the muscles which move the lower jaw, or mandible, during mastication (described in Unit 13).

The vertebral column is made up of 33 small bones, or **vertebrae**, as shown in Figure 27·9*A*. Each vertebra is composed of a thick load-bearing **body**, a **vertebral arch**, and a total of seven projections, or **processes**. The bodies of adjacent vertebrae are held together by tough pads of fibrous cartilage called **intervertebral discs**, and, in life, the spinal cord passes through the large opening, or **vertebral foramen**, which is surrounded by the vertebral arch.

A pair of **transverse processes** extend laterally from each vertebra, while a single **spinous process** projects backwards. In addition, there are two pairs of **articular processes**: the superior articular processes are directed upwards, and the inferior articular processes downwards. In different regions of the vertebral column, these components are modified for different purposes.

The seven **cervical**, or neck, vertebrae are smaller than those of other regions and are easily identified because their transverse processes possess an additional opening, the **transverse foramen**, through which the vertebral arteries and veins can pass. Together with the carotid arteries, which lie in front of the vertebral column, the vertebral arteries supply oxygen and nutrients to the brain, so that this extra protection is clearly important. The spinous process of cervical vertebrae is usually forked to give a double tip.

As shown in Figure 27·9*B*, the first two cervical vertebrae, called respectively the **atlas** and **axis**, are specially shaped to give a wide range of head movement. The large articular surfaces of the atlas allow rocking movements of the head, that is, nodding, or 'yes' movements. A stout peg, called the **odontoid peg**, projects upwards from the axis and is held inside the atlas by a band of ligament. This arrangement enables side to side rotation, or 'no' movements.

Thoracic vertebrae (Fig. 27·9*C*) have well developed spinous and transverse processes and are generally heavier and more robust than cervical vertebrae. Their articular processes fit together in such a way as to limit the possible range of twisting movement. **Facets** for rib articulation are located on the body and transverse processes of the vertebrae. Two facets are needed for each rib. In most cases, one of these is formed between adjacent pairs of vertebrae both of which therefore possess incomplete facets, or **demifacets**.

The largest and strongest vertebrae in the human skeleton are the **lumbar** vertebrae found in the lower back region (Fig. 27·9*D*). They have short wide spinous and transverse processes which provide a secure attachment for back muscles. Their articular processes fit together closely and almost prevent twisting movement. The lumbar vertebrae are supported by the **sacrum**, a triangular wedge of bone formed by the fusion of five sacral vertebrae which is in turn attached firmly to the pelvic girdle. The **coccyx** is a tail remnant consisting of four tiny vertebrae. It has no function and is said to be a vestigial structure, as explained in Unit 42.

2 Appendicular skeleton. This includes the pectoral girdle, the pelvic girdle, and all the bones of the arms and legs.

The pectoral girdle attaches the arms to the axial skeleton. It comprises a shoulder blade, or **scapula**, and a collar bone, or **clavicle** on each side (refer to Fig. 27·8). The scapula is not directly connected to the rest of the skeleton, but is strapped in position by wide bands of muscle which are attached to the scapula over an extended area. The clavicle articulates with the scapula at one end and with the sternum at the other.

The pelvic girdle supports the weight of the upper body and articulates with the legs. It consists

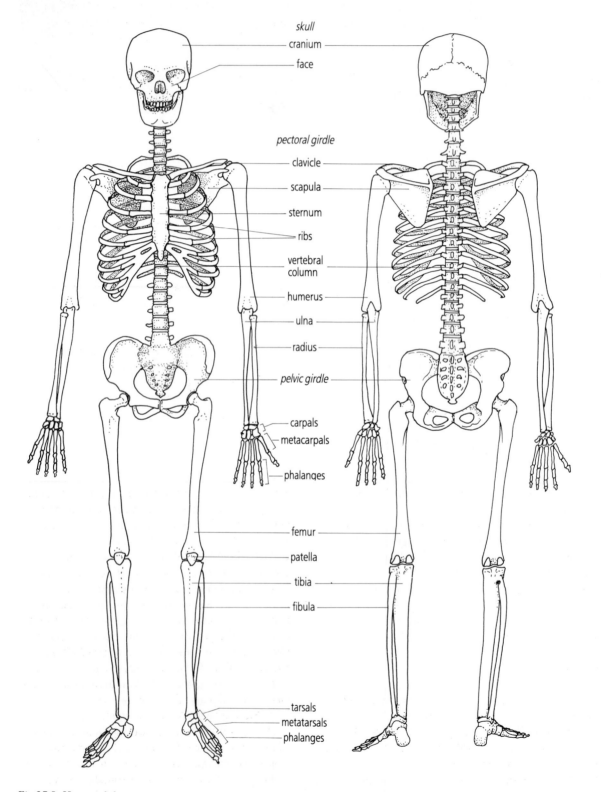

Fig 27·8 Human skeleton

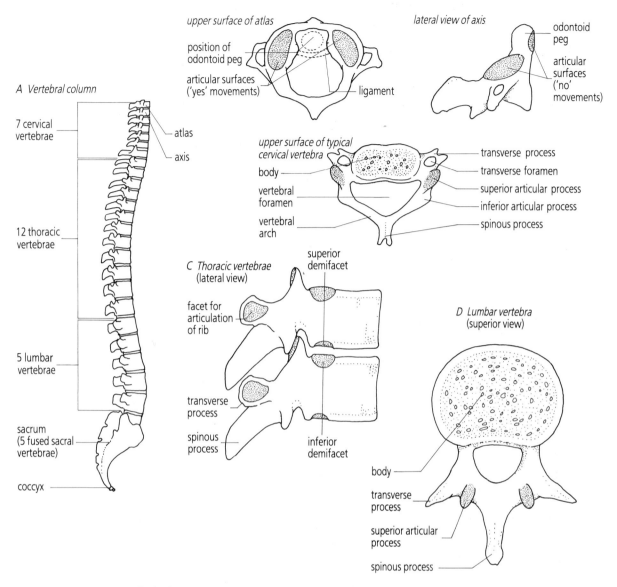

B Cervical vertebrae

upper surface of atlas

position of
odontoid peg

articular surfaces
('yes' movements)

ligament

lateral view of axis

odontoid
peg

articular
surfaces
('no'
movements)

A Vertebral column

7 cervical
vertebrae

atlas

axis

12 thoracic
vertebrae

5 lumbar
vertebrae

sacrum
(5 fused sacral
vertebrae)

coccyx

upper surface of typical
cervical vertebra

body

vertebral
foramen

vertebral
arch

transverse process

transverse foramen

superior articular process

inferior articular process

spinous process

C Thoracic vertebrae
(lateral view)

superior
demifacet

facet for
articulation
of rib

transverse
process

spinous
process

inferior
demifacet

D Lumbar vertebra
(superior view)

body

transverse
process

superior articular
process

spinous process

Fig 27·9 Bones of the vertebral column

of a pair of large bones, the left and right **coxal bones**, which are joined together at the front, and fused to the sacrum and coccyx at the rear. This arrangement of fused bones forms a strong rigid box called the **pelvis**. In females, the pelvis is wider and shallower than in males to provide adequate space for foetal development and childbirth.

The joints between the bones of the arms and legs and their movements are described below.

Joints

Joints occur where bones meet. Some joints hold bones tightly together and permit little or no movement, while others are adapted to provide free movement. Several different types of joints are distinguished according to their structure:

1 Fibrous joints. Bones linked by fibrous joints are held closely together by short fibres embedded in connective tissue. Examples include the jigsaw-like joints called **sutures** which link the bones of the skull, and the joints which fix the teeth into the jaws (described in Unit 13).

2 Cartilaginous joints. Like fibrous joints, cartilaginous joints allow little or no movement. The linking material may be hyaline cartilage, as in the case of the joints between the epiphysis and diaphysis of a growing bone. Alternatively, bones can be held together by fibrous cartilage. Such joints are found between the vertebrae, where the pad of cartilage forms an intervertebral disc, and also at the pubic symphysis where the coxal bones meet at the front of the pelvis.

3 Synovial joints. These joints contain a cavity filled with fluid and are adapted to reduce friction between moving bones. The structure of a typical synovial joint is illustrated in Figure 27·10. As you can see, the joint is surrounded by an **articular capsule** consisting of a thick tubular

221

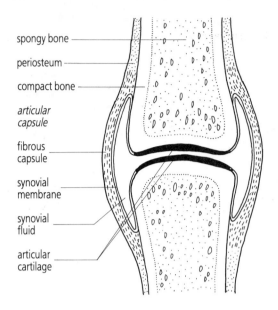

spongy bone

periosteum

compact bone

articular capsule

fibrous capsule

synovial membrane

synovial fluid

articular cartilage

Fig 27·10 Structure of a synovial joint

outer layer of connective tissue, known as the **fibrous capsule**, and a thin inner layer, the **synovial membrane**. The fibrous capsule is extremely strong but remains flexible. Some parts of the capsule may be modified to form distinct **ligaments** holding the bones together, and extra ligaments which do not form part of the capsule

are often present. Inside the joint, the synovial membrane secretes **synovial fluid** which lubricates the joint surfaces and transports food and oxygen to cells in the thin smooth layer of **articular cartilage** which covers the ends of the bones.

In some synovial joints, the ends of the bones are not directly in contact but are separated by an additional pad of fibrous cartilage, known as an **articular disc**, or **meniscus**. These structures reduce joint wear and are found, for example, in the knee joints. **Bursae** (singular bursa) are fluid-filled sacs located between bones and other body tissues. They occur between skin and bone where body movement causes the skin and bone to rub together, and are also found between ligaments and bones, between muscles and bones, and between tendons and bones. The outer covering of each bursa consists of connective tissue lined by a synovial membrane. This produces a lubricating fluid which fills the central cavity, reducing friction and acting as a cushion between the moving parts of the body.

The range of movement possible at individual synovial joints is determined mostly by the shapes of the bones which meet at the joint. Six different types of synovial joint are recognised, as summarised in Figure 27·11. You should study this figure carefully before reading the next section.

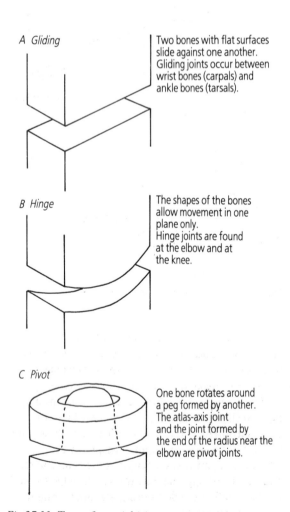

A *Gliding*

Two bones with flat surfaces slide against one another. Gliding joints occur between wrist bones (carpals) and ankle bones (tarsals).

B *Hinge*

The shapes of the bones allow movement in one plane only. Hinge joints are found at the elbow and at the knee.

C *Pivot*

One bone rotates around a peg formed by another. The atlas-axis joint and the joint formed by the end of the radius near the elbow are pivot joints.

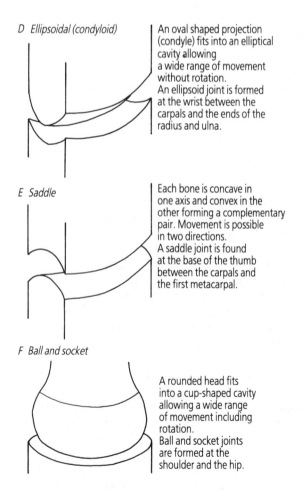

D *Ellipsoidal (condyloid)*

An oval shaped projection (condyle) fits into an elliptical cavity allowing a wide range of movement without rotation. An ellipsoid joint is formed at the wrist between the carpals and the ends of the radius and ulna.

E *Saddle*

Each bone is concave in one axis and convex in the other forming a complementary pair. Movement is possible in two directions. A saddle joint is found at the base of the thumb between the carpals and the first metacarpal.

F *Ball and socket*

A rounded head fits into a cup-shaped cavity allowing a wide range of movement including rotation. Ball and socket joints are formed at the shoulder and the hip.

Fig 27·11 Types of synovial joint

Movement

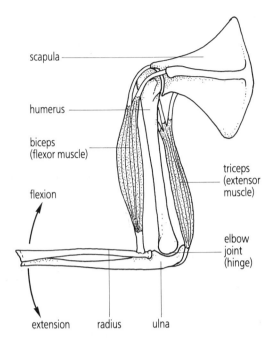

Fig 27·12 Muscles and bones of the elbow joint

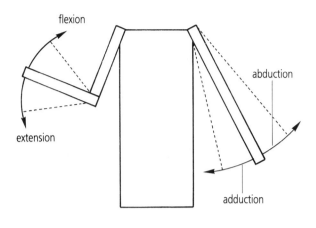

A Angular movements

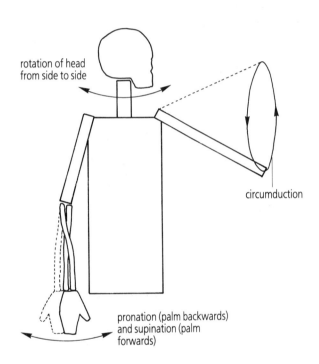

B Rotary movements

Fig 27·13 Movements of the human skeleton

Movement is produced by the action of muscles. As explained in Unit 26, muscles produce force by contraction, that is, a muscle shortens actively, but can be restored to its original length only by pulling its ends. It follows that movement at any joint requires two or more muscles. Frequently, these are arranged in pairs, called **antagonistic pairs**, which act in opposite directions. For example, the biceps and triceps of the human arm make an antagonistic pair (Fig. 27·12). When the biceps contracts, the arm is bent at the elbow and the forearm is raised. At the same time, contraction of the triceps is inhibited by reflex action, and its tendons are pulled apart, lengthening the muscle. If the triceps now contracts, the arm is straightened, contraction of the biceps is prevented, and the biceps is stretched. Reciprocal action of this kind is characteristic of joint movement.

The elbow is a hinge joint which permits movement in one plane, so that a single antagonistic pair is sufficient. The opposing movements produced in this case are called **flexion** and **extension**. At other joints, there may be more than one pair of muscles, and a variety of other movements are possible. **Angular movements** (Fig. 27·13A) include flexion/extension and **abduction/adduction** in which the limb is moved away from the midline of the body (abduction), or towards it (adduction). **Rotary movements** (Fig. 27·13B) are involved between the atlas and axis in turning the head, as already explained. In the arm, rotation of the radius at the elbow allows **pronation** and **supination** of the hand, while rotation at the ball and socket joint of

the shoulder produces a movement called **circumduction**, in which the limb can be moved through a complete circle. All these movements combine to make the arm extremely mobile.

The joints and muscles of the legs produce a more restricted range of movements which are compatible with the role of the legs in walking and support. The knee joint, for example, has additional ligaments to prevent lateral movement, and has a 'locking' mechanism which helps to hold the knee fully extended during standing without undue muscular effort.

Regulation of calcium

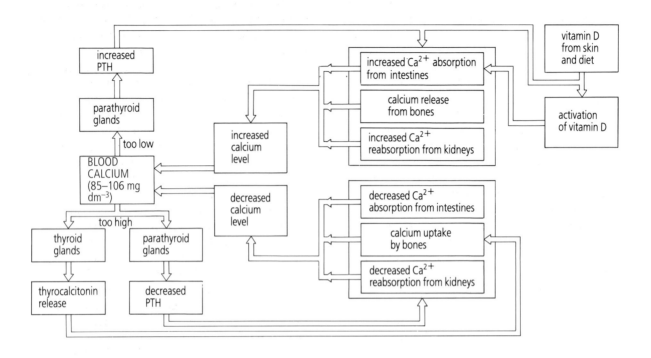

Fig 27·14 Regulation of blood calcium by parathyroid hormone (PTH), vitamin D, and thyrocalcitonin

In addition to its other functions, a person's skeleton acts as an important store of calcium and phosphate in the body. The levels of calcium circulating in the blood are regulated by two hormones, and by vitamin D, as summarised in Figure 27·14. **Parathyroid hormone (PTH)** is released by the parathyroid glands when blood calcium falls below its normal level of 85–106 mg dm^{-3}. Parathyroid hormone has three main effects:

1 PTH causes an increase in the amount of calcium absorbed by the intestine. This is achieved indirectly by the hormone's action on vitamin D, which is chemically altered to its active form. The active form circulates in the blood and is taken up by intestinal cells which are stimulated to produce the calcium carrier proteins responsible for absorption of calcium ions from the gut cavity.

2 PTH acts directly to increase the number and activity of osteoclasts. These cells break down the bone matrix and release calcium and phosphate ions into the blood.

3 PTH increases reabsorption of calcium ions by kidney tubules so that less calcium is lost in the urine.

These effects normally restore calcium to its correct level.

A second hormone, called **thyrocalcitonin**, is released if blood calcium rises above its normal range. Thyrocalcitonin is produced by the thyroid glands and acts by suppressing the activity of osteoclasts, allowing calcium and phosphate to be deposited in the skeleton. At the same time, less PTH is produced, so that absorption from the intestine and reabsorption by the kidneys are both reduced.

Disorders of calcium metabolism can result from shortages of calcium and vitamin D in the diet. In children, vitamin D deficiency gives rise to symptoms of **rickets**. The bones are soft because they are incompletely calcified and they are easily bent, producing serious abnormality. In adults, dietary deficiencies of vitamin D and calcium sometimes cause a similar condition, called **osteomalacia**. This is a rare disease which most often affects women during pregnancy, when large amounts of calcium are absorbed by the developing foetus, or during lactation, when calcium is transferred to milk. Treatment of rickets and osteomalacia usually involves dietary supplements of vitamin D, calcium and phosphate. Additional exposure of the body to ultra-violet light from sunlight or artificial sources allows vitamin D to be manufactured in the skin.

27·4 Locomotion in other vertebrates ■ ■ ■ ■ ■ ■ ■ ■ ■ ■

It is interesting to compare the specializations of the human skeleton with those of other vertebrates. All vertebrates share a common body plan and have skeletons formed of the same basic parts, but there are many differences. Some of these can be related to changes in habitat, for example, support and locomotion in water require adaptations which differ from those needed on land. Similarly, the human skeleton is adapted for bipedal locomotion, while most mammals are four-footed quadrupeds.

Fish gain support from their surroundings and have a streamlined body shape. Most species are propelled forwards by waves of muscle contraction which pass along the body from anterior to posterior, producing a characteristic S-bend motion. The body muscles are organized into separate muscle blocks, or **myotomes**, held together by connective tissue. The number and arrangement of these subdivisions correspond to the arrangement of vertebrae in the vertebral column. Each myotome has its own spinal nerve and can be made to contract individually. The vertebral column is a long rod consisting of incompressible skeletal elements made of cartilage or bone. These prevent shortening of the body, but the joints between them allow the vertebral column to bend easily. Consequently, contraction of the myotomes along one side of the body causes the body to curve towards the contracted side. Alternate contractions on both sides produce lashing movements of the tail which drive the fish forward through the water.

The S-shaped wriggle is seen clearly in cartilaginous fish, Class Chondrichthyes, like dogfish and sharks (Fig. 27·15). These fish have a density slightly greater than that of water and must swim continuously to avoid sinking. The large pectoral and pelvic fins attached to the pectoral and pelvic girdles direct swimming slightly upwards to counteract this. In addition, the tail fin, or caudal fin, has a large lower lobe and a smaller upper lobe. This shape is described as **heterocercal** and helps to produce lift, as well as forward motion. The remaining fins aid stability and help to keep the fish on a straight course through the water.

Bony fish, Class Osteichthyes, regulate their densities by means of a gas-filled bag, or **swim bladder**, so that their overall density exactly matches that of the surroundings. Consequently, all the energy needed for swimming can be directed into forward motion. Bony fish have a **homocercal** tail with equal upper and lower lobes. Lateral motion is usually less pronounced and most of the force of muscle contraction is transmitted to the water by the tail. The pectoral and pelvic fins are held flat against the body during rapid swimming, but are used for steering. In some species, undulations of the pectoral fins can be used to move the fish backwards. This ability is an important adaptation for fish such as the butterfly fish and angel fish of coral reefs, which feed by moving into and out of narrow crevices.

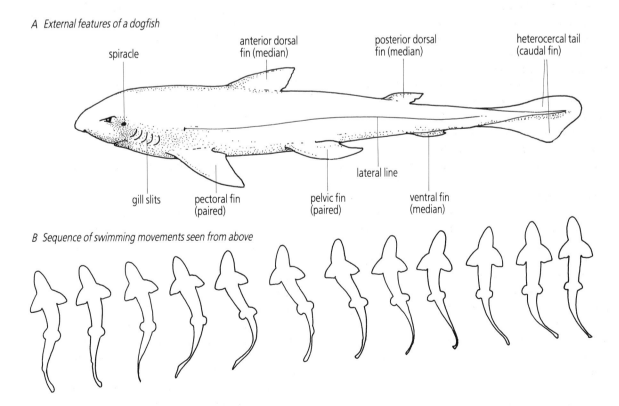

A External features of a dogfish

spiracle

anterior dorsal fin (median)

posterior dorsal fin (median)

heterocercal tail (caudal fin)

gill slits

pectoral fin (paired)

pelvic fin (paired)

lateral line

ventral fin (median)

B Sequence of swimming movements seen from above

Fig 27·15 Swimming in cartilaginous fish

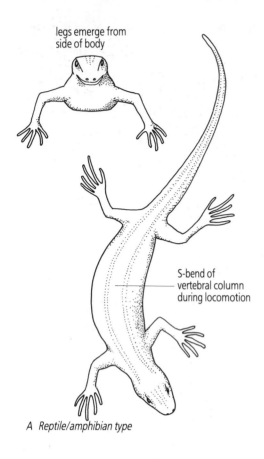

legs emerge from side of body

S-bend of vertebral column during locomotion

A Reptile/amphibian type

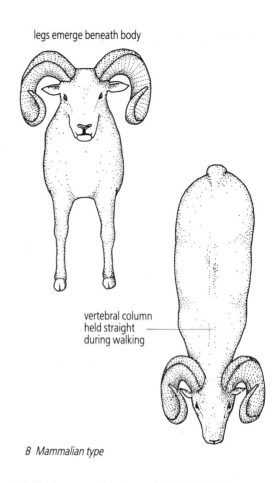

legs emerge beneath body

vertebral column held straight during walking

B Mammalian type

Fig 27·16 Locomotion in quadrupeds

Most land vertebrates are quadrupeds. In four-footed reptiles and amphibians, the legs emerge from the side of the body (Fig. 27·16A), and the S-wriggle is retained as part of the animals locomotion. In mammals (Fig. 27·16B), the legs project beneath the body, providing more effective support. Side to side bending of the body plays little or no part in normal locomotion. Instead, the vertebral column is held more or less rigid during walking, which is carried out using muscles attached to the limbs and limb girdles. In a running mammal, stride length and power are increased by arching the spine first upwards, at the beginning of each stride, and then downwards with the limbs fully extended. In this way, the force produced by the back muscles is transmitted to the ground.

Flight has evolved three times in vertebrates, namely in pterodactyls, birds and bats. It involves far more muscular effort than swimming, walking, or running, because air is a very light medium which is easily displaced. To generate sufficient lift to remain airborne a flying organism must have wings with a large surface area in contact with the air and must beat its wings powerfully.

Feathers are unique to birds. They develop as outgrowths of the epidermis in much the same way as the scales of reptiles or the hair of mammals, and, like these structures, they consist mainly of the fibrous protein keratin. Four types of feathers

are distinguished (Fig 27·17B). **Flight** feathers include the large primary and secondary feathers of the wings, and the large tail feathers. The flat vanes on either side of the feather shaft, or rachis, consist of an interlocking system of barbs and barbules which offers considerable resistance to the flow of air. **Contour** feathers cover most of the remaining area and give the wings and body a streamlined profile. The wings have an aerofoil shape which helps to create lift. Between the contour feathers, **down** feathers, with short shafts and loose barbs, and **filoplume** feathers, with hair-like shafts and a tuft of barbs at the tip, provide effective insulation and help to reduce heat loss. (Note that birds, like mammals, are endothermic.)

The skeleton of a bird (Fig 27·17A) is highly modified for flight. Among the more obvious adaptations are the enlargement of the pectoral girdle and the development of the sternum to form a massive keel for the attachment of the flight muscles. The **pectoralis** muscles provide the power for the downstroke of the wings, while the **supracoracoideus** muscles provide power for the upstroke. This lifting action is possible because the tendon of the supracoracoideus muscle passes through an opening, the **foramen triosseum**, formed between the scapula, coracoid, and clavicle bones, and is attached to the upper surface of the humerus (Fig 27·17C). The number of bones is

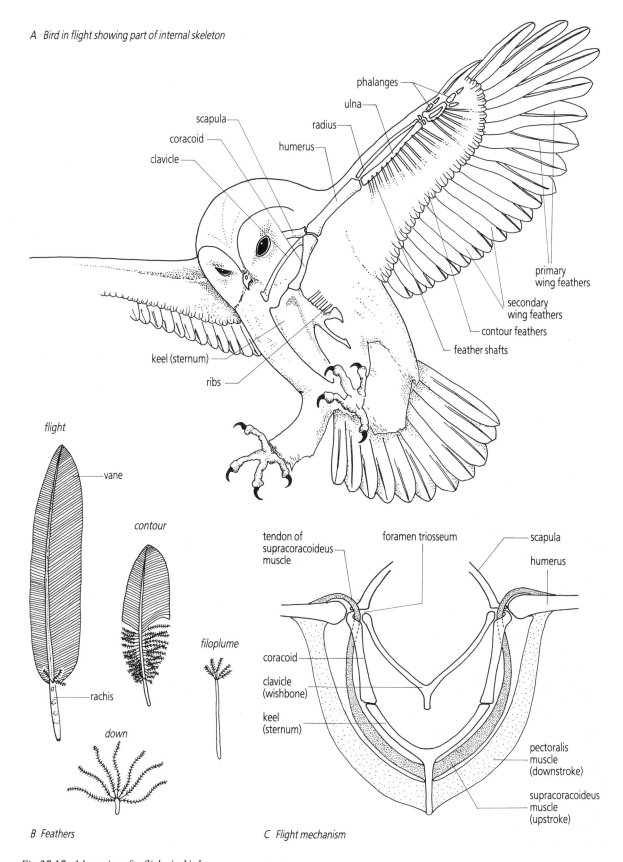

A Bird in flight showing part of internal skeleton

phalanges

ulna

radius

humerus

scapula

coracoid

clavicle

primary wing feathers

secondary wing feathers

contour feathers

feather shafts

keel (sternum)

ribs

flight

vane

contour

filoplume

rachis

down

B Feathers

tendon of supracoracoideus muscle

foramen triosseum

scapula

humerus

coracoid

clavicle (wishbone)

keel (sternum)

pectoralis muscle (downstroke)

supracoracoideus muscle (upstroke)

C Flight mechanism

Fig 27·17 Adaptations for flight in birds

reduced compared with those in the limbs of other vertebrates, and many bones are fused together to increase strength.

The shapes of the wings greatly influence the speed and type of flight which can be achieved. For example, long narrow wings like those of gulls and other seabirds are ideal for gliding into the wind, while short broad wings like those of many garden birds are effective for slow flapping flight.

Bats have a quite different arrangement of wing bones (refer to Fig 42·4), but show a parallel range of adaptations for flight.

28 PATTERNS OF BEHAVIOUR

Objectives

After studying this Unit you should be able to:–

- Draw a diagram to show how receptors and effectors are linked by the nervous system and outline the importance of feedback in controlling behaviour

- Compare 'instinct' and 'learning' and comment on the usefulness of these terms

- Give examples of reflex behaviour

- List and explain the important characteristics of fixed action patterns

- Comment on the importance of sign stimuli and stimulus filtering

- Define motivation and drive

- Give examples of circadian, tidal, and annual rhythms

- Distinguish between return migration, remigration, and removal migration

- List some of the external stimuli used for navigation

- Describe the behavioural interactions between individuals involved in alarm signals, territorial behaviour, agonistic behaviour, courtship, and parental care

- Give an account of social organization and communication in honeybees

- List the main types of learning and comment on their usefulness in the lives of animals

28·1 Introduction ■ ■ ■ ■ ■ ■ ■ ■ ■ ■ ■ ■ ■ ■ ■ ■ ■

Any action taken by an organism is part of its behaviour. For such actions to be possible, organisms must possess **receptors** sensitive to external and internal stimuli, that is, to changes in external and internal conditions. They must also have **effectors**, like muscles or glands, which operate to produce particular responses. Animals are heterotrophic. Typically, they are mobile organisms which search actively for food and shelter. Communication between individuals is needed for successful breeding and care of the young, and sometimes for social behaviour. To cope with all these activities, most animals have a well-developed **nervous system** which coordinates their behaviour, as outlined in Figure 28·1.

An important concept in understanding behaviour patterns is **feedback**. As you can see, the behavioural output can alter the stimuli received, possibly causing further behaviour changes. For example, the presence of water may stimulate drinking behaviour. It is important that drinking should not continue indefinitely but should cease when the animal's needs have been met. This is achieved by feedback involving stimulation of sense organs in the mouth and stomach. Drinking is an important factor in the regulation of body water, helping to maintain a constant osmotic concentration in the body fluids. Many other types of behaviour contribute to homeostasis in a similar way.

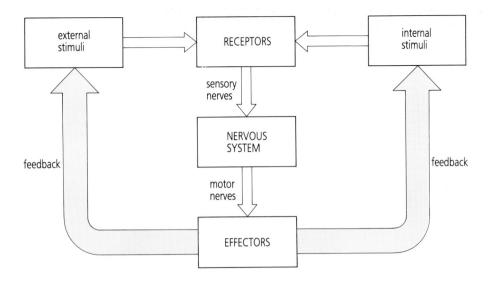

Fig 28·1 Linking role of the nervous system in the control of behaviour

28·2 'Instinct' and 'learning' ■ ■ ■ ■ ■ ■ ■ ■ ■ ■ ■ ■ ■

In the 1920's and 1930's, two very different approaches to the study of behaviour developed. One group of investigators, called **ethologists**, inspired by the work of Lorenz and Tinbergen, studied animals in their natural surroundings and sought to explain the nature of their adaptive behaviour. The second group, known as **behaviourists**, included Watson and Skinner and consisted mainly of experimental psychologists. The behaviourists focused attention on learning processes which were investigated under carefully controlled laboratory conditions. Only a few organisms, notably the white rat, were studied.

Ethologists found it useful to distinguish between **innate** or **instinctive** behaviour, which was inborn and assumed to be genetically determined, and **learned** behaviour, which was acquired as a result of experience. Insects provide many examples of inherited behaviour patterns. Potter wasps, *Eumenes pedunculatus*, are entirely solitary. Males and females emerge in the summer, and, after mating, the female builds a flask-shaped nest formed of fine grains of sand or mud cemented together with saliva and attached to heather or similar vegetation. Inside the nest the female lays a single egg. Then the nest is provisioned with the paralysed bodies of several caterpillars which provide food for the developing larva. Finally, the completed nest is sealed and the female flies off to build another. The entire sequence of mating, nest-building, egg-laying, hunting, and provisioning, is carried out with no training or prior experience whatsoever and it seems reasonable to conclude that this behaviour

must be genetically determined. The life cycles of insects are too short to accomodate long periods of trial and error behaviour leading to learning. Consequently, most complex behaviour patterns are 'pre-programmed' in this way.

On the other hand, young mammals are usually born helpless and must be fed and protected by their parents. Their behaviour develops slowly as they learn by experience and by example. This extended period of development allows complex behaviour patterns to remain flexible so that the animal can adjust its behaviour to suit different conditions.

Based on observations like these, ethologists defined instincts as complex inherited behaviour patterns which would appear in isolated individuals. Instinctive behaviour could be triggered by simple stimuli and was rather inflexible. Learning was defined as the modification of behaviour by experience. The great majority of animals seemed to rely heavily on instinct, while learning was important in the lives of only a few. Behaviourists rejected this interpretation and asserted that learning was the dominant influence in the development of most types of animal behaviour.

Detailed analysis of behaviour patterns does not support either of these views. Mallard ducklings respond selectively to the calls of adult mallards from the moment of hatching and it would be very easy to label this behaviour as 'instinctive' and genetically determined. However, it has been shown that an embryo duckling must be exposed to its own sounds and those of other ducklings while still within the eggshell if this preference is to develop normally. Similarly, the tendency of cats to pursue mice would popularly be described as 'instinct' but involves a whole range of learned and practised body movements and also requires the normal development of the eyes, brain, and muscles. Just what proportion of the animal's hunting behaviour is due to its genetic inheritance and how much should be attributed to its environment is impossible to assess.

Clearly, all behaviour is affected by an organism's genetic make-up and also by its surroundings. Once this is accepted, it becomes pointless and misleading to classify behaviour either as 'instinctive', that is, purely genetic in origin, or as purely learned. Attempts to divide behaviour in this way have hindered rather than helped progress in understanding behavioural mechanisms. It is much more important to discover how particular behavioural patterns develop and to investigate their adaptive functions in helping organisms to survive.

8·3 Reflexes, fixed action patterns, and stimulus filtering ■ ■ ■

It is helpful at this point to think of behaviour as a means of providing economic solutions to survival problems. Like other features of the organism which affect survival, behaviour is subject to evolutionary improvement and it should be no surprise that animals often produce effective behaviour with more or less the minimum equipment in terms of sense organs and the nervous system.

As an example, consider the behaviour of certain kinds of moth. Bats produce pulses of ultrasonic sound which are used for the echolocation of prey such as moths. In some moths, an escape behaviour has evolved – this is as follows: when the bat is some distance away, the moth responds to its cries by flying very irregularly with the result that echolocation becomes more difficult. If it is successful in evading the bat, the moth returns to normal flight. However, if the bat comes nearer and its cries become more intense, the moth folds its wings and drops directly to the ground. This combination of responses often saves a moth from being eaten, but what sensory equipment is needed to make a moth respond in this way?

Noctuid moths have hearing organs, called **tympanal organs**, located between the thorax and abdomen (Fig. 28·2). On each side, there is a tympanum which vibrates for sounds between 3000 Hz and 100 000 Hz. The vibrations of the tympanum are detected by two sensory structures called scolophores, each of which contains a single sensory cell. These send axons to the moth's central nervous system. The two sensory cells differ in their threshold of response. One or both will respond depending on the intensity of sound,

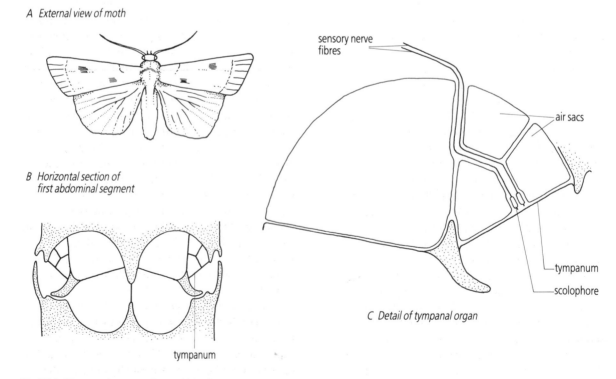

A *External view of moth*

B *Horizontal section of first abdominal segment*

tympanum

sensory nerve fibres

air sacs

tympanum

scolophore

C *Detail of tympanal organ*

Fig 28·2 *Tympanal organs of noctuid moths*

with a maximum sensitivity at about 13 000 – 16 000 Hz. The behavioural output is determined by the number of scolophores which are active and is entirely automatic, or **reflex**, in nature. Thus, an effective response is achieved with just two sensory cells on each side and a minimum of central processing.

Reflexes in other animals often have a similar protective function and are also important in helping the organism to remain in a favourable environment. **Taxes** and **kineses** are simple forms of behaviour based on reflexes. Any movement towards or away from a stimulus where the direction of the stimulus is important is called a taxis. Many organisms show movements towards (positive) or away from (negative) stimuli such as light, temperature, chemicals, and gravity. *Euglena* and other photosynthetic protoctists are positively phototactic, that is, they invariably swim towards a light source of moderate intensity. A kinesis is an increase in the *rate* of movement which depends on the intensity of the stimulus but is unaffected by the direction of the stimulus. For example, a woodlouse moves slowly if at all in damp conditions, but moves about rapidly in dry conditions. The role of reflexes in vertebrates is described in Unit 23.

Observation of animals in their natural surroundings reveals that much of their behaviour consists of sequences of movements called **fixed action patterns**. These resemble reflexes but the responses produced are more complex. Fixed action patterns share the following characteristics.

1 A fixed action pattern consists of a stereotyped sequence of movements which are always performed together. This is seen, for instance, in the courtship display of the male domestic pigeon, which involves a ritual of repeated cooing, strutting and head-bobbing with the tail and wing feathers spread.

2 Once a fixed action pattern has been initiated it is usually completed. When a young cuckoo hatches before the young of its foster parents, it responds to the presence of unhatched eggs by backward pushing movements which result in the eggs being thrown from the nest (Fig. 28·3). These movements begin when an egg is in contact with the young cuckoo's back. The egg is normally retained in this position by the cuckoo's outspread wings, but, if it rolls off back into the nest, pushing movements still continue all the way to the rim of the nest. In a similar way, the adults of many species of ground-nesting birds retrieve an egg outside the nest using their beaks to roll the egg inwards. This response is very stereotyped and the beak is moved fully to the chest even if the egg slips off to one side, in which case the entire sequence of movements will be repeated over and over until the egg is successfully retrieved.

3 An important feature of these stereotyped behaviour patterns is that all of them are produced

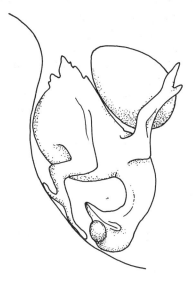

Fig 28·3 Fixed action patterns: A cuckoo removing unhatched eggs from the nest of its foster parents

in response to very simple stimuli, called **sign stimuli**, or **releasers**. This means that the organisms respond to a very restricted part of the total environment. For example, during the breeding season, a male robin defends his territory by attacking other males. The sign stimulus is the red breast of the intruder. A male robin will attack a tuft of red feathers on the end of a wire in just the same way, but will ignore a detailed model bird which lacks the red breast.

Sometimes an artificial stimulus can be found which is more effective than the normal stimulus. Herring gull chicks peck at a red spot near the tip of the parent's beak, an action which triggers the regurgitation of food by the adult bird. Investigation of this response by Tinbergen and others showed that chicks pecked more frequently at a crude cardboard cut out representing a gull's head, provided this had a red spot, than they did at the beak of a complete adult gull with the red spot painted over. Further tests showed that, while red spots were the most effective, other colours which contrasted with the background colour of the beak also initiated the pecking response. Long thin beaks were more attractive than short wide ones. Once he had discovered the important features of the sign stimulus, Tinbergen was able to exaggerate them to create a **supernormal stimulus**. This was a thin red rod with three white stripes near the tip. The chicks pecked more vigorously at the rod than at any other model and consistently preferred it when offered a choice (Fig. 28·4).

4 Careful study of seemingly 'intelligent' behaviour often reveals that sign stimuli are involved so that behaviour can be triggered inappropriately under experimental conditions. Nevertheless, fixed action patterns are adaptive and normally increase the survival chances of the organism or its offspring.

Fig 28·4 Fixed action patterns: pecking response of a herring gull chick to a supernormal stimulus

found in vertebrate nervous systems allow the individual cells to be much smaller so that many more cells can be accomodated, with a corresponding increase in the sophistication of behavioural responses.

The structure and capabilities of the sense organs help to determine the range of stimuli to which an animal can respond. Most animals have sensory capabilities which are different from those of humans. Sometimes, the variations are slight, as in the case of a dog's greater sensitivity to high pitched sounds, but some species have sense organs for which there is no human equivalent. For example, rattlesnakes have pit organs on either side of the head. These are sensitive to infra-red radiation and are used by the snake to locate its prey of small mammals in the darkness. Honeybees respond to polarized light and ultra-violet light both of which are invisible to humans. On the other hand, bees are blind to red light.

The limitations of the sense organs represent one form of **stimulus filtering**, as a result of which organisms respond only to particular features of the environment. Information processing by the sense organs and nervous system further restricts the range of effective stimuli. A frog's eyes, for instance, are specialised as movement detectors in such a way that a moving fly is speedily detected and captured. However, a frog would starve surrounded by freshly killed flies which failed to move. The use of sign stimuli can also be interpreted as a type of stimulus filtering.

Lack of variablility and the use of sign stimuli can both be interpreted as adaptations which promote an efficient use of the nervous system.

The size of the nervous system will always be limited. Therefore, by reducing the number of neurons required for a particular response, the organism can respond appropriately to a wider range of situations. This is particularly true among invertebrates where myelination of nerve cells is poorly developed. The insulating myelin sheaths

28·4 Motivation ■ ■ ■ ■ ■ ■ ■ ■ ■ ■ ■ ■ ■ ■ ■ ■ ■

Suppose you are given a cream bun and that you eat it. Suppose you are given another – will your response to the second be the same as your response to the first? An important observation is that the same stimulus does not always evoke the same response, or, in other words, that the responsiveness of an organism can vary. **Motivation** is the factor which controls responsiveness. A **drive** is motivation for a particular type of activity. Hunger is a drive, thirst is a drive, while sex drive refers to an organism's readiness for sexual activity.

Motivation can arise from internal stimuli. For example, hormones produced by the sex organs

circulate in the blood and act on parts of the brain to stimulate sexual behaviour. However, not all of an animal's activities are initiated by definite external or internal stimuli. Nervous systems are constantly active and contain numerous **'pacemaker'** cells which fire off impulses even in the absence of stimulation. In mammals, such cells determine the basic pattern of breathing movements and are involved in regulating 'waking state'. They may also be important in triggering exploratory behaviour and other forms of behaviour for which there seems to be no immediate stimulus.

28·5 Rhythmic behaviour ■ ■ ■ ■ ■ ■ ■ ■ ■ ■ ■ ■ ■

Almost all organisms show definite cycles or rhythms affecting their metabolism and behaviour. Changes in activity, sleep, feeding and drinking, body temperature and other processes often follow

a cycle of approximately 24 hours and are known as **circadian rhythms** (from the Latin *circa*, approximately, and *dies*, day). It is clearly an advantage for organisms to be active when food is available and predators are scarce, and to be inactive at other times. Some species are **diurnal**, that is, they are most active during daylight hours. Large herbivores like antelope and zebra are typically diurnal. **Nocturnal** animals are active at night and include many desert species, and creatures like bats and owls which have evolved special mechanisms for navigation and hunting in darkness. Still other species, like some small mammals, are **crepuscular**, with maximum activity at dawn and dusk.

Some of these rhythms are triggered by changes in the surroundings, but, very often, rhythms appear to be driven by internal mechanisms known as **biological clocks**. Most organisms kept in constant conditions continue to show repeated cycles of activity more or less indefinitely, but the 'free-running' period of the rhythm is often shorter or longer than 24 hours, so that organisms in the laboratory become progressively out of step with conditions prevailing outside. This is strong evidence for an internal control mechanism. On the other hand, external stimuli, like light and darkness, are needed to keep the rhythms properly synchronised and to permit seasonal adjustments for changes in daylength.

Not all rhythms have cycle times of 24 hours. Intertidal organisms have cycles linked to the tides. Fiddler crabs leave their burrows to feed at low tide, that is, twice every 24 hours, while many species time their spawning to coincide with high spring tides, which occur every 14 days. Sometimes, there are seasonal variations in the spawning pattern, giving an annual cycle as well. The grunion, *Leuresthes tennuis*, is a small fish which spawns on the beaches of California from March to September on nights when the tides are at their highest. At the peak of spawning, thousands of small silver fish appear. Each female, accompanied by several males, swims with a breaking wave as high up onto the beach as possible. The female digs into the sand and deposits 1000–3000 eggs which are fertilized by sperm from the males. When the tide starts to fall, the fish swim back into the sea. The eggs develop in the damp sand until they hatch, and the young grunion are washed away by the next high tide. Annual cycles of breeding behaviour are common in organisms from temperate regions and help to ensure that the young are produced at a time when adequate food is available.

28·6 Migration and navigation ■ ■ ■ ■ ■ ■ ■ ■ ■ ■ ■ ■ ■

Migration is defined as the act of moving from one place to another. Three types of movement can be usefully distinguished, as outlined in Figure 28·5. **Return migration** (Fig. 28·5*A*) involves journeys in two directions by the same individual. The distances covered may be modest, as in the case of the daily movement of plankton from the surface water of the oceans during the day to deeper water at night. Arctic terns, on the other hand, migrate between the North and South poles, a round trip of 40 000 km, and spend up to 8 months of their

annual cycle in non-stop flight. **Remigration** (Fig. 28·5*B*) is a return movement completed by later generations. The best known example is that of North American monarch butterflies, *Danaus plexippus*. Marking experiments indicate that almost 60% of the adults which emerge in the Great Lakes region in the late summer and early autumn migrate 2000–3000 km to overwintering sites in Mexico and Texas. Here the monarchs are relatively inactive and often cluster together in enormous numbers on suitable trees, flying only on

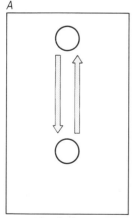
return migration is a to-and-fro movement made by a single individual

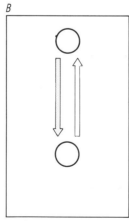
re-migration is a return movement, completed by later generations

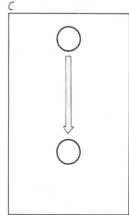
removal migration is a one-way movement made without any intention of returning

Fig 28·5 Different types of migration

the warmer days. Mating starts in February and is followed by a slow northwards migration. Most of the overwintering adults die without travelling far, and it is their offspring which complete the journey to the Great Lakes. Not all monarchs migrate: most regions support small resident populations of individuals which hibernate in hollow trees or under the bark. The third variety of migration, called **removal migration** (Fig. 28·5C), is a one-way movement of the type which enables organisms to colonise new areas. Weak-flying insects like aphids, for example, are dispersed by air currents.

Migrations are often linked to behavioural cycles and to changes in external conditions which trigger migratory behaviour. A bigger problem is to explain how organisms navigate over enormous distances with such precision. European swallows, *Hirundo rustica*, overwinter in central and southern Africa, but individuals are able to return for several years to the same nest site in a barn or under the eaves of an English house. Many migrant birds perform similar feats of navigation. Precisely how these are achieved is not fully understood, but there seems to be no single simple explanation. Some birds orientate with respect to the sun and are able to compensate for its changing position during the day. Night migrants appear to use the moon in a similar way. Larger birds glide on thermal air currents generated over the land and therefore avoid flights over water. Consequently, migrating individuals crossing the Mediterranean from Asia and Europe to Africa are channelled towards one of a number of crossing points including the Straits of Gibraltar in the west, and the Bosphorus in the east. Visual cues like coastlines and mountain ranges may also be important.

Intensive studies of the behaviour of homing pigeons have indicated that the birds are sensitive to magnetic fields and can use the Earth's magnetic field for navigation. For instance, pigeons transported to a release site in aluminium boxes, which do not affect the magnetic field experienced, were able to navigate to the home loft successfully, while individuals transported in iron boxes, which disturb the magnetic field, were unable to do so. Sensitivity to magnetic fields has been demonstrated in many other organsisms including bacteria, beetles, snails, fish, salamanders, and some birds and mammals. In most of these cases, it remains to be discovered to what extent the magnetic sense contributes to orientation and navigation.

In aquatic organisms, stimuli like water temperature, the strength and direction of water currents, and the detailed chemical composition of the water are important. Salmon returning from the sea to the freshwater rivers where they were spawned are able to find their own particular tributary by detecting its unique chemical signature or 'smell', and are unable to navigate if their nostrils are blocked.

28·7 Interaction between individuals ■ ■ ■ ■ ■ ■ ■ ■ ■

Many aspects of an organism's behaviour are directed towards ensuring its own survival, but natural selection also favours behaviour which promotes the survival of the offspring or other related individuals. Such behaviour often involves interaction between individuals.

1 Alarm signals. Alarm signals like the shrill alarm call of a blackbird, the slap of a beaver's tail on the surface of the water, and the display of prominent patches of colour or white fur by mammals, all warn nearby individuals of the approach of predators. The individual giving the alarm signal reduces its own chances of survival even if only slightly, but very often its offspring are likely to benefit. The alarm calls of many species of birds are very similar and individuals often respond to the alarm calls given by a different species.

2 Territorial behaviour. A territory is an area defended against intruders of the same species. The males of many species establish territories prior to mating. The size of the territory and the length of time for which it is defended vary depending upon whether it must provide food for the developing offspring. Blackbirds in woodland have relatively large territories, while herring gulls and many other sea birds compete for suitable nesting sites and establish small territories only a few metres across.

3 Agonistic behaviour. Agonistic behaviour includes threat displays and ritualized fighting. Very often such behaviour is associated with defence of a territory or conflict between rival males. Male herring gulls react to the presence of other male gulls by adopting the upright threat posture (Fig. 28·6A). Once territories have been defined, this posture is usually sufficient to deter an intruder, which may adopt a different posture, known as the submissive posture (Fig. 28·6B), before retreating. However, if the territorial boundaries are not clear, or if the intruding male does not retreat, the males may attack one another. Usually, the attack is short-lived and the weaker individual retreats without sustaining any serious injury. The threat and submissive postures are fixed action patterns which act as sign stimuli triggering the appropriate behaviour in other individuals.

Fig 28·6 Herring gull postures

It is important to appreciate that fighting between individuals is almost always ritualised in this way. Rival red deer stags engage in a trial of strength which looks dramatic, but the protagonists seldom suffer more than minor injuries. The contest is of value in helping to ensure that the strongest and fittest males are the ones most likely to breed. Exceptionally aggressive males are liable to suffer damage themselves and are often less effective in holding together a harem of hinds, so that they father fewer offspring than moderately aggressive males.

4 Courtship. Mating between individuals of the same species is usually essential for the production of healthy, fertile, and well-adapted offspring. Courtship displays have the important effect of making intraspecific mating more certain and preventing the wastage of gametes. The courtship rituals of great crested grebes, *Podiceps cristatus*, are illustrated in Figure 28·7 and provide an example which is typical. Each component of the display represents a fixed action pattern and acts as a trigger or sign stimulus for the next stage. As explained in Unit 41, the displays of closely related species vary in significant ways, with the result that breeding between different species is minimised.

5 Parental care. Adult organisms invest considerable time and effort in feeding and protecting their young. Inevitably this reduces their own chances of survival, for example, by spending more time in the open collecting food, adult birds are more likely to be captured by hawks or other predators. On the other hand, parental care greatly

improves the survival chances of the offspring, so that it is often favoured by natural selection.

The importance of the direct blood relationship between parents and offspring is evident from the behaviour of herring gulls. A pair of adults will incubate their own eggs and feed their own chicks, but invariably attack and eat unguarded eggs and chicks from neighbouring territories.

6 Social behaviour. Social behaviour appears to have developed from parental care with related individuals remaining together first as family groups, and later as larger groups. The evolutionary mechanisms involved are outlined in Unit 42. In this Unit, the social behaviour of honeybees is described.

A honeybee colony consists of a single **queen** and as many as 80 000 sterile female **workers**. Male bees called **drones** are sometimes present – they mate with the young queen but have no other useful function. The queen lays up to 1000 eggs per day into wax cells. Developing larvae are fed for about 3 days on a secretion known as royal jelly, produced by the pharyngeal glands of nurse bees, and for a further 3 days on a mixture of pollen and honey, called beebread. After this, the wax cell is capped and the larva pupates inside, emerging as an adult worker about 12 days later. Workers carry out a definite sequence of activities determined partly by their age and partly by the needs of the hive. At first, they act as nurse bees cleaning empty cells and feeding the developing brood. After 5–7 days, wax secretion starts and the workers begin to cap existing cells and build new

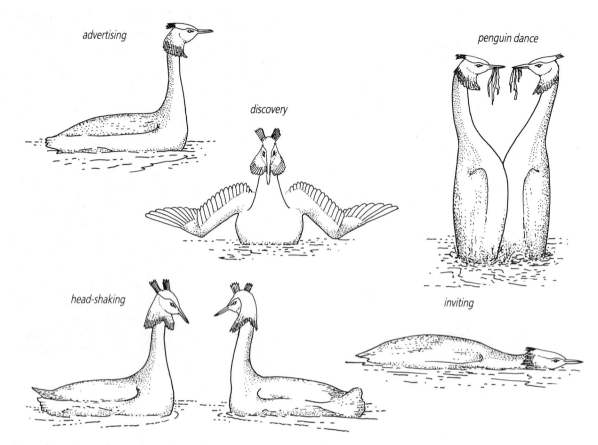

Fig 28·7 Courtship of the great crested grebe

ones. Older workers guard the hive entrance and become foragers, flying out to collect nectar and pollen. Worker bees normally live for 3–5 weeks, while queens may survive for 5 years.

The behaviour of the entire hive is regulated by chemicals, called **pheromones**, which act as signals between individuals. The most important of these is **queen substance** released by the queen and distributed throughout the hive. The workers which attend the queen constantly lick her body and transfer her secretions to other workers by mutual feeding. Queen substance suppresses the development of the ovaries of the workers and helps to trigger the appropriate behaviour patterns. If the colony becomes too large, or if the old queen dies, the supply of pheromone rapidly decreases and some of the workers are stimulated to build specially large queen cells and to continue feeding the larvae inside on a diet of royal jelly. New queens attempt to sting rival queens and queen larvae to death and may found new colonies by swarming.

A fascinating aspect of honeybee behaviour is the ability of foragers to communicate the distance and direction of rich sources of nectar to other workers. This ability had been suspected for hundreds of years, but was first investigated in detail by Karl von Frisch. Like others before him, von Frisch had observed that a dish of sugar solution in the vicinity of a hive might remain

undetected for several hours, but that, once it had been visited by a single forager, it was usually only a few minutes before many other foragers arrived. To find out how information was passed on, von Frisch marked feeding foragers with a coloured dot of paint and observed their behaviour upon return to the hive. His patient and careful experiments carried out over 20 years revealed an astonishing method of communication involving stereotyped body movements performed as a kind of **dance**.

When the food dish is less than 50 m from the hive, a returning forager first contacts a number of other workers on the vertical surface of the comb and shares collected food with them. Then she begins to dance following a roughly circular path alternately to the left and right. This is the **round dance** (Fig. 28·8*A*) and stimulates other bees to begin searching near the hive. Scent stimuli picked up from the body of the dancer may help in locating the food source. Von Frisch discovered that, as the food dish was moved further from the hive, the dance gradually altered to a pattern known as the **waggle dance** (Fig. 28·8*B*), with a rapid forward run incorporated between the turns. During this run, the bee's abdomen vibrates from side to side and bursts of high pitched sound are produced. The waggle dance conveys information in several ways. The angle of the waggle run compared to the vertical is equal to the angle between the food source and the sun and indicates

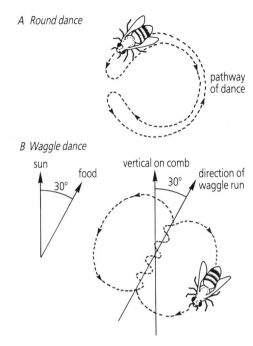

A Round dance

pathway of dance

B Waggle dance

sun

food

30°

vertical on comb

30°

direction of waggle run

Fig 28·8 Communication in honeybees

the direction of the food source. Thus, with a food dish always in the same place, von Frisch observed that the angle of the dance changed during the day, corresponding to the apparent movement of the sun. The interval between dances, the number of waggles, the sounds produced, and the length of the waggle run, all help to indicate distance. The scent of flowers on the body and fur of the forager provide further clues for the identification of the food source. Recent experiments confirm that workers are able to respond to these features of the waggle dance and use them to locate food.

Vertebrate societies differ from insect societies in a number of ways. In particular, there is no parallel for the rigid caste system found in insects. An individual's position in a vertebrate society is not fixed forever but may change depending on the behaviour of other members of the group. Human society depends to a large extent on **culture**, that is, information which is transmitted from generation to generation by non-genetic means. Well-developed language and the use of tools are additional distinguishing features.

28·8 Modification of behaviour ■ ■ ■ ■ ■ ■ ■ ■ ■ ■ ■ ■

The behaviour of most animals alters with time. Often this is the result of developmental processes. For example, newly hatched chickens peck at small objects and their accuracy, measured by the scatter of pecking attempts, improves over the first few days of life. A similar reduction in scatter is observed if the chicks are made to wear hoods with prisms so that they consistently aim wide, or even if the chicks are prevented from pecking for the first few days. Effects of this sort are described as **maturation** and must be attributed to changes in the neuromuscular system of the chick which do not vary with experience. On the other hand, many changes in behaviour result directly from experience and can be properly described as **learning**.

While reading about the different varieties of learning you should remember that an animal's capacity to learn and its ability to learn particular things depend on the structure of its sense organs and nervous system, which are largely determined in turn by the organism's genetic inheritance. Learning introduces a degree of flexibility into the lives of many animals, allowing them to respond more adaptively to their surroundings.

1 Habituation. A sudden shadow causes a ragworm, *Nereis*, to retreat into its burrow. However, if the stimulus is repeated at intervals, the response becomes less pronounced and eventually fades. This is an example of a simple form of learning known as habituation. In general, habituation enables organisms to ignore repeated

stimuli which are not followed by any significant cost or benefit.

2 Imprinting. Shortly after hatching, ducklings and other young birds have a tendency to follow moving objects in their surroundings and show a brief sensitive period during which the shape and form of objects can be 'imprinted', with the result that the young birds will follow them selectively. Normally, of course, the first moving object encountered is the mother bird, and it is obviously adaptive for the young birds to learn her appearance and to follow her. However, if its parents are absent, a young bird may imprint on other species of birds, human beings, or inanimate objects. In later life, such birds will attempt to court and mate with imprinted objects in preference to adults of their own species.

3 Conditioning. One type of conditioning, known as **classical conditioning**, takes place when a new stimulus is presented together with the normal stimulus for a reflex activity. In the early 1900s, the Russian physiologist Pavlov investigated salivation in dogs and discovered that they could easily be trained to salivate in response to the sound of a bell, and produced just as much saliva as they did in response to a small quantity of food placed in the mouth. In conditioning trials, the new stimulus (bell), which Pavlov named the conditioned stimulus, was given immediately before the normal stimulus (food), which he referred to as the unconditioned stimulus. At first, salivation occurred only in response to food, an

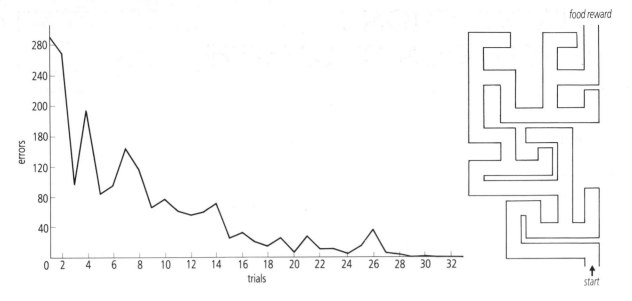

Fig 28·9 *Learning curve for ants learning a maze with a food reward*

unconditioned response, but, after a few trials, the dogs began to salivate before the food was given, and eventually the conditioned stimulus on its own was sufficient to produce a full response, now called the conditioned response.

Operant conditioning occurs when a particular action is either rewarded (positive reinforcement), or punished (negative reinforcement). For example, birds feeding on insects rapidly learn to discriminate between those which are safe to eat and those which are noxious or foul-tasting. In a similar way, the 'trial and error' behaviour of many species leads to useful learning about the environment. Classical and operant conditioning are grouped as **associative learning** because, in each case, the organism forms a new association linking a stimulus to a particular response. The development of such an association can often be represented as a **learning curve**, as shown in Figure 28·9.

4 Exploratory learning. When an animal like a rat is placed in a maze, it will 'explore' the maze even when no food reward is offered. If such an animal is subsequently given a food reward, it learns the correct route through the maze much more quickly than an animal which has no previous experience, but which is given a food reward from the outset. The difference in performance is due to exploratory learning. In the wild, a similar knowledge of the surroundings is often vital in helping animals to evade predators.

5 Insight learning. This involves the ability to recall past experiences and apply them to the solution of new problems. Only a very few animals appear capable of insight learning. Dogs are unable to solve a simple detour problem except by trial and error (Fig. 28·10*A*), but a chimpanzee confronted with a more difficult problem is able to stack boxes to reach a food reward (Fig. 28·10*B*). Humans are unique in the extent to which insight learning is used.

A *Dogs solve a simple detour problem by trial and error.*

B *Chimps are able to stack boxes to reach a food reward*

Fig 28·10 *Insight learning*

238

29 REPRODUCTION 1 PRODUCTION OF GAMETES

Objectives

After studying this Unit you should be able to:–

- List the advantages and disadvantages of asexual and sexual methods of reproduction

- Make labelled drawings of the human male and female reproductive systems

- Describe the process of spermatogenesis, indicating the role of Sertoli cells

- Draw and label a diagram to show the structure of a mature sperm

- State the main functions of the hormone testosterone

- Explain what is meant by secondary sexual characteristics

- Describe the process of oogenesis and make simple drawings to illustrate the development of a follicle

- Discuss the hormonal control of follicular development and preparation of the uterus throughout the menstrual cycle

- Explain the role of the hypothalamus in feedback control of the menstrual cycle

- Give examples of the medical uses of sex hormones

29·1 Asexual and sexual reproduction ■ ■ ■ ■ ■ ■ ■ ■ ■ ■ ■

All organisms die sooner or later, so that, if a particular species is not to become extinct, new organisms must be produced. This process is called **reproduction**.

Asexual (or 'non-sexual') reproduction occurs in bacteria, fungi, protozoa, algae, many plants, and also in a wide range of animals. Only one parent is required and the offspring produced are genetically identical to the parent and to each other. In *Amoeba*, asexual reproduction takes place by **binary fission** after a period of growth during which the chromosomes are duplicated. The organism rounds up and the nucleus divides into two equal parts which pass into the two daughter cells (Fig. 29·1). Daughter cells grow and reproduce in the same way.

Asexual reproduction allows populations of organisms to increase very rapidly to take advantage of favourable conditions. This is discussed further in Unit 56.

Sexual reproduction is used by almost all organisms and is often the only method available for complex organisms such as vertebrates. Typically, it requires the fusion of specialized male and female sex cells, or **gametes**, which are produced by two different parents. The male gamete called a **sperm**, or **spermatozoon**, is small and motile and swims actively towards the larger female gamete, the **ovum**, or **egg cell**. The sperm and egg cell fuse together by a process called **fertilization**, giving a single-celled **zygote**, which finally develops into an adult individual.

Both types of gametes are formed by a process of cell division called **meiosis** which halves the number of chromosomes in the nucleus of the cell. A detailed description of meiosis is given in Unit 36, but, for the purposes of this Unit, it is sufficient to know that the net result is to reduce by half the chromosome complement of each cell. Human body cells, for example, contain 46

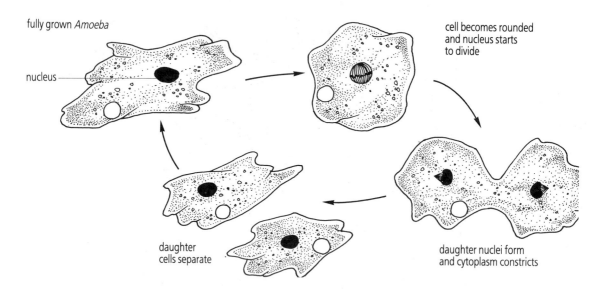

Fig 29·1 Binary fission in Amoeba

chromosomes but, after meiosis, the ova and sperm contain only 23. However, the gamete nuclei fuse together at fertilization so that the original number of chromosomes is restored. In this way, the number of chromosomes in the offspring is the same as in the adult, but each new individual inherits some of its characteristics from each parent.

Sexual reproduction has the enormous advantage that the offspring produced are genetically varied and can be acted on by 'natural selection'. This means that the individuals which are best suited to their surroundings survive to reproduce in their turn, leaving more offspring so that the species as a whole remains adapted to its environment.

In reptiles, birds, and mammals, reproduction involves **internal fertilization**, whereby the sperm and ovum fuse inside the body of the female. For

this to be possible, the male reproductive organs must carry out two essential tasks, namely, the production of sperm (known as **spermatogenesis**), and the transfer of sperm to the female. The female reproductive organs produce egg cells (**oogenesis**), and must be able to receive sperm from the male, at the same time providing the correct conditions for fertilization to occur. In most mammals, the embryo becomes 'implanted' in the uterus of the female so that it can be nourished and protected during its early development. The remainder of this Unit describes the detailed structure of the human male and female reproductive systems and explains how each is specialized for its particular functions. Unit 30 describes copulation, fertilization, and development, leading up to the birth of a new individual.

29·2 Human male reproductive system ■ ■ ■ ■ ■ ■ ■ ■ ■

Figure 29·2 illustrates the human male reproductive system. The male gamete-producing structures, or gonads, called the **testes,** have two functions, the production of sperm, and the manufacture of **testosterone**, an essential male hormone. They are suspended outside the body in a sac called the **scrotum.** A temperature slightly lower than body temperature is optimum for sperm production, which begins at puberty and continues into old age. Each testis contains hundreds of tiny tubes, or **seminiferous tubules**, where sperm are made. These unite to form the coiled tubes of the **epididymis,** which are linked in turn to the **sperm duct,** or **vas deferens,** leading to the urinary tract, and to the outside via the **urethra.** Three types of **accessory glands** are associated

with this region: the **prostate gland, Cowper's glands,** and the **seminal vesicles**. These are responsible for adding enzymatic and nutritive secretions to the sperm, forming a milky liquid known as **semen.** The **penis** is used to transfer semen to the reproductive organs of the female. During sexual arousal, the spongy **erectile tissue** becomes filled with blood, making the penis firm and erect.

Spermatogenesis

Sperm cells are produced by a series of divisions from **germ cells,** or **spermatogonia,** which line the 170 m or so of seminiferous tubules found in each testis. From puberty onwards, about 300

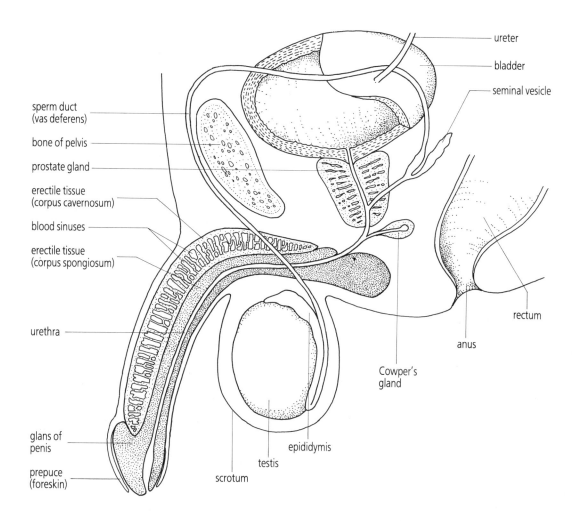

sperm duct
(vas deferens)

bone of pelvis

prostate gland

erectile tissue
(corpus cavernosum)

blood sinuses

erectile tissue
(corpus spongiosum)

urethra

glans of
penis

prepuce
(foreskin)

scrotum

testis

epididymis

Cowper's
gland

anus

rectum

seminal vesicle

bladder

ureter

Fig 29·2 Human male reproductive system

million sperms are produced every day. A
continual supply of spermatozoa is ensured
because the germ cells constantly regenerate by
mitosis, forming a **germ cell layer** near the wall of
each seminiferous tubule (Fig. 29·3).
Spermatogenesis starts when a germ cell migrates
from this layer and enlarges to become a **primary
spermatocyte**. The primary spermatocyte
undergoes two successive cell divisions: the first
produces a pair of cells known as **secondary
spermatocytes,** while the second results in the
formation of four **spermatids**, or immature sperm
cells. This two-stage division takes place by
meiosis.

Throughout spermatogenesis, the dividing cells
are held within the arm-like extensions of
specialised 'nurse cells' or **Sertoli cells**, as
illustrated in Figure 29·3. These cells appear to
exert a controlling influence, acting both as a
barrier and as a transport medium between the
dividing cells and the blood supply, which lies
outside the seminiferous tubules. Spermatids
complete their development into mature
spermatozoa while still closely attached to
Sertoli cells.

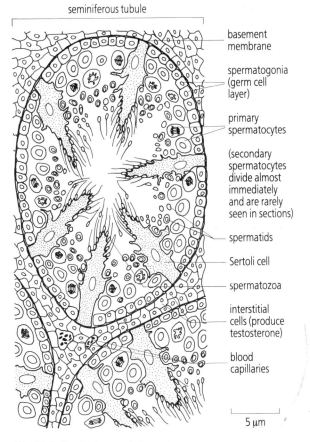

seminiferous tubule

basement
membrane

spermatogonia
(germ cell
layer)

primary
spermatocytes

(secondary
spermatocytes
divide almost
immediately
and are rarely
seen in sections)

spermatids

Sertoli cell

spermatozoa

interstitial
cells (produce
testosterone)

blood
capillaries

5 μm

Fig 29·3 Seminiferous tubules and spermatogenesis

241

A Structure

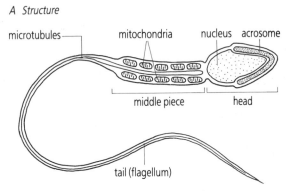

B Scanning electronmicrograph of spermatozoon

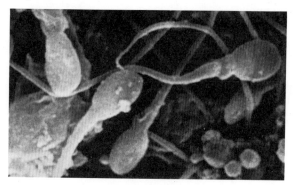

Fig 29·4 Mature spermatozoa (SEM)

The structure of a mature spermatozoon is illustrated in Figure 29·4. As you can see, there are three distinct regions, a head, a middle piece, and a tail. The head is flattened and oval in shape. It contains the nucleus and is capped by an enzyme-filled vesicle, called the **acrosome**, which facilitates the entry of the sperm nucleus into the egg cell at fertilization. The middle piece of the sperm is concerned with energy production, as indicated by the presence of numerous mitochondria. The arrangement of microtubules in the tail region is identical to that found in cilia and flagella (described in Unit 46). Whip-like movements of the tail propel sperm at speeds of up to 4 mm per second.

Spermatozoa are non-motile at first. They are carried passively to the epididymis by a combination of pressure, created by the continual production of new cells and fluid in the seminiferous tubules, and peristaltic movements of the smooth muscle lining the tubules. Spermatozoa are stored in the epididymis where the final maturational changes occur.

The entire development of sperms, from the first division to their presence in the ejaculate, takes about 72 days. In man, spermatogenesis is continuous, but in other mammals it may be seasonal and restricted to certain breeding periods. Sperm formation is inhibited by temperatures above 40°C.

Hormonal control of male sexual activity

The functions of the testes are regulated by hormones called **gonadotrophic hormones**. These are produced by the anterior lobe of the pituitary gland and include **follicle stimulating hormone (FSH)**, and **interstitial cell stimulating hormone (ICSH)**. Together, they promote the growth and activity of the Sertoli cells, and cause the **interstitial cells** located in the spaces between the seminiferous tubules to secrete the male hormone **testosterone**. Testosterone stimulates spermatogenesis and activates the accessory glands to produce secretions collectively called **seminal fluid**. In addition, it is responsible for the more visible features of 'maleness', known as **secondary sexual characteristics**. These include the growth of body and facial hair, the lowering of the voice, and typical male metabolism and muscle/fat ratio.

Testosterone is probably the most important male hormone, but it is not the only one produced by the body. Hormones produced by the adrenal cortex are also known to affect secondary sexual characteristics. The general name for male sex hormones is **androgens**, while the general name for female sex hormones is **oestrogens**. Both groups of hormones are present in male and female mammals, but in different proportions so that the degree of 'maleness' or 'femaleness' is variable depending upon the balance between the levels of androgens and oestrogens in the body. Secondary sexual characteristics may be altered by artificially controlling this balance, as is sometimes done, illegally, to stimulate muscle growth in athletes.

29·3 Human female reproductive system ■ ■ ■ ■ ■ ■ ■ ■

Figure 29·5 illustrates the female reproductive system. Egg cells, or ova, are produced in the **ovaries** and are wafted into the funnel-shaped openings of the **oviducts**, or **Fallopian tubes**. These are fringed by muscular finger-like projections and lined with cilia which beat to create a gentle current. There is no direct connection between the ovary and the oviduct. The paired oviducts are extensions of the **uterus**, or

womb. The uterus has a thick muscular wall, the **myometrium**, and a nutritive blood-filled inner lining, the **endometrium**. At the entrance to the uterus is the **cervix** which forms the upper limit of the tubluar **vagina**. Lining the vagina are secretory gland cells which produce lubricating and nutritive fluids during sexual arousal. Until puberty, the vagina is partially or almost completely closed by a thin membrane, the **hymen**, which perforates to

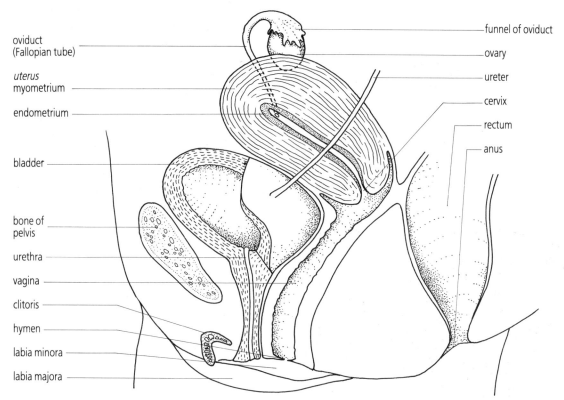

oviduct (Fallopian tube)

uterus
myometrium

endometrium

bladder

bone of
pelvis

urethra

vagina

clitoris

hymen

labia minora

labia majora

funnel of oviduct

ovary

ureter

cervix

rectum

anus

Fig 29·5 Human female reproductive system

allow the flow of menstrual blood. The opening is enlarged by the use of intravaginal tampons. Alternatively, the membrane may tear with slight bleeding during the first act of sexual intercourse. Surrounding the opening of the vagina are two pairs of fleshy folds called **labia**: the inner pair are known as the **labia minora**, while the outer pair are the **labia majora**. The entire region, including the entrances of the urinary and reproductive tracts and the labia, is called the **vulva**. At its anterior end is the **clitoris**, a small erectile organ similar to the male penis, consisting of spongy tissue and containing nerve endings, which heighten sexual arousal on stimulation.

Oogenesis

The ovary, like the testis, has a dual function: the production of gametes, and the secretion of hormones controlling sexual activity. Unlike the testis, the production of ova is not continuous throughout life. In fact, all the potential egg cells, or **primary oocytes**, are present in the ovary at birth. They are produced by the division and enlargement of **germ cells** called **oogonia** during fetal development. There are about 200 000 primary oocytes in each ovary, of which only 400–500 develop to maturity during the active reproductive life of the female.

A further difference between the production of egg cells, or **oogenesis** (see Unit 30), and that of sperm is its cyclical nature. In a human female, an ovum reaches maturity approximately once every 28 days. Its release from the ovary, called

ovulation, is timed to coincide with a number of other events, including the preparation of the uterine endometrium to receive a fertilized egg. This is a monthly cycle, or **menstrual cycle**. Other mammals function according to different time periods. Bitches, for example, are usually 'on heat' twice a year. The general name for these cycles is **oestrous cycles**.

Ovulation in human females occurs at about the mid-point of the menstrual cycle. A mature ovum is the product of a two week developmental process in the ovary, as outlined in Figure 29·6. Development starts when a number of primary oocytes accumulate small clusters of **granular cells** around themselves to become **follicles**. Usually only one follicle develops to maturity to produce an egg cell but on rare occasions, two or more may reach maturity at the same time. When this happens, multiple ovulation and hence multiple fertilization may occur, resulting in twins or multiple births. Twins resulting from fertilization of two different egg cells are called **dizygotic**, or 'non-identical' twins and may be of different sexes. **Monozygotic** or 'identical' twins result from a single zygote which splits after the first cell division to give two separate, but genetically identical, embryos. Monozygotic twins are always of the same sex and always look alike.

During the maturation of a follicle, the primary oocyte enlarges to about one hundred times its original size. It receives nutrients from the cluster of granular secretory cells which proliferate around it. These are separated from the primary oocyte by a thick jelly-like layer, the **zona pellucida** (Fig.

29·6*B*), but they make contact with the primary oocyte through numerous cytoplasmic extensions which penetrate this layer. The granular cells resemble Sertoli cells in their function and are able to regulate the transfer of substances into the primary oocyte from the blood vessels and secretory cells.

As the cluster of granular cells expands, a new layer of cells, called the **theca**, forms around the outside of the follicle (Fig. 29·6*C*). The cells of the theca develop from ovarian connective tissue. Oddly enough, they secrete the male sex hormone, testosterone, which later diffuses into the granular cells surrounding the primary oocyte, where it is converted by enzyme action into the female sex hormone, **oestrogen**.

The last stage in the development of a follicle is the expansion of a central fluid-filled space called the **antrum**, in which the primary oocyte floats suspended by a strand of granular cells (Fig. 29·6*D*). The whole follicle swells like a balloon as the granular cells continue to secrete fluid into the antrum. Just before ovulation, the 'balloon' can be observed as a bump on the surface of the ovary. About this time, the first meiotic division occurs whereby the primary oocyte divides to form two cells called **secondary oocytes**, each with 23 chromosomes. Although the chromosomes are shared equally in this division, the cytoplasm is not. One of the secondary oocytes, destined to become an ovum, retains almost all the cytoplasm, thus keeping its food supply intact, whilst the other consists of little more than a nucleus. This is called a **polar body** and is the first of two polar bodies formed during oogenesis, because the second division of meiosis occurs in a similar way. The ovum thus retains its size through the unequal division of cytoplasm. This second division, the final stage of oogenesis, does not occur until after fertilization. The processes of egg cell and sperm formation, are both examples of **gametogenesis**, and are compared in Figure 29·7.

Ovulation

The egg cell, at this stage more accurately termed the secondary oocyte, is released when the outer lining of the ovary splits along the bulge created by the mature follicle. At the moment of ovulation, antral fluid flows out of the ruptured ovary into the abdominal cavity carrying with it the egg cell, its surrounding jelly (the zona pellucida) and a few granular cells still attached.

The egg cell moves passively into the oviduct by a combination of currents created by the synchronized beating of cilia and muscular contractions of the oviduct wall. It comes to rest about a third of the way along the oviduct, which is the usual site of fertilization.

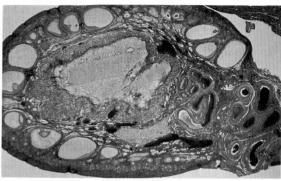

A Light micrograph of ovary section

B Oocyte acquires granular layer

granular cells — — primary oocyte

nucleus — — zona pellucida

C Follicle acquires theca

theca —

D Mature (Graafian) follicle

— theca

— granular cells

— antrum

— secondary oocyte

E Ovulation

ovary —

secondary oocyte discharged from follicle

Fig 29·6 *Development of follicles*

The cells of the mature follicle left behind in the ovary retain an important role in the reproductive process. They enlarge and increase their secretory activity, producing the pregnancy hormone **progesterone**. The whole group of remaining follicle cells becomes a gland-like structure, the **corpus luteum**, or literally 'yellow body', named after the yellow fatty tissue which is deposited to seal the wound after ovulation. If fertilization occurs, the corpus luteum retains its activity throughout pregnancy. If not, it degenerates within ten days.

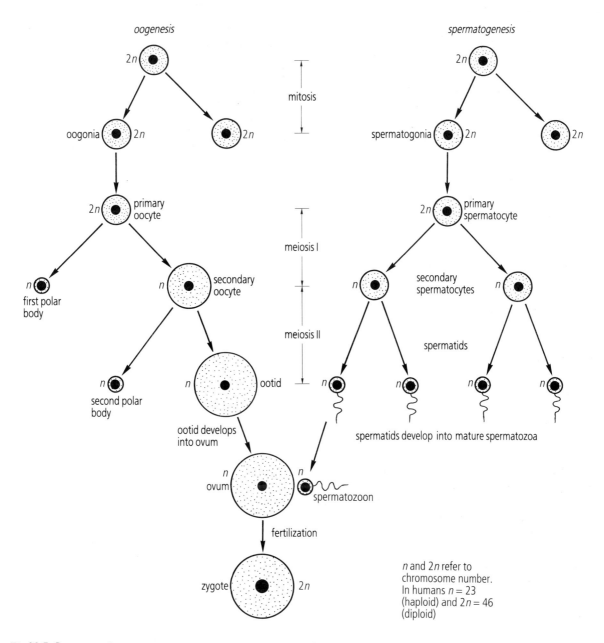

mitosis

spermatogenesis

meiosis I

meiosis II

oogonia 2*n*

2*n* primary oocyte

n first polar body

n secondary oocyte

n second polar body

n ootid

ootid develops into ovum

n ovum

spermatogonia 2*n*

2*n* primary spermatocyte

n secondary spermatocytes *n*

spermatids

spermatids develop into mature spermatozoa

n spermatozoon

fertilization

zygote 2*n*

n and 2*n* refer to chromosome number. In humans *n* = 23 (haploid) and 2*n* = 46 (diploid)

Fig 29·7 Summary of gametogenesis comparing spermatogenesis and oogenesis

29·4 Hormonal control of menstrual cycle ■ ■ ■ ■ ■ ■ ■

As in the male, the activity of the reproductive organs and the production of sex hormones is under the control of the anterior pituitary gland. Some of the same gonadotrophic hormones occur as in males, but in different amounts and with different effects. In females, follicle stimulating hormone (FSH) and **luteinizing hormone (LH)** promote the secretion of sex hormones and the development and release of eggs by the ovary.

The pituitary itself is under the regulating influence of the hypothalamus which triggers the release of gonadotrophic hormones by means of a variety of substances known as **gonadotrophic hormone releasing factors**. The changes which occur during the menstrual cycle are brought about

by cyclic changes in the concentrations of gonadotrophic hormones. The system is self-regulating and incorporates extensive feedback control. For example, the hypothalmus responds to the level of the female hormones oestrogen and progesterone in the blood stream, producing more or less of the corresponding releasing factors, as the situation demands.

In most women, the length of the menstrual cycle varies from about 24 to 34 days, with an average close to 28 days. The average cycle, summarised in Figure 29·8, can be divided into two approximately equal parts, namely the **follicular phase**, that is, the period before ovulation, and the **luteal phase**, the period after

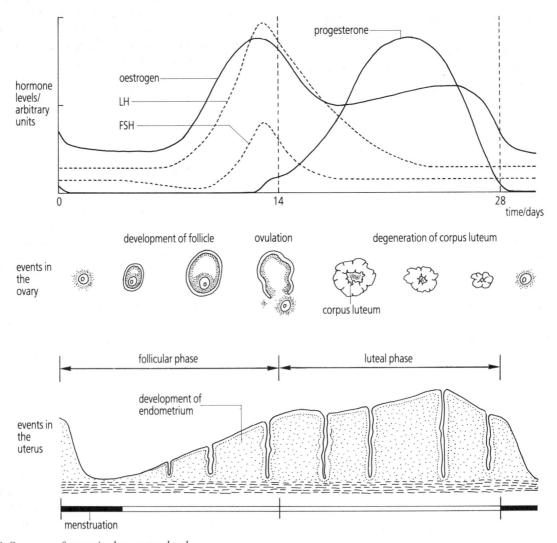

hormone levels/ arbitrary units

oestrogen

LH

FSH

progesterone

0
14
28
time/days

development of follicle

ovulation

degeneration of corpus luteum

events in the ovary

corpus luteum

follicular phase

luteal phase

development of endometrium

events in the uterus

menstruation

Fig 29·8 Summary of events in the menstrual cycle

ovulation. It is easier to understand the sequence of activities if a distinction is made between events which occur in the ovary and those which occur in the uterus.

Events in the ovary

1 Follicular phase. During the first 14 days of the cycle, FSH and LH interact to promote the maturation of an ovum and the secretion of oestrogen by the ovary. In the follicle, LH causes the thecal cells to secrete testosterone and FSH activates enzymes in the granular cells which convert it to oestrogen. Throughout the follicular phase, the level of oestrogen in the blood supply rises. Oestrogen has a **positive feedback effect** on the hypothalamus, that is, it causes increased secretion of releasing factors and hence a surge in LH and FSH production by the anterior pituitary. This surge in LH and FSH production occurs about the 14th day and is the direct cause of ovulation, for the enzymes responsible for splitting the ovary wall at ovulation appear to be stimulated by high concentrations of LH.

2 Luteal phase. After ovulation, the follicular cells take on a new role as the corpus luteum and begin to secrete the hormone progesterone in addition to oestrogen. The presence of both oestrogen *and* progesterone in the blood stream has the opposite effect on the hypothalmus to that of oestrogen alone. The combination of oestrogen and progesterone exerts a **negative feedback effect** on the hypothalamus, inhibiting further secretion of releasing factors and lowering LH and FSH levels drastically. No further follicular development is possible whilst the corpus luteum continues to secrete progesterone. If the egg cell is not fertilized, the corpus luteum degenerates about the 10th day after ovulation and the cycle may be repeated.

In the event of fertilization, the activity of the corpus luteum is maintained throughout pregnancy by a hormone called **human chorionic gonadotrophin (HCG)**, which is secreted by the implanted embryo. The presence of HCG in the urine is a positive indication of pregnancy and provides a simple means of pregnancy testing.

Events in the uterus

1 Follicular phase. During the follicular phase, the rising level of oestrogen stimulates the development of the layer of smooth muscle, the myometrium, and promotes the growth of the glandular epithelial lining of the uterus, the endometrium, in preparation to receive a developing embryo.

2 Luteal phase. After ovulation, progesterone causes the enlarged glandular epithelium to synthesise and store the food reserve glycogen. In addition, there is a great proliferation of blood vessels, making the lining fully prepared to accept an implanting embryo.

In normal circumstances, where pregnancy does not occur, the rapidly falling levels of progesterone and oestrogen cause the myometrium and endometrium to shrink during the last few days of the 28-day cycle. A woman loses up to 50 cm^3 of blood in what is called the **menstrual flow**, as a result of this shrinkage.

The changing levels of oestrogen and progesterone influence the reproductive tract in one further way. Glandular cells in the epithelium of the cervix secrete mucus which lubricates the vagina. Oestrogen stimulates the production of mucus by these cells but progesterone causes the mucus to become viscous and coagulate. In some mammals, like rodents, this has the effect of producing a **mucus plug** which prevents the entry of bacteria into the uterus after ovulation, thus protecting the developing embryo.

Female sex hormones promote the development of female secondary sexual characteristics including breasts, and the typically female distribution of fat and body hair.

Menopause

Towards the late forties, the ovaries of a woman tend not to respond even to high levels of the gonadotrophic hormones FSH and LH. This commencement of infertility is called **menopause**.

As the menstrual cycle ceases, a new balance of hormones is obtained during which period the woman may suffer a range of physiological and psychological discomfort. The onset of infertility is not accompanied by any decrease in the level of sexual arousal and need not affect sexual relationships in later life. Men do not lose fertility in this way; there is no equivalent of the menopause in the male.

Oral contraception

Oral contraceptives, commonly known as 'the pill' contain preparations of manufactured chemicals with similar effects to oestrogen and progesterone. They have a negative feedback effect on the hypothalamus and prevent ovulation by inhibiting the LH and FSH surge. One common form of pill is taken daily for three weeks, commencing the fifth day after bleeding starts. It is then discontinued for one week, during which time the fall in the hormone level induces menstruation. In this way, the menstrual cycle is artificially maintained.

Fertility drugs

Follicle growth and ovulation can sometimes be stimulated in non-productive ovaries by increasing LH and FSH concentrations in the blood. Unlike oestrogen and progesterone, the gonadotrophic hormones cannot yet be imitated by artificial synthesis. Furthermore, LH cannot be extracted in sufficient quantities from natural sources, so human chorionic gonadotrophin is used instead. HCG has similar effects to LH and is easily obtained from placentae after birth. FSH occurs in large quantities in the blood and urine of older women about to enter the menopause. The high levels are produced at this stage in order to stimulate the failing ovaries into action. Supplies of FSH may therefore be extracted from the blood and urine of menopausal women.

The Sheffield College
Hillsborough LRC
Telephone: 0114 260 2254

30 REPRODUCTION 2 FERTILIZATION, PREGNANCY, AND BIRTH

Objectives

After studying this Unit you should be able to:–

- Define copulation and outline the physiological changes which occur during orgasm

- Explain how spermatozoa travel from the vagina to the site of fertilization

- Give a detailed account of the events which occur during fertilization

- Explain the following terms: cleavage, blastocyst, implantation, trophoblastic villi

- Comment on the importance of the chorion, amnion, yolk sac, and allantois

- Describe the structure of the placenta

- List the functions of the placenta

- Describe the events leading to birth

- Draw a simple diagram to illustrate the changes in foetal circulation which occur at birth

- Describe the lactation response and its hormonal control

30·1 Introduction ■ ■ ■ ■ ■ ■ ■ ■ ■ ■ ■ ■ ■ ■ ■ ■

Unit 29 described sperm and egg cell formation and the hormonal control of these processes in the human reproductive system. This Unit describes the methods by which mature gametes are brought together and the subsequent development of the fertilized ovum. It traces the events from fertilization through pregnancy to birth, including parental care.

30·2 Copulation ■ ■ ■ ■ ■ ■ ■ ■ ■ ■ ■ ■ ■ ■ ■ ■ ■

The breeding behaviour of mammals is extremely complex for it involves highly developed courtship behaviour in addition to the physiological mechanisms for uniting sperms and egg cells. The function of courtship is twofold. It reduces the risk of non-profitable mating, restricting it to mature and receptive animals of the opposite sex in the appropriate season of the year. It also serves to select the fittest individuals through competition in mating, so that the new generation are likely to be offspring of parents well adapted to their environment.

The psychological and sociological aspects of courtship are of great importance in human reproduction. However, this section starts with the next stage, that is, **copulation**, the act of introducing sperms into the female reproductive tract, also called sexual intercourse, or **coitus**.

Copulation involves stimulation of touch receptors located in the swollen head, or glans, of the penis and in the clitoris. Such stimulation triggers a variety of autonomic responses including an increase in the heart rate, breathing rate and blood pressure. In the female, the clitoris and breasts become engorged with blood and the nipples and clitoris become erect. The vaginal lining becomes corrugated and begins to secrete mucus. In the male, dilation of arterioles in the spongy tissue of the penis causes it to fill with blood and become firm. At the same time, small quantities of mucus are secreted by glands at the head of the erect organ.

All of these autonomic responses are under the control of the higher brain centres so that thought, sight, sound or memory alone may induce sexual excitement and its associated physiological phenomena. In both males and females, a peak of sexual excitement, called an **orgasm**, may be reached. This may be defined as a state of physiological and emotional release associated with extreme pleasure.

The male orgasm coincides with **ejaculation**. Ejaculation is a spinal reflex response in which the vas deferens and accessory glands (seminal vesicles, prostrate and Cowper's glands) empty their contents in the urethra to be forcibly expelled by a series of rapid muscular contractions. The accessory glands produce **seminal fluid** which contains simple sugars and a group of chemical substances called **prostaglandins**. It mixes with and dilutes the spermatozoa to form the fluid ejaculate called **semen**. The average ejaculate is about 5 cm^3 for the human male but the quantity varies greatly in other mammals. The wild boar probably holds the record at about half a litre of semen for a single ejaculate. Female orgasm is characterised by contractions of the uterus and vagina not dissimilar to those of labour. Unlike male orgasm it is not a prerequisite for fertilization.

30·3 Transport of sperm and fertilization ■ ■ ■ ■ ■ ■ ■ ■ ■

On ejaculation, sperms are thrown up against the cervix where they become lined up parallel to long chain molecules present in the mucus at this point. This helps to ensure that the sperms point in the correct direction. Sperms may arrive at the site of fertilization in the oviduct as quickly as five minutes after ejaculation – much faster than could be accounted for by swimming. Indeed, it is probable that the active swimming movements of spermatozoa are only important in the final penetration of the egg cell, transport through the female reproductive tract being achieved largely by contractions of the uterus and oviduct.

The secretions of the male accessory glands, collectively called seminal fluid, have a number of functions in the transport of spermatozoa. Its alkaline content at pH 7·3 buffers the acid environment (pH 3·8) of the female reproductive tract and the simple sugars it contains provide fuel for sperm cell respiration, necessary for the swimming movements. In addition, it is thought that its prostaglandins induce the contractions of the uterus and oviduct which are responsible for the passive transport of spermatozoa.

Of the hundreds of millions of sperm cells in a single ejaculate, only a few thousand reach the oviduct and only a few hundred reach the ovum. There is no evidence for any chemical attraction between the sperm and ovum. Fertilization thus depends upon random meeting of the two gametes. Sperm cells are 'primed' for fertilization by the acid environment of the uterus and ovary. This causes small pores to open in the head region of the sperm cell through which the enzymes of the acrosome may be released (refer back to Fig. 29·4). The priming process is called **capacitation**.

The acrosome enzymes are lyctic, or splitting, enzymes which enable the sperm to move between the granular cells and penetrate the **zona pellucida** (Fig. 30·1A). When the sperm reaches the ovum, it fuses with the plasma membrane of the egg cell and slowly passes into the cytoplasm, leaving its tail behind. The entry of the sperm into the egg cell in some way stimulates the second and final meiotic division of the egg cell nucleus. Remember that the egg cell is still at this stage a secondary oocyte. This second meiotic division results in an unequal sharing of cytoplasm and the extrusion of a second polar body. Fertilization is completed when the sperm cell nucleus fuses with the nucleus of the ovum restoring the chromosome number to 46 and forming a **zygote** (Fig. 30·1B).

Only one sperm cell can penetrate the ovum. Multiple sperm entry, or **polyspermy**, is prevented in the following way. The sperm cell causes changes in the plasma membrane as it enters, promoting the release of Ca^{2+} ions in the cytoplasm of the ovum. The sudden increase in Ca^{2+} ions causes the **cortical reaction** in which secretory vesicles (cortical granules) in the cytoplasm of the ovum move towards and fuse with the plasma membrane, expelling their contents into the space between the ovum and the zona pellucida. Enzymes present in this secretion alter the zona pellucida so that further entry by sperm cells is prevented.

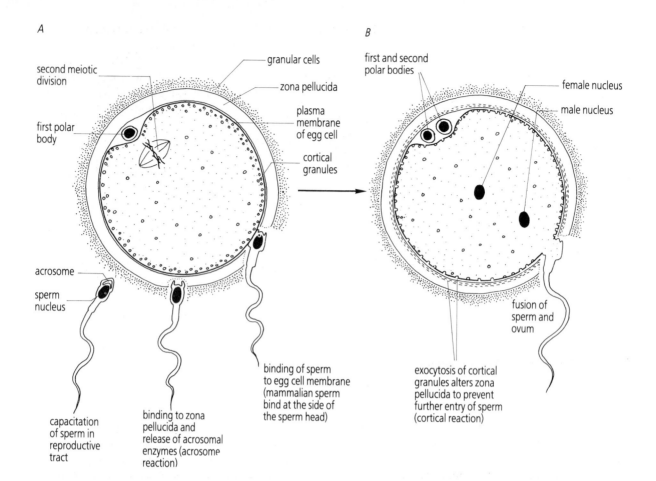

A

second meiotic division
first polar body
granular cells
zona pellucida
plasma membrane of egg cell
cortical granules
acrosome
sperm nucleus
binding of sperm to egg cell membrane (mammalian sperm bind at the side of the sperm head)
capacitation of sperm in reproductive tract
binding to zona pellucida and release of acrosomal enzymes (acrosome reaction)

B

first and second polar bodies
female nucleus
male nucleus
fusion of sperm and ovum
exocytosis of cortical granules alters zona pellucida to prevent further entry of sperm (cortical reaction)

Fig 30·1 Fertilization

Fertilization is possible during a relatively short period. After their release, sperms have a life of approximately 48 hours, while eggs remain fertile for about 10–15 hours. To result in pregnancy, therefore, copulation must occur no more than 48 hours before or 15 hours after ovulation.

30·4 Implantation and fetal development ■ ■ ■ ■ ■ ■ ■

As it travels down the oviduct towards the uterus, the zygote divides several times by a process known as **cleavage**. These divisions produce a ball of cells, but there is no increase in size at this stage.

The process of **differentiation** or cell specialization starts as early as the 20-cell stage, when it is apparent that some of the new cells are larger than others. A distinction can be made between those cells which will form the embryo proper and those which will give rise to the embryonic membranes and the placenta.

The centre of the ball of cells begins to fill with fluid and soon it becomes a hollow ball with a fluid-filled interior. This stage is called a **blastocyst**. On one side of the blastocyst a swelling occurs, formed by the group of cells destined to become the embryo (Fig. 30·2*A*). The remaining thin outer layer, or **chorion**, is composed of cells called **trophoblasts**, whose function is to embed the blastocyst into the uterus wall. All this development takes place within the first seven days after fertilization (days 14–21 of the menstrual cycle). The blastocyst floats freely in the fluid interior of the uterus as the endometrium continues to prepare itself under the influence of progesterone and oestrogen.

Implantation occurs about the 7th day after fertilization, as trophoblast cells stick to the lining of the endometrium and quickly divide and grow between the lining cells, forming an intimate connection between the embryonic and maternal tissues. These **trophoblastic villi** spread out and the blastocyst is drawn deeper into the endometrium, eventually becoming surrounded by the nutrient-rich endometrial cells. For the first few weeks, the developing embryo derives its nourishment by absorbing food and oxygen from these cells but soon a specialized structure, the **placenta**, takes over the function of nutrition and gas exchange.

As the trophoblastic villi burrow deeper into the endometrium, the embryonic area of the blastocyst continues to divide and differentiate. It is disc-shaped and consists of three distinct tissue layers called **germ layers**. Each gives rise to a

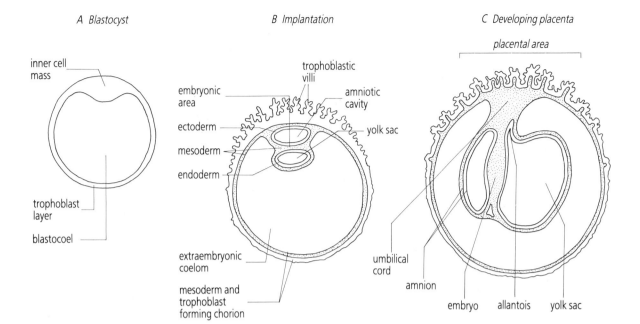

A Blastocyst

inner cell mass

trophoblast layer

blastocoel

B Implantation

trophoblastic villi

embryonic area

ectoderm

mesoderm

endoderm

amniotic cavity

yolk sac

extraembryonic coelom

mesoderm and trophoblast forming chorion

C Developing placenta

placental area

umbilical cord

amnion

embryo   allantois   yolk sac

Fig 30·2 Formation of embryonic membranes

different tissue system: the **ectoderm** develops to form the skin and nervous system, the **endoderm** develops to form the gut, while the central **mesoderm** gives rise to muscle tissue and to connective tissues including bone, cartilage, and blood.

Balloon-like extensions grow out from the embryonic area. These are illustrated in Figures 30·2B and 30·2C. The first to appear is the **amnion** or **amniotic sac** which develops as an extension of the ectoderm, separating the embryo from the overlying trophoblastic cells of the chorion. It encloses a fluid-filled space, called the amniotic cavity, which acts as a shock absorber to protect the young embryo. A second embryonic membrane develops from the embryonic endoderm. This is called the **yolk sac**. Although the yolk sac is for a time the larger of the two cavities, the amnion continues to grow and eventually encloses the yolk sac completely (refer to Fig. 30.3A). The trophoblastic villi and chorion continue to develop, forming the placenta.

The pattern of development of the human placenta is quite different from that observed in many other mammals. In a more typical species, formation of the placenta depends on a second outgrowth from the endoderm, called the **allantois**. This enlarges enormously and fuses with the chorion to replace an earlier primitive placenta formed by the yolk sac.

It is interesting to note a difference here between eutherian, or placental mammals, and metatherial mammals, or marsupials. One reason that the birth of marsupials takes place while they are still only a few centimetres long, is that the allantois does not develop in this way to form a placenta, and they are therefore supplied only by

the short-lived yolk sac placenta.

The human placenta (Fig. 30·3) is an interlocking system of embryonic and maternal tissues formed as the trophoblasts penetrate between the cells of the endometrium, breaking maternal capillaries as they grow. Trophoblastic villi become surrounded by spaces called **lacunae** full of maternal blood. An anticoagulant produced by the trophoblastic cells prevents the blood lacunae from clotting. Five weeks after implantantion the placenta is well established and blood vessels grow out of the middle mesoderm layer of the embryo into the trophoblastic villi, forming **chorionic villi** where exchange of material occurs. These vessels connect with the developing heart and circulatory system of the embryo by means of the **umbilical cord**.

At the end of two months, all the major tissue systems of the embryo are developed. At this stage, it is about 2.5 cm long and is called a **fetus**. During the **gestation** or pregnancy period of 40 weeks, the fetus derives all its food and oxygen from the maternal blood, and discharges waste CO_2 and urea back into it. Transport of materials is made easier by a partial breakdown of the trophoblastic cells of the chorion, which brings the foetal blood closer to the maternal blood, and is also facilitated by microvilli, which are found over the entire surface. There is no direct contact between the two blood systems. Oxygen and CO_2 diffuse between them according to naturally established concentration gradients, oxygen moving into the fetal circulation, and CO_2 outwards from it. The transport of nutrient and waste materials is controlled actively by the placental membranes and requires the expenditure of energy by the trophoblastic cells.

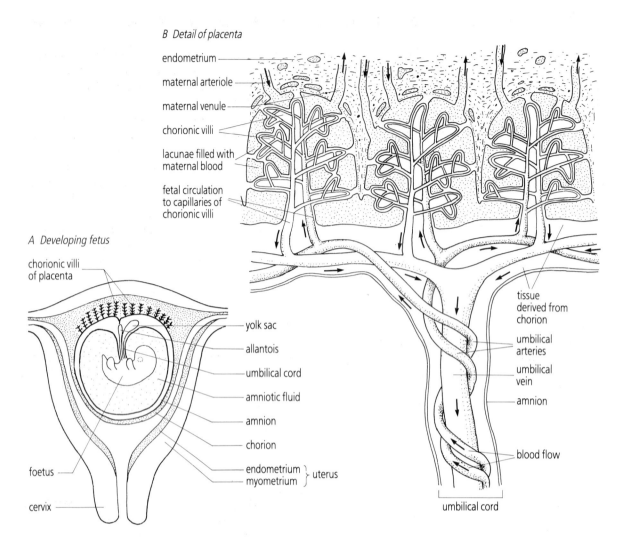

B Detail of placenta

endometrium

maternal arteriole

maternal venule

chorionic villi

lacunae filled with maternal blood

fetal circulation to capillaries of chorionic villi

A Developing fetus

chorionic villi of placenta

yolk sac

allantois

umbilical cord

amniotic fluid

amnion

chorion

endometrium ⎱ uterus
myometrium ⎰

foetus

cervix

tissue derived from chorion

umbilical arteries

umbilical vein

amnion

blood flow

umbilical cord

Fig 30·3 Structure of the placenta

While the placenta is permeable to small molecules, it normally acts as a barrier to proteins and other large molecules which might cross from the fetus into the mother's blood stream. This is important because it prevents the mother's immune system from producing antibodies against the foetus in the same way that her immune system would reject any other foreign material. An exception to this general rule sometimes occurs in the case of a blood group substance known as the Rhesus (Rh) factor. As explained in Unit 19, if Rh antigens from the fetus cross into the blood of a rhesus negative (Rh−) mother, anti-Rh antibodies will be made. Together with other maternal antibodies, anti-Rh antibodies will cross the placenta from the mother to the fetus and may attack the baby's red blood cells, causing **haemolytic disease of the newborn**. On the other hand, the transfer of antibodies is normally desirable because they confer passive immunity against a whole range of disease organisms to which the mother has been exposed. In this way, the baby is protected from infection for the first few months of life, until its own immune system is properly developed.

Although the placenta acts as a barrier to many potentially damaging substances, some harmful chemicals and disease organisms do pass from mother to fetus. These include toxic chemicals, like lead and DDT, nicotine from cigarettes, drugs, and viruses, such as the German measles (rubella) virus. It is not uncommon for babies born to alcoholic and drug-addicted mothers to exhibit 'withdrawal symptoms' at birth.

The nutritive endometrium of the uterus is maintained throughout pregnancy by the hormones progesterone and oestrogen. During the first two months, the corpus luteum of the ovary is stimulated to produce these hormones by human chorionic gonadotrophin (HCG), which is secreted by the embedded trophoblastic cells. After two months, however, the placenta itself begins to secrete oestrogen and progesterone and gradually takes over the dominant role. The placenta has a number of functions: it serves as the organ of nutrition, excretion and gas exchange, as a barrier to disease, and also as a means whereby immunity is acquired. In addition, it acts as an endocrine gland secreting the hormones which control development.

The process by which the mature fetus is expelled from the uterus is called birth, or **parturition** (Fig 30·4). In humans, birth normally occurs from 38 to 42 weeks after conception, with an average around 40 weeks. A series of uterine movements, hours, days, or even weeks before parturition, prepare the fetus for delivery, manipulating it downwards so that its head lies against the cervix. The onset of labour is characterised by a series of involuntary rhythmic contractions of the myometrium which begin at the top of the uterus and sweep downwards. These contractions increase in strength and frequency and cause the cervix to open to a diameter of about 10 cm. During this period, the amniotic sac is ruptured and fluid escapes from the vagina, an occurence commonly termed 'breaking of the waters'. A series of strong uterine contractions, often aided by the mother's conscious abdominal movements, forces the baby through the cervix and vagina. At this stage, the baby remains attached to the placenta by the umbilical cord, but immediately after delivery both the fetal and maternal blood vessels of the placenta completely contract and the placenta separates itself from the wall of the uterus to be expelled by further muscular contractions. This second delivery is called the 'afterbirth'.

The mechanisms which trigger parturition are not fully understood, but it is clear that the fetus plays a major part in timing its own birth. In the period prior to birth, the developing pituitary gland of the fetus secretes adrenocorticotrophic hormone (ACTH), which is believed to act on the fetal adrenal gland, causing the secretion of a second hormone belonging to the glucocorticoid group. This second hormone acts through the placenta to trigger hormonal changes in the mother. Uterine contractions are caused mainly by the hormone **oxytocin**, which is released from the posterior pituitary gland of the mother. Prostaglandins are also present in the blood during parturition, and are known to cause uterine contractions. It is possible that the operation of the hormones which stimulate labour is triggered by the sharp decline in progesterone secretion which occurs a few days before birth.

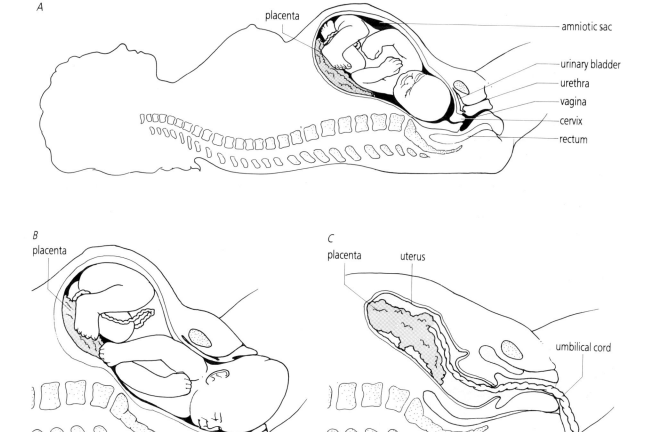

Fig 30·4 Stages in parturition

A remarkable occurrence in the birth of a baby is its immediate adaptation to a new and comparatively hostile environment. In the space of minutes it loses support for its limbs, previously supplied by the amniotic fluid, a direct intravenous food and oxygen supply and the warmth and protection provided by the uterine environment. Its lungs which, until the moment of birth, had been collapsed and full of fluid, take in air, and the body metabolism must begin to use its large fat and glycogen stores in order to generate heat.

Even more remarkable are the changes in the blood and circulatory system which accompany these external signs of independence. As the lungs and digestive system of the baby take over the functions performed by the placenta, two circulatory channels need to be closed, namely, the **foramen ovale**, connecting the two atria of the foetal heart, and the **ductus arteriosus**, which connects the pulmonary artery to the aorta. The reason for these changes is explained in Figure 30·5. During pregnancy, both the foramen ovale and the ductus arteriosus serve to short-circuit the lungs, but after birth they must be closed in order for efficient gas exchange to occur by double circulation.

A further change which occurs at birth affects the pigment haemoglobin found inside red blood cells. As explained in more detail in Unit 18, **fetal haemoglobin** has a higher affinity for oxygen than adult haemoglobin to facilitate transfer of oxygen from the mother. However, it is less suitable as a means of exchange with air and is progressively replaced by adult haemoglobin in the first few weeks of life.

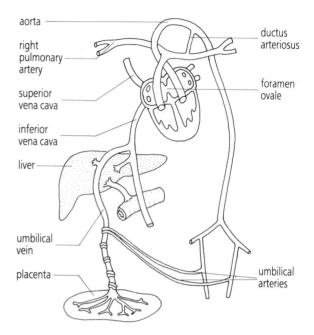

Fig 30·5 Changes in fetal circulation at birth

30·6 Lactation and parental care

All female mammals are able to feed their newborn infants with milk secreted by **mammary glands**. The mammary glands, or breasts of a woman develop under the influence of oestrogen and progesterone. As the levels of these hormones fluctuates during the menstrual cycle, the breasts alter in size slightly. During pregnancy there is a noticeable enlargement, caused by a proliferation of the milk secreting tissue. Milk is produced by gland cells and secreted into groups of tiny sacs called alveoli. The alveoli connect by ducts which lead eventually to a converging point at the nipple.

Milk is released from the nipple by a reflex response stimulated by the sucking infant. Sensory receptors in the nipples connect with autonomic nerves which promote the release of the hormone **oxytocin** by the posterior pituitary gland. Oxytocin causes tiny muscles in the lining of the mammary ducts and alveoli to contract, thus expelling the milk. Although this is an autonomic reflex response, it may be initiated by higher brain centres so that a mother may secrete small droplets of milk even at the sound of her infant crying.

Lactation is inhibited during pregnancy, even though the breasts are well developed, because the production of milk by the secretory cells requires the activity of yet another hormone, called **prolactin**. Prolactin is produced by the anterior pituitary during pregnancy, but its activity is inhibited by the high levels of oestrogen and progesterone at this time. It becomes effective just before the birth when the levels of these hormones drop.

The first milk produced, called **colostrum**, is a rich mixture of protein, lactose, white blood cells, serum, and protective antibodies which provide a useful supplement to the infant's defences against disease. Normal human breast milk is a more watery suspension containing fats, proteins, salts, and lactose. The secretion of milk continues as long as the mother breast feeds her baby, even for several years, but it stops shortly after breast feeding ceases.

Parental care involves more than feeding and protecting the infant. Human infants require a longer period of parental care than other mammals because it is during this time that most learning occurs. Particularly important is the acquisition of language, for it is primarily the skill of communication that enables the knowledge of generations to accumulate, and sets human societies on a level far higher in complexity than those of other animals.

31 DEVELOPMENT OF PLANT STRUCTURE

Objectives

After studying this Unit you should be able to:–

- Outline the life cycle of a broad bean plant

- Describe the structure and germination of a broad bean seed

- Define hypogeal and epigeal germination and give examples of each

- State the importance of apical meristems and name the three primary meristematic tissues

- Give an account of the stages in development of mature plant cells

- Name the three major groups of plant tissues and state the origins of the following tissues: epidermis, cortex, pith, primary phloem, primary xylem.

- Use simple diagrams to explain the structure and adaptations of the following cell types: parenchyma, chlorenchyma, collenchyma, sclerenchyma fibres and sclereids, phloem sieve tube elements and companion cells, xylem tracheids, and xylem vessel elements

- List the functions of stems

- State the role of turgor pressure and vascular bundles in supporting the stem

- Outline the stages of secondary growth and explain the importance of vascular cambium, cork cambium, and lenticels

31·1 Introduction ■ ■ ■ ■ ■ ■ ■ ■ ■ ■ ■ ■ ■ ■ ■ ■ ■

The following five Units describe all aspects of the structure and workings of flowering plants, excluding the biochemistry of photosynthesis and classification, which are described respectively in Units 11 and 54.

The important features of flowering plant organization could be illustrated by any one of thousands of different species. In this Unit, the broad bean, *Vicia faba*, has been chosen as a representative example. The life cycle of the broad bean plant is outlined in Figure 31·1. The bean seed (Fig. 31·1*A*) is formed in the pod of the parent plant and, given favourable conditions, will germinate, producing a young plant, or seedling (Fig. 31·1*B*). The seedling grows to become the adult plant which, in due course, develops flowers, forms new seeds, and dies, thereby completing an annual cycle of events.

The individual cells of the plant do not function independently, but are organized, like animal cells, into **tissues**, that is, groups of similar cells performing the same task. Typically, tissues are arranged together in structures called **organs**. Plants form four basic types of organs, namely, **roots, stems, leaves**, and **flowers**. This 'division of labour' increases the potential complexity of life activities. For example, if one part of the plant body becomes specialized for anchorage or support, then other parts may become adapted to carry out alternative processes, like feeding or reproduction, so that each of these functions is carried out more efficiently.

Some of the features of plant organs are illustrated in Figures 31·1C and 31·1D. However, their detailed structure is best understood by tracing the stages in their development from the unspecialized tissues of the embryo plant. This Unit describes the germination of the broad bean seed and considers the formation and specializations of tissues in the stem. Unit 32 describes the adaptations of roots and leaves.

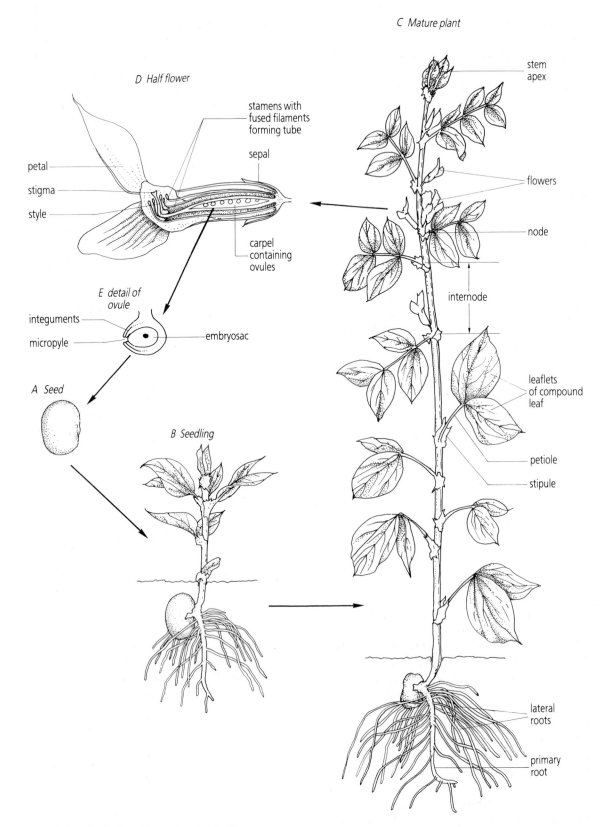

Fig 31·1 Life cycle of a broad bean plant, Vicia faba

256

31·2 Seed structure and germination ■ ■ ■ ■ ■ ■ ■ ■ ■ ■

The broad bean belongs to the plant group known as dicotyledons. Its embryo can be observed by peeling off the outer coat, or **testa**, of a seed which has been soaked in water for 24 hours. It is a simple structure consisting of a single **axis** bearing two large **cotyledons**, or seed leaves. Figure 31·2C shows the embryo with the cotyledons pulled apart. The sections of the axis above and below the junction with the cotyledons are known respectively as the **epicotyl**, and the **hypocotyl**. The tip of the epicotyl, called the **plumule**, grows to become the shoot, while the tip of the hypocotyl, or **radicle**, gives rise to roots.

storage substances by digestive enzymes. For example, amylase enzymes break down starch. Water is taken in mainly because many of the molecules present in the seed, notably cellulose molecules, have a strong electrostatic attraction for polar molecules like water. The physical force which may be developed by imbibition is evident from the practice of the ancient Egyptians, who are believed to have quarried huge stones for building the pyramids by driving wooden wedges (consisting almost entirely of cellulose) into cracks in the rock, and then soaking the wedges with water.

The seed's achievement is less dramatic. The

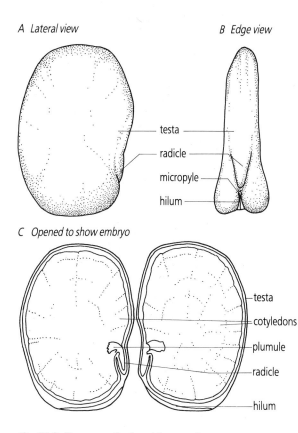

A Lateral view   B Edge view

testa
radicle
micropyle
hilum

C Opened to show embryo

testa
cotyledons
plumule
radicle
hilum

Fig 31·2 Structure of a broad bean seed

Most seeds, including those of the broad bean, become dormant when first formed, that is, they pass through a period when germination cannot take place, even if external conditions are suitable. The mechanisms of dormancy are varied and complex and will be described in Unit 34. Dormancy is broken by exposure to cold, partial drying, or a variety of other environmental changes, and the seeds are then free to germinate. Three basic conditions are required, namely, a supply of **water**, the presence of **oxygen**, and a **favourable temperature**.

In the first stage of germination, a broad bean seed absorbs, or **imbibes**, large amounts of water, swelling to more than twice its original size. This provides suitable conditions for the hydrolysis of

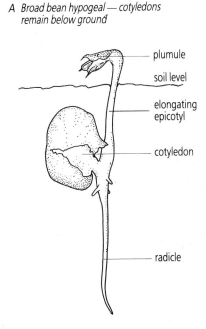

A Broad bean hypogeal — cotyledons remain below ground

plumule
soil level
elongating epicotyl
cotyledon
radicle

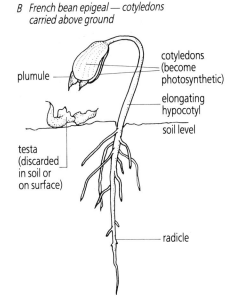

B French bean epigeal — cotyledons carried above ground

plumule
cotyledons (become photosynthetic)
elongating hypocotyl
soil level
testa (discarded in soil or on surface)
radicle

Fig 31·3 A Hypogeal and B Epigeal germination

swelling embryo merely ruptures the seed coat and the radicle emerges and grows to form a structure called the **primary root**. Much of the surface of the primary root is covered by **root hairs**, which increase the surface area in contact with the soil. Later, branches called **lateral roots**, or secondary roots, are produced, and may give rise to further laterals, the whole structure serving to anchor the young plant and absorb water from the soil. The epicotyl extends and grows upwards as shown in Figure 31·3*A*, breaking through the soil surface with the top of the plumule curved over, thereby protecting its delicate tip.

The cotyledons of the broad bean seed remain underground as the shoot develops. This kind of germination is called **hypogeal**. In other seeds, like the French bean, shown in Figure 31·3*B*, the cotyledons are carried above ground during germination, and become the first photosynthetic leaves. This kind of germination is called **epigeal**. Germination is complete when the young plant becomes independent and is able to make all of its own food by photosynthesis.

31·3 Meristems and differentiation

Unlike animals, which can produce new cells for growth and replacement in most parts of their bodies, plants have separate specialized regions, called **meristems**, where the new cells of the plant body are formed. The most important meristems are the **apical meristems**, which occur near the extreme tips of the shoot and root, and at the tips of side branches. Figures 31·4*A* and 31·4*B* show the structures observed at the tip of the shoot of a dicotyledon like the broad bean. As you can see, the cells of the apical meristem, known as **initials**, are relatively small cells. Typically, they have large nuclei and lack a sap vacuole. They divide actively to produce three distinct **primary meristematic tissues**, the **protoderm**, **ground meristem**, and **procambium**. These regions subsequently become the **primary tissues** of the plant, as will be explained shortly.

The apical meristem also gives rise to small leaves which grow to enclose and protect the apex of the shoot, forming an **apical bud**. Similar buds develop in the angles, or **axils**, between leaf branches and the main stem. These are called **axillary buds**, and represent additional meristems, which may, or may not, grow to become lateral branches. Such branches arise from the outer layers of the stem, and are said to be **exogenous** branches. A **node** is any one of the parts where axillary buds are produced (refer to Fig. 31·4*B*), while **internodes** are the intervening sections of straight stem.

The development of mature cells occurs in three stages. The first of these is **cell division**, which takes place in the apical meristem. This is followed by a period of **expansion**, when the newly-formed cells swell and elongate. **Differentiation** occurs as each cell develops into its final specialized form.

A number of changes can be observed during cell expansion (Fig. 31·5). The cells swell by osmotic uptake of water, usually becoming much longer, and small vacuoles appear in the dense cytoplasm, later fusing to give the single large sap vacuole characteristic of mature plant cells. Cell

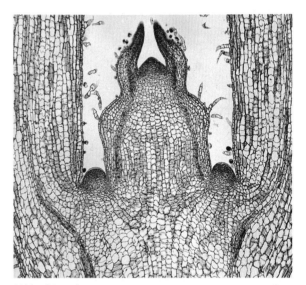

A Light micrograph

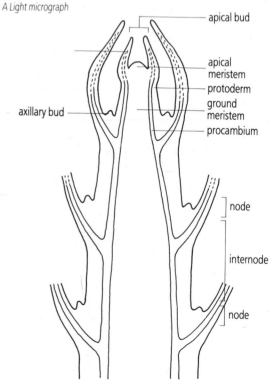

B Diagram

Fig 31·4 Shoot tip of a dicotyledon showing the apical meristem and primary meristematic tissues

258

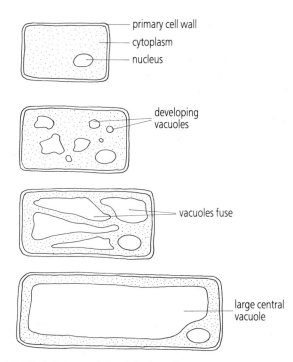

- primary cell wall
- cytoplasm
- nucleus

- developing vacuoles

- vacuoles fuse

- large central vacuole

Fig 31·5 Changes observed during cell expansion

expansion occurs just behind the tip of a growing shoot or root, making these regions the **regions of maximum growth**.

Differentiation is mainly concerned with the growth of **cell walls**. These structures, although non-living, are the most characteristic and visible feature of plant cells. They are secreted by the living part of the cell, called the **protoplast**, while the cell is still increasing in size, and consist mainly of cellulose, pectin and glycoprotein. Cellulose synthesis occurs at, or near, the plasma membrane and cellulose is laid down outside the membrane as a mesh of microfibrils. In the matrix between the microfibrils, pectin and glycoproteins are deposited. These substances are extruded into the wall by **secretory vesicles** from the Golgi body (or dictyosome) of the plant cell, the vesicles fusing with the cell membrane to discharge their contents.

Between the walls of adjoining cells is a thin layer of sticky pectic material, called the **middle lamella**, which binds the cells together. This structure is first formed during division of meristematic cells when vesicles from the Golgi body come to the midline and coalesce. At the same time, the remains of the vesicles fuse to form membranes on either side, thereby isolating the two new protoplasts. In order to separate plant cells in the laboratory, it is common to dissolve the middle lamella by flooding the section with acid – a process called **maceration**.

Although developing cells are bounded by their cell walls and separated by the middle lamella, two features prevent them from being physiologically isolated. Firstly, the cell wall is porous, allowing the free exchange of dissolved materials, and, secondly, there are thin regions where the wall is penetrated by threads of cytoplasm, called **plasmodesmata** (singular **plasmodesma**). The thin areas are called **pits**, and usually occur in opposite pairs between adjacent cells, as illustrated in Figure 31·6*A*.

Almost all plant cells possess a **primary cell wall** laid down during development from meristematic tissues. As part of their differentiation, some cells secrete additional layers of cellulose inside the primary wall to form a **secondary cell wall**. This is true of the water-conducting and supporting tissues, but not of cells which must remain metabolically active. Typically, the microfibrils are laid down in three overlapping layers to form a rigid supporting framework. The pliable pectin of the matrix of both primary and secondary walls may be progressively replaced by **lignin**. This has the effect of toughening the wall, but makes it waterproof, so that the protoplast inside is unable to obtain nutrients and dies, leaving a fluid-filled cavity. Secondary walls are not laid down in the pit regions although they sometimes overhang the edges in the manner illustrated in Figure 31·6*B*, forming **bordered pits**. In gymnosperms, the plant group which includes conifers, bordered pits may have a particularly complex structure.

As a result of differentiation, a variety of plant cells are formed, each specialized to perform a particular function in the life of the plant.

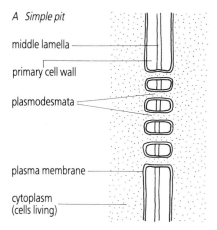

A Simple pit

- middle lamella
- primary cell wall
- plasmodesmata
- plasma membrane
- cytoplasm (cells living)

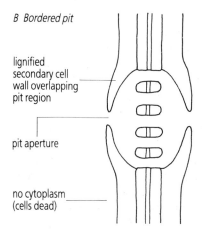

B Bordered pit

- lignified secondary cell wall overlapping pit region
- pit aperture
- no cytoplasm (cells dead)

Fig 31·6 A simple pits and B bordered pits

The primary tissues of a plant belong to three distinct groups, namely **epidermal tissue**, **ground tissue**, and **vascular tissue**. The origins of particular tissues in the stem and their specialized cell types are summarized in Figure 31·7. As you can see, each group develops from one of the three primary meristematic tissues produced by the apical meristem.

molecules, like sugars, other soluble carbohydrates, and amino acids, may be dissolved in the cell sap, while insoluble substances, including starch, proteins, and lipids, are stored as granules or droplets in the cytoplasm. Some parenchyma cells store crystals of inorganic salts, such as calcium oxalate. This may be a method of excretion, allowing excess salts to be conveniently dumped.

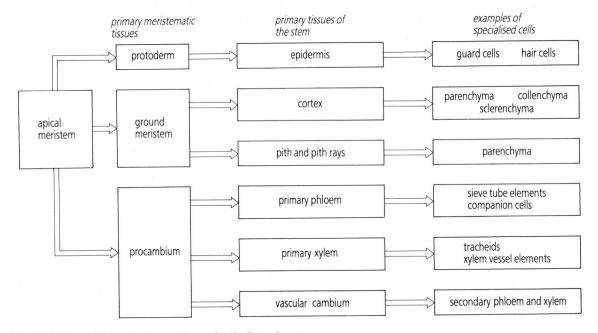

Fig 31·7 Origin of primary tissues and specialized cells in the stem

Epidermis

The epidermis is derived from the protoderm. A plant is protected by the single layer of epidermal cells which cover its surface. In regions of potential water loss, the epidermal cells secrete an external cuticle composed of **cutin**, a polymer of long chain fatty acids which form a cross-linked network impregnated with waxes. The cuticle reduces water loss and helps to protect the plant against mechanical injury and attack by microorganisms like bacteria and fungi. Specialized epidermal cells include guard cells and hair cells, both of which are described in Unit 32.

Cortex

The cortex region is located between the epidermis and the vascular (transporting) tissue of the stem (refer to Fig. 31·8). Differentiation from ground meristem tissue produces cortical cells of several different types.

1 Parenchyma. Parenchyma cells are large, thin-walled, and relatively unspecialized (Fig. 31·9A). They rarely possess secondary cell walls and their protoplasts remain alive. Functions of parenchyma cells include photosynthesis, secretion, support, and the storage of water and food. Food

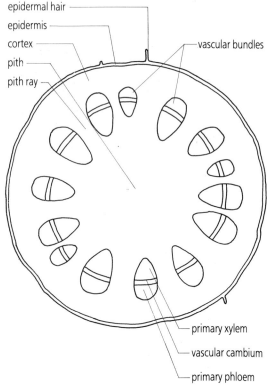

Fig 31·8 Primary structure of the stem of a dicotyledon

Parenchyma tissue is characterized by large intercellular spaces. It is not confined to the cortex regions of the stem and root, but may be found in all plant organs. Parenchyma cells which contain chloroplasts form a tissue called **chlorenchyma**, which is present, for example, in leaves, where it forms the mesophyll layer.

2 Collenchyma. The outer layer of the cortex of a young stem often consists of cells with unevenly thickened cell walls. These are collenchyma cells (Fig. 31·9B). They provide support and protection, but, at the same time, remain flexible enough to allow tissue growth. Collenchyma cells become elongated as the plant develops, often possess chloroplasts, and remain living at maturity.

3 Sclerenchyma. Sclerenchyma cells are specialized for support, and, as differentiation continues, they produce a secondary cell wall in which lignin is deposited. When lignification is complete, the protoplast usually dies. Two types of sclerenchyma cells are distinguished. **Fibres** (Fig 31·9C) are long cells which extend vertically in the spaces between other cells, a process called **intrusive growth**. **Sclereids** (Fig 31·9D) are shorter branched cells often found in the coverings of seeds and in fruits.

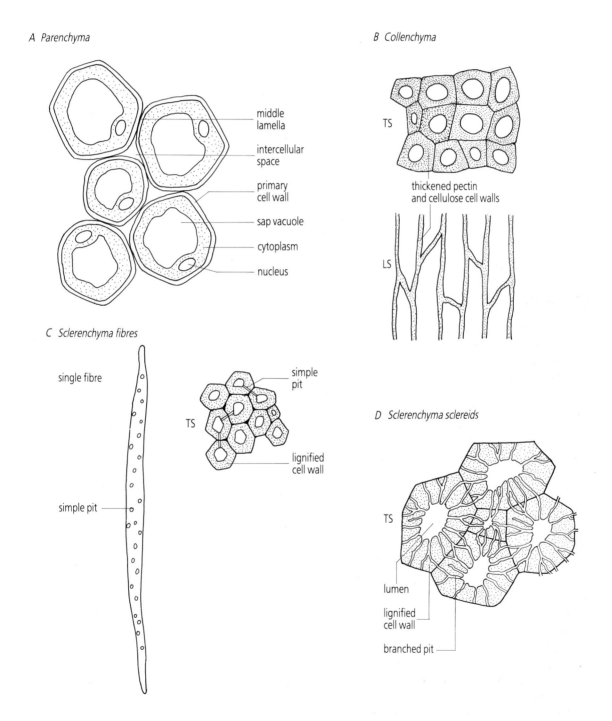

Fig 31·9 Specialized plant cells

Pith

Pith consists of large parenchyma cells with numerous intercellular spaces. Its main function is food storage. The regions of parenchyma extending between the vascular bundles of the stem are known as **pith rays**.

Primary phloem

Phloem and xylem are the vascular tissues of the plant and originate from procambial cells which are arranged in columns in the stem. Differentiation of the procambial cells produces **primary vascular bundles**, as indicated in Figure 31·8. A thin layer of meristematic tissue which will later become the vascular cambium may remain sandwiched between primary phloem and xylem. Phloem is located on the outsides of the vascular bundles and is specialized for transporting the products of photosynthesis, usually sucrose, from the leaves and other photosynthetic tissues to other parts of the plant. In addition to fibres, sclereids, and parenchyma cells, primary phloem contains **sieve tube elements**, and **companion cells**.

1 Sieve tube elements. A sieve tube is a cylindrical column of cells, called sieve tube elements, joined end to end, as illustrated in Figure 31·10A. The end walls of these cells are perforated by enlarged pits, forming **sieve plates**. The protoplast of a sieve tube element remains living, although its nucleus disintegrates as the cell differentiates. The sap vacuole membrane and many cell organelles also disappear, leaving a central region filled with sap directly in contact with a thin layer of cytoplasm close to the cell wall. A mass of strands pass through the sap and penetrate the sieve plates, linking the interior of each sieve tube element directly with the next. The strands contain large amounts of a slimy fibrous protein known as **P-protein**.

2 Companion cells. Each sieve tube element is closely associated with one or more companion cells, which originate by division from the same meristematic initial. The companion cells have normal protoplasts and retain their full complement of cell organelles, and are believed to regulate the metabolic activity of the sieve tube element. The walls separating the two types of cell are thin and uneven, and perforated by numerous plasmodesmata (refer to Fig. 31·10A).

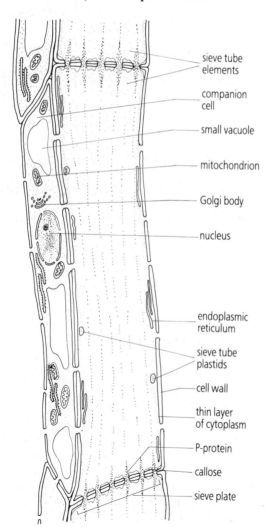

sieve tube
elements

companion
cell

small vacuole

mitochondrion

Golgi body

nucleus

endoplasmic
reticulum

sieve tube
plastids

cell wall

thin layer
of cytoplasm

P-protein

callose

sieve plate

A Diagram of sieve tube and companion cell

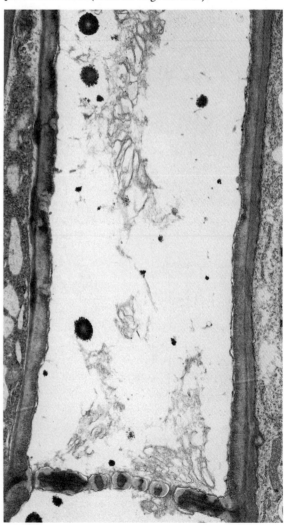

B Electron micrograph

Fig 31·10 Specialized cells of primary phloem

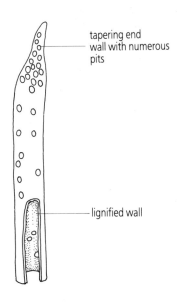

tapering end wall with numerous pits

lignified wall

A Diagram of tracheid

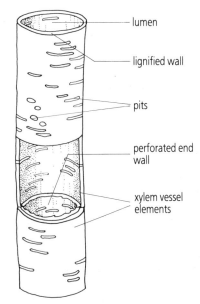

lumen

lignified wall

pits

perforated end wall

xylem vessel elements

B Diagram of xylem vessel elements

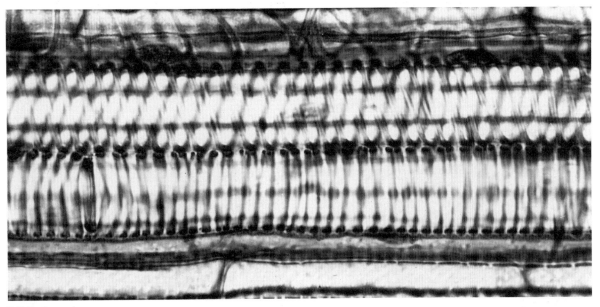

C Light micrograph

Fig 31·11 Specialized cells of primary xylem

Primary xylem

Xylem tissue, located on the inside of the vascular bundles in the stem, is specialized for the transport of water and inorganic salts from the roots to other parts of the plant. The conducting cells of primary xylem, called **tracheids** and **xylem vessel elements**, are often associated with fibres and parenchyma cells. They deposit unusually thick and heavily lignified secondary cell walls, so that their protoplasts die at maturity, leaving a hollow central lumen surrounded by the cell wall. Lignification of xylem vessels prevents collapse under tension as sap pressure changes (see Unit 33.4).

1 Tracheids. Tracheids are elongated cells with tapering oblique end walls and numerous pits

through which water passes freely (Fig. 31·11*A*).

2 Xylem vessel elements. These are typically shorter and fatter than tracheids, and are joined end to end, as indicated in Figure 31·11B. The ends of the vessel elements disappear completely during development, so that they form a continuous tube, called a **xylem vessel**. The first xylem vessels to develop are known collectively as **protoxylem** and have secondary walls in which lignin is deposited in rings or spirals, in a variety of distinctive patterns (Fig. 31·11*C*). This allows the vessels to elongate as growth continues. In mature sections of the stem, **metaxylem** vessels are formed. These are larger and more heavily lignified and have walls which cannot be stretched.

31·5 Stem functions ■ ■ ■ ■ ■ ■ ■ ■ ■ ■ ■ ■ ■ ■ ■

The most important functions of the stem are **support**, **transport**, and **growth**.

In a stem consisting of primary tissues, the function of support is shared between several different types of cells. The living cells of the epidermis, cortex, and pith, all take in water by osmosis. As a result, individual cells develop an internal hydrostatic pressure, called **turgor pressure**, which causes the protoplast to push outwards against the cell wall. This is explained in Unit 33. Turgid cells are rigid and resistant to bending. If the cells lose water, they become limp, or flaccid, and the stem wilts.

Collenchyma and sclerenchyma cells, and the heavily lignified cells of xylem tissue also function in support. In most terrestrial plants, the major mechanical stresses are imposed by the wind, so that the stem must be able to resist bending to remain upright. The vascular bundles, containing the xylem, are tough and inextensible and perform the same function as steel rods in reinforced concrete. Their arrangement as a ring within the stem (Fig. 31·8) provides very effective resistance to wind stress. In the stems of some plants, for example, in the sunflower, *Helianthus*, the vascular

bundles are strengthened by additional sclerenchyma fibres, which form a **bundle cap** (Fig. 31·12).

The sieve tube elements and companion cells of phloem, and the tracheids and vessel elements of xylem, are specialized for transport. The mechanisms involved in their functioning are complex and important, and will be discussed in Unit 33. As already explained, primary growth in stems takes place by differentiation from the apical meristem and the three primary meristematic tissues.

In addition to these essential functions, some stems are specialized to perform other tasks. Photosynthesis and the storage of food and water are among the most common. In cacti, for instance, the stem is capable of storing large volumes of water, and the leaves are reduced to spines, so that the stem becomes the main organ of photosynthesis. Corms, stem tubers, and some types of rhizome are modified stems which act as food storage organs and can be used as a means of asexual reproduction. Most spines and thorns are modified stem branches.

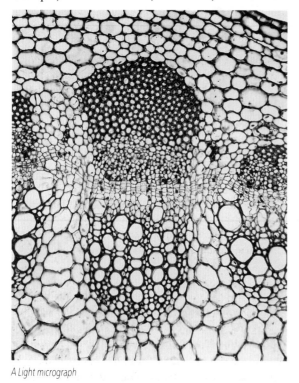

A Light micrograph

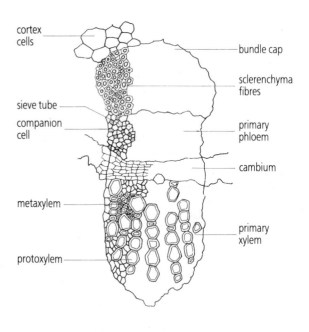

B Diagram

Fig 31·12 Vascular bundle from the stem of the sunflower, Helianthus

31·6 Secondary growth ■ ■ ■ ■ ■ ■ ■ ■ ■ ■ ■ ■ ■ ■ ■

Primary growth is responsible for increases in stem length. Increases in stem diameter, and the continued development of plants from year to year, require a different type of growth, called **secondary growth**. Plants with an annual life

cycle like that of the broad bean usually have a limited capacity for secondary growth. The adult plants simply wither and die as winter approaches, leaving seeds which will germinate in the spring. On the other hand, perennial plants, for example

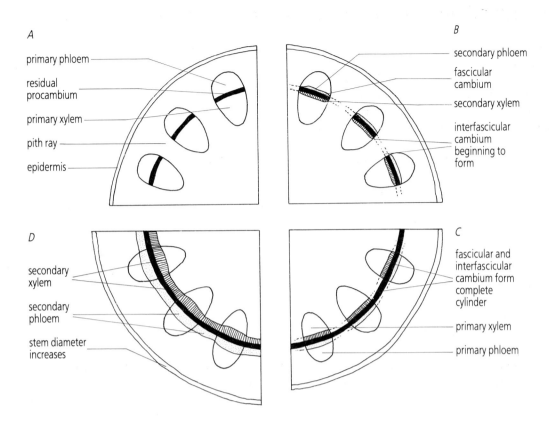

Fig 31·13 Stages in secondary growth of a stem

trees, rely on secondary growth to reach a large size, and may survive for many years, producing seeds in a number of successive seasons.

The main stages of secondary growth are outlined in Figure 31·13. When vascular bundles are first formed, not all of the procambial cells differentiate into primary phloem or xylem. In the early stages of secondary growth (Fig. 31·13B), the remaining cells become active once more, forming regions of **fascicular cambium** inside the bundles. These produce new phloem cells, called **secondary phloem** to the outside, and new xylem cells, called **secondary xylem** towards the inside. At about the same time, some of the parenchyma cells in the pith rays between the bundles are stimulated to divide and start to form a linking region, known as **interfascicular cambium** (Fig. 31·13C). As development continues, the cambial cells form a complete cylinder, which is now referred to as the **vascular cambium**. This produces secondary phloem and xylem in continuous bands (Fig. 31·13D).

As secondary xylem is laid down in layers to the inside, the diameter of the stem increases, and the cambium cylinder and secondary phloem are pushed outwards. The inner core of xylem is known as **wood**. In older stems, **annual rings** can be observed. These appear because the new xylem vessels formed in the spring are large and thin-walled, while those formed later in the autumn are smaller and possess thick walls. Most, but not all,

the xylem cells are arranged in vertical columns. Here and there, horizontal parenchyma cells form structures called **rays**. These are arranged radially, and are living channels which allow metabolic by-products and substances like resin to be deposited in the wood.

The epidermis forms the outermost covering in a young shoot, but, as stems increase in diameter, older tissues split and are lost from the surface. Protection of older stems is achieved by the production of **cork**. Cork tissue consists of small cells with cell walls impregnated with a fatty substance, called **suberin**, which waterproofs the cells and makes them impermeable to gases. Cork is formed by **cork cambium** which is derived initially from the outer layer of cortical cells. As the stem continues to increase in girth, older layers of cork are ruptured and deeper tissues form a new cork cambium. In many trees, cork is formed by cork cambium derived from the parenchyma cells of secondary phloem. The **bark** of the tree consists of all the tissues outside the vascular cambium, that is, cork, cork cambium, cortical cells (if present), and secondary phloem.

Gas exchange by the living tissues of older stems would be impeded except for the presence of structures called **lenticels** which penetrate the outer coverings. These consist of loose parenchyma cells, and allow the inward and outward diffusion of gases, at the expense of some inevitable water loss.

32 ADAPTATIONS OF ROOTS AND LEAVES

Objectives

After studying this Unit you should be able to:–

- Outline the role of the apical meristem of a root

- Describe the distribution of primary tissues in a root

- Comment on the structure and development of root hairs

- Draw and label a diagram showing the structure of cells in the endodermis of a root and state the function of the Casparian strip

- Explain the importance of the pericycle in the formation of lateral roots, and in secondary thickening of roots

- List and describe the functions of roots

- Explain the role of a leaf primordium in development of a leaf

- Distinguish between simple and compound leaves

- Draw and label diagrams to show the external and internal structure of the leaf of a dicotyledon

- Describe the cellular structure and adaptations of the following tissues: upper epidermis, palisade mesophyll, spongy mesophyll, lower epidermis, leaf vascular tissue

- List and describe the functions of leaves

32·1 Root development ■ ■ ■ ■ ■ ■ ■ ■ ■ ■ ■ ■ ■ ■ ■

This Unit describes the structure of roots and leaves and outlines their important functions with the exception of uptake and transport, which are described in Unit 33.

The structures present near the tip of a growing root are illustrated in Figure 32·1. Figure 32·1A shows the appearance of the radicle of a germinating broad bean. Simple experiments in which the radicle is marked with ink reveal that elongation occurs in the region just behind the tip.

The region nearest to the tip consists of meristematic cells, while older cells beyond the region of elongation show signs of differentiation, as in the case of the epidermal cells which develop root hairs (refer to Fig. 32·1B). Unlike the tip of a shoot, which is protected by small leaves inside a bud, the tip of a root is protected by a special structure, called a **root cap**.

The apical meristem within the root tip (Figs 32·1C and 32·1D) gives rise to three primary meristematic tissues, corresponding to those of the shoot. These are the **protoderm, ground meristem**, and **procambium**. In addition, the apical meristem gives rise to cells of the root cap, which are continually renewed to replace cells lost at the surface as the root pushes between the soil particles. At the centre of the apical meristem is a region of cells known as the **quiescent centre**. These cells divide extremely slowly and appear to regulate root growth by limiting the number of rapidly dividing cells, or **initials**, which form the remainder of the meristem.

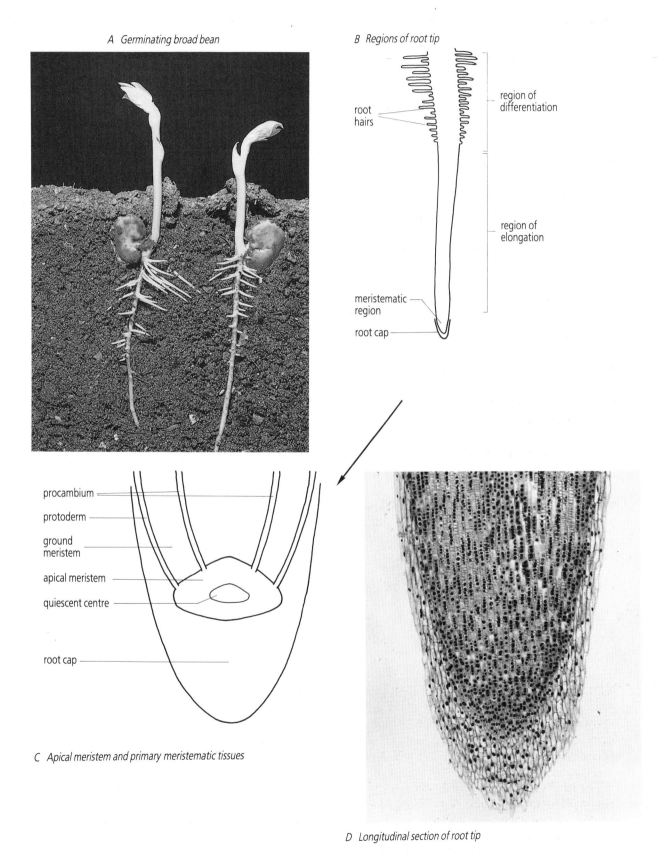

A   *Germinating broad bean*

B   *Regions of root tip*

root hairs

region of differentiation

region of elongation

meristematic region

root cap

procambium

protoderm

ground meristem

apical meristem

quiescent centre

root cap

C   *Apical meristem and primary meristematic tissues*

D   *Longitudinal section of root tip*

Fig 32·1 Root tip of a dicotyledon

32·2 Primary tissues of the root ■ ■ ■ ■ ■ ■ ■ ■ ■ ■ ■ ■

The primary tissues of the root develop from primary meristematic tissues in much the same way as the primary tissues of the stem. However, the distribution of tissues in the roots and some of the specialized cells formed are different, as summarized in Figure 32·2. Figures 32·3*A* and 32·3*B* show the appearance of a typical dicotyledon root in transverse section.

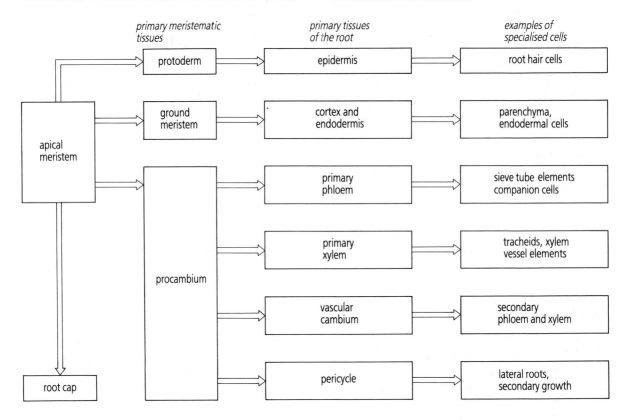

Fig 32·2 Origin of primary tissues and examples of specialized cells in the root of a dicotyledon

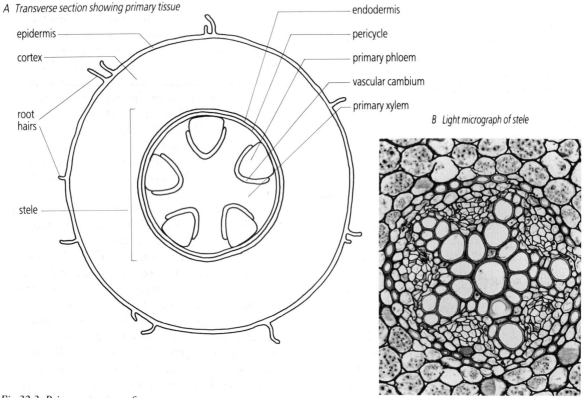

A Transverse section showing primary tissue

B Light micrograph of stele

Fig 32·3 Primary structure of a root

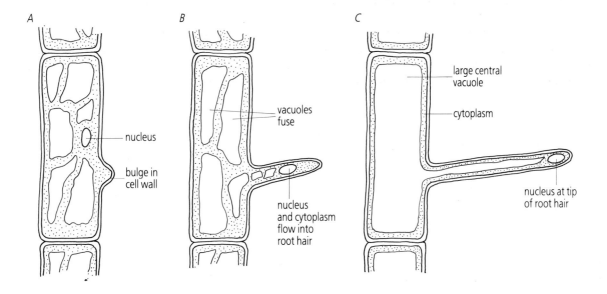

A. nucleus
bulge in cell wall

B. vacuoles fuse
nucleus and cytoplasm flow into root hair

C. large central vacuole
cytoplasm
nucleus at tip of root hair

Fig 32·4 Development of root hairs

Epidermis

The epidermis develops from a single layer of protoderm cells. The protoplasts of these cells are living and their cell walls remain thin. Usually, there is no cuticle. Above the zone of elongation, some of the epidermal cells produce finger-like extensions called **root hairs** (Fig. 32·4). As these increase in length, the nucleus and cytoplasm flow into the root hair and occupy the region nearest to the tip. The root hairs enormously increase the surface area available for absorption of water and inorganic nutrients. For example, the root system of a rye plant 0·5 m tall was estimated to have a total surface area of about 210 m^2. Nevertheless, the useful life of a root hair is short. New hairs are constantly produced in the zone of differentiation, but shrivel and die in more mature sections of the root.

Cortex

In roots, the cortex is a relatively thick layer which originates from ground meristem. It consists mainly of large parenchyma cells separated by large intercellular spaces. The innermost cells of the cortex form the **endodermis**, a cylinder of highly specialised cells surrounding the central vascular tissue. As indicated in Figure 32·5, the protoplasts of the endodermal cells deposit suberin in a continuous strip, called the **Casparian strip**, surrounding each cell. The Casparian strip links each cell with its neighbours, and is completely impermeable to water and solutes. Consequently, water passing through the endodermis from the cortex to the central core of vascular tissue must

pass through the protoplasts of the endodermal cells. This is of great importance in the mechanism of water uptake, as will be explained in the following Unit.

Vascular tissue or stele

The central cylinder of vascular tissue, also called the **stele**, is derived from procambium. The arrangement of phloem and xylem is somewhat different from that in stems, as you can see by comparing Figure 32·3*A* with Figure 31.8. However, specialised phloem and xylem cells differentiate in the same way. The outermost layer of procambial cells forms the **pericycle** of the root. This is a single layer of cells immediately inside the endodermis. The pericycle cells remain capable of division and are involved in the formation of lateral roots. Lateral root branches are said to be **endogenous**, that is, they arise from deep within the root.

Pith is usually absent in roots, and the centre of the stele is occupied by xylem tissue. The largest xylem vessels are heavily lignified and easily recognised in cross section (Fig. 32·3*B*). Typically, the xylem forms a cross or star-shaped region, with phloem tissue located between the extended arms. Undifferentiated procambial cells remain between the xylem and phloem and in the pericycle and may develop to become **vascular cambium**, which allows secondary growth by a process similar to that observed in stems. Pericycle cells normally form the **cork cambium** responsible for producing protective coverings in older roots.

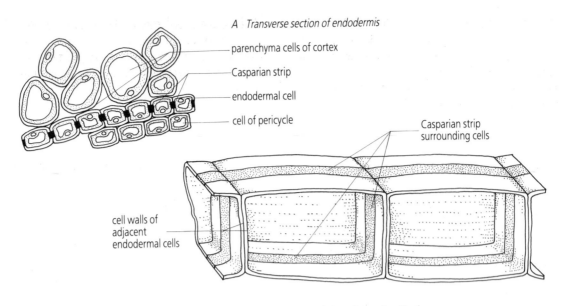

A Transverse section of endodermis

parenchyma cells of cortex

Casparian strip

endodermal cell

cell of pericycle

Casparian strip surrounding cells

cell walls of adjacent endodermal cells

B Three-dimensional view of the cell walls of two endodermal cells to show the barrier formed by the Casparian strip

Fig 32·5 Structure of endodermal cells

32·3 Root functions ■ ■ ■ ■ ■ ■ ■ ■ ■ ■ ■ ■ ■ ■ ■

The roots of plants take in water and inorganic ions and transport these to the stem. In addition, they anchor the plant in the soil.

The mechanical stresses produced by the wind affect the roots quite differently from the stem. Roots are flexible and easily bent, but they must resist stretching forces which would tend to pull the plant out of the soil. This is achieved partly by the arrangement of vascular tissues as a tough central core, and partly by extensive branching of the root system, which traps soil particles and distributes mechanical forces over a wide area.

In many species, specialized roots are found with a great diversity of other functions. Some of these are listed and described below.

1 Adventitious roots. Adventitious roots are produced directly from the tissues of the stem or leaves rather than by branching from primary or lateral roots. They have a structure which is identical to that of true roots. Adventitious roots are important in species which reproduce asexually using horizontal stems, and in climbing plants like ivy, where adventitious roots are modified as organs of attachment.

2 Storage roots. Sugar beet, turnips, carrots, and other commercially important crops have swollen roots containing a mass of parenchymatous storage tissue. Root tubers like those of *Dahlia* plants are used for storage and as a means of asexual reproduction.

3 Root nodules. Some plants, including peas, beans, clover, and other leguminous plants, have roots which are capable of forming a mutualistic association with nitrogen-fixing bacteria. The

bacteria multiply inside small swellings called root nodules. This is explained further in Unit 57.

4 Mycorrhizae. These are similar mutualistic associations between roots and certain types of fungi. Mycorrhizae are described in Unit 52.

5 Haustorial roots. Dodder is a parasitic plant which grows and feeds on the stems of nettles. Its roots penetrate the host's tissues and form swollen absorptive structures called haustoria (singular haustorium) surrounding the vascular bundles of the host stem.

Fig 32·6 Pneumatophores of mangrove

6 Contractile roots. Plants like dandelion, *Taraxacum*, have specialized roots which contract as the plant grows, pulling the stem downwards. New foliage leaves are therefore produced at ground level, where they are less likely to be eaten by herbivores than the surrounding vegetation.

7 Pneumatophores. These are root branches which grow upwards and act as organs of gas exchange for plants such as mangrove which inhabit the oxygen-deficient waters of swamps (Fig. 32·6).

32·4 Leaf development ■ ■ ■ ■ ■ ■ ■ ■ ■ ■ ■ ■ ■ ■ ■

Leaves are produced by the division and differentiation of cells derived from the apical meristem of a shoot. In Unit 31 it was explained that the shoot tip was protected by small leaves forming a bud. Each leaf starts as a small bulge which elongates to beome a **leaf primordium** (Fig. 32·7). On either side of this structure are **marginal meristems** derived from ground meristem tissue, from which the blade of the leaf will grow. These are separated by a thickened middle region. A strand of procambial cells which extends into this region will later become the transport system of the leaf. The primordium is covered by protoderm which develops to form the leaf epidermis. Thus, the three primary meristematic tissues are present in the leaf primordium and contribute to leaf development.

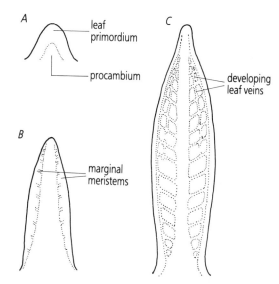

Fig 32·7 Stages in leaf development

32·5 Leaf structure ■ ■ ■ ■ ■ ■ ■ ■ ■ ■ ■ ■ ■ ■

External features

The external features of the leaf of a typical dicotyledon and details of its attachment with the stem are illustrated in Figure 32·8. The flat part of the leaf, called the blade, or **lamina**, provides a large surface area for photosynthesis and is supported partly by turgor pressure, and partly by the leaf **veins**. In dicotyledons, these form a branching network transporting water and inorganic ions into the leaf, and conveying the products of photosynthesis to the stem. The central **midrib** and the leaf stalk, or **petiole**, are composed mainly of vascular tissue. An **axillary**

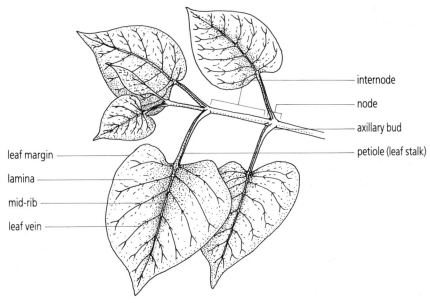

Fig 32·8 External features of a leaf

bud is located in the angle, or **axil**, formed between the petiole and the main stem. Many plants possess outgrowths from the leaf bases known as **stipules**. These are often leaf-like and photosynthetic, but sometimes become tendrils or spines.

(singular **stoma**). The guard cells can change shape to open and close the stomata and help to control gas exchange and water loss. Usually, only a few stomata are found in the upper epidermis. Light passes through the upper epidermis to reach the deeper layers of photosynthetic tissue.

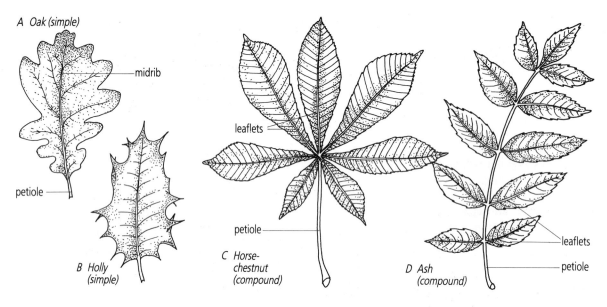

Fig 32·9 Simple and compound leaves

Leaf shape is genetically determined and is characteristic for any particular species. However, leaf shape is also influenced by environmental change, and by the age of the plant. In many trees, leaf position affects the final size and shape. Leaves near the top of the tree are often smaller and narrower than those from lower branches, which are broad and flat with a large area exposed for photosynthesis. **Simple** leaves, like those of oak or holly, have a lamina in a single unit (see Figs 32·9A and 32·9B). **Compound** leaves are composed of separate **leaflets**. These develop from the same leaf primordium and are attached to the main stem by a single petiole. Sometimes, the leaflets are linked at a single point, as in the horse chestnut (Fig. 32·9C), while in leaves like those of ash, the leaflets are arranged on either side of an elongated petiole (Fig. 32·9D).

Internal features

The arrangement of primary tissues in the leaf of a dicotyledon and details of cellular structure are given in Figure 32·10. As you can see, the leaf consists of a number of layers.

1 Upper epidermis. The cells of the upper epidermis produce a thick waxy cuticle which protects the upper surface of the leaf and helps to reduce water loss. The protoplasts of these cells are living, but they do not normally possess chloroplasts. An important exception occurs in the case of the crescent-shaped **guard cells** which lie on either side of small openings called **stomata**

2 Palisade mesophyll. The cells of this layer are chlorenchyma cells which develop from the marginal meristems. They are the most highly adapted cells for photosynthesis and have cytoplasm densely packed with chloroplasts. Typically, they are arranged in rows with the long axis of each cell vertical, thereby trapping light effectively. Large intercellular spaces allow diffusion of gases into and out of the palisade cells. During photosynthesis, carbon dioxide is absorbed and oxygen is released as a waste product. The surfaces of the cell walls are moist to permit these gases to dissolve and represent sites of inevitable water loss.

3 Spongy mesophyll. The cells of this layer are chlorenchyma cells like those of the palisade layer, but are more loosely packed with unusually large intercellular spaces. Spongy mesophyll cells possess fewer chloroplasts than palisade cells and are correspondingly less important in photosynthesis.

4 Lower epidermis. The cuticle secreted by the cells of the lower epidermis is usually much thinner than that covering the upper epidermis. In addition, the lower epidermis is penetrated by numerous stomata. These may be opened and closed by guard cells in the same way as those on the upper surface and elsewhere on the plant.

5 Vascular tissue. The leaf veins and the large central vein of the midrib region contain phloem and xylem. In most species, xylem tissue forms the upper section of each vein and contains tracheids and xylem vessels, as in the stem.

Phloem tissue, consisting of sieve tubes and companion cells, is usually confined to the lower part of the vein. Additional strengthening of the vascular tissue is common. This may involve sclerenchyma and collenchyma cells, formed from mesophyll cells (refer to Fig. 32·10*B*). In some species, collenchyma cells are found at the leaf margins.

A   Distribution of primary tissues (transverse section)

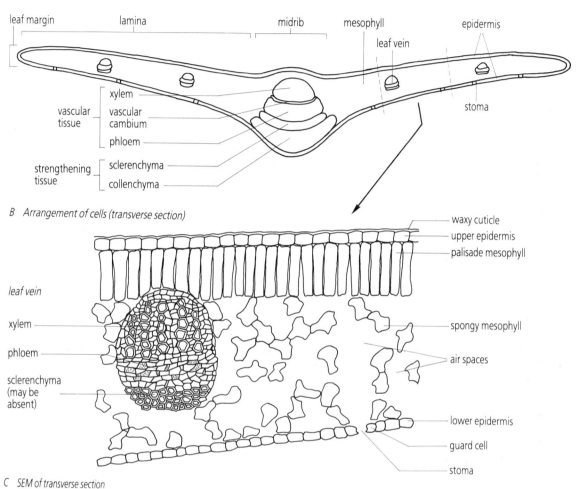

B   Arrangement of cells (transverse section)

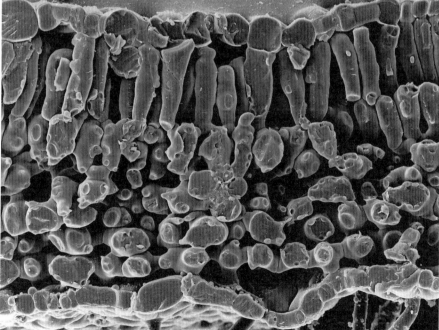

C   SEM of transverse section

Fig 32·10 Internal structure of the leaf of a dicotyledon

Many aspects of leaf structure contribute to effective photosythesis. Leaf responses and growth mechanisms allow leaves to change direction during the day, so that they constantly face the sun, and prevent overlap. The arrangement of foliage leaves which results is known as a **leaf mosaic** (Fig. 32·11). Gas exchange is a related function. During the day, carbon dioxide is absorbed and oxygen is given off, but at night these directions are reversed. The adaptations required for photosynthesis and gas exchange, such as a large surface area, tend to result in the loss of large amounts of water from the leaf surface. Consequently, a whole range of further adaptations have evolved to reduce water loss. Some of these are explained in the next Unit. Unit 11 gives details of the biochemistry of photosynthesis. Some of the specialized functions of leaves are outlined below.

Fig 32·11 Leaf mosaic

1 Protection. It is obviously adaptive for plants to resist attack by herbivores, and a great variety of mechanisms exist for this purpose. Among the most effective is the production of indigestible, poisonous, or unpleasant chemicals which are deposited in the leaves. In the wild cherry and many related species, and also in some grasses, lethal quantities of cyanide can be formed by a process called **cyanogenesis**, in which cyanogenetic glycosides are broken down by enzyme action to yield hydrocyanic acid and a simple sugar. Tannins in oak leaves perform a similar function in deterring insects. An interesting twist in this story involves the caterpillars of some species of butterflies which have become tolerant to the poison produced by a particular species of plant. As they feed, the poison accumulates in the tissues of the young insects, making the caterpillars and the adult butterflies distasteful to birds. This occurs, for example, when monarch butterflies, *Danaus plexippus* feed on the tissues of milkweed, *Asclepias curassavica* (Fig. 32·12).

Fig 32·12 Monarch butterfly, Danaus plexippus, feeding on milkweed, Asclepias curassavica

Other protective devices include spines and prickles, as in holly. Stinging nettles possesses sharp pointed hair cells which penetrate the skin of animals and release an irritant mixture containing histamine and acetylcholine.

2 Food capture. Plants which live in nutrient-deficient habitats benefit by trapping insects and other small invertebrates, whose bodies provide organic nitrogen compounds and other essential substances. The sundew, bladderwort, Venus fly trap, and pitcher plants all have leaves modified as snares. In bladderworts, these take the form of tiny bladders, from which the plant gets its name. The common bladder wort, *Utricularia vulgaris*, is a free-floating aquatic plant (Fig. 32·13*A*). The bladders are like tiny purses with a flap at one end. The pressure of the water inside is normally less than the pressure outside, so that, when a small organism like a *Daphnia* touches the trigger hairs on the flap, it suddenly springs inwards. At the same time, the sides of the bladder spring apart, causing a rush of water which sucks the prey in. The trap closes and the body of the victim slowly decays. Nutrients and water are absorbed to reset the trap. The bladderwort does not secrete digestive enzymes, but these are used by the sundew and other insectivorous plants.

3 Reproduction. Asexual reproduction by leaves is common. In *Bryophyllum* (Fig 32·13*B*), complete new plants are formed at the leaf margins and drop off to become established nearby.

4 Support. Pea tendrils are modified leaflets which wrap round any available structure, holding and supporting the branches of the pea plant. This is why the gardener provides canes with string or netting for his crop.

B Asexual reproduction by leaves of Bryophyllum

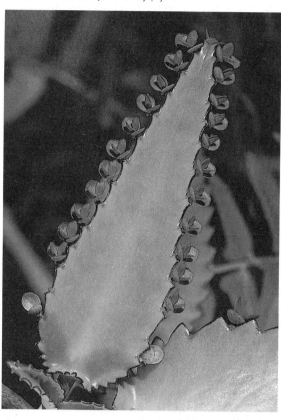

A Bladderwort

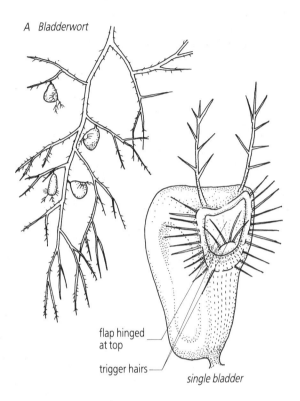

flap hinged at top

trigger hairs

single bladder

Fig 32·13 Specialized functions of leaves

33 UPTAKE AND TRANSPORT IN PLANTS

Objectives

After studying this Unit you should be able to:–

- Define translocation

- Distinguish between passive and active uptake of materials by plant cells

- Describe how osmosis occurs

- Explain each of the following terms as they apply to plant cells: water potential, solute potential, pressure potential, turgid, plasmolysed, incipient plasmolysis

- Distinguish between the symplast and apoplast pathways of water movement through the root cortex

- Define transpiration

- Draw and label a diagram of a simple potometer and comment on its usefulness in estimating transpiration rate

- List the environmental factors which affect transpiration

- Give an account of the cohesion-tension theory of water transport

- Describe the role of guard cells and stomata in limiting water loss through leaves

- Comment on the origin of root pressure

- Describe the uptake of inorganic ions by root hair cells

- List the main inorganic nutrients absorbed and state their importance in plant metabolism

- Outline the stages in conversion of nitrates to amino acids

- Draw a diagram to illustrate the mass flow hypothesis of sucrose transport in phloem sieve tubes

- Outline some of the experimental techniques used to investigate plant transport

33·1 Introduction ■ ■ ■ ■ ■ ■ ■ ■ ■ ■ ■ ■ ■ ■ ■ ■

Plants absorb water and inorganic nutrients through their roots. These substances are needed in the leaves and other parts of the plant, which may be several metres distant. At the same time, carbohydrates produced by photosynthesis in the leaves must be transferred to the stem and roots for respiration, growth, and storage. This Unit describes the structures and mechanisms involved in the uptake and transport, or **translocation**, of these chemicals from place to place. Before studying these processes, you should be familiar with the anatomy of roots, stems, and leaves, as described in Units 31 and 32.

33·2 Uptake by plant cells ■ ■ ■ ■ ■ ■ ■ ■ ■ ■ ■ ■ ■ ■

A plant cell consists of a living region, called the protoplast, surrounded by a non-living cell wall. Dissolved substances move freely through primary cell walls because these have a loose structure with large spaces in the matrix between the mesh of cellulose microfibrils. However, further movement of many chemicals is restricted by the outer membrane which surrounds the protoplast. Only a few substances, including water, dissolved gases, and inorganic ions, have molecules which are small enough to pass directly through the cell membrane. The movement of these substances takes place by diffusion and is described as **passive**, meaning that it does not require energy release by living cells. Diffusion is invariably a 'downhill' process, that is, the molecules move along a concentration gradient from high to low concentration. On the other hand, large molecules and substances which move against their concentration gradients must be transferred by **active** processes. These require a continuous supply of energy, usually obtained by hydrolysis of ATP. In **active transport**, molecules and ions are pumped across the membrane by carrier proteins. Cellulose and other wall materials are extruded from plant cells by exocytosis when secretory vesicles fuse with the outer membrane. The reverse movement, endocytosis, is rarely observed in plants.

33·3 Osmosis and water potential ■ ■ ■ ■ ■ ■ ■ ■ ■ ■ ■ ■

Water is normally drawn into the protoplast by **osmosis**. In biology, this process can be most usefully defined as the net movement of water molecules from a less concentrated solution to a more concentrated solution across a **partially permeable membrane**, that is, across a membrane which allows the small solvent molecules of water to pass but prevents the movement of large solute molecules. The relationship between osmosis and the simple process of **diffusion** is illustrated in Figure 33·1. This shows two containers each of which is set up initially with a strong sucrose solution in the right hand compartment, separated by a barrier from pure water in the left hand compartment.

In Figure 33·1A, the barrier is temporary and is removed, without disturbing the two liquids, at the start of the experiment. In Figure 33·1B, the barrier is a partially permeable membrane which remains in position throughout. After a short time, the sugar and water molecules become completely mixed in A. This is because all the atoms and molecules are in a state of continual random motion and therefore become dispersed throughout the available space. Note that there is a net transfer of sugar molecules from right to left and a net transfer of water molecules in the opposite direction. In B, the movement of sugar molecules is prevented by the membrane, but water molecules continue to diffuse freely, so that there is a net movement of water into the right hand compartment. As a result, the volume of solution in this compartment increases, causing a build up of pressure which pushes liquid up the side arm. The rising column of liquid exerts a hydrostatic pressure on the solution which opposes and finally prevents further entry of water. At equilibrium, the pressure produced can be estimated from the height of the column. The maximum pressure which any solution can develop depends directly on its concentration.

To explain the net transfer of water molecules from one side of the partially permeable membrane to the other, plant physiologists now use the terms **water potential, solute potential,** and **pressure potential**. Water potential is denoted by the symbol psi, ψ, and is measured in pressure units, pascals, Pa (1 Pa $= 1$ Nm^{-2}). The water potential of pure water is equal to 0 Pa. Adding molecules of solute restricts the movement of water molecules and produces a solute potential, ψ_s. This has a negative value which is equal in magnitude to the hydrostatic pressure exerted by the liquid column at equilibrium. Water passes by osmosis from regions of high water potential (zero or slightly negative) to regions of lower water potential (more negative). In this case, water is transferred across the membrane into the sucrose solution.

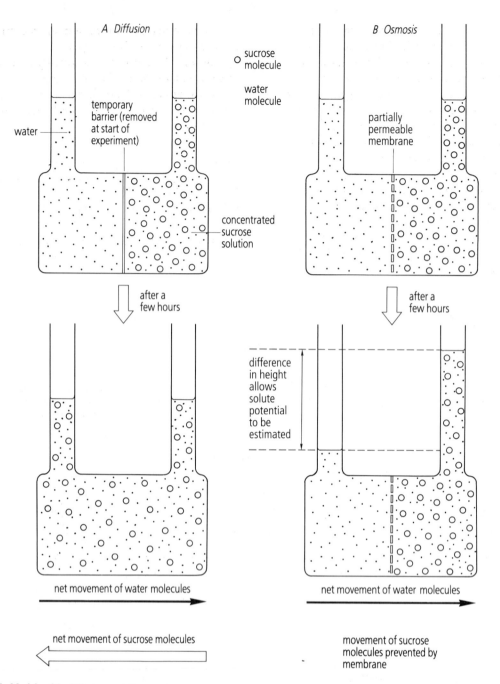

Fig 33·1 Models of A diffusion and B osmosis

When a plant cell is placed in a dilute solution, water tends to be transferred into the interior of the cell (refer to Fig 33·2). The entry of water causes the protoplast to push against the cell wall, resulting in the development of a pressure potential, denoted by ψ_p. The pressure potential has a positive value and is identical to the hydrostatic pressure called **wall pressure** or **turgor pressure**. The water relations of the plant cell are summarized in the following simple equation:

$$\psi_{cell} = \psi_s + \psi_p$$

water potential	=	solute potential	+	pressure potential
		(negative value)		(positive value)

The water potential term ψ_{cell} in this equation can be thought of as the net tendency of a plant cell to take up water by osmosis. The direction of water movement can be determined by comparing the water potential of the cell with that of the surrounding solution. If the water potential of the cell is more negative then water will be absorbed.

In dilute solutions ($\psi_{cell} < \psi_{solution}$), plant cells take up water until the solute potential tending to draw water in is just balanced by the pressure potential tending to oppose water entry. At equilibrium $\psi_s + \psi_p = 0$ and the cells are said to be fully **turgid** (Fig. 33·2A). If, however, a plant cell is placed in a concentrated solution ($\psi_{solution} < \psi_{cell}$), water is withdrawn from the protoplast which therefore shrinks away from the cell wall. A cell in this condition is said to be **plasmolysed** (Fig. 33·2B). The concentration of dissolved substances inside a plant cell can be estimated by finding the external concentration at which the cell just begins to plasmolyse. At this point, called **incipient plasmolysis**, the pressure potential is zero so that the water potential and solute potential of the cell are equal and balanced exactly by the water potential of the external solution, that is, $\psi_{cell} = \psi_s = \psi_{solution}$. Experiments in which plant cells are exposed to solutions of different concentrations allow the solute potential of the cell contents to be estimated. This is done by determining the solute potential of the solution which will result in 50% plasmolysis.

The terminology used to describe the water relations of plant cells has changed in recent syllabuses. Alternative terms which you may find in some older examination questions are summarized below. Note that these terms are regarded as obsolete and will not be used in future examinations.

quantity	symbol	alternative terms
water potential	ψ	
solute potential	ψ_s	osmotic potential, osmotic pressure
pressure potential	ψ_p	wall pressure, turgor pressure

A Turgid cells ($\psi_{cell} < \psi_{solution}$)

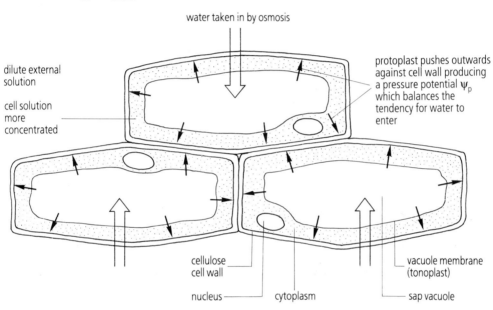

water taken in by osmosis

dilute external solution

cell solution more concentrated

protoplast pushes outwards against cell wall producing a pressure potential ψ_p which balances the tendency for water to enter

cellulose cell wall

nucleus — cytoplasm

vacuole membrane (tonoplast)

sap vacuole

B Plasmolysed cells ($\psi_{solution} < \psi_{cell}$)

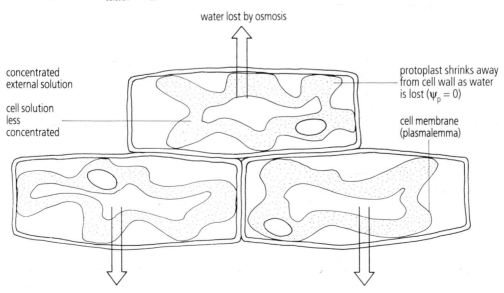

water lost by osmosis

concentrated external solution

cell solution less concentrated

protoplast shrinks away from cell wall as water is lost ($\psi_p = 0$)

cell membrane (plasmalemma)

Fig 33·2 A *turgid and* B *plasmolysed cells*

33·4 Uptake and transport of water and ions ■ ■ ■ ■ ■ ■ ■ ■

Water and ions are transported in xylem vessels. These consist of the lignified cell walls of dead cells and are simple tubes with no active role. Flow occurs from the roots upwards to the stem and leaves, but not in the reverse direction. Figure 33·3 illustrates the probable routes taken by water and ions from the soil to the leaves of a plant.

Uptake by roots

Throughout the root, the protoplasts of adjacent cells are connected by plasmodesmata. As a result, the cytoplasm of the root cells forms a continuous large fluid compartment, which is known as the **symplast**. Materials absorbed by the root hair cells or other epidermal cells can therefore pass by diffusion directly from cell to cell towards the interior of the root. A second large fluid compartment, called the **apoplast**, consists of all the fluid present in the cell walls and in intercellular spaces.

The water potential of the cytoplasm of the root hair cells is normally more negative than the water potential of the soil solution so that water is absorbed by osmosis into the root hairs and may be transferred from cell to cell via the symplast pathway. Water also enters the root through the spaces between the epidermal cells and may be carried inwards by the apoplast pathway, but only as far as the endodermis. As explained in Unit 32, individual endodermal cells are surrounded by a waterproof **Casparian strip** which joins adjacent cells together forming a completely impermeable barrier. Therefore, whichever route is used, water reaching the endodermis must pass through the protoplasts of the endodermal cells before reaching the xylem vessels. Upward movement in the xylem can be explained in either of two ways, that is, by pushing forces originating in the roots or by pulling forces from the leaves. Both types of forces are observed but the leaves play by far the more important part.

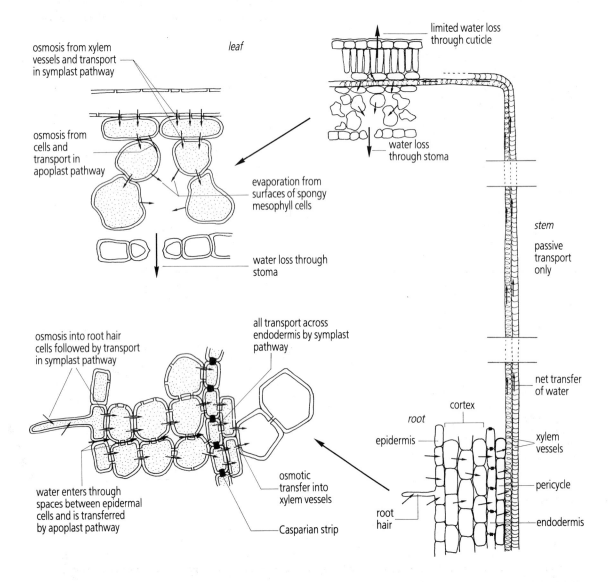

Fig 33·3 Probable routes taken by water and ions from the soil to the leaves of a plant

Transpiration

Transpiration is defined as the loss of water from the leaves and other aerial parts of a plant. On a hot day, a large tree absorbs hundreds of litres of water from the soil. As little as 1% of this is used for photosynthesis, the remaining 99% being lost by evaporation from the surfaces of the leaf mesophyll cells and subsequent diffusion of water vapour through the stomata. The heat energy required for evaporation is absorbed from the surroundings and can have a significant cooling effect. For example, the interior of a forest may be some 5–10°C cooler than the surrounding countryside.

The pulling force developed by transpiration is demonstrated in the experiment illustrated in Figure 33·4A. The apparatus shown is called a **potometer** (literally 'drinking meter'), and can be used to measure the rate of water uptake of the shoot under different environmental conditions. A bubble of air is introduced into the end of the capillary tube and its progress along the tube is followed using the calibrated scale. Note that the potometer does *not* measure transpiration directly and will give an accurate indication only if the rate of water loss is precisely equal to the rate of water uptake. However, for most purposes, the estimate of transpiration obtained is satisfactory. The behaviour of the plant can be compared with that of an **atmometer**, a porous pot filled with water, which shows the effect of physical conditions on the rate of evaporation from its surface (below).

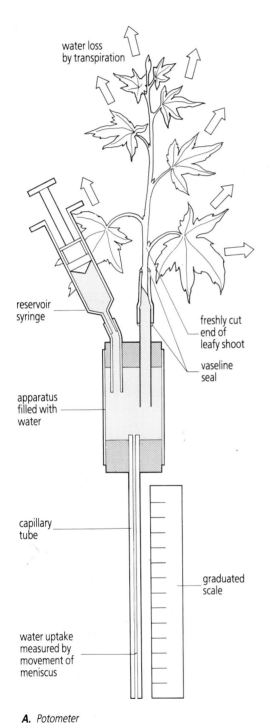

A. Potometer

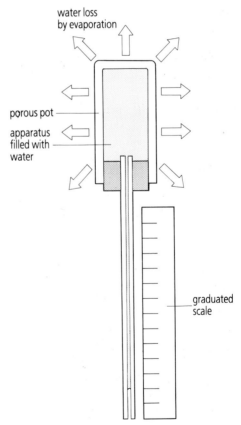

B. Atmometer

Fig 33·4 Diagrams showing A a simple potometer, and B an atmometer used for comparing the rates of evaporation and transpiration under different external conditions

Transpiration is found to vary with the following environmental conditions:

1 Relative humidity. The rate of water loss depends on the difference in concentration of water molecules in the sub-stomatal spaces and in the air outside. High relative humidity reduces water loss.

2 Wind and air currents. The movement of air increases the rate of water loss from the leaves by carrying away molecules of water vapour, thus reducing the relative humidity of the air at the leaf surface.

3 Altitude. At high altitudes, atmospheric pressure is decreased sufficiently to cause an increase in the rate of transpiration.

4 Temperature. Molecules move faster in warm air, hence the rates of evaporation and transpiration are increased.

5 Light. Light has no effect on the rate of evaporation from an atmometer, but, in a plant, it stimulates the opening of stomata and consequently increases transpiration. The effects of light and temperature are complex and interrelated and will be discussed in more detail shortly.

6 Water availability. If soil water is in short supply, the stomata close and transpiration is reduced.

The upward flow of water and dissolved nutrients in the conducting cells of the xylem due to the suction pressure created by transpiration is called the **transpiration stream**. The pulling force is entirely passive in origin, that is, transpiration does not require ATP and is not affected by metabolic inhibitors.

The **cohesion-tension theory** is now generally believed to provide a satisfactory explanation for the magnitude of the transpiration pull, even for large trees. Water molecules are polar and therefore attract each other – this is called **cohesion**. The cohesive forces generated are strong enough to allow water to be pulled upwards under tension for considerable distances without the water column breaking. In addition, water molecules are attracted by **adhesion** forces to the sides of capillary tubes so that taller columns of water can be supported. What is proposed therefore, is that (i) the plant's xylem vessels constitute a series of continuous tubes each supporting a water column which extends from the roots up into leaf veins, (ii) the cohesive and adhesive properties of water molecules ensure that the columns do not break, and (iii) the tension required to raise the water columns against gravity is provided by transpiration at the leaf surfaces.

The site of water evaporation is the interface between the mesophyll cells of the leaf and the intercellular air spaces. Water vapour, resulting from evaporation, diffuses through the gaps between the cells until it accumulates in the sub-stomatal spaces. If the stomatal pores are open, water vapour diffuses out into the atmosphere.

Evaporation at the mesophyll cell surface produces tension in the xylem vessels in the following way. When water leaves the protoplast of a mesophyll cell, the cell contents become relatively more concentrated and the water potential of the cell becomes more negative. This allows the cell to make good its water loss by taking in water directly from neighbouring cells (symplast pathway), or by osmosis from the extracellular fluid (apoplast pathway). In either case, a water potential gradient is set up so that water is drawn from the xylem vessels of the leaf veins. The tension thus created is transmitted back down the stem into the roots and water flows into the leaves along the transpiration stream.

Experimental evidence for the cohesion-tension theory has been obtained in a variety of ways. In one experiment, a device called a dendrograph was used to measure the diameter of trees. It was found that, during the day, the diameter was less, reflecting a reduced pressure in the xylem vessels. In another experiment, sap in xylem tissue was heated and its progress up the stem followed using a thermocouple. It was discovered that the rate of transport varied with the time of day and that rapid transport in small branches started in the morning *before* rapid transport in the main trunk. This result can only be explained if upward movement is the result of a pull originating in the leaves. It is clear that the cohesion-tension theory can hold only for unbroken columns of water. A bubble anywhere in a xylem vessel prevents water transport. Breaks may possibly occur from time to time but it seems likely that they can be repaired by dissolving gases in the cytoplasm of surrounding cells.

Control of water loss

The process of transpiration is very wasteful of water. However, plants control their water loss according to external conditions and internal demands, and are therefore able to osmoregulate to a limited extent. This is achieved mainly by the opening and closing of stomatal pores. Consequently, transpiration is linked to the control of gas exchange, which also occurs through stomata.

The stomata are usually closed at night and open during the day when photosynthesis is in progress. This is made possible by changes in the shape of the **guard cells** which depend in turn upon changes in water potential. The most important factors involved are thought to be the availability of carbon dioxide and ATP. When guard cells are exposed to light, the concentration of carbon dioxide falls and ATP is manufactured by cyclic phosphorylation (explained in Unit 11). These changes activate a K^+ pumping mechanism in the cell membrane of the guard cell, causing the uptake of large numbers of K^+ ions from the surrounding epidermal cells. The cell's water potential becomes more negative and water is absorbed by osmosis, making the guard cell turgid. As the guard cell swells, it changes shape because the cell walls bordering the stoma are particularly thick and resistant to stretching while those on the outer sides are thin and flexible. Consequently, the guard cells become crescent-shaped and hold the stoma open (Fig. 33·5A). At night, carbon dioxide accumulates as a result of respiration and supplies of ATP are used up. The K^+ pump is inactivated and the guard cells lose K^+ ions and become flaccid. As a result, they collapse, thereby closing the stoma (Fig. 33·5B). Under drought conditions, the closing of stomata may involve abscisic acid, but the details of this are not fully understood. In desert plants, including cacti, stomata are opened at night and closed during the day, so that different mechanisms are required.

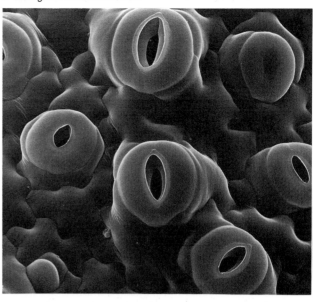

C SEM guard cells and stomata

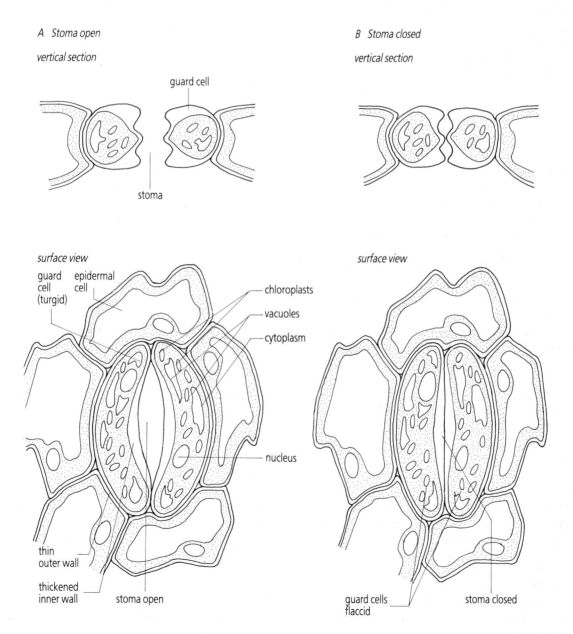

A Stoma open

vertical section

guard cell

stoma

B Stoma closed

vertical section

surface view

guard cell (turgid)   epidermal cell

chloroplasts

vacuoles

cytoplasm

nucleus

thin outer wall

thickened inner wall   stoma open

surface view

guard cells flaccid   stoma closed

Fig 33·5 *Action of guard cells in opening and closing stomata*

283

Root pressure

If a stem is cut just above ground level, sap can be seen to exude from the cut surface under pressure. This pressure is called **root pressure** and can be measured using the apparatus illustrated in Figure 33·6. In the tomato plant, the roots produce an upward force equivalent to an astonishing 8 atmospheres, but in most plants root pressure is found to be much less, no more than about 2 atmospheres. In some plants, including pine trees, root pressure is absent.

Root pressure depends on a pumping process which occurs in the endodermis of the root between the parenchyma cells of the root cortex and the xylem vessels of the vascular tissue. Here, ions of many different kinds are pumped from the cortex into the xylem vessels and are prevented from leaking back by the Casparian strips of the endodermal cells. Consequently, a difference in water potential is established such that water passes by osmosis through the protoplasts of the endodermal cells and flows upwards in the xylem. It is important to appreciate that root pressure is the result of an active process in which energy from ATP is used. If metabolic inhibitors are administered, root pressure disappears completely.

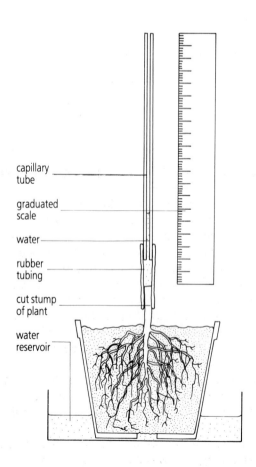

Fig 33·6 Apparatus used to demonstrate root pressure

capillary tube

graduated scale

water

rubber tubing

cut stump of plant

water reservoir

Uptake of ions

The uptake of ions from the soil usually takes place against a concentration gradient and depends on a similar pumping process in epidermal cells at the root surface. Root hair cells absorb nitrates, phosphates, sulphates, and a variety of other essential inorganic nutrients (Table 33·1). For example, they acquire a concentration of K^+ ions up to 70 times greater than that of the soil water. This uptake is assisted by the negative internal electrical potential of the root hair cells which attracts positively charged ions (cations) like K^+ and Ca^{2+}. These are absorbed by a process called **ion exchange** in which H^+ ions are pumped out of the cell at the same time as K^+ and Ca^{2+} ions are pumped in.

Inorganic materials absorbed by the root hairs are transferred across the cortex in the symplast compartment and are then pumped into xylem vessels by the same process which is responsible for root pressure. Most of the absorbed ions are transported in the xylem in an unmodified form but some are converted to other substances. An important example occurs in the case of nitrates (NO_3^-), which are rapidly converted by the root cells, first to nitrites (NO_2^-), and subsequently to ammonia (NH_3) and amino acids. The important steps in this process are:

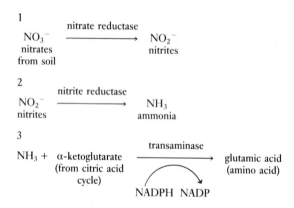

The enzyme **nitrate reductase** responsible for the first stage in the conversion of nitrates to amino acids uses the trace element **molybdenum** as a cofactor. Extremely small quantities of molybdenum are sufficient for the synthesis of this enzyme. Nevertheless, some soils are molybdenum-deficient, as in parts of Australia. Following its manufacture from nitrites, ammonia is converted to the amino acid glutamic acid by a reaction which involves α-ketoglutarate, one of the intermediate substances in Krebs cycle (described in Unit 10). Other amino acids are made by **transamination reactions** involving the transfer of amino groups from glutamic acid. The amino acids produced are transported across the root cortex and can be transported upwards in the xylem to the leaves, where protein synthesis occurs.

Table 33·1 Ions absorbed from the soil and their functions in the plant

ELEMENT	IONS ABSORBED	FUNCTIONS	SYMPTOMS OF DEFICIENCY
Macronutrients			
Nitrogen	NO_3^-, NH_4^+	component of protein, nucleic acids	stunted growth, yellow leaves (chlorosis)
Potassium	K^+	main intracellular cation, water potential changes in guard cells, enzyme cofactor	stunted growth, yellow leaves (chlorosis)
Phosphorus	PO_4^{3-}, $H_2PO_4^-$	component of nucleic acids, ATP, essential for flowering, fruiting, root development	small dark green leaves, abnormal stem colours, reduced root growth
Sulphur	SO_4^{2-}	component of protein	chlorosis in new leaves, reduced root growth
Calcium	Ca^{2+}	formation of middle lamella, enzyme cofactor	small leaves, death of apical buds
Magnesium	Mg^{2+}	component of chlorophyll, enzyme cofactor	chlorosis
Micronutrients (trace elements)			
Boron	complex ion	not known	abnormal leaf colour
Chlorine	Cl^-	maintenance of ionic balance, photosynthesis	small leaves, chlorosis
Copper	Cu^{2+}	component of some cytochromes, enzyme cofactor	abnormal dark green colour of young leaves
Iron	Fe^{2+}, Fe^{3+}	component of cytochromes, cofactor for chlorophyll synthesis	chlorosis of leaf veins
Manganese	Mn^{2+}	cofactor for enzymes of citric acid cycle	reduced growth
Molybdenum	MoO_4^{2-}	metabolism of nitrates	reduced growth
Zinc	Zn^{2+}	cofactor for auxin synthesis	small leaves, short internodes

33·5 Transport of sugars ■ ■ ■ ■ ■ ■ ■ ■ ■ ■ ■ ■ ■

The products of photosynthesis are transported in the form of dissolved sugars, mainly sucrose. This substance is transported by the phloem from its site of manufacture or storage to the areas where it is used. Any region of supply is referred to as a **source**, while a region of utilization is a **sink**.

The leaves of a plant usually act as a source of sugars and the meristematic and growing tissues act as a sink. Storage tissues in structures such as seeds, bulbs or tubers may be a sink during development, and a source when being used. The direction of transport is generally downwards, but depends on the seasonal requirements of different plant structures.

A possible mechanism of sucrose transport is suggested by the **mass flow hypothesis**. According to this hypothesis, there are two sites of active transport of sugars, both involving companion cells, as shown in Figure 33·7. At the source, sucrose is taken up actively by phloem companion cells and passed through plasmodesmata into the adjacent sieve tubes. At the sink, the reverse occurs, the companion cells actively pumping out sugar into the tissues which require it. This establishes a concentration gradient of sugar along a sieve tube from source (concentrated) to sink (less concentrated).

Water is drawn into the sieve tube by osmosis at the source end where the concentration of sugar is high and, as it accumulates, a pressure potential develops in the tube, resulting in the bulk movement of sugar and water towards the sink. At the sink end, the concentration of sugar in the sieve tube decreases and water passes out into the neighbouring tissues.

Among a number of important objections to the mass flow hypothesis is the observation that different substances transported in the phloem are sometimes moved simultaneously in opposite directions, and at different rates.

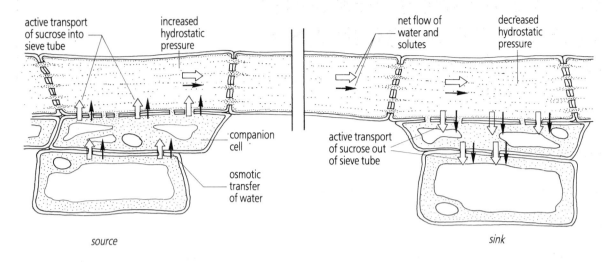

Fig 33·7 *Diagram showing the essential features of the mass flow hypothesis*

33·6 Experimental investigation of plant transport ■ ■ ■ ■ ■

Radioactive tracers

The site and rate of transport of certain substances can be determined using radioactively labelled plant substrates. The radioactive isotope of phosphorus, ^{32}P, for example, can be incorporated into phosphate and given to a plant in its soil water. Another common substrate, CO_2, can be labelled with ^{14}C and will be absorbed by the leaves of a plant enclosed in a perspex box containing a source of $^{14}CO_2$.

The path of these substances can later be traced by the technique of **autoradiography**. After drying and mounting, the whole plant may be placed in contact with X-ray film which reveals the regions of highest radioactivity (Fig. 33·8). Alternatively, the plant may be freeze-dried and then cut into sections suitable for light microscopy. These are covered with a photographic emulsion and later developed, when regions of radioactivity show up as black dots, indicating the exact location of the labelled substances.

Aphids in phloem research

Aphids feed on plant sugars by means of hollow mouthparts called **stylets** with which the insects penetrate the outer tissues of a leaf or stem, often inserting the tips of the stylets into a single phloem sieve tube. The hydrostatic pressure within the tube forces sugar-containing sap outwards into the alimentary canal of the aphid.

The contents of sieve tubes can conveniently be analysed by cutting through the stylet of an aphid and collecting the exuded sap with a micropipette. If the sugar is radioactively labelled, the rate of movement within the sieve tubes can be calculated using a number of aphids at different positions along a plant stem.

As you have seen, the movement of water, inorganic nutrients and sugars in plants depends on a variety of passive and active transport mechanisms. Although different tissues are specialised in quite different ways, it is important to remember that the substances they carry often have interrelated functions. For example, molecules of sucrose made by photosynthesis in a leaf may be translocated in phloem sieve tubes to the roots, where they provide glucose for respiration and intermediates for the synthesis of amino acids. Amino acids are carried upwards in the xylem and may later be used to make new protein in a differentiating cell at the shoot tip. Thus, the mechanisms described in this Unit form a coordinated circulatory system which ensures that all parts of the plant receive the substances they require.

Fig 33·8 *Autoradiography as a method of investigating plant transport*

34 PLANT RESPONSES AND THE ACTION OF PLANT HORMONES

Objectives

After studying this Unit you should be able to:–

- Define a tropism

- Outline some of the experimental evidence for hormonal control of phototropism in coleoptiles

- Describe the role of auxin in positive phototropism in shoots

- Comment on negative phototropism in roots

- Describe geotropism in roots and shoots and state the possible roles of abscisic acid and statocytes

- Briefly describe apical dominance

- List the functions of the following hormones: gibberellins, cytokinins, abscisic acid, ethene

- Define photoperiodism and distinguish between short-day plants, long-day plants, and day-neutral plants

- Outline the role of phytochromes in the flowering response

- Comment on 'florigen' and hormonal control of flowering

- Define and give examples of each of the following: nastic movement, nyctinasty, photonasty, thermonasty, haptonasty

- List some of the economic uses of plant hormones

34·1 Introduction ■ ■ ■ ■ ■ ■ ■ ■ ■ ■ ■ ■ ■ ■ ■ ■ ■

Plants are multicellular and autotrophic, that is, they make their own food by photosynthesis. This mode of life involves only simple responses to changes in the environment. Plants do not move actively from place to place, and do not possess muscles or a nervous system. Nevertheless, they can detect and respond appropriately to stimuli including light, gravity, chemicals, changes in temperature, and contact.

The most frequently observed responses are relatively slow growth movements called **tropisms**. These are directed towards or away from a particular stimulus. Tropisms are controlled by the action of plant hormones, as will be explained shortly. **Nastic movements** are movements of parts of the plant which are not orientated in any particular direction according to the stimulus. They include certain types of growth movements and a variety of rapid responses brought about by changes in the water potential of specialised cells.

This Unit describes the most important plant responses and considers the interaction of plant hormones in regulating processes including growth, dormancy, germination, leaf fall, and flowering.

34·2 Tropisms ■ ■ ■ ■ ■ ■ ■ ■ ■ ■ ■ ■ ■ ■ ■ ■ ■

Positive phototropism in shoots

The shoots of a plant normally grow towards light – this is called positive phototropism. As early as 1880, Charles Darwin and his son Francis investigated this response in grass seedlings, which were grown under different conditions, as indicated in Figure 34·1. Before interpreting their results, it is important to explain that grasses belong to the group of flowering plants known as monocotyledons. Consequently, grass seedlings have a rather different structure from that of the broad bean and other dicotyledons. In particular, the first part of the shoot to emerge from a germinating seed is a tightly rolled bundle of leaves protected by a sheath called the **coleoptile** (Fig. 34·2). The meristematic cells responsible for producing the new cells required for growth are located, not as might be expected near the tip of the coleoptile, but near its base, as is the case in the leaves of all monocotyledons. At the time of germination, the apical meristem at the tip of the stem is protected deep within the coleoptile close to the seed.

What the Darwins discovered was that the tip of the coleoptile was the part which was sensitive to light. Shading the region of cell division near the base of the coleoptile had no effect (Fig. 34·1B), while shading the tips of the seedlings prevented the phototropic response (Fig. 34·1A). Later investigators concluded that the response must

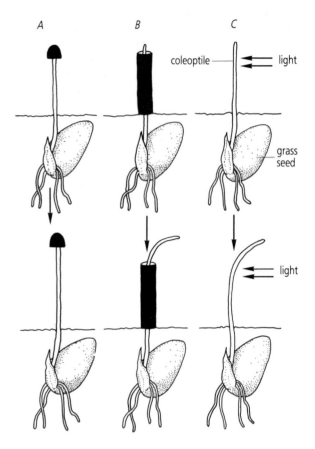

Fig 34·1 Darwin's experiment to discover which part of a coleoptile responds to light

involve cell elongation rather than cell division and that the cells near the tip produced a chemical substance which caused the elongation of other cells lower down in the coleoptile.

Some of the experimental evidence for these conclusions is outlined in Figue 34·3. When the coleoptile tip was removed (Fig. 34·3B), growth stopped. However, if the cut tip was replaced, growth resumed (Fig. 34·3C). A small thin sheet of mica, which was waterproof and impermeable, prevented the growth response (Fig. 34·3D), but a small block of agar jelly through which water soluble substances could diffuse, allowed elongation to continue (Fig. 34·3E). In other experiments, it was discovered that a growth-promoting substance could be transferred to a small agar block which produced curvature of the coleoptile if placed asymmetrically at the tip (Fig. 34·3F). The growth-promoting substance responsible was the first known plant hormone and was named **auxin** (from the Greek word, *auxein*, to grow).

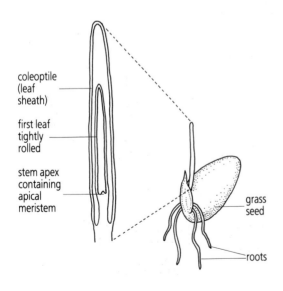

Fig 34·2 Structure of the seedling of a monocotyledon (grass seedling)

Auxin was eventually isolated and chemically identified and was found to be **indoleacetic acid (IAA)**. The degree of curvature produced in the experiment of Figure 34·3F proved to be an extremely sensitive biological test, or **bioassay**, for the presence of indoleacetic acid and was used to estimate the relative amounts of auxin in agar blocks from coleoptiles subjected to different treatments. For example, it was shown that two agar blocks separated by a thin sheet of mica would collect different amounts of auxin if placed beneath a coleoptile illuminated from one side. The agar block from the shaded side contained more auxin and produced more curvature than the block from the illuminated side (Fig 34·3G). Phototropism was therefore the result of an unequal distribution of auxin in the coleoptile, causing increased elongation of the cells on the shaded side.

At first it was believed that light caused the breakdown of auxin, but more recently it has been demonstrated that auxin is actively transported to the shaded side. Consider the experiments illustrated in Figures 34·3H, I, and J. In each case, one end of a length of coleoptile was placed in contact with an agar block containing indoleacetic acid labelled with the radioactive isotope of carbon ^{14}C. At the other end were two small agar blocks separated by mica sheet. In an upright section of coleoptile placed in the dark, the amount of labelled indoleacetic acid detected in the two blocks was equal (Fig. 34·3H), but in a coleoptile illuminated from one side (Fig. 34·3I), or in one placed in a horizontal position (Fig. 34·3J), the auxin was transported from one side of the coleoptile to the other and was distributed unevenly between the two blocks. Measurements of the total quantity of radioactivity in the blocks indicated that the auxin had not been destroyed by light.

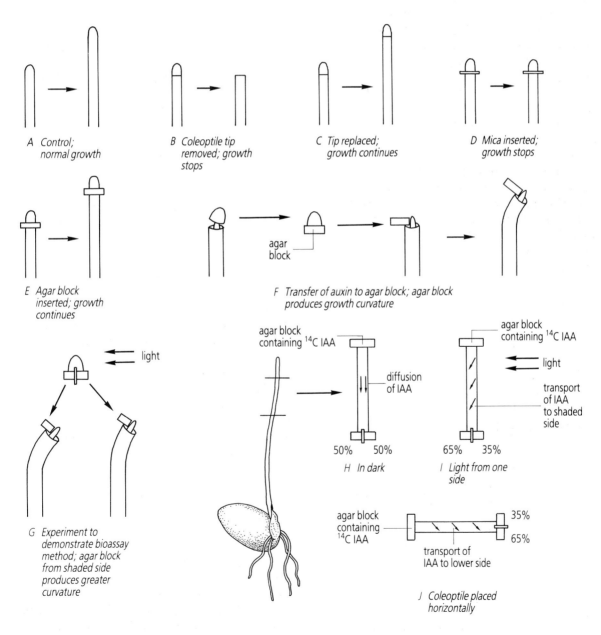

A Control; normal growth

B Coleoptile tip removed; growth stops

C Tip replaced; growth continues

D Mica inserted; growth stops

E Agar block inserted; growth continues

agar block

F Transfer of auxin to agar block; agar block produces growth curvature

light

G Experiment to demonstrate bioassay method; agar block from shaded side produces greater curvature

agar block containing ^{14}C IAA

diffusion of IAA

50%   50%

H In dark

agar block containing ^{14}C IAA

light

transport of IAA to shaded side

65%   35%

I Light from one side

agar block containing ^{14}C IAA

35%

65%

transport of IAA to lower side

J Coleoptile placed horizontally

Fig 34·3 Experiments to investigate tropisms in oat coleoptiles

Although a number of synthetic substances have similar effects, indoleacetic acid appears to be the only naturally-occurring auxin and its presence has been confirmed in many different species. It is produced from the amino acid tryptophan in metabolically active tissues including the apical meristems of most shoots and is normally transported downwards in the stem. Indoleacetic acid promotes the release of H^+ ions into cell walls, reducing pH and providing the appropriate conditions for the activity of enzymes which attack cross links between adjacent cellulose molecules. The result is a loosening of the cell wall which allows the protoplast inside to absorb water by osmosis, stretching and elongating the cell. Precisely how the light stimulus is linked to the transport mechanisms responsible for producing the unequal distribution of auxin which causes growth curvature is not fully understood. However, molecules of a light sensitive pigment concentrated in the cells nearest to the tip of most shoots are likely to be involved.

Negative phototropism in roots

While the shoots of a plant are positively phototropic, the roots are normally negatively phototropic and tend to grow away from a source of light. The possible role of auxin in the tropic responses of roots was investigated by exposing roots and shoots to different concentrations of auxin (Fig. 34·4A). As you can see, low concentrations of auxin which stimulated elongation in roots had no effect on shoots. Higher concentrations stimulated the growth of shoots but inhibited elongation in roots. Based on these observations, it was proposed that negative phototropism resulted from redistribution of auxin to the shaded parts of the root, just as in the shoot, but that the effects of auxin were different (Fig.34·4B). In the shoot, high concentrations of auxin caused elongation of cells and upward curvature, while, in the root, high concentrations of auxin *prevented* elongation and therefore allowed downward curvature.

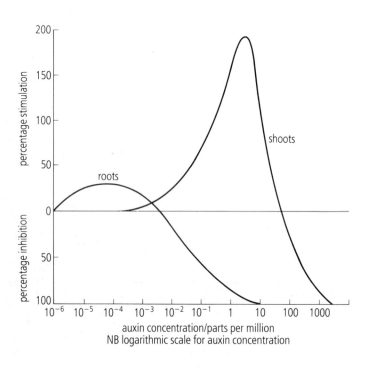

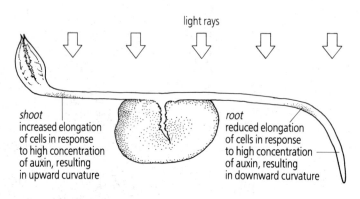

Fig 34·4 Effects of auxin on elongation in roots and shoots

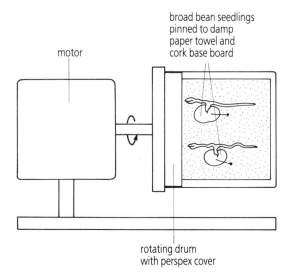

Fig 34·5 Klinostat

Geotropism

A geotropism is a growth movement in a direction determined by gravity. Shoots grow vertically upwards and are described as negatively geotropic, while roots grow downwards and are positively geotropic. The importance of gravity in the geotropic response of the radicles of broad bean seedlings can easily be demonstrated by placing the seedlings in a slowly revolving drum, or **klinostat**, as illustrated in Figure 34·5. This treatment exposes all parts of the seedlings equally to

gravitational stimulation, with the result that the radicles continue to grow horizontally.

For many years, it was believed that geotropism could be explained by redistribution of auxin in exactly the same way as phototropism. This is still thought to be the correct explanation for negative geotropism in shoots, and there is good evidence for the movement of auxin in coleoptiles in response to gravity (Fig. 34·37). However, auxin does not seem to be responsible for the positive geotropism of roots. It has been shown that a different hormone, **abscisic acid (ABA)**, accumulates on the downward sides of roots and it is this hormone which is now thought to inhibit elongation while the cells in the upper regions continue to grow, producing a downward curvature.

The mechanism used by roots and shoots to detect gravity appears to involve cell organelles called **amyloplasts** containing starch grains. These have a density greater than that of cytoplasm and have been shown to sink under the effects of gravity, coming to rest against the cell membrane at the bottom of the cell (refer to Fig. 34·6). The presence of amyloplasts is thought to trigger the transport of auxin and abscisic acid across the membrane. Similar activity by many such cells results in the redistribution of these hormones to the lowest regions of the root or shoot. Cells which contain amyloplasts and which therefore seem to be specialized as gravity detectors are called **statocytes** and have been discovered in all plant organs capable of responding to gravity.

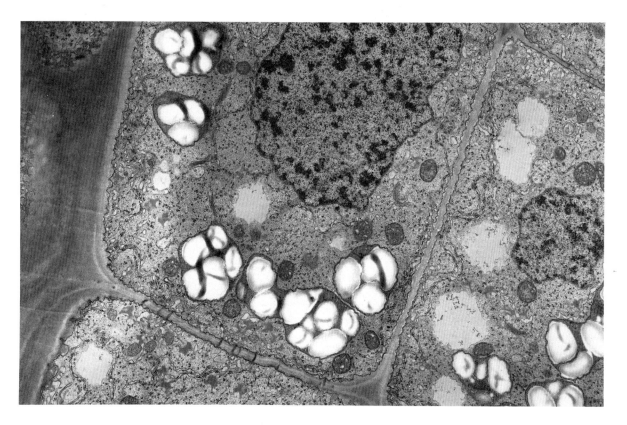

Fig 34·6 Electronmicrograph to show amyloplasts in root cells

Apical dominance

In most plants, the growing tip of the stem inhibits the growth of lateral buds and delays the development of lateral branches. This effect is called **apical dominance** and involves the production of auxin by the apical meristem within the apical bud. Sometimes, the inhibition produced by the apical bud is weak, giving a plant with a wide branching pattern (Fig 34·7A). In other species, the apical bud has a strong inhibiting effect and the plants become tall and thin with little lateral branching (Fig. 34·7B).

The mechanisms of apical dominance are not fully understood. The effect of an apical bud can be mimicked by replacing the apical bud with auxin in a lanolin paste, but it is clear that other plant hormones, including abscisic acid and **cytokinins**, also play a part. In some studies, it has been found that auxin would suppress growth only if the lateral buds had not been allowed to start growing. Once sprouting of the lateral buds had taken place, auxin applied at the tip did not arrest their development.

A

B

Fig 34·7 Patterns of growth in trees with different degrees of apical dominance

34·3 Additional plant hormones ■ ■ ■ ■ ■ ■ ■ ■ ■ ■ ■

The main types of plant hormones and their important functions are listed in Table 34·1. This section describes the different groups of hormones and comments on the interactions which produce particular plant responses.

1 Gibberellins. These were first isolated from a parasitic fungus, *Gibberella fujikuroi*, which causes a damaging disease of rice. Infected seedlings grow abnormally tall and their stems are weakened. Fungal extracts yielded a number of substances, one of which was named **gibberellic acid**. Analysis of plant tissues by the same methods soon revealed that smaller amounts of gibberellic acid and related compounds were normally present, and led to the discovery of their role as plant hormones. At least 50 different naturally-occurring gibberellins have been identified.

Gibberellins stimulate the elongation of cells in the stem and are known to be deficient in some dwarf varieties. They are also involved in leaf growth, fruit development, germination and flowering. In germinating barley seeds, gibberellins produced by the embryo plant are known to stimulate the production of amylase enzymes, leading to the mobilization of stored food reserves in the early stages of germination (see Fig. 34·8).

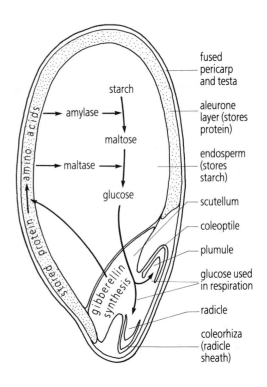

Fig 34·8 Role of gibberellins in germination of barley grains

292

Table 34·1 Summary of plant hormones

HORMONE GROUP	SITES OF PRODUCTION	PARTS AFFECTED	EFFECTS
AUXINS indoleacetic acid 2,4-D (synthetic)	dividing cells in apical and leaf meristems, exported from shoot tip	shoot tip and stem	decreased cell division, increased cell elongation
		lateral buds	growth inhibited (apical dominance)
		vascular cambium	promotes secondary growth
		leaf	inhibits leaf abscission
		fruit	promotes development, inhibits fruit abscission
		seed	no effect on dormancy
		wounds	cell differentiation, growth of adventitious roots
GIBBERELLINS gibberellic acid	chloroplasts, seed embryo, root tip, exported from young leaves	meristems	increased cell division
		stem	increased elongation
		fruit	promotes development
		seed	initiates germination
CYTOKININS	dividing cells in roots, seeds, fruits, exported from roots	meristems	increased cell division, decreased cell elongation
		lateral buds	promotes development
		leaf	delays ageing
		fruit	promotes development
ABSCISIC ACID	leaves, especially ageing leaves, stems, fruits, seeds	root	decreased cell elongation
		apical and lateral buds	induces dormancy
		leaf	promotes leaf abscission
		fruit	promotes fruit abscission
		stomata	promotes closure in drought conditions
ETHENE (ethylene)	most plant organs	fruit	promotes ripening

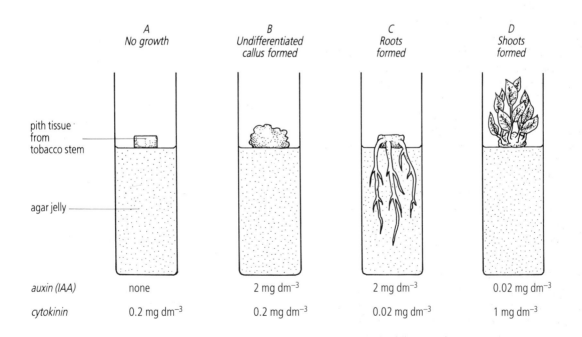

	A No growth	B Undifferentiated callus formed	C Roots formed	D Shoots formed
auxin (IAA)	none	2 mg dm⁻³	2 mg dm⁻³	0.02 mg dm⁻³
cytokinin	0.2 mg dm⁻³	0.2 mg dm⁻³	0.02 mg dm⁻³	1 mg dm⁻³

pith tissue from tobacco stem

agar jelly

Fig 34·9 Auxin, cytokinins and differentiation

2 Cytokinins. Cytokinins are chemically related to the organic base adenine found in nucleic acids and ATP. They interact with auxin to promote cell division in apical meristems and differentiation of plant tissues. For example, depending on the relative levels of auxin and cytokinin, small pieces of pith tissue from tobacco stems may be stimulated to form a mass of undifferentiated cells, called a **callus** (refer to Fig. 34·9). Alternatively, roots or shoots may develop. Whether auxin and cytokinins produce differentiation in intact plants in the same way is not yet understood. It has also been found that cytokinins delay ageing processes including leaf fall.

3 Abscisic acid. The action of abscisic acid in geotropism and a possible role in closing stomata have already been mentioned. Abscisic acid suppresses the activity of auxin, gibberellins, and cytokinins, thereby preventing growth. In the autumn, it is produced in the mature leaves of some deciduous trees and becomes concentrated into buds causing the differentiation of bud scales and the development of **bud dormancy**. Buds remain dormant until the spring when rising levels of gibberellins and cytokinins stimulate growth.

Abscisic acid probably has a similar inhibitory role in some dormant seeds.

The precise role of abscisic acid in leaf fall, called **leaf abscission**, is not known. In this process, illustrated in Figure 34·10, usable nutrients are reabsorbed from the leaf and transported to other parts of the plant. A thin abscission layer is formed at the base of the petiole and a protective layer of corky tissue seals the wound, forming a leaf scar. The loss of the leaves greatly reduces transpiration and is an important adaptation for reducing water loss in winter. High concentrations of abscisic acid applied directly to the leaves are capable of triggering abscission, but this may be due to the inhibiting effects of abscisic acid on auxin and cytokinins. Auxin production normally decreases in ageing leaves, and the resulting decline seems to be a major factor in the timing of abscission. Replacing the lamina of an ageing leaf with auxin in lanolin paste delays or prevents abscission of the petiole.

4 Ethene (ethylene). Ethene gas is almost the only known example of a gaseous hormone. It is produced by fruits in the early stages of ripening and stimulates further ripening.

A

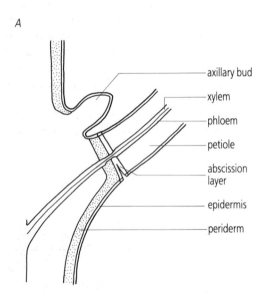

axillary bud
xylem
phloem
petiole
abscission layer
epidermis
periderm

B

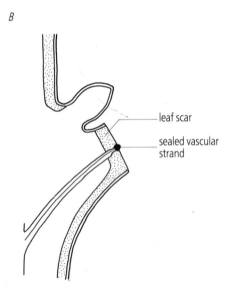

leaf scar
sealed vascular strand

Fig 34·10 Leaf abscission

34·4 Photoperiodism and phytochrome ■ ■ ■ ■ ■ ■ ■ ■

Seasonal activities in plants include flowering, development of fruits and dormancy. In most plants, these activities are timed to coincide with the most favourable time of year, for example, when insect pollinators are available, or when conditions are suitable for seed dispersal. Often, the most important environmental stimulus triggering a change in development is the **photoperiod** associated with a particular season, that is, the number of hours of daylight in each 24 hour cycle. A response affected by changes in the photoperiod is said to show **photoperiodism**.

In many species, flowering is a photoperiodic response. **Short-day plants**, like chrysanthemum, potato, and strawberry, produce flowers when the photoperiod is *less* than some critical value, that is, flowering is prevented if the plants experience more than a certain number of hours of daylight. Such plants usually flower early in the spring, or in late summer or early autumn. In other plants, called **long-day plants**, flowering will not occur until the photoperiod *exceeds* a threshold value, so that they tend to flower as the hours of daylight increase towards the summer. Examples of long-day plants include clover, maize, gladiolus, and spinach. Photoperiod is not the only factor determining the flowering of short-day and long-day plants and it is important to appreciate that factors like temperature and the availability of moisture and soil nutrients are also involved. In **day-neutral plants** the photoperiod is unimportant and these other factors become the main external influence affecting flowering time. Many tropical plants and some temperate species, including dandelions, sunflowers, and tomatoes, are day-neutral.

The measurement of photoperiod by short-day and long-day plants is now known to depend on a light-sensitive pigment, called **phytochrome**, which is present in the leaves. Phytochrome exists in two forms each of which can be converted to the other by light of the appropriate wavelength. The inactive form of phytochrome, P_R, is converted to the active form, P_{FR}, following exposure to red light with peak sensitivity at 660 nm. P_{FR} is slowly converted back to the inactive form in darkness, or may be rapidly converted to P_R by exposure to far-red light, with peak sensitivity at 730 nm:

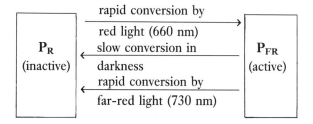

Daylight is a mixture of wavelengths containing both red and far-red light, but red wavelengths predominate, so that, during the day, P_R is converted to P_{FR} which accumulates in the leaves. The onset of flowering seems to be determined by the balance between the two forms of phytochrome pigment, which in some way switches on the genes responsible for flowering.

Flowering may also be initiated by hormones transported from place to place around the plant. The experiment of Figure 34·11 provides clear evidence for some type of chemical influence. In *A*, two cocklebur plants separated by a light-tight partition have been grown with different photoperiods. Cocklebur is a short-day plant with a critical photoperiod of around 15.5 h. With 12 h of light per day, flowers are produced, but with 18 h of light, flowering is inhibited. In *B*, a flowering branch from the 12 h plant has been grafted on to the 18 h plant through a hole in the light-tight partition. If the two parts of the plant continue to receive 12 h and 18 h of light respectively, the 18 h plant subsequently produces flowers although it would not normally have done so. Experiments of this sort have been taken as evidence for the existence of a special flowering hormone, named **'florigen'**. However, such a substance has never been isolated and nothing at all is known about its possible chemical composition. Indeed, it seems more likely that the effect observed is due to the interaction of two or more hormones from the groups already described.

The timing of a whole range of other light-dependent plant responses involves phytochrome as

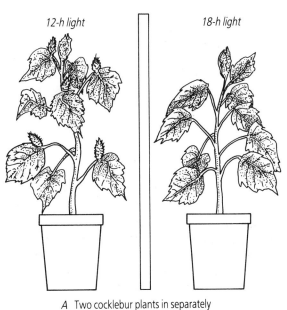

A Two cocklebur plants in separately controlled environmental chambers are exposed to different light/dark cycles. The plant illuminated for 12 h in each 24-h period develops flowers while one illuminated for 18 h does not.

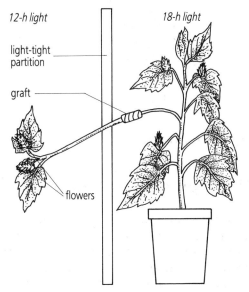

B When a shoot from a plant exposed to 12 h light in each 24-h period is grafted onto one exposed to 18 h, the host plant is stimulated to flower although it would not otherwise have done so.

Fig 34·11 An experiment to demonstrate the possible existence of a flowering hormone ('florigen')

the photoreceptor molecule. These include the germination of some seeds and the normal development of seedlings, leaf abscission, and fruit formation. The effect of far-red light seems curious in view of the normal spectral composition of daylight. However, light filtering to the floor of a forest contains a greater proportion of far-red light, which is absorbed less effectively by leaves. Consequently, it seems possible that far-red conversion may be functional and adaptive for some types of response which are not yet understood.

In continuous darkness, development of plants is severely disrupted, and the production of chlorophyll is prevented. The plants become yellow or white with elongated stems and small leaves. Such plants are said to be **etiolated**. Some of the changes observed may be linked with phytochrome, but light is directly involved in the manufacture of chlorophyll from precursor molecules. This is important in normal circumstances, for it allows the production of chlorophyll in different regions of the developing plant, called **greening**, to be precisely controlled. Photosynthetic pigments are limited to the parts of the plant which actually receive light.

34·5 Nastic movements ■ ■ ■ ■ ■ ■ ■ ■ ■ ■ ■ ■ ■ ■ ■

Nastic movements include responses to a variety of external stimuli including light, changes in temperature, or contact. **Nyctinasty** is the name given to the 'sleep movements' of many flowers and leaves. Sometimes, these movements are triggered by changes in light intensity (**photonasty**), or by changes in temperature (**thermonasty**). Most flowers close at night, as in the case of dandelions, crocus, and daisies, but a few flowers, like the evening primrose, open at night. These are usually pollinated by moths or other night-flying insects.

Haptonastic movements occur in response to contact. Examples include the action of the Venus fly trap and the spectacular wilting response of the sensitive plant, *Mimosa pudica* (Fig. 34·12). At the base of the petiole of a Mimosa leaf, and at the base of each leaflet, there is a swollen region of specialized cells, called the **pulvinus**. Normally, these cells are turgid, but contact or vibration stimuli cause a sudden change in water potential which results in rapid water loss by osmosis. The pulvini become flaccid so that the petioles droop and the leaflets fold together. The collapse of the specialized cells is associated with a change in electrical potential across the cell membranes and active pumping of ions. The adaptive value of the response to the plant is not fully understood, but it may deter herbivores, or reduce water loss in windy conditions.

34.12 Behaviour of the sensitive plant, Mimosa pudica

34·6 Economic importance of plant hormones ■ ■ ■ ■ ■ ■

1 Control of dormancy. Abscisic acid is sometimes used to maintain seeds or vegetative organs in a state of dormancy, allowing cereal foods and potatoes to be stored for long periods. In the brewing industries, barley grains may be stimulated by applying gibberellins to induce simultaneous germination.

2 Ripening and harvesting fruit. Ripening may be delayed or accelerated using hormones. Spraying with auxin delays abscission, keeping the fruit on the trees, while ethene or abscisic acid may be used to make the fruit fall. This degree of control is sometimes required for the mechanical harvesting of fruit crops like cherries or grapes. Fruits including grapes, walnuts, and tomatoes are usually picked and transported in an unripe state and brought to maturity using ethene to make them ready for sale.

3 Selective weedkillers. Synthetic auxins like 2,4-D are used as weedkillers and are thought to work by stimulating a dramatic increase in metabolic rate which causes the plants to use up their food reserves completely and starve. Broad-leaved dicotyledons absorb 2,4-D more effectively than monocotyledons and can therefore be killed selectively, a result which gardeners find very desirable on lawns.

4 Rooting hormones. Application of indoleacetic acid or an equivalent synthetic auxin to the cut end of a stem stimulates the development of adventitious roots and is a useful technique in the propagation of cuttings. In commercial preparations, the auxin is absorbed by an inert material and sold as 'rooting powder'.

35 REPRODUCTION IN FLOWERING PLANTS

Objectives

After studying this Unit you should be able to:–

- List and describe some of the structures used by plants for vegetative propagation and perennation

- Comment on cutting and grafting as methods of artificial propagation

- Draw a diagram to show the half flower of a buttercup and label the following structures: pedicel, receptacle, calyx, sepal, corolla, petal, perianth, androecium, stamen, filament, anther, gynaecium, carpel, ovary, ovule, style, stigma

- Give an account of the development of pollen grains

- Give an account of the development of the embryosac and ovum

- Distinguish between self-pollination and cross-pollination

- List and compare the adaptations required for insect pollination and wind pollination with reference to the structures of broad bean flowers and oat flowers

- Define the following terms: polypetalous, sympetalous, actinomorphic, zygomorphic

- Give examples of other agents of pollination

- List and explain some of the adaptations which result in cross-pollination

- Describe the development of pollen grains leading to double fertilization

- Comment on formation of seeds and fruits in the broad bean

- List and describe the main methods of seed dispersal

35·1 Asexual reproduction ■ ■ ■ ■ ■ ■ ■ ■ ■ ■ ■ ■

This Unit describes the main methods of asexual and sexual reproduction used by flowering plants. The evolutionary significance of sexual mechanisms and relationships with other plant groups are discussed in Unit 54.

In asexual reproduction a single parent organism produces offspring which are genetically identical to each other and to the parent. In plants, this may occur by **vegetative propagation**, that is, by development from a non-sexual part of the plant, such as the root, stem, or leaf, which gives rise to new differentiated tissues, and later separates to become a new individual.

Typical methods of vegetative propagation include the development of horizontal stems, or **stolons**, like those of the strawberry (Fig. 35·1A),

298

which grow along the surface of the ground. Adventitious roots anchor the stolon at nodes, where complete new plants are formed. **Rhizomes** are horizontal underground stems which spread in a similar way. Generally, they are thickened and swollen with food reserves, as in couch grass, a common garden weed (Fig. 35·1B). Couch grass is very difficult to eradicate because any small section of rhizome with one or more lateral buds left in the ground is likely to produce a new plant.

Parts of a plant which are specialised to give rise to new individuals are called **propagules**. Sometimes, these are separated from the parent at an early stage, as in the case of plantlets produced at the leaf margins of *Bryophyllum* (described and illustrated in Unit 32). In other species, propagules may have an additional function in helping the plant to survive unfavourable conditions. Overwintering structures, like bulbs, corms, tubers, and swollen rhizomes, are called **organs of perennation**. A crocus corm (Fig 35·1C) is formed by translocation of food from the leaves and subsequent storage in a short underground stem. Apical and lateral buds within the corm may give rise to new individuals in the following year. At the base of the corm, adventitious roots anchor the plant and absorb water and nutrients. In addition, some of them become contractile and pull the base of the plant downwards, crushing the remains of last season's corm.

Artificial methods of vegetative propagation are of great importance because of the genetic uniformity of the new plants produced. Houseplants are often propagated using small parts of the leaves or shoots, called **cuttings**, which develop roots and grow to become new plants, often with the aid of rooting hormones. Favourable varieties of apples, oranges, and other fruit crops are propagated by **grafting**, in which a branch or bud, called the **scion**, is attached to the stem and roots of a host plant, called the **stock**. Provided the meristematic regions of the two plants are in contact, they will often fuse together to form a single plant. Most fruit trees and roses are produced commercially in this way. Popular varieties originating from a single individual plant have been propagated repeatedly to produce millions of new individuals with the same desirable features. For example, all McIntosh Red apples are genetically identical.

Most plants are capable of vegetative reproduction and gain the advantages of rapid spread in suitable conditions, and, with perennation, occupation of the same favourable habitat from one year to the next. However, long term survival and the production of offspring able to colonize new environments usually require sexual reproduction, as described in the next section.

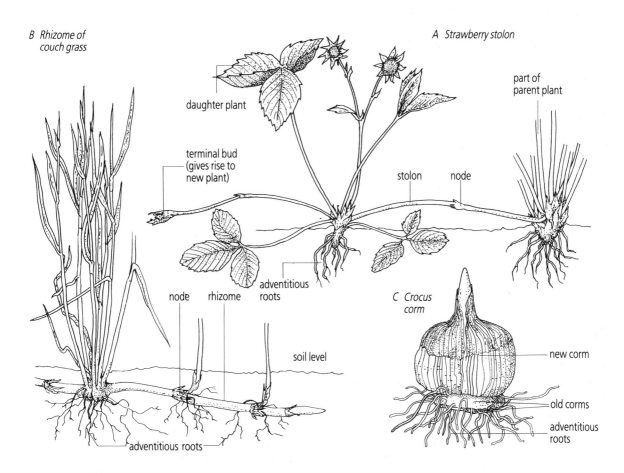

Fig 35·1 Methods of vegetative propagation

35·2 Flower structure ■ ■ ■ ■ ■ ■ ■ ■ ■ ■ ■ ■ ■ ■ ■

Flowers are the reproductive organs formed by angiosperms, that is, dicotyledons and monocotyledons. Typically, they consist of four concentric rings, or **whorls**, of structures attached to a swollen region, or **receptacle**, at the tip of the flower stalk, or **pedicel**.

1 Calyx. The outermost whorl, or calyx, consists of **sepals**, which are normally green and leaf-like and serve to protect the flower bud before the flower opens.

2 Corolla. The corolla consists of **petals**. These are often large and brightly coloured and may possess **nectaries** to attract insects or other organisms. Together, the calyx and the corolla constitute the **perianth** of the flower.

3 Androecium. This is formed by the male reproductive parts of the flower, called **stamens**. Each stamen consists of a stalk-like **filament** supporting an **anther** inside which pollen is produced.

4 Gynaecium. The gynaecium is composed of the female reproductive parts of the flower, called **carpels**. These may be attached separately to the receptacle, or may be fused to make a more complex structure. The swollen base of each carpel is called the **ovary** and may contain one or more **ovules**, while the upper region consists of a stalk-like **style** with a swollen region, or **stigma**, at its tip.

All these structures can be identified in the diagram of Figure 35·2, which shows a half flower of the meadow buttercup, *Ranunculus acris*. You should study this diagram and re-read the description given above before proceeding to the next section.

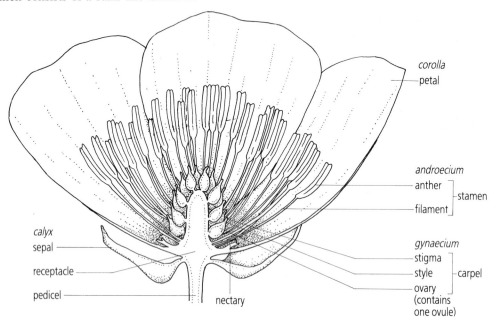

Fig 35·2 Half flower of meadow buttercup, Ranunculus acris

35·3 Formation of gametes ■ ■ ■ ■ ■ ■ ■ ■ ■ ■ ■ ■ ■ ■

Pollen grains and male gametes

The process of male gamete formation which occurs in the anther is summarized in Figure 35·3A. During growth of the anther in the flower bud, four regions of fertile **sporogenous** tissue are formed. Each of these is surrounded by a sterile layer, or **tapetum**, consisting of nutritive cells. A great number of mitotic divisions occur within the sporogenous tissue, producing a swollen mass of cells known as **microspore mother cells**.

At this stage, mitotic activity ceases and each microspore mother cell divides, this time by meiosis, to produce four haploid **microspores**, each of which gives rise to a **pollen grain**. As illustrated in Figure 35·3A, these develop a hard resistant outer wall, the **exine**, and an inner

cellulose wall, called the **intine**. The single haploid nucleus inside divides to form a **pollen tube nucleus**, and a **generative nucleus**, which will later divide again to form two male gametes. It is important to realise that the pollen grain is not itself a male gamete. It is a more complex structure which serves to contain and transport the male gametes, which are naked nuclei, to the female parts of a flower.

In a mature anther, the tapetum shrinks so that the pollen grains are suspended in a hollow sac, while the outer sterile cells become fibrous. Strips of uneven cellulose thickening are laid down in the epidermal cells at each side of the anther in preparation for its **dehiscence**, or splitting, to release the pollen grains (refer to Figs 35·3A and 35·3B).

The outer surface of a pollen grain is often beautifully sculptured with pits and projections forming an intricate pattern which is characteristic for each individual species of flowering plant (Fig. 35·3C). In addition, the material from which the exine is made, called **sapropollenin**, is extremely resistant, allowing pollen grains embedded in sediments or peat to remain unchanged in appearance for thousands or millions of years. The science of plant identification from pollen grains is used to determine which species were present in the vegetation at particular times in the past and gives an indication of the likely climate, even in the absence of other plant remains.

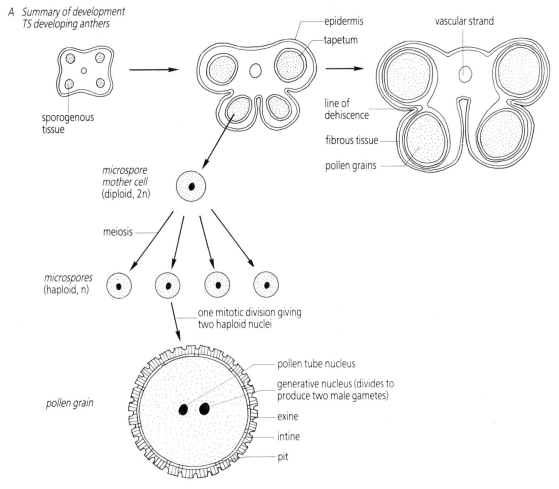

A Summary of development
   TS developing anthers

epidermis
tapetum
vascular strand

sporogenous tissue

line of dehiscence
fibrous tissue
pollen grains

microspore mother cell (diploid, 2n)

meiosis

microspores (haploid, n)

one mitotic division giving two haploid nuclei

pollen grain

pollen tube nucleus
generative nucleus (divides to produce two male gametes)
exine
intine
pit

B Light micrograph TS anther

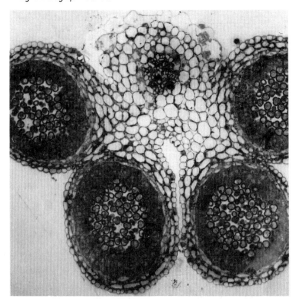

C SEM pollen grains

Fig 35·3 Development of pollen grains

Embryosac and ovum

The process leading to the formation of a female gamete or **ovum** is summarized in Figure 35·4*A*. As the carpel develops in the flower, a small bulge, or **nucellus**, grows out from a region of tissue known as the **placenta**. The nucellus consists of a group of unspecialized nutrient-rich cells which divide mitotically to form an oval mass. As it swells, a pair of covering **integuments** grow around the outside, eventually leaving a single tiny opening, the **micropyle**. The whole structure is called an **ovule** and is lifted clear of the placenta by a stalk or **funicle**.

During this process, a single large cell, called the **megaspore mother cell**, begins to divide by meiosis inside the nucellus. This division results in four **megaspores**, but only one survives: the other three are broken down and absorbed back into the nucellus. The single remaining megaspore enlarges and its nucleus divides mitotically three more times to produce 8 haploid nuclei inside a structure which is now called the **embryosac**. In terms of its development, this is equivalent to the pollen grain of the male. An important difference is that some of the nuclei within the embryosac are enclosed within a membrane and cytoplasm to form separate cells.

The ovum is one of three such cells situated at the micropyle end of the embryosac. It is flanked by two other cells called **synergids**, forming an arrangement known as the 'egg apparatus'. Of the remaining five nuclei, three are located in the **antipodal cells** at the opposite end of the embryosac, and two, called the **polar nuclei**, lie close together near the centre of the embryosac. The polar nuclei are not normally enclosed by membranes and usually fuse to form a single diploid nucleus. Thus, a mature embryosac contains a total of 7 nuclei comprising the ovum, two synergids, and three antipodal cells in addition to the central diploid nucleus.

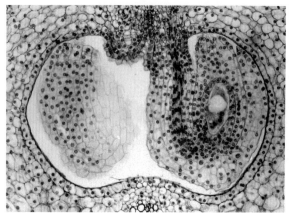

B   *Light micrograph LS ovule*

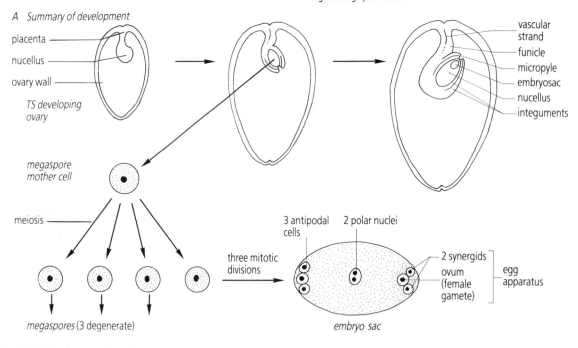

A   *Summary of development*

placenta
nucellus
ovary wall

TS developing ovary

megaspore mother cell

meiosis

megaspores (3 degenerate)

three mitotic divisions

vascular strand
funicle
micropyle
embryosac
nucellus
integuments

3 antipodal cells
2 polar nuclei
2 synergids
ovum (female gamete)
egg apparatus

embryo sac

Fig 35·4 Development of embryosac and ovum

35·4   Pollination mechanisms ■ ■ ■ ■ ■ ■ ■ ■ ■ ■ ■ ■

The transfer of pollen grains from an anther to a receptive stigma is called **pollination**. In **self-pollination**, this transfer takes place within a single flower or between two flowers on the same individual. In **cross-pollination**, the flowers are located on different individuals. Cross-pollination tends to result in greater variability among the offspring and is often favoured by natural selection. Consequently, many species of flowering plants possess adaptations which make cross-pollination more likely.

Insect pollination

Insect pollination is one of the most reliable methods of transferring pollen. Insect-pollinated, or **entomophilous** flowers share the following characteristics:

1 Individual flowers are large and brightly coloured, while small flowers are usually found together in a group, or **inflorescence**, which presents a broad expanse of colour.

2 The flowers are usually scented and provide pollen and nectar as food for insects.

3 The surface of the stigma is coated with a sticky secretion which traps pollen grains, and the flower parts are arranged so as to contact the bodies of visiting insects.

4 The pollen grains are large and thick-walled with projecting spikes which become entangled so that the pollen grains clump together and are easily attached to the body of an insect pollinator.

The buttercup, illustrated in Figure 35·2, is a comparatively unspecialized flower and can be pollinated by a variety of insects. Its petals are separate and the flower is radially symmetrical. Consequently, it is described as **polypetalous** and **actinomorphic**. Other flowers are specialised for pollination by particular types of insects, as in the case of the broad bean (Fig 35·5) and the white dead nettle (Fig 35·6). Both these flowers are adapted for pollination by long-tongued insects, such as bees, and cross-pollination is made much more probable by the structure of the flower. The flowers of these species are bilaterally symmetrical, or **zygomorphic**. The petals of the broad bean are separate, so that the flower is polypetalous. On the other hand, the petals of the white dead nettle are fused to form a tube near the base of the flower, making the flower **sympetalous**.

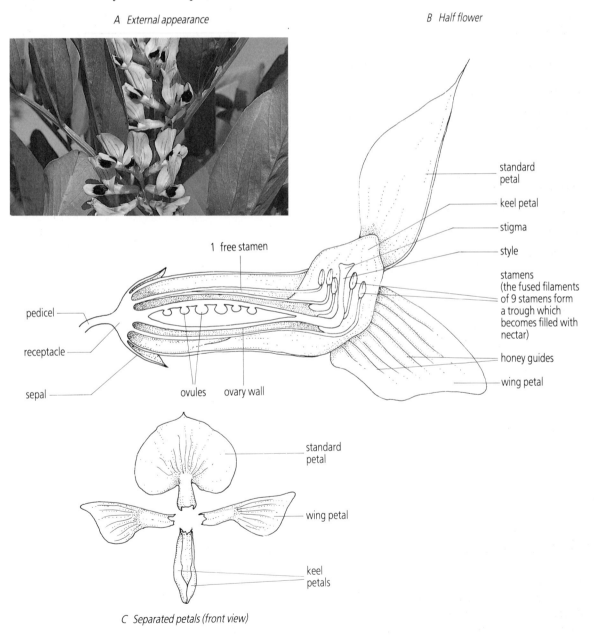

A External appearance

B Half flower

C Separated petals (front view)

Fig 35·5 Structure of the broad bean flower, Vicia faba

303

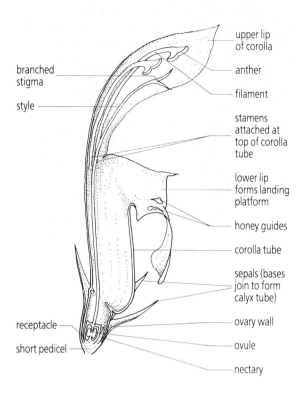

upper lip
of corolla

branched
stigma

anther

style

filament

stamens
attached at
top of corolla
tube

lower lip
forms landing
platform

honey guides

corolla tube

sepals (bases
join to form
calyx tube)

receptacle

ovary wall

short pedicel

ovule

nectary

Fig 35·6 Half flower of white dead nettle, Lamium album

The corolla of the broad bean flower consists of five petals, as shown in Figure 35·5C. These comprise a single large upper petal, called the **standard**, two lateral **wing** petals, and two **keel** petals, which are fused together to form a lower sac-like structure enclosing the male and female reproductive parts. The standard and wings are marked with **honey guides**, which direct visiting bees to the nectaries deep within the flower. In many flowers, such markings reflect or absorb ultraviolet light to which bees are sensitive. As a bee alights on the wing and keel petals, they open downwards with a clicking action, exposing the structures inside.

Nine of the ten stamens are united by their filaments to make a cylinder surrounding the ovary (Fig. 35·5B). The remaining free stamen is located at the top, leaving a groove, or trough, into which the bee inserts its mouthparts to obtain nectar. The anthers ripen while the flower is still in bud and pollen is transferred to the hairs of the style, which form a pollen brush. When a bee lands on the flower, the stigma emerges first and may receive pollen from another flower by contact with the underside of the bee's abdomen. During feeding, the bee becomes dusted with more pollen, but by now the stigma is out of reach, so that self-pollination does not normally occur.

Insect pollination provides many examples of **coadaptation**, in which different species evolve to become dependent on each other for their continued survival. Bees benefit by obtaining food, while the flowers benefit by the transport of pollen from one flower to another. Sometimes, the mutual rewards are different. For example, the female yucca moth of Central America lays its eggs inside the ovary of the yucca flower. Before doing so, it ensures fertilization of the ovules by gathering a ball of sticky pollen from the anthers of another flower and depositing it between the folds of the receptive stigma. The moth larvae and developing seeds mature simultaneously and the larvae grow to full size, feeding exclusively on the yucca seeds. After this, they burrow their way out of the fruit, fall to the ground, and pupate, hatching into adult moths when the next flowering season starts. The yucca plant is pollinated very effectively at the cost of about 20% of its fertile seeds, but the moth and the plant are utterly dependent on each other for continued survival. Other species, like the orchid illustrated in Figure 35·7, achieve pollination by sexual impersonation, producing flowers which resemble the female mating partners of bees or wasps in size and colour, and sometimes in odour. As the male insect attempts to copulate with the flower, pollen is transferred.

Fig 35·7 Sexual impersonation by the orchid Ophrys apifera

Wind pollination

Wind-pollinated, or **anemophilous**, flowers have the same basic structure as insect-pollinated flowers, but are very different in appearance. Their important characteristics are as follows:

1 The flowers lack colour, scent, and nectar and may be small and inconspicuous.

2 In many species, the flowers are located above the surrounding vegetation to gain maximum advantage from wind currents. Pollination often occurs early in the year before the surrounding foliage develops.

3 The stigmas are large and feathery and hang outside the flower to catch airborne pollen.

4 Vast numbers of small, light, smooth-walled pollen grains are produced and released into the air from large anthers suspended outside the flower.

Figure 35·8*A* shows the structure of an **inflorescence** of oats, *Avena sativa*. This consists of 20–30 **spikelets** enclosed by large leaf-like **bracts**, called **glumes** (Fig. 35·8*B*). The spikelets are supported by a stalk, or **rachis**, and each contains several individual flowers, or **florets**, attached to a central axis, or **rachilla**. A floret is enclosed, not by petals, but by two small bracts, known as the **lemma** and **palea**. Long protruding bristles called **awns** arise from the lemma bracts. Three anthers are borne on long delicate filaments and the single carpel has long feathery stigmas (Fig. 35·8*C*).

Although there are many species which produce wind-pollinated flowers, wind pollination is a spectacularly wasteful process. Individual plants produce millions of pollen grains, but these have only a minute chance of reaching a receptive stigma.

movements. On the other hand, very specific relationships may be formed between animals and particular plants. Flowers, such as the trumpet vine, *Campsis radicans*, and *Hibiscus*, which are pollinated by humming birds produce large quantities of nectar to satisfy the high metabolic requirements of these creatures. The flowers are large and usually red or yellow in colour.

Tropical bats are important for the pollination of some types of fruit trees. In some species, the flowers are carried on rope-like branches which dangle down from the canopy and open at night to emit a strong fruity odour. The pollen produced contains much more protein than that of most other types of flower. The bats rely on the pollen as their major source of protein and their jaws are modified to gather it, usually lacking teeth so that the long tongue with its brush-like tip can whip in and out freely. Pollen is transferred as they fly from flower to flower.

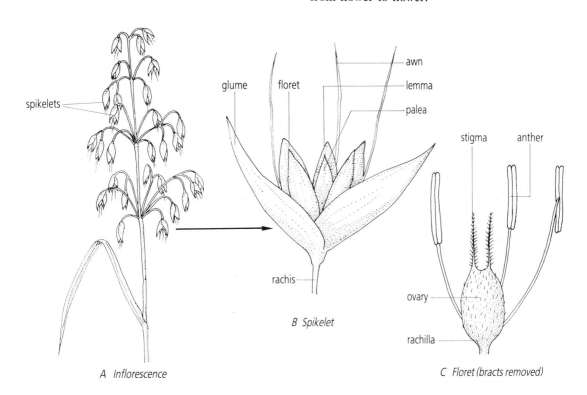

Fig 35·8 Structure of the inflorescence of oats, Avena sativa

Other agents of pollination

In aquatic habitats, plants may use alternative environmental agents to pollinate their flowers. The grass-wrack, *Zostera*, for example, has long thread-like pollen grains which have a density very similar to that of the surrounding seawater. Consequently, they are dispersed by water currents and can be trapped by submerged feathery stigmas.

A number of land plants are pollinated by animals other than insects. This may be a casual, non-specific relationship, as for example, when browsing animals like snails and slugs pick up pollen and distribute it in the course of their daily

Adaptations for cross-pollination

1 Structure. The great majority of flowers are **hermaphrodite**, or bisexual, that is, they possess both male and female structures within the same flower. Such flowers can be adapted for cross-pollination by special features of structure which make self-pollination difficult, as in the case of the broad bean flower described above. These methods depend on the particular relationship between the flower and its pollinating partner.

An interesting example of structural incompatibility, known as **heterostyly**, occurs in the primrose, *Primula vulgaris*. Approximately half the flowers are of the 'pin-eyed' variety (Fig. 35·9A), with a long style holding the stigma near the mouth of the flower and short filaments attaching the anthers lower down inside the corolla tube. The remaining flowers are of the 'thrum-eyed' variety (Fig. 35·9B), in which these positions are reversed. Visiting bees can hardly avoid depositing the pollen they receive from the anthers of one type of flower onto the stigmas of the other. The 50:50 ratio of 'pin' and 'thrum-eyed' flowers is maintained by a genetic mechanism.

2 Protandry and protogyny. In most species, self-pollination is made less likely by **protandry**, in which the anthers ripen first, or, less commonly, by **protogyny**, in which the carpels ripen first. Protandry is widespread in insect-pollinated flowers and occurs, for example, in white dead nettle and rosebay willowherb. Protogyny occurs in arum lilies, and many wind-pollinated plants, including grasses and wild varieties of oats, wheat, and barley. In either case, there is often a short period during which both the male and female parts are ripe, so that self-pollination may take place if cross-pollination has failed.

3 Monoecious species. These are species in which individual plants bear separate male and female flowers. This situation is most common in wind-pollinated plants, such as maize, oak, and sycamore. Self-pollination is possible, but its likelihood is reduced.

4 Dioecious species. In these species male and female flowers occur on separate individuals, so that there are different sexes of plant. Self-pollination is impossible. Many dioecious species are wind-pollinated. They include willow, yew, asparagus, and date palms.

5 Chemical self-incompatibility. Pollen transferred from the anthers to the stigma of the same flower in hermaphrodite species sometimes fails to develop. This is called self-incompatibility and depends on chemical substances within the plant. The process is controlled by self-incompatibility genes. The mechanisms involved are complex, but the net result is a dramatic increase in the proportion of offspring produced by cross-pollination. When self-incompatibility is well developed, as in clover flowers, other features for reducing self-pollination are usually absent.

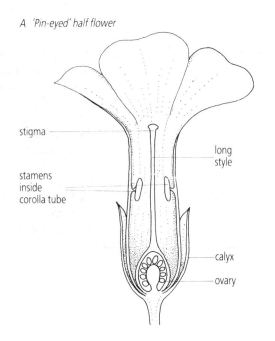

A 'Pin-eyed' half flower

stigma

stamens inside corolla tube

long style

calyx

ovary

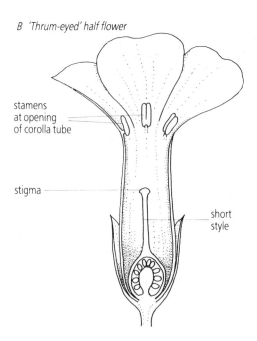

B 'Thrum-eyed' half flower

stamens at opening of corolla tube

stigma

short style

Fig 35·9 Structural incompatibility in the primrose, Primula vulgaris

35·5 Fertilization ■ ■ ■ ■ ■ ■ ■ ■ ■ ■ ■ ■ ■ ■ ■ ■ ■

Pollination is only the first step in the process of sexual reproduction. In contrast to the sperm cells of less advanced plants, the male gametes contained within pollen grains are non-motile and do not require external water to reach the female gamete. They are transported to the embryosac through an outgrowth of the pollen grain called a **pollen tube**, which penetrates the tissues of the stigma, style, ovary wall and ovule.

The pollen tube starts as an outgrowth of the intine which emerges through a pore in the exine. Energy for its germination is supplied initially from food reserves within the pollen grain. Later, the pollen tube secretes enzymes which digest the surrounding cells of the style and ovary to provide the nutrients needed for growth. This activity is controlled by the pollen tube nucleus, present in the advancing tip of the pollen tube.

The pollen tube eventually breaks into the cavity of the ovary and enters an ovule, usually through the micropyle. The pollen tube nucleus and the two male gametes are released. One of the male gametes then fuses with the nucleus of the ovum to form the diploid **zygote** which will subsequently become a new plant embryo.

The pollen tube nucleus degenerates, but the second 'unsuccessful' male gamete does not. Instead, it migrates to the centre of the embryo sac where it fuses with the two polar nuclei, to form the triploid **endosperm nucleus**. This completes a process of **double fertilization**, unique to flowering plants, in which both male nuclei are used (refer to Fig. 35·10).

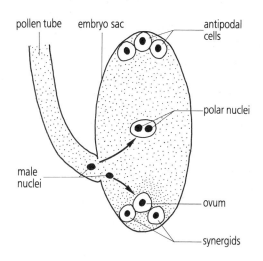

Fig 35·10 Summary of fertilization

35·6 Development of seeds and fruit ■ ■ ■ ■ ■ ■ ■ ■ ■

After fertilization, the ovule develops to form a seed. In this process, the embryo enlarges, forming the embryo shoot, or plumule, and the embryo root, or radicle. In the seeds of monocotyledons, as their name suggests, the embryo gives rise to a single seed leaf, or cotyledon, while in dicotyledons, two seed leaves are formed. In **endospermous** seeds, including castor oil and maize seeds, the endosperm nucleus divides repeatedly by mitosis to form endosperm tissue consisting of large food storage cells. These accumulate food translocated via the vascular strands of the funicle and swell to surround the young embryo, filling the space between the integuments and the remnants of the embryosac. In other seeds, called **non-endospermous** seeds, like the broad bean, the cotyledons are the major storage regions and the development of endosperm

tissue is very limited. The outer seed coat is formed from the integuments and consists of a tough outer layer, the **testa**, and a slightly softer inner layer, called the **tegmen**.

As the seeds mature, the ovary swells to accomodate them, forming a **fruit**. In the case of the broad bean, this is a long pod, called a **legume** (see Fig. 35·11). Dispersal of the seeds is effected when the ovary wall, now referred to as the **pericarp**, dries and splits along its two sides.

The broad bean belongs to a large family of legume-producing plants called the **leguminosae**. However, as a crop plant and the product of intensive genetic selection, its fruit is far too large for the effective dispersal of its seeds. More typical of the group are clover, vetch, gorse, and lupin plants, whose seeds are flicked out from the pod as its walls dry out and spring back.

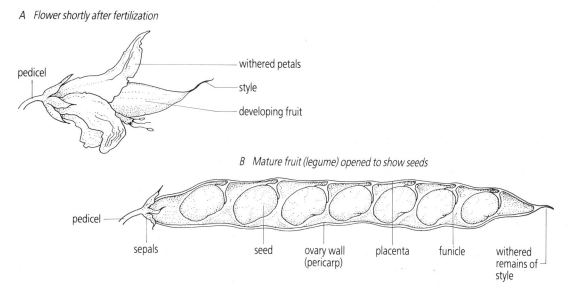

Fig 35·11 Development of fruits in the broad bean

35·7 Dispersal of seeds ■ ■ ■ ■ ■ ■ ■ ■ ■ ■ ■ ■ ■

The function of any fruit is to ensure dispersal of the seeds from the parent plant. This is essential to reduce competition between members of the same species, and for the colonization of new habitats. Several types of dispersal are distinguished.

1 Self-dispersal (Fig. 35·12*A*). The seeds of some plants are expelled explosively from the fruit by rapid movements caused by drying of the pericarp or, more rarely, by turgor pressure. These are sufficient to propel the seeds far away from the parent plant. The gorse fruit is a legume in which the pericarp consists of layers of fibrous tissue set obliquely to each other. As it dries, the two halves curl and suddenly spring back, flicking out the seeds.

2 Wind-dispersal (Fig. 35·12*B*). The poppy produces spore-like seeds which are so light that they may be carried large distances on the wind. These are produced in a dry hollow fruit, called a **capsule**, formed from a number of fused carpels. When mature, the seeds emerge through pores in the capsule as it is shaken by the wind.

Wind dispersal of other fruits and seeds is aided by broad extensions or feathery plumes which delay their descent to the ground. The pericarp of the sycamore fruit extends into a wing. In the dandelion, the calyx becomes a parachute made of hairs. The fruit of the rosebay willowherb is a dry capsule, but its seeds develop feathery hairs which keep them airborne for large distances and give the plant a fluffy appearance.

3 Water-dispersal (Fig. 35·12*C*). This form of dispersal is less common. The fruit or seed becomes hollow, or possesses special bouyant structures enabling it to float on water. A notable example is the coconut fruit which has a loose fibrous outer pericarp and an air space inside the seed itself.

4 Animal dispersal (Fig. 35·12*D*). Some dry fruits and seeds have sticky or spiny projections designed to catch on the fur of passing animals. The wood avens produces a cluster of dried fruit, called **achenes**, each with a barbed hook formed from the style. The entire cluster may stick closely to a passing animal and be carried a long way before the individual achenes break off and disperse.

Fleshy fruits are attractive as food to a variety of animals, mainly birds. They advertise their maturity by changing colour from an inconspicuous green to bright orange, red, blue or black. Internally, the tissues are softened and sweetened by the combined activity of hormones and enzymes to make the flesh tasty.

The edible parts of **true fruits**, such as plums or tomatoes, are formed from the middle layer, or **mesocarp**, and the inner layer, or **endocarp**, of the ovary wall, while the outer layer, called the **epicarp**, forms the fruit 'skin'. In **false fruits**, or **pseudocarps**, like the apple, strawberry and rosehip, the flesh derives from floral parts other than the ovary, often the receptacle.

The seeds of fleshy fruits are protected by a hard testa, sometimes reinforced by a fused stony layer made up of endocarp tissue. They are dispersed, depending on their size, either by being discarded or by being swallowed. In the latter case, the outer seed coat resists enzyme attack in the alimentary canal and the seeds are deposited in the faeces, often at a point remote from the parent plant, and in suitably nutrient-rich surroundings!

Fig 35·12C Water dispersal

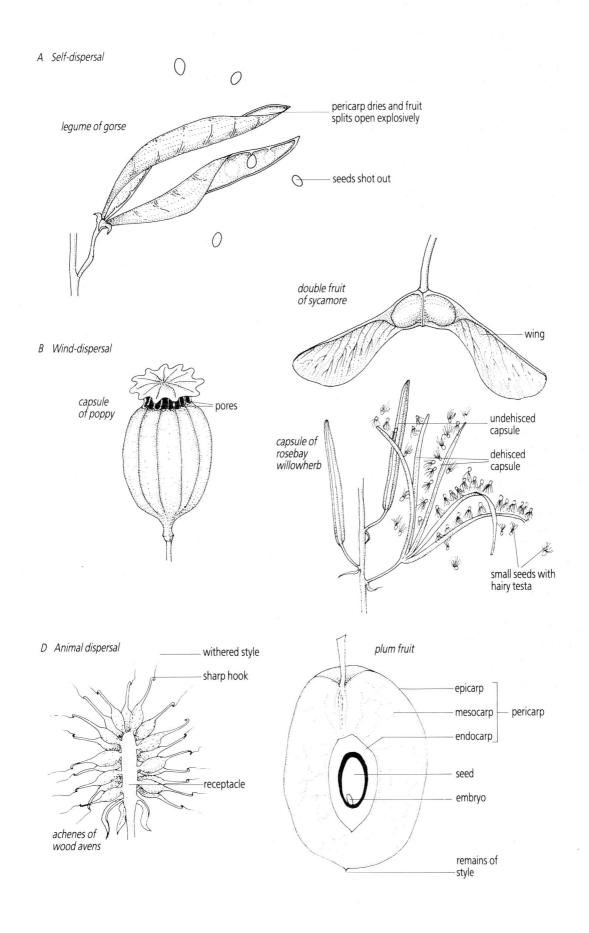

A Self-dispersal

legume of gorse

pericarp dries and fruit
splits open explosively

seeds shot out

double fruit
of sycamore

wing

B Wind-dispersal

capsule
of poppy

pores

capsule of
rosebay
willowherb

undehisced
capsule

dehisced
capsule

small seeds with
hairy testa

D Animal dispersal

withered style

sharp hook

receptacle

achenes of
wood avens

plum fruit

epicarp

mesocarp

pericarp

endocarp

seed

embryo

remains of
style

Fig 35·12 Mechanisms of seed dispersal

Section B Questions ▪ ▪ ▪ ▪ ▪ ▪ ▪ ▪ ▪ ▪ ▪ ▪ ▪

Short-answers and interpretation

1 a Explain the importance of the following types of nutrition in the lives of named organisms:

    i autotrophic nutrition, [3]

    ii saprotrophic nutrition [3]

    iii parasitic nutrition [3]

    iv holozoic nutrition [3]

  b Which of these types are grouped together as 'heterotrophic nutrition'? [1]

2 The diagram below represents part of a transverse section of mammalian ileum.

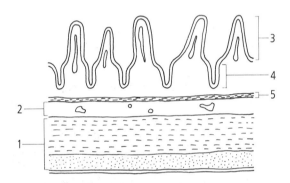

  a What type of muscle is found in region 1? [1]

  b Why is this type of muscle particularly appropriate for the ileum? [1]

  c Name the structures labelled 2–5. [2]

  d List three ways in which structure 3 is suited for the absorption of the products of digestion. [3]

  e Name one substance which is actively transported across the ileum. [1]

  f Give two pieces of evidence which you would require to confirm the existence of active transport in this region of the gut. [2]

University of London School Examinations Board

3 The diagram below represents the stomach of a sheep.

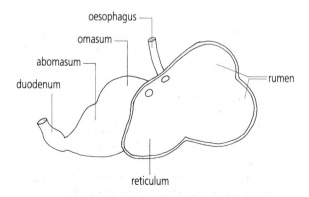

  a Describe the treatment of food which occurs in the rumen. [3]

b Explain what is meant by chewing the cud and explain its significance. [3]

c The abomasum can be described as the true stomach. Explain what this means. [3]

d Indicate the ways in which the alimentary canal of a dog differs in structure from that of a sheep. Your answer should include reference to teeth. [4]

University of London School Examinations Board

4 a i Draw a diagram to show the relationships between the following structures: liver, gall bladder, hepatic duct, cystic duct, common bile duct, duodenum, pancreatic duct. [5]

    ii Explain the mechanism by which secretion of bile into the digestive tract is controlled. [3]

  b State two functions of the liver associated with red blood cell formation. [2]

  c Describe the circulatory pathway by which a molecule of urea, produced by the liver, is eliminated from the blood. [3]

  d State the liver function which has a requirement for vitamin K. [1]

Associated Examining Board

5 Graph *A* below shows the ventilation rate (volume breathed per minute) of a healthy male when at rest and when undertaking different amounts of activity (measured as momentum). Graph *B* shows the tidal volume (shaded) in relation to total lung capacity and residual volume, as recorded by a spirometer, over the same range of activity as in graph *A*.

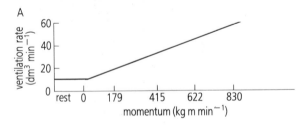

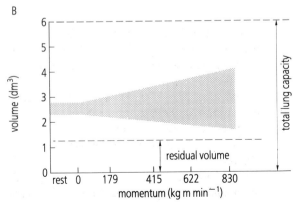

a i What is meant by the term 'tidal volume'? [1]
ii What is the tidal volume of this individual when maintaining a momentum of $830 \, kg \, m \, min^{-1}$? [1]
iii What is the vital capacity of this individual? [1]
iv What is the average number of breaths per minute when maintaining a momentum of $622 \, kg \, m \, min^{-1}$? Show your working. [2]
v What effect does an increase in ventilation rate have on the inspiratory reserve volume? [1]
b What effect would an increase in the carbon dioxide concentration of the inspired air have on the ventilation rate? Describe the physiological mechanism by which this effect is brought about. [6]
c i State what is meant by the term 'respiratory quotient'. [2]
ii Describe an experiment which could be carried out to determine the respiratory quotient of a named animal. [4]
iii Identify two problems which might be encountered in carrying out this experiment on the animal you have named. [2]

Associated Examining Board

6 The diagram below shows some of the structures visible in a ventral view of a section through a mammalian heart.

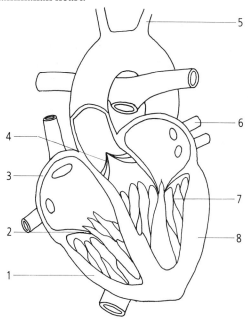

b Explain the significance of the difference in thickness between parts **i** 1 and 3, **ii** 1 and 8. [4]
c Explain how the heart beat is **i** initiated, **ii** controlled [6]
d State and give brief explanation of the effect of the following on the rate of heart beat
i increased adrenalin secretion, [2]
ii increased carbon dioxide levels in the blood, [2]
iii athletic training. [2]

Oxford and Cambridge Schools Examination Board

7 In fish, oxygen is transported in the blood in the form of oxyhaemoglobin. The table below shows the percentage saturation of blood with oxygen of a bony fish (Class Osteichthyes) after equilibrating with oxygen of different partial pressures. The experiment was carried out at two different partial pressures of carbon dioxide.

Partial pressure of oxygen /Pa*	Percentage saturation of blood with oxygen	
	Partial pressure of carbon dioxide at 500 Pa	Partial pressure of carbon dioxide at 2600 Pa
500	30	5
1 000	70	13
2 000	90	24
3 000	96	33
4 000	98	41
5 000	99	48
7 000	100	60
9 000	100	69
11 000	100	76
13 000	100	81

* A pascal (Pa) is a unit of pressure. A pressure of 100 000 Pa is approximately equal to atmospheric pressure (760 mm Hg).

a Present the data in a suitable graphical form. [5]
b Calculate the difference in percentage saturation of blood with oxygen at the two different partial pressures of carbon dioxide at an oxygen partial pressure of 5500 Pa. [1]
c With reference to the graph, describe the effects of different partial pressures of carbon dioxide on the percentage saturation of blood with oxygen. [4]
d Explain how the properties of the haemoglobin molecule are affected by changes in the oxygen and carbon dioxide partial pressures. [4]
e Explain how changes in the oxygen content of the blood at different partial pressures of carbon dioxide are important in the release of oxygen to the tissues of the fish. [3]
f What information do experiments of this type give about the environmental conditions in which fish would maintain a high level of growth as required in commercial fish farming? [3]

University of London School Examinations Board

8 **a i** Make a large labelled diagram to show the microscopic structure of compact bone. [6]
ii Explain how the structure of complact bone makes it a suitable skeletal structure. [4]
b Give four functions of the mammaliam skeleton and one example of each. [4]
c Give one advantage and one disadvantage of exoskeletons compared with internal skeletons. [2]
d Cartilage is another skeletal material. Name two types of cartilage and explain the function of each. [4]

Associated Examining Board

9 The table gives information about substances filtered from blood in normal human kidneys and excreted in urine.

Substance	Approximate daily quantities	
	Filtered through glomeruli (g)	Excreted in urine (g)
Water	180×10^3	1.5×10^3
Sodium (as Na^+ ions)	550.0	2.5
Glucose	187.0	0.0
Amino acids	8.5	0.1
Urea	51.0	30.0

a What do the data in the table indicate about the form in which nitrogen is excreted from the body? [1]

b i It is evident from the table that all the glucose is normally reabsorbed. Explain under what conditions glucose may be found in the urine.

ii Describe a biochemical test which can be used to identify a reducing sugar, such as glucose, in the urine. [3]

c Calculate to one decimal place the percentages of water and sodium ions absorbed in these kidneys. [2]

d Compare the processes by which water and sodium ions are absorbed in the kidneys. [7]

e 80% of the filtered water is reabsorbed in the proximal convoluted tubule and in the descending limb of the loop of Henle. Using this information, together with that in the table and your own knowledge, construct a bar chart to show the mass of water reabsorbed daily in

i the proximal convoluted tubule and the descending limb of the loop of Henle,

ii the ascending limb of the loop of Henle,

iii the distal convoluted tubule and collecting ducts. [4]

f To what extent is the chemical composition of the urine altered in the ureter? [1]

Associated Examining Board

10 A special calorimeter was developed into which a volunteer human being could be placed so that:
(i) measurements of surface temperature (usually taken on the skin) and core temperature (temperature of the blood, estimated by placing a thermocouple adjacent to the tympanic membrane) could be taken and recorded.
(ii) estimates could be made of body heat lost due to evaporation of sweat from the skin surface.

a An adult, nude male volunteer was kept at a constant temperature of 45°C inside the calorimeter. Both surface and core temperatures rose initially but stabilized within 15 minutes. How did this stabilization occur? [5]

b Forty minutes after entering the calorimeter, he was given a quantity of iced water to drink. This was repeated thirty minutes later.

Measurements of skin and core temperature, and of body heat lost due to sweat evaporation were taken at 5-minute intervals after the initial cold drink and the results are given in the Table below.

Time in minutes, after entering calorimeter	Temperature°C		Energy loss through evaporation $J\,s^{-1}$
	skin	core	
iced water given			
40	36.8	37.5	220
45	37.1	37.2	136
50	37.5	36.9	100
55	37.3	37.2	146
60	37.1	37.4	210
65	37.0	37.5	220
iced water given			
70	36.9	37.4	175
75	37.3	37.1	126
80	37.2	37.2	165
85	37.1	37.4	210

Plot these data on graph paper. Construct two graphs, one of skin and core temperature and one of energy loss. [6]

c By reference to your graphs, explain the relationship between skin temperature and evaporation. [3]

d Account for the relationship between the three sets of figures in the Table at 65 minutes. [3]

e Explain why the skin temperature rose after the intake of iced water. [3]

Oxford and Cambridge Schools Examination Board

11 The diagram below illustrates a type of nerve cell found in a mammal.

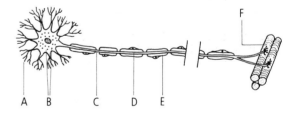

a Name the parts labelled *A–E*. [5]

b What functional type of neurone is shown in the diagram? [1]

c State one function of each of the parts labelled *A*, *C*, and *E*. [3]

d Explain in detail how a nerve impulse is transmitted from part *F* to the muscle fibre. [6]

University of London School Examinations Board

12 a Explain how the autonomic nervous system can be classified into sympathetic and parasympathetic systems on the basis of

i anatomical layout, [4]
ii chemical transmitters. [2]
b i Describe two examples of antagonistic action by the sympathetic and parasympathetic nervous systems. [4]
ii Give one example of non-antagonistic innervation (sympathetic or parasympathetic). [1]
c The diagram below shows the left side of the human cerebral cortex.

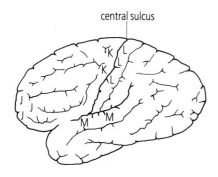

central sulcus

i Of the areas *J*, *K*, *L* and *M*, which is associated with hearing, which with intelligence and which with skin sensory perception? [3]
ii Copy the outline of the cerebral cortex and add the cerebellum, medulla and cervical region of the spinal cord. Label these parts and indicate which part receives impulses from the body's proprioceptors and regulates posture. [4]
d Explain why you may have a sensation of flashing lights if struck on the back of the head. [2]

Associated Examining Board

13 The diagram below shows a section through a vertebrate retina.

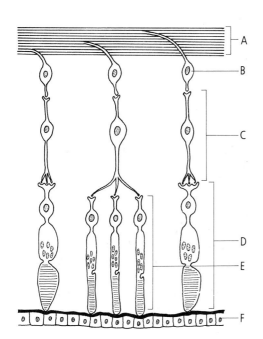

a i Name the layer indicated by *A* [1]
ii Name the cells labelled *B*, *C*, *D*, *E* and *F*. [5]
b State the direction, relative to the diagram, from which light enters the retina. [1]
c Fill in the blanks in the passage below:
Cones provide (i) vision, using three pigments that have absorption maxima spread throughout the visible region. Light of any wavelength from about (ii) nm to (iii) nm will be absorbed in varying degrees by one of these pigments. Rods, on the other hand, have a single pigment named (iv), containing an aldehyde derivative of vitamin (v). Rod cells are sensitive to very (vi) light levels. Therefore rods supply (vii) vision, which is virtually (viii) because it is dominated by a single pigment.
d When a person enters a dimly lit room from bright sunlight, the room seems pitch dark to begin with, but gradually objects become visible. Describe the changes occurring in the photosensitive cells. [3]

Cambridge University Local Examinations Syndicate

14 The diagrams below represent (*A*) a sarcomere of a single myofibril from a striated (striped) muscle fibre in the relaxed state, and (*B*) the comparable region of a myofibril in the same state as seen in greater detail.

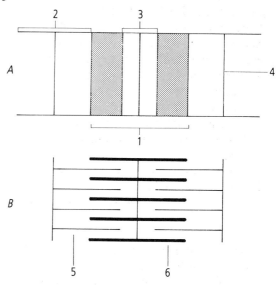

a Name the regions numbered 1 to 4 in diagram *A*. [4]
b Name the filaments numbered 5 and 6 in diagram *B*. [2]
c Explain the appearance of the myofibril shown in *A* in terms of the detailed filament structure shown in *B*. [3]
d i Draw a sketch of the sarcomere and of its constituent filaments as they appear when the myofibril is in the contracted state. [2]
ii Indicate briefly any significant differences from the appearance of these structures in the relaxed state shown in *A* and *B*. [3]

University of London School Examinations Board

15 The diagram below represents a mammalian spermatozoon.

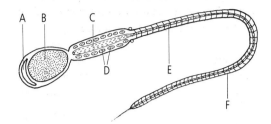

a Name the parts labelled *A–F* [6]
b Explain briefly the functions of the parts *A*, *B* and *D*. [6]
c What events occur in the egg immediately following the entry of the spermatozoon? [4]

University of London School Examinations Board

16 The graphs below show the changes in the concentrations in the blood of four hormones associated with the menstrual cycle.

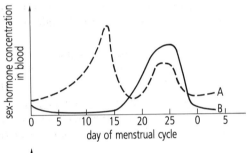

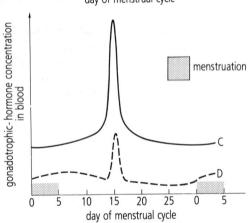

a Which graph represents each of the following hormones? Give one site of production and one effect of each hormone in a mature female.
　i follicle-stimulating hormone (FSH),
　ii luteinizing hormone (LH),
　iii oestradiol (oestrogen),
　iv progesterone. [12]
b What changes occur in the levels of oestradiol and progesterone in the maternal blood in the early stages of pregnancy? [2]
c **i** Explain negative feedback. What is its significance in biological systems? [2]
　ii Explain how negative feedback operates in the control of the hormones in the menstrual cycle. [4]

Associated Examining Board

17 **a** There are wide differences in the rates at which the leaves of various plant species lose water. Such differences may be attributable to structural features of the leaf. The following drawings represent sections of a leaf *Fagus* sp. which loses water relatively quickly and another leaf *Hakea* sp. which loses water slowly.

A Fagus sp.

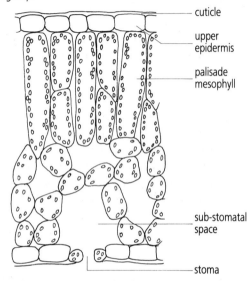

B Hakea sp.

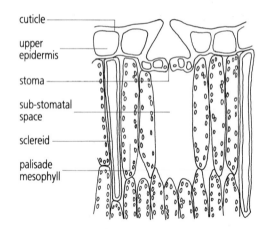

　i Explain fully how the stomata of *Hakea* sp. allow far less water to pass out of the leaf than do the stomata of *Fagus* sp. [3]
　ii Suggest how one other feature, visible in the diagrams, might act to cut down the rate of transpiration. [1]
b The sclereids are thick-walled, lignified elements. Suggest an explanation for their presence in *Hakea* sp. and their absence in *Fagus* sp. [3]
c Which of these two leaves is likely to have the smaller surface area/volume ratio? Give a reason for your choice. [2]
d The following figure shows the variation in the rate of water movement in *Fagus* sp.

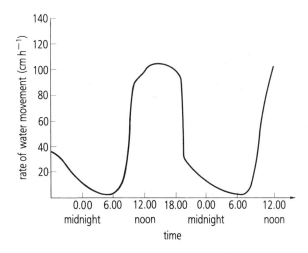

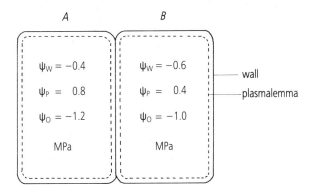

i Describe the variation in the rate of water movement over a 24 hour period. [3]
ii Give an explanation for this variation. [3]
iii Suggest two environmental factors, in addition to any given in your answer to **ii**, which will influence the rate of transpiration. [2]

Welsh Joint Education Committee

18 The water potential (Ψ_w) of a plant cell is determined by two factors, its pressure potential (Ψ_p) and its osmotic potential (Ψ_o). Thus $\Psi_w = \Psi_p + \Psi_o$.

a What does the water potential of a cell measure? [1]
b Consider the two cells *A* and *B* below.

i State which cell has the higher water potential value. [1]
ii State the direction in which water will move by osmosis. Give the reason for your answer. [2]

c If cell *A* was allowed to equilibrate by being placed in pure water, what would be its expected pressure potential? Show how you derived your answer. [2]

d The following experiment was set up to determine the water potential of plant cells. Comparable discs, about 1 mm thick, of potato tuber were cut out and weighed. Replicate samples were taken and placed in a series of sucrose solutions ranging from 0.0 to 0.6 M in

covered dishes at a constant temperature of 20°C. After 1 hour the samples were blotted rapidly between sheets of filter paper and reweighed. The results are shown in the Table below, as the mean percentage change in mass.

Concentration of sucrose solution/M	Mean percentage change in mass
0.0	+22
0.1	+17
0.2	+9
0.3	+3
0.4	−3
0.5	−10
0.6	−15

i Plot these figures on graph paper. [2]
ii Use your graph to work out which concentration of sucrose solution has an osmotic potential equal to the mean water potential of the cells of the tissue. [1]
iii What further information would you need to express the water potential in units of MPa? [1]
iv Why were the dishes containing the discs and sucrose solution covered during the experiment? [1]
v Suggest the reason for expressing the change in mass as a percentage change. [1]

Cambridge University Local Examinations Syndicate

19 Two leaf blades were removed from a healthy intact plant. One of the petioles was left exposed and the other was treated with a hormone. Two weeks later a longitudinal section through the petiole and stem was taken. The tissue distribution of this section is shown in the diagram below.

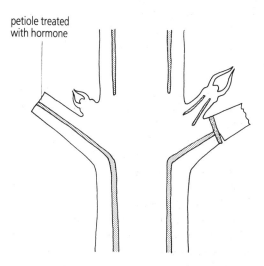

petiole treated with hormone

a i Describe three differences between the treated and untreated parts of the plant which are apparent from the diagram. [3]

ii Identify two areas where actively dividing cells would be present. [2]

b i What method should be used to apply the hormone? [1]

ii Describe a suitable control for this investigation. [1]

c The effect of two hormones *A* and *B* on abscission of leaves in similar plants was measured. The results are shown in the following graphs.

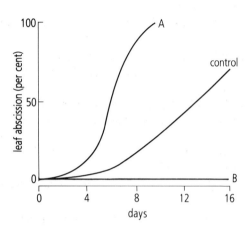

i From these results, which of the two hormones do you think was applied to the cut petiole of the first plant? Explain your answer. [2]

ii Give the name of a hormone which could be hormone *A*. [1]

Joint Matriculation Board

20 a The following diagram shows a half flower.

i Name the parts 1 to 10. [5]

ii Describe five visible features in the flower which you would expect to be different or absent in wind-pollinated flowers. [5]

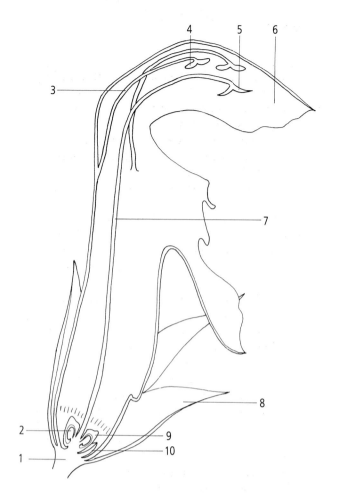

b Explain where each of the following four parts of an angiosperm flower develops.

i microspore, [2]

ii megaspore, [2]

iii male gamete, [2]

iv female gamete. [2]

c *Zostera* is a marine angiosperm whose flowers are submerged under water. Give two ways in which you would expect its pollen grains to differ from those of terrestrial insect-pollinated angiosperms. [2]

Associated Examining Board

Essays

1 **a** List the main constituents of a balanced diet for human beings. [3]
b State the purposes of mastication and digestion in a mammal. [4]
c Describe the characteristic features of the skull, teeth and gut of three mammals with distinctly different diets. [18]
Cambridge University Local Examinations Syndicate

2 Describe the structure of the tissues found in the wall of the mammalian ileum, and discuss their contribution to its functioning. [20]
University of London School Examinations Board

3 **a** Which blood vessels supply materials to the liver? [2]
b Explain the role of the liver in **i** metabolism of lipids and amino acids, and **ii** the regulation of blood sugar levels. [9]
c The liver carries out numerous other activities. Describe briefly examples of such activities and explain why each is important to the organism. [9]
Joint Matriculation Board

4 **a** Describe the structural organization of the breathing system in man. [10]
b Distinguish between abdominal and chest breathing. [3]
c Discuss the ways in which inhalation–exhalation cycles are controlled and maintained. [7]
Oxford and Cambridge Schools Examination Board

5 **a** State the similarities and differences between arteries, capillaries and veins. [6]
b Describe the action of the heart of a mammal. [9]
c Explain the effect of fear on the rate of heart beat. [5]
University of London School Examinations Board

6 Give an illustrated account of the transport of oxygen and carbon dioxide by mammalian blood. [20]
Welsh Joint Education Committee

7 **a** Describe the essential features of the immune system in mammals. [10]
b Give an account of the ABO blood group system in humans and explain why certain ABO group donations cause agglutination in the recipient, while others do not. [6]
c Besides blood, other tissues can be transplanted from one individual to another. Discuss the problems associated with such procedures and the steps taken to minimize transplant failure. [4]
Joint Matriculation Board

8 **a** By means of a labelled diagram, illustrate the structure of a mammalian kidney tubule. [6]
b How is the water content of the body fluids regulated? [9]
c Why is osmoregulation important to the animal? [3]
Cambridge University Local Examinations Syndicate

9 **a** Define homeostasis. [3]
b How do animals regulate the compensation of the blood? [12]
c Discuss the extent to which plants carry out homeostasis. [3]
Cambridge University Local Examinations Syndicate

10 **a** What are the benefits and costs of maintaining a constant, high body temperature in mammals? [6]
b Describe the responses of a mammal to cold temperatures, in both the short term and the long term. [8]
c Why is prolonged exposure to cold frequently fatal? [4]
Cambridge University Local Examinations Syndicate

11 **a** How is the nerve impulse generated and conducted? [14]
b Why is the 'all or none' law of nerve conduction important in understanding how nerves communicate within the nervous system? [6]
Oxford and Cambridge Schools Examination Board

12 Describe the difference between
a hormonal and nervous coordination. [6]
b a simple reflex action and a voluntary action. [6]
c the sympathetic system and the parasympathetic system. [8]
Joint Matriculation Board

13 Discuss the functioning and importance of eyes in the lives of animals. [20]
University of London School Examinations Board

14 Explain the role of
a the cochlea in the detection of sound and the discrimination of volume and pitch. [10]
b the retina in the detection of colour and the discrimination of colour. [10]
Joint Matriculation Board

15 **a** Define the term hormone. [3]
b Describe how hormones released by the hypothalamus and pituitary gland affect the functions of the thyroid gland, the ovaries and the kidney. [12]
c Suggest why hormonal rather than nervous stimuli are used to control these processes. [3]
Cambridge University Local Examinations Syndicate

16 With reference to a mammal:
a i distinguish between the appearance of striated, cardiac and smooth muscle as seen using a light microscope; [8]
ii briefly state the functions of each of the three muscle types [6]
b i What further details of the structure of striated muscle can be resolved using an electron microscope? [6]
ii Explain the contraction of striated muscle in the light of present knowledge. [8]
University of Oxford Delegacy of Local Examinations

17 Describe the macroscopic and microscopic structure of a long bone and discuss the ways in which this structure is related to its functioning. [25]
Associated Examining Board

18 a Illustrate what is meant by the terms instinctive (or innate) behaviour and learned behaviour, using specific, non-human examples. [12]
b Describe the advantages of a social existence over a solitary existence, with reference to named examples. [8]
Cambridge University Local Examinations Syndicate

19 This question refers to the mammal.
a Describe
i spermatogenesis,
ii the production of semen,
iii the fertilization of the egg,
iv the development of the fertilized egg up to its implantation. [14]
b What hormonal changes take place during gestation and what are their effects [6]
Joint Matriculation Board

20 a Describe the structure and functions of the human placenta. [10]
b Discuss the problems which may arise as a result of the exchange of materials across the placenta, between the blood of the mother and the blood of the foetus. [10]
Cambridge University Local Examinations Syndicate

21 Write an essay on 'Structural Adaptations for Photosynthesis'. [24]
Associated Examining Board

22 Write an essay on 'Respiratory Surfaces'. [20]
Associated Examining Board

23 a Describe and account for the circulatory and respiratory changes which occur in skeletal muscle during a session of strenuous exercise. How does the muscle recover after such exercise? [12]
b Describe the mechanisms concerned with the transport of carbon dioxide in the blood and its release at the alveolar surface. [8]
Joint Matriculation Board

24 a Describe the structure and distribution of xylem in a herbaceous plant. [10]
b Describe how the structure and distribution of xylem within the plant are related to the functions performed by this tissue. [10]
University of London School Examinations Board

25 a What is meant by the term water potential?[4]
b Describe the path taken by water through an angiosperm plant, from the soil to the atmosphere. [7]
c Explain the mechanisms involved in this movement in terms of water potential. [7]
Cambridge University Local Examinations Syndicate

26 Compare transport in flowering plants with circulation in mammals. [20]
University of London School Examinations Board

27 a What are the effects of each of the following plant hormones on plant tissue?
i auxin,
ii gibberellin,
iii cytokinin,
iv abscisic acid. [10]
b Explain how the balance between two or more of the hormones listed in a controls
i seed dormancy and germination,
ii leaf senescence and abscission. [5]
c What is a bioassay? Describe how the auxin concentration of a solution could be determined by bioassay. [5]
Joint Matriculation Board

28 a With reference to named examples in each case, show how flowers are adapted for
i wind pollination, [4]
ii cross pollination. [4]
b Describe the events which occur in the flower from pollination to the formation of a seed. [10]
Cambridge University Local Examinations Syndicate

Section C

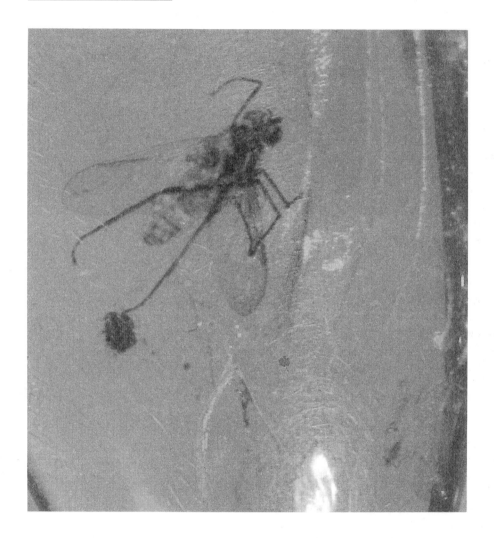

GENETICS
AND EVOLUTION

36 CHROMOSOMES AND CELL DIVISION

Objectives

After studying this Unit you should be able to:–

● Explain what is meant by the cell cycle

● Describe interphase and explain its division into G1, S, and G2 stages

● Draw diagrams to illustrate the following stages of mitotic cell division: prophase, prometaphase, metaphase, anaphase, telophase, cytokinesis

● Give an account of the important events taking place during each stage

● Explain the terms diploid and haploid

● Describe the main features of meiosis leading to the production of gametes

● Explain synapsis and crossing-over

● Comment briefly on the evolution of meiosis

36·1 Introduction ■

Cell division is essential for the survival of all living organisms. Asexual reproduction and the growth of body tissues both involve **mitosis**, a form of division in which the daughter cells produced are genetically identical to the parent cell. Sexual reproduction involves a different type of division, called **meiosis**, which results in the production of genetically varied sex cells, or gametes. This Unit describes these processes in detail as they occur in the cells of eukaryotic organisms.

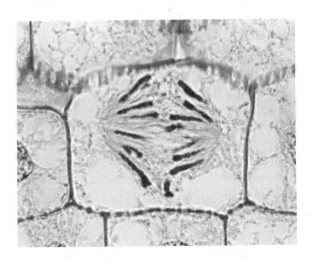

Mitosis in an onion root tip cell

36·2 The cell cycle ■ ■ ■ ■ ■ ■ ■ ■ ■ ■ ■ ■ ■ ■ ■

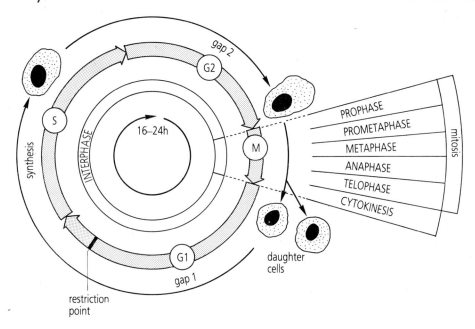

Fig 36·1 The cell cycle

Many body cells in a multicellular organism pass through a well-defined sequence of stages culminating in division and the formation of new cells. This is the **cell cycle**, as outlined in Figure 36·1. The individual stages in the cycle are as follows.

Interphase

During interphase the cell is metabolically active and able to synthesize new proteins and DNA. Interphase involves cell growth, measured by an increase in dry mass, as new organic compounds are manufactured.

At the start of interphase, the nucleus of the new cell contains a fixed number of **chromosomes**. Each of these is a very long, single molecule of DNA which is associated with **histone** protein to form a darkly-staining material called **chromatin**. The number of chromosomes present in each body cell of a particular species is fixed and characteristic for that species. For example, human cells contain 46 chromosomes, mouse cells contain 40, and the cells of a broad bean plant contain 12.

In the interphase nucleus the chromosomes are uncoiled, or **uncondensed**, forming fine threads which are just visible at the resolution of the electron microscope. The threads consist mainly of **nucleosomes**, that is, short lengths of DNA, 200 base-pairs long, each wound on a 'bead' of histone protein (illustrated in Unit 8). At any one time, only a small proportion of the cell's DNA is copied to messenger RNA for protein synthesis. In these regions, the DNA is fully unwound so that transcription can take place. One or more dense spherical structures, called **nucleoli**, are present inside the nucleus, and these are responsible for making ribosomal RNA, transfer RNA and some ribosomal proteins. Each nucleolus is formed around a specialized chromosome region, called a **nucleolar organizer**.

Interphase is divided into three shorter stages, respectively **G1 (gap 1)**, **S (synthesis)**, and **G2 (gap 2)**, as shown in Figure 36·1.

The duration of G1 is very variable and determines the time taken for cell multiplication in different tissues. Near the end of G1, a proportion of body cells pass a critical point, called the **restriction point**. When this happens, the cells become **committed** and will complete their cycles and divide to form new cells, regardless of external conditions. Any cell which does not pass the restriction point is unable to divide. In most organisms, differentiated tissue cells lose the ability to divide and remain locked in G1. As these cells die, they are replaced by division from undifferentiated cells called **stem cells**.

The durations of S and G2 are far less variable. Two important events occur during the S, or **synthesis**, stage. First, all the cell's DNA undergoes **replication** by the semi-conservative process described in detail in Unit 8. When replication is complete, each chromosome consists of two identical molecules of DNA lying side by side. This copying process provides the two complete sets of genetic instructions which will eventually pass into the daughter cells. The second event is duplication of the **centrioles** – these structures will later migrate towards opposite poles of the cell and may help in the formation of the spindle, although their precise function is not understood.

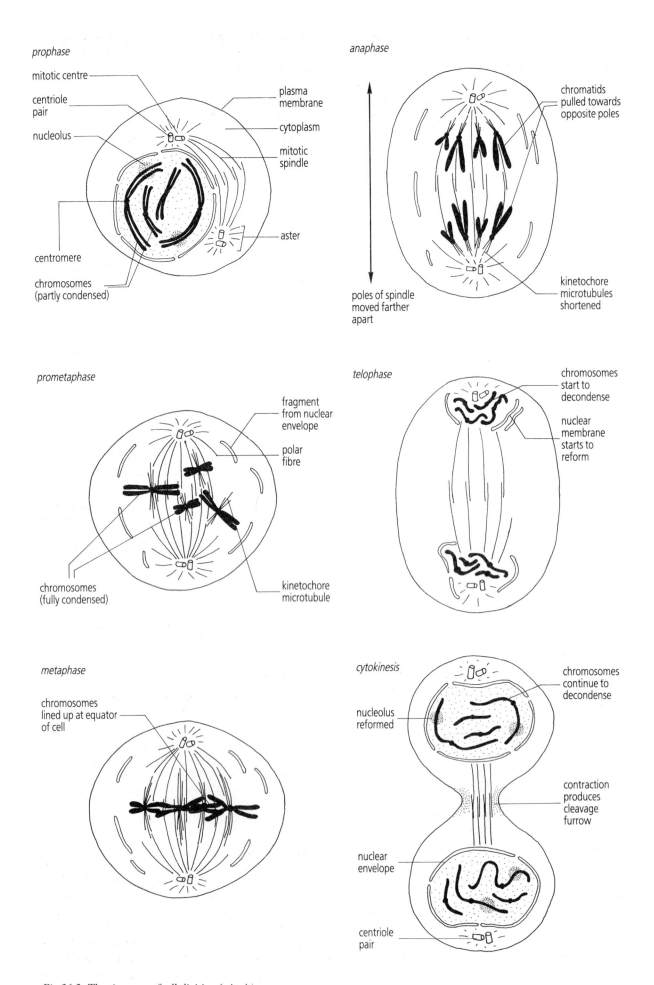

prophase

mitotic centre

centriole pair

nucleolus

plasma membrane

cytoplasm

mitotic spindle

aster

centromere

chromosomes (partly condensed)

anaphase

chromatids pulled towards opposite poles

kinetochore microtubules shortened

poles of spindle moved farther apart

prometaphase

fragment from nuclear envelope

polar fibre

chromosomes (fully condensed)

kinetochore microtubule

telophase

chromosomes start to decondense

nuclear membrane starts to reform

metaphase

chromosomes lined up at equator of cell

cytokinesis

chromosomes continue to decondense

nucleolus reformed

contraction produces cleavage furrow

nuclear envelope

centriole pair

Fig 36·2 The six stages of cell division (mitosis)

The stage called G2 represents the cell's final preparations for division and involves the synthesis of **histone** and **spindle** proteins.

Interphase normally accounts for at least 90% of the total cell cycle. The remaining events are concentrated into a comparatively short **division**, or **M phase**, typically lasting 1–2 hours. This short period may be further subdivided into six stages, namely **prophase, prometaphase, metaphase, anaphase, telophase**, and **cytokinesis**. Strictly speaking, **mitosis** includes only the first five of these stages, corresponding to the process of **nuclear division**. The sixth stage, called cytokinesis, refers to **cytoplasmic division** and separation of the daughter cells. It is important to take note of this distinction, for there are many organisms in which mitotic division of the nucleus is not directly followed by cytoplasmic division, as in many fungi (Unit 46).

The events taking place in each of the six stages are summarized in Figure 36·2.

Prophase

The transition from G2 to prophase is not a sharply defined event. The chromosomes slowly **condense**, that is, they shorten and thicken and finally become visible under the light microscope. Each chromosome can then be seen to consist of sister **chromatids** linked together by a structure known as a **centromere**. The two chromatids correspond to two identical molecules of DNA produced earlier, during the S stage. Each chromatid contains one molecule of DNA and each is an exact copy of the other.

As condensation continues, the nucleoli shrink and finally disappear, although the **nucleolar organizers**, that is, the chromosome regions responsible for nucleolus formation, may remain visible throughout division.

Prophase is also marked by the development of an arrangement of microtubules known as the **mitotic spindle** (see Fig. 36·2). At the beginning of prophase, the cytoplasmic microtubules which form the cytoskeleton (described in Unit 7) are broken down, allowing the cell to round up into a spherical shape. This breakdown provides a large 'pool' of **tubulin** molecules which are gradually incorporated into new microtubules. The first of these to appear are called **polar fibres** and they radiate from a pair of **mitotic centres** which move slowly towards opposite poles of the cell. In animal cells, a structure called an **aster** is formed at each end of the spindle, with fibres radiating in all directions. A **centriole pair** is located inside each mitotic centre. The exact function of the centrioles is in some doubt as, in experimental cultures, division continues normally even when they are destroyed using a laser microbeam. In plant cells, the mitotic centres are less well defined, there is

no aster and centrioles are completely absent. Thus, centrioles are not directly essential for spindle formation.

Prometaphase

Prometaphase begins with the sudden disintegration of the nuclear envelope. The spindle moves to occupy the central region of the cell and the centromere of each chromosome forms **kinetochores** (Fig. 36·3). These structures lie on either side of the chromosome so that the **kinetochore microtubules**, or **kinetochore fibres**, project in opposite directions. The fibres interact with the spindle, often resulting in agitated chromosome movement.

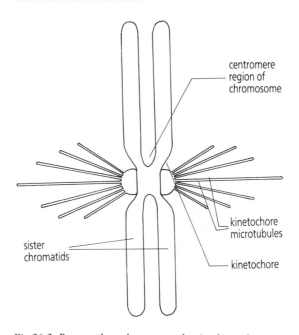

Fig 36·3 Prometaphase chromosome showing kinetochores

Metaphase

All the chromosomes are arranged with their centromeres near the **equator** of the cell. The plane of alignment is called the metaphase plate. The kinetochore microtubules lie parallel to the spindle and pull towards opposite poles, holding the chromosomes in place under tension. Compared with other stages, metaphase may last for a long time.

Anaphase

Anaphase begins when the kinetochores at each centromere suddenly split. The kinetochore fibres now slide towards opposite poles of the spindle, actively pulling the chromatids apart. At the same time, the spindle lengthens. The biochemical basis for these movements is not fully understood, but the net effect is to form two daughter nuclei, each genetically identical to the other. Note that both of these contain one chromatid from each parent chromosome: the two nuclei are identical because the chromatids themselves are identical molecules of DNA produced by replication.

Telophase

The nuclear membrane reforms, the chromosomes start to uncoil, or **decondense**, and nucleoli reappear.

Cytokinesis

This is the process of cytoplasmic division, sometimes also called **cleavage**. It usually begins before nuclear division is complete, that is, during the later stages of anaphase, or in telophase. A **cleavage furrow** appears at the equator of the cell and deepens progressively until the daughter cells separate (Fig. 36·2). The forces required are produced by actin-myosin interactions similar to those involved in muscle contraction (Unit 26).

36·3 Meiosis ■ ■ ■ ■ ■ ■ ■ ■ ■ ■ ■ ■ ■ ■ ■ ■ ■

To understand the process of meiosis, you must have a clear idea of its function. While mitosis produces genetically identical cells for growth and asexual reproduction, meiosis is only of value to organisms which reproduce sexually. With some exceptions to be discussed later, the cells produced following meiosis are the sex cells, or **gametes**.

In all cells about to undergo meiosis, the chromosome complement is described as **diploid (2n)**. Human spermatogonia and oogonia contain 46 chromosomes, as do normal human body cells. Detailed study of the chromosome complement reveals that the chromosomes can be arranged as 23 **homologous pairs** (Fig. 36·4). In 22 of the pairs, the chromosomes match one another exactly in appearance: these chromosomes are called **autosomes**. The remaining chromosome pair are the **sex chromosomes**, designated **X** and **Y**. It is important to appreciate that, even although they look alike, the two chromosomes forming a pair need *not* be genetically identical.

Gametes contain the **haploid** number of chromosomes, called *n*, that is, exactly half the diploid number. In human gametes there are 23 chromosomes, comprising one from each of the 22 autosomal pairs together with one sex chromosome. Fusion of male and female gametes at fertilization restores the diploid number, so that all the cells in the body of the offspring again contain 46 chromosomes.

You can see from Figure 36·5, which compares meiosis and mitosis, that meiosis involves two separate cell divisions. Many features of the two types of division are similar, but, during **prophase I** of meiosis, two additional processes occur which dramatically affect the outcome:

1 Synapsis. This is the pairing together of homologous chromosomes to form structures called **bivalents**. When synapsis is complete, the chromosomes are linked together throughout their length.

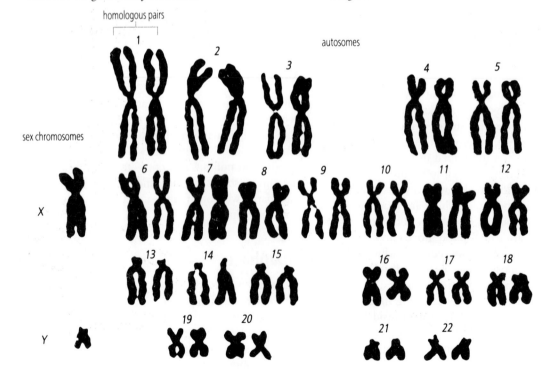

Fig 36·4 Human male chromosomes (karyogram)

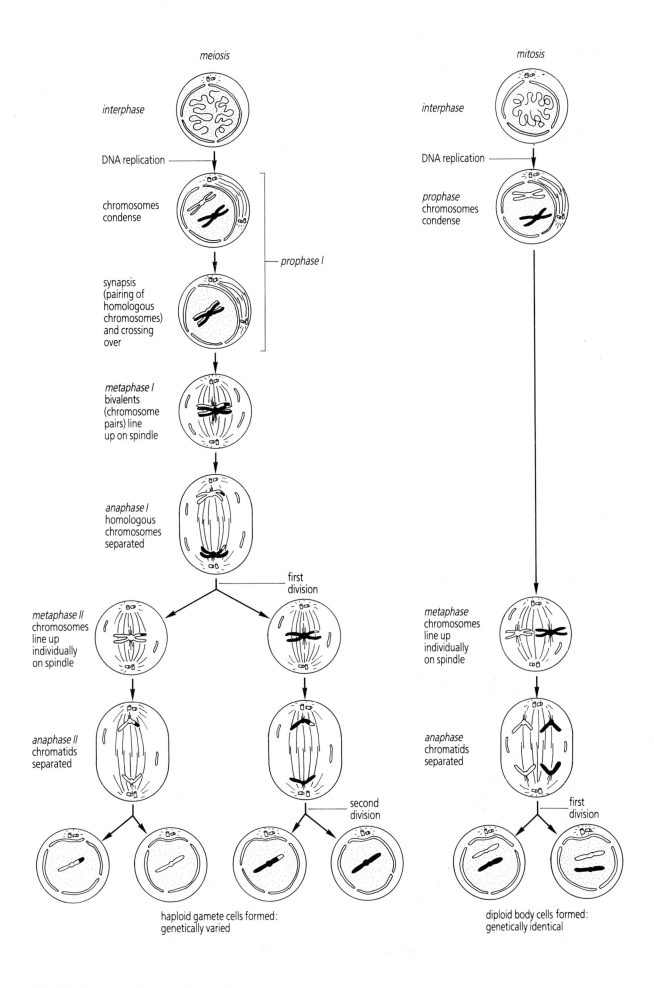

meiosis

mitosis

interphase

interphase

DNA replication

DNA replication

chromosomes
condense

prophase
chromosomes
condense

⎤
⎥ prophase I
⎦

synapsis
(pairing of
homologous
chromosomes)
and crossing
over

metaphase I
bivalents
(chromosome
pairs) line
up on spindle

anaphase I
homologous
chromosomes
separated

first
division

metaphase II
chromosomes
line up
individually
on spindle

metaphase
chromosomes
line up
individually
on spindle

anaphase II
chromatids
separated

anaphase
chromatids
separated

second
division

first
division

haploid gamete cells formed:
genetically varied

diploid body cells formed:
genetically identical

Fig 36·5 Comparison between meiosis and mitosis

2 Crossing-over. Crossing-over or **chiasma formation** takes place following synapsis. In this process, homologous chromatids break and rejoin at precisely corresponding points so that genetic material is exchanged between them (Fig. 36·6). Crossing-over is a random event leading to **genetic recombination**, that is, it generates new combinations of genes. The cross-overs, also called **chiasmata** (singular **chiasma**), can occur anywhere along the length of the chromosome. Typically, 2 or 3 chiasmata are formed on each pair of chromosomes in the production of human gametes. Gametes which contain new arrangements of genes are called **recombinant** types.

Prophase I normally lasts for several days. In its later stages the chromosomes become fully condensed and undergo **desynapsis**, remaining linked only where chiasmata have been formed. The subsequent stages of meiosis I separate the members of each homologous pair. Note carefully that it is entire chromosomes, each consisting of two daughter chromatids, which are segregated into the daughter cells of the first division. The second division, called **meiosis II**, is very similar to mitosis and serves merely to separate the chromatids, resulting in the production of four haploid gamete cells (refer to Fig. 36·5).

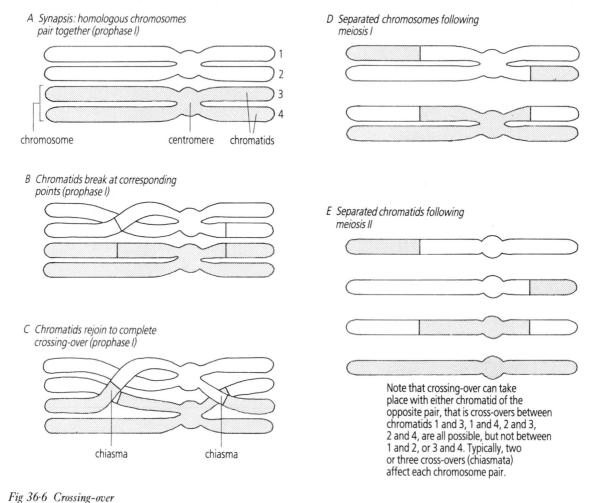

A Synapsis: homologous chromosomes pair together (prophase I)

chromosome   centromere   chromatids

B Chromatids break at corresponding points (prophase I)

C Chromatids rejoin to complete crossing-over (prophase I)

chiasma   chiasma

D Separated chromosomes following meiosis I

E Separated chromatids following meiosis II

Note that crossing-over can take place with either chromatid of the opposite pair, that is cross-overs between chromatids 1 and 3, 1 and 4, 2 and 3, 2 and 4, are all possible, but not between 1 and 2, or 3 and 4. Typically, two or three cross-overs (chiasmata) affect each chromosome pair.

Fig 36·6 Crossing-over

36·4 Evolution of meiosis ■ ■ ■ ■ ■ ■ ■ ■ ■ ■ ■ ■

In the context of evolutionary development, the mechanisms of meiosis represent a series of elegant adaptations to the problems of rearranging and transmitting genetic information in sexual reproduction.

The first eukaryotic cells reproduced asexually and were undoubtedly haploid, having only one gene for each of their characteristics. In such organisms, a change, or **mutation**, in any

particular gene will directly affect survival. Unfavourable mutations may be lethal, while favourable ones confer an immediate advantage. Populations of organisms with different favourable characteristics compete with one another but, except by repeated mutation, there is no way of combining favourable characteristics in the same individual. Evolution is slow because mutation is a random process and beneficial mutations are rare.

Fusion between haploid cells may have given rise to new types of organism, each with a pair of genetically different nuclei. These individuals would combine the favourable characteristics of the 'parent' lines and would have an obvious selective advantage. Furthermore, because they now have two genes for each of their characteristics, the new organisms are protected against the most damaging effects of additional mutations. Thus, one copy of each gene is free to mutate without loss of viability and may ultimately come to serve a new and useful function. Similar processes still occur in fungi, many of which exist as **dikaryotic** individuals (Unit 46).

Initially the two nuclei in each cell might have divided separately, but, in later evolution, the nuclei fused so that the two complete sets of chromosomes became attached to the same spindle during mitosis, forming a true diploid organism. Meiosis, which separates sets of chromosomes, may have evolved firstly as an adaptation producing genetically varied haploid offspring capable of rapid proliferation. Subsequent fusion of haploid cells, called **fertilization** or **syngamy**, completes a cycle of events observed in all sexually-reproducing organisms. Variations in the relative timing of meiosis and syngamy give rise to different types of life cycle, as explained in Unit 51.

The remaining Units in this Section describe the consequences of meiosis for inheritance, leading in turn to a consideration of the important topics of population genetics and evolution.

37 PATTERNS OF INHERITANCE

Objectives

After studying this Unit you should be able to:–

- Describe the experimental technique used by Mendel to cross between different varieties of garden peas

- State the Law of Segregation and, with the aid of diagrams, show how segregation of alleles helps to explain the results of a monohybrid cross

- Comment on the relationship between genes and chromosomes

- Explain the need for the test cross method and state how this might be applied

- Distinguish between complete dominance and incomplete dominance

- Using the inheritance of a person's ABO blood group as an example, explain what is meant by multiple alleles

- Define a mutant allele and comment on the inheritance of mutant characteristics

- State the Law of Independent Assortment and explain in detail the results of a dihybrid cross

- Briefly define the following terms as used in the text: pure-breeding, contrasted character, dominant, recessive, allele, phenotype, genotype, homozygous, heterozygous, locus, co-dominant, mutagen, carrier, recombinant type

37·1 Gregor Mendel ■ ■ ■ ■ ■ ■ ■ ■ ■ ■ ■ ■ ■ ■ ■ ■

The study of patterns of inheritance, called **heredity**, or more commonly **genetics**, has been important for hundreds of years. From the earliest days of agriculture, plant and animal breeders have tried to improve their stocks by **selective breeding**, attempting to combine the best features of different **strains**, or varieties, of organism. This process is known as **domestication**. Until the late 19th century, this was a rather hit or miss affair because the mechanisms underlying inheritance were not understood.

The main rules governing inheritance were first discovered by **Gregor Mendel**, a monk at the Augustinian monastery in Brunn (now Brno in Czechoslovakia). Mendel chose to study garden peas, *Pisum sativum*, and, starting in 1856, he carried out many experiments, crossing different varieties of pea plant. He succeeded brilliantly where many previous investigators had failed because of his insight in asking precisely the right questions and because of his patient and logical approach.

Pea plants are normally self-fertilizing and can be maintained as **pure-breeding** lines, meaning that the characteristics of the parent plants are transmitted unaltered to the offspring, generation after generation. Mendel cross-fertilized between pure-breeding lines showing **contrasted characters**. For example, he crossed together red-flowered and white-flowered varieties, tall and dwarf plants, and plants with smooth and wrinkled seeds. To prevent self-fertilization, he removed the stamens from the developing flowers of one variety. Later he transferred pollen to the stigma of each flower from the ripe anthers of a second variety. The seeds produced, called **hybrid** seeds, were collected and used to grow new plants in subsequent years.

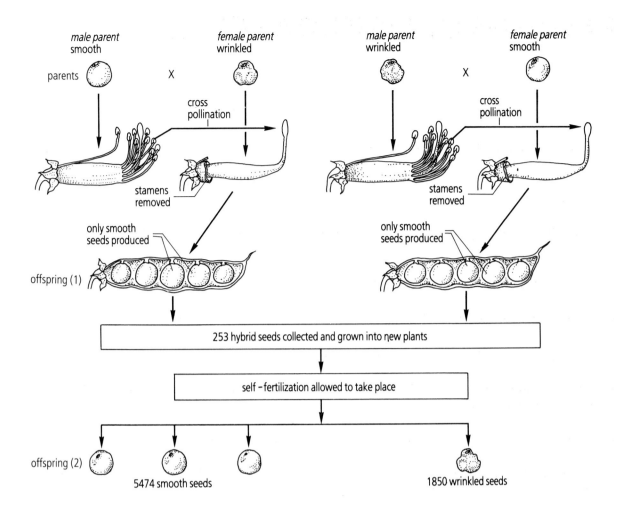

Fig 37·1 Summary of one of Mendel's experiments

In one particular experiment, summarized in Figure 37·1, Mendel crossed, or **hybridized**, two pure-breeding varieties, one with smooth rounded seeds, and the other with wrinkled seeds. In the next generation, called the **offspring (1) generation**, all the seeds were smooth irrespective of whether they developed on a parent plant belonging to the wrinkled strain fertilized with pollen from the smooth strain or vice versa. Mendel collected 253 of the hybrid seeds and grew them into new plants which he allowed to flower and set seed by self-fertilization in the normal way. The new seeds were the grandchildren of the original cross, called the **offspring (2) generation**. There were 7324 seeds, of which 5474, almost exactly three-quarters were smooth, and 1850 were wrinkled.

From a whole range of similar results, Mendel reached an important conclusion which is now known as **Mendel's First Law**, the **Law of Segregation**, or, less commonly, the **Law of Particulate Inheritance**. This states that **an organism's characteristics are controlled by 'factors' which are normally carried in pairs, but which occur singly in the gametes.** (It is now known that this occurs because the cells of most sexually-reproducing organisms are diploid, that is, they carry two sets of homologous chromosomes, whereas the gamete cells are haploid and carry only one.) The character which appeared in the **offspring (1)** generation, smooth seeds in the example above, Mendel called **dominant**, while the character which was temporarily obscured, wrinkled seeds, was called **recessive**.

Mendel's 'factors' were purely hypothetical and their physical form remained undiscovered for 100 years. The modern equivalents are called **genes**, each of which is a short length of DNA carrying the coded instructions for the development of just one characteristic. As described in Units 8 and 9, a typical gene codes for the formation of messenger RNA molecules from which individual molecules of a particular protein can be produced. A pea plant contains thousands of different proteins, ranging from the pigment molecules which regulate flower colour, to enzymes which affect food storage and hence shape in the seeds.

Representing the factor for smooth seeds as 'S' and the factor for wrinkled seeds as 's', the results of Mendel's experiment can be explained with the aid of Figure 37·2. The dominant 'smooth' factor and the recessive 'wrinkled' factor are called **allemorphs**, or **alleles**, because they are **alternative forms of the gene for one characteristic**. The external appearance, or **phenotype**, of the male parent is 'smooth'. The

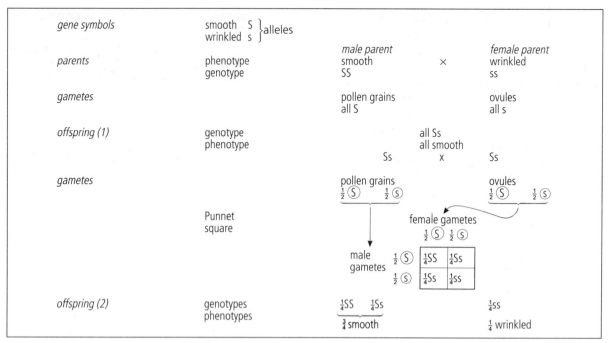

gene symbols	smooth S wrinkled s	alleles		
		male parent		*female parent*
parents	phenotype genotype	smooth SS	×	wrinkled ss
gametes		pollen grains all S		ovules all s
offspring (1)	genotype phenotype		all Ss all smooth	

Ss   x   Ss

gametes   pollen grains ½ Ⓢ   ½ ⓢ   ovules ½ Ⓢ   ½ ⓢ

Punnet square

female gametes
½ Ⓢ ½ ⓢ

male gametes

	½ Ⓢ	½ ⓢ
½ Ⓢ	¼SS	¼Ss
½ ⓢ	¼Ss	¼ss

offspring (2)   genotypes   ¼SS ¼Ss   ¼ss
phenotypes   ¾ smooth   ¼ wrinkled

Fig 37·2 Explanation of Mendel's experiment

alleles are carried in pairs and, since this is a pure-breeding line, the corresponding **genotype** is **SS**. This is the **homozygous dominant** genotype. The genotype of the pure-breeding wrinkled female is **homozygous recessive, ss**.

The gametes formed by these parents contain only one allele from each pair: pollen grains contain the allele **S** and ovules **s**. The genotype of the offspring resulting from fertilization is therefore **Ss**. This is called the **heterozygous** genotype. Because the smooth allele is dominant, the seeds produced are smooth and outwardly indistinguishable from those of the homozygous dominant parent line. The recessive wrinkled allele is completely masked. However, when the heterozygous offspring (1) seeds are grown to maturity it is clear than a number of different types of gametes can be formed. Following meiosis, exactly half of the pollen grains will contain the **S** allele, and half the **s** allele. Similarly, half the ovules are of each type.

Fertilization is assumed to be random between the different types of gametes so that the offspring (2) generation can have any of three different genotypes. Some will be homozygous dominant, **SS**, some heterozygous, **Ss**, and some homozygous recessive, **ss**. The proportions of each type to be expected are best determined using the

Punnett square method, as illustrated (Fig. 37·2). The 'list' of male gametes is transferred to one side of a table, while the 'list' of female gametes is copied at the top. For this cross, the table consists of four boxes, which are now filled. Considering the first entry, you can see that, when two gametes fuse, there is a chance, or **probability** of 1/2 that the male gamete will carry the S allele, and a chance of 1/2 that the female gamete will carry the same allele. The probability of the genotype SS is therefore 1/2 x 1/2 = 1/4. The remaining entries are arrived at in the same way. Consequently, the probabilities of the different genotypes are 1/4 **SS**, 1/2 **Ss** and 1/4 **ss**. Three quarters of the offspring (2) seeds are expected to be smooth and 1/4 wrinkled. In other words, there should be an approximately **3 : 1 ratio** of phenotypes, as Mendel observed.

The terms F_1, or **first filial generation**, and F_2, **second filial generation** are often used as if they were synonymous with **offspring (1)** and **offspring (2)**. Strictly, F_1 refers *only* to the offspring of a cross between homozygous individuals, and F_2 *only* to offspring obtained by crossing the F_1 individuals. You should be aware of these narrow definitions, although they are not universally observed.

37·2   Genes and chromosomes ■ ■ ■ ■ ■ ■ ■ ■ ■ ■ ■

Each cell of a pea plant contains 14 chromosomes, comprising 7 homologous pairs. One of these

pairs, represented in Figure 37·3*A*, carries the alleles for smooth or wrinkled seeds. These alleles are always found in a definite position or, **locus**, along the length of a particular chromosome. In this case, either **S** or **s** can be present. The arrangement shown gives the heterozygous **Ss** genotype. Alleles have already been defined as 'alternative forms of the gene for one characteristic'. You should now extend this definition to include the idea that alleles are **alternative forms of the gene at a particular locus**. The remaining regions of these chromosomes and the other six chromosome pairs contain hundreds of additional loci, all arranged in a definite order and each with its own alleles.

Figure 37·3*B* outlines what happens to the homologous chromosome pair and the genes they carry when meiosis occurs, as it would, for example, in a pollen mother cell. In the interphase preceding meiosis, the cell's DNA replicates, producing new copies of both the **S** and **s** alleles. Meiosis I separates pairs of homologous chromosomes into different cells, while meiosis II separates chromatids so that the gametes eventually formed contain only one allele from each original pair. It should be obvious that the result of meiosis is to ensure that exactly half the gametes produced by a heterozygous **Ss** individual will contain the **S** allele, and half the **s** allele. The observed behaviour of the chromosomes is, therefore, entirely consistent with Mendel's First Law.

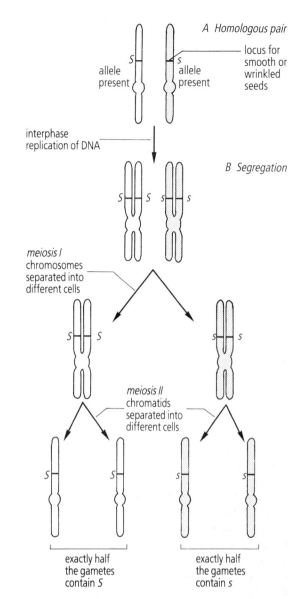

Fig 37·3 Chromosome explanation of segregation

37·3 Examples of monohybrid inheritance ■ ■ ■ ■ ■ ■ ■ ■

In monohybrid crosses only one characteristic, controlled by alleles at a single locus, is considered.

Test crosses

A test cross is carried out to determine the genotype of a particular individual when this is not known. For example, in guinea pigs, the allele for black coat colour, **B**, is dominant to the allele for brown, **b**. The homozygous recessive type, **bb**, is brown, but the homozygous dominant, **BB**, and heterozygous, **Bb**, individuals are both black and quite indistinguishable in appearance. For a guinea pig breeder wishing to maintain a pure-breeding black line, a simple method of separating these genotypes is invaluable. This is done by crossing the black individuals with the homozygous

recessive genotype, **bb**. Note that **all test crosses involve a cross with the homozygous recessive**. There are two possible outcomes, as outlined in Figure 37·4.

If a black individual is homozygous, then all the offspring of the cross must be black. However, if the black individual is heterozygous, half the offspring are expected to be brown. You should appreciate that some care is needed in interpreting the results of a test cross when the number of offspring is small. Even a single brown individual among the offspring gives a definite indication that the test individual was heterozygous. On the other hand, an all black litter does not necessarily indicate that the test individual was homozygous. In the heterozygous cross, the probability of each black offspring is 1/2. It follows that an all black litter of 4 will arise by chance on 1/2 x 1/2 x 1/2 x 1/2 = 1/16 of occasions. Where there is an

| | | gene symbol | black B | alleles |
| | | | brown b | |

A. *black individual homozygous*			B. *black individual heterozygous*		
	male	*female*		*male*	*female*
parents	black	brown	*parents*	black	brown
	BB	bb		Bb	bb
gametes	all ⓑ	all ⓑ	*gametes*	½Ⓑ ½ⓑ	all ⓑ
offspring (1)	all Bb		*offspring (1)*	½Bb	½bb
	all black			½ black	½ brown

Fig 37·4 Example of a test cross

uncertainty of this kind, the test cross can be repeated. Larger numbers of offspring, 6 or more, make misleading results much less likely.

Codominance

In the examples discussed so far, the dominant allele of a pair has completely masked the effect of the non-dominant, or recessive, allele. This relationship is called **complete dominance** and applies to the great majority of genes. In a few cases, however, one of the alleles is not completely masked so that both alleles are expressed in the heterozygous individual.

Special symbols are used for codominant alleles. For example, when the alleles controlling flower colour in *Antirrhinum* (snapdragon) flowers are written as C^R and C^W; C^R represents the allele for red flowers and C^W the allele for white flowers. Two common varieties of snapdragon differ in flower colour – one is pure-breeding red, $C^R C^R$,

the other is pure-breeding white $C^W C^W$. Surprisingly, when these varieties are crossed, the offspring (1) flowers, corresponding to the genotype $C^R C^W$, are all pink. Neither the red allele, C^R, nor the white allele, C^W, is fully dominant over the other. Self-pollination gives rise to an offspring (2) generation in which the results of Mendelian segregation are clearly seen. Because each gamete contains only a single allele from any pair, chance combination of the offspring (1) gametes results in approximately 1/4 of the offspring (2) individuals giving red flowers, 1/2 with pink flowers, and 1/4 with white flowers. You should compare this outcome with the expected proportions of smooth and wrinkled seeds obtained in Figure 37·2. Other examples of codominance include the inheritance of coat colour in cattle where a cross between certain breeds of red and white cattle produces an intermediate coat colour, called roan.

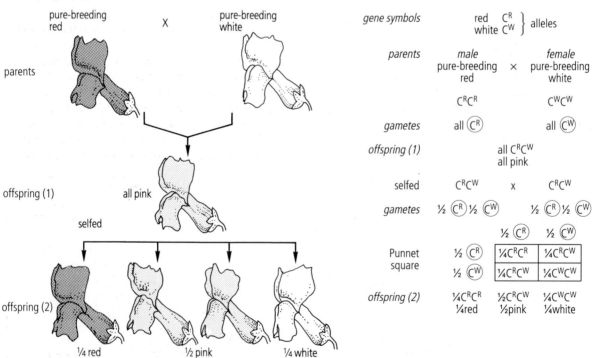

Fig 37·5 Incomplete dominance

Multiple alleles

A person's blood group is controlled by alleles at a single locus, so that each individual carries only two blood group alleles. Nevertheless, there are three rather than two alternative forms of this gene: these are designated I^A, I^B, and I^O. The I^A and I^B alleles are described as **codominant**. In other words, I^A and I^B are both dominant to I^O. This arrangement gives rise to four different blood groups which are listed, together with the corresponding genotypes, in Table 37·1. As you can see, blood groups A and B can be produced by two different genotypes, whereas there is only one possible genotype for groups AB and O. Blood group information is often used in attempting to resolve paternity disputes and you should try to ascertain for yourself which blood groups can arise among the children of different crosses.

The human ABO system provides one example of multiple alleles, but there are other cases where 3 or more alleles can occupy the same locus. These include eye colour in *Drosophila* fruit flies and the inheritance of coat colour in some breeds of rabbit.

Table 37.1 ABO blood groups

gene symbols	A ⎫ alleles	I^A
	B ⎬	I^B
	O ⎭	I^O

BLOOD GROUP (phenotype)	POSSIBLE GENOTYPES
A	$I^A I^A$ $I^A I^O$
B	$I^B I^B$ $I^B I^O$
AB	$I^A I^B$
O	$I^O I^O$

Mutant alleles

The genetic material, DNA, is extremely stable and is normally transmitted from generation to generation without alteration. A **gene mutation** is defined as a chance alteration in the DNA coding for a particular protein. In nature, such mutations occur only very rarely, but they can be made to occur more often by exposing the organism to mutagenic agents, or **mutagens**. These include radiation, such as X-rays, and UV light, and certain chemicals, like formaldehyde, colchicine, and some pesticides. Natural or artificially induced mutations are always random events, so that it cannot be predicted which individual genes within the DNA will be affected. An altered gene, called a **mutant allele**, is replicated along with the normal DNA and may be transmitted to future generations.

As you will recall, genes regulate protein synthesis by determining the sequence of amino acids in each protein molecule. Any change in the DNA is liable to affect this sequence and hence the structure and properties of the resulting protein. The most likely result is that the protein produced will be defective in some way, or may not function at all. Only very occasionally is the protein formed by a mutant allele in any way superior to its normal counterpart. In most cases, mutant alleles simply fail to code for an effective protein. The consequence of this is that many mutant alleles are recessive to their normal alleles. They are damaging or lethal to the individual only if carried in the homozygous condition.

For the human population, it is estimated that new mutations at a typical gene locus will arise once in every 50 000 sperm or egg cells. Taking the imaginary case of a recessive mutant allele, **a**, the allele might be transmitted to the offspring which then have the genotype **Aa**. Such individuals are called **carriers**; they are phenotypically normal but carry the mutant allele in heterozygous form. From known mutation rates, it can be calculated that the average person carries recessive mutant alleles at three or more loci. However, because there are thousands of different loci, there is only a small chance of parents having the mutant forms of the same genes. Consequently, most crosses involving a mutant allele are like that shown in Figure 37·6A, One parent is heterozygous, **Aa**, while the other is homozygous dominant, **AA**, with two normal alleles at this locus. All the offspring have the normal phenotype, but, on average, half of them will be carriers. The adverse effects of the mutation can become evident only if both parents

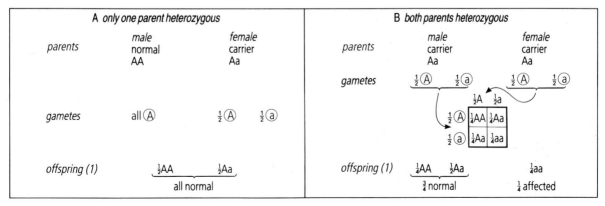

Fig 37·6 Inheritance of mutant alleles

are heterozygous for the same mutant allele (Fig. 37·6*B*). When this happens, one quarter of the offspring are expected to be homozygous recessive, **aa**, and may suffer from any one of a number of distressing conditions. Human diseases inherited in this way include **galactosaemia** (discussed briefly in Unit 19), **phenylketonuria**, and **cystic fibrosis**. As many as 1 in 2000 children are affected by cystic fibrosis. The disease increases the activity of the glands producing mucus, sweat,

and digestive juices, resulting in impaired digestion, and obstruction of the intestine and bronchi, leading in turn to respiratory infection. The disease becomes progressively more debilitating so that in the past few sufferers survived far into adult life.

Some rare diseases are caused by dominant mutant alleles. These include **Huntington's chorea** and **achondroplasia**, which causes one kind of dwarfism.

37·4 Dihybrid inheritance ■ ■ ■ ■ ■ ■ ■ ■ ■ ■ ■ ■ ■

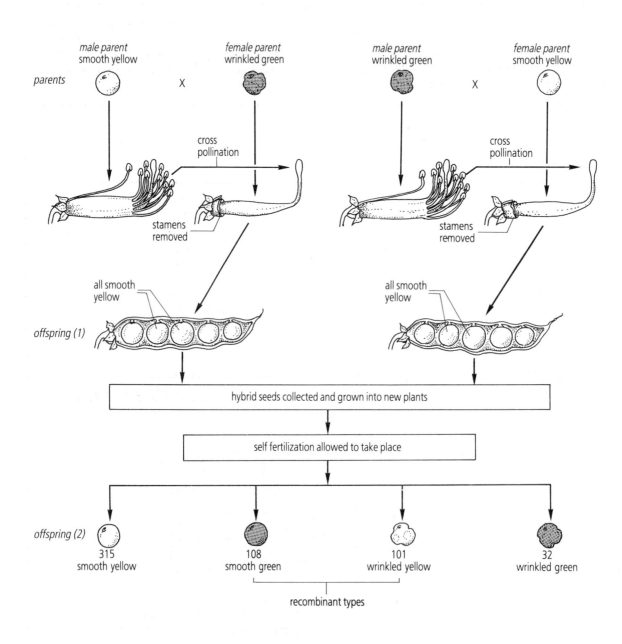

Fig 37·7 Dihybrid inheritance

In dihybrid crosses, two characteristics controlled by genes at different loci are considered. In one experiment, outlined in Figure 37·7, Mendel investigated dihybrid inheritance by crossing plants from two pure-breeding strains, one with smooth yellow seeds, and the other with wrinkled green seeds. The offspring (1) were all smooth yellow, these being the dominant characteristics. In the following year the offspring (1) seeds were planted and allowed to self-fertilize. A total of 556 offspring (2) seeds were collected from these plants, comprising 315 smooth yellow, 108 smooth green, 101 wrinkled yellow, and 32 wrinkled green. Smooth green and wrinkled yellow are new combinations of characters different from those shown by the original parent lines. They are therefore described as **recombinant types**.

Figure 37·8 provides the genetic explanation for these results. As you can see, there are two separate pairs of alleles, **S** and **s** controlling seed shape, and **Y**, **y** for seed colour. The genotype for pure-breeding smooth yellow is written **SSYY**, while pure-breeding wrinkled green is **ssyy**. Only one allele from each pair is represented in the gametes so that these must be **SY** and **sy** respectively. All the offspring (1) are smooth yellow and have the genotype **SsYy**, heterozygous for both characteristics.

The range of expected offspring (2) genotype is determined by the Punnet square method (refer to Fig. 38·7). It shows offspring (1) phenotypes in the proportion 9 : 3 : 3 : 1, almost exactly as Mendel observed. From a range of similar results, Mendel proposed a second generalization which is now known as **Mendel's Second Law or the Law of Independent Assortment. This states that the alleles of genes at different loci segregate independently of one another during the formation of gametes.**

In the next Unit, the significance of Mendel's Laws is assessed and some important exceptions discussed. However, before continuing, you should review this Unit together with Unit 36. The topics covered provide an essential introduction to the study of genetics and evolution and should be known thoroughly.

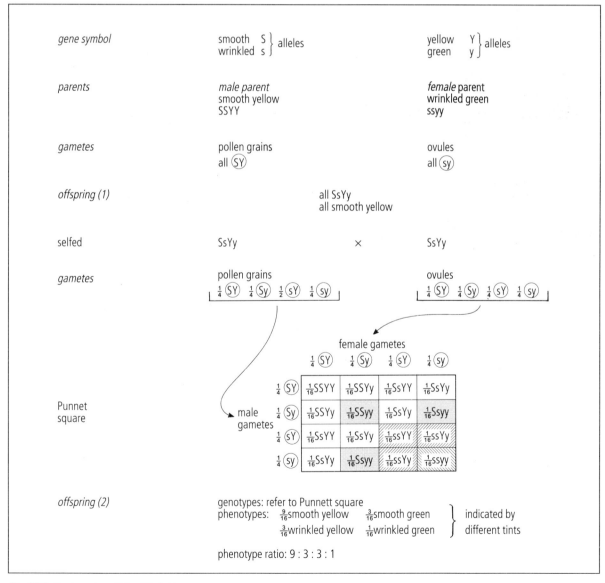

Fig 37·8 Explanation of the dihybrid cross

38 EXCEPTIONS TO MENDEL'S LAWS [   ]

Objectives

After studying this Unit you should be able to:–

● Describe the inheritance of sex in *Drosophila*, humans, and birds, and explain the terms homogametic and heterogametic

● Define sex-linked genes and describe the inheritance of eye colour in *Drosophila* and haemophilia in humans

● Define non-disjunction of chromosomes and outline the nature of Down's syndrome, Klinefelter's syndrome, and Turner's syndrome

● Comment on the origins and importance of polyploidy

● Distinguish between deletion, inversion, translocation, and duplication as forms of chromosome mutation

● Give examples of extra-nuclear genes

● Draw diagrams to show the chromosomal basis for independent assortment and linkage of genes

● Describe how linkage can be identified using the test cross method and write the formula for calculating cross-over values

● Comment briefly on chromosome mapping

● Give examples of pleiotropy, simple interaction, epistasis, and polygenic inheritance

38·1 Introduction ■ ■ ■ ■ ■ ■ ■ ■ ■ ■ ■ ■ ■ ■ ■ ■

Mendel's Laws, explained in the previous Unit, are as follows:

1 Segregation. An organism's characteristics are controlled by factors (genes) which are normally carried in pairs, but which occur singly in the gametes.

2 Independent assortment. Alleles of genes at different loci segregate independently of one another during the formation of gametes.

These statements are not 'laws' in the same sense as those of the physical sciences. Strictly speaking, they are generalizations which hold true for the majority of cases, but not for all. This Unit describes exceptions to Mendel's Laws – some of these occur only rarely, but others play an essential part in the life of all sexually-reproducing organisms.

In some species, the chromosomal mechanisms of sex determination give rise to First Law exceptions because they result in some genes being carried singly in the adult, instead of in allelic pairs. Chromosome abnormalities and the presence of extra-nuclear genes can also lead to the inheritance of unpaired genes.

Inheritance of sex

The use of the fruit fly, *Drosophila melanogaster*, for genetic experiments was pioneered by the American geneticist, **T.H. Morgan**. His work from 1907 onwards confirmed Mendel's results and led to a whole range of new discoveries, for which he received the Nobel Prize in 1933. The flies, which are kept in small specimen tubes, breed very rapidly (at 20°C a new generation is produced every 10–15 days). Eggs laid in the culture medium develop into larvae which then pupate to emerge finally as adult flies. Mating is delayed for 12 hours after the adults appear and, during this time, male and female virgin flies can be collected.

The appearance of male and female *Drosophila* and the normal chromosome complements of male and female cells are shown in Figure 38·1. The male flies are distinguished by their rounded dark-tipped abdomen and by the possession of sex combs. As you can see, their cells contain four pairs of chromosomes. Three pairs, called **autosomes**, are exactly alike in appearance; the fourth pair are the **sex chromosomes**. In male *Drosophila*, one of these, called the **X-chromosome**, is rod-shaped, while the other, called the **Y-chromosome**, has a distinctive hooked shape. Consequently, males are said to be the **XY**, or **heterogametic** sex. Cells of female *Drosophila* contain the same total number of chromosomes (Fig. 38·1B), but both the sex chromosomes are rod-shaped. In other words, females have an **XX** arrangement, described as **homogametic**.

A *Male: external features*

sex combs

rounded dark-tipped abdomen

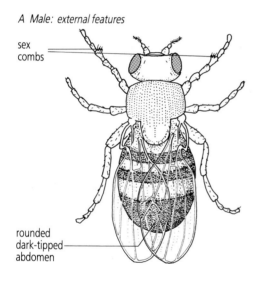

chromosome complement

sex chromosomes

X

Y

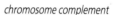
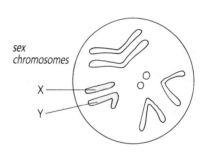

B *Female: external features*

sex combs absent

pointed abdomen

chromosome complement

sex chromosomes

X

X

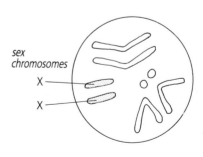

Fig 38·1 Male and female Drosophila melanogaster

Consider what happens when male and female *Drosophila* form gametes by meiosis (Fig. 38·2). In the case of the sex chromosomes, separation of homologous pairs produces two different kinds of male gamete, half containing an X-chromosome and half a Y-chromosome. Female cells form only one variety of gamete, all of which contain a single X-chromosome. Fertilization therefore provides an approximately equal number of males and females in the next generation. Note that the sex of the offspring is determined by a particular arrangement of entire chromosomes and not by individual genes. Similar mechanisms have been found in most other sexual species, including man (refer to Fig. 36·4). Surprisingly, it appears that human Y-bearing sperms have a slightly better chance of fertilizing an ovum than the X-bearing sperms, with the result that more boys are born. However, in many countries, women tend to live for longer than men so that females outnumber males in the population as a whole.

It must be noted that the male is not always the heterogametic sex. In butterflies and moths, and in birds, the males are homogametic XX, while the females possess both X and Y-chromosomes.

Sex-linked genes

Genes carried on the sex chromosomes are said to be **sex-linked**. A most interesting finding is that in many sexual species (although not in *Drosophila*) the Y-chromosome is much smaller than the X, or may be absent altogether. Even when it is present and similar in size, the Y-chromosome very often seems to be **genetically inert**, carrying few if any functional genes. By contrast, the X-chromosome has many loci carrying a range of important genes. XX individuals have two alleles for these characteristics, but XY individuals have only one.

Figure 38·3 shows the results of two different crosses involving red-eyed and white-eyed *Drosophila*. This characteristic is controlled by a gene carried at a locus on the X-chromosome; X^R, the allele for red eyes, is dominant to X^r, the allele for white eyes. In the first cross (Fig. 38·3*A*), the red-eyed female parent is homozygous dominant X^RX^R. All of her gametes carry the X^R allele, which is thereby transmitted to all of the offspring. The white-eyed male parent, with both X and Y-chromosomes, possesses just one eye colour allele located on the X-chromosome and passed to the X-bearing sperm. As you can see, only the female

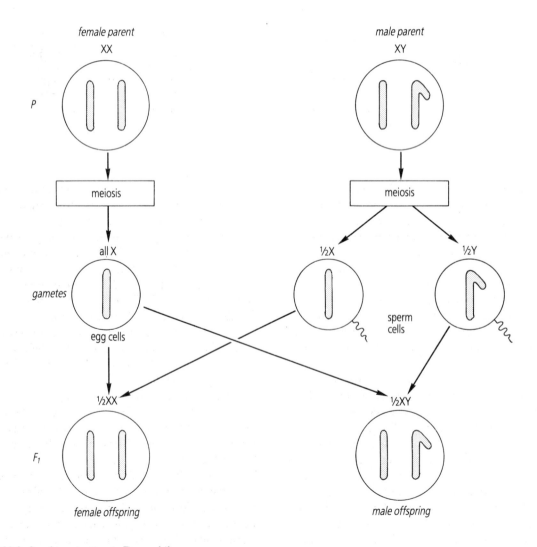

Fig 38·2 Sex determination in Drosophila

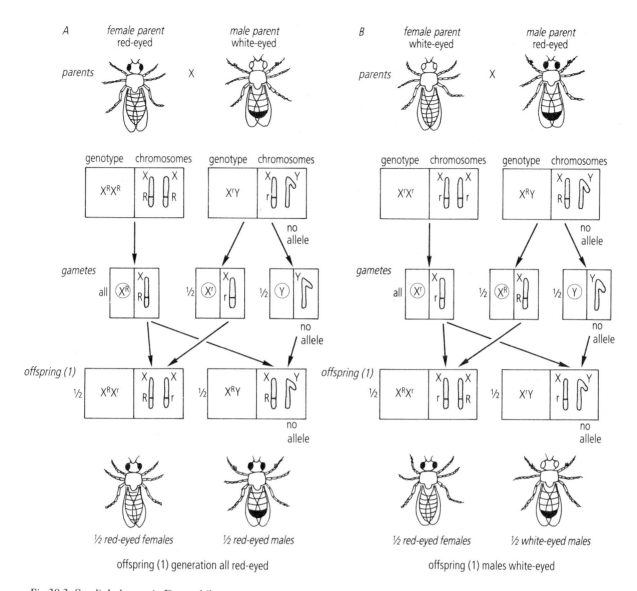

A female parent red-eyed X male parent white-eyed

parents

genotype chromosomes   genotype chromosomes

$X^R X^R$ $X^r Y$ no allele

gametes

all X^R ½ X^r ½ Y   no allele

offspring (1)

½ $X^R X^r$ ½ $X^R Y$ no allele

½ red-eyed females   ½ red-eyed males

offspring (1) generation all red-eyed

B female parent white-eyed X male parent red-eyed

parents

genotype chromosomes   genotype chromosomes

$X^r X^r$ $X^R Y$ no allele

gametes

all X^r ½ X^R ½ Y   no allele

offspring (1)

½ $X^R X^r$ ½ $X^r Y$ no allele

½ red-eyed females   ½ white-eyed males

offspring (1) males white-eyed

Fig 38·3 *Sex-linked genes in* Drosophila

offspring receive X^r, the allele for white eyes, making them heterozygous $X^R X^r$. Since X^R is dominant, all the offspring(1) females show the red-eyed phenotype. Male offspring receive only one X-chromosome and therefore only one gene. Their genotype, written $X^R Y$, makes them red-eyed like the females. In the reciprocal cross (Fig. 38·3B), the offspring (1) females remain heterozygous $X^R X^r$ but this time, the males receive a single **r** allele from their mother. As a result, the offspring (1) genotype is $X^r Y$ and the phenotype is white-eyed.

A number of mutant sex-linked genes are known to affect the human population. Examples include genes for certain forms of **colour blindness**, and for **haemophilia**, a serious condition in which the blood fails to clot properly. Sufferers are unable to produce one of a number of blood clotting factors (described in Unit 18) and are therefore prone to prolonged bleeding from cuts and internal haemorrhages. The normal form of the gene, X^H, is carried on the X-chromosome and is dominant to its mutant allele, represented by X^h. As outlined in Figure 38·4, females have

two alleles for this characteristic so that there are three possible genotypes. Fortunately, haemophilia is a rare disease, so that the vast majority of women are homozygous $X^H X^H$, and only a small proportion are carriers, $X^H X^h$. These women are unaffected themselves but are able to transmit the mutant allele to their children. Three possible crosses are set out in Figure 38·4. Affected females, genotype $X^h X^h$, are exceedingly uncommon, there being only a handful of recorded cases. Males have only one functional allele and so are either completely normal $X^H Y$, or have the disease, genotype $X^h Y$.

A famous family history involving haemophilia is illustrated in Figure 38·5. This kind of chart is called a **pedigree**. It shows members of the British Royal Family descended from Queen Victoria. Queen Victoria herself was a carrier of the haemophilia allele which most probably arose as a fresh mutation affecting the germ cells of one of her parents. You should follow the chart trying to assign the appropriate genotypes to individual descendants. The present line, descended from King Edward VII, is unaffected by the disease.

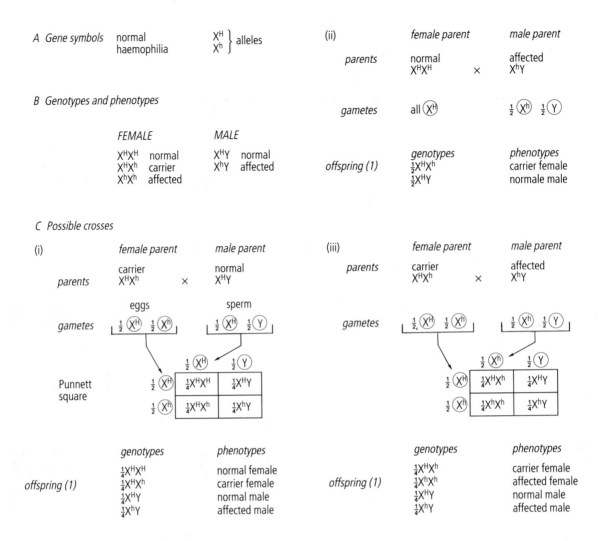

A *Gene symbols* normal X^H } alleles
                  haemophilia X^h

B *Genotypes and phenotypes*

FEMALE		MALE	
X^HX^H	normal	X^HY	normal
X^HX^h	carrier	X^hY	affected
X^hX^h	affected		

C *Possible crosses*

(i)

	female parent		*male parent*
	carrier		normal
parents	X^HX^h	×	X^HY

          eggs   sperm

gametes $\frac{1}{2}(X^H)$ $\frac{1}{2}(X^h)$ $\frac{1}{2}(X^H)$ $\frac{1}{2}(Y)$

Punnett square

	$\frac{1}{2}(X^H)$	$\frac{1}{2}(Y)$
$\frac{1}{2}(X^H)$	$\frac{1}{4}X^HX^H$	$\frac{1}{4}X^HY$
$\frac{1}{2}(X^h)$	$\frac{1}{4}X^HX^h$	$\frac{1}{4}X^hY$

offspring (1)

genotypes	phenotypes
$\frac{1}{4}X^HX^H$	normal female
$\frac{1}{4}X^HX^h$	carrier female
$\frac{1}{4}X^HY$	normal male
$\frac{1}{4}X^hY$	affected male

(ii)

	female parent		*male parent*
	normal		affected
parents	X^HX^H	×	X^hY

gametes   all (X^H) $\frac{1}{2}(X^h)$ $\frac{1}{2}(Y)$

offspring (1)

genotypes	phenotypes
$\frac{1}{2}X^HX^h$	carrier female
$\frac{1}{2}X^HY$	normale male

(iii)

	female parent		*male parent*
	carrier		affected
parents	X^HX^h	×	X^hY

gametes $\frac{1}{2}(X^H)$ $\frac{1}{2}(X^h)$ $\frac{1}{2}(X^h)$ $\frac{1}{2}(Y)$

	$\frac{1}{2}(X^h)$	$\frac{1}{2}(Y)$
$\frac{1}{2}(X^H)$	$\frac{1}{4}X^HX^h$	$\frac{1}{4}X^HY$
$\frac{1}{2}(X^h)$	$\frac{1}{4}X^hX^h$	$\frac{1}{4}X^hY$

offspring (1)

genotypes	phenotypes
$\frac{1}{4}X^HX^h$	carrier female
$\frac{1}{4}X^hX^h$	affected female
$\frac{1}{4}X^HY$	normal male
$\frac{1}{4}X^hY$	affected male

Fig 38·4 Inheritance of haemophilia

Chromosome abnormalities

1 Non-disjunction. The normal process of meiosis ensures that each gamete will contain a haploid set of chromosomes, for example, human gametes contain 23. Sometimes, however, one or more pairs of homologous chromosomes fail to separate so that some gametes contain extra chromosomes, while others contain too few. This is called **non-disjunction** and accounts for a number of abnormalities, among which **Down's syndrome** is the most important. The symptoms of this condition, which affects as many as 2 per 1000 live-born children in Europe and North America, include a characteristic facial and bodily appearance and limited mental development. It is caused by non-disjunction of one of the smallest pairs of chromosomes, designated **chromosome 21** (refer to Fig. 36·4). Individuals with Down's syndrome possess three of these chromosomes, an arrangement called **trisomy 21**. Their cells contain a total of 47 chromosomes instead of the usual 46. Down's syndrome is increasingly common in children born to mothers over 40 and can sometimes be detected following **amniocentesis**.

In this procedure, a small sample of amniotic fluid, which contains cells from the developing foetus, is withdrawn using a long fine needle. If microscopic examination of the fluid reveals Down's syndrome, the parents may be given the option of terminating the pregnancy.

Non-disjunction of the sex chromosomes causes two principal types of genetic disorder, known respectively as **Klinefelter's syndrome**, and **Turner's syndrome**. Klinefelter's syndrome occurs in about 2 out of every 1000 live male births. Sufferers have one or more extra sex chromosomes. An XXY formation is the most frequent, but XXYY, XXXY, and XXXXY have also been recorded. Generally, there are no outward physical symptoms but affected individuals are unable to produce sperm and are consequently sterile. Men with an extra Y-chromosome, XYY, are taller than average and, at one time, were thought to be more aggressive, although this now seems to be incorrect. People with one X-chromosome, written as XO, or with three, suffer from Turner's syndrome. They are female in appearance but ovaries fail to develop, and there may be other physical and mental abnormalities.

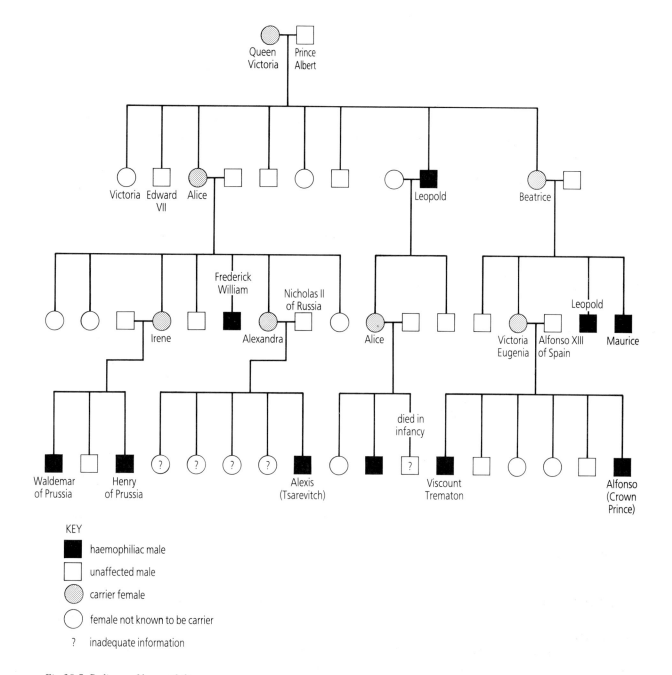

KEY

■ haemophiliac male

□ unaffected male

◉ carrier female

○ female not known to be carrier

? inadequate information

Fig 38·5 Pedigree of haemophilia

In plants, non-disjunction sometimes affects not just a single pair of homologous chromosomes, but all pairs simultaneously, with the result that gametes containing a full diploid set of chromosomes are formed. Fertilization involving such gametes can then produce **triploid** (3n) or **tetraploid** (4n) individuals. This phenomenon is called **polyploidy**. Triploid individuals may be able to reproduce asexually, but are generally infertile because homologous chromosomes fail to pair correctly during prophase I of meiosis. However, in a tetraploid cell, the four homologous chromosomes of any particular type can join together to make two pairs, so that normal meiosis can still take place, producing more diploid gametes. This means that tetraploid individuals can reproduce amongst themselves and, in effect, they form a new species distinct from the original parental type. Polyploidy has played an important part in the evolution of many flowering plant species including important crop species such as wheat, and is discussed further in Unit 40.

2 Chromosome mutation. Chromosome mutations involve changes in the structure of chromosomes and in the sequence of genes carried, although not in the genes themselves. (Gene mutation is a separate process, as described in Unit 37.) Changes are most likely to occur when chromatids break and rejoin during crossing-over in prophase I of meiosis. As a result, sections of one chromosome may be lost altogether, or incorporated into other chromosomes. A number of specific abnormalities are possible as outlined in Figure 38·6. **Deletion**, involving a loss of genes, is often fatal. However, **inversion, translocation,** and **duplication** all lead to a reshuffling of genes

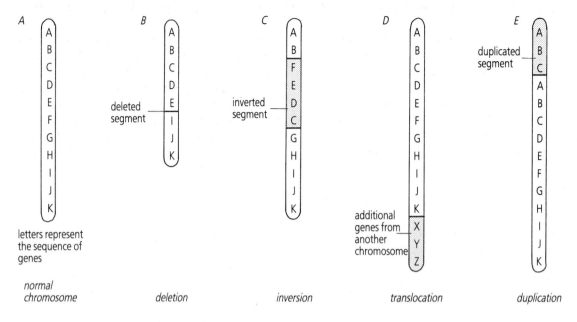

Fig 38·6 Chromosome mutations

along the length of the chromosomes. Such changes can be advantageous, as you will discover later in this Unit.

Extra-nuclear genes

Mendelian inheritance mechanisms operate when the genetic material is carried on homologous pairs of chromosomes in the nucleus of eukaryotic cells. However, not all of the cell's DNA is confined to the nucleus. A small but significant proportion is present in the cytoplasm inside cellular organelles. In mouse liver cells, for example, mitochondrial DNA represents about 1% of the total, while, in the leaves of maize, as much as 15% of the total cellular DNA may be contained in the chloroplasts. It is now known that both mitochondrial DNA and chloroplast DNA include many functional genes from which there is transcription of mRNA leading to protein synthesis. In other words, mitochondria and chloroplasts make some of their own proteins, including some which are needed for respiration and photosynthesis. These genes are not duplicated in the cell nucleus so that they are essential for normal metabolism.

Organelles and their genes are transmitted to the offspring in the cytoplasm of the gametes (Fig. 38·7). Obviously, if the gametes are similar in size, both parents will contribute extra-nuclear genes but, if they are very dissimilar, such genes may come from only one parent. Mammalian sperm cells and pollen grains from flowering plants contribute very little cytoplasm to the zygote so that the inheritance of organelles is often completely maternal.

A Gametes similar in size (e.g. yeast)

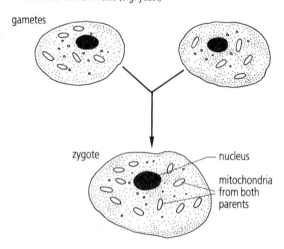

B Gametes dissimilar in size (e.g. mammals)

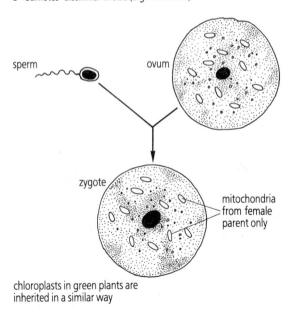

chloroplasts in green plants are inherited in a similar way

Fig 38·7 Inheritance of cytoplasmic genes

342

38·3 Second Law exceptions ▪ ▪ ▪ ▪ ▪ ▪ ▪ ▪ ▪ ▪ ▪

The reasons for exceptions to Mendel's Second Law are best explained by considering the chromosomal basis for independent assortment of alleles. Figure 38·8*A* shows two pairs of alleles at loci on different chromosomes. As you can see, separation of the chromosomes at meiosis I can occur in either of two ways so that a heterozygous individual, **AaBb**, will produce four different sorts of gametes, **AB**, **ab**, **Ab**, and **aB**, in equal proportions. A cross between two such individuals gives the 9:3:3:1 ratio of phenotypes characteristic of a dihybrid cross (refer to Fig. 37·8). On the other hand, each chromosome contains hundreds of genes so that it must often happen that two pairs of alleles are located on the same chromosome. This situation is illustrated in Figure 38·8*B*. Assuming that crossing-over does not take place, only two different types of gametes can be produced. These have the same arrangement of genes as in the parental cells.

Genes on the same chromosome are said to be **linked** and will tend to be inherited together. Only when crossing-over takes place can **recombinant gametes**, with new combinations of genes, be produced (Fig. 38·8*C*). The number of such gametes depends essentially on two things: the first is the distance apart of the genes on the

chromosome: the second is the probability that a cross-over will occur between them. For any particular species, the average number of cross-overs remains approximately constant so that the distance between genes becomes the most important variable. If the genes are far apart, crossing-over between them is frequent, but, if they are close together, crossing-over is unlikely and recombinant gametes are rare.

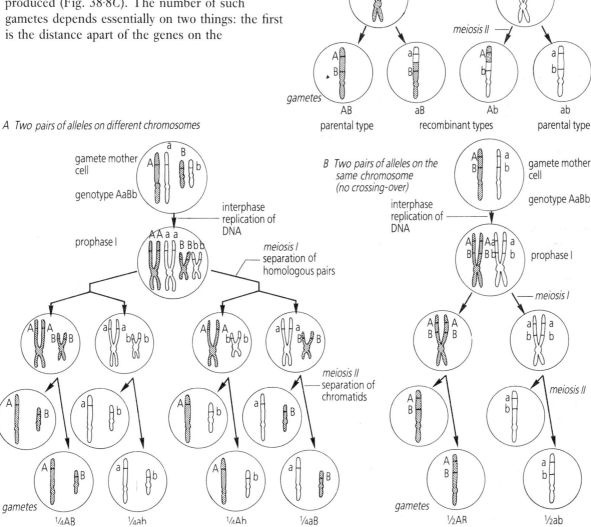

Fig 38·8 *Chromosomal basis for independent assortment and linkage*

In genetic experiments, linkage between genes is suspected if the phenotype ratios obtained from a dihybrid cross differ markedly from those expected. The situation can be investigated further by a special application of the **test cross** method (explained first in Unit 37). For example, in *Drosophila*, long wings are dominant to a mutant form called vestigial wings and normal red eyes are dominant to purple eyes. Flies heterozygous for both characteristics can be obtained by crossing pure-breeding lines as indicated in Figure 38·9A. The test cross is carried out between a heterozygous female **LlNn**, and a male with the homozygous recessive genotype, **llnn**. (Remember that a test cross always involves a cross with the homozygous recessive.) Without linkage, the female is expected to produce four different types of gametes in equal proportions. These combine with male gametes, all of which have the genotype **ln**, such that four different phenotypes are expected amongst the offspring in the ratio 1:1:1:1 (refer to Fig. 38·9B). Two of the phenotypes, long/purple and vestigial/normal show new combinations of characteristics not shown by either parent.

When this cross was actually carried out by Morgan's co-workers, using several pairs of flies, the results summarized below were obtained:

phenotype	number of offspring	
long/normal	1339	parental type
long/purple	154 ⎫	recombinant
vestigial/normal	151 ⎭	types
vestigial/purple	1195	parental type
Total	2839	

As you can see, the recombinant phenotypes appear much less often than is expected. This is a clear indication that the genes are carried on the same chromosome and can only be separated by crossing-over. The **cross-over frequency**, or **cross-over value (COV)** is calculated according to the formula:

$$COV = \frac{\text{number of recombinants}}{\text{total number of offspring}} \times 100\%$$

Note that, without linkage, the COV obtained from the test cross would be 50%
Here

$$COV = \frac{154 + 151}{2839} \times 100\%$$
$$= 10\cdot7\%$$

COVs like this are extremely useful because they allow the sequence of genes carried on the individual chromosomes to be determined – this process is called **chromosome mapping**. To illustrate the principle, consider the COVs between **L** and **N** and an imaginary third gene, called **X**:

gene pair	COV
L – N	10·7
L – X	18·5
N – X	7·8

It is easy to see that only one arrangement of genes satisfies these values, as follows:

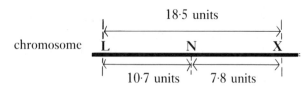

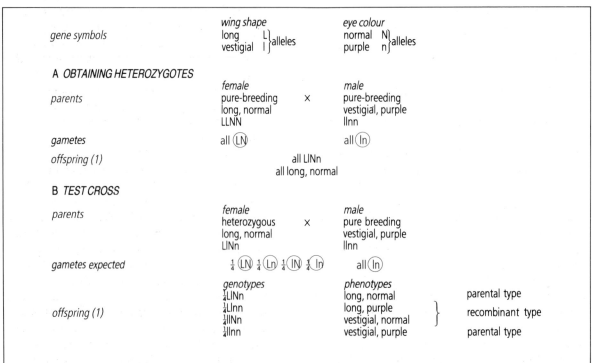

Fig 38·9 Test cross method of investigating linkage

In *Drosophila*, this principle has been used to map the distribution of literally hundreds of genes. An interesting observation is that all of the known genes fall into one or other of four large **linkage groups**, corresponding exactly to the number of homologous pairs of chromosomes. Similar results have been obtained for other organisms, providing good evidence for the central role of chromosomes in the transmission of hereditary information.

An important effect of linkage is to keep together advantageous combinations of genes, or genes with related functions. Chromosome mutations, referred to earlier, may promote linkage of this sort. Very closely linked groups of genes are sometimes called **supergenes**. At least some of the many genes responsible for producing human blood clotting factors are linked in this way so that they are invariably inherited together.

38·4 Genotypes and phenotypes ■ ■ ■ ■ ■ ■ ■ ■ ■ ■ ■

Organisms inherit hundreds of genes from their parents. These contribute to the total genotype, sometimes called the organism's **genome**. Most of the genes are inherited in a simple Mendelian fashion, but this pattern is often obscured in the phenotypes of the offspring, partly because individual genes have many effects, and partly because genes at different loci interact. The concluding section of this Unit outlines some of these processes.

1 Pleiotropic effects. A gene which controls more than one character is said to be **pleiotropic**. For example, in one study of laboratory rats, a whole range of abnormalities including an enlarged heart, closed nostrils, tracheal obstruction, and lack of elasticity in the lungs were all found to be caused by a single mutant allele. The normal form of this gene was later shown to produce a protein essential for the formation of cartilage. Such widespread damage arises because cartilage is a vital structural component present in many different body organs. In view of the complex pathways of normal metabolism, it is not surprising that effects like this are extremely common – so much so that most genes are now thought to be pleiotropic.

2 Simple interaction. Interaction is the converse of pleiotropy. It occurs when two or more genes affect the same phenotypic character. Two pairs of genes control the inheritance of comb shape in poultry (Fig. 38·10). Both pea and rose-combed individuals possess dominant alleles but when pure-breeding pea and rose poultry are crossed, the genes interact to produce a new phenotype, called walnut, among the offspring (1). Crossing the offspring (1) individuals in turn produces a typical 9:3:3:1 ratio and a fourth comb shape, called single. To improve your understanding of this result, identify and write down the possible genotypes for the four different comb shapes from the Punnett square of Figure 38·10. How would you establish the genotype of a walnut hen of unknown parentage?

3 Epistasis. Epistasis occurs when an allele of one gene suppresses the action of an allele of

Table 38·1 Inheritance of coat colour in mice

agouti	A	alleles
non-agouti	a	
black	B	alleles
brown	b	
coloured	C	alleles
albino	c	

genotype	phenotype
A* B* C*	agouti (grey)
aa B* C*	black
aa bb C*	brown
A* bb C*	cinnamon
** ** cc	albino

*, either allele may be present

another gene. Coat colour in mice is controlled by three different loci, as outlined in Table 38·1. One locus carries a dominant allele **A**, called agouti, or the recessive allele, **a**, non-agouti. An agouti individual has a band of yellow near the tip of each hair which modifies the overall appearance of the fur. The second locus is occupied by the alleles **B** or **b**, responsible for black or brown pigmentation. If **C**, the dominant allele for coloured coat is present at the third locus, then the phenotype of the mouse is determined by the alleles at the other two loci. However, the homozygous recessive genotype **cc** at this locus produces a white, or **albino** coat regardless of the alleles present. In other words, the albino allele suppresses the action of the other genes.

4 Polygenic inheritance. In many organisms, characteristics like height and mass do not vary in an obviously Mendelian fashion. Often, there is no separation into sharply distinct categories, such as red and white flowers, smooth and wrinkled seeds, or tall and dwarf individuals. Instead there is a complete range of intermediate phenotypes, as shown in Figure 38·11. This is known as **continuous variation**. The observed phenotypes usually conform to a bell-shaped curve, called a **normal distribution curve**. Most males have heights close to the average, or **mean** value, with fewer either very short or very tall. This pattern is typical of **polygenic inheritance** in which many genes interact. Sometimes the

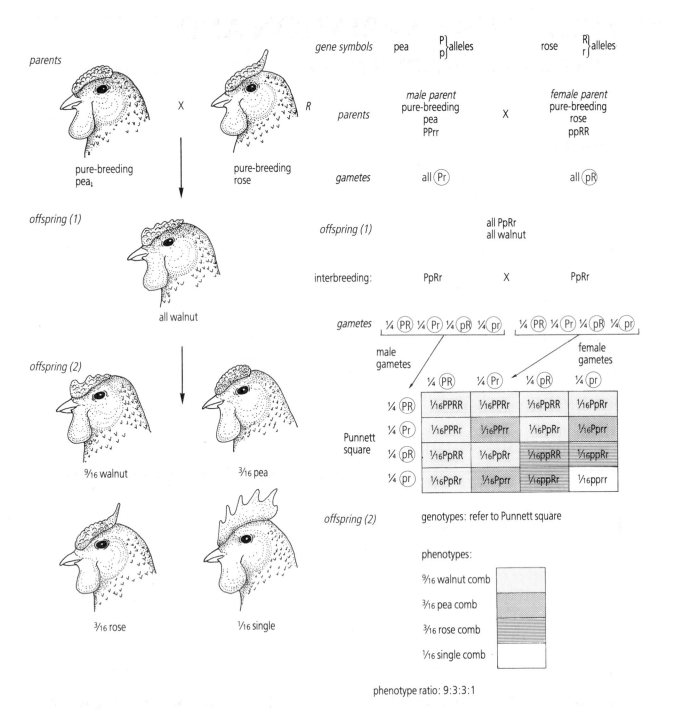

Fig 38·10 Inheritance of comb shape in poultry

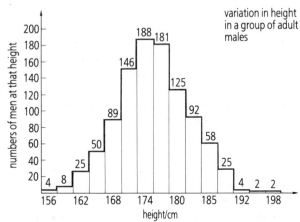

variation in height in a group of adult males

size of men in cm (based on heights of 1000 students aged 18–23 years)

Fig 38·11 Distribution of human heights

existence of the genes responsible can be demonstrated, but, in many cases, this is impossible. A further problem is that many normally-distributed characteristics are influenced by the environment, so that it becomes difficult to separate genetic from environmental effects. Nevertheless, polygenic inheritance plays an essential part in extending the range of variation, providing raw material for the action of evolutionary forces.

39 EVOLUTION, VARIATION AND NATURAL SELECTION

Objectives

After studying this Unit you should be able to:–

- Define evolution

- Explain what is meant by adaptation

- Comment on special creation and the inheritance of acquired characteristics as theories of evolution

- Give a simple account of the neo-Darwinian theory of evolution

- List the sources of genetic variation in natural populations

- Describe natural selection and industrial melanism in the moth *Biston betularia*

- Explain the Hardy-Weinberg method for calculating gene frequencies

- Show how the effects of selection on genotype frequencies from generation to generation can be calculated

- State the Hardy-Weinberg Law

- Give examples of transient polymorphism and balanced polymorphism

- Describe the effects of stabilizing selection, directional selection, and disruptive selection, and give examples of each type

39·1 Evolution and adaptation ■ ■ ■ ■ ■ ■ ■ ■ ■ ■ ■ ■

Biologists define evolution in two different ways:
1 Evolution is a process of gradual change whereby organisms of one species remain suited to, or **adapted** to, their environment.
2 Evolution is the process which results in the development of new species from previous forms. This is called **speciation** and is assumed to explain the appearance of new kinds of organisms during the history of life on Earth.

These types of evolutionary change are not alternatives but are complementary, and both must be explained by any satisfactory theory of evolution. This Unit outlines the **neo-Darwinian theory** and describes the main mechanisms which contribute to adaptation. Unit 40 considers the concept of biological species in detail.

The idea of adaptation can be illustrated with reference to almost any part of an organism's structure. For example, a bird's wing is suited to flight both in its general and detailed structure. The overall shape of the wing provides lift and can be altered appropriately for take-off or for level flight. The primary feathers are light and strong, while the smaller contour feathers give a smooth aerodynamic profile. In addition, the supporting bones have internal struts which improve their mechanical properties without increasing their mass. In most birds, the flight muscles are large and powerful, with the largest part of their bulk concentrated close to the base of the wings where the muscles are attached to a specially modified sternum. It is important to ask how all these features first developed and how they came to be shared by almost all birds.

1 Special creation. Many religions teach that species were created by God and have existed on the Earth unaltered from generation to generation. According to this view, an organism's adaptive features are inherited from its ancestors and were present fully formed at the moment of creation. The essential elements of the theory are not open to scientific investigation and cannot be tested. Other evolutionary theories accept that species characteristics do change and that all present day species are derived from a few very simple types of organism which first arose from non-living material more than 3000 million years ago.

2 Inheritance of acquired characteristics. In 1809, the French zoologist Jean-Baptiste Lamarck published his *Philosophie Zoologique* in which he proposed a mechanism to explain the evolution of more advanced species of organisms from simpler ones. He suggested: (1) that body organs can be developed and improved by repeated use but are weakened by disuse, and (2) that changes in structure acquired during an organism's lifetime are transmitted to its offspring. For example, Lamarck believed that the long neck and forelegs of giraffes arose from their habit of browsing from trees. He argued that the short-necked ancestors of giraffes must have succeeded in lengthening their necks by continual stretching so that, over many generations, they produced offspring with longer and longer necks.

In some respects this theory is an attractive one but there is no experimental evidence to support it. An organism's inherited characteristics are controlled by its genes and there is no known way in which the structure of the DNA forming the genes could be altered during its life to reflect an individual's activities.

3 Neo-Darwinian theory. A coherent and plausible theory of evolutionary change was first described in detail by Charles Darwin (1809– 1882). As a young man, Darwin was sent to study medicine at Edinburgh University and later to study divinity at Cambridge. He was unsuccessful as a student but developed an absorbing interest in natural history. In 1831, he obtained a position as naturalist on the survey ship HMS *Beagle*, and, over the next five years, he visited Patagonia, Tierra del Fuego, Chile, Peru, and a number of Pacific islands, including the Galapagos islands. His extensive studies of the geology and natural history of these areas convinced him that evolutionary change was possible, and later led to his proposal of **natural selection** as the main mechanism of evolution. His theory, published in the *Origin of Species*, 1859, triggered a revolution in human thinking and transformed the study of biology.

Its modern equivalent, called the **neo-Darwinian theory**, can be stated simply as follows:

(a) **Organisms produce far more offspring than can possibly survive**. An adult female frog, *Rana temporaria*, typically produces about 2000 eggs, many of which are successfully fertilized and develop into tadpoles. However, in most places, the population of adult frogs remains approximately constant. This means that the individual offspring have a very low average chance of survival, probably 0.01% or less. All sexual species produce surplus offspring most of which die before reaching sexual maturity.

(b) **Organisms differ in ways which affect their survival**. Any investigation of a species reveals numerous differences of size, colour, shape, detailed structure, and physiology, many of which must affect the survival of individual organisms. The sources of this variation must be considered.

(c) **Much variation is controlled by genes**. There are thousands of examples of variation with a known genetic basis. Many of these characters are inherited in a simple Mendelian fashion, while others, like height, or body mass, depend on interaction between genes, although the same features can also be affected by the environment (refer to Unit 38).

(d) **Natural selection keeps species adapted**. From the observations above it follows that the organisms which are best adapted are the ones which are most likely to survive and leave offspring in the next generation. Provided the variation is controlled by genes, these offspring are likely to share the favourable characteristics of their parents and will transmit their favourable genes in turn to subsequent generations. This process is called **natural selection** and will normally ensure that individuals of one species remain adapted to their environment.

(e) **New species can arise only by isolation**. Natural selection on its own is not enough to explain how new species are formed. This can happen only if populations of a species are separated so that they do not interbreed. The separated populations adapt to their own particular environments and may diverge, eventually forming new species. There are two possible methods of isolation, namely **geographical isolation** and **ecological isolation**. Their relative importance will be discussed in Unit 40.

39·3 Sources of variation ■ ■ ■ ■ ■ ■ ■ ■ ■ ■ ■ ■

For evolution to occur, there must be sources of variation on which natural selection can act. The most important sources are **gene mutation** and **chromosome mutation**. As outlined in Unit 37, gene mutations are chance alterations in the sequence of DNA coding for proteins. Most are damaging, but a few are beneficial, leading to the synthesis of new proteins or proteins with improved characteristics.

Chromosome mutation, described in Unit 38, is believed to have played a vital part in the evolution of advanced organisms. Simple organisms like bacteria contain far less DNA and far fewer genes than eukaryotic organisms. A human cell, for example, has 800 times more DNA than the bacterium *E. coli*. Increases in the quantity of DNA per cell seem to have come about by duplication and translocation affecting the chromosomes and possibly also as a result of viral infections. Surprisingly, only a small fraction of the total DNA in a eukaryotic cell codes for proteins. Some of the remaining non-coding regions are thought to make successful chromosome mutations more likely and therefore to increase the rate at which new proteins evolve.

Given a source of new genes and increased amounts of DNA, independent assortment of chromosomes and crossing over during meiosis help to produce genetic recombination, leading to the formation of genetically varied gametes and thereby maintaining variation among the offspring. Most species have adaptations for cross fertilization, which further increases genetic variability.

39·4 Natural selection in action ■ ■ ■ ■ ■ ■ ■ ■ ■ ■ ■

In the following example quoted from the *Origin of Species*, Darwin explains how insect pollinated flowers might have arisen from self-pollinated ones:

'It may be worth while to give another and more complex illustration of the action of natural selection. Certain plants excrete sweet juice, apparently for the sake of eliminating something injurious from the sap: this is effected, for instance, by glands at the base of the stipules in some Leguminosae, and at the backs of the leaves of the common laurel. This juice, though small in quantity, is greedily sought by insects; but their visits do not in any way benefit the plant. Now, let us suppose that the juice or nectar was excreted from the inside of the flowers of a certain number of plants of any species. Insects in seeking the nectar would get dusted with pollen, and would often transport it from one flower to another. The flowers of two distinct individuals of the same species would thus get crossed; and the act of crossing, as can be fully proved, gives rise to vigorous seedlings which consequently would have the best chance of flourishing and surviving. The plants which produced flowers with the largest glands or nectaries, excreting most nectar, would oftenest be crossed; and so in the long-run would gain the upper hand and form a local variety. The flowers, also, which had their stamens and pistils placed, in relation to the size and habits of the particular insects which visited them, so as to favour in any degree the transportal of the pollen would likewise be favoured.'

Arguments of this kind illustrate the explanatory power of natural selection but are inevitably weakened by a lack of experimental evidence. However, since Darwin's time, there have been many detailed studies of natural selection which confirm its importance as a mechanism of change.

The normal form of the moth *Biston betularia*, has speckled greyish white wings. The colour and pattern are a very effective camouflage against the lichen-covered branches and trunks of trees, where the moths usually rest during the day. In 1848, a single specimen of a previously unrecorded black, or **melanic**, form of the moth was captured near Manchester. The black form appears to have been better camouflaged against the soot-blackened trees in industrial areas and spread rapidly until, by 1895, 98% of the population of *Biston betularia* in the Manchester area were black. This phenomenon, called **industrial melanism**, affects several other species of moths and is now common in industrial areas and in rural areas affected by pollution. It has been investigated over many years by H.B.D. Kettlewell and his co-workers.

By breeding experiments, it was discovered that the melanic phenotype was controlled by a dominant mutant gene. To investigate the role of natural selection in the spread of the gene, it was necessary to identify the predators involved and to show that they were selective in taking the moths which were less well camouflaged. Kettlewell built up a large stock of moths and carried out a series of experiments, comparing results from unpolluted rural woodland (Dean End Wood, Dorset) with those from heavily polluted woodland (Rubery, Birmingham). In both locations, equal numbers of normal and melanic moths were released onto tree trunks where they could be observed from hides. Birds hunting by sight, including robins, song thrushes, and other insectivorous species, preyed on the moths. In the rural situation, the melanic moths appeared conspicuous to the human eye, while the normal moths were very difficult to see (Fig. 39·1A). The birds captured a total of 190 moths, of which 164 were melanic and only 26 normal, a striking difference giving the normal moths a clear survival advantage. On the other hand, in industrial woodland, it is the normal moths which appear conspicuous (Fig. 39·1B). Here, the survival advantage is reversed and the bird predators take a higher proportion of the normal moths.

In other experiments, Kettlewell released large numbers of males of both colour varieties marked on the underside with a dot of cellulose paint for identification. Surviving moths were recaptured on subsequent nights using mercury-vapour light traps and traps containing newly-emerged virgin females (which release a scent, or pheromone, strongly attractive to males). The proportions of moths recaptured are summarized in Table 39·1. As you can see, a larger proportion of moths were recaptured from the industrial site, possibly due to

Table 39·1 Recapture of normal and melanic male moths, Biston betularia, from unpolluted and polluted woodland

| | % recaptures | |
	Normal	Melanic
Dorset (unpolluted)	12·5	6·3
Birmingham (polluted)	13·1	27·5

a more efficient arrangement of traps. However, in both locations, approximately twice as many of the better camouflaged variety were recaptured, confirming their survival advantage.

All these results provide a convincing explanation based on natural selection for the observed frequencies of normal and melanic moths in the natural populations of these areas – Dorset: normal 94·6%, melanic 5·4%, Birmingham: normal 10·1%, melanic 89·9%. Interestingly, there is evidence that the numbers of melanic moths in industrial areas have declined slightly since legislation was introduced in 1956 and 1964 to restrict atmospheric pollution.

Fig 39·1 Industrial melanism in Biston betularia

39·5 Calculating allele frequencies ■ ■ ■ ■ ■ ■ ■ ■ ■ ■

In studying natural populations, it might seem difficult at first to estimate the relative proportions of the two alleles forming a pair. For example, in humans, the disease cystic fibrosis is caused by a recessive mutant allele and affects about 1 in every 2000 live-born children. If the normal form of the gene is referred to as **C**, and the mutant allele as **c**, then affected individuals must have the genotype

cc, but what are the proportions of **CC** and **Cc** genotypes in the remainder of the population? This question can be answered using a simple mathematical technique, known as the **Hardy-Weinberg** method.

To explain the technique, suppose that a gene **A** occurs in the population with a frequency p, while its recessive allele, **a**, occurs with the frequency q. The **allele frequencies** are calculated as fractions of the total number of alleles so that, provided **A** and **a** are the only alleles for this locus, $p + q = 1$. Assuming **A** was nine times more common that **a**, then $p = 0.9$ and $q = 0.1$. Now consider what happens when individual members of the population form gametes. Individuals with the genotype **AA** will form only **A** gametes, others with the genotype **Aa** will form both **A** and **a** gametes, while those with the genotype **aa** will form only **a** gametes. However, if **A** alleles are nine times as common in the population as a whole, then it follows that nine tenths of all the gametes produced will contain **A** and only one tenth **a**. Therefore, when fertilization occurs, there is a chance, or probability, of 0.9 that any individual offspring will receive an **A** allele from its father. Provided mating is random, there is an equal probability that the **A** allele will be received from the mother. The probability of the genotype **AA** is $0.9 \times 0.9 = 0.81$, that is $p \times p$ or p^2. By a similar argument, the probability of the genotype **aa** is $q \times q$ or q^2, while the probability of **Aa**, which can arise in two different ways according to which parent provides the **A** allele, is $2pq$. Like the allele frequencies, these **genotype frequencies** represent fractions of the total population so that $p^2 + 2pq + q^2 = 1$.

Returning to the example given at the beginning of this section, you can see that, for cystic fibrosis, $q^2 = 0.0005$, that is, $1/2000$, the proportion of **cc** genotypes. It follows that the allele frequency $q = \sqrt{(0.0005)}$, or approximately 0.022. Since $p + q = 1$, $p = 0.978$. The expected genotype frequencies can now be calculated:

frequency of **C** allele	p	$= 0.978$
frequency of **c** allele	q	$= 0.022$
probability of **CC** genotype	p^2	$= 0.9565$
probability of **Cc** genotype	$2pq$	$= 0.0430$
probability of **cc** genotype	q^2	$= 0.0005$

This application of the Hardy-Weinberg method shows that, although cystic fibrosis is a relatively rare disease, as many as 4.3% of the population must be carriers.

The Hardy-Weinberg method is also used to predict the effects of natural selection, as will now be explained:

Suppose the initial frequencies of two alleles **B** and **b** are $p = 0.010$ and $q = 0.990$, and that individuals with the dominant phenotype, genotypes **BB** and **Bb**, are at an advantage compared with those showing the recessive phenotype, genotype **bb**. (This is precisely the situation which would arise with the first appearance of melanic moths in an area of polluted woodland.) If it is further assumed that 0.2, or 20%, of the **BB** and **Bb** individuals survive to reproduce, as against 0.1, or 10%, of the **bb** individuals, then the change in allele frequencies is easily calculated:

frequency of **B** allele	$p = 0.010$
frequency of **b** allele	$q = 0.990$

genotypes in first generation before selection:

BB	**Bb**	**bb**
p^2	$2pq$	q^2
0.0001	0.0198	0.9801

proportion surviving to reproductive age:

0.20	0.20	0.10

genotypes in first generation after selection:

BB	**Bb**	**bb**
0.000 02	0.003 96	0.098 01

proportion of remaining **B** alleles:
$$2 \times 0.000\,02 + 0.003\,96 = 0.004\,00$$

proportion of remaining **b** alleles:
$$0.003\,96 + 2 \times 0.9801 = 0.199\,98$$

total remaining alleles:

$$2 \times 0.000\,02 + 2 \times 0.003\,96 + 2 \times 0.098\,01 = 0.203\,98$$

new allele frequencies after selection:

$$p = \frac{0.004\,00}{0.203\,98} = 0.020$$

$$q = \frac{0.199\,98}{0.203\,98} = 0.980$$

These are the allele frequencies before selection in the second generation and can be used to calculate the frequencies to be expected in the third and fourth generations and so on. This is exactly what was done to produce the graph given in Figure 39·2. As you can see, the frequency of **B** in the population rises slowly at first, then much more rapidly and finally slowly again, giving the curve a characteristic sigmoid, or S-shape. If selection against **bb** was continued indefinitely, the **b** allele would ultimately disappear from the population altogether.

It is important to appreciate that the change in allele frequency described above is *caused* by selection. If the same method is repeated without assuming selection in favour of one of the genotypes, then it is found that the allele frequencies do not change. This result leads to an important conclusion which is known as the **Hardy-Weinberg Law**. The Law states that **in a large random mating population, there will be no change in allele frequency from generation to generation except when there is selection, mutation, or migration**. In effect, Hardy-Weinberg calculations provide a mathematical proof that natural selection, mutation, and migration must be the main agents of evolutionary change.

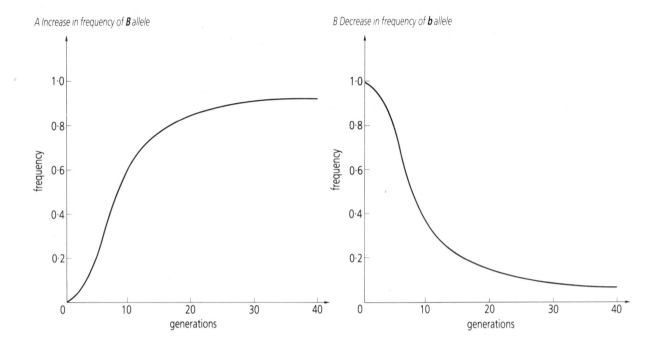

A Increase in frequency of **B** allele

B Decrease in frequency of **b** allele

Fig 39·2 Effects of selection in favour of a rare dominant allele (calculated using the Hardy–Weinberg method)

39·6 Transient and balanced polymorphism ■ ■ ■ ■ ■ ■ ■

Polymorphism is said to occur when individuals of one species in the same population exist in two or more distinct forms. These different forms, called **morphs**, must be phenotypic variations of the same character. For example, normal and melanic moths are alternative morphs for body colour. **Transient polymorphism** arises when there is strong selection pressure against one of the morphs so that it will eventually disappear from the population. In **balanced polymorphism**, on the other hand, selection acts to maintain the different morphs indefinitely.

The Hardy-Weinberg method can be applied to show that balanced polymorphism is likely to arise for characteristics controlled by a single pair of alleles whenever individuals with the heterozygous genotype are at a selective advantage compared with homozygous dominant and homozygous recessive individuals. An important example concerns the human disease **sickle cell anaemia** (described previously in Unit 18).

The disease is caused by a recessive mutant allele which codes for an abnormal haemoglobin,

haemoglobin S. Homozygous recessive individuals, **ss**, have red blood cells which contain only haemoglobin S – this becomes insoluble at low levels of dissolved oxygen, causing a change in shape (see Fig. 39·3). The deformed red cells block the body capillaries resulting in tissue

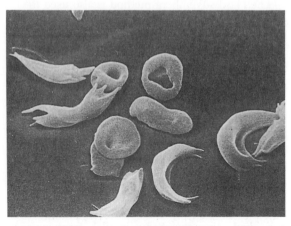

Fig 39·3 SEM to show sickled red blood cells in the blood of a person suffering from sickle cell anaemia

damage, so that affected individuals rarely survive to reproductive age. Heterozygous individuals, **Ss**, have red cells which contain both normal haemoglobin and haemoglobin S and are said to show **sickle cell trait**. Normally, their red cells are able to transport oxygen, but, under conditions of oxygen shortage, the red cells collapse into the sickle shape characteristic of the disease. A very interesting observation is that the distribution of the sickle cell allele confers resistance to the distribution of the malarial parasite, *Plasmodium falciparum* (Fig. 39·4), and it has been shown that the sickle cell gene confers resistance to the disease, possibly because infected red cells become sickled and are removed by phagocytosis before the parasites reach maturity. In areas where malaria is endemic, heterozygous individuals are at an advantage compared with homozygous, **SS**, individuals, so that the sickle cell allele is maintained at a high level in spite of its disadvantages in the homozygous condition.

Selection in favour of heterozygous individuals, called **heterozygous advantage**, is thought to be widespread in natural populations and is believed to account for a large proportion of the polymorphism which is observed.

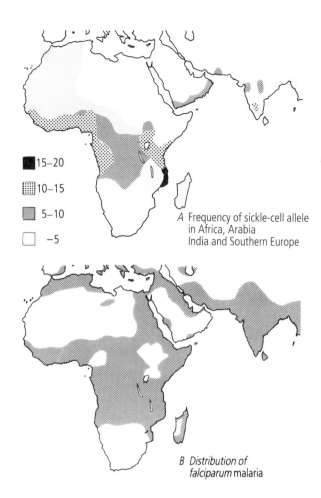

15–20

10–15

5–10

–5

A Frequency of sickle-cell allele in Africa, Arabia India and Southern Europe

B Distribution of falciparum malaria

Fig 39·4 Distribution of the allele for sickle cell anaemia and malaria caused by Plasmodium falciparum

39·7 Selection and polygenic inheritance ■ ■ ■ ■ ■ ■ ■ ■ ■

It has already been pointed out (Unit 38) that many of an organism's characteristics, like height or body mass, vary continuously rather than discontinuously, giving a normal distribution of phenotypes. Natural selection can affect such variation in three different ways (see Fig. 39·5).

1 Stabilizing selection. This occurs when there is selection against the extremes of the range (Fig. 39·5*A*). Many species of plants produce flowers at a particular time of the year triggered by changes in day length. Individuals which flower earlier or later than the average are often far apart and may fail to attract pollinating insects. Bees, for example, collect nectar selectively from flowers which are abundant. Therefore, individuals which flower at other times are less likely to be pollinated and will leave fewer offspring in the next generation. Stabilizing selection ensures that the majority of individuals of a particular species will continue to flower at approximately the same time.

Human infant mortality provides another example of stabilizing selection. The optimum body mass at birth is 3.6 kg. Babies which are heavier than this, as well as those which are

lighter, show an increased mortality. Thus, babies of average size are selected for, and babies at either extreme are selected against, although improvements in medical treatment make survival much more likely.

2 Directional selection. This results in a change in phenotype in one particular direction (Fig. 39·5*B*). Many predator-prey relationships are likely to have evolved in this way. The cheetah which is faster is more likely to be successful in killing prey, while the antelope which is faster is more likely to evade capture. Selection will always act in the direction of increased speed.

3 Disruptive selection. This involves selection against intermediate phenotypes in favour of the extremes (Fig. 39·5*C*) and appears to be less common, although it may play an important role in speciation, to be discussed in the next Unit.

A fascinating recent investigation of spawning behaviour in Pacific salmon, which breed only once in their lifetimes, shows that spawning males can be of two types. Most males are large and aggressive and fight vigorously amongst themselves for a position close to a female. However, in many populations, a proportion of much smaller males, called 'jacks', ascend the river and, capitalizing on their small size, sneak between the rocks of the river bed until they are close to a spawning female. When the female releases her eggs, these males release sperm and may be successful in fertilizing up to 40% of the offspring. Males of an intermediate size could not compete effectively with either of the existing types.

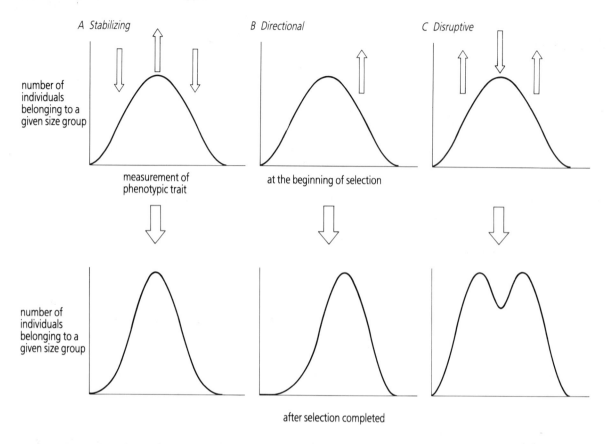

Fig 39·5 Selection and polygenic inheritance

40 ISOLATION AND THE ORIGIN OF SPECIES

Objectives

After studying this Unit you should be able to:–

- Define the terms 'gene pool' and 'species'

- Comment on the likely role of geographical isolation in the evolution of Darwin's finches and explain what is meant by adaptive radiation

- Describe in simple terms how continental drift can explain the present day distribution of placental and marsupial mammals

- Give examples of convergent evolution

- Distinguish between allopatric, parapatric, and sympatric speciation

- Describe these processes in detail and comment on their relative importance

40·1 What is a species? ■ ■ ■ ■ ■ ■ ■ ■ ■ ■ ■ ■ ■ ■ ■

It is easy to distinguish a cat from a dog or a mouse from an elephant, and simple to explain that they are members of different **species**, or kinds, of organism. Not all differences between species are equally obvious and, although species are normally distinct and separate in nature, it is surprisingly difficult to provide a completely satisfactory species definition.

Differences in structure and shape or colour are often used by field biologists to distinguish species but are not useful in developing a general species concept. A tadpole and a frog or a caterpillar and a butterfly would be recognised as individuals belonging to the same species, as would the brightly-coloured male pheasant and the drab female, but the visible differences involved are much larger than between some types of organisms which are classified as separate species.

Early attempts to control malaria by spraying insecticides were often unsuccessful because much of the effort was directed against the wrong mosquitoes. What was first identified as a single species, *Anopheles maculipennis*, was eventually separated into no fewer than six different species, only three of which transmitted malaria. The adults of these species are very difficult to distinguish, although there are differences between

their eggs and larvae, and in their breeding habits and behaviour.

A much more useful approach is based, not on structural features, but on the idea of a **reproductive community** sharing access to the same genes. Collectively, the genes available are referred to as the species **gene pool**. Two species definitions have been suggested:

1 A species is a group of actually or potentially interbreeding populations which are reproductively isolated from other such groups.

2 A species is the largest and most inclusive reproductive community of sexual and cross-fertilizing individuals which share in a common gene pool.

The test of whether two types of organism belong to the same species is whether they normally interbreed in nature and are capable of producing fertile offspring.

The mallard, *Anas platyrhynchos*, and the pintail, *Anas acuta*, are among the commonest fresh water ducks in the Northern Hemisphere. In captivity, these ducks mate and produce fully fertile offspring, with no loss of viability even in the second and third generations. In nature, such interbreeding is extremely rare, although the ducks frequently nest side by side. In contrast, many types of toad, *Bufo*, breed together, or **hybridize**, regularly but the offspring produced are sterile. In both these cases, the organisms involved do not meet the species definitions and are therefore classified as separate. On the other hand, melanic and normal *Biston betularia*, which look very different, interbreed freely to produce fertile offspring and are therefore grouped as one species.

Some difficulties arise with geographically separated populations. For example, European house sparrows, *Passer domesticus*, clearly do not interbreed with North American house sparrows, which were introduced there between 1852 and 1860. Nevertheless, the two populations are very similar and would not be thought of as separate species. When the time interval extends to thousands or tens of thousands of years, the situation is much less clear cut. The species

definitions do not apply at all to organisms like *Amoeba* which reproduce only by asexual methods. However, exclusively asexual organisms are very much the exception rather than the rule and even bacteria have some method of genetic exchange (Unit 45).

You should appreciate that, unlike other groups used for classifying organisms (Unit 44), species are real biological entities. The individuals which form a species share a relationship which is quite different in type from that between different genera or classes. The fact that no one definition of a species can be universally applied does not reflect a weakness in the species concept, but indicates that new species have evolved and are still evolving. The mechanisms required for this process of speciation form the subject matter of the remainder of this Unit.

40·2 Geographical isolation ■ ■ ■ ■ ■ ■ ■ ■ ■ ■ ■ ■ ■ ■

The islands of the Galapagos archipelago (Fig. 40·1), visited by Darwin in 1835, lie on the equator about 1000 km off the coast of Ecuador, and almost 5000 km from the nearest land to the west. The islands are of volcanic origin and are believed to have been thrust up out of the sea about 2 000 000 years ago, so that no direct connection with the South American mainland has ever existed. Indeed, the submarine platform on which the islands stand is separated from the continental shelf by a deep trench. Therefore, with the exception of recent introductions by man, the ancestors of all the animals and plants which inhabit the islands must have first arrived by accident.

How could this happen? In the case of terrestrial organisms, the most likely explanation stems from the fact that the islands lie in the path of the Humboldt current which flows from south to north along almost the entire coastline of South America. The continent contains huge regions of rainforest and it is not uncommon for mats of vegetation, trees, and other debris to be carried down rivers and out to sea. Perhaps, in the past, it could have happened that animals and plants were ferried from the mainland, and that some of them became the first colonists of these remote islands. This view is supported by the very restricted range of species found on the Galapagos, reflecting the rarity of successful ocean crossings, and by the nature of the groups represented. Amphibians, that is, frogs, newts, salamanders, and toads, are killed by contact with sea water and do not occur. Similarly, large mammals, which would be unlikely to survive a voyage of several weeks or months are completely absent. On the other hand, reptiles, which can survive without food and water for long periods, are common and have become the

dominant land animals. Birds, insects, and bats, which could have been carried by storms, are also present.

What fascinated Darwin was that, while all these organisms resembled the mainland forms in a general way, many of them appeared to have diverged, forming new species which were **endemic** to the islands, that is, species which were found nowhere else. For example, among the 700 species or so of flowering plants which have now been described, 250 are endemic. Endemic animals include the giant tortoise, *Geochelone elephantopus*, and two species of iguana lizard, the land iguana, *Conolophus subcristatus*, and the marine iguana, *Ambylrhynchus cristatus*.

Of particular interest are **Darwin's finches**, a group of small short-tailed birds with greyish brown or black plumage. Fourteen species are recognized (Fig. 40·2). These resemble one another in breeding behaviour and internal anatomy, but differ considerably in the shapes of their beaks and feeding habits. Darwin reasoned that these different forms must have developed by descent from a common ancestor. Following an initial successful colonization, the finches would have been likely to spread to neighbouring islands, building up a series of isolated populations, each of which would become adapted to its own particular environment. Freed of competition from mainland forms, individuals which began to exploit different food sources from the parental forms could have been at an advantage and are likely to have multiplied, forming, first, new local varieties, and later separate species.

Intensive research has tended to confirm Darwin's interpretation of this process, called **adaptive radiation**. A summary of the probable evolutionary history of the group is given in Figure

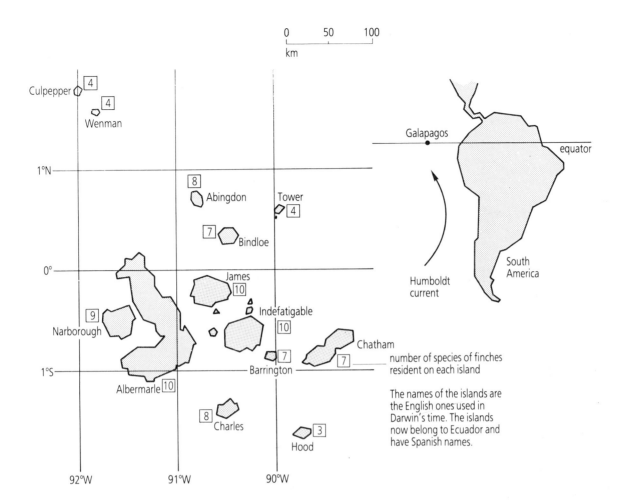

Fig 40·1 The Galapagos Archipelago

40·3. The ancestral type is believed to have been a seed-eating ground finch which gave rise to two separate lines, namely ground finches (genus *Geospiza*) inhabiting arid lowland and coastal regions, and tree finches (genus *Camarynchus*) inhabiting humid inland forests. Among present day ground finches, differences in beak size have been shown to allow the various species to cope effectively with different sizes of seeds. Among the tree finches, one species remains a seed eater, while the remainder are specialized for feeding on insects. The woodpecker finch, *Camarynchus pallidus*, resembles a European woodpecker in many of its habits, but, lacking the long tongue of a true woodpecker, uses a cactus spine to probe crevices in the bark, dropping the spine when an insect emerges.

The number of finch species occupying each island is given in Figure 40·1. As you can see, the larger islands, which offer a wider range of habitats, support more different species than the smaller more remote islands. An interesting observation is that remote islands, although having fewer species, have a higher proportion of endemic species. For example, of four species found on Culpepper and Wenman, three are unique to these islands, confirming the importance of geographical isolation in their evolution.

Darwin's findings from the Galapagos are far from unique. Many oceanic islands support endemic species: examples include Aldabra, Ascenscion, the Cape Verde Islands, the Falklands, Hawaii, Madagascar, Tristan da Cunha, the Seychelles, and many more. In every case, these islands are widely separated from the nearest land mass and have remained isolated for long periods. However, similar patterns of variation can be detected in island populations which are far less remote. Many of the islands surrounding Britain were formed as recently as 4000–10000 years ago as a result of a general rise in sea level after the last Ice Age. There are no endemic species but many islands have distinctive populations, particularly of species unlikely to breed with mainland forms. For example, there are distinct varieties of the long-tailed field mouse, *Apodemus sylvaticus*, on St Kilda, the Outer Hebrides, Shetland, and twelve other islands off the Scottish coast. The St Kilda wren is larger than the mainland wren and shows slight differences in colouration and song.

On a quite different scale, geographical isolation helps to explain the distribution of living organisms on the different continents of the earth. Geological evidence indicates that the great land masses of the earth are not fixed in position but are attached to

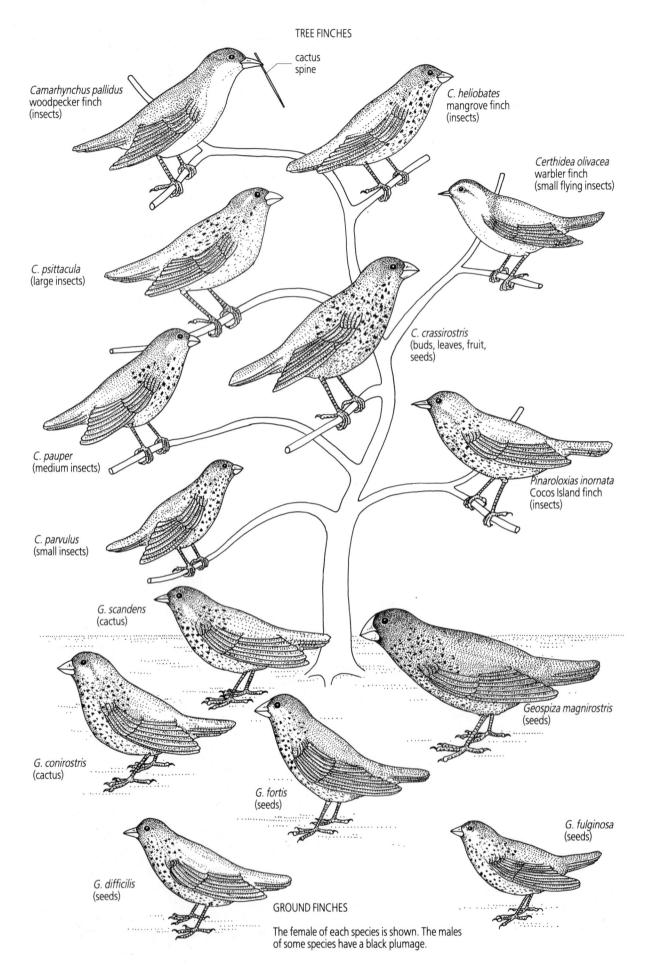

TREE FINCHES

Camarhynchus pallidus
woodpecker finch
(insects)

cactus
spine

C. heliobates
mangrove finch
(insects)

Certhidea olivacea
warbler finch
(small flying insects)

C. psittacula
(large insects)

C. crassirostris
(buds, leaves, fruit,
seeds)

C. pauper
(medium insects)

Pinaroloxias inornata
Cocos Island finch
(insects)

C. parvulus
(small insects)

G. scandens
(cactus)

G. conirostris
(cactus)

Geospiza magnirostris
(seeds)

G. fortis
(seeds)

G. fulginosa
(seeds)

G. difficilis
(seeds)

GROUND FINCHES

The female of each species is shown. The males
of some species have a black plumage.

Fig 40·2 Darwin's Finches

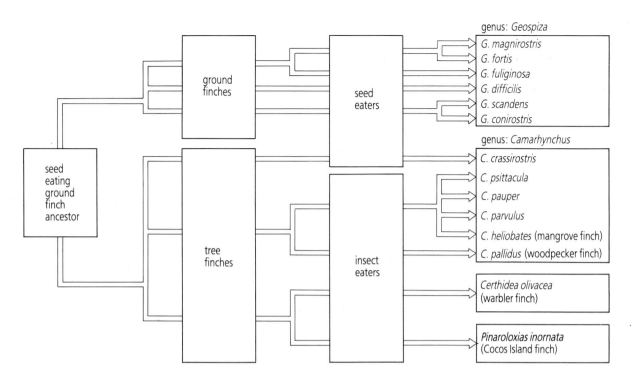

Fig 40·3 *Evolutionary history of Darwin's finches*

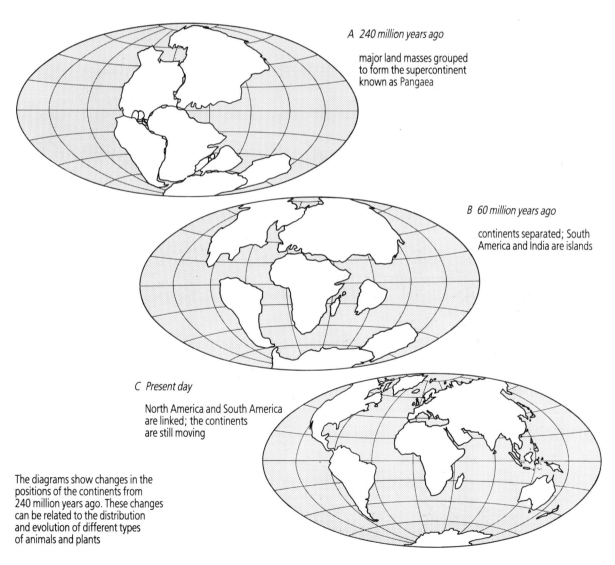

A *240 million years ago*

major land masses grouped to form the supercontinent known as Pangaea

B *60 million years ago*

continents separated; South America and India are islands

C *Present day*

North America and South America are linked; the continents are still moving

The diagrams show changes in the positions of the continents from 240 million years ago. These changes can be related to the distribution and evolution of different types of animals and plants

Fig 40·4 *Continental drift*

colossal plates, called **tectonic plates**, which move slowly but inexorably. The process is called **continental drift**. The positions of the continents have changed during the history of the Earth and are changing today (Fig. 40·4). The boundaries between the plates are marked by chains of volcanoes, and earthquakes and volcanic activity are concentrated in regions of slippage.

The importance of these movements for the theory of evolution is that the distribution of fossils and present day organisms can be precisely related to the positions of the continents in the past. For example, the earliest mammals appeared about 250 million years ago at a time when the major land masses had collided to form one supercontinent, known as **Pangaea**. (Fig. 40·4*A*). They were evidently successful and spread widely, so that, as the continents drifted apart, some populations were isolated from others. Pouched, or **marsupial mammals**, are believed to have developed first in North America, when this was still part of Pangaea, and to have spread to the areas which were to become South America, Australia, and New Zealand. In the north, the marsupials were later eclipsed by the more advanced **placental mammals**, but, in isolation, adaptive radiation produced a wide range of specialized organisms. Many of the South American forms became extinct in competition with placental mammals when a new land connection, the isthmus of Panama, was formed about 2 000 000 years ago, but in Australia, marsupials and primitive egg-laying mammals, called **monotremes**, still exist.

It is interesting that the specializations evolved by the marsupial mammals (see Fig. 40·5) are often very similar to those found in placental mammals. This is an example of **convergent evolution**, which is said to occur when the similarities between organisms are the result of natural selection and adaptation to the same ecological conditions and not directly the result of descent from a common ancestor. The teeth of kangaroos are modified for grinding in much the same way as those of placental herbivores, but, much more striking are the similarities in body form between the Tasmanian wolf, a marsupial carnivore, and its placental equivalents, or between the marsupial and placental moles. These similarities can be explained in the sense that particular structures are needed for a particular mode of life. On the other hand, different solutions to the same problem are often possible. Kangaroos rely on bipedal locomotion and a balancing tail to escape predators, while horses have evolved a completely different set of adaptations.

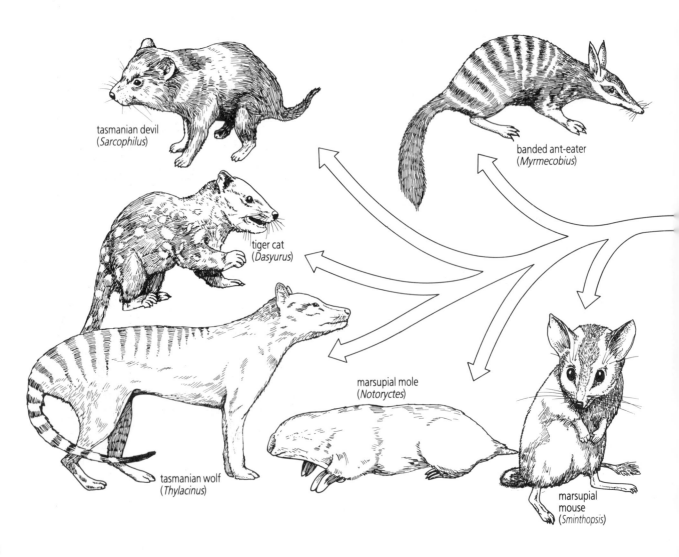

tasmanian devil
(*Sarcophilus*)

banded ant-eater
(*Myrmecobius*)

tiger cat
(*Dasyurus*)

tasmanian wolf
(*Thylacinus*)

marsupial mole
(*Notoryctes*)

marsupial
mouse
(*Sminthopsis*)

koala bear
(*Phascolarctos*)

flying phalanger
(*Petaurus*)

kangaroo
(*Macropus*)

wombat
(*Vombatus*)

opossum-like
ancestor

bandicoot
(*Perameles*)

Fig 40·5 *Adaptive radiation of marsupial mammals*

The two species definitions given earlier in this Unit describe species as reproductively isolated communities consisting of individuals which share in the same gene pool. The definitions apply to all the individual members of a single species and state what they have in common. In the context of evolutionary change, it is just as important to ask how species differ from one another. The answer is that, while two species may have access to some of the same genes, their total gene pools are never identical – each species has its own unique gene pool different from that of any other species. Thus, explaining the origin of new species becomes a matter of explaining the circumstances under which the gene pools of two populations can diverge. According to these circumstances, three types of speciation are distinguished, called **allopatric**, **parapatric**, and **sympatric** speciation.

Allopatric speciation

Allopatric speciation ('allo', different; 'patric', place) occurs between geographically isolated populations and depends on the following mechanisms.

1 Adaptation. No two geographically distinct regions will be exactly alike in climate or ecology, so that separated populations will be subject to different selective pressures. Through natural selection, each will become adapted to its own environment and may diverge from the parental type, especially if there is competition with other organisms. Beneficial mutations appearing in one area will not be available to individuals in the other.

2 Founder effect. The first colonists of an oceanic island, or the first individuals to spill over into a previously unoccupied area, are the **founders** of a new population. If relatively few individuals are involved, they may carry an unrepresentative selection of genes from the parent population. When this happens, the new population, descended from the founders, will have a divergent gene pool from the outset. This is called the **founder effect**. Natural selection may act to increase divergence, producing a distinct group of organisms in the new region.

3 Genetic drift. This is the name given to random changes in gene frequency in small populations. A rare gene in a colonizing population may be lost altogether, or may increase to become common, by chance alone, provided the change is not of selective value. Genetic drift can also occur in larger populations if mating between individuals is non-random. Some organisms, like snails, have a very limited mobility, so that each group of interbreeding individuals is confined to just one small part of the available habitat. The effective

population size is much smaller than the total number of individuals. Like the founder effect, genetic drift may help to produce divergence between geographically isolated populations.

For extreme instances of isolation, like the Galapagos islands, or the drifting apart of continents, it is easy to see how these mechanisms could lead to the development of new species. However, not all of the millions of different species of organisms which inhabit the earth can have been isolated to this extent. Does allopatric speciation explain the origin of most species, or that of only a few?

Part of the answer to this question lies in a more detailed consideration of the natural features which act as barriers to reproduction. Mountain ranges and deserts are obvious examples, but it is not so clear at first that many populations are isolated by their adaptations to a particular habitat. Plant species adapted for life at high altitudes cannot compete with lowland forms and may be effectively isolated from similar species elsewhere. Equally, organisms inhabiting different river valleys, forests, marshes, or grassland regions, may be unable to interbreed because the intervening habitat is unsuitable. In fact, this kind of geographical isolation is extremely common. Sometimes, isolated populations are only a few metres apart.

An interesting example of speciation resulting from geographical isolation on a small scale has been described in the case of the desert pupfish. These small fish live in hot springs in the desert surrounding Death Valley on the California-Nevada border. During the last Ice Age, aquatic connections existed between the various habitats, but these gradually dried up leaving isolated populations of fish. These have evolved into four separate species. One species, *Cyprinodon diabolis*, is entirely confined to Devil's Hole, a hot spring 50 m deep. The total number of individuals fluctuates between about 200 and 800.

In trying to assess the overall importance of geographical isolation, it is vital to appreciate that the present day distribution of species may be very different from their distribution in the past. The fact that closely related species are now found together in the same place does not necessarily indicate that they were formed there rather than in isolation. Significantly, the appearance of new species can often be linked to the changes in climate and conditions associated with repeated glaciations. The slow advance and retreat of the ice separated some populations from others, but, equally important, it sometimes removed natural barriers, allowing previously isolated populations to mix and bringing newly evolved species together. With a few exceptions to be discussed shortly, the evidence available indicates that most if not all species were formed as a result of natural selection acting on geographically isolated populations.

Parapatric speciation

Parapatric speciation ('para', next to; 'patric', place) occurs between populations in directly adjacent areas. It differs from allopatric speciation because the populations exchange genes freely.

Some indications of how parapatric speciation might occur come from studies of the wind pollinated grass, *Agrostis tenuis*, growing on the waste material from copper mines. Copper-tolerant individuals of this species are among the very few plants able to grow on mine soil, but they are less successful than normal *Agrostis tenuis* on pasture soil. Normal pasture individuals are killed by exposure to the levels of copper in mine soil. Crosses between tolerant and pasture individuals produce offspring which are less successful in both habitats. In these circumstances, any tendency which leads to more frequent crosses with the same type of plant will be favoured by natural selection. This is precisely what has happened at the Drws-y-Coed copper mine in Wales. *Agrostis tenuis* growing on the mine soil flower earlier in the season than pasture plants. Thus, reproductive isolation has developed without any actual barrier to interbreeding at the border between the populations. In time, the two populations would be expected to evolve into new species.

Sympatric speciation

Sympatric speciation ('sym', same; 'patric', place) is said to occur when organisms inhabiting the same area become separated into two or more reproductively isolated groups, without the need for geographical separation. There are two ways in which this might happen:

1 Polyploidy. This is the condition in which the cells of an organism contain additional sets of chromosomes, giving **triploid** or **tetraploid** individuals, or individuals with even higher multiples of the haploid chromosome number (refer to Unit 38). It has played a vital part in the evolution of flowering plants. The main form of polyploidy leading to speciation, called **allopolyploidy**, occurs when individuals of two

different but related species are accidentally hybridized. The offspring of such a cross may be vigorous plants, although they are usually sterile because their chromosomes are unable to pair up correctly at meiosis. However, new plants with double the hybrid chromosome number sometimes appear. Often, these are fully fertile and they form a new species distinct from either of the original parent types. Bread wheat, *Triticum aestivum*, is a hybrid polyploid resulting from a cross between a wild grass and an early cultivated variety. Less commonly, new types of plants may appear by **autopolyploidy** in which diploid gametes formed by individuals of the same species fuse to produce tetraploid offspring. Artificially induced autopolyploidy has been used to produce new and more vigorous varieties of crop plants such as sugar beet and tomatoes.

2 Ecological isolation. When a new adaptation arises in a few members of an interbreeding population, one of two things is likely to happen. Either the new feature will be generally favoured by natural selection so that it spreads, eventually to all individuals, or it will not be favoured and is likely to disappear. If divergence is to occur, the new adaptation must lead directly to reproductive isolation for the individuals which possess it. A possible example concerns insects feeding on different host plants. Many species are confined to just one type of host plant, although females occasionally lay eggs on different plants. Experiments have shown that females which develop successfully on a new host often return to lay their own eggs, preferring the new plant to the normal host species. The offspring produced are exposed to different conditions and, after several generations, they are likely to have acquired a range of adaptations to cope with the new host. Such individuals are said to be ecologically isolated from the rest of the population – they inhabit the same area but are separated by their feeding behaviour. Any tendency for them to breed with each other rather than with the parent species will be favoured, leading to further divergence. Similar processes could have taken place among parasites, many of which are host specific.

In these limited circumstances, sympatric speciation is possible. However, it is clear that geographical separation must have been by far the more important trigger for evolutionary change.

41 INTEGRITY OF SPECIES

Objectives

After studying this Unit you should be able to:–

- Define an isolating mechanism

- Distinguish between reproductive isolation and geographical isolation

- Describe the circumstances in which isolating mechanisms are likely to evolve

- Give examples of behavioural isolation

- Briefly explain what is meant by sexual selection

- Comment on the importance of other premating mechanisms

- Explain why postmating mechanisms are less likely to be improved by natural selection except in the case of prevention of fertilization

- Compile a summary table listing the different possible types of isolating mechanism

41·1 Isolating mechanisms ■ ■ ■ ■ ■ ■ ■ ■ ■ ■ ■ ■ ■

Units 39 and 40 were concerned with explaining the appearance of new species. It is less obvious, but equally important, to ask how existing species are kept separate and distinct from each other. Hundreds of different species live side by side in any habitat. Why is it that the individual members of these species normally breed with other members of the same species and not with members of a different species?

Investigations of natural communities show that species identity is almost always maintained by one or more **isolating mechanisms**. These are defined as the **biological properties of individuals which prevent successful interbreeding with members of different species.**

You must appreciate from the outset that isolating mechanisms are relevant only for species capable of inhabiting the same area. It is unfortunate that the term 'isolation' is used to refer to two quite different types of separation, namely **reproductive isolation** and **geographical isolation**. The important distinctions can be explained with the aid of Figure 41·1, which charts stages in the evolution of new species. As you can

used

see, a single original population divided by a natural barrier gives rise to two separated populations which are free to evolve differently. When the barrier is removed, Figure 41·1D, the outcome depends upon how far the two populations have diverged. Initially, interbreeding is very likely to occur. If the populations have remained similar and the hybrids are viable and successful as well as fertile, then the populations will fuse and accumulated differences between them will be lost. On the other hand, following a long period of separation, it is more probable that the differences will be considerable and that hybrid forms will be at a disadvantage compared with either parent population. In these circumstances, there will be strong selection in favour of any characteristic which reduces breeding between the populations, making it more likely that individuals will breed with their own kind. These characteristics will be rapidly improved by natural selection and will become the isolating mechanisms which keep the populations distinct. Reproductively isolated in this way, the two populations are now separate species each able to keep its own particular adaptations.

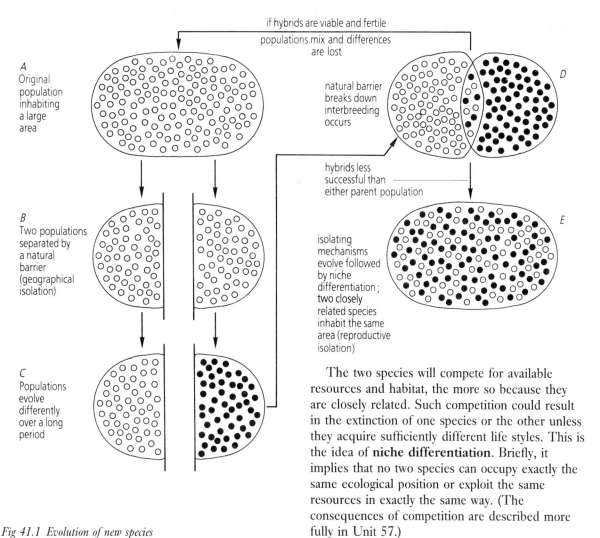

if hybrids are viable and fertile
populations mix and differences
are lost

A
Original population inhabiting a large area

natural barrier breaks down interbreeding occurs

D

B
Two populations separated by a natural barrier (geographical isolation)

hybrids less successful than either parent population

C
Populations evolve differently over a long period

isolating mechanisms evolve followed by niche differentiation; two closely related species inhabit the same area (reproductive isolation)

E

Fig 41.1 Evolution of new species

The two species will compete for available resources and habitat, the more so because they are closely related. Such competition could result in the extinction of one species or the other unless they acquire sufficiently different life styles. This is the idea of **niche differentiation**. Briefly, it implies that no two species can occupy exactly the same ecological position or exploit the same resources in exactly the same way. (The consequences of competition are described more fully in Unit 57.)

41·2 Premating mechanisms ■ ■ ■ ■ ■ ■ ■ ■ ■ ■ ■ ■ ■

A summary classification of isolating mechanisms is given in Table 41·1. Premating mechanisms are

Table 41·1 Classification of isolating mechanisms

PREMATING MECHANISMS

1 *Behavioural isolation:* courtship behaviour ensures mating with members of the same species.
2 *Seasonal isolation:* mating seasons do not overlap.
3 *Habitat isolation:* habitat preferences keep members of different species apart.
4 *Mechanical isolation:* the physical structure of the organism prevents successful transfer of gametes from individuals of a different species.

POSTMATING MECHANISMS

1 *Prevention of fertilization:* gamete transfer takes place but there is no fertilization.
2 *Hybrid inviability:* hybrid organisms die or survive poorly in competition with parental types.
3 *Hybrid infertility:* hybrid organisms may be vigorous but do not produce functional gametes.
4 *Hybrid breakdown:* first generation hybrids are viable and fertile but later generations are not.

those which act to prevent crosses between individuals from different species before the transfer of gametes. They are favoured by natural selection because they prevent wastage of gametes.

Behavioural isolation

A whole range of behavioural mechanisms have evolved which help to keep species separate.

Some organisms rely on chemical stimuli to bring the mating partners together. Newly-emerged female silkworm moths release minute quantities of a sexual attractant or sex **pheromone**. The secretion, called **bombykol**, from the name of the moth, *Bombyx mori*, is dispersed into the surrounding air. Male moths (Fig. 41·2) possess elaborate feathery antennae with sensory cells capable of detecting and responding to a single molecule of pheromone. Astonishingly small changes in concentration guide the male moth to the female. Similar chemical signals are used by many other female moths but each species produces its own distinctive odour

Fig 41·2 Antennae of a male silkworm moth

which only attracts males of the same species. Pheromones are so effective that caged virgin females can be used to estimate the population sizes of moths in different areas, as in the case of *Biston betularia* (Unit 39).

Auditory and visual signals can be equally important in maintaining reproductive isolation. In grasshoppers, crickets, and cicadas, females actively seek out males, which produce characteristic species-specific songs. Sound production occurs by **stridulation** in which one part of the body is rapidly vibrated against another. For example, crickets stridulate by rubbing together specially modified veins on their forewings. Bird songs serve similar functions in courtship but may also be involved in defending territory and social behaviour.

The importance of behaviour as an isolating mechanism is seen particularly clearly in the sexual displays of many birds. Figure 41·3 shows ten different poses which are used by male mallard ducks, *Anas platyrhynchos*, in their breeding displays. It is interesting that the same poses are used by male ducks of other species including teal, *Anas crecca*, and gadwall, *Anas strepera*. However, males of these species have their own distinctive brightly-coloured plumage and repeat the poses in quite different sequences, so that females normally respond only to males of the same species.

In many species, reproductive isolation depends partly upon **sexual dimorphism**, each sex developing its own special characteristics. This is the result of **sexual selection**, which is said to occur when natural selection acts differently on the two sexes. The peacock's tail and the beautiful but bizarre plumage of male birds of paradise are hardly likely to increase their individual chance of survival or of evading predators. However, selection will always favour those characteristics which enable individuals to leave the largest possible number of offspring in the next generation. The males of many species compete with other males for attention and those with a

Fig 41·3 Courtship poses of male mallard

lack-lustre appearance leave few if any descendents. For females the selective pressures are different: their chances of leaving offspring in the next generation are likely to be improved by good camouflage and effective maternal behaviour.

Seasonal and habitat isolation

These are not very common in animals but are widespread among plants. Closely related species often differ most noticeably in their flowering seasons, or in their requirements for soil type or climate. The sea campion, *Silene maritima*, grows on rocky beaches, while the bladder campion, *Silene vulgaris*, grows in fields and meadows. As a result, breeding between the species is rare, although the hybrids formed are fully fertile. Factors like these reduce the number of interspecific crosses, but often they are not the only isolating mechanisms used.

Mechanical isolation

When the detailed anatomy of insects is studied it is quite common to find that the biggest differences between species involve the structure of the external sex organs. This observation led to the 'lock and key' theory, according to which the sex organs of a male will fit correctly into those of a female of the same species, but not into those of another species. There is little direct evidence to support the theory and it seems that courtship displays are almost invariably the main isolating mechanism.

On the other hand, the mechanical structure of flowers is often adapted to take advantage of a particular species of pollinating insect. The white dead nettle, *Lamium album*, can only be pollinated by a long-tongued insect, such as a bee, and is therefore isolated from flowers pollinated by different insects. The behaviour of insects also helps to maintain reproductive isolation. For example, honey bees usually collect nectar from many flowers of the same shape, colour, or scent, ignoring other flowers, although these may be visited on a different day. Some flowers are pollinated by humming birds or bats, while others, like the bee orchids, resemble the females of various species of insect and are pollinated when males are deceived into attempting copulation. In every case, the structure of the flower favours transfer of pollen from another flower of the same species. The mechanisms of pollination are discussed in more detail in Unit 35.

41·3 Postmating mechanisms ■ ■ ■ ■ ■ ■ ■ ■ ■ ■ ■

Postmating isolating mechanisms reduce the success of interspecific crosses. Unlike premating mechanisms, they do not prevent wastage of gametes. With few exceptions, postmating mechanisms are likely to be difficult or impossible to improve by natural selection.

Prevention of fertilization

External fertilization is very common in aquatic organisms. In many species, behavioural, seasonal and habitat isolation help to prevent interspecific crosses, but it must sometimes happen that the eggs and sperm of different species are present together. Fertilization may still be prevented because the eggs do not produce the correct chemical signals to attract the sperm, or because the sperm fail to penetrate. A similar situation arises when the stigma of a flower receives pollen from a different species. Such pollen grains often fail to germinate, or grow only slowly.

At least in plants, the mechanisms which prevent fertilization must be capable of improvement by natural selection because they can prevent the production of non-viable seeds. Individuals which possess these characteristics must normally leave more offspring in the next generation.

Reduced hybrid success

In this context 'success' means the ability to produce a normal number of healthy fertile offspring. If the hybrids have a reduced viability or are infertile, they will leave fewer offspring. These mechanisms are difficult to improve by natural selection because the reproductive effort of the original parents is wasted. Indeed it is better to view reduced hybrid fertility and hybrid breakdown simply as the result of accumulated genetic differences between the species. Few species are kept separate only by such mechanisms. Premating mechanisms are strongly favoured by selection and it is very probable that they will develop in addition to postmating mechanisms if interbreeding between the species is at all widespread.

42 EVICENCE FOR EVOLUTION

Objectives

After studying this Unit you should be able to:–

- Distinguish between direct and indirect evidence for evolution

- Describe how fossils are formed and summarize the most important conclusions from the fossil record

- Outline the main adaptations observed in the evolution of the horse

- Explain with examples how each of the following provides evidence of shared ancestry among living organisms:
  - taxonomy
  - molecular biology
  - comparative anatomy
  - embryology
  - vestigial organs
  - biogeographical evidence
  - artificial selection

- Give examples of selfish, altruistic, and social behaviour

- Explain how advanced forms of social behaviour might have evolved

42·1 Introduction ■ ■ ■ ■ ■ ■ ■ ■ ■ ■ ■ ■ ■ ■ ■ ■ ■ ■

Evolution is the single most important concept in biological science because it provides a linking theme without which a vast number of observed facts would simply fail to make sense. Evolution is controversial not so much because the scientific evidence is in any doubt, but because some people do not like or refuse to accept its implications in a religious or philosophical context. This Unit presents a summary of the major lines of evidence in favour of evolutionary theory. You should consider for yourself whether any alternative explanation could be supported so effectively.

The distinctive fossils of ammonites are abundant in parts of Dorset. A successful group in the Mesozoic, they became extinct at the same time as the dinosaurs.

42·2 What is evidence? ■ ■ ■ ■ ■ ■ ■ ■ ■ ■ ■ ■ ■ ■ ■

Two main categories of evidence can be used in evaluating any hypothesis. The private detective of crime fiction investigating a murder would ask many questions: How was the murder committed? Why? Who are the likely suspects? And so on. The corpse and the murder weapon are the inescapable facts, the **direct evidence** that a crime has taken place. **Indirect evidence** would include observations relating to motive and opportunity.

The evaluation of a scientific hypothesis is of course much more rigorous, but both direct and indirect evidence remain essential. Equally, the

hypothesis must generate predictions which are testable and experiments which are carried out must not give conflicting results. The direct evidence for evolution comes mainly from the study of fossils. The indirect evidence is concerned with the mechanisms of evolution: How could it have taken place? And also with anticipated outcomes: If evolution has occurred, what relationships are to be expected among the organisms alive today? It is simply untrue to suggest that evolutionary ideas cannot be tested.

42·3 Fossils ■ ■ ■ ■ ■ ■ ■ ■ ■ ■ ■ ■ ■ ■ ■ ■ ■

Fossils are the remains of organisms found in rocks. Dead organisms usually rot and decompose, or are eaten by other organisms, with the result that the valuable chemical substances contained in their bodies are recycled and can be used again. On the other hand, it sometimes happens that a carcass is buried almost immediately by shifting sands or volcanic ash. Alternatively, dead organisms may be trapped in swamps or marshes, or sink to the bottom of lakes or oceans. In these circumstances, decay is likely to be prevented by the absence of oxygen. Later, accumulated sediments compress the remains and the original hard parts are slowly replaced by inorganic salts deposited from water seeping through. This process is called **petrification**. After many millions of years such fossils may be exposed on the surface by geological movements and erosion.

Occasionally, organisms are preserved in other ways, as in the case of insects trapped in tree resin fossilised to form **amber**. Mammoth remains discovered in Siberia were so well preserved in ice that their meat was eaten by dogs. **Trace fossils** include footprints and fossilized droppings, or **coprolites**.

The most important conclusions from the fossil record are:

1 The simplest organisms appeared first and survived for hundreds of millions of years before the appearance of more advanced forms.
2 When fossils are arranged in chronological order, clear patterns of evolutionary descent normally emerge. Changes in structure can be interpreted as adaptations to new conditions.
3 Extinction is a frequent event so that only a very few fossil forms are represented among the organisms living today.

A simplified evolutionary history of horses is shown in Figure 42·1. Each branch of the tree

represents a different genus of horses, but only the more important ones are named. The illustrations show reconstructions of fossil forms together with details of limb and tooth structure. The ancestral horse, *Hyracotherium*, was a relatively small animal about 0·3 m tall at the shoulder. It was a woodland creature and probably ate soft vegetation and fruit. The changes observed, including increased size, development of high-crowned molar teeth, and loss of limb bones, can all be related to a change in habitat from woodland browsing to grazing in open country. Alterations in limb structure contributed to increased speed, important for escape from predators. Horses which retained the earlier adaptations eventually became extinct through failure to compete successfully with other herbivores.

The evolution of horses is an important example because many of the intermediate stages in development appear in the fossil record. Also important is the fact that the transitions from one limb type to another occurred fairly quickly and were separated by long periods with little change. Many similar instances of accelerated development in other fossil groups suggest that evolution is not always a gradual process, but quite often occurs as a series of 'jumps', in much the same way as a new technological innovation, such as electric lighting, quickly replaces older less efficient technology.

The occurrence of rapid transitions helps to explain why intermediate forms are sometimes 'missing' from the fossil record. In some cases, the fossils of intermediate forms have been found. For example, the suspected evolutionary link between birds and reptiles was positively established with the discovery of *Archaeopteryx*, a fossil form showing both avian and reptilian characteristics (see Fig. 42·2). The existence of such 'missing links' is strong direct evidence for evolution.

369

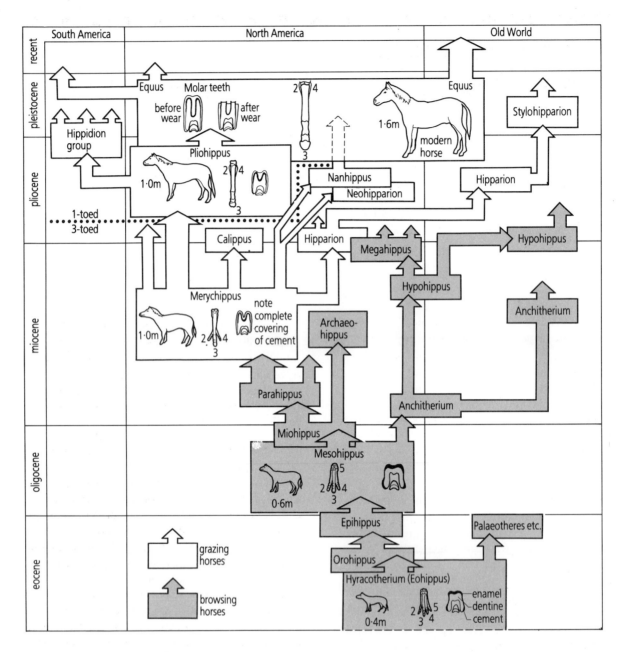

Fig 42·1 *Evolution of the horse*

Fig 42·2 *Fossil of the primitive bird* Archaeopteryx

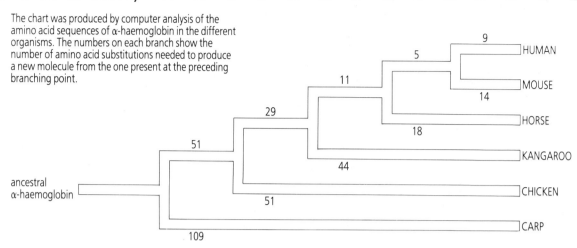

The chart was produced by computer analysis of the amino acid sequences of α-haemoglobin in the different organisms. The numbers on each branch show the number of amino acid substitutions needed to produce a new molecule from the one present at the preceding branching point.

Fig 42·3 Probable evolutionary history of α-haemoglobin

This section summarizes the evidence indicating shared ancestry among living organisms. The evidence considered is indirect in the sense that evolution is not observed at first hand. However, if modern organisms were to possess any features which failed to conform with evolutionary principles, or could not be explained by processes of mutation and natural selection, then the theory could be definitely rejected.

Taxonomy

Taxonomy is the science of classification. Although the details of classification schemes vary, it is a remarkable fact that the great variety of living organisms *can* be arranged in orderly groups on the basis of shared features. Most biologists believe that such groupings reflect real evolutionary relationships. For example, all insects possess a body divided into three sections, the head, thorax, and abdomen, a tracheal system, three pairs of jointed legs, one pair of antennae, and many other aspects of detailed structure, physiology, and behaviour. It is difficult to avoid the conclusion that all these characteristics are shared by virtue of descent from a common ancestor. Unit 44 considers the methods of taxonomy and develops this argument at greater length.

Molecular biology

All organisms use nucleic acids, DNA and RNA, as coding devices for the manufacture of proteins. The genetic code which controls protein synthesis (see Unit 9) is almost universal. In a similar way, virtually all organisms are capable of releasing energy from glucose by the process of glycolysis. However, these mechanisms are not the only ones possible, nor even necessarily the best. Why should they be preserved? It seems probable that certain modes of metabolism were arrived at by chance

among the first living organisms, and remained fixed thereafter. For example, once a particular version of the genetic code became established, it is difficult to see how substantially different versions could evolve. Even a single change would be likely to cause numerous mistakes in protein assembly. Very few deviant codes have been discovered, notably in mitochondria and in *Paramecium*, but none differs by more than two or three codons from the 'universal' code.

The development of rapid techniques for analysing the amino acid sequences of proteins gives a new opportunity for investigating evolution. Figure 42·3 shows a scheme describing the probable history of the α-haemoglobin molecules of several vertebrate species. This was produced by computer analysis of sequence information. The numbers on the branches of the tree are the minimum number of amino acid substitutions required to produce a new molecule from the one at the previous branching point, each substitution being the result of a single gene mutation, that is, a change affecting just one base pair in the DNA sequence coding for the protein. The important point is that the relationships revealed are precisely those expected from taxonomic and fossil evidence. Studies of other proteins, such as cytochromes, have given essentially the same results with other organisms.

Comparative anatomy

It is a feature of evolution that the same structure is often adapted and developed for a number of distinct functions. The basic components of the vertebrate skeleton are common to thousands of different species, but the sizes and shapes of the individual bones can vary enormously. For as some instance, some of the **visceral arches** supporting the gills of primitive fish act as jaws in bony fish, or as ear ossicles in mammals. Others survive as cartilage supports for the tongue and larynx.

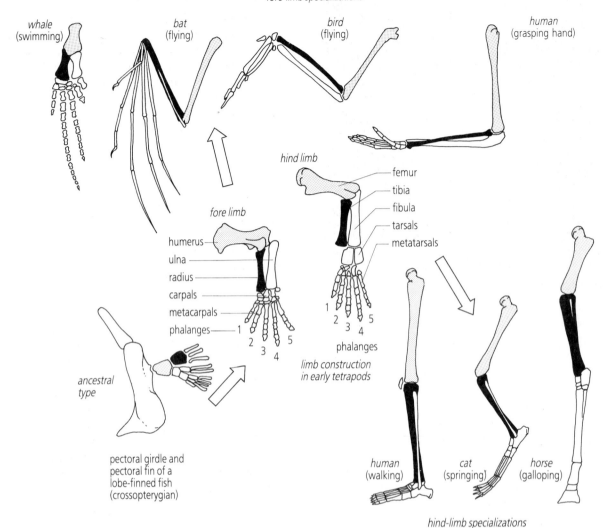

fore-limb specializations

whale (swimming)   *bat (flying)*   *bird (flying)*   *human (grasping hand)*

hind limb

femur
tibia
fibula
tarsals
metatarsals

fore limb

humerus
ulna
radius
carpals
metacarpals
phalanges — 1
2
3 4 5

1
2 3 5
4
phalanges

limb construction in early tetrapods

ancestral type

pectoral girdle and pectoral fin of a lobe-finned fish (crossopterygian)

human (walking)   *cat (springing)*   *horse (galloping)*

hind-limb specializations

Fig 42·4 Specializations of vertebrate limbs

Figure 42·4 shows various specializations of vertebrate limbs. The first land vertebrates are thought to have evolved from a group of lobe-finned, or **crossopterygian** fish, which possess pelvic and pectoral fins with a similar arrangement of skeletal elements. The pattern of limb construction in early terrestrial amphibians and reptiles can be followed in all later forms, including birds and mammals. A very wide range of different adaptations occur, but the same structures are evident throughout.

A fascinating variation in the five-fingered, or **pentadactyl**, pattern is found in the giant panda. Pandas are related to bears, but, unlike bears, pandas have adopted a herbivorous diet consisting entirely of bamboo. They eat only the bamboo shoots and strip off the leaves using an apparently flexible thumb. However, the panda's 'thumb' is in addition to its five normal digits and is developed from the **radial sesamoid** bone, normally a small component of the wrist. There are many similar examples of strange compromises in structure: these are easily explained by mutation and natural selection.

Embryology

The embryonic development of vertebrates is extremely uniform. For example, all early embryos have branchial grooves and internal gill pouches. In fish, these meet to form gill slits and later become an important part of the respiratory system. On the other hand, in terrestrial vertebrates, the grooves and pouches almost disappear, so that the only traces which remain in an adult mammal are the Eustachian tube and the auditory canal. Human embryos also pass through stages in which they possess a two-chambered heart and a tail. The differences between vertebrate species are best interpreted as variations on a theme. Certain mechanisms are inherited from a common ancestor: these can be modified but not scrapped completely.

Vestigial organs

Vestigial organs are present both in embryo and in adult organisms. They have no known function, although they seem to have had a function in the

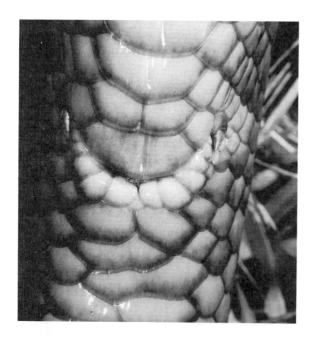

Fig 42·5 *Vestigial leg bones are visible as external spurs in this python.*

Biogeographical evidence

The distribution of living organisms has already been referred to in Unit 40 as important supporting evidence for evolution. Briefly, the land masses of the earth can be divided into several main regions, each with its own characteristic life forms. The differences between regions cannot be explained by climate alone but are the result of continental drift, which isolated primitive ancestral populations.

Artificial selection

Deliberate selection of organisms with certain characteristics has been the main method used by plant and animal breeders for centuries, and has transformed the appearance of many domestic species. Modern cereal varieties yield several times the amount of grain produced by the wild grasses from which they are descended, and livestock yields are also tremendously improved. The huge diversity of breeds of domestic dogs from Chihuahuas to Great Danes illustrates the extent of variation which can be produced. In evolutionary terms, these changes are astonishingly rapid, typically on a time scale of a few tens or hundreds of years, instead of thousands or millions of years. Artificial selection provides a useful model for natural selection, indicating that the key mechanisms of evolution can indeed produce dramatic changes in form.

past. The **coccyx** of the human skeleton is a useless tail remnant, while the **appendix**, which corresponds to the caecum of herbivorous animals, is too small to have any useful role in digestion. Vestigial leg bones are found in snakes (Fig. 42·5), but make no contribution to locomotion, which depends exclusively on movements of the rib cage. Such structures seem to represent an evolutionary link with previous species. Ostrich wings and the partial development of an egg tooth in the embryos of marsupial mammals are additional examples.

42·5 Selfishness, altruism and social behaviour ■ ■ ■ ■ ■▭▭

If the theory of evolution is to be accepted then it must be possible to explain the origins of *all* the observed characteristics of organisms. A single exception, which could not be properly accounted for, would require the theory to be modified or rejected. The final section of this Unit considers how animal behaviour is interpreted in evolutionary terms. Some forms of behaviour are easy to explain, but others appear, at least superficially, to contradict the normal view of evolution by natural selection.

Selfish acts are a common feature of animal behaviour. Many sea birds, like blackheaded gulls, breed in large colonies, with nests only a metre or so apart. Although the colony provides protection from predators of other species, newly-hatched chicks are quite liable to be eaten by adult birds from nearby nests. Similarly, a male lion which is successful in taking over a pride will often kill any young lions, which are the offspring of the previous dominant male. Such acts are likely to be favoured by natural selection because they increase the reproductive potential of the individual, allowing him to leave more offspring.

Much less obviously adaptive are behavioural acts which involve individual sacrifice. Many small birds give characteristic alarm calls when a predator such as a hawk first appears. Such an action warns other birds of danger but must at least slightly reduce the individual's chances of escape. Similarly, adult male baboons take up exposed positions on the outside of their troop and will threaten approaching predators or rival troops. Worker honeybees possess barbed stings which may remain embedded in the skin of their victims. Stinging a predator provides effective protection for the hive, but the bee itself is often fatally injured. These examples illustrate **altruism**, which is defined in biological terms as **behaviour which reduces an individual's chance of survival while at the same time increasing the probability of survival for a different individual**.

It is very important to ask how altruism could evolve because at first sight it seems to conflict with the fundamental principles of natural selection. The answer to this puzzle lies in the concept of **kin selection**, that is, selection in

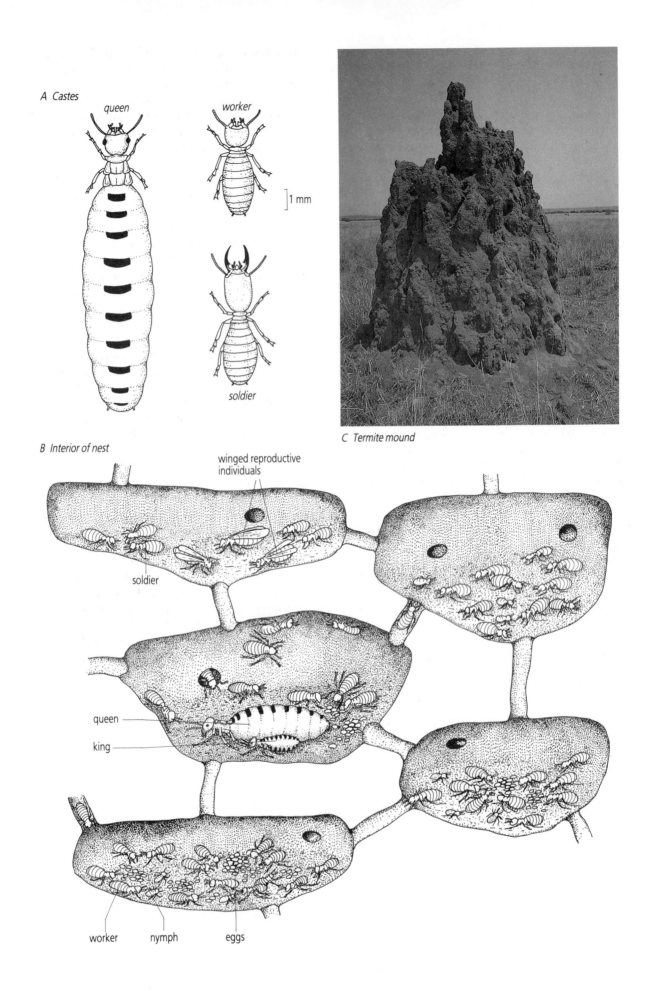

A Castes

queen

worker

] 1 mm

soldier

C Termite mound

B Interior of nest

winged reproductive
individuals

soldier

queen

king

worker   nymph   eggs

Fig 42·6 Nest and castes of the termite Amitermes hastatus

favour of blood relatives. For instance, in the context of parental care, it is easy to see that a gene which contributes to more effective parental care will be likely to increase in frequency, even at the expense of reducing individual viability. In precisely the same way, alarm calls and the protective behaviour of baboons will be favoured by selection provided the organisms which benefit normally include the offspring of the individual producing the call.

More dramatic examples of altruistic behaviour are seen in the social behaviour of insects. The termite, *Amitermes hastatus* (Fig. 42·6), has a well developed social system in which reproduction is normally limited to a single male and female, the primary **king** and **queen**, often the original founders of the colony. Brood care, food gathering, and the defence of the colony are the function of separate **castes** of **workers** and **soldiers**. The behaviour of these individuals is entirely altruistic: all of their lives are spent in supporting the reproductive effort of the king and queen. Indeed, the development of reproductive organs in workers and soldiers is prevented so that they are unable to produce their own offspring. At intervals, some of the larvae from eggs laid by the queen develop into winged male and female reproductive forms which erupt from the colony, mate, and fly off to found new colonies.

It is now known that the cooperative effort involved in social behaviour is possible because the individuals which comprise a colony are invariably members of an extended family, and therefore have many of their genes in common. The reproductive success of the colony as a whole ensures that many copies of these genes will be passed on to the next generation. Each individual which is part of the colony therefore contributes to the survival of many of its own genes whether or not these are represented in its own offspring. If selfish individuals appear which reproduce on their own account, without the support of the colony, they tend to leave fewer copies of their genes. In other words, in the special circumstances of social behaviour, the genes which control altruistic behaviour can sometimes be favoured by selection, while genes for selfish behaviour are selected against.

Other social insects include ants, wasps, and bees. Ants are enormously successful and, in tropical ecosystems, they replace earthworms as the main organisms processing soil and leaf litter. It is no coincidence that all these insects belong to the same group, the Order Hymenoptera, so that they share a rather unusual method of sex determination. Males are haploid and are produced from unfertilized eggs, while females are diploid and develop from fertilized eggs. An important consequence is that the sterile female workers in any colony each have about three-quarters of their genes in common with their mother, the queen. The thresholds which must be passed for the cooperative success of the colony to exceed the potential individual success are correspondingly lower than in termites, where individuals share on average only about half their genes with their parents. This helps to explain why social organization has evolved independently several times among the hymenoptera, but only once in the termites. The social behaviour of honeybees is described in Unit 28.

In recent years there has been a tendency to apply the discoveries of 'sociobiology' rather uncritically to human behaviour, for example to 'explain' differences in the social roles of men and women. However, the evolutionary study of social behaviour gives only limited insights because so much of human experience depends on cultural factors accumulated over thousands of years of civilisation.

Section C Questions ■ ■ ■ ■ ■ ■ ■ ■ ■ ■ ■ ■ ■

Short-answers and interpretation

1 a Make large labelled drawings to show one pair of chromosomes at
  i metaphase of mitosis, [4]
  ii metaphase I of meiosis, [4]
  iii metaphase II of meiosis. [4]
  b How is sex genetically determined in man? [8]
  c In humans, a gene responsible for clotting blood is carried on the **X** chromosome. People who carry only the recessive gene bleed easily and are called haemophiliacs. The diagram below shows the occurrence of haemophiliacs in a certain family.

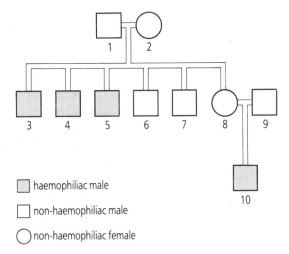

☒ haemophiliac male

☐ non-haemophiliac male

◯ non-haemophiliac female

Indicate the genotype for each individual numbered in the diagram. [10]
University of London School Examinations Board

2 Erminette fowls have mostly light-coloured feathers with an occasional black one, giving a flecked appearance. A cross of two erminettes produced a total of 48 progeny consisting of 22 erminettes, 14 blacks and 12 pure whites.
  a From these results, suggest a genetic basis for the erminette pattern. [4]
  b How would you test your hypothesis? [2]
Cambridge University Local Examinations Syndicate

3 Wild type *Drosophila melanogaster* (a species of fruit fly) has long straight wings which lie over the thorax and abdomen. One **mutant** has wings of normal length curling up over the thorax; another **mutant** has only vestigial wings too small to be either straight or curled. The **locus** for this wing-shape gene and the **locus** for this wing-size gene are **not linked**. The **allele** for curled wing is **recessive** to the **allele** for straight wing and the **allele** for vestigial wing is **recessive** to the **allele** for long wing.

a Explain the meanings of the genetic terms in bold type in the passage above: **mutant, locus, not linked, allele** and **recessive**. [5]
b A female *D. melanogaster* with curled wings was crossed with a male with vestigial wings. All the offspring had long straight wings.
  i Why is it reasonable to infer that the female parent in this cross was homozygous for the alleles for straight wing? [1]
  ii Give a genetic diagram of this cross. [4]
  iii When two of the offspring (1) were mated with each other, the resulting phenotypes in the offspring (2) generation were:

> 82 long straight wings
> 27 curled wings
> 37 vestigial wings

Give a genetic diagram of this cross and comment on the relationship between the observed and the 'expected' results. [5]
c Suppose the genes for wing shape and wing size had been close to one another on a chromosome.
  i What would have happened to the offspring (1) phenotype in **b**? [1]
  ii How would the ratio of phenotypes in **b iii** have been affected. Explain your answer. [4]
Associated Examining Board

4 In poultry, comb shape is controlled by two pairs of alleles. **P** represents the dominant allele for a pea-shaped comb and **R** represents the dominant allele for a rose-shaped comb. These alleles are carried on different chromosomes. A cross between a homozygous pea-combed fowl and a homozygous rose-combed fowl produced an offspring (1) generation in which all birds had walnut combs. In the offspring (2) generation, four types of combs appeared in the ratios: 9 walnut: 3 pea: 3 rose: 1 single.

Indicate the type of comb in birds
  a with the genotype **PPRR**
  b with the genotype **pprr**
  c with the genotype **Pprr**
  d resulting from a cross **PPRR** × **pprr**
  e resulting from a cross **PPrr** × **pprr** [5]
Welsh Joint Education Committee

5 A certain species of flowering plant shows variation in two characters: height of the stem and the distribution of chlorophyll in the leaves. The plants may be tall or short, whilst chlorophyll distribution may be normal, variegated (partial) or absent.

In an experiment, plants with tall stems and green leaves which were known to be pure breeding for both characters were crossed with plants with short stems and variegated leaves. The offspring (1) generation from this cross consisted of plants with tall stems and green leaves and plants with tall stems and variegated leaves in the ratio of 1:1.

a i Construct a genetic diagram to show this cross, and the offspring (1) plants produced. [3]

ii Explain in words the result of this cross. [3]

b The offspring (1) plants with tall stems and variegated leaves were self-pollinated and 650 seeds were sown. The offspring (2) plants were observed and counted six weeks after germination. The following results were obtained:

Plants with tall stems and green leaves   124
Plants with tall stems and variegated leaves   251
Plants with short stems and green leaves   40
Plants with short stems and variegated leaves   82

i Construct a genetic diagram to show the results of the offspring (2) self-pollination. Relate the genotypes of the offspring (2) shown to the phenotypes produced. [6]

ii Explain in words as fully as possible the results obtained in this offspring (2) cross. [4]

c Suggest ways by which a gardener could successfully maintain a supply of plants of the variety with variegated leaves. [4]

University of London School Examinations Board

6 Figure 39.1 shows two varieties of the peppered moth *Biston betularia* in natural surroundings.

a *Biston betularia* is said to show polymorphism. What do you understand by this term? [2]

b In a mark, release and recapture experiment carried out in an industrial area, the following results were obtained:

variety	number marked and released	number recaptured
melanic	154	82
non-melanic	64	16

i Calculate the percentage of each variety recaptured. [2]

ii Suggest three possible reasons for the significant difference between the recapture figures for the two varieties. [3]

c A similar experiment was carried out in a rural area and produced the following results:

variety	number marked and released	number recaptured
melanic	473	30
non-melanic	496	62

i Calculate the percentage of each variety recaptured. [2]

ii Comment on the difference between the results from the industrial and rural areas. [2]

d How can the concept of natural selection be supported by these results? [3]

University of London School Examinations Board

7 The diagrams below show the heads of four of the ten species of finch inhabiting a volcanic island in the Galapagos group, some 600 miles from the mainland of Ecuador, South America.

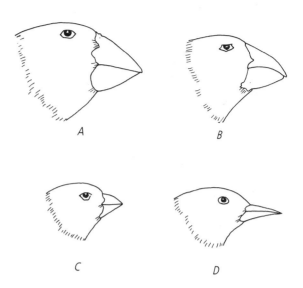

a What major difference between the four species is shown in the diagram? [1]

b How might the difference you have given in **a** be related to the way of life of the finches? [2]

c These finches probably descended from a common ancestral stock of finches. Suggest how these ancestors may have reached the island. [2]

d What evidence to support theories of how evolution occurs may be deduced from the information provided by these finches? [3]

e There are more species of finch than other birds on the island. Suggest an explanation for this. [2]

f Why is the diversity of finch types on the South American mainland less than on the island? [2]

g The different finch species on the island do not interbreed. What might this indicate? [2]

h All the species of finch on the island are dull-coloured. Give two ways in which recognition might occur between sexes of the same species of finch. [2]

University of London School Examinations Board

8 a What is a fossil? [2]

b Indicate two ways in which fossils may be formed.

c Briefly describe a method used for estimating the age of a fossil. [2]

d Describe an example of fossil evidence which has helped establish the possible evolutionary development of a group of organisms. [4]

e Suggest two possible reasons for gaps in the fossil record. [4]

f Describe briefly one further type of supporting evidence for evolution other than fossils, giving a named example. [4]

University of London School Examinations Board

377

Essays

1 a Give an illustrated account of meiosis. [10]
b What does meiosis have in common with mitosis and in what way does it differ? [4]
c Discuss the significance of the differences. [6]
Oxford and Cambridge Schools Examination Board

2 Describe the principles of genetics formulated by Mendel. Relate these principles to chromosomes and their behaviour during nuclear division and fertilization. [20]
University of London School Examinations Board

3 a With the aid of an example, distinguish between the terms phenotype and genotype. [4]
b Write brief notes on each of the following (where appropriate using specific examples).
   i sex-linked inheritance,
   ii polygenic inheritance,
   iii sex determination. [12]
c Show how non-disjunction during spermatogenesis could result in offspring suffering from Klinefelter's syndrome or Turner's syndrome. [4]
Cambridge University Local Examinations Syndicate

4 a Explain how you would carry out and record the results of a dihybrid cross (to obtain offspring (1) and offspring (2) generations) in a named organism, emphasizing the reasons for the various procedures. [5]
b Consider the offspring (1) generation of a dihybrid cross. Explain how the results in the offspring (2) are dependent on the behaviour of chromosomes during meiosis in the offspring (1). [10]
c Typically, a 9:3:3:1 phenotypic ratio is obtained in the offspring (2) of a dihybrid cross. What effect does i linkage and ii co-dominance have on this ratio. Explain your answer. [5]
Joint Matriculation Board

5 a Describe the main features of the synthetic (neo-Darwinian) theory of evolution. [9]
b Discuss the evidence from the geographical distribution of organisms that supports this theory. [9]
Cambridge University Local Examinations Syndicate

6 a Mutation is a source of genetic variation in a population. How does such variation arise, how is it maintained within the population, and what is its importance? [14]
b Asexual reproduction reduces the level of genetic variation in a population. Outline the mechanisms by which asexual reproduction is achieved and under what circumstances might a lack of genetic variation be advantageous? [6]
Joint Matriculation Board

7 a What is a species? [3]
b Describe how new species may arise. Illustrate your answer with named examples. [11]
c Explain why closely-related species may be unable to interbreed successfully. [4]
Cambridge University Local Examinations Syndicate

8 Explain how genetic variations, natural selection and isolation mechanisms may be involved in the process of evolution. [20]
University of London School Examinations Board

9 'Genes mutate, organisms are selected and groups evolve.' Discuss. [20]
Associated Examining Board

10 Show how evidence from two of the following topics has been used to provide support for the theory of evolution of organisms.
   a distribution of animals and plants,
   b comparative anatomy,
   c palaeontology. [9,9]
Cambridge University Local Examinations Syndicate

Section D

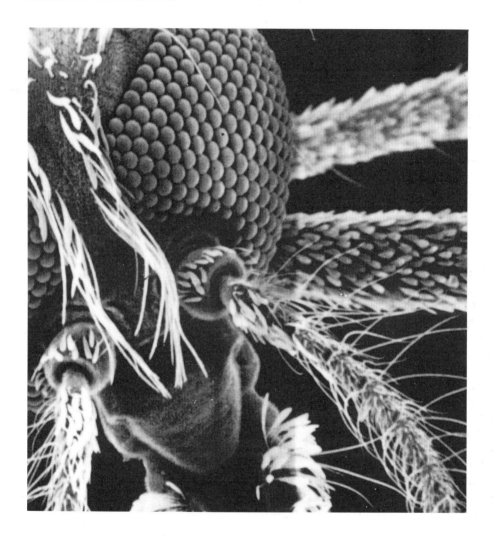

ORIGIN AND
DIVERSITY OF LIFE

43 ORIGIN OF LIFE

Objectives

After studying this Unit you should be able to:–

- List the gases present in the Earth's atmosphere before life evolved

- Describe the experiment carried out by Stanley Miller to demonstrate the synthesis of organic molecules from inorganic substances

- Explain what is meant by coacervate droplets

- Explain the importance of glycolysis for the first living organisms

- Briefly describe the evolution of photosynthesis

- State two reasons why atmospheric oxygen was damaging to early life forms

- Draw and label diagrams to illustrate the structure of prokaryotic and eukaryotic cells and list their characteristic features

- Briefly describe the symbiotic theory of the origin of eukaryotes

- Outline a method for dating rocks

- Comment on the importance of microfossils

43·1 Formation of the Earth ■ ■ ■ ■ ■ ■ ■ ■ ■ ■ ■ ■ ■ ■

This Unit describes one of the most probable sequences of events from the earliest beginnings of life on Earth to the present day. It is not intended to answer all the possible objections or to explore alternatives, but simply to present a summary along with some of the more important pieces of supporting evidence.

The Earth was formed between 4500 and 5000 million years ago from a cloud of dust particles surrounding the Sun. As the mass of the planet increased, heat generated by gravitational compression and radioactive decay caused the interior to melt, producing a dense molten iron and nickel **core** (Fig. 43·1). This was surrounded by a cooler liquid **mantle** of iron and magnesium silicates. The outermost layer, or **crust**, consisting of lighter silicate compounds, solidified to form the continents and the sea floor. As the cooling process continued, gases from the hot interior escaped through volcanoes, forming an **atmosphere** consisting of hydrogen, water vapour, methane, ammonia, nitrogen, and hydrogen sulphide. This mixture of gases had strong chemical **reducing** properties and was quite unlike the atmosphere today. In particular, oxygen was missing. At about the same time, large quantities of water vapour condensed to form the oceans.

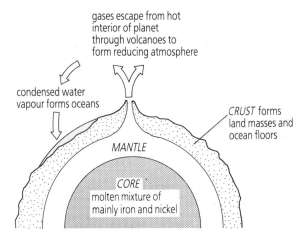

gases escape from hot interior of planet through volcanoes to form reducing atmosphere

condensed water vapour forms oceans

CRUST forms land masses and ocean floors

MANTLE

CORE molten mixture of mainly iron and nickel

Fig 43·1 Structure of the planet Earth

43·2 Synthesis of organic chemicals ■ ■ ■ ■ ■ ■ ■ ■ ■ ■ ■

Many of the chemicals needed as subunits for making biologically useful molecules are thought to have been formed in the shallow waters of the oceans as the products of chemical reactions between simple inorganic substances. The probable nature of these reactions was first demonstrated by an American scientist, Stanley Miller. In 1953, he set up the experiment illustrated in Figure 43·2, intending to recreate conditions on Earth before life evolved. The essential components of his model system were a mixture of gases simulating the primitive atmosphere, water, and an external energy source. A number of possible natural energy sources existed, such as ultra-violet light, lightning, and radiation. In Miller's apparatus, gases and water vapour were exposed to spark discharges to represent lightning. Following this, the water was condensed, then boiled and recirculated. After a week, the solution in the apparatus was found to contain a variety of organic compounds including the amino acids glycine and alanine, lactic acid, and urea.

In more recent experiments, other amino acids, sugars including ribose and deoxyribose, purines, pyrimidines, and nucleotides have all been produced. In the laboratory, one of the easiest molecules to make is the purine base adenine. This combines with ribose and phosphate

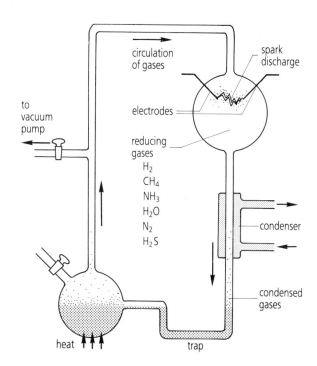

Fig 43·2 Stanley Miller's apparatus

compounds under the influence of ultra-violet light to form ATP. In the absence of oxygen, all of these compounds accumulated in the waters of the primitive planet, forming a mixture sometimes called the **'primordial soup'**.

43·3 Organization into chemical systems ■ ■ ■ ■ ■ ■ ■ ■

Over millions of years, the complexity of organic compounds in the 'soup' gradually increased. In warm concentrated solutions, these substances are able to join together, or **polymerize**, to form long chain molecules. The process can occur spontaneously, but takes place more rapidly in the presence of a catalyst. It is probable that chains of amino acids, polysaccharides, and long chains of nucleotides, called **polynucleotides**, were all produced in shallow rock pools, particularly where organic compounds had accumulated by adsorption onto the surfaces of clay particles. The polynucleotide molecules had the capacity to produce new similar molecules by replication (described in Unit 8).

A parallel development was the formation of large aggregations of molecules called **coacervate droplets**. The potential importance of these structures was first recognised by the Russian scientist, Oparin, who investigated the properties of artificial droplet systems. Droplets made by

dissolving gelatin protein and a polysaccharide material, called gum arabic, in water showed many similarities to living cells. For example, they acquired a membrane-like lipid coating, capable of absorbing a range of different substances from the surrounding solution. Droplets containing enzymes could take up substrate molecules and release products back into the external solution. In later experiments, photosynthetic systems were produced by making droplets which contained chlorophyll. Using light energy, these tiny chemical systems were able to absorb an oxidized dye and return it in a reduced form back to the surrounding solution. Stable droplet systems tended to 'grow' by absorbing chemicals from those which were less stable. The composition of coacervate droplets formed in the oceans before life evolved would have been quite different but they may have shared some of the properties of Oparin's model systems.

43·4 The earliest living cells ■ ■ ■ ■ ■ ■ ■ ■ ■ ■ ■ ■ ■ ■

The first living cells are thought to have arisen from coacervate droplets which contained polynucleotides, and in particular molecules of RNA. In time, some RNA molecules became able to direct the assembly of chains of amino acids and so formed proteins. Precisely how this occurred is not yet fully understood but, once the template system for protein synthesis was available, droplet systems could carry out the major processes characteristic of living cells. With the manufacture of enzymes and structural proteins the transition to living organisms became complete. The enzymes were used in new synthetic pathways leading to the production of vital substances previously obtained by absorption, while the ability to make structural proteins permitted growth and asexual reproduction by a fission process – the parent 'cell' splitting to form two or more smaller daughter cells.

All these activities required increasing quantities of ATP. However, supplies of ATP in the surrounding water were limited, so that many primitive organisms must have run out of fuel and broken up. Those which survived possessed a synthetic pathway for the manufacture of ATP. The most important pathway of this kind, called **glycolysis**, is common to almost every organism alive today. As described in Unit 10, it involves the breakdown of glucose to provide energy for the production of ATP. Supplies of glucose in the surroundings outlasted supplies of ATP, so that organisms capable of glycolysis gained a considerable advantage. Thus, the first true organisms were heterotrophic and anaerobic. Raw materials were taken in through the outer membrane and new substances synthesized using energy derived from glucose.

The nucleic acids inside these first cells were organized into coding systems for protein synthesis, but they were not enclosed within a nuclear membrane. Such cells are described as **prokaryotic**. Present day prokaryotes include bacteria and blue-green algae, which are described in detail in Unit 45. Other modern cells possess a true nucleus enclosed within a pair of membranes, forming the nuclear envelope: these cells are described as **eukaryotic**.

43·5 Evolution of photosynthesis ■ ■ ■ ■ ■ ■ ■ ■ ■ ■ ■

Oxygen in the atmosphere was first produced by the action of ultra-violet light on water vapour. Subsequently, many organic molecules were oxidized and the natural supply of organic substances diminished. Atmospheric oxygen was also damaging because it reacted with water to form hydrogen peroxide (H_2O_2). This substance attacks RNA and DNA and is capable of destroying a cell's genetic instructions. Consequently, the accumulation of oxygen in the atmosphere favoured organisms which were able to manufacture the enzyme **catalase**. This enzyme protects against the effects of hydrogen peroxide by converting it back to oxygen and water.

At about the same time, a group of chemicals called **porphyrins** became important. The molecules of these substances contain atoms of iron or magnesium which can be harmlessly oxidized, thereby acting as a defensive buffer against oxygen. A major breakthrough in evolution occurred when certain organisms began to use a particular porphyrin molecule, **chlorophyll a**, to absorb light energy and harness it for the synthesis of ATP. In this way, photosynthesis began. Many of the photosynthetic organisms required only simple inorganic substances and energy from the Sun to manufacture all their own body chemicals, that is, they were 'self-feeding', or **autotrophic**. Their success ensured the survival of the remaining heterotrophic organisms, which were now able to use a new and inexhaustible supply of food.

43·6 Accumulation of oxygen in the atmosphere ■ ■ ■ ■ ■ ■

In addition to light energy, photosynthesis requires a source of hydrogen ions (H^+) and electrons (e^-) which can be combined with carbon dioxide to make carbohydrates. Some of the early organisms used hydrogen sulphide (H_2S) as a source – much more importantly, a second group developed the ability to use water. The process of **photolysis**, or 'light-splitting', of water, summarised in the equation below, provided hydrogen ions and electrons, and, at the same time, produced oxygen gas as a waste product:

$$H_2O \rightarrow 2H^+ + 2e^- + \tfrac{1}{2}(O_2)$$

The release of oxygen by photosynthetic organisms and its rapid accumulation in the atmosphere had an irreversible effect on the subsequent evolution of life. The remaining organic chemicals in the 'primordial soup' were broken down into carbon dioxide and oxidized sediments, and, secondly, a layer of **ozone**, that is, oxygen with the molecular formula O_3 rather than O_2, began to appear in the upper atmosphere. This acted as a barrier to the penetration of ultra-violet light and prevented the production of new organic chemicals. Consequently, conditions changed so that they were no longer suitable for the development of living organisms from non-living material.

43·7 Evolution of eukaryotic cells ■ ■ ■ ■ ■ ■ ■ ■ ■ ■ ■

The differences in structure between prokaryotic and eukaryotic cells, summarized in Table 43·1, are so striking that it is not immediately clear how eukaryotes could have evolved from prokaryotes. Nevertheless, several different lines of evidence indicate that an evolutionary link exists between them. Eukaryotes have a more complex structure and appear much later in the fossil record but, despite the structural differences, the two types of cell share numerous vital biochemical pathways, notably glycolysis and the light reactions of photosynthesis.

The **symbiotic theory** suggests that important cell organelles, like mitochondria, chloroplasts, cilia, flagella, and centrioles were at one time independent prokaryotic organisms which became engulfed by larger cells and continued to survive inside their cytoplasm, eventually becoming essential cell components.

Supporting evidence for this theory is not hard to find: mitochondria resemble aerobic bacteria in size and structure, while chloroplasts are very similar to certain kinds of prokaryotic photosynthetic organisms. Mitochondria and chloroplasts both possess their own DNA which is in the form of a circular 'loop' like that of

Table 43·1 Prokaryotic and Eukaryotic Cells

PROKARYOTES	EUKARYOTES
Structure	*Structure*

Features	*Features*
1. Small calls (1–10 μm)	1. Large cells (10–100 μm)
2. DNA not asociated with histones and usually in the form of a loop	2. DNA divided into lengths (chromosomes) and combined with histone protein to form chromatin
3. Nuclear material not enclosed by membranes	3. True nucleus bounded by membranes (nuclear envelope)
4. No membrane-bound cell organelles, no centrioles, no microtubules	4. Mitochondria, chloroplasts, centrioles, endoplasmic reticulum and microtubules all present
5. Small ribosomes (70 S)*	5. Large ribosomes (80 S)*
6. Cell walls present but chemically different from eukaryotes	6. Cell walls absent except in plant cells (cellulose) and fungal cells
7. Simple bacterial flagella composed of flagellin protein	7. Complex 9+2 cilia and flagella consisting of tubulin and other proteins
8. Cell division by simple fission, no spindle	8. Cell division by mitosis, spindle formed
9. Sexual systems rare; genetic material passes from donor to recipient	9. Sexual reproduction by meiosis and fertilization common
10. Very varied metabolic pathways – unusual energy sources exploited, many anaerobic forms	10. Many shared metabolic pathways – glycolysis, Krebs cycle, almost all aerobic

*S – Svedberg unit see index-glossary

prokaryotes. Even in human cells, these organelles possess their own ribosomes which are smaller and quite distinct from those in the remainder of the cytoplasm. They make some of their own proteins and reproduce independently of the host cell.

Some antibiotics inhibit the growth of prokaryotic cells by preventing protein synthesis. These same antibiotics prevent protein synthesis by ribosomes inside mitochondria and chloroplasts, but leave cytoplasmic ribosomes unaffected. Conversely, some drugs inhibit protein synthesis in the cytoplasm of eukaryotic cells, but allow it to continue within the cell's organelles.

Both mitochondria and chloroplasts have a double membrane – the inner membrane contains electron transport proteins corresponding to those of prokaryotes. The outer membrane resembles the membranes of the host cell in structure and composition and can be thought of as equivalent to an engulfing vacuole.

It is assumed that cilia, flagella, and centrioles arose from associations between some types of bacteria and other cells.

Whatever the details of their evolution, the appearance of eukaryotes greatly increased the potential for further development. In due course, true multicellular organisms evolved with complex tissue and organ systems. The range of diversity of these forms is the main subject of the remaining Units in this Section.

43·8 The geological time chart ■ ■ ■ ■ ■ ■ ■ ■ ■ ■ ■ ■ ■ ■

Figure 43·3 charts the history of living organisms from their earliest beginnings, with time divisions corresponding to the main rock strata found in the Earth's crust. Sedimentary rocks were originally formed from accumulated sand and mud at the bottom of lakes and oceans. In many areas of the World, geological movements have lifted and tilted whole sections of the Earth's crust so that the thickness of the different layers can be measured. An impressive example is the Grand Canyon in Arizona, where erosion by the Colorado River has exposed sedimentary rocks millions of years old.

The age of rocks can be estimated by measuring their depth below the original surface, but this is relatively inaccurate for it assumes a constant rate of sedimentation. Furthermore, it does not allow for the process of erosion that must have occurred when the rock face was exposed. A much more accurate method of dating makes use of the fact that a radioactive isotope of uranium, ^{238}U, which is usually present, decays into lead at a very slow, but known, rate. With appropriate techniques, the amount of this decay can be estimated and the age of the rock determined to within about 10%. Other isotopes, such as those of thorium or rubidium, are sometimes used. (The radioactive isotope of carbon, ^{14}C, is used to date many historical relics, but is of no use in dating rocks because its half life is less than 6000 years.)

Organisms can be located on the geological time scale only if remains of them can be found in the rocks. The fossils of invertebrates with hard shells or exoskeletons are found in sedimentary rocks up to 600 million years old and the first vertebrate fossils appear in rocks about 500 million years old (refer to Fig. 43·3). Traditionally, geologists recognized the Cambrian period at the beginning of the Palaeozoic era as one of the most important markers in the study of evolution, simply because no obvious fossils could be found in older rocks.

More recently, the 'Precambrian' history of organisms has been explored using new techniques of fossil hunting. Very thin sections of rocks have been cut to reveal microscopic fossils, or **microfossils**, resembling bacteria, in flint-like black chert rocks as old as 3100 million years. Figure 43·4 shows examples of prokaryotic organisms from the Gunflint Chert formation, Ontario, estimated to be 1900 million years old. Such fossils reveal outlines of the structure of primitive organisms and can indicate whether particular molecular structures were present. However, many details of the early history of life are still uncertain.

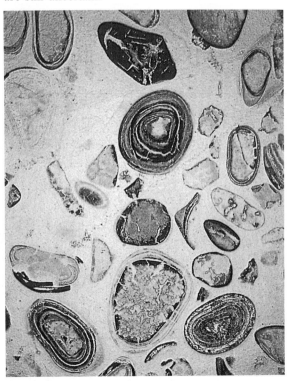

Fig 43·4 Microfossils from the Gunflint Chert formation, Ontario

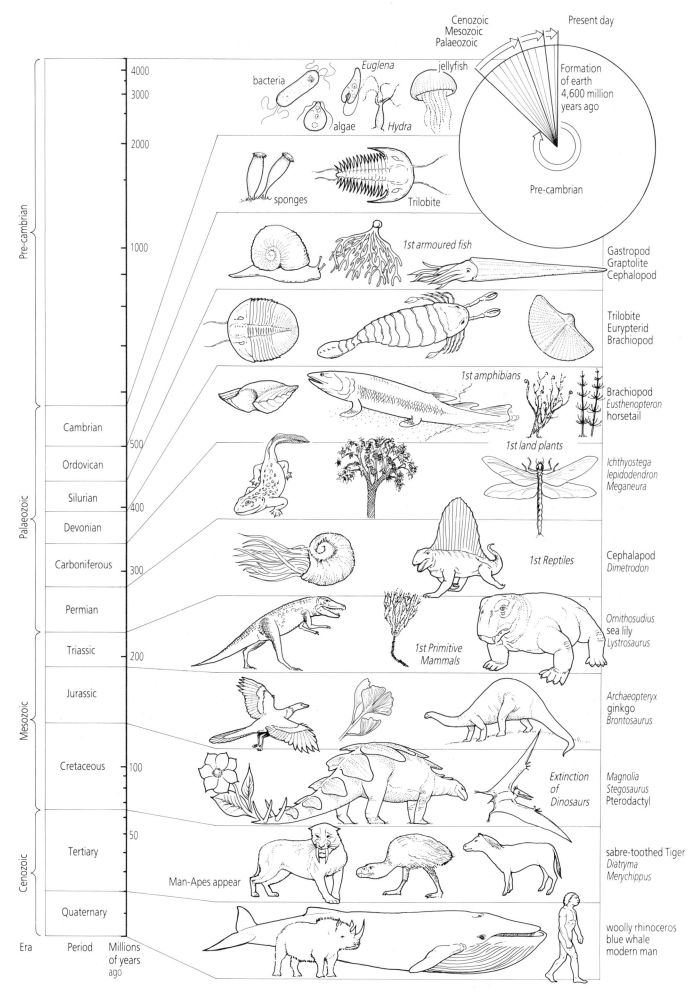

Fig 43.3 *Geological eras and the history of life on Earth*

44 CLASSIFICATION OF LIVING ORGANISMS

Objectives

After studying this Unit you should be able to:–

- Explain what is meant by phylogenetic classification

- Construct and use simple identification keys

- Distinguish between homologous and analogous structures

- Outline the Five-Kingdom system of classification and state the distinguishing features of organisms belonging to each Kingdom

- Draw a diagram summarizing the evolutionary relationships between these Kingdoms

- Name the major classification groups in order of size

- Explain the binomial system devised by Linnaeus

44·1 Introducing classification ■ ■ ■ ■ ■ ■ ■ ■ ■ ■ ■ ■ ■

The study of classification is called **systematics**. This is essential in biology to understand the complex variety of form observed in living organisms. Any group used for classification purposes is called a **taxon** (plural **taxa**). **Taxonomy** is the process of ordering these groups into a larger scheme. Although there are many possible strategies for classifying the several millions of different organisms so far identified, in practice some are more useful than others. The best taxonomic schemes are based on natural relationships brought about by common ancestry, rather than incidental similarities of size or colour. Thus, biological classification is said to be **phylogenetic**, meaning that it reflects the evolution of organisms and is not just an arbitrary system developed for convenience.

One practical way of attempting to classify organisms is to place them into groups according to visible external features. You can try this for yourself by examining the diagrams given in Figure 44·1. These show a variety of organisms which are often found inhabiting fresh leaf litter. To help you carry out the exercise a number of questions are listed in Table 44·1 and you should answer these, after copying the Table.

A method like this can be a useful beginning in the process of classification and often reveals a number of natural groupings. You will have discovered, for example, that all the organisms with three pairs of legs also possess antennae and it seems reasonable to suppose that they must be related. On the other hand, some problems are obvious. In particular, which features should be regarded as more important in cases of conflict? For example, specimens D and I both have tail appendages but, while D has many pairs of legs, I has only three pairs. Should they be classified together or apart? Equally, are all the organisms in their adult form or could some of them be in their larval stage?

In attempting to solve problems like these, taxonomists would study many additional features, including the life cycles, internal anatomy, biochemistry, development and behaviour of the organisms. Taxonomy has become a complex science. Nevertheless, much classification still depends upon subjective interpretation of the facts.

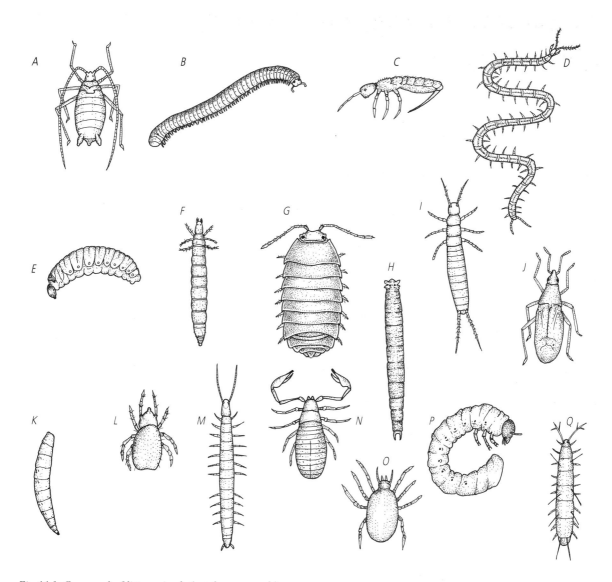

Fig 44·1 Common leaf litter animals (not drawn to scale)

Table 44·1 External features of organisms illustrated in Figure 44·1

QUESTIONS	ANSWERS (Use the letter which identifies the specimen)
1. How many pairs of legs?	none
	3 pairs
	4 pairs
	many pairs
2. Is the body divided into segments?	yes
	no
3. Are wings present?	yes
	no
4. Are there visible tail appendages?	yes
	no
5. Are antennae present?	yes
	no
6. Describe the antennae where present	branched
	longer than body

Copy this table onto a separate sheet of paper before completion

44·2 Identification keys ■ ■ ■ ■ ■ ■ ■ ■ ■ ■ ■ ■ ■

Biology students encounter a rather different set of problems in field work when they are required to identify specimens from a particular habitat. Field guides contain **identification keys** for this purpose, usually consisting of a series of alternative clues based on the external features of different organisms. The first step in preparing a key is often to compile a table of similarities and differences very much like Table 44·1. From this table, it is relatively simple to construct a key (one example is given in Fig. 44·2) and, using it, you should be able to identify specimens *A–Q*. Check your identifications with the answers printed at the end of the Unit.

1.	Jointed legs present	. . . . EITHER go to clue number	2
	No legs	. . . . OR go to clue number	15
2.	3 pairs of legs	. . . . (INSECTS)	4
	More than 3 pairs of legs	. . . .	3
3.	4 pairs of legs	. . . . (ARACHNIDS)	9
	More than 4 pairs of legs	. . . .	11
4.	Wings present	. . . . plant bug (Order Hemiptera)	
	Wings absent	. . . .	5
5.	Tail appendages present	. . . .	6
	No tail appendages	. . . .	8
6.	Antennae longer than body	. . . . soil aphid (Family Aphididae)	
	Antennae shorter than body	. . . .	7
7.	Two tail appendages	. . . . bristletail (Order Diplura)	
	Single tail appendage doubled back under body	. . . . springtail (Order Collembola)	
8.	Last body segment shorter than others	. . . . proturan (Order Protura)	
	Last body segment similar in size to others	. . . . cockchafer larva (*Melolontha*)	
9.	Large crab-like pincers	. . . . pseudoscorpion (Order Chelonethi)	
	No large pincers	. . . .	10
10.	Body in two sections	. . . . beetle mite (Order Acarina)	
	No clear body division	. . . . gamasid mite (Order Acarina)	
11.	Tail appendages present	. . . .	12
	No tail appendages	. . . .	14
12.	Branched bristle-like antennae	. . . . *Pauropus* (Class Pauropoda)	
	Antennae unbranched	. . . .	13
13.	Long tail appendages resembling antennae	. . . . *Geophilus* Centipede (Class Chilopoda)	
	Tail appendages short, not like antennae	. . . . *Scutigerella* (Class Symphyla)	
14.	Oval body	. . . . *Armadillidium* woodlouse (Class Crustacea)	
	Elongated body	. . . . *Blaniulus* millipede (Class Diplopoda)	
15.	Tail appendages present	. . . . crane-fly larva (*Tipula*)	
	No tail appendages	. . . .	16
16.	Body tapering towards tail	. . . . blowfly larva (*Calliphora*)	
	Body not tapering	. . . . weevil larva (Family Curculionidae)	

Fig 44·2 Identification key for the organisms illustrated in Figure 44·1

44·3 Homologues and analogues ■ ■ ■ ■ ■ ■ ■ ■ ■ ■ ■

An identification key can be made up using almost any characteristics of the organisms concerned; its only criterion of success is its effectiveness in enabling the observer to make a correct identification. Taxonomists have to be more selective in their choice of distinguishing characteristics when devising a comprehensive classification scheme, choosing only those which are believed to reveal true relationships between organisms. Characteristics of this nature are called **homologues**, where a homologue is defined as any structural or biochemical feature which is shared between two or more organisms by virtue of a common ancestral link. Examples of homologous structures in vertebrates would be backbones, feathers, or a five-fingered limb.

Organisms may share characteristics which are not homologues. These cannot be used in classification because they do not necessarily reflect common ancestry: wings, for example, are common to bats, birds, and insects, but the different types of wings are not homologues. Features such as these which are shared by unrelated organisms are called **analogues**, meaning that they have the same function, although their basic structure may be quite different. The development of such features in different groups is called **convergent evolution** (Unit 42).

The taxonomist has to distinguish between homologues and analogues in order to group organisms together in such a way as to express their evolutionary links. One method of achieving this is called **cladistics**, as outlined in the next section.

44·4 Cladistics ■ ■ ■ ■ ■ ■ ■ ■ ■ ■ ■ ■ ■ ■ ■ ■

Cladistics is a method of taxonomy based on constructing groups, or **clades**, comprising organisms which share a **unique homologue**. Birds, for example, form a clade sharing the unique homologue, feathers, and mammals form a clade because they possess mammary glands and suckle their young. Both birds and mammals belong to a larger clade, vertebrates, through their common possession of a backbone. Surprisingly, fish do not form a clear cut clade on their own because, although they have many features in common, such as gills, scales, fins, and a tail, none of these features is unique to fish.

The relationships of clades to each other may be expressed in a branching diagram called a **cladogram**. The simple cladogram shown in Figure 44·3 is based on the list of shared features given in Table 44·2. The joints, or **nodes**, of the branching system represent shared homologues. As you proceed from left to right, the groups, or clades, become smaller, and the smaller they become, the more homologous characteristics they have in common. On the cladogram illustrated, the sheep and the long-eared bat have four features in common (jaws, lungs, fur, mammary glands); the sheep and the blackbird have two features in common (jaws, lungs), while the sheep and the salmon have only one feature in common (jaws).

Clearly, it is possible to make more than one cladogram for a group of organisms like this, but the criterion for choosing one above all others is simple: the best cladogram provides the maximum possible number of shared homologues. The first step in cladistic classification is the construction of a list of characteristics such as that shown in Tables 44·1 and 44·2. This is often a laborious task. To be as effective as possible, a whole range of approaches must be used, comparing the structural, physiological, and biochemical features of the organisms. Computers are used to process the data and can be programmed to produce cladograms which can then be assessed and compared.

The validity of this method of classification depends upon the assumption that, if two organisms share a homologue, they must be related. There are very good theoretical reasons for thinking that this is so, and also for assuming that, the more homologues shared, the more closely related two groups must be. The nodes of a cladogram should therefore represent the common ancestry, or **phylogeny**, of the organisms, and the entire branching system should reflect their divergent evolutionary history.

Taxonomists have often tried to express phylogeny in schemes of grouping, but their attempts have sometimes failed because they have concentrated on the search for common ancestral types in the fossil record. In the absence of such evidence, there has been too great a reliance on subjective interpretation. The advantage of the cladistic method is that it produces possible schemes from readily available information. These can be used in helping to evaluate existing schemes built on traditional methods. In this way, the overall approach to classification becomes more objective and scientific. However, if too few features are used in making cladograms, the results are liable to be misleading, so that caution is required.

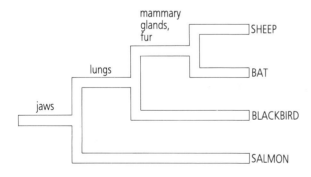

Fig 44·3 A simple cladogram for four vertebrate organisms

Table 44·2 Shared homologues of vertebrate animals

STRUCTURE	ANIMALS			
	bat	blackbird	sheep	salmon
fur	√		√	
jaws	√	√	√	√
lungs	√	√	√	
mammary glands	√		√	

44·5 How many Kingdoms? ■ ■ ■ ■ ■ ■ ■ ■ ■ ■ ■ ■ ■

If all living organisms could be represented on a single cladogram, this would be like a branching tree. The tips of the branches would be the fundamental units of classification, called **species**, where each species is defined as a group of similar organisms which share a common gene pool. (The species concept is described in detail in Units 40 and 41.) The branching tree shows the major pathways of evolution, while the trunk comprises primitive organisms similar to those which first existed on Earth. Within this structure, living organisms can be divided into major groups, called **Kingdoms**, in several different ways.

In this book, the Five-Kingdom system proposed by the American biologist R. H. Whittaker and subsequently modified by L. Margulis and K. V. Schwartz has been followed in classifying all groups of organisms. The names and important characteristics of the five Kingdoms are as follows.

1 Kingdom Prokaryotae. All prokaryotic organisms are included in Kingdom Prokaryotae. Most are unicellular and relatively simple in structure. However, many prokaryotic cells have specialized biochemical pathways which enable them to exploit unusual energy sources, such as hydrogen sulphide (H_2S), or methane (CH_4). The Kingdom includes numerous types of **bacteria**, and **blue-green bacteria** (Unit 45).

2 Kingdom Protoctista. This includes eukaryotic organisms with a unicellular or simple multicellular structure. The most important groups are **protozoa**, that is, single-celled heterotrophic protoctists, and **algae**, or photosynthetic protoctists. These are described in Units 46 and 51 respectively. The Kingdom also includes slime moulds and a variety of other aquatic and parasitic organisms.

Earlier versions of the Five-Kingdom system incorporated an alternative kingdom, called Kingdom Protista, which included single-celled eukaryotes only. Defining the Kingdom in this way presents a problem in classifying groups like green algae with both unicellular and multicellular representatives. The broader definition of Kingdom Protoctista overcomes some of these problems and allows the remaining three Kingdoms to be defined clearly. However, you should appreciate that Kingdom Protoctista includes some groups traditionally thought of as plants.

3 Kingdom Plantae. The members of Kingdom Plantae are multicellular and autotrophic, with chloroplasts containing chlorophylls a and b and other photosynthetic pigments. They are distinguished from photosynthetic protoctists by their life cycles which include a diploid embryo stage. Plant structure and physiology are described

in detail in Units 31–35, and the evolution of plant groups is discussed in Units 53 and 54.

4 Kingdom Fungi. Fungi are eukaryotic organisms which reproduce by forming spores and lack cilia and flagella at all stages of their life cycles. Their bodies consist of fine threads, called hyphae, in which dividing walls are often absent. Some fungi feed saprotrophically by releasing enzymes and absorbing the soluble products of digestion, while others are parasitic, as explained in Unit 52.

5 Kingdom Animalia. Animals are multicellular heterotrophic eukaryotes. The nuclei of their body cells are diploid, and they reproduce by means of small motile male gametes (sperm) and large non-motile female gametes (ova). Cilia and flagella are normally present. The range of animal types and their evolutionary relationships are described in Units 47–50. The evolutionary relationships between these five Kingdoms are shown in outline in Figure 44·4A, and a summary of the most important groups is given in Figure 44·5.

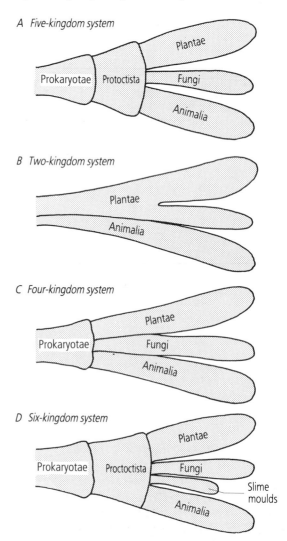

A Five-kingdom system

B Two-kingdom system

C Four-kingdom system

D Six-kingdom system

Fig 44·4 Outline of the Five-Kingdom system and comparisons with other possible classification schemes

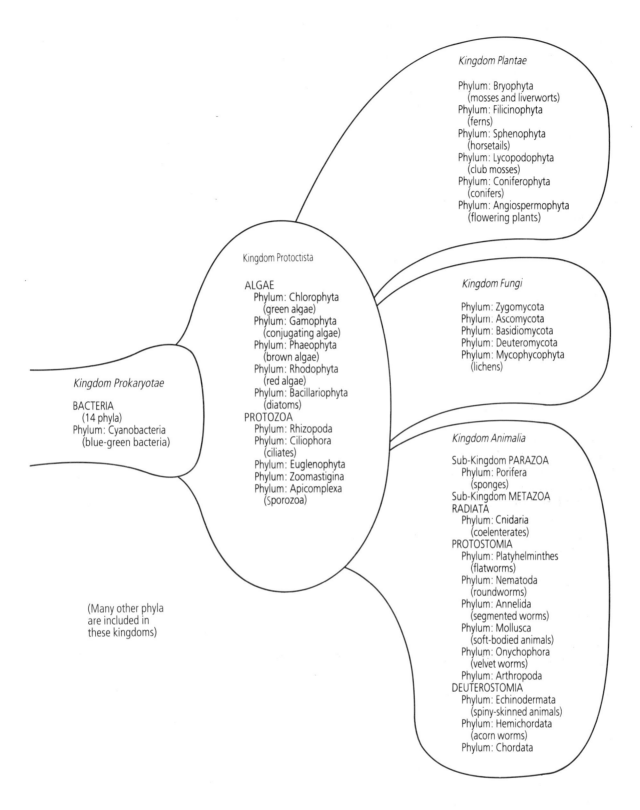

Fig 44·5 The Five-Kingdom System of classification

Traditionally, only two Kingdoms were distinguished. Four-Kingdom and Six-Kingdom schemes have also been suggested. Comparison with the Five-Kingdom system shows that these schemes are consistent with the same general pattern of evolutionary descent, as illustrated in Figure 44·4B, C, and D. Most biologists would now agree that there is an important discontinuity between the prokaryotes and eukaryotes which should be recognized by a separate prokaryote Kingdom, usually called **Prokaryotae**. However, at the present time, there is no general agreement about the separation of the other groups. No one scheme can be considered as final or correct.

44·6 Classification groups ■ ■ ■ ■ ■ ■ ■ ■ ■ ■ ■ ■

No matter how they vary, all classification schemes recognize a hierarchy of groups or clades between the largest units, the Kingdoms, and the smallest units, species. In descending order of size, these groups are **Kingdoms, Phyla** (singular **Phylum**), **Classes, Orders, Families, Genera** (singular **Genus**), and **species**. Three examples of organisms classified into these groups are given in Table 44·3. You will notice that, in some cases, taxonomists have found it useful to subdivide the major clades. For example, Phylum Chordata is divided into several **Sub-phyla**.

Table 44·3 Three examples of classification

	cabbage white butterfly	man	white dead nettle
KINGDOM	Animalia	Animalia	Plantae
PHYLUM	Arthropoda	Chordata	Angiospermophyta
SUB-PHYLUM		Vertebrata	
CLASS	Insecta	Mammalia	Dicotyledoneae
SUB-CLASS	Pterygota		
ORDER	Lepidoptera	Primates	Tubiflorae
FAMILY	Pieridae	Hominidae	Labiatae
GENUS	*Pieris*	*Homo*	*Lamium*
SPECIES	*brassicae*	*sapiens*	*album*

In Table 44·3, the terms Athropoda, Pterygota, Insecta, and so on are the names of taxons, that is, they are the names of the groups actually used for classification. It is important to appreciate that there are other commonly used terms which are *not* names for distinct taxonomic groups. **Protozoa** and **algae** refer to different types of protoctistan cells but are not taxon names. Protozoa are single-celled heterotrophic protoctists. This description encompasses several different phyla and is a convenient way of referring to cells at a particular level of organization and with certain types of specialization. In a similar way, the term 'algae' is used to describe photosynthetic protoctists with a range of plant-like characteristics. Some protoctistan phyla do not fit either description.

Other useful but strictly non-taxonomic groupings include **radiata, protostomia,** and **deuterostomia** in Kingdom Animalia (see Unit 48). In Kingdom Plantae, terms such as **tracheophyte** and **gymnosperm** remain useful for explaining the evolutionary development of plant groups (Units 53, 54).

44·7 Binomial system ■ ■ ■ ■ ■ ■ ■ ■ ■ ■ ■ ■ ■

Organisms are referred to not just by their species name, but also by their genus, rather like a forename and a surname. It is customary to use a capital letter for the genus and small letters for the species. The lion, *Felis leo*, and the tiger, *Felis tigris*, belong to the same genus and are immediately recognised as being closely related. You should adopt the habit of underlining names like these when you write them. (In books they are printed in italics.) This **binomial** ('two name') system was first devised by the Swedish botanist **Carl von Linne**, or **Linnaeus**, (1707–78) who devoted most of his life to the naming and classification of organisms in a way which could be universally understood and copied. He adopted Latin as the language for his 'great alphabet of nature' and translated his own name into its Latin equivalent, *Carolus Linnaeus*. The letter L in brackets after a binomial name indicates that the species was named by Linnaeus.

Table 44·4 Answers to identification key

SPECIMEN	IDENTIFICATION
A	soil aphid (Family Aphididae)
B	*Blaniulus* millipede (Class Diplopoda)
C	springtail (Order Collembola)
D	*Geophilus* centipede (Class Chilopoda)
E	weevil larva (Family Curculionidae)
F	proturan (Order Protura)
G	*Armadillidium* woodlouse (Class Crustacea)
H	crane-fly larva (*Tipula*)
I	bristletail (Order Diplura)
J	plant bug (Order Hemiptera)
K	blowfly larva (*Calliphora*)
L	beetle mite (Order Acarina)
M	*Scutigerella* (Class Symphyla)
N	pseudoscorpion (Order Chelonethi)
O	gamasid mite (Order Acarina)
P	cockchafer larva (*Melolontha*)
Q	*Pauropus* (Class Pauropoda)

45 KINGDOM PROKARYOTAE AND VIRUSES

Objectives

After studying this Unit you should be able to:–

- List the important characteristics of organisms classified in Kingdom Prokaryotae

- Distinguish between the coccus, bacillus, vibrio, and spirillum forms of bacterial cells

- Draw and label a diagram to show the main structural features of a bacterial cell

- Describe the structure of bacterial cell walls and distinguish between gram-positive and gram-negative bacteria

- List the different types of bacterial nutrition and give a specific example of each type

- Distinguish between obligate anaerobes, facultative anaerobes, and obligate aerobes in terms of their ability to use oxygen

- Describe binary fission and the production of endospores as methods of asexual reproduction in bacteria

- Draw and label a diagram to show the main structural features of the cells of blue-green bacteria

- List the main characteristics of blue-green bacteria and comment on their evolutionary links with the chloroplasts of eukaryotic cells

- Describe the structure and variety of virus particles

- Give an illustrated account of the life cycle of a bacteriophage virus

- List the important ecological effects of bacteria

- Outline the economic importance of bacteria and viruses and give a simple explanation of 'genetic engineering'

45·1 Introduction ■ ■ ■ ■ ■ ■ ■ ■ ■ ■ ■ ■ ■ ■ ■ ■ ■ ■

CLASSIFICATION

Kingdom: Prokaryotae
prokaryotic
widespread, many types thrive in extreme
 conditions
autotrophic, heterotrophic, diverse metabolic
 pathways

Bacteria and **blue-green bacteria** are the two main groups of organisms included in the Kingdom Prokaryotae. They are the simplest of living organisms but play a vital role in the ecology of the Earth, particularly in the recycling of bioelements. Certain types of bacteria, e.g. blue-green bacteria are among the only organisms able to 'fix' atmospheric nitrogen: ultimately they provide most of the nitrogen-containing compounds in the bodies of other living organisms. A single gram of garden soil contains at least 2000 million bacteria.

The structure of prokaryotic cells is simple but, unlike eukaryotes, they show an astonishing range of metabolic pathways. The examples described in the remainder of this Unit illustrate some of this diversity and help to place bacteria and blue-green bacteria in their proper evolutionary perspective.

45·2 Bacteria ■ ■ ■ ■ ■ ■ ■ ■ ■ ■ ■ ■ ■ ■ ■ ■ ■

Structure

The overall size and shape of bacterial cells varies enormously but four main groups can be identified (Fig. 45·1). Round or spherical cells are called **cocci**, rod-shaped cells are called **bacilli**, comma-shaped ones **vibrios**, and spiral ones **spirilla**. Many genera of bacteria are named after these shapes, e.g. *Streptococcus*, *Staphylococcus*, *Lactobacillus*, but shape is not now used as the sole basis for classification: metabolic differences are much more important. Taking these differences into account, as many as 16 separate prokaryotic phyla are recognized. Fourteen of these can be usefully grouped as 'bacteria'.

A Cocci

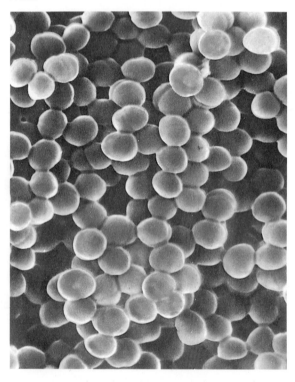

C Vibrios

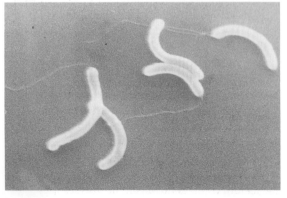

B Bacilli

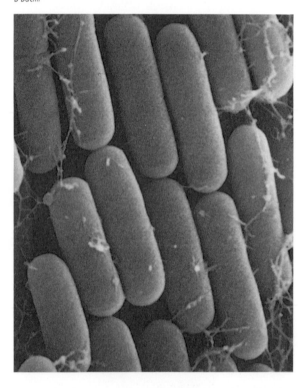

D Spirilla

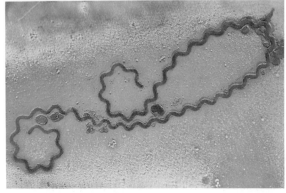

Fig 45·1 Shapes of bacterial cells

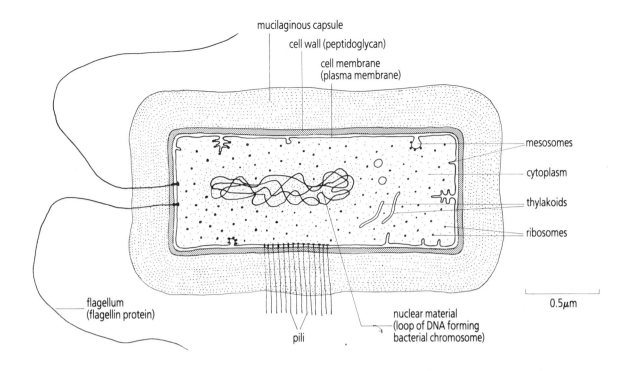

mucilaginous capsule

cell wall (peptidoglycan)

cell membrane (plasma membrane)

mesosomes

cytoplasm

thylakoids

ribosomes

flagellum (flagellin protein)

pili

nuclear material (loop of DNA forming bacterial chromosome)

0.5μm

Fig 45.2 Structure of a typical bacterial cell

The structure of a typical rod-shaped bacterial cell is illustrated in Figure 45·2. Virtually all bacteria have a rigid **cell wall** consisting of molecules of polysaccharide cross-linked by short chains of amino acids. This rather complex **peptidoglycan** structure gives strength and performs the same function as cellulose in the cell walls of green plants. Sometimes, an additional layer of **lipopolysaccharide** material is deposited. In the absence of this layer, the cell wall reacts with a stain called **crystal violet**, producing a purple colour, but, if the lipopolysaccharide layer is present, the reaction is prevented so that the cell walls remain colourless. **Gram staining**, named after the Danish microbiologist who developed this technique, distinguishes between two major groups of bacteria according to the structure of their cell walls: **gram-positive** bacteria, which take up the stain, are more susceptible to lysozyme, an enzyme in saliva and nasal secretions, than **gram-negative** bacteria, and are also much more likely to be affected by antibiotics like penicillin. Lysozyme attacks the cell wall directly, while penicillin is effective because it resembles the substrate of an enzyme responsible for forming cross-links. Molecules of penicillin inhibit the enzyme by combining with the active centres of its molecules. The net result is that the cell wall splits, allowing the cell contents to swell by osmosis. The bacterium finally bursts open, or **lyses**, because the water potential of the cytoplasm is more negative that the water potential of the surrounding medium. Many bacteria which cause disease have a mucilaginous **capsule** as their outermost layer; its function is to give protection against the host's body defences.

The **cell membrane** of bacteria is a selectively permeable barrier between the inside and outside of the cell and has a fluid-mosaic structure very like the membranes of eukaryotic cells. Sometimes it is invaginated to form **mesosomes**. In most cases no definite function for these structures is known; some types are involved in cell division and spore formation (described later). Tubular or flattened **thylakoids** containing chlorophyll and other pigments are present in the cytoplasm of photosynthetic bacteria.

Motile bacteria are propelled by **flagella** which are anchored in the cell membrane and pass through the cell wall. The projecting part, or **filament**, of each flagellum is constructed entirely from **flagellin** protein subunits and has a fixed spiral or corkscrew shape. By a remarkable mechanism the base of the flagellum is rotated about 100 times per second by a tiny 'motor' located in the cell membrane. The energy needed to produce rotation is obtained from H^+ ions which are pumped out of the cell by membrane carrier proteins and then re-enter. (As explained in Unit 10, a similar H^+ gradient drives the production of ATP in mitochondria.) Many motile bacteria show **chemotaxis**, that is, they swim towards a source of food, or away from noxious chemicals. One recently discovered species contains small particles of an iron compound (magnetite) and is able to detect and use the Earth's magnetic field. In the northern hemisphere, these bacteria are North seeking with the result that they normally swim downwards until they reach the bottom sediments on which they feed. In the southern hemisphere, similar species are consistently South seeking.

Numerous **pili** cover the surface of some bacteria. They are fine threads of protein; some of them may help the bacterium to stick to a solid surface but, in many cases, it is not known what function they serve.

Nutrition and metabolism

Some aspects of bacterial metabolism are shared. For example, all bacteria have DNA as their genetic material and use the same 20 amino acids for making proteins. They use glucose as a subunit for cell walls and synthesize ATP using energy from glycolysis. The great diversity of metabolism is seen in the ways in which these essential molecules are obtained.

1 Autotrophic forms. These bacteria have the simplest nutritional requirements. Using carbon dioxide as their only carbon source, and ammonium compounds as their only nitrogen source, they make all their own vitamins, sugars, amino acids, and nucleotides.

Chemosynthetic, or **chemoautotrophic** bacteria derive their energy from the oxidation of inorganic compounds such as ammonia (NH_3), or hydrogen sulphide (H_2S). The most important examples are the **nitrifying bacteria**, including *Nitrosomonas* and *Nitrobacter*, which help to recycle nitrogen by converting ammonium compounds to nitrites (NO_2^-) and nitrates (NO_3^-) (refer to Unit 62). Sulphur oxidizers, like *Thiobacillus*, have a similar role in the sulphur cycle, converting hydrogen sulphide and other sulphur compounds to sulphates directly useful to plants, by means of the reaction:

$$H_2S + 2O_2 \longrightarrow SO_4^{2-} + 2H^+$$

Some related bacteria thrive in the acid waters of hot springs and one species, *Sulpholobus acidocaldarius*, is killed by temperatures below 55 °C. It grows best at 70–75 °C with pH in the range 2–3.

There are three groups of **photosynthetic** bacteria, called **green sulphur bacteria**, **purple sulphur bacteria**, and **purple non-sulphur bacteria**. All of them are poisoned by oxygen and carry out photosynthesis according to the equation:

$$CO_2 + \underset{\text{hydrogen donor}}{2H_2X} \overset{\text{light}}{\longrightarrow} \underset{\text{carbohydrate}}{(CH_2O)} + H_2O + 2X$$

The **hydrogen donor** substance, represented by H_2X, is never water: in the green sulphur bacteria and the purple sulphur bacteria it is hydrogen sulphide, while in the purple non-sulphur bacteria, it is a small organic molecule such as lactic acid, pyruvic acid, or ethanol. The main photosynthetic pigment in green sulphur bacteria is **chlorobium chlorophyll**, which is chemically very similar to the chlorophyll a of green plants. A rather different pigment, **bacteriochlorophyll**, is found in the other two groups.

2 Heterotrophic forms. Most bacteria are heterotrophic, that is, they obtain energy from the breakdown of preformed organic chemicals. Usually, they are **saprotrophic**, feeding on dead organic material by releasing enzymes and absorbing the soluble products of enzyme action. Different types of heterotrophic bacteria exploit different carbon compounds. The normal energy source may be as simple as methane (CH_4), or it may be a complex polysaccharide like cellulose. More commonly, intermediate compounds like lactic acid, pyruvic acid, or glucose are used. A few species are able to break down organic ring compounds including those found in petroleum. In general, saprotrophic bacteria play a vital ecological role as decomposers, maintaining supplies of carbon, nitrogen and other elements in forms useful to living organisms.

In the laboratory, many of these bacterial species can be isolated and characterized by their nutritional requirements and by their sensitivity to oxygen. **Obligate anaerobes** are believed to be similar to some of the first living organisms. Oxygen inhibits their growth so that they are restricted to places deep in the soil or in ocean and freshwater sediments where oxygen is absent or in short supply. A few strictly anaerobic species, such as the sulphate reducers, *Desulphovibrio*, are able to use oxygen in a combined form. **Aerobic** bacteria derive energy from the breakdown of food chemicals using oxygen. Some species are **obligate aerobes**, but more often such bacteria are **facultative anaerobes**, that is, when oxygen is in short supply, they derive energy anaerobically or may exploit alternative combined forms of oxygen. For example, **denitrifying bacteria** in the soil normally respire aerobically but are able to break down nitrates (NO_3^-), or nitrites (NO_2^-) when oxygen is not available.

Some species of bacteria form associations with other organisms. Sometimes, the outcome is mutually beneficial, as in the case of **root nodule bacteria** in leguminous plants (described in Unit 57), or the cellulose-digesting bacteria of ruminant animals. Relatively few bacteria are **pathogenic**, although the diseases they cause may be important. More often their effects are positive, like those of vitamin K-producing bacteria in the human intestine, or acid-producing skin bacteria which help to protect against infection by pathogenic forms. Some examples of bacterial diseases and their methods of transmission are discussed in Unit 19.

In most instances, the symptoms of disease are caused by the release of substances called **toxins** which block essential metabolic pathways in the host cells. The toxins produced by *Clostridium* *botulinum*, which causes botulism, and *Clostridium tetani*, which causes tetanus or 'lock-jaw', are among the most potent poisons yet discovered. Such toxins are only incidentally harmful to humans as *Clostridium* species normally live saprotrophically in the soil where it is believed that the toxins act to suppress the activity of other bacteria in competition for the same food supply. Other types of bacteria and some fungi produce substances called **antibiotics** for precisely the same purpose. Typically, these are effective only against bacterial metabolism and many, including **penicillin**, **streptomycin**, and **actinomycin**, have become vital in the treatment of bacterial diseases.

Growth and reproduction

The **nuclear material** of a bacterium consists of a closed loop of DNA, the **bacterial chromosome**, which is usually concentrated into a small area of the cell. There is no nuclear membrane and the DNA is not combined with histone protein as in eukaryotic cells. *Escherichia coli (E. coli)*, a bacterium found in enormous numbers in the human large intestine, contains a loop of DNA 1100 μm in length, that is, about enough DNA for 2500 genes. Only a small number of these are transcribed at any one time, most gene expression being controlled by operator genes (refer to Unit 9).

In suitable conditions, protein synthesis is extremely rapid and each bacterium divides by **binary fission** as often as once every 20 minutes. Replication of DNA produces new loops each of which becomes attached to a mesosome. As the cell grows, the loops are moved farther apart, so that each daughter cell receives its own copy of the cell's genetic instructions (Fig. 45·3). In fast growing cultures, replication of the DNA starts again before division is complete, helping to shorten the generation time. Rapid reproduction is

favoured by natural selection for it allows bacteria to exploit any available food supply effectively in competition with other organisms. In ideal conditions, the increase in numbers can be very dramatic indeed (see Unit 56).

A Bacterial chromosome held in position by mesosome

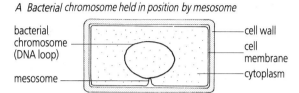

B New DNA loop formed by replication

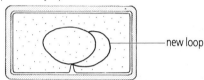

C Loops separate as cell elongates

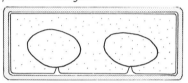

D Constriction of cell membrane and cell wall – daughter cells form

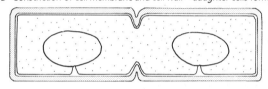

Fig 45·3 Binary fission in bacteria

A few genera of bacteria, including *Bacillus* and *Clostridium*, form structures called **spores**, or **endospores**, as a normal part of their life cycles (Fig. 45·4). Endospore formation begins with the invagination of the cell membrane to form a small compartment containing at least one loop of DNA.

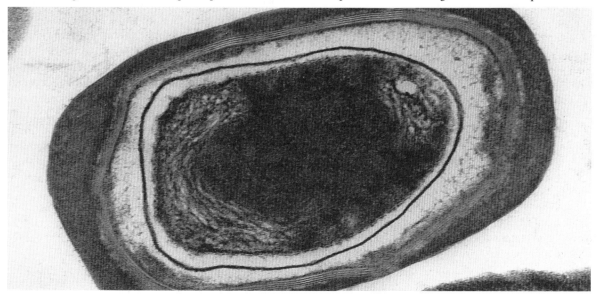

Fig 45·4 Bacterial endospore

Around this, a thick wall comprising several distinct layers is deposited. Endospores have a very low rate of metabolism; they are resistant to dehydration, toxic chemicals and extremes of temperature and can survive for many years without any external source of nutrients. Often they are small enough and light enough to be carried around by air currents. In suitable conditions, the endospore 'germinates', releasing a normal bacterial cell capable of growth and reproduction.

Pasteurization, the treatment of milk by heating to 63 °C for 30 minutes, or to 72 °C for 15 seconds, kills most pathogenic bacteria, including those which cause tuberculosis and typhoid, but does not kill spores. Heat **sterilization**, which destroys all bacteria and their spores requires temperatures as high as 115–125 °C and is achieved by steam heating under pressure (autoclaving). Alternatively, sterilization is carried out using antiseptic chemicals or gamma-ray irradiation.

Reproduction in bacteria is almost invariably asexual but a few species have evolved a primitive method of sexual reproduction, which leads to **genetic recombination**. **Conjugation** in *E. coli* occurs when two cells become linked by means of tubular **sex pili** through which DNA may be transferred (Fig. 45·5). Sometimes, the DNA becomes incorporated into the chromosome loop of the recipient cell so that it will be replicated along

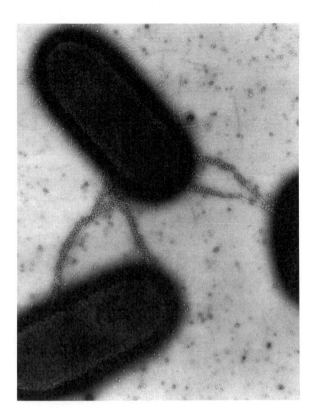

Fig 45·5 Conjugation in E. coli

with the the rest of the cell's genetic instructions and transmitted to the daughter cells of future generations. In this way, the genotypes of some cells are permanently altered.

45·3 Blue-green bacteria ■ ■ ■ ■ ■ ■ ■ ■ ■ ■ ■ ■ ■ ■ ■

CLASSIFICATION

Phylum: Cyanobacteria (blue-green bacteria)
prokaryotic
widespread: marine, freshwater, terrestrial
photosynthesis producing oxygen, chlorophyll a
Anabaena, Nostoc

The cells of blue-green bacteria are larger than most bacteria and have a more complex structure (Fig. 45·6). All of them carry out photosynthesis using water as the hydrogen acceptor and produce oxygen, exactly like eukaryotic algae and green plants:

$$6CO_2 + 6H_2O \xrightarrow{\text{light}} C_6H_{12}O_6 + 6O_2$$

carbon dioxide / water / glucose / oxygen

They contain chlorophyll a and a variety of other pigments fixed to specialized photosynthetic membranes, called **thylakoids**, formed initially by invagination from the cell membrane.

In some species new cells separate following binary fission, while in others they remain attached forming chains or filaments, as in *Anabaena* and *Nostoc*. In suitable conditions, blue-green bacteria multiply rapidly, producing dense masses which may trap inorganic particles. Circular markings in sedimentary rocks called **stromatolites** are trace fossils reflecting the activity of blue-green bacteria in the past. They are common in rocks from 600 to 2500 million years old. Similar communities of blue-green bacteria exist today in Western Australia and the Bahamas, in warm shallow bays where evaporation from the sea produces unusually salty conditions.

Like nitrogen-fixing bacteria, many blue-green bacteria convert atmospheric nitrogen into nitrates which are then incorporated into amino acids and proteins. In *Nostoc*, nitrogen fixation occurs only inside special thick-walled cells called **heterocysts**. The thick walls are necessary to exclude oxygen which disrupts the structure of **nitrogenase**, the most important enzyme involved. Sometimes, resistant spore-like structures also called heterocysts are formed; these survive adverse conditions and may later 'germinate' to produce new filaments.

The close similarities between blue-green bacteria and the chloroplasts of eukaryotic cells have already been mentioned (Unit 43). However,

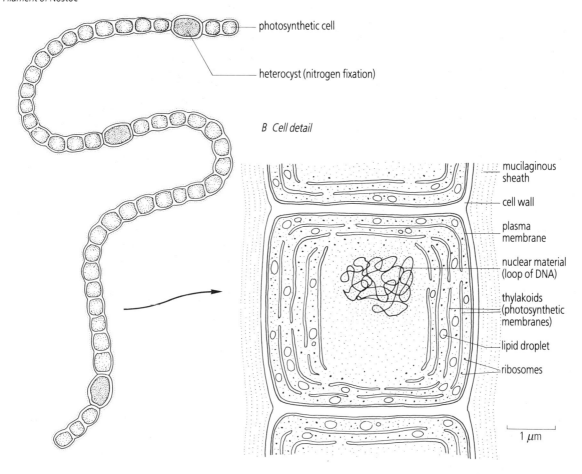

A *Filament of* Nostoc

photosynthetic cell

heterocyst (nitrogen fixation)

B *Cell detail*

mucilaginous sheath

cell wall

plasma membrane

nuclear material (loop of DNA)

thylakoids (photosynthetic membranes)

lipid droplet

ribosomes

1 μm

Fig 45.6 Structure of blue-green bacteria

it is clear that chloroplasts are not the direct descendants of blue-green bacteria because they have a different group of photosensitive pigments. Chloroplasts contain molecules of chlorophyll a and chlorophyll b, whereas blue-green bacteria possess only chlorophyll a, with further differences in the range of accessory pigments present. This seemed to present a problem for the symbiotic theory of the origin of eukaryotic cells but, in the late 1960s, a related group of organisms, now called **Phylum Chloroxybacteria**, was discovered. These are very similar in structure to blue-green bacteria but contain both chlorophyll a and chlorophyll b, making them the free-living prokaryotic counterparts of chloroplasts and removing an important objection to the theory.

45·4 Viruses ■ ■ ■ ■ ■ ■ ■ ■ ■ ■ ■ ■ ■ ■ ■ ■ ■

Viruses are not prokaryotes. They are not classified with living organisms because they lack many vital cellular structures and are incapable of independent activity. They exist only as parasites, taking over the metabolic pathways of a host cell. Viruses themselves are not cells and should not be thought of as forerunners of the first organisms, not least because they depend entirely on living cells for their reproduction. Instead, viruses appear to have originated from small fragments of RNA or DNA which replicate separately from the main chromosome loop inside many bacteria. These acquired a protective protein coat making transmission from cell to cell more likely.

Virus infections are responsible for many human diseases including smallpox, chicken pox, measles, rubella (German measles), polio, mumps, influenza, colds, yellow fever, infectious hepatitis, rabies, and HIV. Some types are implicated as contributory factors in cancer and autoimmune diseases like multiple sclerosis. Other viruses are plant and animal parasites affecting many important crop species. One example is **tobacco mosaic virus (TMV)** which attacks the leaves of tobacco plants. The structure of this virus is illustrated in Figure 45·7. As you can see, each

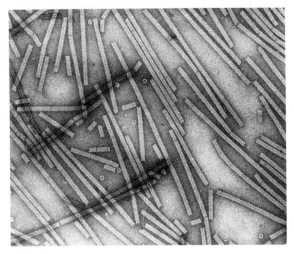

A Electronmicrograph

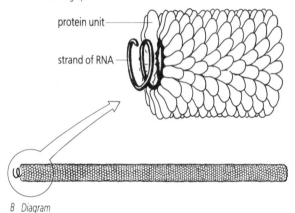

protein unit

strand of RNA

B Diagram

Fig 45·7 Tobacco mosaic virus (TMV)

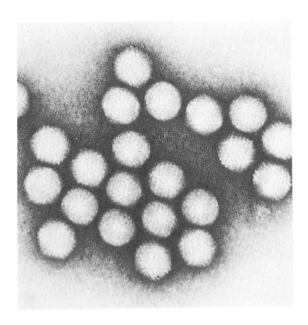

Fig 45·8 Adenovirus

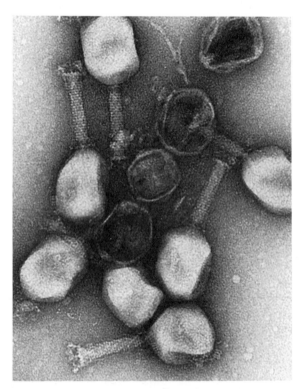

Fig 45·9 Bacteriophage

virus particle consists of a spiral coil of RNA surrounded by a protein covering, or **capsid**. In TMV this consists of 2200 identical protein molecules, or **capsomeres**, fitted together to form a cylindrical column with an internal groove which exactly accommodates the RNA molecule. This regular structure allows TMV to be crystallized and it can be kept in this form for many years without losing its infective properties. An alternative structure is shown by **adenoviruses**, a group which includes some of the many viruses that cause colds in humans (Fig. 45·8). The particles have an icosohedral shape, with a total of 20 triangular faces, each formed from repeating protein subunits. This shape gives close to the maximum possible internal volume for a given surface area. The space inside is filled by DNA.

A rather more complex structure is found in **bacteriophage** viruses which attack bacterial cells. Figure 45·9 shows one type, called a *lambda* bacteriophage, which includes *E. coli* among its normal hosts. The capsid is formed from at least five different types of protein subunit. When a *lambda* bacteriophage contacts a bacterium it becomes attached and the DNA it contains is literally injected into the cytoplasm. There are two possible outcomes, as illustrated in Figure 45·10. In the **lytic pathway** the host DNA is disrupted

and the viral DNA replicates repeatedly. New viral proteins are made, allowing the spontaneous assembly of new virus particles which are liberated when the host cell bursts open. In the **lysogenic pathway**, the viral DNA becomes incorporated in the bacterial chromosome and is replicated along with it, so that copies of the viral DNA pass to all the cell's descendants. Very occasionally, the viral DNA is released, causing the bacterium to enter the lytic pathway. Note that the *lambda* DNA is able to form a small closed loop, called a **plasmid**.

Retroviruses have RNA as their genetic material but differ from other RNA viruses by

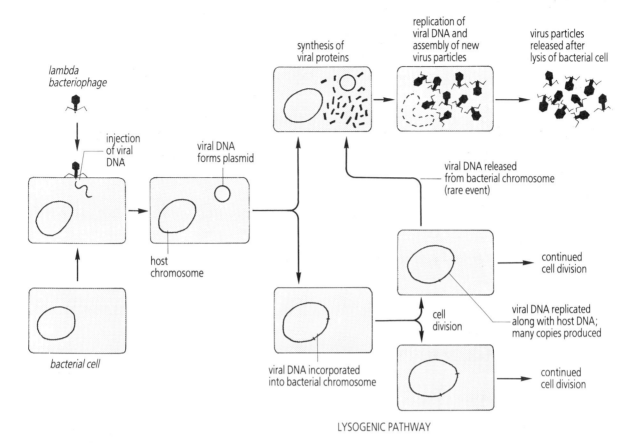

synthesis of
viral proteins

replication of
viral DNA and
assembly of new
virus particles

virus particles
released after
lysis of bacterial cell

lambda
bacteriophage

injection
of viral
DNA

viral DNA
forms plasmid

viral DNA released
from bacterial chromosome
(rare event)

host
chromosome

continued
cell division

viral DNA replicated
along with host DNA;
many copies produced

cell
division

bacterial cell

viral DNA incorporated
into bacterial chromosome

continued
cell division

LYSOGENIC PATHWAY

Fig 45·10 *Life cycle of* lambda *bacteriophage*

possessing an enzyme known as **reverse transcriptase**. Following infection of a host cell, molecules of the enzyme are released and cause synthesis of DNA from the viral RNA template. In this way, viral genes are converted from RNA into DNA form. Typically, the new DNA segment is incorporated into the host DNA and is replicated along with it. Unlike DNA incorporated from other types of virus, DNA derived from a retrovirus may sometimes become active without killing the host cell. The reasons for this are complex, but the result is that new viruses can be continually produced without premature cell death.

Human immunodeficiency virus (HIV), the virus which causes AIDS, is an important member of the retrovirus group (Fig. 45·11). HIV is transmitted during sexual intercourse and by direct transfer of blood. Once inside the body, the virus attacks a special type of T-lymphocyte which is vital for normal functioning of the immune system. The virus may remain dormant for some time, possibly even for years, but is liable to be triggered into activity when T cells start to multiply to combat infection by bacteria or by another virus. If this happens, new HIV particles are produced and the T cells are unable to carry out their normal role. Once the disease takes hold, victims may develop tumours and become less and less able to fight off minor infections. Not everyone who is

exposed to the virus will get the disease, but AIDS itself is always fatal. Certain drugs can slow the progress of the disease and alleviate symptoms, but, at the moment, there is no cure. The use of condoms reduces the chance of transferring HIV infection during sexual intercourse.

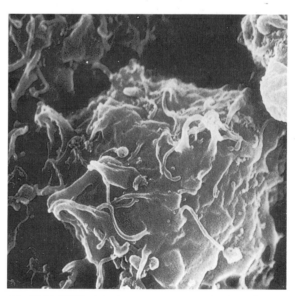

Fig 45·11 *SEM of the HIV virus budding from a T lymphocyte*

45·5 Economic importance of bacteria and viruses ▪ ▪ ▪ ▪ ▪

The essential ecological roles of bacteria in nutrient cycles like the nitrogen and carbon cycles, their beneficial associations with other organisms, and the importance of pathogenic forms have all been mentioned previously. This section describes some additional ways in which bacteria and viruses directly affect human activities.

1 Fermentation. This is a form of anaerobic respiration in which organic compounds are broken down into new and potentially useful substances. In the dairy industry, fermenting bacteria including *Lactobacillus* and *Streptococcus* are used in the production of cheeses and yoghurt. These organisms can carry out glycolysis but do not have the necessary enzymes for the Krebs cycle. Consequently, although they can use lactose as an energy source, they are unable to break it down completely, so that lactate is released as a waste product. The lactate accumulates, lowering the pH and explaining the 'sharp' taste. Other fermentation products include vinegar, sauerkraut and the food additive monosodium glutamate. Fermented grass, called **silage**, is an important winter feed for cattle.

2 Sewage treatment. Effective sewage treatment is essential to prevent the spread of faeces-borne diseases. Both anaerobic and aerobic bacteria are involved. In the **activated sludge process**, solids contained in the raw sewage are allowed to settle. The sediment, or sludge, is acted on by anaerobic bacteria. These produce methane which is often used as a fuel to power the sewage plant. Treated sludge may be dried and used as a fertilizer or pumped into ships for dumping at sea. The liquid from the sewage is passed to holding tanks where air is bubbled through under pressure to encourage the growth of aerobic bacteria and other microorganisms. These digest soluble organic compounds to produce carbon dioxide and water. After filtering, the effluent can be safely discharged into rivers.

3 Biological control. This is the use of predatory or pathogenic organisms to control the population numbers of pest species affecting crops. A well known example is the introduction of ladybird beetles into greenhouses to control the numbers of greenfly (aphids) infesting tomato plants. Bacteria and virsuses can sometimes be used in a similar way. For example, the viral disease **myxomatosis** is a natural disease of some rabbit species in South America but was deliberately introduced to reduce population numbers of rabbits in Australia and Western Europe. The caterpillars of butterflies and the larvae of other insects are attacked by bacteria and several species of *Bacillus* have proved effective in treating crops.

4 Industrial microbiology. The metabolic accomplishments of bacteria are increasingly important for the food, chemical, and drug industries. Many different enzymes are extracted and purified for biological and medical research. Others are used in the manufacture of soaps, cheeses, as meat tenderizers, and in 'biological' washing powders. Protein made by microorganisms provides a cheaper protein source than many traditional animal feeds. For example, 'Pruteen', a dried product from *Methylophilus* bacteria, is produced in quantity by ICI, using methanol and ammonia as starting materials. The production of antibiotics has already been mentioned.

5 Genetic engineering. This is a rapidly developing technology whereby individual genes are inserted into bacterial cells with the intention of manufacturing the gene products in quantity. Insulin and the anti-viral protein interferon have already been produced in this way and there is tremendous potential for making other human substances including hormones, antibodies, enzymes, and blood clotting factors, all of which seem likely to become invaluable in the treatment of disease. In many cases, the genetically engineered product will give a dramatic improvement in therapy. For example, until now the only insulin available for the treatment of diabetes was extracted from the pancreas tissue of slaughtered animals. Unfortunately, insulin from animal sources is not identical to human insulin so that many diabetics develop antibodies against it after prolonged treatment. Sometimes drugs are needed to suppress the patient's immune responses. With human insulin made by bacteria none of these problems are likely to arise.

The processes involved in introducing genes into bacteria are complex but the essential principles of one method can be understood from Figure 45·12. Many bacteria protect themselves against viral infection by producing special enzymes, called **restriction enzymes**, which recognize and attack specific sequences of nucleotides in DNA. The action of a particular restriction enzyme, called Eco R1, is shown in Figure 45·12*A*. This produces a 'staggered' cut, leaving complementary sequences on the loose ends of the DNA molecule. When Eco R1 is used to treat chromosomes from human cells, these are split into many short fragments some of which may contain single genes coding for a particular human protein. As outlined in Figure 45·12*B*, such fragments can be incorporated into bacterial plasmids and then reintroduced into bacterial cells, where they replicate as independent units. Culturing the host cells allows many copies of the original human genes to be produced. The greatest problem in genetic engineering is to identify and isolate those bacteria which form a useful product.

A   The action of the restriction enzyme Eco R1

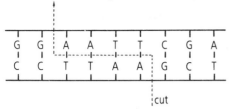

B   The principle used to incorporate chromosomal genes into bacterial plasmids

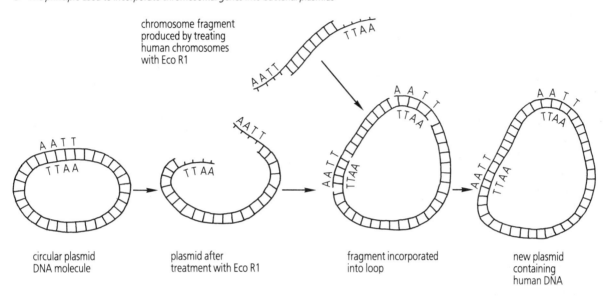

chromosome fragment
produced by treating
human chromosomes
with Eco R1

circular plasmid
DNA molecule

plasmid after
treatment with Eco R1

fragment incorporated
into loop

new plasmid
containing
human DNA

Fig 45·12 Genetic engineering

As explained in Unit 9, gene expression in bacteria is controlled by an operon system in which the structural gene responsible for producing mRNA is switched on or off by an operator gene, which is in turn controlled by a repressor gene (refer to Fig. 9·7). Fragments of DNA obtained from whole chromosomes normally contain structural genes together with the promoter regions which initiate transcription, but often lack the corresponding operator and repressor genes. When such fragments are incorporated into plasmids or into the bacterial chromsome, they are almost always expressed and, since there is no mechanism for switching them off, the host cells often synthesize large quantities of the desired protein.

A different problem arises with DNA obtained from other sources. For example, the DNA coding for a particular protein can sometimes be made artificially from purified mRNA using an enzyme called **reverse transcriptase**. This accurately reproduces the original coding sequence, but the DNA formed lacks the promoter region required for mRNA synthesis. This problem can be overcome by attaching the required sequences, or by joining the synthesized DNA to a known bacterial gene which is complete, so that the two

genes are expressed together. The bacterial protein and the introduced protein are linked when first formed, but are separated by chemical means.

A similar approach has been used to insert genes into eukaryotic cells. In one study, the human gene coding for the production of growth hormone was linked to the promoter sequence for a mouse protein called metallothionein, and successfully introduced into the DNA of a fertilized mouse ovum. The metallothionein protein is normally produced when cells are exposed to small amounts of heavy metals like zinc. However, genetically-engineered embryos produced growth hormone instead and grew to more than twice the normal size. In addition, they were able to transmit their potential for increased growth to their offspring as part of their DNA. Giant mice are of doubtful value, but there are many possible applications for similar techniques in developing new breeds of cattle, sheep, or poultry.

There is no doubt that, as techniques improve, genetic engineering, or 'biotechnology' as it is sometimes called, will have a revolutionary impact. A further example would be to introduce genes for nitrogen fixation into crop plants, thereby avoiding the need for expensive fertilizer.

46 KINGDOM PROTOCTISTA 1 PROTOZOA

Objectives

After studying this Unit you should be able to:–

- Name some of the organisms included in Kingdom Protoctista

- Outline the distinguishing features of organisms belonging to Phylum Rhizopoda, Phylum Ciliophora, Phylum Euglenophyta, Phylum Zoomastigina, and Phylum Apicomplexa

- Describe the structure and life activities of *Amoeba proteus*

- Discuss the mechanisms of amoeboid movement

- Describe the structure and life activities of *Paramecium caudatum*

- Explain the important mechanisms of ciliary movement

- Comment on the structure of *Euglena gracilis*

- Describe the life cycles of the malarial parasites *Plasmodium vivax* and *Plasmodium falciparum*

- List some of the possible control measures against malaria

46·1 Introduction ■ ■ ■ ■ ■ ■ ■ ■ ■ ■ ■ ■ ■ ■ ■ ■

CLASSIFICATION

Kingdom: Protoctista
eukaryotic: unicellular, colonial and simple
  multicellular forms
widespread in aquatic and damp habitats
autotrophic, heterotrophic, parasitic

The Kingdom Protoctista comprises an immense variety of organisms characterized by eukaryotic organization and a relatively simple body structure. Most protoctists are single-celled, but there are a number of important multicellular species. The Kingdom includes heterotrophic microorganisms,

such as *Amoeba* and *Paramecium*, and autotrophic forms like *Euglena*, as well as *Trypanosoma* and *Plasmodium*, the parasites which cause sleeping-sickness and malaria. Also included are the green algae, red algae, brown algae, diatoms, and some fungus-like forms.

Protoctists can be divided into several not strictly taxonomic groups, including **protozoa** and **algae**. The protozoa are single-celled organisms which are typically heterotrophic and lack a cellulose cell wall. The examples described in this Unit illustrate the possible diversity of protozoan organization. The algae show specializations of their own and are described separately in Unit 51.

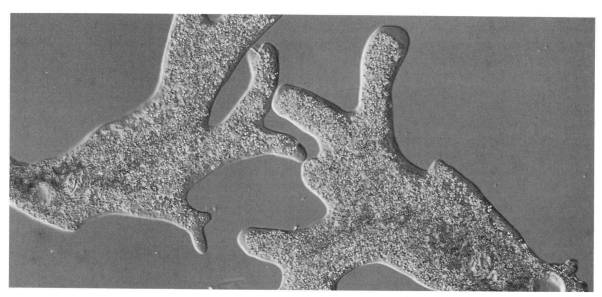

Amoeba proteus. *The prominent pseudopodia are used for locomotion and food capture.*

CLASSIFICATION

Phylum: Rhizopoda
eukaryotic, unicellular
widespread: marine, fresh water, soil
heterotrophic, parasitic, feeding and locomotion by
  pseudopodia
asexual reproduction only
Amoeba, Entamoeba, Arcella

Structure

The rhizopods are common water and soil organisms. Most species are microscopic, but a few are visible to the naked eye as small white specks. One of the larger species, *Amoeba proteus*, inhabits the bottom sediments of fresh water ponds and may grow to 500 μm (0.5 mm) in length. Its main features are illustrated in Figure 46·1.

The cell membrane, or **plasmalemma**, which forms the external surface of the organism has a fluid mosaic structure like that of other biological membranes (refer to Unit 6). The living substance inside, called **protoplasm**, is subdivided into a **nucleus** and **cytoplasm**. The nucleus of *Amoeba proteus* is roughly bun-shaped and has no fixed position. Two layers of cytoplasm are distinguished. The outer layer, called **ectoplasm** or **plasmagel**, is transparent and jelly-like. The inner layer, called **endoplasm**, or **plasmasol**, contains numerous organelles, such as

mitochondria and food vacuoles, as well as other cell inclusions, like crystals and granules. Endoplasm is fluid and flows freely through the body of the organism during movement.

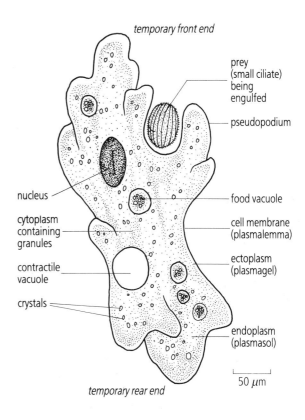

Fig 46·1 Amoeba proteus

Life activities

Amoeba has a large surface area to volume ratio and carries out gas exchange and excretion by simple diffusion across the body surface. Ammonia is the main nitrogenous waste product. Crystals in the cytoplasm consist of **carbonyl diurea**, another excretory product.

Although the outer membrane must be permeable to allow gas exchange and excretion to occur, one disadvantage is that, at least in fresh water species, water tends to enter the organism by osmosis. This is inevitable because the water potential of the cytoplasm is invariably more negative than that of the external environment. Water enters continuously, but is pumped out by the action of the **contractile vacuole**.

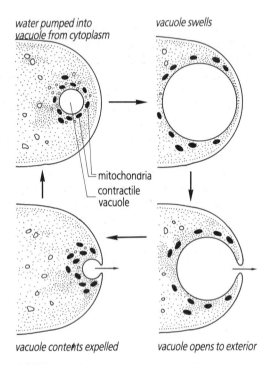

water pumped into vacuole from cytoplasm

vacuole swells

mitochondria
contractile vacuole

vacuole contents expelled

vacuole opens to exterior

Fig 46·2 Action of the contractile vacuole in Amoeba

The vacuole swells slowly as water is pumped in from the cytoplasm, and then collapses suddenly, expelling its contents to the outside (see Fig. 46·2). In *Amoeba proteus*, the time taken for one complete cycle of filling and emptying varies from 2 to 15 minutes in normal conditions, but can be altered by placing the organism in more dilute or more concentrated solutions. In this way, the water content of the cytoplasm is kept constant: the process is called **osmoregulation**. The mechanisms which control vacuole activity are not yet fully understood, but the numerous mitochondria clustered round are thought to supply energy for the pumping process. Marine species have an internal concentration of dissolved substances similar to the external concentration and usually lack contractile vacuoles.

Locomotion and food capture in *Amoeba* take place by means of blunt projections, called 'false feet' or **pseudopodia** (singular **pseudopodium**). In normal locomotion, these appear to be thrust out more or less at random, but the organism can respond to its surroundings by moving away from some stimuli, such as light, or towards others. *Amoeba* feeds by **phagocytosis**. Pseudopodia spread out widely to surround the prey organism or food particle, and then engulf it completely, forming a **food vacuole**. Experiments using food dyed with a harmless indicator show that the contents of the vacuole become first acid and later alkaline, presumably to facilitate the action of different digestive enzymes. Simple substances, like glucose and amino acids, are absorbed into the cytoplasm, and any solid remains are left behind when the food vacuole finally fuses with the cell membrane.

Asexual reproduction in *Amoeba proteus* occurs by means of **binary fission**. The nucleus contains from 500 to 600 very small chromosomes and divides by mitosis. Each daughter cell receives approximately half the cytoplasm and will feed and grow until large enough to reproduce itself. Many rhizopods survive unfavourable conditions by forming protective **cysts**, but this does not occur in *Amoeba proteus*. No method of sexual reproduction has been observed.

Molecular basis of amoeboid movement

Movement by means of pseudopodia, or **amoeboid movement** as it is usually called, has been studied intensively. It occurs not just in protozoa, but in many of the cells of multicellular organisms, such as white blood cells. Figure 46·3 shows stages in the formation of a pseudopodium. Initial development depends on a reduction in the thickness of the plasmagel, possibly in response to appropriate internal or external stimulation. The underlying plasmasol, which is under pressure, flows into the weakened area, forming first a bulge and then a tube. The movement is sustained by contraction of the outer gel layer which squeezes inwards, causing cytoplasmic streaming towards the tip of the pseudopodium. Within the advancing tip, in a region called the **fountain zone**, plasmasol is converted to plasmagel which is then deposited on the sides of the tube. At the temporary rear end of the organism, the reverse transition from gel to sol occurs, allowing movement to continue in any direction.

Several lines of evidence suggest that amoeboid movement involves **actin** protein filaments, which are found abundantly in the cytoplasm. For example, when the actin-binding drug phalloidin is injected into an amoeba, movement stops instantly. In cell extracts, the filaments show sol to gel transitions and can contract in combination with other proteins. They can be cross-linked by

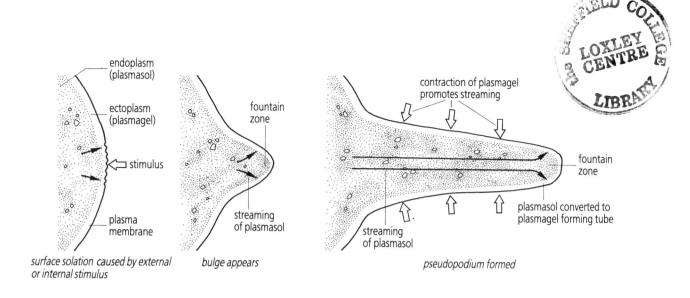

Fig 46·3 *Formation of pseudopodia in* Amoeba

fimbrin to form a solid gel, or broken into short lengths by **gelsolin** to make a sol. Contraction depends upon the presence of **myosin** protein and is triggered by calcium ions, Ca^{2+}, as in the case of muscle contraction (Unit 26). Precisely how these changes are linked to particular stimuli remains a mystery. It is not known how *Amoeba* can distinguish between food particles and inorganic debris, or how it can detect and move away from unfavourable stimuli.

Other rhizopods

Amoeba proteus is a free-living heterotroph. The Phylum Rhizopoda also includes many parasitic forms. Some of these are transmitted directly from host to host, while others are transmitted in contaminated food and water. In tropical regions, *Entamoeba histolytica*, which is responsible for amoebic dysentery, is an important human parasite. The spread of the disease can be prevented by purification of water supplies and proper treatment of sewage. The mouth amoeba, *Entamoeba gingivalis*, is associated with periodontal disease and is transmitted by kissing.

Other rhizopods differ from *Amoeba* by possessing tiny shells, called **tests**. *Arcella* (Fig. 46·4) constructs its test by glueing together sand grains and other inorganic particles. Such structures have been preserved in sedimentary rocks and give the Rhizopoda a fossil record which extends to the beginning of the Palaeozoic Era.

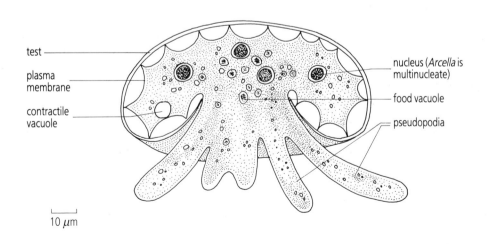

Fig 46·4 Arcella, *a rhizopod which builds a test*

CLASSIFICATION

Phylum: Ciliophora (ciliates)
eukaryotic, unicellular
widespread: marine, fresh water
heterotrophic, feeding and locomotion by cilia,
  rarely parasitic
sexual and asexual reproduction
Didinium, Paramecium, Stentor, Vorticella

Structure

All the organisms which belong to the Phylum
Ciliophora have numerous tiny hairs, called **cilia**
(singular **cilium**). In most species, these beat with
a coordinated rhythm to drive the organism rapidly
through the water. Members of the Phylum are
often referred to simply as 'ciliates'.

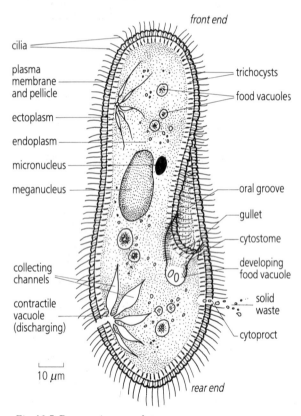

Fig 46·5 Paramecium caudatum

The structure of a large ciliate, *Paramecium
caudatum*, is illustrated in Figure 46·5. As you can
see, the bases of the cilia are embedded in
ectoplasm and pass through a stiff outer covering,
the **pellicle**. This is located immediately inside the
cell membrane and gives the organism a fixed
slipper-like shape. The internal structure of
Paramecium is more complex than that of *Amoeba*
and the organism is said to show a greater degree
of **protoplasmic differentiation**. Thus, in
Paramecium, many functions are carried out by
separate structures specially developed for the
purpose.

Life activities

Like *Amoeba*, *Paramecium* has a large surface area
to volume ratio and is able to carry out gas
exchange and excretion by diffusion across the cell
membrane. However, *Paramecium* has two
contractile vacuoles for osmoregulation, and each
of these is associated with a permanent system of
collecting channels which empty into the main
vacuole.

Paramecium feeds on bacteria. These are
collected from the surrounding water by the
beating of the cilia lining the **oral groove**, which
produce a whirlpool effect. Captured bacteria pass
via the **gullet** to a modified region of the pellicle,
called the **cytostome**, where food vacuoles are
formed. The contents of the food vacuoles are
digested as they circulate in the endoplasm. Solid
wastes are expelled through a second modified
region of the pellicle, called the **cytoproct**.

The behaviour of *Paramecium* involves a simple
but effective **avoiding reaction** (Fig. 46·6). The
organism swims forward in a spiral path until it
encounters an obstacle, or a stimulus such as a
noxious chemical or a change in temperature.
When this happens, the ciliary beat is reversed to
move the organism back a short distance, after
which it changes direction and promptly sets off
again. Although entirely a trial and error process,
the avoiding reaction enables *Paramecium* to stay in
favourable surroundings.

Unpleasant stimulation or mechanical contact
can also trigger the discharge of very fine hairs
from flask-shaped **trichocysts** located in the
pellicle. These are believed to give some protection
against protozoan predators, or may possibly
anchor the organism during feeding.

Asexual reproduction takes place by **binary
fission**. As you will see from Figure 46·5,
Paramecium has two nuclei, comprising a large

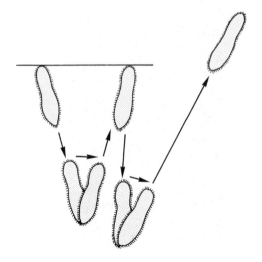

Fig 46·6 Avoiding reaction of Paramecium

meganucleus and a small **micronucleus**. In binary fission, only the micronucleus divides by mitosis. The daughter meganuclei are formed without a mitotic spindle and it seems unlikely that there is an exact separation of replicated copies of DNA. The relationship between the two types of nuclei is not fully understood, but it appears that the meganucleus is responsible for directing the majority of protein synthesis, although it may in turn be controlled by the micronucleus.

Sexual reproduction in *Paramecium* occurs by **conjugation**. In this process, two individuals of different mating types join at their oral grooves. The meganuclei degenerate completely and genetic exchange is carried out solely with the micronuclei. The details of conjugation are complex, but the net effect is to produce genetically varied offspring.

Ciliary movement

The body of *Paramecium* is completely covered by cilia which are arranged in parallel rows (Fig. 46·7). As the cilia beat, each cilium goes through a definite sequence of movements, as illustrated in Figure 46·8*A*. During the active stroke, the cilium is fully extended and exerts the maximum force on the surrounding water. In the recovery phase which follows, the cilium returns to its original position with an unrolling motion which reduces drag. The cycles of adjacent cilia are slightly out of phase, so that they do not bend at exactly the same moment (Fig. 46·8*B*). This gives rise to a **metachronal rhythm** in which waves of ciliary activity pass along the organism from front to rear.

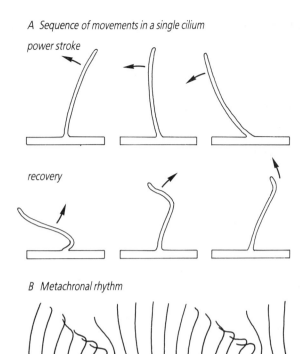

A Sequence of movements in a single cilium

B Metachronal rhythm

Fig 46·8 *Movements of cilia*

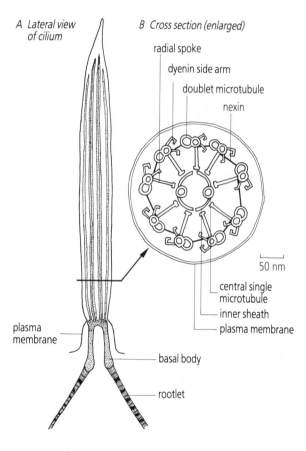

A Lateral view of cilium

B Cross section (enlarged)

radial spoke
dyenin side arm
doublet microtubule
nexin
50 nm
central single microtubule
inner sheath
plasma membrane
plasma membrane
basal body
rootlet

Fig 46·9 *Ultrastructure of a cilium*

The detailed structure of cilia in *Paramecium* and other organisms has been revealed by electron microscope studies and the molecular interactions responsible for movement are now partly understood. Figure 46·9 shows the structure of a typical cilium. The central core or **axoneme**

Fig 46·7 *External surface of* Paramecium *showing rows of cilia (SEM)*

consists of a characteristic '9+2' arrangement of microtubules. The two microtubules located at the centre are simple tubes, but the nine outer microtubules have a doublet structure consisting of one complete and one incomplete tubule, giving a figure 8 appearance. Associated with the microtubules are many additional proteins. Some of these form radial spokes and an inner sheath, while linking molecules of an elastic protein called **nexin** hold the outer ring together. The most important accessory proteins are the **dyenin** side-arms which project at intervals. Experiments in which the cross-linking proteins are removed by detergent (see Fig. 46.10A) show that the dyenin side-arms 'walk' along the adjacent doublet when magnesium ions, Mg^{2+}, and ATP are added. This is very similar to the basic mechanism of muscle contraction in which myosin molecules 'walk' along adjacent actin filaments (Unit 26). When the doublets are held in position by cross links, the same action produces an energetic bending movement (Fig. 46.10B).

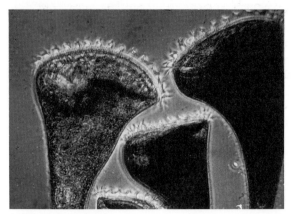

A Light micrograph   Stentor

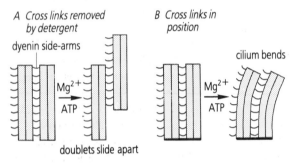

Fig 46·10 Action of dyenin side arms

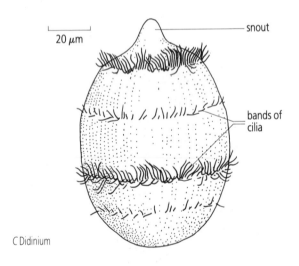

B SEM Vorticella

The mechanisms which trigger ciliary movement are not yet understood. Coordinated action of many cilia is partly the result of viscous forces in which a bending cilium pulls upon its neighbours. However, in *Paramecium* and many other ciliates, the basal bodies are linked by a network of additional microtubules, which may also have a coordinating effect. It is not known how the ciliary beat of *Paramecium* can be reversed during the avoiding reaction, although this may be related to changes in electrical potential across the cell membrane.

Cilia and flagella, which have an identical structure, are found in many other groups of organisms. Cilia often act in groups for the transport of food or fluids. It is interesting to note that **centrioles** have an identical structure to the basal bodies of cilia.

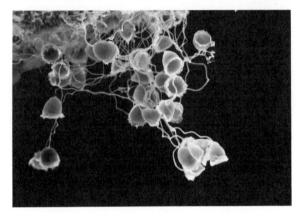

Fig 46·11 A Stentor, B Vorticella, C Didiniur

Other ciliates

Most ciliates feed on bacteria and are motile, like *Paramecium*. *Stentor* and *Vorticella* (Fig. 46·11A, 46·11B), are exceptional by remaining attached to water plants or other solid surfaces. *Stentor* is a very large protozoan, up to 2000 μm (2 mm) in

length and roughly trumpet shaped. Specially modified cilia surrounding the rim create a powerful current of water which draws bacteria and other protozoans inwards. *Vorticella* feeds in a similar way, but has a long thread-like stalk which contracts like a spring in response to vibration or other stimuli. *Didinium* is a predator. It has a projecting snout armed with trichocysts and swims around vigorously until it strikes a suitable target, usually *Paramecium*. Once penetrated, the prey is engulfed bodily (Fig. 46.11C).

CLASSIFICATION

Phylum: Euglenophyta
eukaryotic, unicellular
widespread: marine, fresh water
mainly autotrophic using chloroplasts for
    photosynthesis, food storage by paramylon
pellicle made of protein, locomotion by single large
    flagellum
asexual reproduction only
Euglena

Phylum: Zoomastigina
eukaryotic, unicellular
widespread:
heterotrophic, chloroplasts absent, many parasitic
    forms
locomotion by one or more flagella
sexual and asexual reproduction
Trichomonas, Trypanosoma

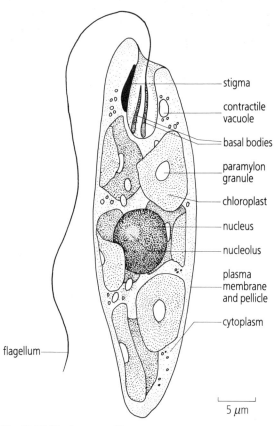

stigma
contractile vacuole
basal bodies
paramylon granule
chloroplast
nucleus
nucleolus
plasma membrane and pellicle
cytoplasm
flagellum

5 μm

Fig 46·12 Euglena gracilis

Flagella have the same structure as cilia but are
normally larger and are carried singly or in small
groups. Flagellate protozoans include *Euglena
gracilis*, illustrated in Figure 46·12. *Euglena* is a
common pond organism and is often so plentiful
that the surface water is tinged green. There are
two basal bodies but only a single flagellum is long
enough to be used for locomotion. Near its base is
a prominent red granule or **stigma**, sometimes
rather misleadingly referred to as an 'eyespot'. The

stigma absorbs light and, depending on the
direction of light rays, it casts a shadow on the
basal region. In some way this triggers **postive
phototaxis**, that is, a movement towards the light
source which helps the organism to maintain a
favourable position for photosynthesis.

In two-kingdom schemes of classification, the
position of *Euglena* is a source of controversy since
the organism shows a combination of plant-like
and animal-like features. The chloroplasts of
Euglena are very similar to those of plants and
contain many of the same pigments. On the other
hand, the organism lacks a cellulose cell wall and
stores glucose in the form of **paramylon**, a
polysaccharide with $1\beta-3$ glycoside linkages
between subunits. Furthermore, some euglenoid
species lack chloroplasts altogether and are able to
feed as heterotrophs. Most biologists would now
agree that *Euglena*'s evolutionary links with plants
and animals are rather remote and that the
organism is better classified as part of a broad
eukaryotic kingdom showing unicellular or simple
multicellular organization.

Phylum Zoomastigina includes a great variety of
flagellate forms all of which are heterotrophic and
lack chloroplasts. Among the most important are
the parasites *Trypanosoma gambiense* and
Trypanosoma rhodesiense which cause different types
of sleeping sickness (Fig. 46·13). The organisms
are transmitted by the bite of the tsetse fly and
thrive in the blood of domestic and wild animals as
well as in man. Control is difficult and large areas
of Africa are virtually uninhabitable because of the
disease.

Trichomonas and related species are mutualistic
protozoans which live in the guts of termites. They
digest cellulose and so enable their hosts to survive
on a diet consisting entirely of wood.

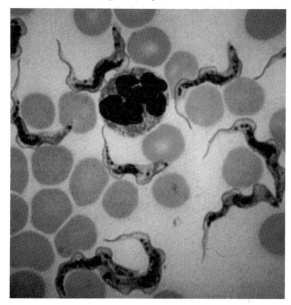

*Fig 46·13 Trypanosomes in the blood of a person with
sleeping sickness (light micrograph)*

CLASSIFICATION

Phylum: Apicomplexa (Sporozoa)
eukaryotic, unicellular
parasitic forms only
sexual reproduction and asexual reproduction
  involving production of spores by schizogony
Plasmodium

Life cycle

Worldwide at least 200 million peolple suffer from
malaria, with an estimated 2 million deaths
annually. Four species of *Plasmodium* can infect
man. The two most important are *Plasmodium
vivax* and *Plasmodium falciparum. Plasmodium vivax*
is prevalent in temperate and subtropical areas,
while *Plasmodium falciparum* is confined to tropical
regions. The disease is transmitted by the females
of several species of *Anopheles* mosquito. Females
require several blood meals before laying eggs,
while males feed harmlessly by sucking fruit juices.
The distribution of malaria is determined partly by
the availability of suitable vectors and partly by the
fact that development of the parasite in the
mosquito requires a certain minimum temperature.
In Africa, *Anopheles gambiae* is the most effective
vector, being long-lived and usually a human-biter.

The life cycle of *Plasmodium vivax* is shown in
Figure 46·14. The cycle in the human host begins
with the bite of an infected female mosquito.
Before feeding, the mosquito injects saliva
containing an anticoagulant, and this carries with it
sporozoites, which are the infective form of the
parasite. As you can see, these are transported in
the blood to the liver where they enter liver cells
and reproduce asexually by a process of multiple
fission called **schizogony**. Most of the **merozoites**
released pass out of the liver in blood vessels, but,
in *Plasmodium vivax*, a small proportion remain to
infect other liver cells. A similar phase of asexual
reproduction occurs in red blood cells. After about
48 hours, the red cells rupture to release new
merozoites. Repeated cycles produce the periodic
bouts of fever characteristic of the disease in its
early stages. Eventually, some parasites develop
into male and female **gametocytes** which remain
circulating in the blood. New vectors are infected
when these sexual forms are drawn up in a blood
meal. Inside the mosquito's stomach, eggs and
sperm develop and fertilization takes place. The
zygote burrows through the wall of the stomach
and forms a small cyst on the outside. Each cyst
can produce as many as 10 000 sporozoites which
migrate to the salivary glands, ready to infect a new
human host. The cycle in the mosquito takes a
minimum of 12 days.

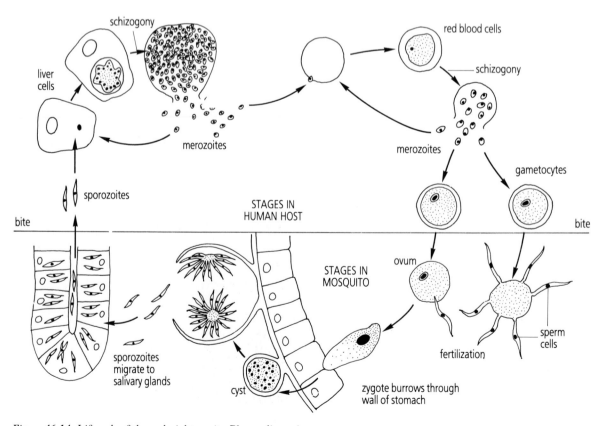

Figure 46·14 Life cycle of the malarial parasite, Plasmodium vivax

Control of malaria

In principle, the disease can be controlled by breaking the life cycle of the parasite at any point. Control measures are taken against the parasite in the human population and against the vector organism. Some of the possible methods are listed below:

1 Treatment and isolation of infected people. Antimalarial drugs, including quinine, chloroquine, and mepacrine, can be used either to prevent infection or against established parasites. Although parasites can be eliminated from the blood, the disease is difficult to cure because a reservoir of infection often remains in the liver cells. Isolation using mosquito nets or similar devices is effective in preventing the parasite from passing to new vectors but is not practicable in areas where a large proportion of the population is affected.

2 Drainage of breeding grounds. The larval stages of the mosquito require static or stagnant water in which to develop. Filling ponds and draining marshes close to human habitation reduces the number of adult mosquitoes able to transmit the disease.

3 Destruction of larvae. As an alternative to drainage, standing water can be sprayed with oil or insecticide to kill the larvae. This has the disadvantage that it must be repeated frequently and is damaging to other aquatic organisms.

4 Spraying houses with insecticide. This is the most effective method of control. The walls of houses are sprayed with a persistent insecticide to kill adult mosquitoes. Initially, DDT was used but in many areas DDT-resistant strains of mosquito have developed, resulting in the use of progressively more toxic insecticides, like malathion and dieldrin. Regrettably, eradication campaigns have been only partially successful and it may well be that long-term control of the disease will depend on the development of an effective vaccine. So far, intensive efforts on this front have failed.

47 SIMPLE MULTICELLULAR ORGANISMS ☐

Objectives

After studying this Unit you should be able to:–

- Explain the advantages of large size and multicellular organization

- Comment on the evolution of multicellular forms

- Using the structure of the sponge *Leucosolenia* as an example, explain what is meant by the cellular level of organization

- List the important characteristics of organisms belonging to Phylum Cnidaria

- Draw diagrams to illustrate the structure of *Hydra* and explain what is meant by the tissue level of organization

- Explain how *Hydra* carries out the following life activities: feeding and digestion, movement and nervous coordination, reproduction

- Outline the importance of polyp and medusa stages in the life cycles of other coelenterates

- Distinguish between diploblastic and triploblastic organization

- List the important characteristics of organisms belonging to Phylum Platyhelminthes

- Draw diagrams to illustrate the structure of *Dendrocoelom* and explain what is meant by organ systems

- Describe the parasitic adaptations of the liver fluke *Fasciola* and the tapeworm *Taenia*

47·1 Development of multicellular organisms ■ ■ ■ ■ ■ ■ ■ ■

Tiny organisms are very vulnerable to changes in their surroundings but larger organisms are able to maintain their position and shape and can protect themselves against predators and extremes of climate. Increased size allows organisms to exploit new habitats and new life styles, so that it is often favoured by natural selection.

The maximum size of any organism is affected by its ability to obtain raw materials from the environment and to dispose of wastes. In single-celled species, such processes normally occur by diffusion across the outer body surface and can be carried out effectively only when the organism possesses a sufficiently large surface area to volume ratio. Because diffusion of dissolved

chemicals is slow, most moneran and protoctist cells are extremely small, with dimensions which rarely exceed 50–100 μm. Large size cannot be achieved by a simple increase in cell mass and it follows that the body of a large organism must be subdivided into smaller units, each of which can obtain its own raw materials. This **multicellular** pattern of construction makes the total surface area available for exchange many times greater than the area of the external body surface. It has a number of other major advantages which have played a vital part in the evolution of more complex organisms.

1 Multicellular organisms have a tremendous potential for **cell specialization**, leading to the development of tissues and organs which allow

different parts of the body to fulfil different functions. This 'division of labour' promotes an efficient organization and is fundamental in explaining the origins of advanced animals and plants. A solitary protoctist cell must remain capable of a full range of life activities and is therefore unable to specialize in the same way.

2   The simplest multicellular organisms are aquatic so that the fluid in the spaces between the cells can be derived directly from the surrounding medium. More complex aquatic and terrestrial animals develop transport systems and the body cells are bathed in **tissue fluid** derived from blood. This arrangement makes possible a whole range of homeostatic mechanisms whereby the composition of tissue fluid is regulated, giving a dramatic improvement in overall efficiency.

3   Only multicellular organisms can become fully adapted for life in terrestrial environments. Single-celled organisms carry out gas exchange and excretion across the entire body surface, which must remain moist and permeable for the purpose. In dry conditions water evaporates from the moist surface and the organisms rapidly dehydrate. In a large multicellular organism this problem is overcome by the development of specialized organs for functions such as gas exchange and excretion. Because these organs are internal, the body surface can be made waterproof and impermeable, reducing water loss to such an extent that an active existence is possible even in desert conditions. Some moneran and protoctist cells can survive dry seasons by **encystment**, that is, by forming a temporary waterproof layer at the body surface, but they are never capable of sustained activity without an abundant external supply of water.

It is not known exactly how the first multicellular organisms developed. One suggested mechanism is that, as organisms increased in size, additional nuclei were formed by repeated nuclear division, with dividing membranes appearing later. An alternative view is that the many-celled condition first appeared when certain kinds of protoctist cells failed to separate following normal cell division, giving rise to various types of colonial organism. There are many examples of this in present day organisms, especially among the green algae (Fig. 47·1). Some colonies, like those of *Gonium*, are very simple, while others show increasing degrees of complexity. Colonies of *Volvox* are small green spheres which spiral through the water. They consist of thousands of individuals which show the earliest beginnings of cell specialization. Individuals at the anterior end are unable to reproduce but are specialized for locomotion, having large 'eyespots' and powerful flagella. Only a few individuals near the posterior of the colony are able to reproduce; some of these give rise to male gametes, or sperm cells, while others form ova. The zygotes produced by fertilization are retained within the colony and each

forms a resistant wall which enables it to survive freezing or drying. When favourable conditions return daughter colonies are liberated.

There is no sharp distinction between colonial and multicellular organisms. However, it has proved convenient to keep the term 'multicellular' for organisms in which the non-reproductive cells are differentiated into several types. On this basis, *Volvox* is regarded as colonial, while the other groups described in the remainder of this Unit are truly multicellular. These groups, sponges, coelenterates, and flatworms, show increasing degrees of specialization and exploit a wide variety of different life styles.

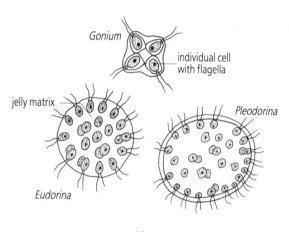

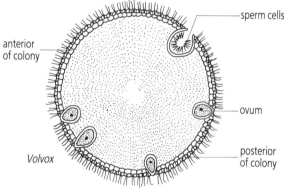

Volvox (light micrograph)

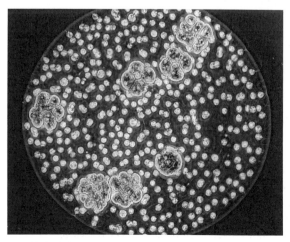

Fig 47·1 Colonial green algae

47·2 Phylum Porifera (Sponges) ■ ■ ■ ■ ■ ■ ■ ■ ■ ■

CLASSIFICATION

Phylum: Porifera (Sponges)
cellular level of organisation
marine, fresh water
sessile, heterotrophic, feeding by collar cells
  (choanocytes)
Leucosolenia

Almost all sponges live in the sea and are **sessile** organisms, that is, they are permanently attached to rocks, or to the sea bed, where they obtain food by extracting plankton from water taken in through thousands of tiny pores. The body of a simple sponge such as *Leucosolenia* is a hollow tube with a large opening at the top, called the **excurrent opening**, or **osculum** (Fig. 47·2). Sponges have no specialized tissues or organs but are constructed from just a few main types of cell – they are said to exist at the **cellular level of organization**.

The outer surface of the sponge is covered by **epithelial cells** while the inner layer consists mainly of collar cells, or **choanocytes**, each of which possesses a single large flagellum. The beating of the flagella draws a current of water into the interior through the tubular openings of pore cells, or **porocytes**. Sandwiched between the inner and outer layers of cells, there is a thin layer of jelly-like material, called **mesoglea**. This contains amoeboid **mesenchyme cells** capable of developing into any of the more specialized types. The mesenchyme cells secrete needles of calcium carbonate called **spicules** which give skeletal support and help to protect the sponge against being eaten by predators.

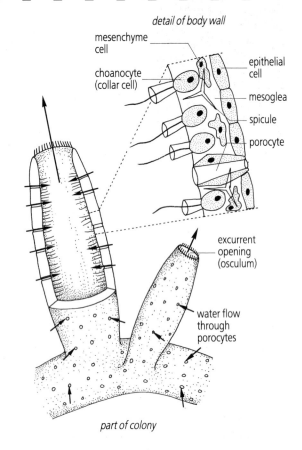

Fig 47·2 *Structure of* Leucosolenia, *a simple sponge*

Most sponges reproduce sexually. The male and female gametes develop from mesenchyme cells and clouds of sperm are released into the water. After fertilization, the zygotes develop into multicellular free-swimming larvae which eventually settle to develop into new individuals.

47·3 Phylum Cnidaria (Coelenterates) ■ ■ ■ ■ ■ ■ ■ ■ ■ ■

CLASSIFICATION

Phylum: Cnidaria (Coelenterates)
diploblastic, radially symmetrical
marine, fresh water
sessile, free-living, heterotrophic, feeding by
  stinging cells (cnidoblasts)
life cycle including separate polyp and medusa
  stages
Class: Hydrozoa
    both polyp and medusa normally represented
    Hydra, Obelia

Class: Scyphozoa (large jellyfish)
    medusa dominant
    Aurelia

Class: Anthozoa (sea anemones, corals)
    polyp dominant
    Actinia

General features

Coelenterates are almost all marine and are most abundant in warm shallow seas. They include jellyfish, sea anemones and corals. The name 'coelenterate' means literally 'hollow gut' and is an apt description of the structure of these organisms. The main body cavity, called the **gastrovascular cavity** or **enteron**, is used for the digestion of food and connects to the exterior by way of a single opening, the **mouth**. Coelenterates are all carnivores but they do not actively pursue their prey. Their method of food capture helps to explain why they are **radially symmetrical**, with bodies which can be divided into mirror-image parts about any vertical plane. Surrounding the mouth are a series of **tentacles** bearing stinging cells, called **cnidoblasts**, which sting and immobilize any small organism which comes into contact with them. Captured prey pass into the

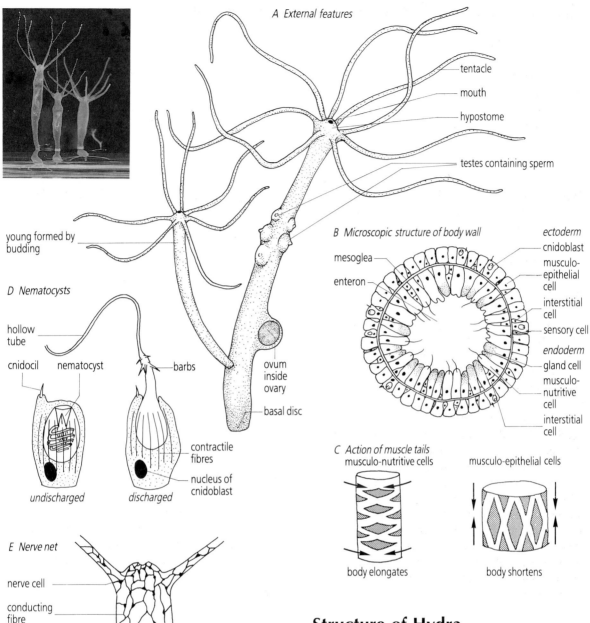

A External features

tentacle
mouth
hypostome
testes containing sperm

young formed by budding

ovum inside ovary

basal disc

B Microscopic structure of body wall

mesoglea
enteron

ectoderm
cnidoblast
musculo-epithelial cell
interstitial cell
sensory cell
endoderm
gland cell
musculo-nutritive cell
interstitial cell

D Nematocysts

hollow tube

cnidocil   nematocyst   barbs

contractile fibres

nucleus of cnidoblast

undischarged   discharged

C Action of muscle tails
musculo-nutritive cells

musculo-epithelial cells

body elongates   body shortens

E Nerve net

nerve cell

conducting fibre

Fig 47·3 Structure of Hydra viridis

Structure of Hydra

The green hydra, *Hydra viridis*, lives in ponds, lakes and rivers attached to water plants or rocks. Its external features and detailed structure are shown in the diagrams given in Figure 47·3. The body of *Hydra* is **diploblastic**, that is, it consists of two cell layers separated by a jelly-like **mesoglea**. The activity of the cells forming each layer can be coordinated to a much greater extent than in sponges, so that they act together as functional groups. Such groups of cells are called **tissues**, and *Hydra* and other coelenterates are therefore said to have reached the **tissue level of organization**. The outer cell layer, called **ectoderm**, is a protective epithelium, while the inner layer, or **endoderm** is a digestive epithelium. Within each of these layers there may be several specialized cell types (refer to Fig. 47·3B). The least specialized cells, called **interstitial cells**, are capable of developing into any of the other types, giving *Hydra* a remarkable ability to replace lost or damaged parts, a process called **regeneration**.

interior through the mouth, which is also used for the removal of solid waste.

The phylum is divided into three main classes – Hydrozoa, Scyphozoa (large jellyfish), and Anthozoa (sea anemones, corals). As will be explained in more detail later, many coelenterates include both **polyp** and **medusa** stages in their life cycles. Polyps are normally sessile and attached, with mouths which face upwards, while medusae are free-swimming umbrella or bell-shaped animals, with a mouth and tentacles which trail downwards. There are several species of fresh water *Hydra*. Although these have no medusa stage and exist only as polyps, they are easily obtained and illustrate the essential features of coelenterate organization extremely well.

Feeding and digestion

Hydra traps its prey by means of **cnidoblasts** which are concentrated in large numbers in the tentacles. As illustrated in Figure 47·3D, each of these cells surrounds a stinging capsule, or **nematocyst**, containing a coiled tube and a small quantity of poison. When a short external barb, the **cnidocil**, is stimulated, the tube is suddenly and explosively turned inside out, piercing the surface of the prey and injecting the poison. The mechanism of discharge is not fully understood, but a sudden increase in fluid pressure inside the capsule and contractile fibres in the cytoplasm of the cnidoblast appear to be involved. Discharge is not triggered unless chemical and mechanical stimulation occur simultaneously. Several other kinds of nematocyst are formed, with threads which stick to or wrap round the prey. Each nematocyst can be used only once and must be replaced by a new one: these develop inside interstitial cells.

Once the prey is paralysed, the tentacles contract, pulling it slowly towards the mouth. Inside the body cavity, *Hydra's* victim is digested by protease and lipase enzymes secreted by **gland cells**. Small particles are ingested by **musculo-nutritive cells** and digestion is completed in food vacuoles inside the cytoplasm of these cells. Continual beating of the flagella keeps the contents of the enteron well mixed.

The colour of a green *Hydra* is due to numerous mutualistic algal cells, called **zoochlorellae**, which live within the cytoplasm particularly in the larger endodermal cells. In light conditions, they carry out photosynthesis and produce oxygen and food materials which are directly useful to the host.

Movement and nervous coordination

Body movement in *Hydra* is possible because many ectoderm and endoderm cells possess long extensions, called **muscle tails**, which contain contractile fibres, or **myonemes**. The muscle tails of the ectodermal **musculo-epithelial** cells are arranged longitudinally so that the body shortens when they contract. On the other hand, the muscle tails originating from the musculo-nutritive cells project laterally and cause the body to lengthen (refer to Fig. 47·3C).

The coordination of these movements depends on two kinds of **nerve cells**. These are **sensory cells**, found mainly in the ectoderm, and **conducting cells**, which form a network, or **nerve net**, extending throughout the mesoglea (Fig. 47·3E). Stimulation of the sensory cells generates nerve impulses which are transmitted through the net to the muscle cells. Weak stimuli result in a local contraction while strong ones may cause the

animal to contract completely, forming an inconspicuous little blob.

Hydra can alter its position of attachment by slow **'gliding'** movements on its base, and can also move much more quickly by a kind of **'somersaulting'** motion in which the tentacles are fixed to the bottom before allowing the base to 'flip over' to a new position. *Hydra viridis* is positively phototactic, that is, it tends to move towards a source of light. Occasionally, bubbles of gas are secreted by the cells of the basal disc, allowing the organism to float passively at the water surface.

Reproduction

Hydra is capable of reproducing sexually and asexually. The asexual method is called **budding**. In the summer months, new individuals develop as outgrowths from the body wall and eventually separate from the parent (refer to Fig. 47·3A). Sexual reproduction occurs in the autumn. Ova and sperm are formed by divisions of the interstitial cells inside swellings in the ectoderm. Ovaries and testes often develop on a single individual. The egg cells remain in their ovaries and are fertilized by sperms released from the testes of nearby animals. After fertilization, each zygote divides and forms a resistant **cyst** which is able to survive the extremes of winter. Although its life cycle is different, *Hydra* is classified as a hydrozoan because of its close similarities in structure with other members of this group. The absence of the medusa stage in *Hydra* is thought to be an adaptation for life in fresh water.

Life cycles of other coelenterates

Obelia is a marine coelenterate which, like *Hydra*, belongs to Class Hydrozoa. However, the life cycle of *Obelia* is much more representative of the group. As you can see from Figure 47·4, it is a colonial organism which grows by asexual budding into a characteristic plant-like form. Individual polyps are similar to *Hydra* but they remain attached and share nutrients through a common enteron. The colony is supported by a chitinous outer covering called the **perisarc** which is secreted by the ectoderm.

The colony is unable to reproduce sexually but some buds develop into structures called **blastostyles**. These produce tiny free-swimming sexual medusae complete with tentacles and gonads, which release egg cells and sperm into the water. Fertilization results in a zygote which develops to form a **planula** larva consisting of a ball of cells with a ciliated outer layer. This is dispersed by tidal currents before settling onto a suitable rock or piece of seaweed where it becomes the founder of a new colony.

In scyphozoan jellyfish, the medusa stage is

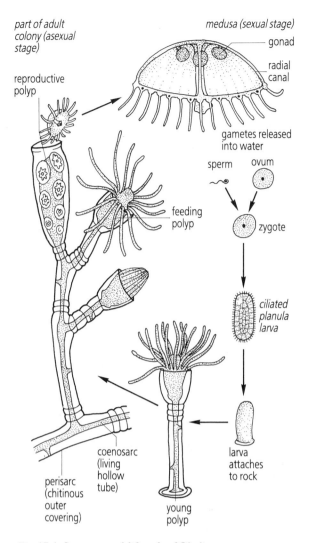

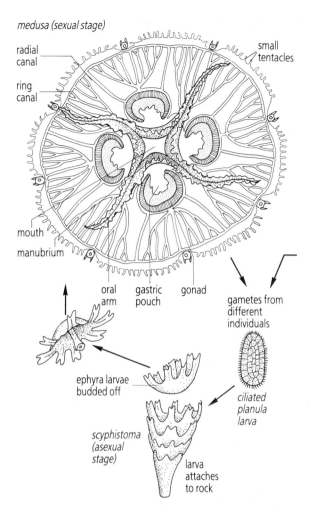

Fig 47·4 Structure and life cycle of Obelia

Fig 47·5 Structure and life cycle of a scyphozoan jellyfish, Aurelia

dominant and the polyp stage is reduced or absent. *Aurelia* (Fig. 47·5) is one of the commonest jellyfish. Male and female medusae occur separately and the four purple or pink horseshoe-shaped structures near the centre of the bell are either testes or ovaries. Eggs are released into the gastrovascular cavity of the female and are fertilized by sperm, which the males release directly into the sea. Ciliated planula larvae escape and settle to form small polyps which feed and may reproduce asexually by budding. In the autumn and winter months, the polyps develop to a **scyphistoma** stage in which horizontal divisions appear and tiny medusae, called **ephyra** larvae, are budded off. Polyps may survive for several years, each year producing a crop of new medusae.

The third group of coelenterates, Class Anthozoa, includes sea anemones and corals. These lack a medusa stage and exist only as polyps. Figure 47·6 shows the external features of the beadlet anemone, *Actinia equina*, one of the commonest species on rocky shores around the British Isles. Stony corals are abundant in warm seas and secrete calcareous exoskeletons which have accumulated over millions of years to form immense reefs and numerous coral atolls.

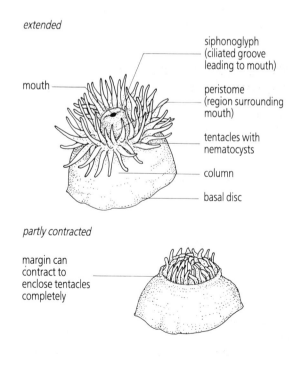

Fig 47·6 External features of the sea anemone Actinia equina

419

Phylum Platyhelminthes (Flatworms) ■ ■ ■ ■ ■ ■ ■ ■

CLASSIFICATION

Phylum: Platyhelminthes (flatworms)
triploblastic, acoelomate, bilaterally symmetrical
marine, freshwater
heterotrophic, many parasitic forms
digestive system with single opening, flame cells
  present

Class: Turbellaria
    free-living flatworms
    Dendrocoelom, Polycelis

Class: Trematoda (flukes)
    parasitic, digestive system present
    Fasciola

Class: Cestoda (tapeworms)
    parasitic, no digestive system, reproduction by
      strobilization
    Taenia

Body plan

The platyhelminth body plan differs from the
coelenterate body plan in two major ways.
Flatworms are bilaterally rather than radially
symmetrical and possess a third layer of body cells,
called **mesoderm**, giving a three-layered, or
triploblastic organization.

Bilateral symmetry, that is, mirror-image
structure on either side of a central plane, is
characteristic of organisms which move actively in
search of food. Platyhelminths show the first clear
signs of a related evolutionary process called
cephalization, in which the organism's sense
organs and nervous system are progressively
concentrated in a definite head region. The
triploblastic body plan gives additional scope for
tissue differentiation and has allowed flatworms to
develop **organ systems** for particular functions. A
free-living flatworm, like *Dendrocoelom lacteum*,
illustrated in Figure 47·7, possesses relatively well
developed digestive, excretory, nervous, and
reproductive systems, and is said to show an **organ
system level of organization**. Internal parasites
like tapeworms often lack one or more of these
organ systems, but are highly specialized for
reproduction.

Platyhelminths do not have a circulatory system
and carry out gas exchange by diffusion. Their
flattened body shape is essential to preserve a large
surface area to volume ratio and to keep diffusion
distances short.

Structure of *Dendrocoelom*

Free-living flatworms, Class Turbellaria, are found
in marine and freshwater habitats. *Dendrocoelom*

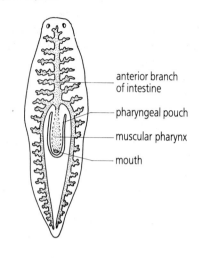

A Digestive system

- anterior branch of intestine
- pharyngeal pouch
- muscular pharynx
- mouth

B Transverse section

- flame cell
- ectoderm
- testis
- circular muscle
- mesoderm
- longitudinal muscle
- endoderm
- cilia on ventral surface
- nerve cord
- oviduct
- mesenchyme tissue

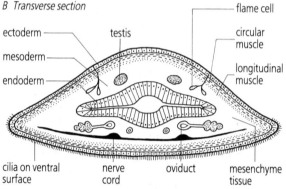

C Excretory system

- flame cell
- excretory pore
- tuft of cilia
- flame cells

D Nervous system

- ganglia forming 'brain'
- nerve cord

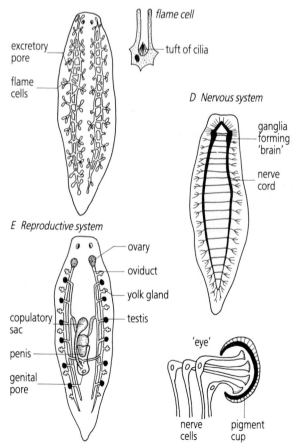

E Reproductive system

- ovary
- oviduct
- yolk gland
- copulatory sac
- testis
- penis
- genital pore

- 'eye'
- nerve cells
- pigment cup

Fig 47·7 Structure and organ systems of Dendrocoelom lacteum

lacteum is common in freshwater lakes and ponds. Its body is white, making its internal structures more easily visible than in pigmented species. The animal glides over the bottom sediments and water plants by waves of muscular contraction and by beating of the cilia on its lower surface. It feeds mainly on dead animal tissue by protruding its muscular pharynx. Particles of food taken into the branching gut (Fig. 47·7*A*) are partially digested before being engulfed by pseudopodia to form food vacuoles in the endodermal cells lining the cavity. Digested food is transferred by diffusion to the rest of the body.

The mesoderm consists of unspecialized **mesenchyme** cells together with differentiated cells which form the circular and longitudinal muscles and certain organ systems (see Fig. 47·7*B*). *Dendrocoelom* has an excretory system (Fig. 47.7*C*) consisting of numerous **flame cells** and a network of tubes connected to the exterior via excretory pores. Each flame cell has a hollow centre containing a tuft of cilia which beat continually to drive a current of liquid outwards. It is likely that osmoregulation is the most important function of the flame cells as most of the metabolic wastes produced can be removed by diffusion.

Free-living flatworms respond to chemical stimuli, light and water currents and show more varied behaviour than *Hydra*. *Dendrocoelom* has specialized sense organs ('eyes') and its nervous system (Fig. 47·7*D*) consists of a pair of ventral nerve cords and groups of neuron cell bodies, called ganglia.

Reproduction is sexual. Individual organisms are hermaphrodite (Fig. 47·7*E*) but self-fertilization is prevented by the structure of the reproductive organs.

Adaptations of parasitic forms

While turbellarians are free-living, the two remaining groups of Platyhelminths, Class Trematoda (flukes) and Class Cestoda (tapeworms), are entirely parasitic and show important adaptations for the parasitic mode of life.

Fasciola hepatica, the liver fluke, is a parasite of sheep, cattle and other mammals. Its structure and life cycle are illustrated in Figure 47·8. Flukes are recognisably similar to free-living flatworms, but differ by having an external cuticle and suckers for attachment. *Fasciola* feeds directly on the tissue fluids and liver cells of its host. An individual host is often parasitised by many flukes and sexual reproduction involving cross-fertilization occurs. Egg capsules pass out in the faeces. As you can see, the remainder of the life cycle involves several stages and an intermediate host, frequently the pond snail, *Limnaea truncatula*. New mammalian hosts are infected by swallowing food or drink containing encysted **cercaria** larvae. Humans are occasionally infected by eating wild watercress growing in water contaminated with the faeces of sheep or cattle.

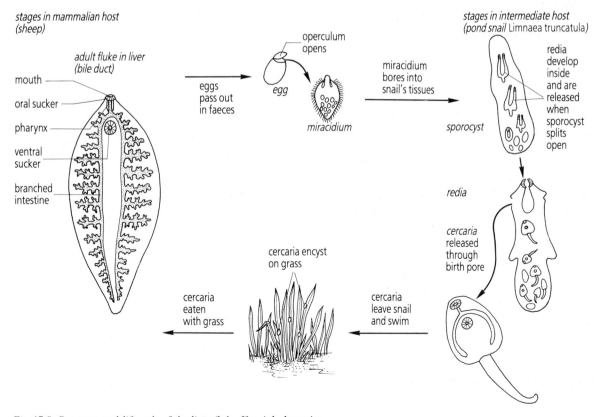

Fig 47·8 Structure and life cycle of the liver fluke, Fasciola hepatica

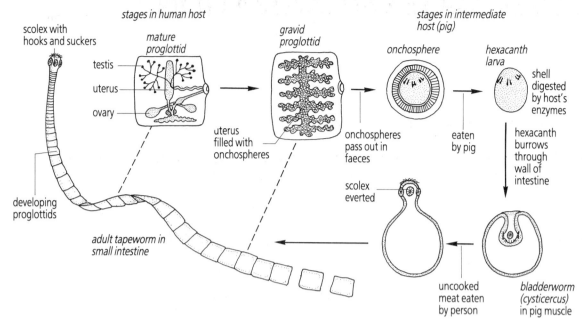

stages in human host

scolex with hooks and suckers

mature proglottid

testis

uterus

ovary

developing proglottids

adult tapeworm in small intestine

uterus filled with onchospheres

gravid proglottid

stages in intermediate host (pig)

onchosphere

hexacanth larva

shell digested by host's enzymes

onchospheres pass out in faeces

eaten by pig

hexacanth burrows through wall of intestine

scolex everted

uncooked meat eaten by person

bladderworm (cysticercus) in pig muscle

Fig 47·9 Structure and life cycle of the pork tapeworm, Taenia solium

The life cycle of the pork tapeworm, *Taenia solium*, also involves an intermediate host (see Fig. 47·9), but the parasite itself is more highly adapted. Adult tapeworms can live only in the human intestine. The **scolex** at the anterior end is attached to the host by hooks and suckers, while, at the posterior end, a series of sections, called **proglottids**, are budded off. The budding process is called **strobilization** and it allows mature tapeworms to reach a length of 2–3 m. A thin cuticle protects the tapeworm from the host's digestive enzymes, but allows simple food materials, like glucose and amino acids, to be absorbed directly from the host's intestine. Inside each segment, there is no trace of a digestive system. Instead, most of the space is occupied by reproductive organs. Self-fertilization within one segment can occur, but cross-fertilization with other segments or with other tapeworms in the same host is possible. A ripe proglottid typically contains 30,000–40,000 eggs or **onchospheres**, which pass to the outside in the faeces. Each egg is surrounded by a protective shell. The intermediate host, in this case a pig, becomes infected by consuming food contaminated with human sewage. Inside the pig's alimentary canal, the shell of the egg is digested away and the six-hooked, or **hexacanth**, larva burrows through the wall of the intestine into a blood vessel. Eventually, the larva comes to rest in a muscle and grows to form a bladderworm, or **cysticercus**. The cycle is completed when raw or partially cooked pork containing bladderworms is eaten. The beef tapeworm, *Taenia saginatum*, has an essentially similar cycle.

To varying degrees the life cycles of *Fasciola* and *Taenia* illustrate the range of adaptations needed by a successful parasite. Some of these adaptations are listed here:

1 Increased reproductive potential. The probability of a single egg reaching an appropriate host is often extremely small so that many eggs must be produced. A mature beef tapeworm produces more than a million eggs per day. In a typical life cycle, asexual and sexual phases of reproduction alternate to produce enormous numbers of offspring, without loss of genetic variability.

2 Reduced structure. Loss of non-essential structures is a very common feature of parasitic forms. It is advantageous and likely to be favoured by natural selection since it allows more resources to be directed towards reproduction.

3 Protection from host defence mechanisms. Many organisms are killed by the stomach acid and digestive enzymes of mammals. Liver flukes and tapeworms survive by means of a protective cuticle. In a similar way, the malarial parasite, described in Unit 46, escapes attack by the human host's immune system by spending most of its time as an intracellular parasite, safe inside the host's own cells.

4 Transmission by an intermediate host. The most successful parasites have a life cycle involving two or more species of host organism. The advantages of such an arrangement are obvious for vector-borne organisms like the protozoans causing malaria or sleeping sickness, but are equally important for other parasites. The intermediate host is usually used for an additional phase of reproduction, and energy gained from its tissues enables the parasites to survive for a longer period away from their primary host. Furthermore, the behaviour of the intermediate host often makes transmission more likely, for example, by depositing parasites where they are liable to be eaten.

48 COELOMATE ORGANIZATION

Objectives

After studying this Unit you should be able to:–

- Draw and label simple diagrams to illustrate diploblastic, triploblastic, and coelomate organization

- Explain the main advantages of coelomate organization

- List the distinguishing features of organisms belonging to Phylum Annelida, Phylum Mollusca, Phylum Echinodermata, and Phylum Nematoda

- Outline the potential advantages of metameric segmentation

- Compare the structure and life activities of the ragworm, *Nereis diversicolor*, with those of the earthworm *Lumbricus terrestris*

- Draw and label a diagram to illustrate the main features of the molluscan body plan and comment on the variations observed in Class Pelecypoda, Class Gastropoda, and Class Cephalopoda

- Comment on the evolution of animal phyla and explain the meanings of the following terms: Parazoa, Metazoa, Radiata, Protostomia, Deuterostomia

48·1 What is a coelom? ■ ■ ■ ■ ■ ■ ■ ■ ■ ■ ■ ■ ■

The organisms discussed in this Unit show further developments in body structure and tissue organization compared with the phyla described in Unit 47. As outlined in Figure 48·1, coelenterates have a diploblastic structure, while platyhelminths are triploblastic. More advanced organisms have an additional body cavity called a **coelom** (Fig. 48·1C). This is a fluid-filled space in the mesoderm dividing the muscles of the gut from those of the body wall. Nematode worms, which are described briefly at the end of this Unit, possess one kind of cavity called a pseudocoelom. Such a cavity is merely an irregular space between cells: a true coelom has a definite shape and is lined by a thin layer of epithelial cells, known as the **peritoneum**. The outer layer of mesoderm in a coelomate animal is called **somatic mesoderm**, while the inner layer is called **splanchnic mesoderm**.

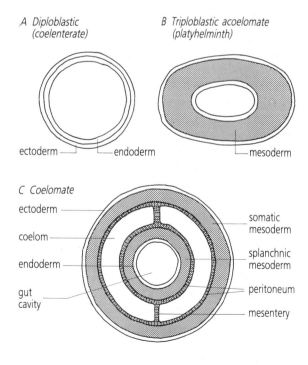

Fig 48·1 Diploblastic, triploblastic and coelomate organization

This new body plan gives a number of important advantages:

1 Separation of gut muscles and body wall muscles. In an acoelomate organism, such as the flatworm *Dendrocoelom*, food is dispersed throughout the branching digestive system by movements of the whole body. All coelomate organisms have a digestive system with two openings, the mouth and anus, and gut muscles which function independently of the body wall. These features allow digestion to become much more efficient because muscular movements like peristalsis can be developed, and also because the alimentary canal can be increased in length to provide a greater surface area for absorption of digested substances. The alimentary canal becomes more fully differentiated as an organ system, with separate regions specialized to perform different digestive functions.

2 Locomotion. The coelom can aid locomotion in two ways. In the earthworm, the coelomic fluid provides a 'hydrostatic skeleton'. The circular and longitudinal muscles of the body wall act on the fluid enclosed within each segment and, because the fluid is incompressible, the shape of the segments can be controlled. In other organisms, efficient systems of locomotion have developed based on hard skeletal materials, like chitin or bone, and the muscles of the body wall are specialized to produce limb movement.

3 Development of internal organs. The coelom provides a protected place for specialization of organ systems. Sometimes, the coelomic fluid acts as a medium for exchange, providing food and oxygen and removing waste, but, in more advanced forms, specialized respiratory organs and a circulatory system have developed. These structures are essential for the evolution of large active organisms because, with the loss of a flattened body shape, exchange of materials between tissue cells and the external environment cannot be accomplished by simple diffusion.

Within the coelom, body organs are held in position by sheet-like extensions of the peritoneum, called **mesenteries** (refer to Fig. 48·1C). The organs are supported and lubricated by the coelomic fluid so that they slide over each other during digestive or respiratory movements.

There are many possible variations in coelomate organization some of which are illustrated by the groups of organisms considered in the remainder of this Unit, namely Phylum Annelida (segmented worms), Phylum Mollusca (soft-bodied animals), and Phylum Echinodermata (spiny-skinned animals). These groups are not closely related and each has its own particular specializations. As you read, you should try to assess to what extent the coelom has contributed to the success of these three groups.

48·2 Phylum Annelida (segmented worms) ■ ■ ■ ■ ■ ■ ■

CLASSIFICATION

Phylum: Annelida (segmented worms)
triploblastic, coelomate, bilaterally symmetrical, body consisting of many similar segments, nephridia present
marine, freshwater, terrestrial
heterotrophic: sessile, free-living, parasitic

Class: Polychaeta
  marine worms with parapodia and numerous chaetae
  Arenicola, *Nereis*, *Pomatoceros*

Class: Oligochaeta
  freshwater and terrestrial forms with relatively few chaetae
  Lumbricus

Class: Hirudinea (leeches)
  parasitic forms without parapodia and chaetae
  Hirudo

Segmented body plan

Apart from the coelom, the most important feature of annelid structure is **metamerism**, that is, the subdivision of the body into many similar segments. Typically, an individual segment, or **metamere**, contains a number of complete organs, which are therefore duplicated many times along the length of the organism. Such a body plan is extremely versatile because it allows further specialization of organ systems. Segments at the anterior become adapted for feeding and, through cephalization, sense organs and nervous tissue are concentrated in the head region. At the same time, structures duplicated further back can be reduced in size or may disappear altogether. In a similar way, segments in other parts of the body become specialized for locomotion, digestion, or reproduction. As you will see shortly, annelids have exploited this body plan to good effect. Two other major groups of organisms show comparable forms of metameric segmentation, namely arthropods and chordates (described respectively in Units 49 and 50). The success of a segmented body plan is an important factor in helping to explain the tremendous diversity of forms observed among these organisms.

Polychaete adaptations

Polychaetes are the least specialized group of annelids. They are marine or estuarine. British representatives include ragworms, *Nereis*, and lugworms, *Arenicola*, as well as sessile tube-building worms like *Pomatoceros*, all of which are found abundantly in seashore habitats.

The structure of *Nereis diversicolor* is illustrated in Figure 48·2. Typically, the animal is 7–10 cm long and consists of 90–120 segments. With the exception of the head and the last segment, every segment carries a pair of fleshy projections, called **parapodia**, which are used as paddles during swimming and for gas exchange. Tufts of sharp bristles, or **chaetae**, sticking out from the tips of each parapodium help to grip smooth surfaces and may give some protection against predators.

In transverse section, Figure 48·2B, the body wall is seen to consist of several layers. The outer **cuticle** is a tough but flexible covering secreted by the epidermis. Beneath this lie layers of **circular** and **longitudinal muscle** and large strands of oblique muscle attached to the bases of the parapodia. The cavity of the coelom is completely enclosed by the peritoneum and is separated from adjacent segments by an internal partition, or **septum**, which helps to hold the internal organs in place. Coelomic fluid contains dissolved substances and amoeboid cells.

Many *Nereis* species are active carnivores pursuing smaller organisms with their powerful jaws. *Nereis diversicolor* has a more sedate method of food capture and lives in soft sand or mud where it forms a conical net of mucus extending around the opening of its burrow. Movements of the parapodia maintain a current of water which draws in small organisms and other food particles. From time to time, the net is eaten together with its contents, and the animal makes another.

The digestive system (Fig. 48·2C) has two openings. At the anterior end, the muscular **pharynx** turns inside out as it is extended through the mouth, exposing a pair of jaws which grasp the food. Extension is achieved by a sudden contraction of the circular muscles of the body wall which increases the hydrostatic pressure of the coelomic fluid. Direct muscles pull the pharynx inwards. There is a short oesophagus leading to the intestine where chemical digestion and absorption occur. Solid waste is voided via the anus.

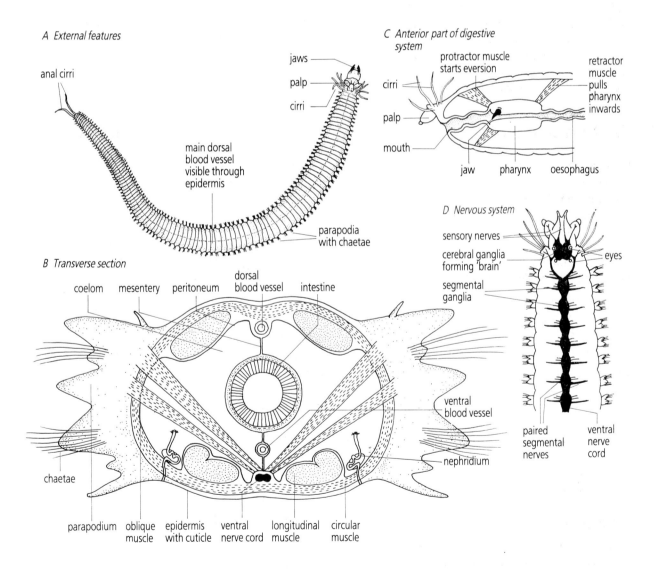

Fig 48·2 Structure and organ systems of Nereis diversicolor

Unlike the simple multicellular organisms described in Unit 47, many annelids have well developed circulatory systems containing blood with an oxygen-carrying pigment. In *Nereis*, haemoglobin is used, but in some tube-dwelling forms, this is replaced by **chlorocruorin**, a green copper-based pigment. Blood is circulated by muscular contractions of the larger vessels and flows forwards in the main dorsal vessel, which can usually be seen as a thin red line. Gas exchange occurs in capillary networks close to the surface of the parapodia, assisted by movements which maintain a flow of oxygenated water.

Excretion and osmoregulation appear to be carried out by the pair of **nephridia**, found inside almost every segment. These have a tubular structure reminiscent of vertebrate kidney tubules and it is tempting to suggest that they function in a similar way. However, their precise mode of action remains in doubt.

The nervous system of *Nereis* (Fig. 48·2D) is clearly centralized and consists of a definite **ventral nerve cord** with segmental ganglia and paired segmental nerves. **Giant fibres** present in the ventral nerve cord provide a means of rapid communication. At the anterior end, the **cerebral ganglia** form a primitive 'brain' with connections to well-developed sense organs, which include chemical, tactile, and light-sensitive structures. Each of the four eyes has a rudimentary lens capable of gathering light from different directions. Such a high degree of cephalization is explained by the animal's carnivorous habits.

Swimming movements are coordinated by the segmental ganglia. Interaction between the ganglia triggers waves of muscle contraction which throw the body into S-shaped undulations, and contraction of parapodial muscles assists their paddle-like action. In some polychaetes, parapodia are modified for other functions, for example, as feathery gills in *Arenicola*.

The sexes are separate and spawning occurs in early spring, when gametes are formed inside most segments. Ripe females burst open to release their eggs and males discharge sperm into the sea. Fertilization is external and zygotes develop into ciliated **trochophore** larvae.

Oligochaete adaptations

Oligochaetes are terrestrial or freshwater annelids thought to have evolved from a polychaete ancestor. Many species, including the earthworm *Lumbricus terrestris* (see Fig. 48·3) are adapted for burrowing. The numerous differences between such an organism and a marine polychaete like *Nereis* can be attributed almost entirely to their different life styles. Earthworms are adapted for a terrestrial existence and are harmless herbivores munching their way slowly through the soil. Like all oligochaetes, they lack parapodia and have a greatly reduced number of chaetae, so that they present a relatively smooth profile.

The body of *Lumbricus* can be up to about 25 cm long and consists of about 150 segments. At the anterior end, a sensory lobe, called the **prostomium**, overlaps the mouth, which is located in the first true segment. With the exception of this segment and the final segment, or **pygidium**, which carries the anus, every body segment possesses four pairs of chaetae (refer to Fig. 48·3A

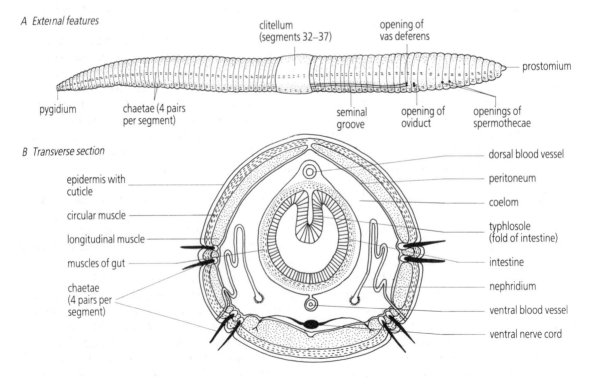

Fig 48·3 Structure of Lumbricus terrestris

426

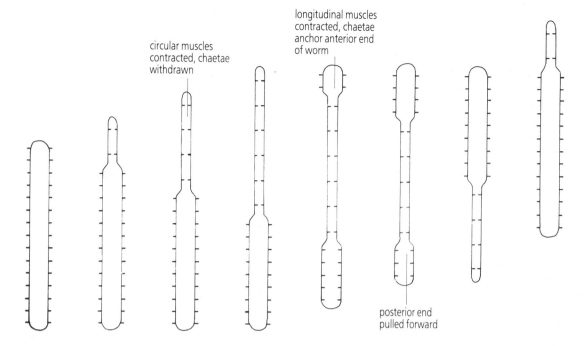

circular muscles contracted, chaetae withdrawn

longitudinal muscles contracted, chaetae anchor anterior end of worm

posterior end pulled forward

Fig 48·4 Locomotion in Lumbricus terrestris

and 48·3*B*). There is little visible specialization of the head region, but a thickened region, called the **clitellum**, is found from segments 32–37. Shallow grooves run along the ventral surface from the clitellum to the external openings of the reproductive organs.

Earthworms feed on soil and on vegetation gathered from the surface. Their activities aerate and mix the soil and promote the formation of humus, so that they play an essential part in improving soil fertility. Coping with a diet which is often poor in nutrients requires a number of modifications in the structure of the alimentary canal, which now includes a **crop** for storing ingested material, and a muscular **gizzard** for grinding. The **typhlosole** projecting from the dorsal surface of the intestine (see Fig. 48·3*B*) provides increased surface area for absorption. *Lumbricus terrestris* voids its faeces within the soil, unlike several other common species, which deposit their casts on the surface.

The remaining organ systems of *Lumbricus* are fairly similar to their counterparts in *Nereis*. However, the method of locomotion used by *Lumbricus* is very different and does not involve the lateral undulations of the body characteristic of polychaete swimming. Instead, movement results from changes in the shape of individual segments produced by contraction of the body wall muscles. The fluid enclosed in the coelom of each segment cannot be compressed and acts as a 'hydrostatic skeleton'. Thus, when the circular muscles contract, the segments tend to become longer and thinner, while contraction of the longitudinal muscles causes them to become shorter and fatter. During crawling, shown diagrammatically in Figure

48·4, alternate waves of contraction in the circular and longitudinal muscles, are combined with movements of the chaetae which anchor one part of the animal while the remainder moves along.

Sexual reproduction in terrestrial organisms is frequently more complex than in aquatic forms, many of which discharge their gametes directly into the surroundings. On land, copulation and internal fertilization are essential, but can only be achieved if potential mating partners are able to find one another and interact appropriately. Although earthworms are hermaphrodite, self-fertilization is prevented by the physical arrangement of the reproductive organs, which are formed in a number of anterior segments. When a suitable mate has been identified, the worms become linked by sticky secretions from the clitellum and mutual transfer of sperm takes place. Sperm cells are stored in the body and used for fertilization during the formation of egg cocoons, which may continue for several months.

Hirudinea

The leeches, Class Hirudinea, are specialised as external parasites, many of which suck the blood of vertebrates. They are closely related to oligochaetes but have a reduced number of internal partitions and no chaetae. Large suckers at the anterior and posterior ends are used for attachment. During feeding, the medicinal leech, *Hirudo medicinalis*, once in popular use for blood-letting, makes a Y-shaped wound and injects anaesthetic and anticoagulant substances. The leech gorges itself with blood and finally drops off. Lateral pouches in the walls of the digestive system hold enough blood to last for months.

427

48·3 Phylum Mollusca (soft-bodied animals) ■ ■ ■ ■ ■ ■ ■

CLASSIFICATION

Phylum: Mollusca (soft-bodied animals)
triploblastic, coelomate, unsegmented body
consisting of head, muscular foot and visceral
mass, radula and shell usually present
marine, freshwater, terrestrial
heterotrophic: sessile, free-living

Class: Pelecypoda (Lamellibranchiata)
body laterally flattened, shell in two sections
hinged together
Mytilus

Class: Gastropoda
bilateral symmetry obscured by torsion
Helix, Limnaea, Patella

Class: Cephalopoda
head well developed and surrounded by
tentacles
Loligo, Octopus

Molluscs include snails and slugs, clams, mussels,

limpets, and large free-swimming forms like octopus and squid. More than 100,000 species have been described, making molluscs one of the most successful invertebrate groups. They are abundant in aquatic and damp terrestrial habitats and a tremendous variety of forms has evolved.

The essential features of the molluscan body plan are illustrated in Figure 48·5A. As you can see, the body consists of three main sections, a **head** bearing tentacles, a muscular **foot** responsible for locomotion, and a **visceral mass**, which contains the remainder of the body organs. Above the visceral mass, a fold of tissue, called the **mantle**, extends like an umbrella to cover the rest of the body. Its upper surface secretes a hard calcareous **shell**, helping to protect the delicate gills located inside the mantle cavity. The digestive system is well developed and includes a distinct stomach and digestive glands. A unique rasping organ, the **radula**, can be protruded through the mouth and is used by herbivores for breaking off small fragments of plant material, or by carnivores for boring into the flesh of their prey.

A Basic molluscan body plan

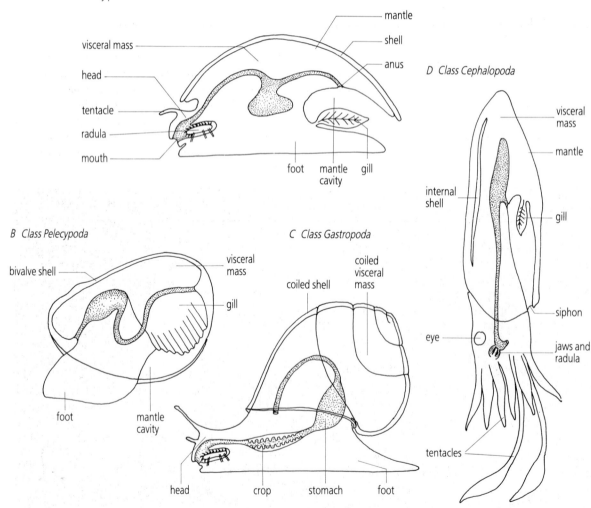

Fig 48·5 Molluscan body plan and its variations in the three main classes of molluscs

In its simplest form, the molluscan body plan occurs in only a very few living molluscs, and each of the three main classes of modern molluscs has developed its own particular modifications (see Fig. 48·5*B*, *C*, and *D*). Class Pelecypoda includes many familiar marine molluscs like clams, scallops, and mussels, which possess a hinged or **bivalve** shell. In the common mussel, *Mytilus edulis*, as in most other pelecypods, the gills are enlarged and specialized for filter feeding. Each gill is penetrated by thousands of tiny holes lined with cilia which beat to produce a steady current of water. The gill acts as a sieve and trapped particles of food are transported to the mouth by a flow of mucus. *Mytilus* normally lives attached to rocks by means of **byssus threads** secreted by byssal glands in the foot, but, in many other pelecypods, the foot is adapted for burrowing and holds the animal in position in the sand or mud.

Gastropod molluscs are distinguished by a curious process, called **torsion**, which takes place during early development. Typically, the visceral mass is rotated through 180°. In addition, the organs on one side of the body degenerate and the visceral mass becomes coiled inside a spiral shell. Most shells have a clockwise twist. The advantages of torsion are not fully understood. Terrestrial gastropods like *Helix aspersa*, the common garden snail, belong to a group called pulmonates, in which the moist lining of the mantle cavity is used for gas exchange and forms a simple lung. Many freshwater snails are pulmonates which have reverted to an aquatic existence, as in the case of *Limnaea*.

Squid, cuttlefish, and octopus, Class Cephalopoda, have abandoned the sedentary habits of other molluscs and are active predators. They show a remarkable degree of cephalization, possessing complex brains and highly specialized sense organs. Experimental studies show that cephalopods are intelligent animals which learn quickly. Giant squid are the largest of all invertebrates and may be as much as 20 m in length, including tentacles.

48·4   Phylum Echinodermata (spiny-skinned animals)   ■ ■ ■ ■ ■

CLASSIFICATION

Phylum: Echinodermata (spiny-skinned animals)
triploblastic, coelomate, unsegmented, bilateral
  symmetry in larval forms replaced by pentameric
  symmetry in adults, water vascular system with
  tube feet used for locomotion, calcareous ossicles
  in dermal layer
marine forms only
heterotrophic: free-living

Class: Asteroidea (starfish)
  Asterias

Class: Echinoidea (sea urchins)
  Echinus

Starfish and sea urchins are the most familiar echinoderms and share a body plan which is radically different from that of any other invertebrate group. Their most striking features are five-pointed, or **pentameric**, radial symmetry, and spiny skins due to the possession of calcareous ossicles.

Sea urchins are one of the two main classes of echinoderms.

Asterias rubens, illustrated in Figure 48.6, is the commonest British starfish, found on the lower shore and in rock pools. It moves using rows of **tube feet** connected to an internal **water vascular system** (see Fig. 48·6*B*). This is a unique echinoderm feature and consists of radial canals in each arm linked to a central ring canal, which opens to the exterior via a short tube, known as the stone canal, and a sieve plate called the **madreporite**. Each tube foot is connected to a bulb-shaped **ampulla** with walls containing muscle fibres. Contraction of the ampulla closes a valve inside the tube linking the ampulla with the rest of the system, so that fluid is pushed into the tube foot, causing it to be extended. The tube feet end in suckers which grip the substrate and are used to pry open the shells of bivalve molluscs like scallops and cockles, on which starfish feed. The mouth of the starfish is located on its underside, or **oral** surface, while the anus is located near the centre of the upper, or **aboral** surface. During feeding,

parts of the digestive system are extruded through the mouth and ingested liquid passes to a branching system of tubes inside each arm.

A transverse section of one arm (Fig. 48·6*C*) reveals the internal structures responsible for most other body processes. Gas exchange between the coelomic fluid and the external medium occurs through delicate tubular extensions of the peritoneum called **skin gills**. These are protected by the calcareous ossicles embedded in the dermal layer and projecting as fixed or movable spines. Tiny pincer-like **pedicellaria** formed from modified spines are able to open and close, trapping small organisms and helping to keep the external surface clean.

Sea urchins have a complete shell formed by fusion of ossicles and a body structure similar to that of starfish, but with the five arms folded upwards and joined at the top. They feed on vegetation scraped from the rocks.

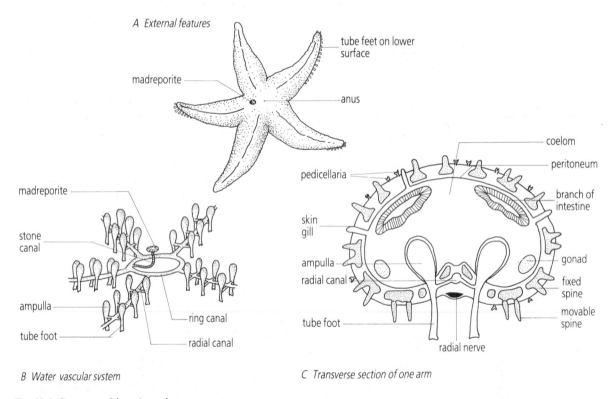

Fig 48·6 Structure of Asterias rubens

48·5   Phylum Nematoda (roundworms) ■ ■ ■ ■ ■ ■ ■ ■

CLASSIFICATION

Phylum: Nematoda (roundworms)
triploblastic, pseudocoelomate, unsegmented,
 locomotion by contraction of longitudinal muscle
 fibres (circular muscles absent)
widespread: marine, freshwater, terrestrial
heterotrophic: free-living, many parasitic forms
Ascaris, Enterobius, Toxocara, Trichinella, Wuchereria

Nematodes are very common aquatic and soil organisms which live in millions in almost every available habitat. Most species are microscopic, but a few reach 30 cm in length. Although successful, they are significantly less advanced in structure than the other animal phyla described in this Unit. The body cavity of a nematode (see Fig. 48·7*C*) is a **pseudocoelom** formed by fusing of vacuoles in the mesoderm layer. The pseudocoelom confers some of the advantages of a true coelom, but the intestine is a single layer of cells without its own muscles, so that food must be agitated by movements of the whole body.

Typical nematodes, Figure 48·7A, are small white or transparent worms easily recognized by their continual twisting and bending movements. These result from contraction of the four separate groups of longitudinal muscles, which flex the body in opposite directions. There are no circular muscle fibres.

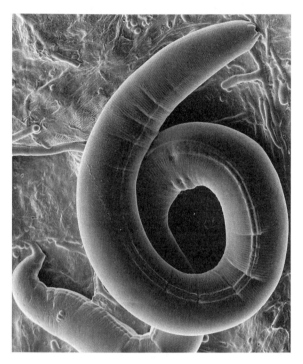

A Typical nematode (SEM)

Parasitic nematodes are found in a tremendous variety of animals and plants and many are economically important agents of disease in livestock and crops. About 50 species affect the human population. *Ascaris lumbricoides* lives in the intestine and is one of the largest nematodes. Although rare in Britain, it is one of the commonest of all human parasites. Adult worms cause relatively little damage unless they become numerous enough to block the intestine. Another common parasite is the pin worm, *Enterobius vermicularis*, which lives in the colon. Ripe females deposit their eggs around the anus, often producing intense itching. *Ascaris* and *Enterobius* have similar life histories to the extent that no intermediate hosts are involved and that eggs produced are directly infective to the human host. Safe drugs are available for the treatment of both infections.

More serious diseases include toxocariasis caused by parasites transmitted from domestic dogs and cats. People are not the normal hosts but are liable to become infected by handling their pets. The young *Toxocara* worms migrate around the body causing extensive damage and may lodge in the retina to produce blindness, especially in children. In many countries, pork is likely to contain the cysts of *Trichinella* worms and must be properly cooked to avoid the debilitating effects of trichiniasis, which result from the encystment of new larvae in human tissue. *Wuchereria bancrofti* and related nematodes are transmitted by mosquitoes and prolonged infections give rise to the horrifying symptoms of elephantiasis.

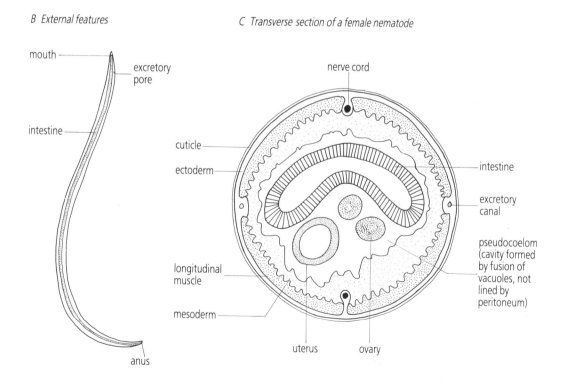

B External features

C Transverse section of a female nematode

Fig 48·7 Nematodes

Depending on the classification scheme used, about 30 different phyla are recognized in the Kingdom Animalia. Eleven of these, chosen to illustrate the most important variations in body plan, are described in this book. There is little doubt that animals are derived from protoctist ancestors, but the precise relationships between the various groups remain in doubt. The coelom, for example, develops in several distinct ways, and it is not clear which method represents the ancestral type, or whether coelomate organization evolved independently on more than one occasion. Similar problems exist in explaining the origins of segmented and unsegmented forms.

One of several possible evolutionary schemes which overcome the major difficulties is given in Figure 48·8. The first animals are thought to have arisen from heterotrophic flagellated protoctist organisms. The **Parazoa**, often considered as a Subkingdom, includes organisms with no definite shape and no organized tissues, or, in other words, the sponges, Phylum Porifera. The remaining phyla, grouped as Subkingdom **Metazoa**, have a definite shape, organized tissues, and possess a digestive cavity. The simplest metazoans, called **Radiata**, are diploblastic and radially symmetrical, like the coelenterates, Phylum Cnidaria.

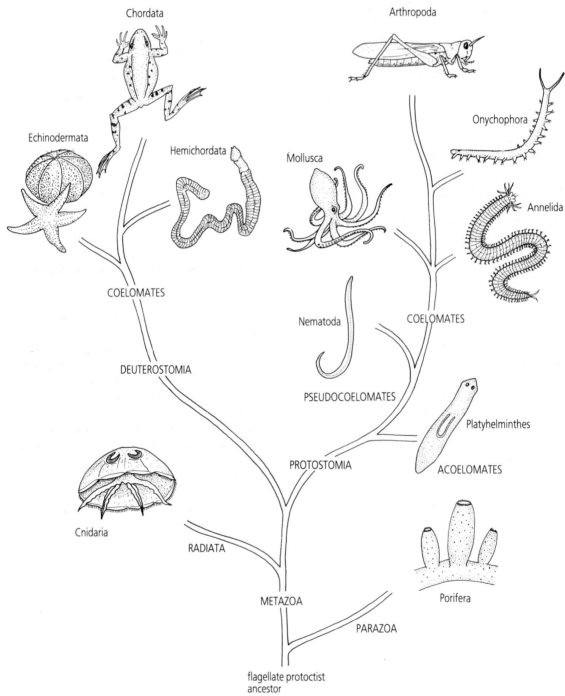

Fig 48·8 One possible scheme of evolutionary relationships of animal phyla

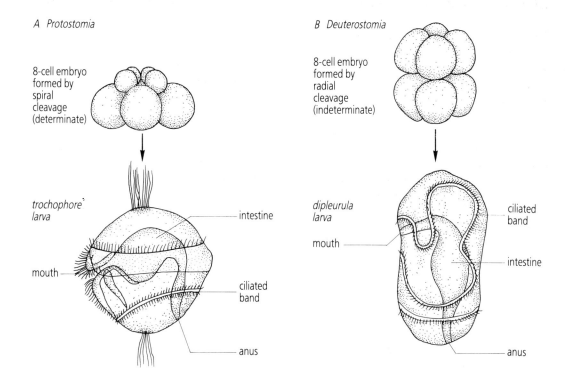

A Protostomia

8-cell embryo
formed by
spiral
cleavage
(determinate)

trochophore
larva

mouth

intestine

ciliated
band

anus

B Deuterostomia

8-cell embryo
formed by
radial
cleavage
(indeterminate)

dipleurula
larva

mouth

ciliated
band

intestine

anus

Fig 48·9 Embryonic development of protostomes and deuterostomes

A major division is recognized between **Protostomia** and **Deuterostomia**, now widely believed to represent two quite separate lines of animal evolution. The evidence for this distinction rests on studies of embryonic development, as outlined in Figure 48·9. Protostomia, that is, Platyhelminthes, Nematoda, Annelida, Mollusca, Onychophora, and Arthropoda, are characterized by **determinate** cleavage following a spiral pattern. The third division is unequal so that the top layer of cells in an 8-cell embryo are smaller and lie in the angles between the cells below. Each cell is destined to become a particular part of the adult organism and its removal prevents normal development. Later, the mesoderm arises from a group of cells located near the junction of the ectoderm and endoderm. Typically, a **trochophore** larva is formed in which the embryonic opening, or blastopore, becomes the mouth.

In sharp contrast, deuterostomes, comprising Echinodermata, Hemichordata, and Chordata, show **indeterminate** radial cleavage. Individual cells removed at an early stage are able to develop into small but complete embryos and their loss can be tolerated. The mesoderm is formed by evagination from the gut cavity, or archenteron, and the blastopore becomes the anus. A structurally different larva, called a **dipleurula** is formed.

The acorn worms, Phylum Hemichordata, are marine worm-like organisms closely related to the chordates, as indicated by their embryological development, and in some species, by the presence of pharyngeal gill slits and a hollow dorsal nerve cord. A structure similar to the notochord may also be present, although this is now known to develop in a different way.

Without doubt, the most successful and sophisticated protostomes are the arthropods, while the most advanced deuterostomes are the chordates. As you read the descriptions of these phyla given in the next two Units, you will find it useful to recall their very different evolutionary backgrounds.

49 ARdTHROPODS

Objectives

After studying this Unit, you should be able to:–

- List the characteristics of organisms belonging to the Phylum Arthropoda

- Comment on differences and similarities between the body plans of arthropods and annelids

- Identify organisms belonging to the five main classes of modern arthropods

- Draw and label a diagram showing the structures present in the body wall of an arthropod and comment on their functions

- Describe the structure of a typical arthropod hinge joint

- Give examples of the versatility of the arthropod exoskeleton as a building material

- Describe the process of ecdysis and comment on its importance

- Give an account of the structure, organ systems, and life activities of the locust

- List the important adaptations of insects for life in terrestrial habitats

- Distinguish between the mechanisms of flight using direct and indirect muscles

- Describe the life cycles of hemimetabolous and holometabolous insects and explain the advantages of complete metamorphosis

- Give examples of the economic importance of arthropods

49·1 Origin of arthropods ■ ■ ■ ■ ■ ■ ■ ■ ■ ■ ■ ■ ■ ■

Arthropods have clear evolutionary links with annelids. Organisms of both groups are protostomes, showing spiral cleavage during early development, and both types have a segmented body plan. Their common anatomical features include a dorsal blood system, a ventral nerve cord with paired segmental ganglia, and paired appendages on at least some of the body segments.

Additional evidence for the arthropod-annelid link is provided by the **velvet worms**, Phylum Onychophora, a small group of tropical organisms which includes *Peripatus* (see Fig. 49·1). Like other members of its phylum, *Peripatus* possesses a combination of arthropod and annelid features. For example, the animal has an arthropod-like blood system and its coelom is reduced by the growth of

connective tissue in which a new cavity, called the **haemocoel**, develops. Gas exchange occurs by means of a system of tubes similar to the tracheal system in arthropods. On the other hand, the excretory system consists of nephridia, typically an annelid structure.

Locomotion in *Peripatus* depends on hydrostatic pressure which helps to keep the hollow legs rigid and effective as supports. In all arthropods the same function is carried out by a rigid outer casing, or **exoskeleton**, with joints to allow movement. The exoskeleton provides mechanical protection and acts as a framework for muscle attachment.

The three essential components of the arthropod body plan, a segmented body, jointed appendages, and a hard exoskeleton, first occurred together in several groups of organisms which developed from annelid-like ancestors more than 550 million years ago. These groups included the **trilobites** which dominated the seas for hundreds of millions of years, but subsequently became less abundant and finally died out about 220 million years ago. Crustaceans, which first appeared in the fossil record at about the same time, continued to thrive and are represented today by numerous marine and freshwater species. Arachnid fossils are known from 400 million years ago, while fossil insects and other terrestrial forms are found in rocks 350 million years old.

Like their ancestors, modern arthropods are spectacularly successful organisms. Together, they account for more than 80% of all known animal species (Fig. 49·2). This Unit outlines some of the variety of arthropod structure and discusses the important factors which contribute to their success.

Fig 49.1 The velvet worm, Peripatus

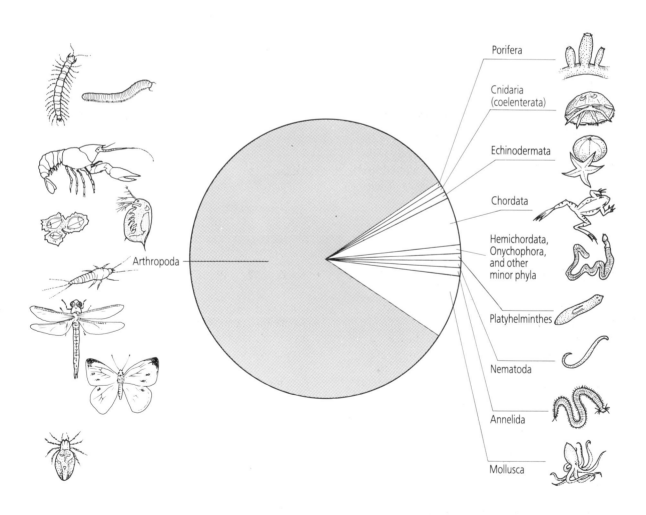

Fig 49·2 Numbers of animal species in different groups

435

49·2 Classification of arthropods ■ ■ ■ ■ ■ ■ ■ ■ ■ ■

Figure 49·3 shows the typical appearance of organisms belonging to the five main classes of modern arthropods. A formal classification scheme is given below:

CLASSIFICATION

Phylum: Arthropoda
triploblastic protostomia, coelom replaced by
  haemocoel, segmented body with jointed legs
  and hard external skeleton
marine, freshwater, terrestrial
heterotrophic: free-living, parasitic

Class: Chilopoda (centipedes)
  elongated body with many walking legs, one
    pair from each adult leg-bearing segment
  Lithobius

Class: Diplopoda (millipedes)
  elongated body with many walking legs, two
    pairs from each adult leg-bearing segment
  Blaniulus

Class: Crustacea
  marine and freshwater arthropods with two
    pairs of antennae, most segments with
    appendages
  Astacus, Balanus, Carcinus, Daphnia, Oniscus

Class: Insecta
  freshwater and terrestrial arthropods with
    one pair of antennae, clearly-defined head,
    thorax, and abdomen, three pairs of legs

  Sub-class: APTERYGOTA
    wingless insects
    Lepisma

  Sub-class: PTERYGOTA
    insects with wings

    Division: Hemimetabola (Exopterygota)
      life cycle without pupa stage, young
        stages resemble adults, wings develop
        externally
      Anax, Schistocerca

    Division: Holometabola (Endopterygota)
      life cycle including pupa stage, young
        stages unlike adults, wings develop
        internally
      Musca, Pieris

Class: Arachnida
  terrestrial arthropods with chelicerae, body
    divided into prosoma and opisthosoma, four
    pairs of legs
  Epeira, Ixodes, Scorpio

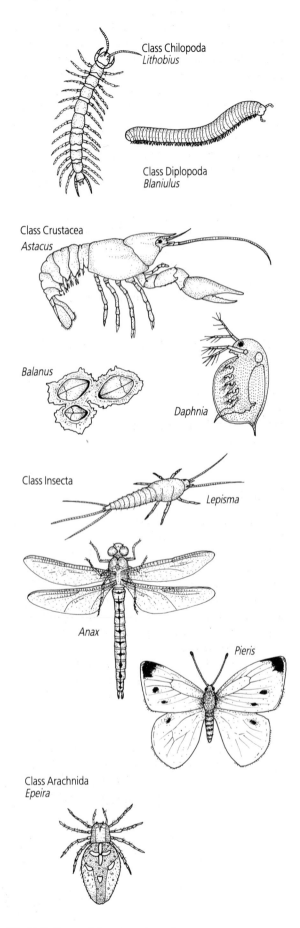

Class Chilopoda
Lithobius

Class Diplopoda
Blaniulus

Class Crustacea
Astacus

Balanus

Daphnia

Class Insecta

Lepisma

Anax

Pieris

Class Arachnida
Epeira

Fig 49·3 Range of form in arthropods

Centipedes (Class Chilopoda) and millipedes (Class Diplopoda) are distinguished by the number of pairs of legs which arise from each adult body segment. The body shape and the number of legs are adaptations to rather different modes of life. Millipedes are herbivores which live in leaf litter and soil – their extra legs provide additional thrust for burrowing. Centipedes are found in similar habitats but are specialized as predators. Their legs are longer and more widely spaced, with the result that the animal can move rapidly in pursuit of prey.

Crustaceans are successful marine and freshwater organisms with two pairs of antennae and appendages from most body segments. Many species are small or microscopic, like freshwater *Daphnia*, or marine plankton species, which occur in immense numbers in the surface waters of the oceans. Larger species include crabs and lobsters, some of which grow to considerable size, as in the case of giant spider crabs, which sometimes exceed 45 cm in body width, with legs spanning 3·5 m or more. Very few crustaceans are terrestrial; woodlice are among the best adapted forms, but are confined to moist habitats.

Insects, on the other hand, are predominantly terrestrial. Their most characteristic features are a body clearly divided into three regions, the head, thorax, and abdomen, a single pair of antennae, and three pairs of legs. Some insects, like springtails and silverfish, *Lepisma*, are wingless (Sub-class Apterygota), but the majority have two pairs of wings arising from the second and third thoracic segments (Sub-class Pterygota). Within this group there are important variations in life cycle and development.

Arachnids are also terrestrial. They include spiders, scorpions, ticks and mites. All these organisms have a body in two sections, the prosoma and opisthosoma, and possess four pairs of legs. There are no antennae – instead the first segment of the prosoma carries a pair of jointed **chelicerae** armed with poison fangs.

49·3 Advantages and disadvantages of the exoskeleton ■ ■ ■ ■

All arthropods have an exoskeleton. The structure and properties of this hard outer layer are of major importance in explaining the success of the group, but, as you will discover, there are also some disadvantages.

Structure

The typical structure of part of the exoskeleton, or **cuticle**, is illustrated in Figure 49·4. The outermost layer, called the **epicuticle**, is extremely thin, often as little as 1–2 μm. In crustaceans, centipedes and millipedes, this layer is permeable, but, in insects and spiders, it is impregnated with waxes which make it completely impermeable, so that it forms a vital waterproof barrier which prevents drying out.

The second, inner layer, known as the **procuticle**, is 30–40 μm thick and is composed of a tough nitrogen-containing polysaccharide, called **chitin**, together with fibrous proteins. It forms a continuous flexible covering, or **endocuticle**, reinforced at intervals by rigid sections of hardened **exocuticle**, shaped into skeletal plates called **sclerites**. Typically, the endocuticle is folded over in the regions between sclerites making **articular membranes** which allow movement between adjacent tubular sections of the exoskeleton. Internal notches or ridges of cuticle called **apodemes** provide surfaces for muscle attachment.

The entire cuticle is a non-living structure produced by the cells of the **epidermis**, which also secretes a tough **basement membrane**. The whole structure resists bacterial attack. Note that the body wall of an arthropod lacks the alternate layers of circular and longitudinal muscles so characteristic of annelid organization. The peritoneum is also absent because the coelom has been replaced as the main body cavity by a new cavity, called the **haemocoel**.

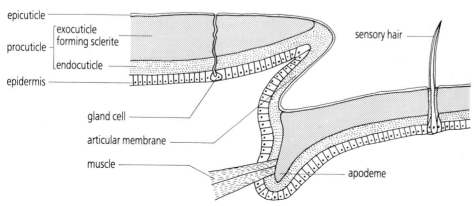

Fig 49·4 Structure of arthropod cuticle

Movement

An arthropod limb consists of a series of hardened tubular sections linked together. As you can see from Figure 49·5A, one of the tubes at each joint is equipped with a pair of small projecting pegs. These fit into corresponding sockets in the second tube so that the two tubes are hinged together. The range of movement possible is determined by the shapes of the tubes. Usually, the joint can be moved through about 60° from its fully straightened, or **extended** position, to its folded, or **flexed** position.

valve between the prosoma and opisthosoma, an action which unavoidably prevents the circulation of blood. Unfortunately, the spider's gas exchange organs are located in the rear body section, so that its burst of high speed is rapidly terminated by a lack of oxygen.

Versatility

The exoskeleton has excellent structural properties which have allowed body appendages to become specialized for a great variety of functions. In an

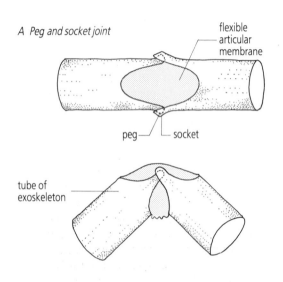

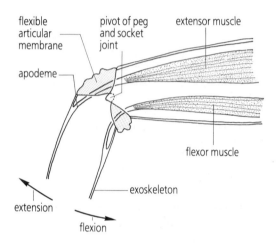

Fig 49·5 *Structure of an arthropod hinge joint*

The action of the muscles which operate the limb can be understood only when the three-dimensional structure of the joint is taken into account. Look carefully at Figure 49·5B and note that the extensor muscle is connected to its apodeme above the peg and socket pivot, so that it straightens the joint when it contracts. The flexor muscle, on the other hand, is connected below the pivot and bends the joint.

It is important to appreciate that hinge joints are the only mechanically satisfactory form of linkage which can be formed between tubes of similar sizes. Ball and socket joints are sometimes found where appendages meet the body, but free limb movement is only possible when the limb itself is subdivided into several sections linked by joints which hinge in different planes. This is why an arthropod's leg has many more joints than that of a mammal.

An interesting variation in the mechanism of limb movement occurs in spiders, where some of the leg joints lack extensor muscles, extension being achieved instead by fluid pressure from inside the front body section, or prosoma. The effect is similar to blowing out the finger of a rubber glove, and probably contributes to rapid movement. Pressure is maintained by sealing a

organism like a lobster, for example, different pairs of appendages serve as antennae, mouthparts, pincers, walking legs, gills, and reproductive structures. The sizes and shapes vary enormously, but the basic pattern of construction and the building material remain the same. Similarly, in insects, the mouthparts of different species are specialized to cope with many different diets. The mandibles of beetles can cut through copper sheet and give a painful bite. In mosquitoes, the same structures are modified to form needle-like stylets which pierce the skin, while in honey bees they are reduced to small spoons used for moulding wax.

Much of this versatility can be attributed to the hardening process which converts the soft material of the endocuticle into the tough rigid sclerites of the exocuticle (refer to Fig. 49·4). In insects, this process is called **sclerotization** and depends on the deposition of a tough insoluble protein, **sclerotin**, in the spaces between chitin molecules. The proteins become cross-linked by **tanning**, a chemical reaction in which phenol compounds are used as bridges between adjacent protein chains. In crustaceans, the exoskeleton is hardened by an alternative series of reactions which result in the deposition of calcium carbonate and calcium phosphate. This is called **calcification**.

Ecdysis

The greatest single disadvantage of the exoskeleton is that its size is fixed, so that it must be cast off at intervals to allow for growth. The process of shedding the old cuticle is called moulting, or **ecdysis**. A new cuticle is formed underneath, and, following ecdysis, the organism inflates its body to a new larger size before the cuticle hardens. These changes are reflected in the growth curves of arthropods (Fig. 49·6). At each ecdysis, the length of the organism changes abruptly, giving a characteristic step-like appearance. In contrast, the dry mass of the organism increases steadily between moults as new body materials are accumulated.

Some of the changes preceding ecdysis are outlined in Figure 49·7. In the first stage, the epidermal cells separate from the old cuticle and begin to divide laterally to produce a folded surface. At the same time, the epidermal cells release moulting fluid. This contains **protease** and **chitinase** enzymes which do not become active until the epidermal cells have secreted a new layer of epicuticle. The new epicuticle protects the epidermis from enzyme attack. However, the old endocuticle is completely digested and materials from it are reabsorbed and used to build the new cuticle. As much as 90% of the old cuticle may be recycled in this way. The only regions immune from digestion are the hardened sclerites of the exocuticle and the old epicuticle. When the new cuticle is completely formed, the organism swells by taking in air or water, splitting the remains of the old cuticle and allowing the organism inside to emerge. Until the new cuticle hardens, the organism is very vulnerable to predators or mechanical injury.

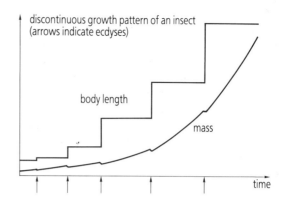

Fig 49·6 Growth curves of arthropods

The timing of ecdysis is controlled by the hormone **ecdysone** and is triggered by increasing concentrations of the hormone circulating in the blood.

Size

The exoskeleton is very efficient as a means of support for a small organism. However, as arthropods become larger, their volume and body mass increase in proportion to the cube of their linear dimensions, so that the exoskeleton must become very much heavier and thicker if it is to remain effective. There is no solution to this mechanical problem which limits the size of terrestrial arthropods, usually to no more than 10–15 cm in any direction. The most massive insects are goliath beetles which may reach 100 g. Aquatic arthropods gain support from the surrounding water and can be correspondingly bigger.

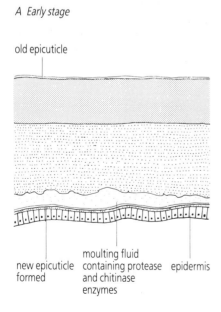

A Early stage

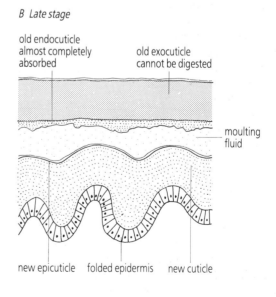

B Late stage

49·4 Insect adaptations ■ ■ ■ ■ ■ ■ ■ ■ ■ ■ ■ ■ ■ ■ ■

Insects are the most successful group of arthropods because many of them are supremely well adapted for life on land. This section describes the structure and life activities of insects, using the locust as an example to explain the important adaptations which enable insect species to thrive even in the driest terrestrial habitats.

External features

Figure 49·8*A* shows the appearance of the desert locust, *Schistocerca gregaria*. As you can see, each body region is specialized for particular functions. Cephalization is well advanced, with the nervous system, sense organs, and feeding apparatus all concentrated in the head. The most prominent sense organs are the antennae and the large compound eyes, although simple eyes, called ocelli, are also present.

Locusts have a large abdomen which contains the digestive, excretory and reproductive organs. Each segment is protected by two curved plates, the **tergum**, which covers the dorsal surface, and the **sternum**, covering the ventral surface. These are linked by folds of soft cuticle to allow for movement and expansion. The **spiracles** are the external openings of the tracheal system.

Feeding and digestion

Locusts have mouthparts specialized for biting and chewing plant material. As illustrated in Figure 49·8*B*, these consist of a flap-like upper lip, or **labrum**, and a lower lip, or **labium**, between which lie a pair of shiny black jaws, or **mandibles**, with large cutting and grinding edges, and a pair of secondary jaws, called **maxillae**. The maxillae and labium have jointed appendages called **palps**

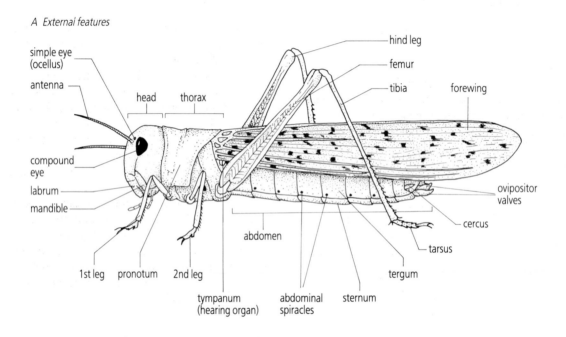

A *External features*

Fig 49·8 Structure and organ systems of Schistocerca gregaria

The thorax is specialized for locomotion and consists of three segments, each with a pair of jointed legs. All three pairs of legs have spines and claws to improve grip during walking and climbing, while the large hind legs are adapted for jumping. There are two pairs of wings, the long narrow forewings being kept folded over the larger hindwings except during flight. The dorsal sclerite, or tergum, of the first thoracic segment is enlarged to form a structure called the **pronotum** which extends backwards to cover the bases of the forewings.

which are used to manipulate and taste food before it is chopped into smaller pieces. In the later stages of chewing the food is moistened with saliva before being passed to the mouth.

The locust's gut consists of three main regions (see Fig. 49·8*C*). The **fore-gut** is lined with cuticle and comprises the oesophagus, crop, and gizzard. Ingested food material is stored in the crop where it may be partly digested by amylase enzymes in the saliva. The gizzard is an active grinding organ with internal ridges of cuticle covered by tiny teeth. From here, food enters the

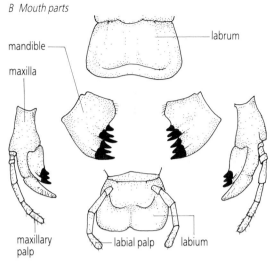

Fig 49·8 (cont.)

mid-gut which consists of a short tube lined by endodermal cells. This is the most important region for digestion and absorption. At its anterior end, tubular extensions, called **mesenteric caecae** produce enzymes and increase the surface area available for absorption. As food passes from the gizzard into the mid-gut region it is enclosed in a sieve-like tube of chitin, called the **peritrophic membrane**. This is permeable to enzymes and digested nutrients, but protects the delicate endodermal cells from abrasion. The peritrophic membrane is secreted continuously and forms an outer covering on the faecal pellets. Water is extracted from the faeces as they pass through the colon and rectum of the **hind-gut**, so that the pellets produced are quite dry.

C Digestive system

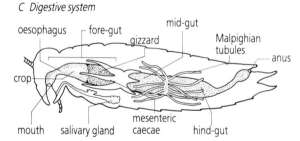

D Tracheal system

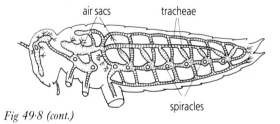

Fig 49·8 (cont.)

Excretion

The excretory system of the locust consists of **Malpighian tubules**. In dissection these appear as fine pink threads attached to the posterior end of the mid-gut at its junction with the hind-gut (refer to Fig. 49·8C). The tubules extend throughout the body cavity. Their ends are closed but they absorb

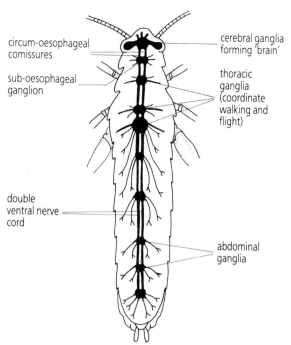

E Nervous system

Fig 49·8 (cont.)

nitrogenous waste products such as sodium and potassium urate from the blood. Muscles in the tubule wall agitate the blood and produce peristaltic movements. As the liquid inside the tubule travels towards the gut, carbon dioxide is secreted, producing acid conditions which favour the precipitation of solid **uric acid**. A paste containing uric acid crystals is emptied into the hind-gut where water is removed.

The locust's ability to carry out digestion and excretion without loss of water are important adaptations for terrestrial life. Locusts do not drink and obtain their water from the food they eat, or as 'metabolic water' produced by respiration. Some insects, like wood-boring beetles, which eat very dry food, survive almost entirely on metabolic water, making effective adaptations to prevent water loss essential.

Gas exchange and circulation

Another important adaptation for terrestrial life is an efficient gas exchange system. In insects this is a branching network of tubes called the **tracheal system** (Fig. 49·8D). The larger tubes, or **tracheae**, are lined with cuticle and have thickened rings or spirals of chitin to prevent collapse. From their external openings, air is piped through a system of air sacs to much narrower tubes, called **tracheoles**. These have permeable walls not lined by cuticle and are at least partly filled with fluid. They penetrate to every region of the body. Oxygen from the air dissolves in the fluid and diffuses into the tissue cells, while carbon dioxide produced by respiration diffuses in the opposite direction.

The system is ventilated by muscular movements of the abdomen coordinated with opening and closing of the spiracles. Air is drawn in through the anterior spiracles and expelled through those at the rear. At other times, the spiracle valves are closed to prevent loss of water.

The mechanisms of gas exchange used by insects eliminate the need for a well developed circulatory system. Insects have an **open circulation** in which the blood bathes the body organs directly (see Unit 17). The main structure is the tubular heart which collects blood from the haemocoel and pumps it forwards into the head. The blood percolates through the spaces between the tissues and slowly returns to the abdomen.

The action of indirect muscles is explained in Figure 49·9. The wings are raised by contraction of the vertical **elevator muscles** which stretch between the floor of the thorax, or **sternum**, and its movable roof, or **tergum** (Fig. 49·9A). The action of the **depressor muscles** is less obvious. These muscles run from front to back inside the thorax, and their contraction causes the roof of the thorax to arch upwards lifting the base of the wing (Fig. 49·9B). In many insects, the wings move downwards against increasing resistance until a definite 'click' point is reached. Some of the energy provided by muscle contraction is stored in pads of a rubber-like protein, called **resilin**, located in the wing pivots, and, as the wing passes

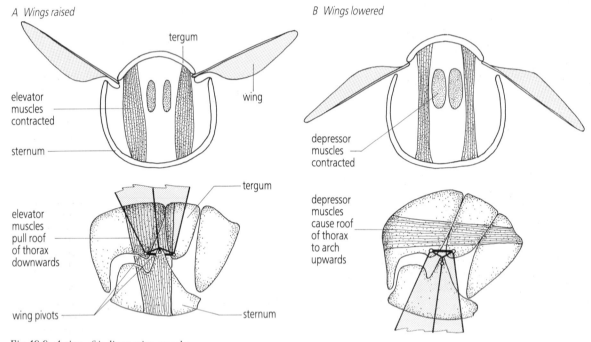

Fig 49·9 Action of indirect wing muscles

Flight

The ability to fly is clearly important. It allows insects to escape from their enemies, to exploit new food resources, to colonize new areas, and to find their mates more easily.

Two rather different mechanisms of flight have evolved. In locusts, and in insects like dragonflies, **direct muscles** attached to the bases of the wings provide the main power for propulsion. The wings beat relatively slowly, 20 times per second in the locust, and the forewings and hindwings tend to beat at different times. In the second group, which includes the majority of insects, the direct muscles are used for changing the angle of the wings during flight, or for folding the wings when at rest, but do not provide much power. Instead, the wings are made to move by **indirect muscles** which work by changing the shape of the thorax. Insects with indirect flight muscles often beat their wings rapidly, for example, 250 times per second in honey bees.

the click point, the stored energy is released, producing a powerful downstroke.

The click mechanism allows the thorax and its muscles to become an oscillating system and was essential for the evolution of small insect species. Such insects must beat their wings rapidly in order to fly effectively and require special flight muscles, called **fibrillar muscles**, which can be stimulated to vibrate at the appropriate frequencies. The wing tips follow a figure-of-eight path and the angle of the wings is automatically adjusted so that they provide lift for up to 85% cent of the cycle. In many groups, the second pair of wings is reduced in size, as in wasps and bees (Order Hymenoptera). In flies, midges, and mosquitoes (Order Diptera), the hind wings are reduced to small club-shaped balancing organs called **halteres**, while in beetles (Order Coleoptera) the forewings are modified to hardened protective **elytra** so that only the hindwings flap in flight.

Sense organs and behaviour

To coordinate complex activities, sophisticated sense organs and a well developed nervous system are needed. In locusts, the nervous system consists of a double ventral nerve cord with many neuron cell bodies concentrated into ganglia (Fig. 49·8E). Several ganglia are fused to form the brain which serves to collect information from the sense organs and to direct body activity accordingly.

Locusts are well equipped with sense organs. There are two large **compound eyes**, each composed of thousands of hexagonal **ommatidia**. As you can see from Figure 49·10, a single ommatidium is a complete visual unit isolated from its neighbours by pigment cells. Light rays are normally received over a very narrow angle (1–2°) so that a composite or 'mosaic' image is built up. The focus of the eyes is fixed but they are very sensitive movement detectors. In dim light, the pigment inside the pigment cells migrates to the ends of the cells so that light from adjacent units can stimulate the retinal cells. The image formed is now blurred but its brightness is increased. The simple eyes, or **ocelli**, have a structure similar to that of individual ommatidia. Simple eyes are present along with compound eyes in many adult insects, but their function is not fully understood.

The locust's antennae, mouthparts, and feet are covered with tiny bristles called **sensilla**, which are also found distributed over the body surface. Different types of bristles are sensitive to touch, air movements, or to chemical stimuli. Flying movements are initiated when a locust tethered to a wire is gradually lifted until its legs lose contact with the surface. If kept suspended, the locust stops flying after a time but can be made to start by blowing a jet of air towards its head. Both responses are reflexes triggered by stimulation of touch sensilla.

Some sensilla can detect sound vibrations, but locusts have an additional structure for hearing, called the **tympanum**. This is a thin membrane located in the body wall on either side of the first abdominal segment. Sound waves vibrate the membrane and stimulate sensory cells to produce nerve impulses. The most important function of the tympanum is to detect courtship songs which male locusts produce by stridulation, rubbing toothed ridges on the inside of the legs against the margins of the wings.

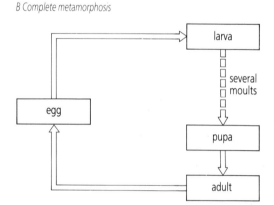

Fig 49·11 Life cycles of insects

Life cycles

Insects belonging to the two Divisions of Sub-class Pterygota have significantly different life cycles. Locusts show **incomplete metamorphosis** and are said to be **hemimetabolous**. Their young stages, called **nymphs**, resemble the adults and have compound eyes, jointed legs, and wing buds which develop externally (Fig. 49·11A). The stages between successive moults are called **instars**. First instar nymphs hatch from the eggs and must undergo ecdysis to become second instar nymphs, and so on. Sexually mature adults are the fifth instar stage. After mating, the female burrows into the sand using her abdomen and lays up to 100 eggs in a frothy secretion which hardens to form an egg pod. Depending on the temperature, the whole life cycle can be completed in 50–70 days.

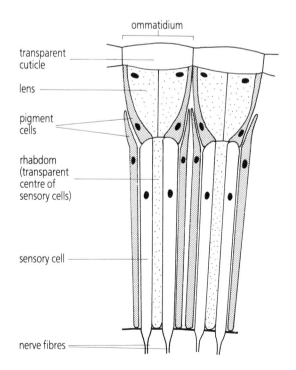

Fig 49·10 Structure of the compound eye

Holometabolous insects have a life cycle which includes a **pupa** stage during which **complete metamorphosis** takes place. They include butterflies and moths, beetles, flies, and most other familiar insect groups. The young stages are called **larvae** and are usually quite unlike the adult in appearance. They lack many of the structures present in adults, such as jointed legs and compound eyes, and are often reduced to feeding machines whose main function is to search for and consume food. Once it has reached a certain size, the larva stops feeding and encloses itself within a protective case, thereby forming a pupa. Inside, the tissues of the larva are broken down and reorganised to form the adult body. The wing buds develop internally and do not become visible until the pupa stage.

This pattern of life cycle has two major advantages – firstly, the larval and adult stages may become highly specialized for particular functions, that is, feeding, in the case of the larva, and reproduction in the adults. Secondly, the two stages can exploit different food resources and occupy quite different ecological niches, so that competition between them is avoided.

49·5　Economic importance of arthropods　■　■　■　■　■　■　■　■

Arthropods are so numerous that it is inevitable that many species directly or indirectly affect human activities. Some examples are given below:

1 Damage to crops. Every year 10–15% of the World's food production is destroyed by insects. A large locust swarm can contain 40 000 million insects each capable of consuming its own mass of food per day and it is hard to imagine the scale of their devastating effect. Less dramatic, but equally important, is the damage done by sap-sucking aphids, wood-boring beetles, and other plant-eating insects.

2 Vectors of disease. Insects transmit malaria, sleeping-sickness, yellow fever, typhus, and elephantiasis, as well as many diseases of crops and livestock. Blood-sucking arachnids called ticks transmit one type of viral encephalitis and are important as vectors of diseases which affect sheep. Dutch elm disease is caused by the fungus *Ceratocystis ulmi* and is carried from tree to tree by bark beetles, *Scolytus multistriatus*.

3 Biological control. The principle of biological control is to introduce a predator or parasite to reduce the numbers of a damaging pest species. In California, the citrus crop was seriously threatened by the scale insect, *Icerya purchasi*, accidentally introduced from Australia, until its natural enemy, the ladybird beetle, *Vedalia cardinalis*, was deliberately introduced. The ladybird keeps the scale insects in check, limiting losses to an acceptable level.

4 Pollination. While some insects are damaging, most are harmless or positively beneficial. Cereal crops like wheat are wind-pollinated, but many other food plants rely on insects for the transfer of pollen necessary to ensure normal seed and fruit development.

5 Food. Crabs, lobsters, and shrimps are economically important as food sources for the human population. Terrestrial arthropods are not exploited to the same extent, although some species are eaten locally. For example, African and Australian native foods include termites and large maggots, sometimes toasted and covered with sugar or chocolate. The most widely accepted food product is honey. Cochineal, obtained from beetles, is sometimes used as a food colouring, but the tendency is to replace it with synthetic alternatives.

6 Silk. Silk is the most important non-food product obtained from insects. In commercial silk farming, the caterpillars of silkworm moths, *Bombyx mori*, are reared on trays provided with mulberry leaves until they spin cocoons. At this stage, the larvae are killed by steam or hot air. Each cocoon provides 600–900 m of usable fibre which is unwound, cleaned, and used to make thread and luxury fabrics. China, Japan, and the USSR are the leading producers.

50 CHORDATES

Objectives

After studying this Unit you should be able to:–

- List the characteristics of organisms belonging to the Phylum Chordata

- Write a summary classification for the Phylum and identify organisms belonging to all the main groups

- Comment on the evolutionary origins of chordates and explain the importance of paedomorphosis in the evolution of free-swimming forms

- Draw and label a diagram to illustrate the important features of the lancelet, *Branchiostoma*

- Outline the events leading to the evolution of cartilaginous fish (Class Chondrichthyes) and bony fish (Class Osteichthyes)

- List some of the problems associated with colonization of terrestrial habitats and explain how these have been overcome during the evolution of terrestrial vertebrates

50·1 Origin of chordates ■ ■ ■ ■ ■ ■ ■ ■ ■ ■ ■ ■ ■

Chordates show a tremendous range of adaptations. Their body plan is very different from that of arthropods, but just as versatile. Its special features have resulted in the evolution of large active organisms, allowing chordates to become the dominant animals in the food chains of most marine, freshwater, and terrestrial ecosystems.

Together with echinoderms and hemichordates, chordates are **deuterostomes**, showing indeterminate radial cleavage during development (refer to Fig. 48·9). In echinoderms and hemichordates, and in primitive chordates, the embryos develop into a free-swimming larva, called a **dipleurula**. Although the adults of these groups are very dissimilar, the embryos and larvae are strikingly alike, so much so that it is difficult to avoid the conclusion that the groups are related. The fossil evidence is patchy, but it appears that the first chordates arose more than 600 million years ago, most probably from sessile echinoderm-like organisms feeding by means of ciliated tentacles. Precisely when and how the transition to a free-swimming fish-like form took place is not known, but some of the stages in the process can be deduced from the structure and life-styles of modern organisms.

50·2 Characteristics and classification ■ ■ ■ ■ ■ ■ ■ ■ ■ ■

At some time in their lives, all chordates possess the following structures:

1 Notochord. This is a rigid dorsal rod consisting of vacuolated cells surrounded by a tough outer coat. In primitive chordates its function is to prevent shortening of the body so that most of the force of muscle contraction is translated into bending movements useful for swimming.

2 Hollow dorsal nerve cord. The nerve cord is formed by invagination from the outer cell layer, or ectoderm, of the embryo and develops as a groove which is later closed off at the top (refer to Unit 23). In vertebrates, the nerve cord is enclosed and protected by the bone or cartilage of the vertebral column.

3 Pharyngeal gill slits. In the simplest chordates, the pharynx is a sieve-like structure with numerous slits used to strain food particles from the water. In vertebrates, the number of slits is greatly reduced and they may be modified for different purposes. In fish and larval amphibians their walls are lined with feathery gills used for gas exchange, while in reptiles, birds, and mammals, the only opening which remains is the Eustachian tube.

In addition to these features, most chordates possess a closed circulatory system and a true tail, that is, a post-anal extension of the body. The segmentation of chordates is best interpreted as a secondary adaptation for swimming. The S-shaped wriggle characteristic of swimming is produced by separated muscle blocks, or **myotomes**, which contract one after another in a serial pattern. The vertebral column and the nerves controlling the muscles show a corresponding segmented organization, but the body organs located ventrally show no trace of segmentation, even during embryonic development.

A formal classification of Phylum Chordata is given below. Acorn worms and related species are sometimes included as chordates, but lack a proper notochord and have other unusual features, making them distinct enough to deserve a phylum of their own, called Phylum Hemichordata. Hemichordates show an interesting mixture of echinoderm-like and chordate-like features, but are not on the main line of chordate evolution.

CLASSIFICATION

Phylum: Chordata
triploblastic coelomate deuterostomia, notochord, hollow dorsal nerve cord, pharyngeal gill slits present at some stage during development, secondary segmentation of dorsal body regions

ACRANIATA: chordates with no skull or vertebral column

Sub-phylum: Tunicata (sea squirts)
marine chordates with free-swimming larva and sessile adult stage, adults with external tunic and pharynx modified for filter feeding
Ciona

Sub-phylum: Cephalochordata (lancelets)
marine chordates with free-swimming larva and burrowing adult stage, adults with pharynx modified for filter feeding
Branchiostoma

CRANIATA (Vertebrata): chordates with skull and vertebral column of cartilage or bone

Sub-phylum: Agnatha – craniates without jaws

Class: Cyclostomata (lampreys and hagfish)
freshwater and marine, parasitic feeding by rasping suckers, well-developed notochord in adult
Lampetra

Sub-phylum: Gnathostomata – craniates with jaws

Class: Chondrichthyes
marine fish with cartilaginous endoskeleton and skin with denticles
Scyliorhinus

Class: Osteichthyes
marine and freshwater fish with bony endoskeleton, swim bladder, and skin with scales
Salmo

Class: Crossopterygii
lobe-finned fish and lung-fish
Latimeria, Neoceratodus

Class: Amphibia
freshwater and terrestrial vertebrates with moist permeable skin, external fertilization and aquatic larval stage
Rana, Triturus

Class: Reptilia
freshwater, marine, and terrestrial vertebrates with impermeable scaly skin, internal fertilization and soft-shelled eggs
Lacerta, Vipera

Class: Aves
homoiothermic terrestrial vertebrates with feathers, internal fertilization and hard-shelled eggs
Columba

Class: Mammalia
homoiothermic freshwater, marine and terrestrial vertebrates with hair, mammary glands and internal fertilization

Sub-class: Prototheria (egg-laying mammals)
Ornithorhyncus, Tachyglossus

Sub-class: Theria (mammals with internal development)
Infraclass 1: Metatheria (marsupial mammals)
Macropus
Infraclass 2: Eutheria (placental mammals)
Oryctolagus

Diagrams showing the external features of representative members of all these groups are given in Figure 50·1. Where possible, British species have been chosen.

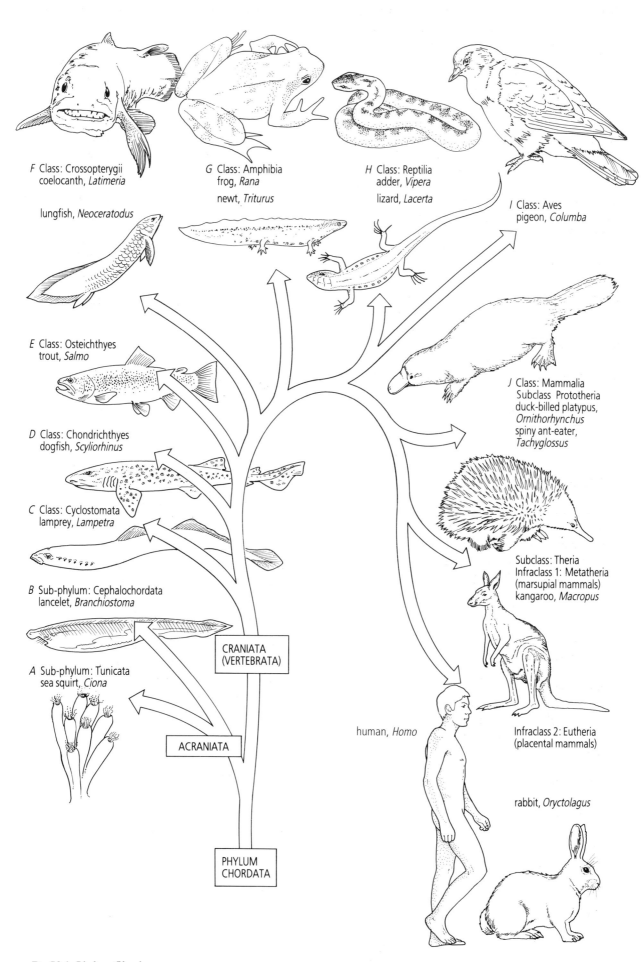

F Class: Crossopterygii
coelocanth, *Latimeria*

G Class: Amphibia
frog, *Rana*

newt, *Triturus*

H Class: Reptilia
adder, *Vipera*

lizard, *Lacerta*

I Class: Aves
pigeon, *Columba*

lungfish, *Neoceratodus*

E Class: Osteichthyes
trout, *Salmo*

D Class: Chondrichthyes
dogfish, *Scyliorhinus*

C Class: Cyclostomata
lamprey, *Lampetra*

B Sub-phylum: Cephalochordata
lancelet, *Branchiostoma*

CRANIATA
(VERTEBRATA)

A Sub-phylum: Tunicata
sea squirt, *Ciona*

ACRANIATA

J Class: Mammalia
Subclass Prototheria
duck-billed platypus,
Ornithorhynchus
spiny ant-eater,
Tachyglossus

Subclass: Theria
Infraclass 1: Metatheria
(marsupial mammals)
kangaroo, *Macropus*

human, *Homo*

Infraclass 2: Eutheria
(placental mammals)

rabbit, *Oryctolagus*

PHYLUM
CHORDATA

Fig 50·1 Phylum Chordata

447

Tunicates differ from ancestral echinoderms in two important ways. First, tentacle feeding is replaced by filter feeding using the pharynx, and, secondly, the duration of the larval stages is increased. The potential advantages of both changes are easy to imagine. Trailing tentacles are exposed and delicate. Pharyngeal slits initially improved water flow over the feeding surfaces, allowing the tentacles to be smaller. Later, the pharynx became enlarged with numerous slits and the tentacles disappeared. Development of the free-swimming larval stages improved dispersal and may have assisted in evading predators.

An adult sea squirt, Figure 50·2A, lives attached to the sea bed and uses its enormous pharynx to filter food particles from the water. Its outer surface is covered by a coat, or tunic, consisting mainly of a cellulose-like carbohydrate. Water is drawn in through the open mouth at the top, while water and wastes are expelled through a second opening, called the atriopore. Such an organism looks improbable as an ancestor for other chordates, but its larva, Figure 50·2B, possesses a notochord, hollow dorsal nerve cord, and pharyngeal slits, and is unquestionably chordate-like in its organization.

Modern chordates are thought to have arisen from tunicates in which the free-swimming larval stages were further favoured by natural selection. Metamorphosis tended to occur later and later in the life cycle, until, after many generations, the adult stage was lost altogether. At the same time, the development of sexual organs was accelerated so that larval individuals became sexually mature and able to breed. This process, called **paedomorphosis**, made it possible for the descendents of tunicates to adopt different life-styles and to exploit different resources.

Cephalochordates lack the sessile adult stage and remain bilaterally symmetrical and capable of swimming throughout their lives. The structure of the lancelet, *Branchiostoma*, is illustrated in Figure 50·3. Lancelets are fairly common marine organisms living in soft sand around the coast. During feeding, adults bury themselves with their open mouths facing upwards. A current of water is drawn in by ciliary action and filtered food particles are collected by the pharynx. The buccal cirri prevent the entry of large particles which might be damaging.

Branchiostoma possesses all the essential features of chordate organization in a simple form and is thought to be very similar to the ancestral type from which the more advanced craniate groups evolved. Look carefully at Figure 50·3 and note the positions of the nerve cord, notochord, and the segmented arrangement of the muscle blocks, or myotomes. This general pattern differs very little from that observed in vertebrates. The circulatory system of *Branchiostoma* and the peripheral connections of the nervous system also follow the vertebrate plan.

There are eight classes of living vertebrates. The most primitive are the lampreys and hagfish, Class Cyclostomata (refer to Fig. 50·1C). These organisms are the only living representatives of the Sub-phylum Agnatha, or jawless fish, and differ considerably from their extinct ancestors. The extinct forms, called **ostracoderms**, were small fish protected by external bony plates. They were sluggish animals, probably spending most of their time on the bottom of freshwater lakes or rivers, or on the sea bed, feeding on decayed organic matter. Lampreys and hagfish are parasites which attach themselves to the bodies of other fish and feed by means of rasping suckers, slowly chewing away at

A Adult

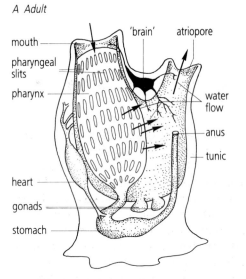

B Free-swimming larva (not to scale)

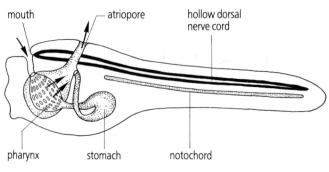

Fig 50·2 Structure of the sea squirt, Ciona

the host's tissues. They have lost all trace of a bony covering and have an endoskeleton consisting entirely of cartilage. There are far fewer gill slits than in *Branchiostoma* and, between the slits, the pharynx is supported by cartilaginous gill arches, or visceral arches

A second group of fossil fish, called **placoderms**, appeared about 400 million years ago and were successful throughout the Devonian period. They differed from ostracoderms in having the first pair of supporting gill arches modified to form jaws. In addition, they had paired fins which improved stability in the water and made active swimming possible. For reasons which are not clear, the placoderms were particularly successful in colonising freshwater habitats. However, towards the end of the Devonian, about 350 million years ago, major changes in climate produced warm conditions which gradually dried up many lakes and rivers. At about this time, one group of placoderms became adapted to salt water and colonised the sea, where their descendents gave rise to the cartilaginous fish, Class Chondrichthyes,

that is, sharks, skates, and rays. A second group remained in freshwater and gave rise to the bony fish, Class Osteichthyes. Much later, the bony fish successfully invaded the oceans.

An important obstacle to colonization of salt water habitats is the need for osmoregulation. Many freshwater organisms are unable to survive in sea water because the concentration of dissolved substances in sea water is greater than that in their body fluids, so that water is lost by osmosis. Cartilaginous fish overcome this problem by retaining urea in the blood until the total concentration of dissolved substances is slightly higher than in sea water. Bony fish have evolved a quite different mechanism of osmoregulation which involves drinking sea water and actively pumping out excess salts through the gills. Nitrogenous waste is converted to non-toxic trimethylamine oxide and only small quantities of urine are produced (see Unit 20). Some bony fish, like the salmon, migrate from freshwater to sea water and back again, and show a remarkable ability to adjust their physiology appropriately.

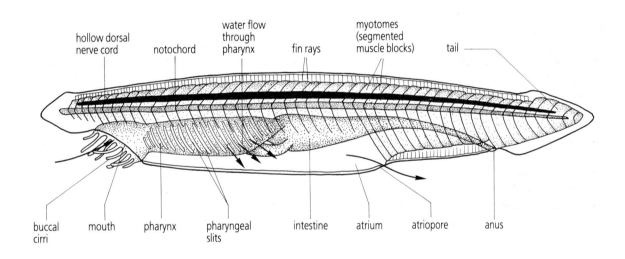

Fig 50·3 Structure of the lancelet, Branchiostoma

50·4 Conquest of land ■ ■ ■ ■ ■ ■ ■ ■ ■ ■ ■ ■

Life on land is very different from life in water and requires effective solutions to many different problems. This section outlines the most important of these problems and describes how they have been progressively overcome during the evolution of chordate groups.

1 Gas exchange. Feathery gills work well in water but, when a fish is removed from water, the gill filaments stick together, drastically reducing the surface area available for gas exchange. On the other hand, oxygen is not very soluble and is at least 30 times more plentiful in air so that

mechanisms for air-breathing are potentially valuable for aquatic organisms living in conditions of oxygen shortage. The first air-breathing structures evolved among the freshwater ancestors of the bony fish and were simple pouches which developed as outgrowths from the pharynx. These were filled with air and allowed fish to survive stagnant conditions and seasonal drying out of ponds. Modern lung-fish exploit a similar mode of life, but, in other bony fish, the pouches became transformed into the **swim bladder**, used for balance and to adjust buoyancy.

2 Support. Water is a dense medium which provides support. Terrestrial organisms gain very little support from air and require stronger skeletons and different mechanisms for locomotion. A major step in evolution occurred when lobe-finned fish, Class Crossopterygii, the group which includes lung-fish, started to use their fins as a means of support. At first, the fins were probably used for paddling in shallow water or for short journeys between ponds, but there were many advantages to be gained by invasion of terrestrial habitats, including freedom from predators, and increased food supply. Primitive amphibians differed comparatively little from lobe-finned fish (Fig. 50·4) except in the increased development of leg bones and limb girdles. The arrangement of these bones is remarkably constant throughout the vertebrates, although the limbs are modified for many different purposes. (The range of possible adaptations is illustrated in Fig. 42.4.)

3 Water loss. Exposure to air results in evaporation of water from respiratory organs and from the general body surface. The skin of amphibians is moist and permeable, although mucus secretion from skin glands helps to reduce water loss to some extent. Gas exchange occurs partly through the skin and through the mucous membranes of the mouth as well as through the lungs, making water loss inevitable in dry conditions. Consequently, most amphibians are confined to damp habitats. Reptiles, birds, and mammals have an impermeable surface layer and their excretory and gas exchange organs are modified to conserve water, making these groups more completely adapted for life on land.

4 Reproduction. Amphibians have external fertilization and aquatic larval stages and must return to water to breed. The larvae are often vulnerable to predators and many strange mechanisms of giving them greater protection have arisen, as in the case of the midwife toad, which transfers fertilized eggs onto her back. The eggs stick to the skin which grows rapidly to enclose them completely. Larval development takes place within the skin and young frogs are eventually released. Internal fertilization and the evolution of shelled eggs freed reptiles from the water and allowed them to colonise much drier environments. Internal development of the young as in placental mammals represents a further advance.

5 Homeostasis. Terrestrial habitats are much more variable than aquatic ones, especially in relation to temperature. Terrestrial vertebrates have adapted to this in various ways. Reptiles are poikilothermic and must regulate their body temperatures by their behaviour, lying sideways in the early morning to gain heat and facing the sun, or resting in the shade at mid-day. Birds and mammals are homoiothermic, that is, they generate heat within their own tissues and maintain a constant body temperature independent of external conditions. This provides optimum conditions for enzyme action and makes possible a tremendous range of tissue and organ specializations. To support increasingly active life-styles major changes in circulation, complex sense organs and sophisticated hormonal and nervous control mechanisms were all necessary. These systems are most advanced in mammals and are described at length elsewhere in this book.

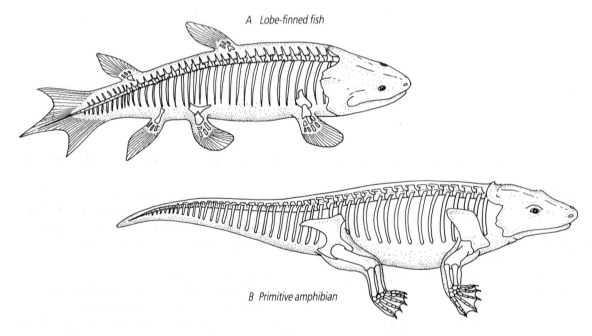

A   Lobe-finned fish

B   Primitive amphibian

Fig 50·4 Skeletons of a lobe-finned fish and a primitive amphibian

51 KINGDOM PROTOCTISTA 2
ALGAE

Objectives

After studying this Unit you should be able to:–

- List the important characteristics of algae

- Outline the distinguishing features of organisms belonging to Phylum Chlorophyta, Phylum Gamophyta, Phylum Phaeophyta, Phylum Rhodophyta, and Phylum Bacillariophyta

- Describe the structure and life cycle of the unicellular green alga *Chlamydomonas.*

- Describe the structure and life cycle of the filamentous green alga *Spirogyra.*

- Distinguish between isogamous, anisogamous and oogamous species

- Draw diagrams to illustrate zygotic, gametic, and sporic life cycles and comment on the differences between these types

- Explain the meanings of the following terms: alternation of generations, gametophyte, gametangium, sporophyte, sporangium

- Describe the life cycle of *Ulva lactuca* as an example of isomorphic alternation of generations

- Describe the structure and life cycle of *Fucus vesiculosus*

- Give examples of the ecological and economic importance of algae

51·1 Introduction ■ ■ ■ ■ ■ ■ ■ ■ ■ ■ ■ ■ ■ ■ ■ ■ ■

Algae are photosynthetic protoctists with a unicellular, colonial, or simple multicellular structure. In this context the term algae is a convenient way of referring to several phyla. However, you will recall 'algae' is not, strictly, a taxonomic term. Algae share the following plant-like characteristics:

1 Their cells are surrounded by **cell walls**, separate from and lying outside the cell membrane.
2 Their cytoplasm usually contains one or more large **vacuoles**.
3 They possess photosynthetic pigments enclosed in special organelles called **plastids**, of which

chloroplasts are the commonest type.
4 Reproduction is usually sexual and may involve an **alternation of generations** between haploid and diploid individuals in the life cycle.

Many different algal groups can be distinguished according to their structure and life-styles, and by the presence of particular photosynthetic pigments or storage materials. This Unit outlines the most important adaptations of the five phyla classified below. The green algae, Phylum Chlorophyta, have many features in common with higher plants, and a close evolutionary link is presumed.

Phylum: Chlorophyta (green algae)
eukaryotic, unicellular, colonial, simple
  multicellular, life cycle including flagellate
  zoospore stage
marine, freshwater, terrestrial
photosynthetic, chlorophylls a, b, starch food
  reserve
Chlamydomonas, Gonium, Ulva, Volvox

Phylum: Gamophyta (conjugating algae)
eukaryotic, unicellular, colonial, flagella absent,
  sexual reproduction by conjugation
freshwater forms only
photosynthetic, chlorophylls a, b, starch food
  reserve
Spirogyra

Phylum: Phaeophyta (brown algae)
eukaryotic, multicellular, sexual reproduction
  involving small bi-flagellate male gametes
  (sperm) and large immobile female gametes
  (eggs)

marine
photosynthetic, chlorophylls a, c, fucoxanthin,
  laminarin food reserve
Fucus, Laminaria

Phylum: Rhodophyta (red algae)
eukaryotic, multicellular, flagella absent, sexual
  reproduction
marine
photosynthetic, chlorophyll a only, starch-like
  food reserve (floridean starch)
Porphyra

Phylum: Bacillariophyta (diatoms)
eukaryotic, unicellular, colonial, each cell with a
  protective shell consisting of two valves
  impregnated with silica, sexual reproduction
marine, freshwater
photosynthetic, chlorophylls a, c,
  chrysolaminarin food reserve
Melosira

51·2 *Chlamydomonas* ■ ■ ■ ■ ■ ■ ■ ■ ■ ■ ■ ■ ■ ■

Structure

Chlamydomonas is a common genus of unicellular
green algae. Most species are found in freshwater
ponds, lakes, and ditches, where they are
sometimes so numerous that the water is coloured
green.

A single *Chlamydomonas* cell (Fig. 51·1*A*) is
approximately pear-shaped, about 10 μm in
diameter, and about 20 μm long. The cell wall
consists mainly of cellulose and may be covered by
a thin layer of mucilage. The remainder of the cell,
often referred to as the **protoplast**, lies inside the
cell wall and is surrounded by the cell membrane.

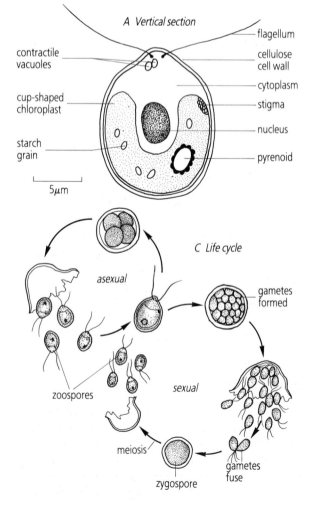

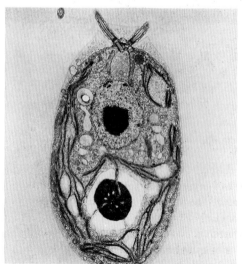

B EM Chlamydomonas *cell*

Fig 51·1 Structure and life cycle of Chlamydomonas

At the anterior end, a pair of flagella extend through the cell wall and lash the water to produce swimming movements. The red 'stigma' detects light and influences the activity of the flagella. *Chlamydomonas* swims towards light and individuals tend to gather in regions of optimum light intensity. This response is called **positive phototaxis** and depends on the presence of the stigma, although the precise mechanism is not understood.

Most of the space inside the protoplast is occupied by the large cup-shaped chloroplast. This contains an additional structure, the **pyrenoid**, which is the major site of starch formation. The pyrenoid consists mainly of proteins, including the enzyme **ribulose bisphosphate carboxylase**, responsible for fixing carbon dioxide during the dark reactions of photosynthesis (see Unit 11). The nucleus is surrounded by the chloroplast and is located at the centre of the cell. In freshwater species, water enters the cell by osmosis and is actively pumped out by the contractile vacuoles, which appear to function in much the same way as those of *Amoeba* or *Paramecium*.

Reproduction

Asexual reproduction occurs when the protoplast divides to form 2, 4, 8, or sometimes more, new protoplasts inside the original cell wall. Each of these develops its own cell wall, flagella, stigma, and contractile vacuoles, and becomes a miniature *Chlamydomonas* cell. The new individuals, called **zoospores**, are released when the parent cell wall ruptures (see Fig. 51·1C).

Sexual reproduction occurs following the production of gametes. Each parent cell gives rise to 8, 16, or 32 gametes. These lack cell walls, but have flagella and are able to swim. Most species of *Chlamydomonas* are **heterothallic**, so that fusion takes place only between gametes produced by individuals of different **mating strains** (often called + and − strains). Gametes formed by different strains are usually equal in size and are called **isogametes**. However, in some species, fusion occurs between differently-sized gametes, or **anisogametes**.

The zygote formed by fusion develops a thick wall and becomes a zygospore. This is a resistant structure which may remain dormant for long periods. The zygote is the only diploid stage in the life cycle of *Chlamydomonas*. It divides by meiosis to produce four haploid daughter cells which are liberated when the zygospore bursts (refer to Fig. 51·1C). Like the new individuals produced by asexual reproduction, these new cells are called zoospores.

51·3 *Spirogyra* ■ ■ ■ ■ ■ ■ ■ ■ ■ ■ ■ ■ ■ ■ ■ ■

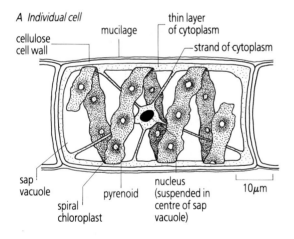

A *Individual cell*
cellulose cell wall
mucilage
thin layer of cytoplasm
strand of cytoplasm
sap vacuole
spiral chloroplast
pyrenoid
nucleus (suspended in centre of sap vacuole)
10 μm

B *Conjugation*
swellings from adjacent cells
zygospore germinates
new filament
zygote
zygospore
conjugation tube
gametes formed
male gamete passes through tube

Fig 51·2 Structure and conjugation in Spirogyra

Structure

Tangled floating masses of *Spirogyra* are common in shallow stagnant freshwater ponds. Each strand is an unbranched filament up to 30 cm long consisting of hundreds of cylindrical cells joined end to end, as illustrated in Figure 51·2A. The cells are about 30 μm in diameter and 60−80 μm long and their cellulose cell walls are covered by a layer of mucilage which gives the filaments a characteristically slimy feel.

Inside each cell, the cell membrane and a thin outer layer of cytoplasm surround a large sap vacuole at the centre of which lies the cell nucleus held in position by cytoplasmic threads. Depending on the species, the peripheral cytoplasm may contain one or more spiral chloroplasts studded with pyrenoids. All the cells in the filament have a similar structure and there is no division of labour leading to cell specialization.

Reproduction

Filaments of *Spirogyra* increase in length by mitotic cell division which may take place in any cell. Asexual reproduction occurs when the filament breaks to form two or more new individuals – this is called **fragmentation**.

Sexual reproduction involves **conjugation** between cells (Fig. 51·2*B*). Usually, the cells belong to different filaments, but in some species **homothallic** conjugation, that is, conjugation between different regions of the same filament, has been observed, indicating that genetically different mating strains are not essential. Swellings from adjacent cells fuse to form a **conjugation tube** through which the protoplast from one cell is able to migrate. The moving protoplast is often referred to as the male gamete, while the stationary protoplast is called the female gamete. After fusion, the zygote secretes a thick resistant wall and becomes a zygospore. As in *Chlamydomonas*, the zygote is the only diploid stage in the life cycle. Meiosis takes place before germination of the zygospore. Three of the nuclei formed degenerate, leaving a single haploid cell which gives rise to a new filament.

51·4 Life cycles of algae ▪ ▪ ▪ ▪ ▪ ▪ ▪ ▪ ▪ ▪ ▪ ▪ ▪ ▪

Any life cycle which includes sexual reproduction involves diploid (2n) cells which undergo meiosis, and haploid (n) cells called gametes which fuse by a process known as **syngamy**, or **fertilization**, to produce a diploid zygote.

When the gametes are of equal size and resemble each other, as in most species of *Chlamydomonas*, they are called **isogametes**, and the species is said to be **isogamous**, or to exhibit **isogamy** (Fig. 51·3*A*). In some species, the gametes differ slightly in size – such gametes are called **anisogametes** (heterogametes) and the species is **anisogamous** (Fig. 51·3*B*). More commonly, the gametes differ in activity as well as in size. Sperm are small motile gametes usually produced in large numbers, while egg cells are large and immobile. Species which produce eggs and sperm exhibit **oogamy** (Fig. 51·3*C*). Many brown algae, like *Laminaria* and *Fucus*, described later in this Unit, are oogamous.

The life cycles of algae also vary in the relative timing of meiosis and syngamy. In *Chlamydomonas* and *Spirogyra*, meiosis occurs during germination of the zygote, which is thus the only diploid stage in the life cycle (Fig. 51·4*A*). This is called a **zygotic life cycle** and is probably the most primitive type in the evolutionary sense. Diatoms and certain green algae have a **gametic life cycle** like that of most animals (Fig. 51·4*B*). In this case, meiosis results in the production of gametes, so that all the vegetative cells are diploid.

In the third type of life cycle, called a **sporic life cycle**, meiosis and syngamy are separated by additional stages so that there are two distinct generations of organisms (Fig. 51·4*C*). In the **gametophyte** generation, the individuals are haploid and produce gametes. In the **sporophyte** generation, the individuals are diploid and produce reproductive cells called **meiospores** following meiosis. The meiospores are haploid and give rise

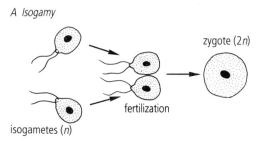

A Isogamy

zygote (2n)

isogametes (n)

fertilization

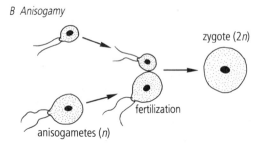

B Anisogamy

zygote (2n)

anisogametes (n)

fertilization

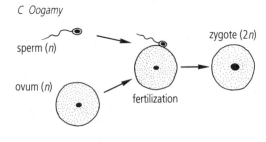

C Oogamy

sperm (n)

ovum (n)

zygote (2n)

fertilization

Fig 51·3 Isogamy, anisogamy, and oogamy

to new gametophyte individuals. Thus, the sporic life cycle exhibits a clear cut **alternation of generations** between haploid and diploid individuals.

In some algae, like the sea lettuce, *Ulva*, the gametophyte and sporophyte individuals are

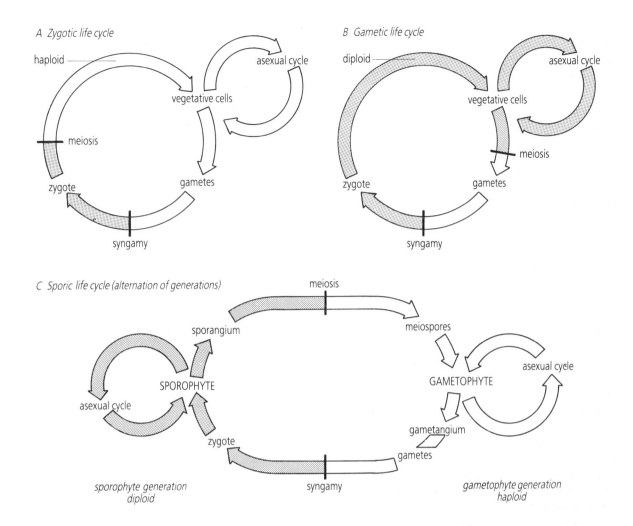

A *Zygotic life cycle*

haploid

asexual cycle

vegetative cells

meiosis

zygote

gametes

syngamy

B *Gametic life cycle*

diploid

asexual cycle

vegetative cells

meiosis

zygote

gametes

syngamy

C *Sporic life cycle (alternation of generations)*

meiosis

sporangium

meiospores

SPOROPHYTE

asexual cycle

GAMETOPHYTE

asexual cycle

gametangium

zygote

gametes

*sporophyte generation
diploid*

syngamy

*gametophyte generation
haploid*

Fig 51·4 Life cycles in algae

identical in appearance. Consequently, there is an alternation of **isomorphic** generations. More often, the gametophyte and sporophyte individuals are very different, so that there is an alternation of **heteromorphic** generations. Sometimes, as in *Fucus*, one generation becomes totally dependent upon the other and is no longer capable of a separate existence.

The life cycles of *Ulva* and *Fucus* are described in the next two sections. To help you understand the structures and processes involved, a summary of important terms is given in Table 51·1.

51·5 *Ulva* ■ ■ ■ ■ ■ ■ ■ ■ ■ ■ ■ ■ ■ ■ ■ ■ ■ ■ ■

Ulva lactuca, the sea lettuce, is a common intertidal species on rocky seashores. The cells of the organism divide mainly in two planes, producing a flat sheet-like expanse, or **thallus**, two cells in thickness. This is anchored by a small **holdfast** (Fig. 51·5*A*). Each cell contains one nucleus and a cup-shaped chloroplast with a single pyrenoid.

All vegetative individuals appear similar but are in fact of two different types which can be distinguished by their methods of reproduction. Any cell in the haploid thallus, or gametophyte, can become a **gametangium** or gamete-producing cell. Typically, 8 or 16 isogametes are formed, each with two flagella (refer to Fig. 51·5*B*). These can fuse only with gametes from another thallus, so that *Ulva* is heterothallic. The diploid zygote

develops to form a complete diploid thallus, or sporophyte. Zoospores are produced when sporophyte cells develop into structures called **sporangia** (singular **sporangium**) inside which meiosis takes place. The zoospores are normally larger than gametes and can be identified by the presence of four flagella. They are haploid and give rise directly to new gametophyte individuals. This cycle of events shows alternation of generations in one of its simplest forms. Because the gametophyte and sporophyte generations look so similar and are equally important in the life cycle, *Ulva* provides an example of isomorphic alternation of generations ('iso' – same; 'morph' – form).

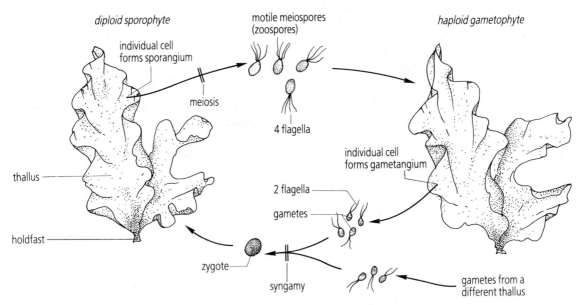

Fig 51·5 Structure and life cycle of Ulva lactuca

51·6 *Fucus* ■ ■ ■ ■ ■ ■ ■ ■ ■ ■ ■ ■ ■ ■ ■ ■ ■

Fucus vesiculosus, or bladder wrack, is a common British seaweed found abundantly in the intertidal region on all rocky shores. For part of the day it is battered by the waves and for the remainder it is exposed to the combined drying effects of the wind and the Sun. Few organisms can tolerate such extreme conditions and it is interesting to note the structural features of *Fucus* which enable it to thrive in this habitat.

A *Fucus* thallus may reach 1 m or more in length and consists of fronds which show a characteristic two-forked, or **dichotomous**, branching pattern (Fig. 51·6*A*). At each branching point the fronds divide equally. They are supported by a thickened **midrib** region consisting of elongated fibrous cells, and by gas-filled bladders, or **vesicles**, which increase buoyancy so that the fronds float freely in the water, exposing the maximum area for photosynthesis. The entire surface is covered by a thick layer of mucilage which reduces colonization by epiphytic organisms and helps to prevent desiccation at low tide. The thallus is connected by a short stalk, or **stipe**, to the disc-shaped holdfast, or **hapteron**. This secretes a specially adhesive form of mucilage and remains firmly fixed to the rock even in the roughest weather.

The photosynthetic pigments present include chlorophylls a and c. These pigments are green, but their colour is masked by large quantities of the brown pigment **fucoxanthin**. Fucoxanthin absorbs blue and green light. These wavelengths penetrate seawater much more effectively than red and yellow, and it follows that brown algae, which possess fucoxanthin, can survive and continue to carry out photosynthesis at greater depths than other groups of algae.

Reproduction

Fucus reproduces sexually. *Fucus vesiculosus* is **dioecious**, that is, male and female gametes are produced by separate individuals. The species does show alternation of generations, but the male and female gametophytes are very much reduced and develop within the tissues of sporophyte individuals, upon which they are completely dependent.

The reproductive structures, called **conceptacles**, are located at the tips of the fronds (Fig. 51·6*B*). Male conceptacles can be identified by their orange colour. They are lined with branching hairs upon which male gametangia, or **antheridia**, may develop. Meiosis occurs inside each antheridium and is followed by a series of four mitotic divisions resulting in the production of 64 nuclei. After division of the cytoplasm, 64 tiny bi-flagellate sperm, or **antherozoids**, are formed. Female conceptacles are greenish in colour and contain female gametangia, called **oogonia** (singular **oogonium**). Each of these produces 8 egg cells, or **oospheres**.

Partial drying out of the conceptacles causes them to shrink and squeezes out their contents, so that gametes are usually released following exposure of the thallus between the tides. Fertilization occurs in the surrounding water and the zygote formed eventually settles into a crevice and becomes attached to the rock. Rapid mitotic divisions give rise to a new sporophyte individual.

Table 51·1 Summary of terms used in describing plant life cycles

GAMETES AND FERTILIZATION

gamete: specialized haploid cell which fuses with another similar cell to produce a new diploid individual

syngamy: fusion of gametes (= fertilization)

isogamy: fusion of gametes of equal size. Both gametes may be motile

anisogamy: fusion of gametes of unequal sizes. Both gametes may be motile

oogamy: fusion of a small motile male gamete (sperm) with a large non-motile female gamete (ovum)

zygote: diploid cell formed by fusion of gametes

LIFE CYCLES

zygotic life cycle: life cycle in which the zygote is the only diploid cell and all other cells are haploid (Fig. 51·4A)

gametic life cycle: life cycle in which the gametes are the only haploid cells and all other cells are diploid (Fig. 51·4B)

sporic life cycle: life cycle with *alternation of generations*, that is, with distinct generations of haploid and diploid individuals. Both types of individual possess vegetative (non-reproductive) cells (Fig. 51·4C)

ALTERNATION OF GENERATIONS

gametophyte: haploid individual producing gametes by mitosis

gametangium: cell in which gametes are formed

antheridium: gametangium in which male gametes are formed

antherozoid: male gamete (= sperm)

oogonium gametangium in which a female gamete is formed

oosphere: female gamete (= ovum)

archegonium: female gametangium with a protective jacket of sterile cells (not found in algae)

sporophyte: diploid individual which develops from a zygote and produces reproductive spores by meiosis (meiospores)

sporangium: cell in which spores are produced

meiospore: haploid spore which gives rise to a new gametophyte individual

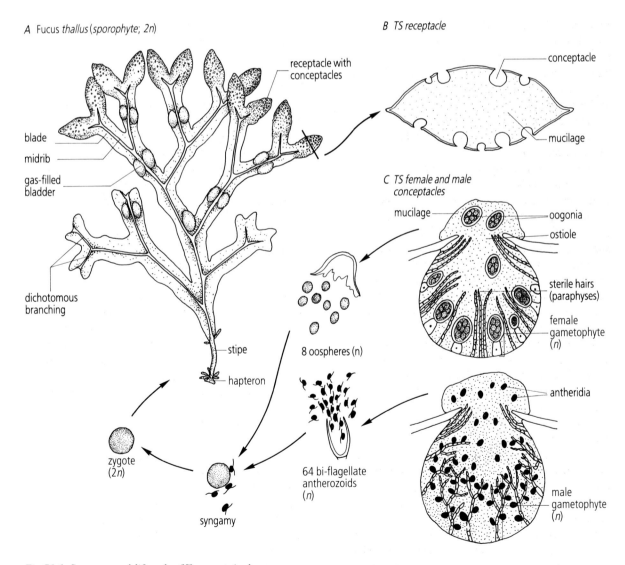

Fig 51·6 Structure and life cycle of Fucus vesiculosus

51·7 Ecological and economic importance of algae ▪ ▪ ▪ ▪ ▪

1 Photosynthesis. More than half the world's photosynthesis is carried out by algae living in the surface waters of the oceans. Huge amounts of carbon dioxide are removed annually from the atmosphere, and supplies of oxygen are replenished. As the major producers in marine ecosystems, algae provide food for zooplankton, and indirectly for most other marine organisms.

2 Food. Without exception, the world's fisheries depend on algal production. In addition, algae are sometimes harvested directly for food. The red seaweed *Porphyra* is common on exposed rocky coasts and in parts of Wales it is cooked and made into a sort of pancake called laver bread. Another species of *Porphyra* is eaten extensively in Japan, where it is called nori. A substantial industry has developed for its cultivation, collection, and processing.

3 Extracted substances. Alginic acid is a polysaccharide which can be extracted in quantity from brown seaweeds including *Fucus* and *Laminaria*. Its derivatives, called **alginates**, are used as gelling or emulsifying agents in the manufacture of ice cream, cosmetics, car polishes, paints, and pharmaceuticals. **Agar** and **carrageenan** are polysaccharide materials obtained from red seaweeds. They have gelling properties similar to those of alginates. Agar is used in the preparation of growth media for culturing microorganisms. A mixture of agar and nutrients is dissolved in hot water, sterilized by heating under pressure, and then cooled so that it sets as a jelly.

4 Fertilizer. Seaweed is sometimes used as a fertilizer in coastal regions, as in northwest France. Liquid extracts of brown seaweeds are used in growing vegetables and glasshouse crops and are a valuable source of potassium and trace elements.

5 Geological deposits. Diatoms, Phylum Bacillariophyta, are an important group of planktonic algae, easily identified by their protective shells (Fig. 51·7). These are rigid structures impregnated with silica. The compacted remains of diatoms form important geological deposits on the sea bed. Fossil diatom deposits, called keiselghur, or diatomite, can be processed to give a hard powder used in filtration processes and as an abrasive. Dynamite is made by adding

Fig 51·7 Diatoms

nitroglycerin to kieselghur to produce a granular material which can be handled safely.

6 Algal 'blooms'. Unusually high levels of nutrients arising from pollution of water by sewage, or leaching of fertilizers from agricultural land, can result in an explosive growth of algae, particularly if the the water is warm. Excessive populations of algae constitute an algal bloom and may encourage the growth of aerobic bacteria. These multiply rapidly and can use up all or most of the oxygen dissolved in the water, causing the death of fish and other organisms. This process is difficult to reverse and is a serious problem in many lakes and rivers.

Certain kinds of marine algae produce toxins which accumulate in the bodies of shellfish. Blooms of these algae cause the shellfish to become a potential source of food poisoning, making them unfit for human consumption.

52 KINGDOM FUNGI

Objectives

After studying this Unit you should be able to:–

- List the important characteristics of fungi

- Draw a diagram to illustrate a typical fungal life cycle and explain the terms dikaryotic, plasmogamy, and karyogamy

- Outline the distinguishing features of organisms belonging to Phylum Zygomycota, Phylum Ascomycota, Phylum Basidiomycota, Phylum Deuteromycota, and Phylum Mycophycophyta

- Describe the structure and life cycle of the bread mould, *Rhizopus stolonifer*

- Describe the structure and life cycle of the parasitic fungus, *Claviceps purpurea*

- Describe the structure and life cycle of the mushroom, *Agaricus campestris*

- Explain what is meant by a mycorrhiza

- Give examples of the ecological and economic importance of fungi

52·1 Introduction ■ ■ ■ ■ ■ ■ ■ ■ ■ ■ ■ ■ ■ ■ ■ ■ ■ ■ ■

Familiar fungi include moulds, toadstools, mushrooms, and yeasts. All these organisms lack chlorophyll and are heterotrophic. Fungi do not ingest their food but release enzymes into their surroundings, causing the break down of complex molecules into simple soluble substances which can be absorbed. Many fungi are **saprotrophic**, meaning that they feed on dead organic remains. Others are **parasitic** and obtain nutrients directly from living organisms. Among the parasitic forms, some are **obligate parasites** which can survive only in the tissues of a living host. **Facultative parasites** are able to live saprotrophically, often after causing the death of their hosts, and include a number of important plant pathogens.

A typical fungus consists of fine threads or filaments called **hyphae** (singular **hypha**) forming a tangled mass, or **mycelium**. Each hypha is surrounded by a cell wall containing the nitrogenous polysaccharide **chitin**, the same substance as found in the exoskeletons of insects. Cross walls, or **septa** may sub-divide the filament, but seldom separate the 'cells' completely, so that cytoplasm can flow more or less freely throughout the mycelium. Growth occurs only at the tips of the hyphae.

The nuclei of the hyphae are normally haploid. In many species, spores are produced asexually at the tips of specialized reproductive hyphae. Sexual reproduction takes place by conjugation between different mating strains. Often, the parental nuclei do not fuse immediately, but remain separate and may divide repeatedly, giving rise to a **dikaryotic** (heterokaryotic) mycelium, containing paired

Fungi play a vital role in many ecosystems as saprotrophs, breaking down organic material and so recycling its components.

459

haploid nuclei. Eventually, the nuclei fuse to form diploid zygotes. These divide by meiosis to produce haploid meiospores each capable of germinating to form a new haploid mycelium. This cycle of events is summarized in Figure 52·1. The life cycle of fungi is essentially zygotic (cf Fig. 51.4*A*), but the situation is complicated by the delay between **plasmogamy** (cytoplasmic fusion) and **karyogamy** (nuclear fusion). Cilia and flagella are absent from all stages of the life cycle.

As defined above, Kingdom Fungi includes four phyla, namely Phylum Zygomycota, Phylum Ascomycota, Phylum Basidiomycota, and Phylum Deuteromycota. In addition, the kingdom includes the lichens, Phylum Mycophycophyta. A lichen is the result of a mutualistic association between a fungus and a photosynthetic organism, either a green alga, or a cyanobacterium. The fungus partner is always dominant.

A number of fungus-like forms are classified in the Kingdom Protoctista. These include *Phytophthora infestans*, the parasite which causes potato blight, and other important pathogenic species. Such organisms have cell walls which lack chitin and are made of cellulose. In addition, their life cycles involve a flagellated zoospore stage, making them significantly different from the organisms described in the remainder of this Unit.

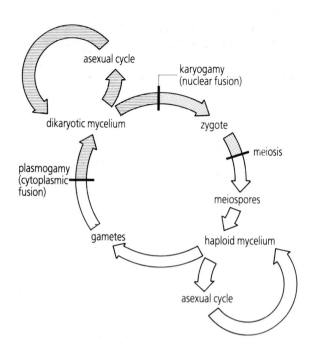

Fig 52·1 Diagram to show the sequence of events in a typical fungal life cycle

CLASSIFICATION

Kingdom: Fungi
eukaryotic, mycelium consisting of hyphae with
  cell walls containing chitin, cilia and flagella
  absent
sexual and asexual reproduction by spores
terrestrial
heterotrophic, saprotrophic, parasitic

Phylum: Zygomycota
hyphae aseptate, sexual reproduction by
  conjugation producing resistant zygospores
terrestrial, widespread
saprotrophic, some parasitic forms
Rhizopus

Phylum: Ascomycota
hyphae septate, asexual reproduction by conidia,
  sexual reproduction producing an ascus
  containing ascospores
terrestrial, widespread
saprotrophic, parasitic
Claviceps, Saccharomyces

Phylum: Basidiomycota
hyphae septate, reproduction usually sexual
  leading to the production of club-shaped
  basidia bearing basidiospores
terrestrial, widespread
saprotrophic, parasitic
Agaricus

Phylum: Deuteromycota
fungi which lack specialized structures for
  sexual reproduction
terrestrial, widespread
saprotrophic, parasitic
Penicillium, Dactylaria

Phylum: Mycophycophyta (lichens)
mutualistic associations between a fungus and a
  photosynthetic organism, either a green alga or
  a cyanobacterium
terrestrial, extreme habitats
predominantly autotrophic
Cladonia

52·2 *Rhizopus* – a saprotrophic fungus ■ ■ ■ ■ ■ ■ ■ ■ ■

Structure

The bread mould *Rhizopus stolonifer* is a common representative of the Phylum Zygomycota. Spores of *Rhizopus* are carried by air currents and will germinate on any suitable organic material. The fungus grows luxuriantly on bread, but will also attack and damage berries and other fruit while in transit or storage.

The hyphae of *Rhizopus* do not have cross walls or dividing membranes and are said to be **aseptate**, or **coenocytic**. Nuclei are distributed at intervals throughout the cytoplasm which forms one continuous mass (Fig. 52·2*B*). Organelles like mitochondria are free to move within the hypha and may be concentrated in actively growing regions. Once a mycelium is established, large hyphae called **stolons** grow out laterally. Where they touch the substrate new hyphae develop (Fig. 52·2*A*). *Rhizopus* feeds saprotrophically by means of fine branching hyphae, or **rhizoids**, which penetrate the substrate and are responsible for releasing enzymes and absorbing nutrients.

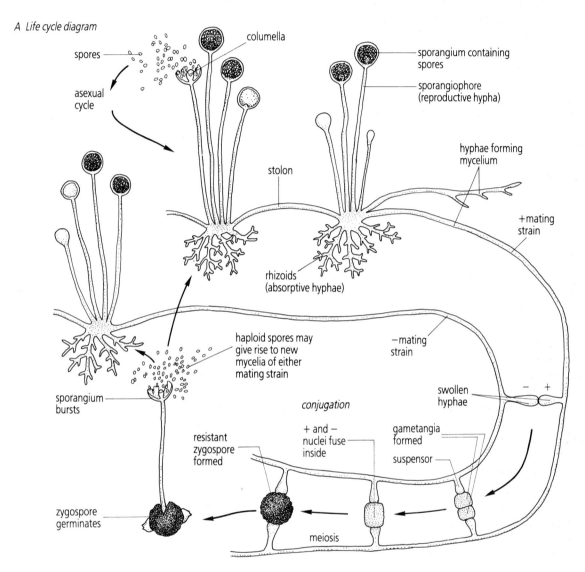

A Life cycle diagram

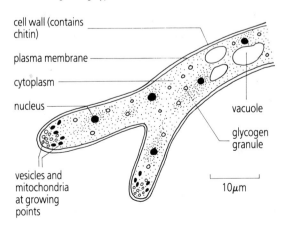

B Detail of growing hypha

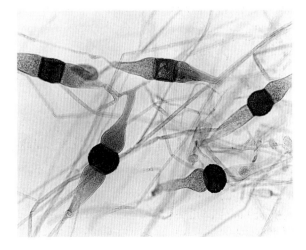

C Photograph of mycelium

Fig 52·2 Structure and life cycle of Rhizopus stolonifer

461

Reproduction

Asexual reproduction involves reproductive hyphae called **sporangiophores** which grow vertically. The tip of each sporangiophore swells and a partition develops separating off a section which becomes a **sporangium**. The dome-shaped tip of the hypha inside the sporangium is called the **columella**. Immature sporangia are whitish, but, as development continues, crystals of calcium oxalate are deposited in the sporangium wall, producing a dense black colour (see Fig. 52·2C). When the sporangium is ripe, its outer wall dries and splits to release numerous resistant spores which can be dispersed by air currents. The pin head appearance of the sporangia explains the name 'pin moulds' used to describe *Rhizopus* and other members of the Phylum Zygomycota.

Conjugation, the method of sexual reproduction observed in *Rhizopus*, takes place between parents of genetically different mating strains. It occurs much less frequently than asexual reproduction because there is a limited chance that spores from + and − strains will be carried to the same substrate. When the mycelia of + and − strains meet, short club-shaped branches develop and swell (Fig. 52·2A). Next, the tips of the branches are walled off to form **gametangia** each of which contains many nuclei from the parent hypha. When the gametangia fuse, the nuclei inside fuse in pairs, one from each parent, resulting in the production of many diploid nuclei. All but one of these degenerates, leaving a single diploid nucleus which immediately undergoes meiosis to yield four haploid nuclei. Three of these degenerate, so that just one haploid nucleus remains. The structure formed, called a **zygospore**, secretes a thick protective wall impregnated with crystals of calcium oxalate and may remain dormant for long periods. In favourable conditions, the zygospore germinates, producing a sporangium which bursts to liberate numerous haploid spores.

52·3 *Claviceps* – a parasitic fungus ■ ■ ■ ■ ■ ■ ■ ■ ■

Fungi belonging to the Phylum Ascomycota are distinguished from other fungi by the possession of structures called **asci** (singular **ascus**) formed as a result of sexual reproduction. An ascus is a capsule inside which haploid nuclei from different parental strains fuse to give a diploid zygote nucleus. The zygote nucleus divides by meiosis and, following one or more mitotic divisions, **ascospores** are produced. Each ascospore consists of a nucleus and cytoplasm surrounded by a membrane and a protective spore wall. Typically, they are dispersed by wind currents. Asexual reproduction in ascomycotes occurs by means of spores called **conidia**, which are produced at the tips of specialized reproductive hyphae.

Yeasts, mildews, cup fungi, and truffles are all ascomycotes. *Claviceps purpurea*, described here, is an important parasitic species which causes the disease **ergot** of rye and other grasses. The life cycle of *Claviceps* is illustrated in Figure 52·3. Ascospores are produced in the spring when rye is in flower and are carried by the wind. Those which reach young flowers germinate and give rise to a mycelium which spreads throughout the ovary of the flower. At the surface of the mycelium, short hyphae called **conidiophores** are formed and asexual spores, or conidia, are budded off. These are embedded in a sweet, sticky secretion which provides food for insects and can be transported to other flowers. The mycelium

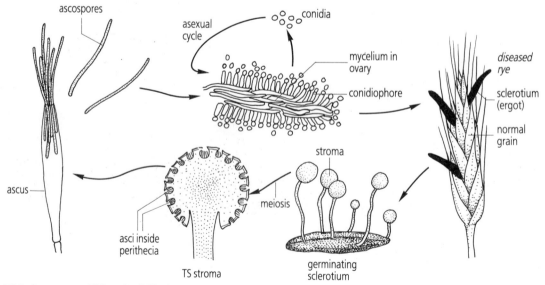

Fig 52·3 Structure and life cycle of Claviceps purpurea

continues to grow, then begins to harden, and is finally transformed into a hard purple structure, called a **sclerotium**, or ergot. This resembles a normal grain in shape, but is somewhat larger. In the autumn, or during harvesting, the sclerotia are easily dislodged and fall to the ground, where they pass the winter.

When conditions become suitable, the sclerotium germinates, producing several small mushroom-like bodies. In the head, or **stroma**, of these, male and female gametangia develop and fuse. The asci formed are thin and cylindrical and each contains eight long ascospores lying side by side. The asci are grouped inside flask-shaped **perithecia**, and, as they mature, the ascospores are extruded and shot upwards under pressure.

The sclerotia of *Claviceps* contain a variety of poisonous alkaloid substances. In the past, ergot was often milled along with grains of rye, producing flour which contained as much as 10% powdered mycelium. Baked into bread, contaminated flour caused abortions, madness, hallucination and death among the human population. A burning sensation in the extremities is one symptom of ergot poisoning and, in the Middle Ages, the disease was known as St Anthony's fire. Today, cases of ergotism in people are rare, but the fungus is still a common cause of poisoning in cattle grazing on infected grasses. Extracted alkaloids are used as drugs to control haemorrhage, especially during childbirth.

52·4   *Agaricus* – a mushroom   ■ ■ ■ ■ ■ ■ ■ ■ ■ ■ ■ ■ ■

Familiar examples of fungi belonging to the Phylum Basidiomycota include mushrooms, toadstools, bracket fungi and puffballs. Most species are saprotrophic, but the group also includes wheat rust and other fungal diseases of crops. The major characteristics of the Phylum can be illustrated by describing the structure and life cycle of the common field mushroom *Agaricus campestris*.

The spores of *Agaricus* are haploid and germinate to produce mycelia which grow underground (see Fig. 52·4). Fusion of hyphae from different mating strains produces a new dikaryotic mycelium which develops extensively without fusion of nuclei and eventually gives rise to one or more spore-bearing bodies, each called a **basidiocarp**. These are mushrooms and are usually the only structures visible above ground level. Each consists of an upright stalk, or **stipe**, supporting a domed cap, or **pileus**. As the mushroom develops, the edges of the pileus break away from a ring of tissue, the **velum**, and the cap flattens to expose numerous sheet-like **gills** radiating from the centre. The surfaces of the gills are covered by a spore-bearing layer known as the **hymenium**. Within this layer, nuclear fusion occurs and is promptly followed by meiosis, producing haploid nuclei which are then isolated into **basidiospores**. These develop at the tips of characteristic club-shaped structures called **basidia** (singular **basidium**). The spores are discharged by an explosive mechanism and are shot out into the space between the gills. Millions of spores are released and transported away by air currents.

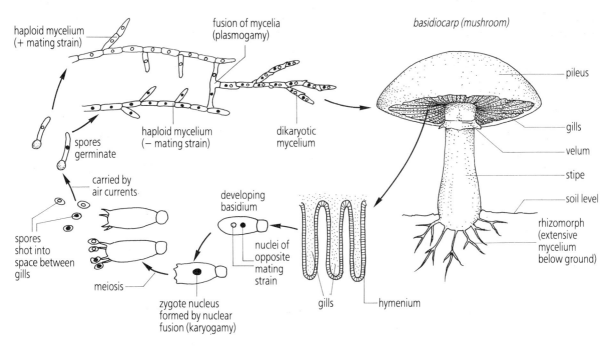

Fig 52·4 Structure and life cycle of Agaricus campestris

52·5 Phylum Deuteromycota ■ ■ ■ ■ ■ ■ ■ ■ ■ ■

Phylum Deuteromycota, or **Fungi Imperfecti**, is
an artificial grouping comprising about 20 000
species which are known only by their asexual
stages. The sexual stages have either not been
observed or have been lost altogether. Most
deuteromycotes are thought to be derived from
ascomycotes because of similarities in the structure
of the mycelium and in the way asexual spores, or
conidia, are formed. The phylum includes the
mould species *Penicillium chrysogenum* from which
the antibiotic **penicillin** is extracted (see section
52·7). Although vast quantities of mycelium have
been cultured, no sexual stages have ever been
seen. Nevertheless, genetic recombination does
occur and is believed to be the result of a
parasexual process in which genetically different
nuclei fuse. Meiosis occurs and new haploid nuclei
are formed which give rise directly to a new
mycelium.

Another member of the phylum is the soil
fungus *Dactylaria* which traps and feeds on
nematode worms. Special loops attached to the
hyphae can be triggered to close by contact with a
nematode passing through (Fig. 52·5). The
mechanism is osmotic and relies on the rapid
uptake of water. The body of the trapped worm is
penetrated by hyphae and its tissues slowly
digested.

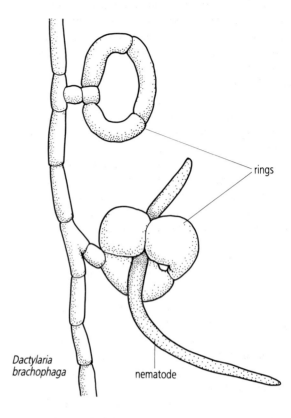

Fig 52·5 Nematode trapping loops of Dactylaria

52·6 Fungal associations ■ ■ ■ ■ ■ ■ ■ ■ ■ ■ ■ ■ ■

Lichens are formed as the result of a close
association between a fungus and a photosynthetic
partner. Their structure provides an excellent
example of **mutualism** and is described in Unit
57. Lichens are pioneer organisms and are often
the first to appear in exposed habitats such as bare
rock or cooled volcanic lava. They are
extraordinarily well adapted to survive in harsh
conditions and can tolerate prolonged periods of
cold or desiccation. Lichens grow slowly but are
abundant in Arctic and Antarctic regions. Solid
mats of *Cladonia* and other lichens provide forage
for reindeer in the far north.

A **mycorrhiza** is a fungal association with the

roots of a higher plant. Endomycorrhizal fungi are
unicellular parasites living inside root cells.
Ectomycorrhizal fungi have hyphae which form a
dense covering surrounding the root tips and
penetrate between the root cells. The fungus
obtains organic nutrients from the root cells, and
the host plant benefits by taking up calcium,
phosphate, and potassium ions and other ions
absorbed by the fungus from the soil. At least 80%
of flowering plants form mycorrhizae and, in some
species, including orchids, citrus trees, and pine
trees, they are essential for normal growth and
development. Some fungi promote growth by
secreting plant hormones such as IAA (Unit 34).

52·7 Ecological and economic importance of fungi ■ ■ ■ ■ ■

1 Decomposition. Like bacteria, saprotrophic
fungi act as decomposers in ecosystems. Many
feed on dead and decaying material in the soil
and help to recycle nutrients, such as phosphates
and sulphates, which are taken up by plants.

The diversity of digestive enzymes in
saprotrophic fungi allows them to exploit unusual

substances for food. Fungi may attack cloth, paint,
leather, waxes, some kinds of insulation on wires
and cables, photographic film, the coating on
camera lenses, even aviation fuel. More
conventionally, they cause widespread rotting of
stored food, often producing toxic waste products
harmful to man. The basidiomycote *Merulius*

lachrymans causes dry rot in the timber frames of houses, causing millions of pounds worth of damage each year. Its fruiting bodies are like wrinkled pancakes and grow to 30 cm in diameter, each one producing enough spores to infect a city. The hyphae of the fungus absorb water from the atmosphere and may thrive in relatively dry timber, reducing it to a rotting pulp.

2 Food. Fungi are vital for a whole range of **fermentation industries**. Yeast species, such as *Saccharomyces cerevisiae*, are used to oxidize sugars to ethanol and carbon dioxide. The process is called alcoholic fermentation (see Unit 10) and is exploited commercially in making wines, beer, and bread. Cheeses are produced mainly by bacterial fermentation, but fungi are often added to improve flavour and texture, as in blue cheeses, where the blue veining consists of the fungal mycelium and spores. Different species of the mould *Penicillium* are used in making Camembert and Roquefort cheeses.

Agaricus bisporus is the commercially grown species of mushroom. A great variety of other fungi are collected and eaten, but care is needed in distinguishing those which are edible from those which are not. Truffles are the underground spore producing structures of various species of the genus *Tuber* which grow in association with the roots of trees, usually oaks. Their spores are dispersed by small mammals attracted to the truffle by its distinctive odour. Truffles are prized ingredients in French cookery and are collected with the aid of dogs and pigs trained to sniff them out.

3 Extracted substances. Antibiotics are the most important substances extracted from fungi. Penicillin was first discovered and named by Sir Alexander Fleming in 1928, and later developed as a medical treatment against bacterial infection by Howard Florey and Ernst Chain. Penicillin is effective against most types of bacteria which cause diphtheria, pneumonia, meningitis, gas gangrene, syphilis and gonorrhea, and also against the *Staphylococcus* bacteria often responsible for sepsis of wounds. The drug revolutionised the treatment of these diseases and dramatically improved post-surgical recovery rates. Penicillin is not effective against *Mycobacterium tuberculosis*, but the search for other antibiotic substances resulted in the discovery of the bacterial antibiotic **streptomycin**, which, used in combination with other drugs, virtually eliminated tuberculosis from the developed world.

Penicillin is still the safest and most useful antibiotic and can be chemically modified to produce more drugs. Unfortunately, like many other antibiotics, it has been over-used, both in the treatment of minor human infections and as a supplement to animal feeds. Exposed to strong selective pressures, many species of bacteria have evolved resistant strains capable of making enzymes which attack penicillin and other antibiotics. Multi-resistant strains of *Staphylococcus aureus* (MRSA) now pose a real threat to health in many hospitals.

Other substances which can be extracted in commercial quantities from cultures of fungi include citric acid and vitamins.

4 Disease. Very few fungi are human parasites. The skin diseases athlete's foot and ringworm are caused by closely related species of *Tinea* which digest the protein keratin, while *Candida albicans*, a type of yeast, causes a mouth or vaginal infection known as thrush.

Fungal infections of plants have huge ecological and economic effects. Most cereal crops, like wheat and corn, are susceptible to fungal diseases called **smuts** and **rusts**. Wheat rust, *Puccinia graminis*, has a life cycle involving two different hosts and is therefore difficult to control using chemical fungicides. The most effective strategy has been to develop more resistant crop species by selective breeding. Ergot is an important parasite of rye. Apples, grapes, cherries, and roses are affected by obligate plant parasites called **powdery mildews**. **Dutch elm disease** is caused by the fungus *Ceratocystis ulmi* which is carried from tree to tree mainly by bark beetles of the genus *Scolytus*. The fungal hyphae kill the tree by blocking its water-conducting vessels and by releasing toxins.

53 KINGDOM PLANTAE 1 PRIMITIVE LAND PLANTS

Objectives

After studying this Unit you should be able to:–

- List the important characteristics of plants

- Outline the distinguishing features of organisms belonging to Phylum Bryophyta, Phylum Filicinophyta, Phylum Sphenophyta, and Phylum Lycopodophyta

- Comment on the major problems which had to be overcome in the evolution of terrestrial plants

- Describe the structure and life cycle of the liverwort, *Marchantia polymorpha*

- Describe the structure and life cycle of the moss, *Funaria hygrometrica*

- Explain the importance of tracheids

- Distinguish between microphyllous and megaphyllous leaves

- Describe the structure and life cycle of the male fern, *Dryopteris filix-mas*

- Briefly describe the structure of the horsetail, *Equisetum arvense*

- Describe the structure and life cycle of the clubmoss, *Selaginella kraussiana*

- Comment on the importance of heterospory

- Compare the adaptations to terrestrial life of all the species listed above

53·1 Introduction ■ ■ ■ ■ ■ ■ ■ ■ ■ ■ ■ ■ ■ ■ ■ ■ ■ ■

Plants are multicellular eukaryotes with cell walls made of cellulose. They are autotrophic and derive energy by photosynthesis, trapping light energy by means of chlorophylls a and b, and other pigments, such as xanthophylls and carotenoids. These pigments are enclosed in chloroplasts which give plants their green colour. The food storage material is starch. Plants reproduce sexually, and are adapted for life on land, although external water may be required for the transfer of gametes. All plant life cycles involve alternation of generations.

The evolutionary history of plants is fairly well understood and the relationships between the different groups are summarised in Figure 53·1. A close link with green algae (Phylum Chlorophyta)

is evident from the similarities in structure and physiology between chlorophytes and primitive plants. The development of multicellular organization in chlorophytes is described in Unit 47.

In this Unit, the structure and adaptive features of four phyla are described. Mosses and liverworts, Phylum Bryophyta, are the most primitive group, and lack fully-developed water conducting tissues. They are said to be **non-vascular** plants. Ferns, Phylum Filicinophyta, horsetails, Phylum Sphenophyta, and clubmosses, Phylum Lycopodophyta, all possess water-conducting cells, or **tracheids**, and are called **vascular plants**, or **tracheophytes**. They show increasingly effective adaptations for life on land.

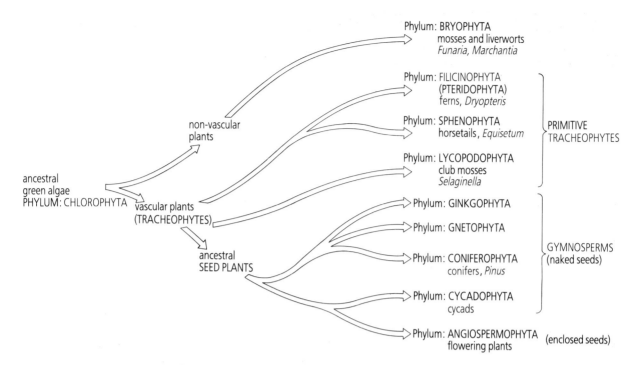

Fig 53·1 Evolutionary relationships in Kingdom Plantae

CLASSIFICATION

Kingdom: Plantae
eukaryotic, multicellular, cell walls containing
  cellulose, life cycle involving alternation of
  generations
terrestrial, freshwater
photosynthetic, chlorophylls a, b, starch food
  reserve

Phylum: Bryophyta
plants which lack fully-developed water
  conducting tissues
gametophyte generation dominant
damp terrestrial habitats, freshwater, external
  water required for gamete transfer

Class: Hepaticae (liverworts)
  flattened ribbon-like or 'leafy' gametophyte
    with single-celled rhizoids
  Marchantia

Class: Musci (mosses)
  upright gametophyte with 'stem' and 'leaves',
    rhizoids multicellular
  Funaria, Sphagnum

Phylum: Filicinophyta (ferns)
plants with vascular tissues (tracheophytes),

megaphyllous leaves arising directly from
underground rhizome
sporophyte generation dominant, usually
homosporous
terrestrial, freshwater, external water required
for gamete transfer
Dryopteris

Phylum: Sphenophyta (horsetails)
plants with vascular tissues (tracheophytes),
microphyllous leaves, underground rhizome,
stem with nodes
sporophyte generation dominant, homosporous
terrestrial, external water required for gamete
transfer
Equisetum

Phylum: Lycopodophyta (clubmosses)
plants with vascular tissue (tracheophytes),
microphyllous leaves, horizontal stem with
upright branches, true roots
sporophyte generation dominant, some species
heterosporous
terrestrial, external water required for gamete
transfer
Selaginella

53·2 Evolution of land plants ■ ■ ■ ■ ■ ■ ■ ■ ■ ■ ■ ■

The first land plants appeared 400–450 million
years ago. Their cellular structure was very similar
to that of green algae, but additional specializations
were necessary to overcome several major
problems.
1 Water loss. Water evaporates from moist
surfaces exposed to the air. Land plants must

prevent drying out or remain confined to damp
habitats. Many seaweeds which inhabit the
intertidal zone survive temporary exposure by
producing mucilage. A similar, but much more
important mechanism, involves the secretion of a
wax-like mixture of fatty substances called **cutin**.
This impregnates the cell walls of the epidermis of

land-adapted species and often forms a separate external coating, or **cuticle**. The cuticle is waterproof and impermeable and forms a highly effective barrier against water loss. Bryophytes usually lack a cuticular layer and live in wet places, but all tracheophytes restrict water loss in this way.

2 Gas exchange. Carbon dioxide is required for photosynthesis. In algae, this is obtained by diffusion across the entire body surface, but in land plants, tiny pores, called **stomata**, are present. These give access to a loosely packed mesophyll tissue inside the plant, where gas exchange takes place. The surfaces of the mesophyll cells can remain moist without causing excessive water loss.

3 Support and transport. All terrestrial organisms require support. Plants must expose a large surface area for photosynthesis and there is an obvious advantage in spreading this surface at some distance above ground level, where it will not be shaded by the surrounding terrain, or by other plants. Such an arrangement requires flat photosynthetic organs (**leaves**) supported by an upright **stem** which is anchored in the ground by **roots**. A major advance in the evolution of these organs was the development of water-conducting cells called **tracheids** with strengthened cell walls containing lignin. These cells are part of the plant tissue known as **xylem**. Together with **phloem** cells, they form tough rigid vascular bundles and leaf veins which act as a supporting framework for the rest of the plant (see Units 31 and 32). Water and minerals are absorbed by the roots and transported upwards by the xylem, while sugars and other nutrients are transported from the leaves to other parts of the plant by the phloem.

4 Reproduction. A typical feature of sexual reproduction in algae is the formation of motile male gametes. The sperm swim towards the female gametes and an external supply of water is essential. This remains true in bryophytes and primitive tracheophytes, but there is a tendency for the sporophyte generation, which produces aerial spores, to become dominant in the life cycle. In more advanced plants, the gametophyte generation becomes further reduced, and various mechanisms have evolved which avoid the need for external water. These advanced forms will be described in Unit 54.

53·3 Phylum Bryophyta ■ ■ ■ ■ ■ ■ ■ ■ ■ ■ ■ ■ ■ ■

Bryophytes are restricted in distribution mainly to damp terrestrial habitats, and the margins of freshwater rivers and lakes. They form a significant proportion of the vegetation of tundra and mountain regions, and have an important ecological role in the colonization of bare land, for example burnt heathland, or volcanic ash.

The Phylum includes two main classes, examples of which are given below.

1 Class Hepaticae – liverworts

Liverworts are so named because they were used by mediaeval herbalists to treat liver disease, a practice which arose from the 'doctrine of signatures', according to which a resemblance in shape between a plant and a human body organ was thought to indicate curative powers.

Marchantia polymorpha, illustrated in Figure 53·2, is commonly found along damp garden paths and in other moist places. The main plant body is the flattened dark green gametophyte, referred to as a **thallus** because it lacks differentiated roots, stems, and leaves. It has a prominent **midrib** and shows a two-forked, or dichotomous, branching pattern. The upper surface of the thallus is specialized for photosynthesis, with pores and air chambers containing loosely packed photosynthetic cells with numerous chloroplasts. The cells which form the lower surface lack chloroplasts and some are modified as **rhizoids** which anchor the thallus and absorb water and minerals.

Asexual reproduction occurs by means of cup-shaped structures which develop on the upper surface. These contain tiny balls of cells, called **gemmae**, which may be dispersed by raindrops and give rise to new gametophytes. Sexual reproduction involves specialized male and female gametangia which are borne on umbrella-shaped outgrowths from the gametophyte thallus. *Marchantia polymorpha* is **dioecious**, in other words, the sexes are separate. The male gametangia, or **antheridia**, are flask-shaped structures filled with sperm. Ripe female gametangia, called **archegonia**, consist of a single egg cell, or **oosphere**, surrounded by a jacket of sterile cells. Sperms are transferred from the male to the female gametophyte by splashing raindrops. After fertilization, the zygote becomes an embryo which develops to form a simple sporophyte. This remains attached to the gametophyte and is completely dependent upon it. A mature sporophyte consists of a **foot** region, a short stalk, or **seta**, and a sporangium, or **capsule**, from which haploid spores are released in dry weather.

2 Class Musci – mosses

Funaria hygrometrica is a common yellowish-green moss which grows in damp shady places and often invades burnt areas, like the site of an old bonfire. It is a common greenhouse species.

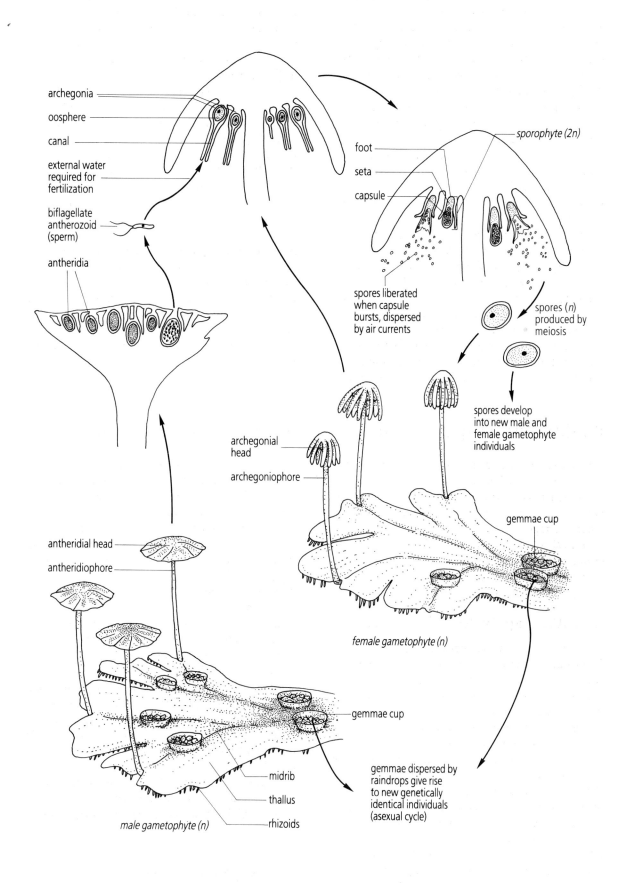

archegonia
oosphere
canal
external water required for fertilization
biflagellate antherozoid (sperm)
antheridia

foot
seta
capsule
sporophyte (2n)

spores liberated when capsule bursts, dispersed by air currents

spores (n) produced by meiosis

spores develop into new male and female gametophyte individuals

archegonial head
archegoniophore

gemmae cup

female gametophyte (n)

antheridial head
antheridiophore

gemmae cup

midrib
thallus
rhizoids

male gametophyte (n)

gemmae dispersed by raindrops give rise to new genetically identical individuals (asexual cycle)

Fig 53·2 Life cycle of the liverwort, Marchantia polymorpha

The gametophyte of *Funaria* develops from a spore as a branching filamentous structure called a **protonema** (see Fig. 53·3). At intervals, multicellular 'buds' are formed which grow vertically, forming a small upright plant with spirally arranged 'leaves'. The blade of each 'leaf' consists of a single layer of cells, except for the supporting midrib region. The 'stem' contains a central strand of tubular cells which may help to conduct water, but tracheids are absent. The whole structure is anchored by rhizoids, which are usually multicellular.

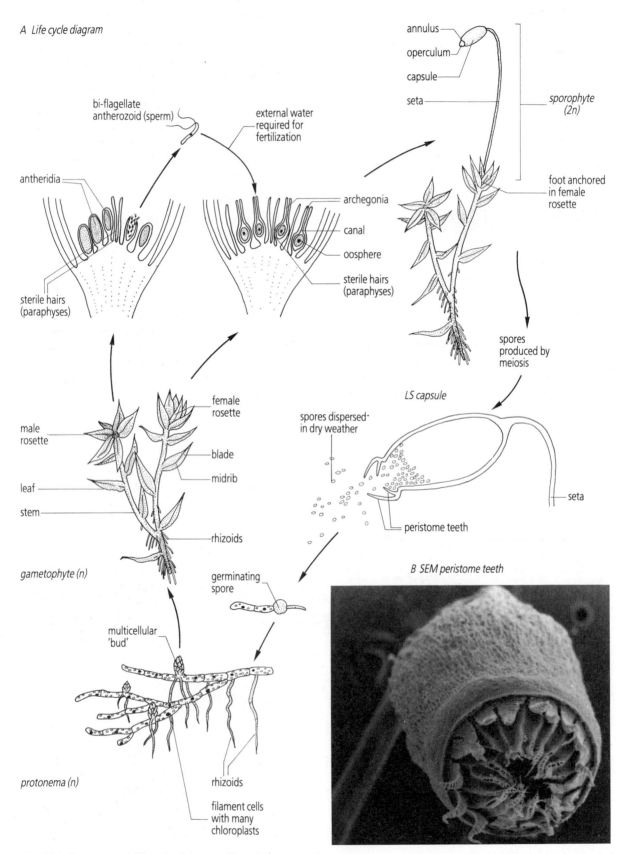

A *Life cycle diagram*

annulus
operculum
capsule
seta
sporophyte (2n)

bi-flagellate antherozoid (sperm)
external water required for fertilization

foot anchored in female rosette

antheridia
archegonia
canal
oosphere
sterile hairs (paraphyses)

sterile hairs (paraphyses)

spores produced by meiosis

female rosette
male rosette
blade
midrib
leaf
stem
rhizoids

LS capsule

spores dispersed in dry weather

seta

peristome teeth

gametophyte (n)

germinating spore

B *SEM peristome teeth*

multicellular 'bud'

protonema (n)
rhizoids
filament cells with many chloroplasts

Fig 53·3 Structure and life cycle of the moss, Funaria hygrometrica

Continued growth and branching of the protonema constitutes one form of asexual reproduction. Some mosses, but not *Funaria*, reproduce asexually by means of gemmae, like those of liverworts. *Funaria* is **monoecious** and gametes for sexual reproduction are produced in antheridia and archegonia, which occur in clusters, called **rosettes**, at the tips of fertile branches. Male and female rosettes can be distinguished by their shape. When a mature antheridium is covered by rain water, a large mucilaginous cell in its wall swells and ruptures, releasing the sperm. These swim towards the egg, and appear to be attracted by a chemical secretion, possibly malic acid.

After fertilization, the diploid zygote divides and elongates rapidly to form the sporophyte. This is larger than the comparable structure in liverworts. Initially, it is dependent on the gametophyte for nutrition, but later the seta and the top section of the sporophyte, called the **capsule**, become photosynthetic. Nevertheless, the sporophyte remains attached to the gametophyte and never becomes independent.

Meiosis and spore production take place within the capsule. At first, the capsule lid, or **operculum**, is held in position by a ring of mucilaginous cells known as the **annulus**. When the capsule is ripe, these cells take up water and burst, and the operculum drops off. In damp conditions, the release of spores is prevented by rows of **peristome teeth** (see Fig. 53·3B). However, in dry weather, the teeth lose water and flex outwards, allowing the spores to be carried off by air currents. This remarkable adaptation ensures that dispersal occurs only at the most favourable times.

53·4 Primitive tracheophytes ■ ■ ■ ■ ■ ■ ■ ■ ■ ■ ■

The primitive tracheophytes possess vascular tissue but have motile sperm, and remain dependent on an external supply of water for reproduction. They include ferns, Phylum Filicinophyta, horsetails, Phylum Sphenophyta, and clubmosses, Phylum Lycopodophyta. Modern-day members of these phyla represent only a small proportion of the species which once existed. The reconstruction illustrated in Figure 53·4 shows a variety of tree-like forms which flourished in the warm conditions of the Carboniferous period. With the evolution of seed plants, almost all of them became extinct, but their compressed remains are mined today as coal.

Significantly, the sporophyte generation with its resistant aerial spores dominated the life history of these early terrestrial plants, while the gametophyte was reduced in size and was relatively short-lived. The presence of tracheids allowed sporophytes to become larger and more specialized as the vascular tissue gave support to tree-like stems and side branches.

Fig 53·4 Reconstruction of Carboniferous swamp

The first true leaves appeared in this group and were of two distinct ancestral types. **Microphyllous** leaves (Fig. 53·5A), like those of clubmosses and horsetails, arose as small outgrowths of the stem, each with only one vascular strand or vein derived from the vascular cylinder of the stem. **Megaphyllous** leaves (Fig. 53·5B) are found in ferns and seed plants, and originated from flattened lateral branches. The spaces between the branches became filled with leaf tissue, called mesophyll, which later differentiated into palisade and spongy layers.

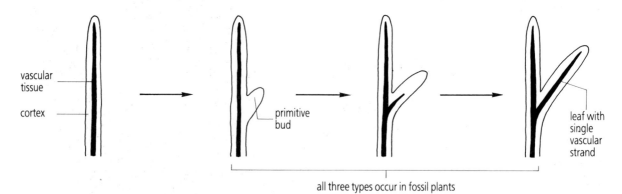

A Evolution of microphyllous leaf

vascular tissue

cortex

primitive bud

all three types occur in fossil plants

leaf with single vascular strand

B Evolution of megaphyllous leaf

flat region of photosynthetic tissue (lamina)

end branches form leaf veins

dichotomously branching stem

reduced dichotomy

Fig 53·5 *Hypothetical stages in the evolution of leaf structure*

53·5 Phylum Filicinophyta (ferns) ■ ■ ■ ■ ■ ■ ■ ■ ■ ■

Ferns are the most successful descendants of these early forms, with about 12 000 living species. Most of these are found in tropical regions, but many species grow abundantly in damp habitats in temperate regions. *Dryopteris filix-mas*, the male fern, is a common woodland species throughout Britain, and is tolerant of a wide range of conditions and soil types. Its leaves, or **fronds**, are greatly divided and arise from a horizontal underground rootstock, or **rhizome**, by a characteristic uncoiling growth, called **vernation** (see Fig. 53·6). Sporangia occur in horseshoe-shaped clusters, or **sori** (singular, sorus), located on the undersides of the fronds. Each sorus is covered and protected by an umbrella-like flap, the **indusium**, which eventually shrinks to expose the underlying sporangia.

Spore dispersal depends on turgor movements which are initiated in dry conditions. The **annulus** is a line of specialised cells which extends around the sporangium, each cell having a thin outer wall and a thick inner wall. The forces produced as these cells dry out cause the sporangium to split open at a weak region, called the **stomium**, consisting of thin-walled cells. As drying continues, the annulus curls farther and farther backwards until a critical point is reached when its outer walls can no longer withstand the tension placed upon them and suddenly fail, allowing a bubble of air to enter each annulus cell. The annulus whips back towards its original position, catapulting the spores outwards.

A spore released in this way can germinate in suitably damp conditions, forming a small heart-shaped structure, called a **prothallus**, which represents the gametophyte generation of *Dryopteris*. The prothallus is anchored by single-celled rhizoids and contains chloroplasts so that it is self-supporting and independent. Antheridia and archegonia develop as shown in Figure 53·6 and motile sperms are released when the prothallus is covered by water. They are attracted to the egg cell by chemotaxis. Only one zygote develops from each gametophyte, eventually growing into a new sporophyte plant.

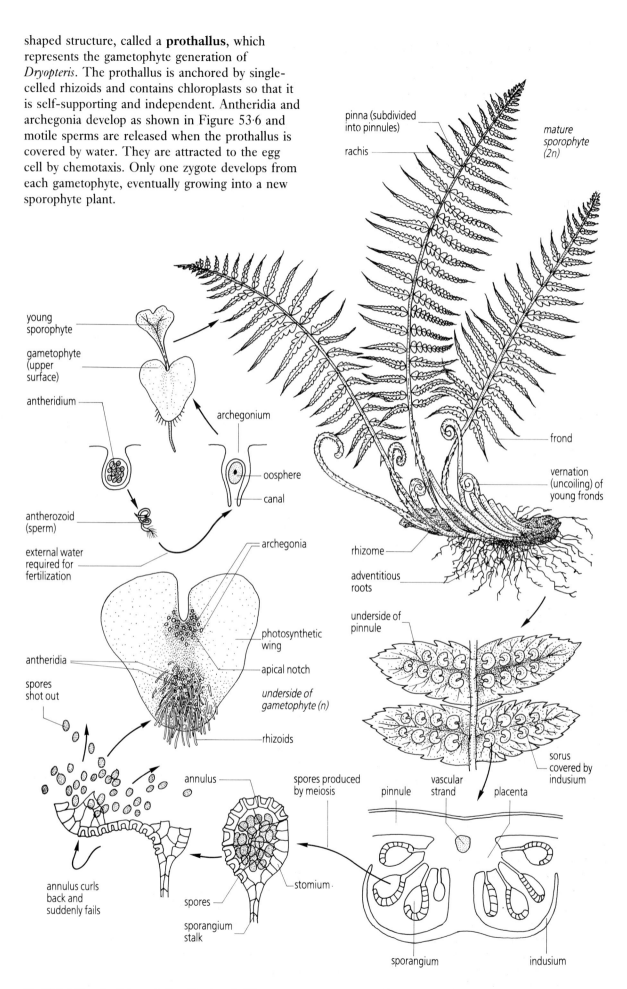

pinna (subdivided into pinnules)

rachis

mature sporophyte (2n)

young sporophyte

gametophyte (upper surface)

antheridium

archegonium

oosphere

canal

antherozoid (sperm)

external water required for fertilization

archegonia

photosynthetic wing

apical notch

underside of gametophyte (n)

antheridia

spores shot out

rhizoids

frond

vernation (uncoiling) of young fronds

rhizome

adventitious roots

underside of pinnule

sorus covered by indusium

annulus

spores produced by meiosis

pinnule

vascular strand

placenta

annulus curls back and suddenly fails

spores

stomium

sporangium stalk

sporangium

indusium

Fig 53·6 *Life cycle of the male fern*, Dryopteris filix-mas

473

53·6 Phylum Sphenophyta (horsetails) ■ ■ ■ ■ ■ ■ ■ ■ ■ ■

Horsetails grew abundantly in the swamps of Carboniferous times, but there are only about 25 living species, all belonging to the genus *Equisetum*. Extinct species grew to 15 m or more in height. Today, horsetails are no more than 2 m tall and most species inhabit cool moist places. Silica deposited in the epidermis of the stem gives them an abrasive texture. The early American settlers used them for cleaning pots and pans and gave them the common name 'scouring rushes'.

The structure of the sporophyte of *Equisetum arvense*, a common British species which grows on waste ground and in gardens, is illustrated in Figure 53·7A. A creeping underground stem, or rhizome, allows *Equisetum* to invade and colonize suitable areas quickly, often to the exclusion of other plants. Storage organs, or **tubers**, attached to the rhizome help the plant to survive through the winter. Upright green ridged photosynthetic stems arise at intervals. The sterile vegetative shoots possess joints, or **nodes** from which numerous lateral branches may arise. In addition, each node is surrounded by a whorl of small microphyllous leaves. Fertile shoots appear in spring and develop club-shaped **strobili**, (singular, strobilus) or cones.

Following meiosis, spores are produced. These have coiled flaps, or **elaters**, which straighten out in dry conditions and act as tiny sails to catch the wind (see Figure 53·7B). When a spore lands in a damp place, the elaters coil up and the spore germinates. A tiny gametophyte about the size of a pin head is produced and the life cycle is completed in much the same way as in *Dryopteris*.

A *Sporophyte* (2n)

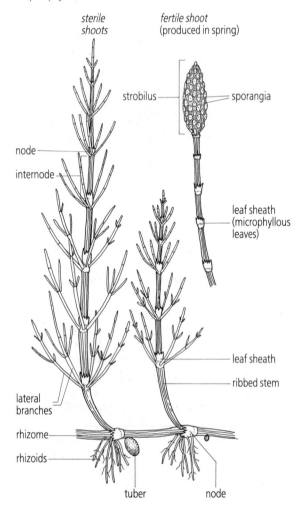

B *Spores*

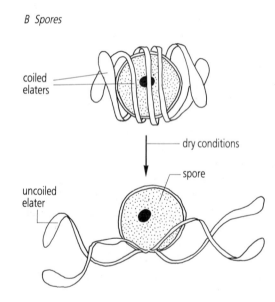

Fig 53·7 A *Structure, and* B *spores of the horsetail*, Equisetum arvense

53·7 Phylum Lycopodophyta (clubmosses) ■ ■ ■ ■ ■ ■ ■ ■ ■

The phylum is represented by about 1000 living species, most of which are tropical. Among the relatively few British species is *Selaginella selaginoides* which is found in damp mossy or grassy places usually on mountains. An introduced species, *Selaginella kraussiana*, is a common greenhouse plant and is described here.

The sporophyte of *Selaginella kraussiana*, illustrated in Figure 53·8, has a moss-like appearance and consists of a series of creeping branches at ground level. Microphyllous leaves are produced in opposite pairs, resulting in an arrangement of four rows along the length of the stem. The upper leaf in each pair is small, while the lower leaf is larger. At the base of young leaves there is a small membranous outgrowth, or **ligule**, which later withers and disappears. The sporophyte is anchored by **rhizophores** which originate just behind branching points in the stem and grow downwards.

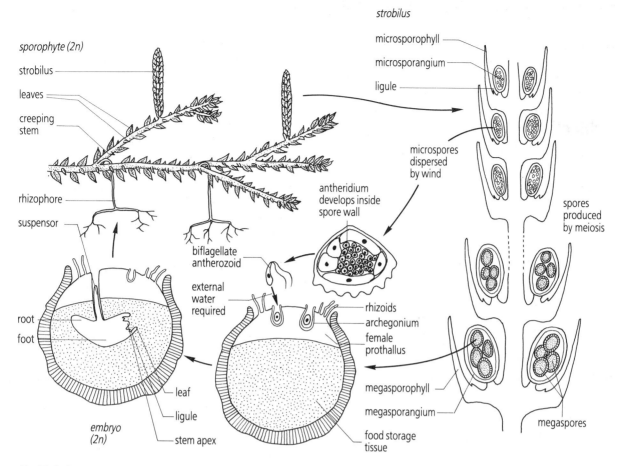

Fig 53·8 Structure and life cycle of Selaginella kraussiana

A mature plant has upright reproductive structures called **strobili** (cones) which consist of numerous sporangia formed in the axils of leaves, or **sporophylls**, on vertically growing shoots. *Selaginella* exhibits **heterospory**, that is, it produces spores of two distinct kinds. **Megasporangia** are usually found in the lower part of the strobilus, and each gives rise to four large haploid spores, called **megaspores**. These germinate to produce female gametophytes. On the other hand, **microsporangia** are found in the upper part of the strobilus and produce numerous **microspores** which form male gametophytes.

Selaginella is more completely land-adapted than any of the plants described so far by virtue of the way in which its gametophytes develop. The male gametophyte develops entirely within the wall of the microspore, which contains stored food materials. Thus, it is no longer an independent plant, but depends entirely on the activities of the sporophyte for its nutrition. When the microspore germinates, two cells are formed. One of these, called the **prothallial cell**, is a small cell which does not divide again. It represents the last remnant of the vegetative part of the male gametophyte. The other cell becomes an antheridium and division continues to produce numerous biflagellate sperm inside a jacket of sterile cells. Microspores are released about midway through their development and may be

transported to other plants, or rain down onto the megasporangia below.

Development of female gametophytes takes place inside the megasporangia. These are much larger than microsporangia and remain attached to the strobilus throughout their development. A female prothallus and a separate food storage region are formed within the spore wall, and archegonia are exposed when the spore wall splits. In wet weather, sperms released from the microspores swim to the oosphere and fertilization is achieved. The zygote develops, still within the remains of the megaspore, and an embryo is formed which grows initially by absorbing food from the storage region. Later, the embryo becomes photosynthetic and a new independent sporophyte individual is formed.

Selaginella is not a direct descendant of the forms which gave rise to seed plants, but its life cycle gives an important insight into the mechanisms which must have been involved. Heterospory, the reduction in size of the male gametophyte, and the nutritional dependence of the gametophytes, can all be seen as adaptations to terrestrial life. Avoiding the need for external water as a medium for fertilization represented the most important remaining obstacle in the evolution of fully adapted land plants. The ways in which present day plants overcome the problem are explained in the next Unit.

475

54 KINGDOM PLANTAE 2 ADAPTATIONS OF SEED PLANTS

Objectives

After studying this Unit you should be able to:–

- Draw a flow diagram to show the important stages in the life cycle of a typical seed plant

- List and explain the evolutionary advances of seed plants compared with other plant groups

- List the distinguishing features of Phylum Coniferophyta

- Describe the structure and life cycle of the Scots pine, *Pinus sylvestris*

- Comment on the economic importance of conifers

- List the distinguishing features of Phylum Angiospermophyta

- List and explain the adaptations which contribute to the success of angiosperms

- Draw and label a diagram to show the main structural features of the grass, *Festuca pratensis*

- Draw and label a diagram to show the main structural features of a shoot of horse chestnut, *Aesculus hippocastanum*

- Compile a table to compare the structure of monocotyledons and dicotyledons

- Comment on the economic importance of angiosperms

54·1 Evolution of seed plants ■ ■ ■ ■ ■ ■ ■ ■ ■ ■ ■ ■ ■

The life cycle of *Selaginella*, described in Unit 53, shows four major evolutionary advances compared with the life cycles of liverworts, mosses, and ferns. These advances are (1) **heterospory**, in which different sizes of spores give rise to separate male and female gametophytes, (2) reduction in size of the male gametophyte and its development to maturity inside the spore wall of the microspore, (3) transfer of intact microspores by air currents, and (4) nutritional dependence of the embryo, which remains inside the megaspore wall and attached to the sporophyte during early development.

The significance of these features can be explained with the aid of Figure 54·1, which summarizes the type of life cycle shared by *Selaginella* and all more advanced plants. As you can see, the sporophyte generation has become dominant and forms both microsporangia and megasporangia. Inside these structures, haploid microspores and megaspores are produced by meiosis.

The microspores develop a resistant wall and are small enough and light enough to be carried to female structures elsewhere on the same plant, or on other plants. In *Selaginella*, the microspores are relatively unspecialized, but in more advanced groups, they are modified to form **pollen grains**. Various adaptations have evolved for pollen transport, for example, by wind or insects, as will

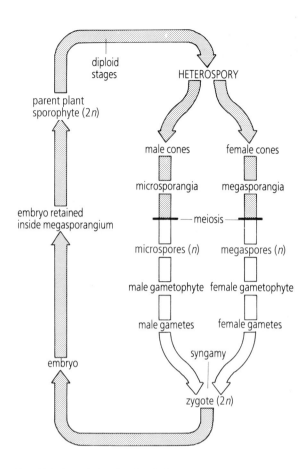

Fig 54·1 Basic life cycle of seed plants

be explained later. The male gametophyte develops inside the microspore and gives rise to male gametes which are retained for some time within the microspore wall. In the case of *Selaginella*, the male gametes are motile antherozoids which must be released before fertilization can occur. Thus, *Selaginella* requires an external supply of water in which the male gametes can swim. In the plant groups described in this Unit, the need for a

covering film of water is removed by the growth of a **pollen tube** through which the male gametes, now naked nuclei, travel directly to the female gamete.

The megaspores of *Selaginella* remain inside the megasporangia attached to the sporophyte. Fertilization and early development of the embryo take place within the megaspore wall. A structure called a **seed** is formed when additional protective layers, called **integuments**, grow to surround the megaspore. Inside the seed, development of the embryo is delayed during dispersal from the parent plant. In suitable conditions, the seed germinates and the embryo gives rise to a new sporophyte individual.

With the evolution of pollen grains and resistant seeds, seed plants became well adapted for colonization of terrestrial habitats and a great variety of forms developed. Traditionally, two main groups of seed plants were recognized, **gymnosperms**, with naked seeds, and **angiosperms**, with seeds contained inside an ovary wall, forming a fruit. In the classification system used in this book, four separate phyla of gymnosperms are distinguished, namely Phylum Coniferophyta (conifers), Phylum Cycadophyta (cycads), Phylum Gnetophyta, and Phylum Ginkgophyta.

The great majority of living gymnosperms are trees or shrubs belonging to Phylum Coniferophyta. There are about 550 species. Compared with gymnosperms, angiosperms are spectacularly successful plants, with more than 230 000 species classified in a single phylum, Phylum Angispermophyta. The remainder of this Unit outlines the life cycles and characteristic features of representative members of these groups. The structure and physiology of angiosperms is described in more detail in Units 31–35.

4·2 *Pinus sylvestris* – a conifer ■ ■ ■ ■ ■ ■ ■ ■ ■ ■

CLASSIFICATION

Phylum: Coniferophyta (conifers)
plants with naked seeds (not protected by an ovary wall) vascular tissues containing xylem tracheids and phloem sieve cells, xylem vessels and phloem companion cells absent, well-developed secondary thickening, leaves typically needle-like with xerophytic adaptations, often evergreen
sporophyte generation dominant, heterosporous, usually bearing cones in which sporangia develop, gametophyte generation reduced, microspores form pollen grains transported by wind, no external water needed for fertilization
Pinus, Larix, Sequoia, Taxus

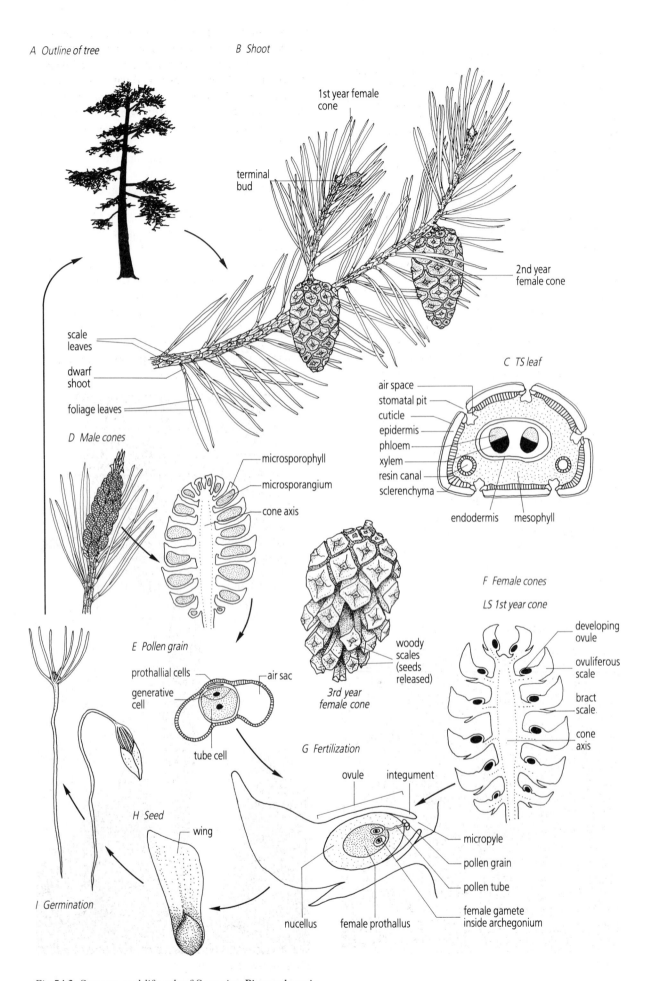

A *Outline of tree*

B *Shoot*

1st year female cone

terminal bud

2nd year female cone

scale leaves

dwarf shoot

foliage leaves

C *TS leaf*

air space
stomatal pit
cuticle
epidermis
phloem
xylem
resin canal
sclerenchyma

endodermis   mesophyll

D *Male cones*

microsporophyll

microsporangium

cone axis

E *Pollen grain*

prothallial cells

generative cell

air sac

tube cell

woody scales (seeds released)

3rd year female cone

G *Fertilization*

F *Female cones*

LS 1st year cone

developing ovule

ovuliferous scale

bract scale

cone axis

ovule   integument

micropyle

pollen grain

pollen tube

female gamete inside archegonium

H *Seed*

wing

nucellus   female prothallus

I Germination

Fig 54·2 Structure and life cycle of Scots pine, Pinus sylvestris

Structure

Scots pine, *Pinus sylvestris*, is often used as an example to illustrate the typical life cycle and main structural features of conifers. *Pinus sylvestris* is native to western and northern Europe and Russia and has been extensively planted in Europe and North America. Individual trees grow to about 35 m and are recognised by their reddish brown or orange flaking bark and characteristic irregular outline (Fig. 54·2*A*).

A young shoot possesses two types of leaves (Fig. 54·2*B*). As you can see, the surface of the stem is covered by small **scale leaves** arranged in a spiral pattern. Small buds in the axils of the scale leaves give rise to **dwarf shoots** from which the **foliage leaves** develop in pairs. Each leaf is needle-like and semi-circular in cross section. The shape of the leaf reduces the surface area available for water loss and the detailed anatomy of the leaf reveals further adaptations for water conservation (Fig. 54·2*C*). These include a thick waxy cuticle, small internal air spaces, and the location of the stomata in sunken pits. Such water-conserving, or **xerophytic**, features allow the leaves to be retained for several years. As older leaves are shed new ones are formed, so that the branches of the tree are never bare. Most other conifers are evergreen, but a few species such as larches, *Larix*, are deciduous and show seasonal leaf loss.

The dwarf shoots are shed along with the old foliage leaves and are therefore said to show **limited growth**. On the other hand, the apical bud at the tip of the main shoot is capable of **unlimited growth** and may give rise to new lateral branches.

All species of pines are trees. Their wood consists mainly of elongated secondary xylem cells called **tracheids**. The walls of the tracheids are impregnated with lignin and the empty cavities of adjacent cells are linked by **bordered pits** (Fig. 54·3). In conifers, the region of primary cell wall in the centre of the pit is thickened with a deposit of waxy material, forming a structure known as a **torus**. The phloem of conifers contains sieve cells, fibres, and parenchyma cells, but companion cells are absent.

Life cycle

The tree is the sporophyte generation of the plant and is **monoecious**, that is, separate male and female structures are formed on the same individual. These develop from shortened lateral branches with modified leaves called **sporophylls**. Male cones consist of up to 100 **microsporophylls** each of which carries a pair of **microsporangia**, or **pollen sacs**, attached to its lower surface, while female cones consist of a similar number of **megasporophylls** upon which **megasporangia** are formed.

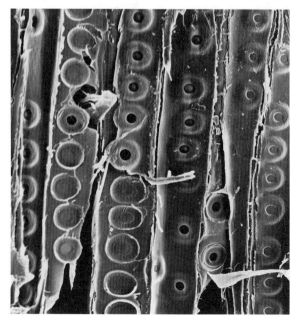

A SEM bordered pits

Fig 54·3 Bordered pits in tracheids of pine wood

B Transverse section (EM)

Inside the microsporangia, numerous haploid microspores are formed following meiosis: these develop into **pollen grains** with lateral air sacs to aid wind dispersal (Fig. 54·2*E*). The male gametophyte is represented by a few prothallial cells within the pollen grain. Large numbers of pollen grains are shed in dry weather in April or May and may be carried long distances by the wind. Successful pollination is completed when a pollen grain becomes trapped in mucilage secreted by a receptive **ovule**. It then germinates, producing a **pollen tube** through which the male gamete travels to fertilize the female gamete situated deep within the ovule (Fig. 54·2*G*).

Ovules are formed in the female cones (Fig. 54·2*F*), and each consists of a food storage region, the **nucellus**, surrounded by a protective layer, called the **integument**. At one end, the integument is perforated by a tiny pore, the **micropyle**. The ovule beomes ready to receive pollen after about a year of development, but at this time the female gametes have not yet been formed. Trapped pollen grains release chemicals which trigger a sequence of changes which takes a further year to complete. A single cell in the nucellus, called the **archesporial cell**, swells and divides by meiosis to produce four haploid **megaspores**. Only one of these survives, growing and dividing to form a female gametophyte (prothallus) which eventually produces three or four gamete-containing vessels called **archegonia**.

The female gamete inside an archegonium is fertilized in the spring following pollination and the development of seeds takes a further year. The embryo formed is surrounded by a food store and protected within a hard seed coat formed by the integument. The seeds are released when the sporophylls of the cone dry and curl upwards and wing-like extensions of the integument enable them to be dispersed by wind (Fig. 54·2*H*). In favourable surroundings, the seed may germinate and ultimately give rise to a new tree.

Economic importance of conifers

Pines, spruces, firs, and other conifers are a major source of softwood timber used in house construction and for fences and telegraph poles. The timber can be pulped to make paper or compressed to make chipboard, the basic material for modern furniture. In addition, pines are a source of resins from which turpentine and pine oil are extracted.

Redwoods, like coast redwood, *Sequoia sempervirens*, and Sierra bigtree, *Sequoiadendron giganteum*, are probably the largest of any plant or animal species. Coast redwoods grow to more than 100 m, while Sierra bigtree grows to about 80 m, but is more massive (Fig. 54·4). The trunk of the General Sherman tree is estimated to weigh 625 000 kg. Because of their excellent timber qualities, redwoods have been extensively felled and 90% of virgin redwood forest has been lost in the last 150 years. Only half the remaining trees are protected from future logging.

Fig 54·4 Sierra bigtree, Sequoiadendron giganteum

54·3 Phylum Angiospermophyta ■ ■ ■ ■ ■ ■ ■ ■ ■ ■

CLASSIFICATION

Phylum: Angiospermophyta (flowering plants)
plants with seeds protected by an ovary wall
forming a fruit, xylem containing vessels and
tracheids, phloem with sieve tubes and
companion cells

sporophyte generation dominant, heterosporous,
sexual reproduction by flowers, gametophyte
generation reduced, no external water needed
for fertilization

Class: Monocotyledoneae
embyro with one seed leaf, leaves parallel-
veined, vascular bundles scattered in stem,
secondary thickening rare

perianth segments similar (not differentiated
into sepals and petals), flower parts usually in
multiples of three, many species wind-
pollinated
Avena, Lilium, Zea

Class: Dicotyledoneae
embryo with two seed leaves, leaves net-veined,
vascular bundles arranged as a ring in stem,
secondary thickening common

perianth consisting of sepals and petals, flower
parts in multiples of four or five, many species
insect-pollinated
Aesculus, Helianthus, Lamium, Ranunculus, Vicia

The flowering plants are the most abundant group
of land plants. Since their first appearance in mid-
Cretaceous times about 110 million years ago they
have replaced gymnosperms as the dominant
vegetation in most terrestrial habitats except those
of the northern temperate region. The enormous
success of angiosperms can be attributed to the
following features.

1 Flowers. Flowers are unique to angiosperms.
The detailed structure of flowers is described in
Unit 35. One of their most important functions is
to protect the floral parts during development.
Many flowers have become adapted to attract
insects or other pollinators, for example, by the
development of large brightly coloured petals,
nectaries and scent glands. The transport of pollen
by insects is much more efficient than the wasteful
process of wind-pollination and **co-evolution** of
flowers and pollinating insects has been a key
factor in the success of both groups.

2 Development of carpels. The ovule of an
angiosperm develops inside a modified leaf called a
carpel which surrounds and encloses the ovule,
forming an **ovary**. The ovary wall protects the
ovule during fertilization and seed formation, and
in most species gives rise to the **fruit** of the plant.
As explained in Unit 35, fruit structure can be
adapted in various ways to ensure effective
dispersal of the seeds.

3 Rapid development of seeds. In
conifers, the production of ovules, the growth of
the pollen tube and the formation of seeds are
extremely slow, requiring up to three years in *Pinus
sylvestris*. In angiosperms, these processes are much
more rapid. Food translocated from the rest of the
plant allows the ovules to grow to maturity in a few
weeks. The pollen grains derive nutrient from the
tissues of the stigma and style and pollen tubes
may elongate at rates of up to 15 cm per day.

The female gametophyte in flowering plants is
reduced to a single multinucleate cell, known as
the **embryosac**. As indicated in Figure 54·5, two
male nuclei are involved in fertilization of the
embryosac. One of these fuses with the female
gamete to form the zygote, while the other fuses
with a central diploid nucleus to form the triploid
endosperm nucleus. In some seeds, this gives rise
to a specialized food storage tissue, the endosperm,
which accumulates food reserves for use by the
embryo during germination. The process of
double fertilization is unique to flowering plants
and may be one factor in enabling seeds to develop
to maturity within a single season. You should
compare Figures 54·5 and 54·1 to identify the
structures in flowering plants which correspond to
microsporangia and megasporangia in *Selaginella*,
and in conifers.

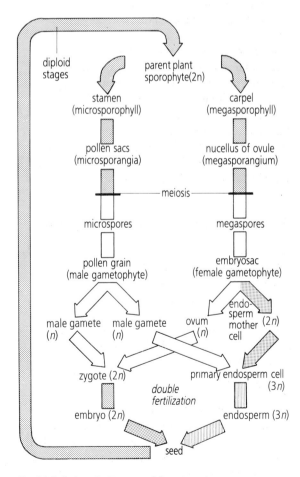

Fig 54·5 Life cycle diagram of flowering plants

4 Adaptations for cross-pollination. The genetic variability of the offspring of angiosperms is increased by mechanisms which promote cross-pollination, while making self-pollination less likely. The range of possible mechanisms is outlined in Unit 35.

5 Adaptations for vegetative reproduction. Conifers have no method of asexual reproduction and are therefore unable to spread vegetatively. In contrast, many flowering plants have special structures which allow them to reproduce vegetatively so that they can increase in numbers and gain a firm hold in favourable habitats. Sexual reproduction continues to produce genetically varied offspring capable of colonizing new areas.

6 Structural and physiological adaptations. In addition to the reproductive adaptations listed above, angiosperms possess many features which are likely to confer an advantage in competition with other groups of plants. For instance, their vascular tissues are more highly developed with specialized cell types including xylem vessels and phloem companion cells. Adaptations for the control of water loss, the development of broad flat leaves for efficient photosynthesis and the production of secondary compounds for protection against herbivores also help to explain the success of the group.

Many of these adaptations are explained at greater length elsewhere in this book.

54·4   Monocotyledons and dicotyledons ■ ■ ■ ■ ■ ■ ■ ■ ■ ■

Monocotyledons include grasses, irises, lilies, orchids, palms, and bamboos. Although there are fewer species than in dicotyledons, monocotyledons are a successful group and dominate the grasslands of the World. Species such as wheat, oats, barley, and maize are important as food crops. Figure 54·6 illustrates the main structural features of *Festuca pratensis*, or meadow fescue, a common species of grass. The floral structure of *Avena sativa* is described in Unit 35.

Dicotyledons are the more numerous class of angiosperms and include many insect-pollinated species and trees and shrubs with secondary thickening. Most fruit, vegetable and nut crops are obtained from dicotyledons and some species of trees, such as oak and walnut, are used for timber. The structure of the horse chestnut, *Aesculus hippocastanum*, is illustrated in Figure 54·7.

You should compare the structure of monocotyledons and dicotyledons as represented by these examples and draw up your own table of comparison using the information provided in the classification summary at the beginning of section 54·3. Each group has its own specializations and monocotyledons should not be considered as more primitive than dicotyledons.

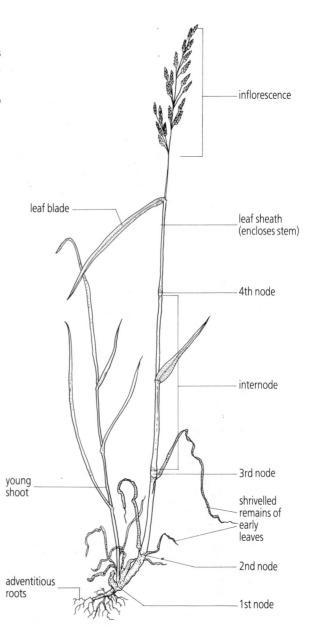

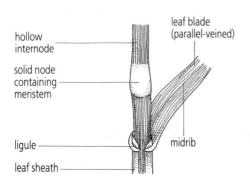

Fig 54·6 Structure of meadow fescue, Festuca pratensis

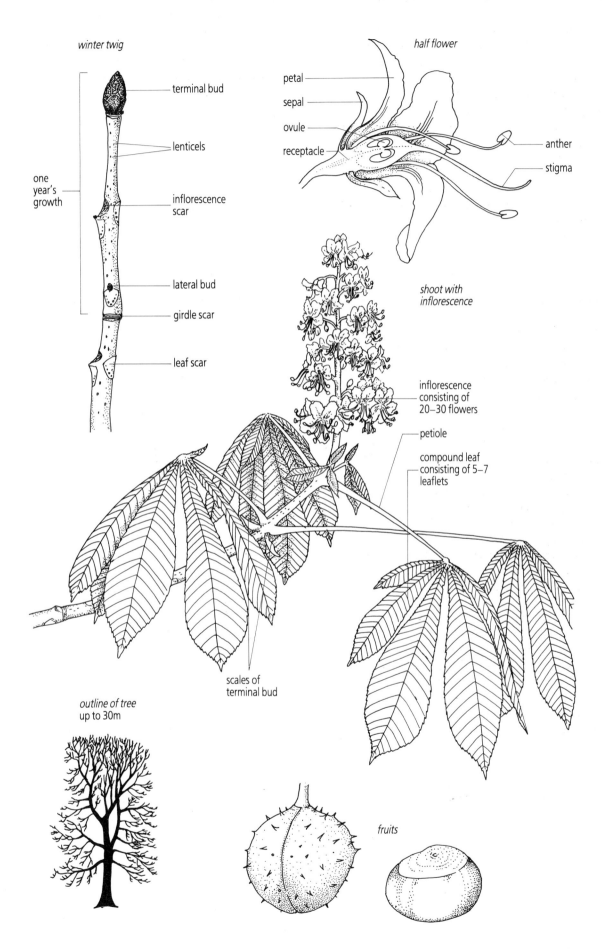

terminal bud

lenticels

one
year's
growth

inflorescence
scar

lateral bud

girdle scar

leaf scar

half flower

petal

sepal

ovule

receptacle

anther

stigma

shoot with
inflorescence

inflorescence
consisting of
20–30 flowers

petiole

compound leaf
consisting of 5–7
leaflets

scales of
terminal bud

outline of tree
up to 30m

fruits

Fig 54·7 Structure of horse chestnut, Aesculus hippocastanum

Section D Questions ■ ■ ■ ■ ■ ■ ■ ■ ■ ■ ■ ■

Short-answers and interpretation

1 Construct a dichotomous key by which someone could identify the seven *Ranunculus* fruits shown in the diagrams below.

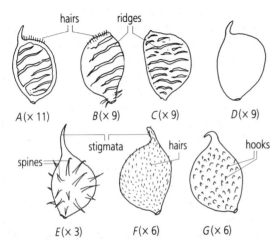

A = *R. cincinatus*   B = *R. trichophyllus*
C = *R. peltatus*   D = *R. acris*
E = *R. arvensis*   F = *R. auricomus*
F = *R. parviflorus*

To save time you can refer to the fruits by the letters **A–G** when writing out your key. [6]

Associated Examining Board

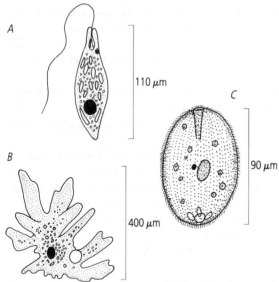

2 **a** The diagrams above show three examples of protozoans belonging to different phyla. Name the phylum to which each belongs. [3]
b Describe briefly how locomotion is achieved by each of these organisms. [6]
c State two ways in which the organism shown in *A* differs from a cell of a filamentous alga. [2]
d How do organisms like those shown assist in sewage treatment? [3]

University of London School Examinations Board

3 An animal has the following features:
   bilateral symmetry
   triploblastic organization
   acoelomate structure
   flame cells
a Briefly describe each of these features [4]
b **i** Name the phylum to which the animal belongs.
ii Name one class from the phylum named in **i**.
iii Describe one characteristic which will distinguish the class you have named in **i** from other classes in this phylum. [3]

Associated Examining Board

4 **a** **i** Name a class of the phylum Annelida. [1]
ii Write down, in the accepted scientific manner, the name of a species of annelid from this class. [2]
b **i** State two characteristics that enable you to classify the organism into the phylum Annelida. [2]
ii State one characteristic of the class of Annelida you have chosen. [1]
c Describe the mechanism of locomotion in a named annelid. [4]

University of Oxford Delegacy of Local Examinations

5 **a** Name an animal with a hydrostatic skeleton and explain how its skeleton helps its locomotion. [6]
b Give four differences between arthropod and mammal skeletons. [4]
c The diagram shows the surface of an arthropod limb joint.

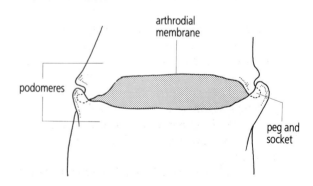

i Explain how this joint allows movement in one plane. [4]
ii Compare and contrast structure and movement in this joint and in a human finger joint. [6]

Associated Examining Board

6 The diagram below is of a fresh water organism as revealed by the electron microscope. The organism is green and has two flagella.

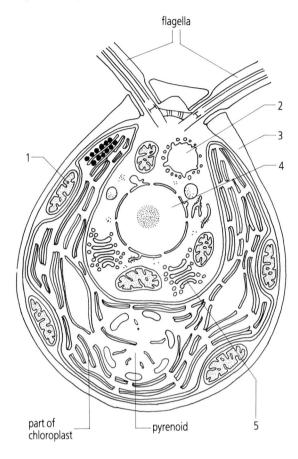

flagella

1

2

3

4

part of chloroplast

pyrenoid

5

a Name the parts numbered 1–5. [5]
b Outline the functions of organelles 1, 4 and 5. [6]
c Draw a diagram to show the structure of a typical flagellum as seen in transverse section in an electron micrograph. [3]
d How would you classify this organism? Give your reasons. [3]

University of London School Examinations Board

7 a State when and where **i** meiosis, **ii** chemotaxis and **iii** spore discharge occur in the life cycle of a bryophyte. [6]
b In bryophytes, the sporophyte is often described as being 'parasitic on the gametophyte'. Why is it so described? [3]
c Indicate the importance of **i** wet conditions and **ii** dry conditions in the life cycle of a bryophyte. [6]

University of London School Examinations Board

Essays

1 a Suggest how life may have originated on Earth. [6]
b With reference to suitable examples discuss:
  i the advantages of eukaryotic organization over prokaryotic organization,

ii the benefits of multicellularity over acellular (unicellular) organization. [12]

Cambridge University Local Examinations Syndicate

2 Devise a key to distinguish between the main animal phyla. For each phylum, you may use all the major structural features that you consider diagnostic. [25]

Cambridge University Local Examinations Syndicate

3 a In what respects do bacteria differ from all other free-living organisms? [4]
b Discuss the suggestion that bacteria are more varied in their mode of life than all the other main groups of organisms. [10]
c Indicate the main steps by which each of these other main groups may have evolved from a bacterial ancestor. [6]

University of Oxford Delegacy of Local Examinations

4 a Describe the structure and mode of replication of a virus. [12]
b Indicate how viruses may be transmitted from host to host. [3]
c Explain how viruses may cause disease. [3]

Cambridge University Local Examinations Syndicate

5 Give an account of the economic importance of viruses, bacteria and fungi. [20]

University of London School Examinations Board

6 Define the term 'division of labour'. Using named examples, give an illustrated account of the way in which changes in the division of labour within the body have characterized invertebrate evolution. [20]

Welsh Joint Education Committee

7 Discuss the importance in biology of the ratio of surface area to volume. [20]

Associated Examining Board

8 a Describe the mechanism of locomotion in **i** a ciliate protozoan, and **ii** *Amoeba*. [5,5]
b Discuss why these mechanisms are restricted to small organisms, and explain why large animals, such as fish, use muscles for locomotion. [8]

Cambridge University Local Examinations Syndicate

9 a What is a parasite? [3]
b Give an illustrated account of the life history of a named endoparasite. [12]
c Comment on the threats to its survival. [5]

Oxford and Cambridge Schools Examination Board

10 What particular biological difficulties arise with increase in size of an organism. How have these difficulties been overcome in the course of evolution? [20]

Oxford and Cambridge Schools Examination Board

11   Survey the various methods used by invertebrate animals to obtain food.   [20]
University of London School Examinations Board

12   With reference to both plants and animals, discuss the problems of being sessile and the adaptations to a sessile habit.   [20]
Associated Examining Board

13   a List the distinctive features of insects.   [4]
     b Describe how a named insect **i** flies and **ii** walks.   [11]
     c Suggest reasons why very large insects have not evolved.   [3]
Cambridge University Local Examinations Syndicate

14   a List the main problems that had to be overcome before animals could successfully colonize the land.
     b Show how insects have overcome these problems.   [13]
Welsh Joint Education Committee

15   a List the Divisions (Phyla) of the Plant Kingdom, indicating the main characteristics of each Division.   [10]
     b Explain why the Spermatophyta (seed plants) are often referred to as the 'higher plants'.   [8]
Cambridge University Local Examinations Syndicate

16   Organisms with a true coelom are regarded as being more advanced than acoelomate organisms. What are the advantages of coelomate organization?   [20]
Oxford and Cambridge Schools Examination Board

17   a List the adaptations that plants required in order to colonize the land.   [7]
     b By comparing *Fucus* with *Funaria*, discuss the extent to which mosses can be said to be successful terrestrial plants.   [13]
Welsh Joint Education Committee

18   a What is meant by 'alternation of generations'?   [4]
     b Give an account of the life cycle of a fern, emphasizing how the life cycle is related to the environment.   [10]
     c What are the major differences between the life cycle of a fern and that of a bryophyte?   [6]
University of London School Examinations Board

19   Discuss the adaptations shown by *Pinus* for life in a terrestrial habitat.   [20]
Welsh Joint Education Committee

20   Discuss the ways in which flowering plants appear to be adapted to life on land.   [20]
University of London School Examinations Board

Section E

ECOLOGY

55 ENERGY AND ECOSYSTEMS [   ]

Objectives

After studying this Unit you should be able to:–

- Define ecology and explain the concept of an ecosystem

- Give examples of food chains and food webs, indicating the importance of different trophic levels

- Outline the relationships between grazing, parasitic, and detritus food webs

- Explain the terms pyramid of numbers, pyramid of biomass, and pyramid of energy

- Comment on the flow of energy through an ecosystem and list the various ways in which energy is gained and lost

- Briefly discuss the application of some of these principles in modern agriculture

55·1 Ecology and ecosystems ■ ■ ■ ■ ■ ■ ■ ■ ■ ■ ■ ■

Ecology is the study of the interactions between living organisms and their environment. The study of a single species is called **autecology** and considers the behaviour of populations and the reasons for the abundance and distribution of particular organisms. **Synecology** is the study of complete **ecosystems** consisting of many different species which interact with each other and with their non-living surroundings.

A good example of an ecosystem is provided by a forest. This contains trees, smaller plants, birds, animals, and soil microorganisms together with the non-living components of the soil, such as rock particles, organic remains and inorganic salts.

Ecosystems are affected by climatic factors including sunlight, rain, wind and humidity.

All these things work together to form a balanced, yet constantly changing, unit of nature. To keep such a system in equilibrium, energy is required. This is supplied directly or indirectly by the Sun and usually enters the ecosystem through the process of photosynthesis carried out by green plants.

Energy is passed from organism to organism in the form of food. Consequently, the study of feeding relationships is a good starting point for the general discussion of energy flow in ecosystems.

55·2 Food chains, food webs and trophic levels ■ ■ ■ ■ ■ ■

An example of a simple **food chain** from a pond ecosystem is illustrated in Figure 55·1. Food chains always start with a self-feeding, or autotrophic, organism called a **producer**. With few exceptions, the producers are green plants which convert a small fraction of the Sun's energy into a stored chemical form. Some of this energy is transferred to herbivores in food and may later be transferred to one or more carnivores. The herbivores in an ecosystem form the **primary consumers** and subsequent organisms in the

various food chains make up secondary and often tertiary levels of consumption. In the pond example, *Daphnia* is a **primary consumer**, *Hydra* is a **secondary consumer** and the dragonfly nymph is a **tertiary consumer**. These feeding levels are known as **trophic levels**. It is very rare to encounter a chain of more than six organisms because the energy initially stored by autotrophic organisms is progressively used up as it passes through the trophic levels.

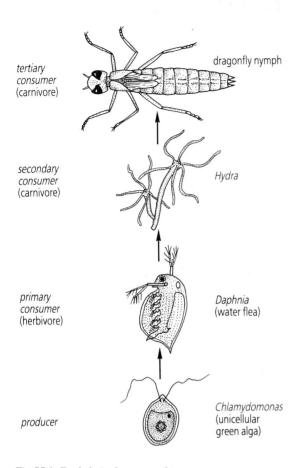

tertiary consumer (carnivore) — dragonfly nymph

secondary consumer (carnivore) — Hydra

primary consumer (herbivore) — Daphnia (water flea)

producer — Chlamydomonas (unicellular green alga)

Fig 55·1 Food chain from a pond ecosystem

Linear food chains offer a very simplistic representation of the actual feeding relationships in an ecosystem. Returning to the pond example, it is obvious that the dragonfly nymph does not depend entirely on *Hydra* for its nutritional requirements. It eats many other small organisms as well and sometimes small leeches and water beetles. A very complex network of inter-related food chains exists in the pond, forming a **food web**. A simplified version of the pond food web is shown in Figure 55·2. Since it is based directly on green plants, it is called a **grazing food web**.

Grazing food webs omit some very important organisms, namely the **decomposers**. These are bacteria and fungi which are abundant in all but very acid regions, feeding on the dead remains of all the other organisms in the ecosystem, regardless of trophic level. The decomposers have a food web of their own called the **detritus food web**. (As a biological term 'detritus' is used to describe fragments of organic remains.)

A further group of organisms present in almost all ecosystems are **parasites**. These feed on the living bodies of organisms at all trophic levels, but the energy obtained is not usually transferred to the next trophic level. Instead it is transferred to a third food web, called the **parasitic food web**. A complete picture of feeding relationships comprises three linked food webs, that is, the grazing, detritus, and parasitic food webs (Fig. 55·3).

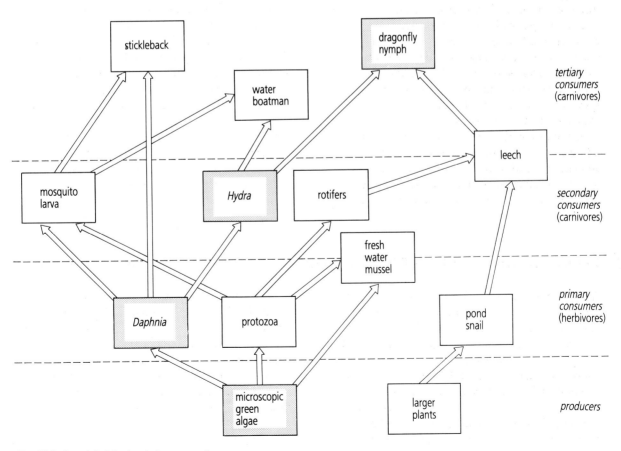

Fig 55·2 Simplified food web from a pond ecosystem

489

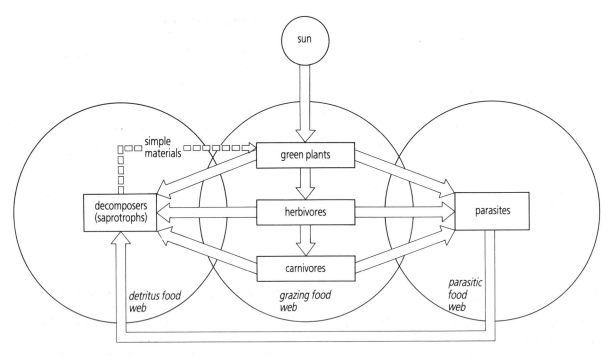

Fig 55·3 *Feeding relationships in a typical ecosystem (grazing food web, detritus food web, parasitic food web)*

55·3 Ecological pyramids ■ ■ ■ ■ ■ ■ ■ ■ ■ ■ ■ ■ ■ ■ ■

A significant feature of grazing food webs like that illustrated in Figure 55·2 has so far not been considered. This is the question of the numbers of individuals occupying each trophic level. Herbivores tend to be less numerous than the plants they eat, while predators tend to be larger than their prey and to consume many of them during their life-span. Consequently, there is often a marked decrease in numbers from green plants to herbivores and from herbivores to carnivores. A **pyramid of numbers** therefore exists, with green plants forming the base, and top consumers, the apex (Fig. 55·4). This generalization does not apply equally well to all ecosystems; an obvious exception is provided by a single tree which can support a large number of smaller herbivores such as greenfly.

This difficulty may be overcome by considering the **pyramid of biomass** of an ecosystem, where 'biomass' is a measure of the amount of organic matter present. Thus, instead of the numbers of organisms per square metre, their mass in grams of dry material per square metre (g m^{-2}) is charted. Usually this results in a pyramid but, once again, as shown in Figure 55·5B, it does not always do so. The reason for this is that biomass has to be measured at one particular point in time and does not take into account the fact that the numbers of organisms may vary over a longer period. In January, the biomass of animals in the English Channel exceeds that of plants. However, this is a temporary winter phenomenon: over the whole

year, the average biomass of autotrophs does in fact greatly exceed that of the consumers in the ecosystem.

To give an accurate representation of the transfer of material from one trophic level to the next, it is necessary to take account of the time taken to accumulate new materials. Reliable comparisons between different organisms and different trophic levels can be made on the basis of **productivity**, that is, the mass of new organic

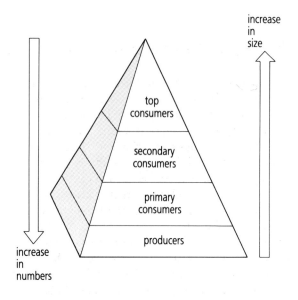

Fig 55·4 *Pyramid of numbers*

490

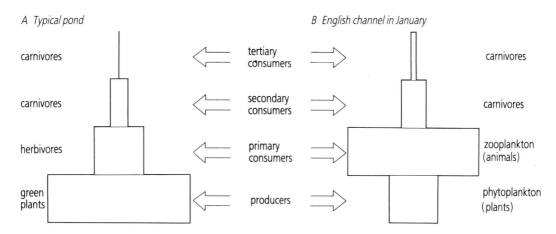

carnivores ← tertiary consumers →

carnivores

carnivores ← secondary consumers → carnivores

herbivores ← primary consumers → zooplankton (animals)

green plants ← producers → phytoplankton (plants)

Fig 55·5 Pyramids of biomass

material formed in grams dry mass per square metre per year (g m^{-2} yr^{-1}).

Because the energy value of materials such as carbohydrates, fats and proteins differ, it is usual to standardise the measurement of productivity by converting 'dry mass' to its energy equivalent measured in units of kJ m^{-2} yr^{-1}. Using these units, an upright **pyramid of energy** is obtained for all grazing food webs. Figure 55·6 shows a pyramid of energy calculated for a food chain based on beef cattle farming. This clearly illustrates the transfer of energy and wastage of energy between trophic levels.

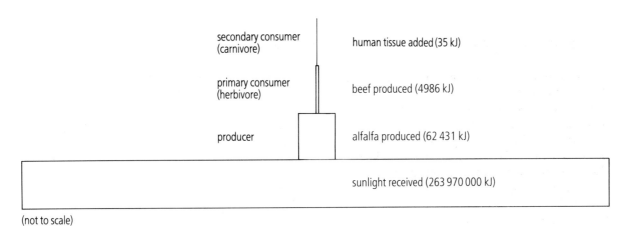

secondary consumer (carnivore) human tissue added (35 kJ)

primary consumer (herbivore) beef produced (4986 kJ)

producer alfalfa produced (62 431 kJ)

sunlight received (263 970 000 kJ)

(not to scale)

Fig 55·6 Pyramid of energy for a food chain involving beef cattle

55·4 Ecological energetics ■ ■ ■ ■ ■ ■ ■ ■ ■ ■ ■ ■ ■

It is important to remember that the transfer of materials represented by ecological pyramids is continuous; energy flowing through the trophic levels is replaced by input from the Sun and is lost, mainly through respiration and radiation of heat. The study of the dynamics of this system of energy flow is called **ecological energetics**.

The amount of solar radiation reaching an ecosystem varies according to the latitude, the climate and the time of year. How efficient the producers are in converting solar energy into a stored chemical form is important for all the members of an ecosystem. This can be determined by collecting and weighing the amount of plant material produced by 1 m^2 of land over a given period of time. The amount of energy stored in this material can be estimated by burning it in a calorimeter which is used to measure the heat produced. Comparisons with the total radiant energy received over the same period show that only about 0·5% of the energy input from the Sun becomes available to consumers in stored chemical form. In fact, plants 'fix' rather more than this amount of energy but the remainder is used up for the plants' own respiration. The **net productivity** of an ecosystem is a measure of the energy fixed by plants which is available to consumers. Net productivity varies widely in different ecosystems as Figure 55·7 shows.

Figure 55·8 shows the overall dynamic flow of

energy in a typical ecosystem. Consider first what happens to the material which constitutes net productivity. A proportion of the energy it contains will remain stored in overground or underground plant structures. The rest will enter the grazing food web by being eaten directly by a herbivore. Alternatively, it may enter the detritus or parasitic food webs.

An additional input available to consumers is provided by migratory organisms or particles of organic matter blown or washed into the ecosystem. However, this import of energy is generally balanced by export through organisms and their remains leaving the system in a similar way. A large proportion of the energy entering each trophic level will be used for maintenance activities such as respiration and is ultimately degraded in value by being converted into heat which is lost from the ecosystem.

The point has already been made that there is a great wastage of energy between successive trophic levels. This energy loss has economic implications. For example, in terms of energy input and output, a plant crop gives far greater productivity than a herd of beef cattle supported by the same area of land. Nevertheless, efforts to minimize the amount of plant material consumed by other herbivores and decomposers and maximize the amount reaching cattle have resulted in increasing beef production per unit area by up to seven times in

some areas. In a similar way, the controversial practice of keeping livestock closely confined in small spaces, instead of letting them run free, saves a considerable amount of energy which would otherwise be wasted in respiration.

Also of importance to the farmer is growth efficiency. Selective breeding produces varieties of crop plants and livestock which grow and mature more quickly. For instance, in Asia strains of rice have been developed which are capable of yeilding three crops in a year instead of only one.

Modern methods of agriculture have increased productivity far in excess of that of natural ecosystems. However, the benefits which arise from the reduction of wastage through pests and disease are offset because additional inputs of energy are required for the production of fertilizers and pesticides, to say nothing of pumping water or heating animal houses. Energy is also expended in ploughing the soil and in sowing and harvesting the crop. The main source of this extra energy is fossil fuels which represent the stored energy from previous ecosystems. In fact, it has been calculated that modern intensive agricultural practices require about nine joules of fossil fuel energy to produce one joule of food on the dinner plate. It may be concluded that the maximum production from land space is not compatible with the most efficient use of energy.

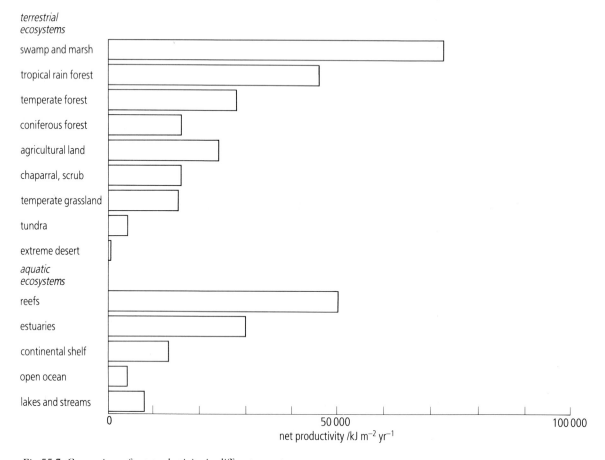

Fig 55·7 Comparison of net productivity in different ecosystems

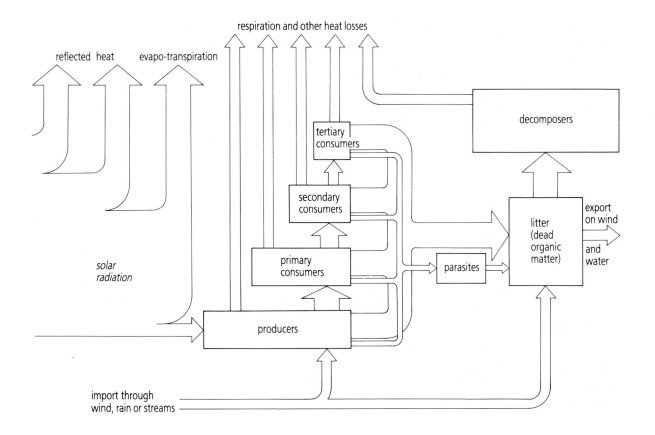

Fig 55·8 Summary of energy flow in a typical ecosystem

493

56 POPULATIONS

Objectives

After studying this Unit you should be able to:–

- Define the following terms: community, population, birth rate, death rate, population density, population dynamics

- Describe exponential growth and relate this to the intrinsic rate of increase of a population

- Discuss the shape of the sigmoid growth curve and define carrying capacity

- Compare the sigmoid growth curve with the J-shaped growth curve

- Draw survival curves for r-strategists and K-strategists and comment on the differences

- Give examples of environmental resistance

56·1 Population dynamics ■ ■ ■ ■ ■ ■ ■ ■ ■ ■ ■ ■ ■ ■

The living organisms in any ecosystem are collectively known as a **community**. The interactions between the various organisms which form the community are best understood by studying and comparing the composition and behaviour of smaller groups called **populations**, that is, groups of individuals of a particular species.

The simplest investigation of a population involves a count of individual organisms. If the numbers are recorded over a period of time, patterns of growth and decline become apparent. The **birth rate** or **natality** of the population is the number of new individuals produced in a given time, usually expressed as a percentage of the total population. A birth rate of 20% per year means

that 20 new individuals are produced each year for each 100 individuals already present in the population. Human birth rates are measured in live births per 1000 head of population per year. The **death rate** or **mortality** is the number dying per unit time measured in a similar way. **Population density** is the number of individuals per unit area in terrestrial ecosystems, or per unit volume in aquatic ecosystems. The study of variations in all these factors is called **population dynamics** and involves mathematical modelling of population growth and the analysis of interactions with other living organisms and with the non-living environment.

56·2 Growth of populations ■ ■ ■ ■ ■ ■ ■ ■ ■ ■ ■ ■ ■ ■

One way of investigating population growth is to start with a new population which can be kept in a laboratory, provide it with adequate food and space and regularly count the numbers of new individuals. Microorganisms such as bacteria and yeasts are very convenient for this purpose because of their rapid rate of reproduction. Figure 56·1 illustrates the curve obtained from growing a colony of yeast in optimum conditions.

In the initial period of the experiment, the

population increases quite slowly. However, the number of cells doubles with each new generation, so that the growth rate progressively increases, causing the curve to rise steeply. This sort of explosive growth is called **exponential growth** and is typical of organisms colonizing a new environment.

A given population has a maximum rate of increase which is called its **intrinsic rate of increase (r)**. this is determined by factors such as the time taken between generations and the life-span and fertility of individuals and is characteristic for each particular species. The intrinsic rate is a constant which determines the slope of the population growth curve in ideal conditions. The actual rate of increase, which depends on the number of individuals present, can be calculated according to the equation:

(1) $$\frac{dN}{dt} = rN$$

where $\frac{dN}{dt}$ is the rate of increase at a particular time, t, r is the intrinsic rate of increase, and N is the number of individuals in the population. A more useful form of this equation enables projections of population growth to be made:

(2) $$N = N_0 e^{rt}$$

where N_0 is the starting population, N is the predicted population after a certain time t, and e is the constant 2.72 (called the natural logarithmic base).

In ideal conditions, yeast cells have an intrinsic rate of increase, r, of approximately 0.5 h^{-1}. In other words, the population increases by half every hour. (This value is obtained for the results of experiments like that of Fig. 56·1). As an example, suppose an initial population of 10 yeast cells was allowed to grow for 6 hours. What would be the expected number of individuals in the final population? As you can see, $N_0 = 10$, $t = 6$, and $r = 0.5$. Substituting these values in equation (2) gives:

$$N = 10 \times 2.72^{(0.5 \times 6)}$$

hence:

$$N = 200.86$$

Thus, after 6 hours, the population is expected to contain about 200 individuals.

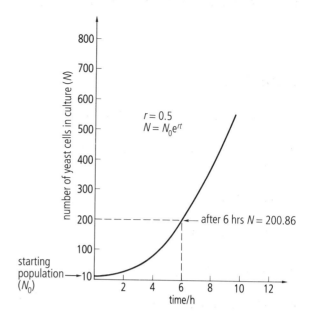

Fig 56·1 Exponential growth of a yeast population

Clearly, exponential growth cannot continue indefinitely. If measurements are continued over a further period, the population curve is found to level off, as shown in Figure 56·2. This happens because the cells grow more slowly as the available nutrients are used up, and also because they begin to poison each other with accumulated waste products such as ethanol. In other words, the birth rate falls and the death rate increases. An equilibrium is reached after about 18 hours, when the death rate exactly balances the birth rate.

The growth of yeast populations usually follows an S-shaped curve, or **sigmoid curve**, sometimes called the **logistic curve** for population growth.

Competition for food and accumulation of waste products are only two examples of **density-dependent** factors which limit exponential growth.

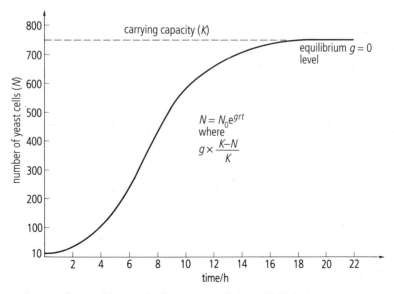

Fig 56·2 Sigmoid growth curve illustrated by growth of a yeast population with limited resources

Many other external factors influence the fertility and survival of the individuals in a population. Collectively, these contribute to **environmental resistance** to population growth. Increasing environmental resistance places an absolute limit on the number of individuals of one species which can be supported in a given habitat. This limit is called the **carrying capacity** of the environment, usually denoted by K.

The effect of environmental resistance on population growth can be modelled by introducing a term known as the **growth realization factor** (g) into the exponential equation, equation (1). The value of the growth realization factor is given by:

(3) $$g = \frac{(K - N)}{K}$$

Adding this to equation (1) gives:

(4) $$\frac{dN}{dt} = \frac{(K - N)rt}{K}$$

Alternatively:

(5) $$N = N_0 e^{\frac{(K - N)rt}{K}}$$

When the number of individuals in the population, N, is much less than the carrying capacity, the growth realization factor has a value close to 1.0. and the population is expected to grow exponentially. However, as N approaches K, the growth realization factor, and the rate of population growth both tend towards zero.

Using equations (4) and (5), population growth of laboratory populations can often be modelled with reasonably accuracy.

56·3 Variations on the sigmoid curve ■ ■ ■ ■ ■ ■ ■ ■ ■

The growth curve obtained for a yeast culture in the laboratory gives an idealized impression of growth patterns in natural ecosystems. A more typical version of the sigmoid curve is shown in Figure 56·3 which illustrates the increase in sheep population following their introduction into Tasmania in 1814. It is clear from this graph that the number of individuals is never static but rises and falls about an average which follows a sigmoid curve.

Although the sigmoid curve is typical for new colonizers, it does not apply to all populations. Figure 56·4 shows the growth pattern for a population of planktonic algae which shows a vigorous acceleration and never levels off for long enough to establish a carrying capacity. Called a J-curve, it is typical of populations which produce many generations of offspring per year. One reason for a curve like this is that the population may have very precise survival limits in terms of its environmental requirements. Unfavourable changes such as the accumulation of toxic wastes or the scarcity of food cause most of the individuals to die off quickly, leaving only a few (usually in spore or zygote form) to survive. In following years, the same pattern of population growth is repeated.

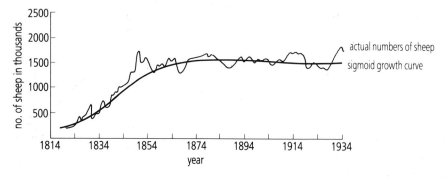

Fig 56·3 Growth curve of sheep following their introduction into Tasmania

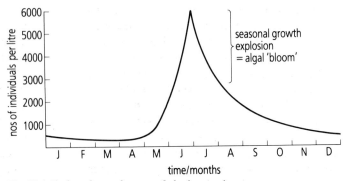

Fig 56·4 J-shaped growth curve of planktonic algae

56·4 Human Populations ■ ■ ■ ■ ■ ■ ■ ■ ■ ■ ■ ■ ■

The human population is unusual in its growth pattern, as can be seen from Figure 56·5, because it remains in a period of exponential growth of frightening proportions. In developed countries, this can be attributed to a dramatic lowering of environmental resistance caused by improvements in agriculture and medicine. Some limit must eventually be reached, leading to speculation about the possibilities of sigmoid or J-type outcomes for the future of mankind. For the present, the net rate of growth is about 1.9%; in other words, between 75 and 80 million extra people per year.

The study of human populations, called **demography**, has become a complex science and involves many more factors than simple birth and death rates.

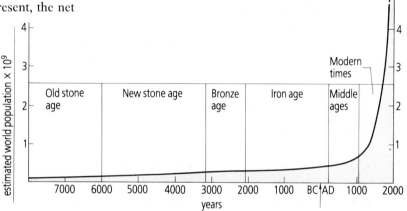

Fig 56·5 Increase of human population

56·5 Strategies for survival ■ ■ ■ ■ ■ ■ ■ ■ ■ ■ ■ ■

Birth rate and death rate figures are crude indicators of population behaviour because they disregard factors such as age and sex. However, the probability of producing offspring or of dying does vary with these factors. The relationship between age and survival is best examined by constructing a **survival curve**. This is a graph of the number of surviving individuals in a population plotted against age. Figure 56·6 shows survival curves of three different types.

Curve A is typical of most invertebrate and plant populations which have a very high mortality rate in the young stages. The oyster provides a good example of this, producing thousands of larvae of which very few are able to attach themselves to a suitable rock for development into the adult stage. Such organisms have a number of things in common. In addition to producing large numbers of offspring they take little or no care of them and have short life spans and generation times. They are called **r-strategists** because their life-style is adapted to give the highest possible value of *r* (the intrinsic rate of increase).

At the other extreme, represented by curve C, are organisms which have stable populations very close to *K* (the carrying capacity of the environment). These are called **K-strategists** and show what are usually considered to be evolutionary advances over r-strategists. Typically, they live in stable surroundings and are closely adapted to a particular and specialized part of the

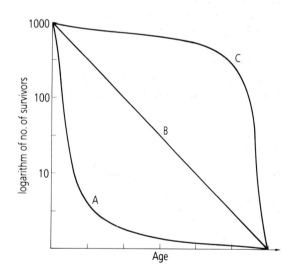

Fig 56·6 Survival curves

environment. They have long life spans and generation times, produce small numbers of offspring and exhibit parental care towards them. Closely regulated societies keep their population in balance by reducing birth rates rather than exposing themselves to increased death rates. Mountain sheep, man, and elephants provide examples of this sort of curve.

Curve B represents the rare situation in which there is an equal chance of dying at all ages. *Hydra* is an example of such an organism.

497

56·6 Factors limiting population growth ■ ■ ■ ■ ■ ■ ■ ■ ■

The factors which produce environmental resistance can be classified as extrinsic or intrinsic. Extrinsic factors are those which affect the population from outside, like climate, available food supply, competition from other species, predation and parasitism. Intrinsic factors involve competition with other members of the same species.

As examples of extrinsic environmental resistance, the effects of temperature and food availability on the carrying capacity of the environment are shown in Figures 56·7 and 56·8. The first graph plots the growth curves for the water flea *Moina macrocopa* at three different temperatures and the second shows the effect of varying the quantity of food supplied to two populations of the flour beetle, *Tribolium confusum*.

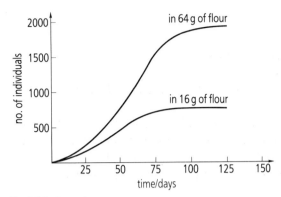

Fig 56·8 *Growth curves of flour beetles,* Tribolium confusum, *in different quantities of flour*

Intrinsic environmental resistance occurs when individuals in a population compete with each other. Such competition is often more intense than competition between species because individuals of the same species compete for exactly the same resources. Many animals defend a **territory** against members of their own species. Such behaviour is necessary if the offspring are to receive sufficient food and resources to ensure their survival.

Factors limiting population growth rarely operate singly and frequently defy measurement or definition. It is relatively easy to study the influence on populations of the physical components of the environment such as temperature, water or light. The ways in which many different populations interact within a community are more difficult to understand and must be approached by first examining the relationships which may exist between individuals of just two different species. Unit 57 discusses a variety of such interactions.

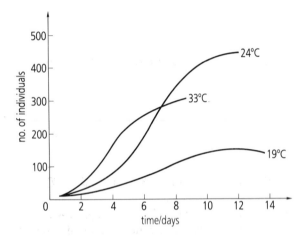

Fig 56·7 *Growth curves of populations of the water flea,* Moina macrocopa, *at different temperatures*

57 ASSOCIATIONS

Objectives

After studying this Unit you should be able to:–

- Define and give examples of each of the following types of association: cooperation, mutualism, commensalism, amensalism, competition, predation, parasitism

- Describe the structure of a lichen as an example of mutualism

- Give an account of nitrogen fixation in root nodules

- Describe experiments to demonstrate competition and give an example of competitive exclusion

- Explain the terms niche and niche differentiation

- Discuss the beneficial effect of predation

- Give examples of parasitism and comment on evolutionary trends in the parasite – host relationship

57·1 Types of association

Organisms almost never exist in isolation from other species so that the growth of a population usually depends on the activities of other members of the plant and animal community. This Unit examines the variety of interactions which occur between one species and another.

Table 57·1 summarizes the possible types of association. As you can see, most associations involve food in one way or another. Some species simply compete for available food supplies or make brief predator-prey encounters. Others interact more intimately over a longer period of time to form lasting associations. The term **symbiosis** (literally 'living together') is now commonly used to describe these longer term associations. Sometimes symbiotic associations benefit both species, as in **mutualism**, but they may benefit only one, as in **commensalism** and **parasitism**. In Table 57·1, beneficial effects are symbolized (+), neutral effects (0) and harmful effects (−). In terms of population growth, harmful effects can be thought of as contributing factors to environmental resistance.

57·2 Cooperation

Cooperation describes a loose association between organisms capable of living independently. However, it is of mutual benefit and increases the growth rates of the populations involved. One example is the early warning system developed by mixed herds of grazing animals in the African savannah. Populations of impala, baboons and giraffes could quite well exist on their own, but gain extra security from mixing with others and responding to the alarm signals given by different species when a potential predator approaches the group.

A further example of cooperation is the association between a crab and a sea anemone. The anemone fixes itself to the crab's back gaining free transport and food debris, while the crab gains a protective camouflage.

Table 57·1 Summary of associations between different species of organisms

ASSOCIATION	DEFINITION	IMMEDIATE EFFECTS	EXAMPLES
cooperation	loose association in which both species benefit	+/+	sea anemone/crab
mutualism	close association in which both species benefit	+/+	lichens (fungus/alga) herbivore/cellulose-digesting bacteria root nodule (*Rhizobium*/ leguminous plant)
commensalism	close association in which one species benefits	+/0	remora fish/shark hermit crab/empty mollusc shell
amensalism	one species inhibits the growth and survival of another species but does not exploit it for food	0/−	*Penicillium*/bacteria
competition	one population tends to eliminate the other and both suffer	−/−	*Paramecium caudatum*/ *Paramecium bursaria*
predation	one species attacks and kills another for food	+/−	lion/zebra lynx/snowshoe hare
parasitism	the parasite feeds on the living body of the host	+/−	tapeworm/man *Claviceps*/rye

key: + beneficial effect   − damaging effect   0 slight effect or no effect

57·3 Mutualism ■ ■ ■ ■ ■ ■ ■ ■ ■ ■ ■ ■ ■ ■

Mutualism is an intimate association which is essential for the growth and survival of the participating organisms.

The body of a **lichen** consists of the hyphae of a fungus and the cells of a photosynthetic alga so closely associated that they behave as a single organism. Figure 57·1 shows the form and internal structure of a lichen called *Sticta*. The fungus contributes to this association by supplying structural support and by absorbing water and minerals. The fungus benefits because most of its organic food materials are manufactured by algal photosynthesis. Lichens reproduce asexually by means of tiny fragments containing a mixture of undifferentiated fungal and algal cells. These break off and blow away from the parent, becoming lodged in cracks and crevices where new individuals can develop. The success of lichens is evident from their ability to colonize uninhabited and inhospitable environments such as barren rocks and brick walls where neither fungi nor algae could survive alone.

Other examples of mutualism involve algae, particularly in association with coelenterates. The common green hydra, *Chlorohydra viridissima*, for instance, contains single-celled green algae called **zoochlorellae** inside its endodermal cells. These provide oxygen and other metabolic products from photosynthesis and receive protection and raw materials like carbon dioxide, nitrates and

phosphates, all of which are formed as excretory products of the host's metabolism.

The occurrence of mutualism in coelenterates can affect their distribution. Reef building corals rarely grow in depths below 30 m because they are dependent on the **algae** they contain. The association between algae and corals is not fully understood, but it is thought that the plants promote the growth of the reef by fixing carbon dioxide and hydrogencarbonate in combination with calcium ions from the sea. This helps the coral to deposit calcium carbonate which, along with other solid remains, forms an integral part of the reef structure.

A rather different type of mutualism involves the 'borrowing' of enzymes by one organism from another. Mammalian herbivores do not possess genes for the synthesis of cellulase enzymes although cellulose is a major constituent of their diet. Consequently, cellulose digestion depends on anaerobic microorganisms like bacteria and protozoa which live in the stomach and intestine of herbivores. The microorganisms produce the appropriate enzymes and release substances which can be absorbed and used as a source of energy by the host. They gain protection, warmth and a constant supply of food, but tend to be digested as they pass backwards down the gut and may supply the herbivore with up to 20% of its nitrogen requirement. Wood-eating termites have similar populations of mutualistic protozoa.

In terrestrial ecosystems, probably the most important example of mutualism is the association of soil bacteria of the genus *Rhizobium* with leguminous plants. Although both *Rhizobium* and legumes such as clover, peas and beans can live

A Appearance of thallus

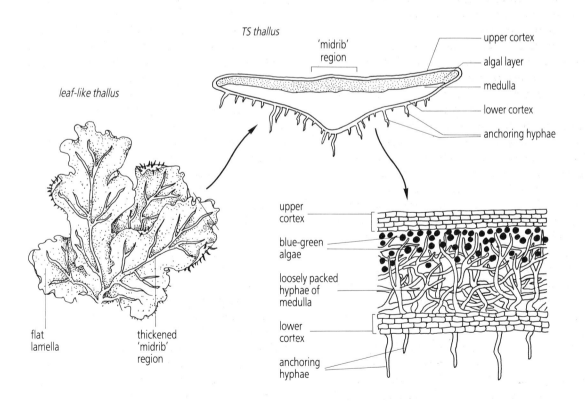

B Structure

Fig 57·1 The lichen Sticta

independently, together they form structures called **root nodules** which carry out nitrogen fixation (Fig. 57·2). In this process, molecules of atmospheric nitrogen are converted into ammonia some of which becomes available to the plant for making amino acids and proteins.

The bacteria are attracted to the legume root by a chemical secretion and their entry into the cells of the cortex takes place by means of tiny intracellular channels called **infection threads**. Once inside the host root they multiply and change shape, becoming non-motile structures known as **bacteroids**. These secrete growth hormones called cytokinins which stimulate cell division, producing externally visible swellings or nodules attached to the roots.

The conversion of nitrogen into ammonia requires a large quantity of energy, because molecules of N_2 gas are very stable. The bacterial enzyme **nitrogenase** which catalyses the reaction is destroyed by contact with oxygen and effective fixation therefore requires relatively anaerobic conditions. Within the root nodule, oxygen is absorbed by an oxygen-carrying pigment called **leghaemoglobin**. This resembles haemoglobin in red blood cells and has a high affinity for oxygen, absorbing and storing it as it enters the root cells. Consequently, the concentration of oxygen in the cytoplasm is reduced and damage to nitrogenase is avoided. Like haemoglobin, leghaemoglobin is red in colour so that nodules appear bright pink when cut open.

The evolutionary implications of mutualistic associations are of great interest. If the partners cannot live apart, they must evolve together. Ultimately, they are likely to become completely integrated, forming a single living unit. Similar processes may have been involved in the evolution of cell organelles, including mitochondria, chloroplasts and centrioles, which are thought to have arisen from mutualistic prokaryotic ancestors. Root nodule bacteroids might be regarded as the fore-runners of nitrogen-fixing cell organelles.

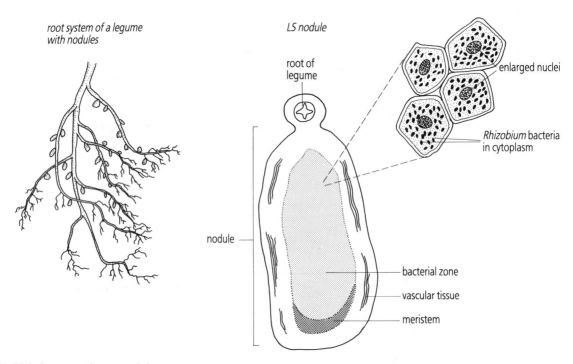

Fig 57·2 Structure of a root nodule

57·4 Commensalism and Amensalism ■ ■ ■ ■ ■ ■ ■ ■ ■ ■

Commensalism and amensalism are looser associations in which one partner remains relatively unaffected.

A commensal is an organism which feeds literally 'at the same table' as another organism. An example occurs in the case of the remora fish sometimes seen attached to the belly of a shark. The shark suffers a small loss in streamlining but the remora fish is dependent on the shark for scraps of food. Its dorsal fin is modified into a suction disc in order to cling onto the shark's inhospitable skin.

Amenalism is the converse of commensalism because the organisms involved cannot exist together. One of the relatively few known examples involves the inhibitory action of moulds such as *Penicillium notatum* on bacterial populations.

57·5 Competition ■ ■ ■ ■ ■ ■ ■ ■ ■ ■ ■ ■ ■ ■ ■

Competition arises when two populations depend on the same limited resources. A common outcome is **competitive exclusion** of one of the populations. This can be illustrated by experiments using two species of the protozoan, *Paramecium*, *P. aurelia* and *P. caudatum*, which are nearly identical in their environmental requirements. If the two species are grown in a mixed culture, *P. caudatum* suffers a marked decline (Fig. 57·3).

A contrasting result is shown in Figure 57·4, which illustrates the growth curves obtained for competing populations of *P. caudatum*, and a third species *P. bursaria*. Although both populations are affected adversely, their association does not result in competitive exclusion. This is possible because their environmental requirements do not overlap so precisely as those of *P. caudatum* and *P. aurelia*.

Such experiments indicate that the intensity of

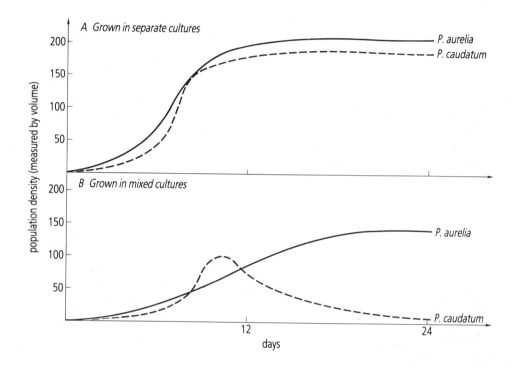

Fig 57·3 *Population growth curves of* P. caudatum *and* P. aurelia *grown separately and in mixed cultures*

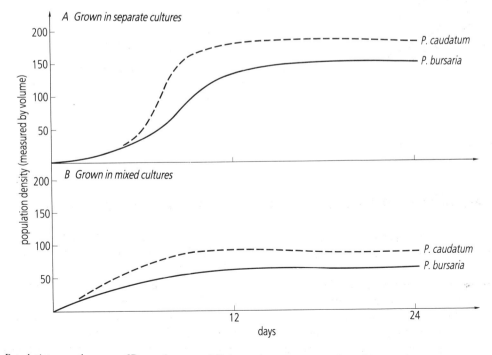

Fig 57·4 *Population growth curves of* P. caudatum *and* P. bursaria *grown separately and in mixed cultures*

the competition between two species depends on the extent to which they share the same environmental resources. The word **niche** is used to describe the precise set of resources, living and non-living, required by a particular organism in order to feed, protect itself and breed successfully. Competition can therefore be thought of in terms of niche-overlap between two species, the greater the overlap, the more intense the competition.

Competitive exclusion of one species does not necessarily result in its extinction. In many cases, species shift their environmental requirements (niche) over successive generations so that overlap and competition are reduced to a minimum. This is called **niche differentiation**. An example is provided by two marine diving birds, the cormorant and the shag, which share the same hunting ground and both feed on fish. The cormorant feeds on bottom-dwelling species like flounders and prawns, while the shag feeds on swimming fish and eels in the upper waters.

Niche differentiation is an important evolutionary process and a major cause of diversity within biological communities.

57·6 Predation ■ ■ ■ ■ ■ ■ ■ ■ ■ ■ ■ ■ ■ ■ ■ ■

Predation is a type of interaction between two populations in which one species called the predator, attacks and kills the other, called the prey. These terms usually apply only to the relationship between carnivores and their food. (Note that herbivores are called grazers, not predators.)

Although the immediate effects of the relationship are beneficial to the predator and harmful to the prey, the dynamics of predator-prey interactions are complex and there may be benefits for the prey population. Predation can improve the genetic composition of the prey population by weeding out its weaker members and can also ensure that the prey population does not exceed the carrying capacity of the environment.

Figure 57·5 illustrates the changing patterns in the growth of a population of moose living on Isle Royale in Lake Superior. Moose colonized the island at the beginning of the century by walking across 20 km of frozen water from Canada. Free from wolves and other predators, their numbers increased exponentially until the mid 1930s, when the carrying capacity of the environment was exceeded. Overpopulation denuded the area of short vegetation and a population crash ensued, followed by another short period of exponential growth in the 1940s, after the vegetation reappeared. In 1949, timber wolves crossed the ice and their effect on the moose population was immediate and beneficial. The wolves keep the moose population between 600 and 1000, somewhat below carrying capacity but controlled by predation and not starvation.

The numbers of predators and prey is never precisely stable. Depending on the lifetimes of the predator and prey species and on their reproductive potential, population numbers

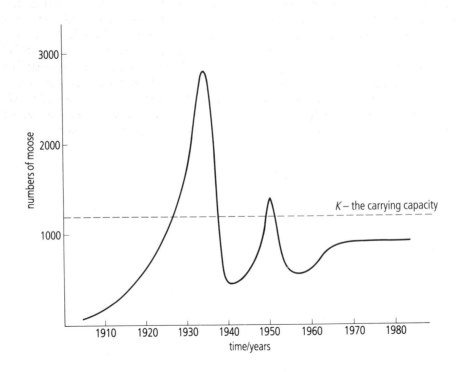

Fig 57·5 Growth of moose population on Isle Royale, Lake Superior

sometimes follow a cycle, as illustrated by the example of Figure 57·6. An increase in the availability of prey is likely to result in a corresponding increase in the numbers of predators. However, if the number of predators increases too rapidly, they may over-exploit the prey population, causing a decline and a subsequent reduction in the food supply. In competition for scarce resources, many predators die and the population falls to a low level. In these circumstances, the prey population starts to recover, and the cycle begins again.

Such predator-prey cycles are liable to occur when a predator species is highly specialized for feeding from one particular type of prey. However, most predators have a range of alternative prey species, and switch from one to another depending on conditions, so that their numbers do not fluctuate widely from year to year.

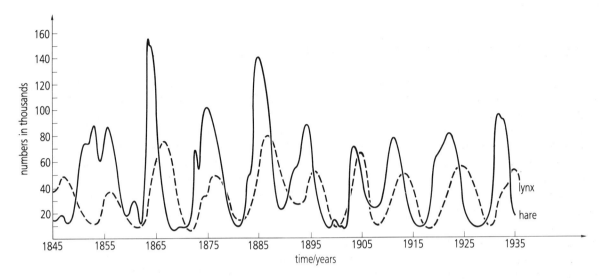

Fig 57·6 *Predator-prey interactions illustrated by the numbers of pelts of lynx and snowshoe hare purchased by the Hudson Bay Company*

57·7 Parasitism ■ ■ ■ ■ ■ ■ ■ ■ ■ ■ ■ ■ ■ ■ ■ ■

A parasite derives food from the living body of another organism, called the **host**. Many parasitic associations are described in Section D of this book.

Although it is common to think of parasites as harmful to their hosts, there is a clear evolutionary tendency for pathogenic effects to be reduced to a minimum. A parasite which invariably killed its host species would rapidly become extinct.

Parasites with less debilitating effects have a larger number of potential hosts and spread more widely. The host population also adapts over many generations, becoming more resistant. When livestock are introduced to new areas their lack of resistance to local diseases is often evident. For example, European cattle exported to Africa suffer and die from sleeping sickness, whereas the native African cattle are relatively unharmed.

58 DIVERSITY OF ECOSYSTEMS

Objectives

After studying this Unit you should be able to:–

- Outline the important features of the following biomes: tundra, taiga, temperate deciduous forest, temperate evergreen woodland, temperate grassland, tropical grassland, tropical rain forest, desert

- Give examples of zonation and comment on the importance of environmental gradients

- Define a climax community

- Describe ecological succession

- Distinguish between primary and secondary succession

58·1 Biomes ■ ■ ■ ■ ■ ■ ■ ■ ■ ■ ■ ■ ■ ■ ■ ■ ■ ■

The place where an organism lives is called its **habitat**. Some habitats, such as a coral reef, are large, while others, such as an oak leaf, are small. This Unit describes the variety of habitats and ecosystems which occur throughout the World and links the distribution of biological communities to non-living, or **abiotic** factors.

The living world, or **biosphere**, can be divided into a number of **biomes**, as shown in Figure 58·1. Each biome is a large community consisting of organisms which share particular temperature, rainfall and humidity requirements. Although their characteristic features are easily recognized, transition from one biome to the next occurs gradually and there are no sharp boundaries between regions.

The important features of some of the major biomes are outlined below. The same biome may occur in widely separated areas of the World wherever climatic and edaphic (soil) conditions are similar. For example, areas of tropical rain forest are found in Malaysia, West Africa and South America. The organisms which inhabit these areas have evolved quite separately, but often show strikingly similar adaptations.

Tundra

Tundra is confined mostly to the Northern Hemisphere where it forms a circumpolar band separating the polar icecap from the coniferous forests to the south. The region is characterized by low temperatures and a short growing season. The lower ground levels are permanently frozen (permafrost) and only the hardiest plant forms can survive. *Sphagnum* moss, sedges, heather and lichens dominate the bogs and marshes which appear in the short summer and provide food for visiting caribou, insects and migratory birds.

Taiga

Taiga is a Siberian word meaning 'coniferous forest'. Large areas of taiga stretch across North America, northern Europe and Asia and are a major source of commercial timber. The evergreen conifers shade the ground below and prevent the development of ground shrubs. Typical animals include hibernating or migratory animals such as bears, moose, squirrels and birds. There are far fewer species than in deciduous forests.

Temperate deciduous forest

Temperate deciduous forest is found in areas with an annual rainfall of 75–150 cm distributed fairly evenly throughout the year. It is dominated by hardwood trees such as oak, beech or maple, which rise to 40 or 50 m. Their leaves shade the lower layers to some extent, but allow stable

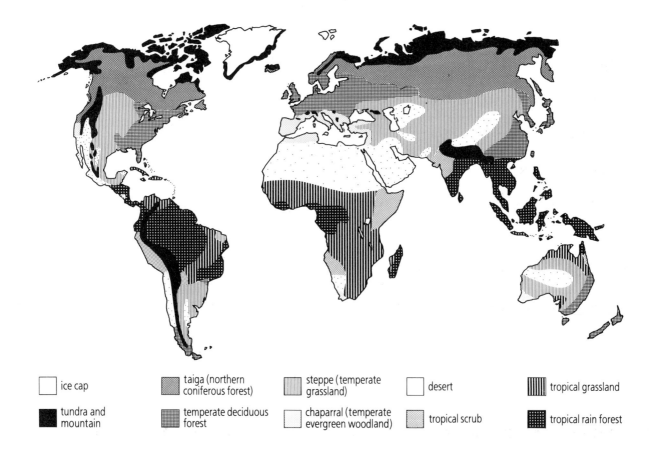

Fig 58·1 World distribution of biomes

communities of shrubs and ground plants to develop. This is the vegetation natural to most of Britain, but much of the original forest has been cleared.

Temperate evergreen woodland (chaparral)

Chaparral is found in mild temperate regions of the World with abundant winter rainfall and relatively dry summers, for example, in areas surrounding the Mediterranean. It is characterized by evergreen thickets, small oaks or eucalyptus trees and often includes man-made vineyards or olive groves.

Temperate grassland (steppe)

Temperate grassland is usually found in the central areas of continents where rainfall is 25–75 cm per year. The prairies of the western United States, and the grasslands of the USSR, Argentina, South Africa and Australia provide examples. The natural community is dominated by grasses and large

herbivores such as bison. Burrowing mammals like ground squirrels and prairie dogs, and ground-nesting birds are common. Much of this region has been exploited for agriculture both for growing cereal crops and for raising beef and dairy cattle.

Tropical grassland

A broad belt of tropical grassland or savannah stretches across Africa and other areas are found in Australia and South America. The annual rainfall may be as much as 125 cm, but a dry season prevents the development of forests. Grasses survive drought by means of extensive underground rhizome systems. The animals typical of these regions are herbivores such as antelope, zebra and wildebeest, and carnivores like lions and cheetahs.

Tropical rain forest

Tropical rain forest is found in areas close to the equator with an annual rainfall of 200 cm or more and comprises a spectacular variety of different

507

organisms each with their own life-styles and adaptations. Most of the vegetation forms a dense layer, called the canopy layer, 25–35 m above ground level. This supports a large community of small animals and plants, including many **epiphytes**, that is, commensal or parasitic plants attached to the trunks and branches of trees. Tall trees extend their branches above the canopy. The forest floor is rich in decomposing vegetation and supports an enormous range of fungi and invertebrates. However, not much light penetrates the upper layers of the forest and the vegetation is dense only where there is a break in the canopy.

Desert

Deserts receive a rainfall of less than 25 cm per year. They support a very limited vegetation composed of xerophytic (drought-adapted) plants, such as cacti and succulents, and fast-growing annuals which appear and disappear with the infrequent rains. Almost all the animals supported by this meagre primary production avoid the extremes of climate by burrowing, feeding only at dusk and dawn.

58·2 Zonation ■ ■ ■ ■ ■ ■ ■ ■ ■ ■ ■ ■ ■ ■ ■ ■

A journey north from Mexico to the Arctic circle, or the ascent of a mountain in the Andes, would reveal a similar sequence of changes from tropical rain forest through to tundra. These changes are caused by an **environmental gradient**, in this case a temperature gradient. A similar **zonation** effect can be seen in a more leisurely way by walking down a rocky beach to the sea at low tide. Some animals and plants are specialized for life on the upper part of the shore where they are exposed for most of the time, while others are found lower on the shore and are exposed only at low tide.

Zonation also occurs in aquatic ecosystems according to depth, light penetration and water temperature. Another example is the layering of vegetation in a forest ecosystem from the tallest light-loving trees above the canopy to the shade-loving ferns and mosses which may receive only 1–5% of the available light.

Boundaries between different habitats are called **transition zones**. Often these contain populations of animals and plants specially adapted for life in the transition zone, as well as species characteristic of the two major communities at the limits of their tolerance range. This is well illustrated by the sea-shore example which can be thought of as a transition zone between the terrestrial and the marine ecosystems and has its own highly adapted organisms.

The brown seaweed *Fucus* is a typical intertidal inhabitant with special adaptations including a holdfast enabling it to cling to rocks despite buffeting waves, a covering of mucilage to prevent the fronds from drying out or sticking together, reproductive and dispersal mechanisms which rely on the tide, and a brown photosynthetic pigment (fucoxanthin) for maximum light absorption while submerged. Even so, as Figure 58·2 shows, minor differences between species ensure uneven distribution along the steep environmental gradient of the shore.

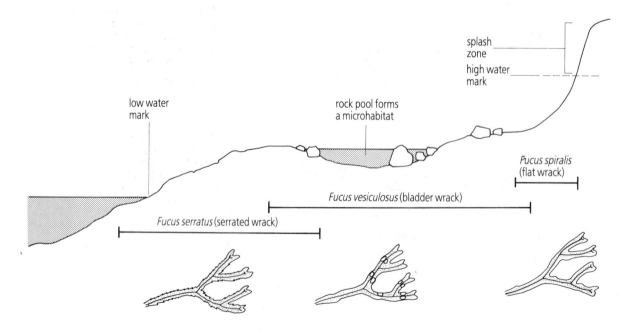

Fig 58·2 Zonation of three species of Fucus *on a rocky shore*

58·3 Succession ■ ■ ■ ■ ■ ■ ■ ■ ■ ■ ■ ■ ■ ■ ■ ■ ■

The first colonists of a bare area such as cooled lava or a newly exposed rock face begin a process of succession which leads eventually to a balanced final ecosystem in which the living organisms form a **climax community**. The living organisms of the different biomes described in the earlier part of this Unit form climax communities.

In many parts of the World, the first plants to colonize a new habitat would be lichens and mosses. Initially, these primary colonizers have little competition from other species, but the lichens and mosses produce acids which weather the rock causing dust particles to accumulate. Eventually, a shallow soil is formed which can support secondary colonizers such as grasses. As the grass roots bind the soil, the decay of organic matter adds more humus thereby improving the soil fertility. New plants appear and begin to gain a foothold. In regions of temperate deciduous forest, these new colonizers will be shrubs like bilberry and crowberry. Later these will be replaced by fast growing birch trees and finally by other broad leaved trees which form the climax forest.

An example of succession in a shallow aquatic ecosystem is illustrated in Figure 58·3. As you can see, the result in this case is a slow process of filling in the lake which will eventually result in dry land. This is not the only possible outcome. A stable aquatic ecosystem can be formed if the lake is deep enough or if currents, waves or outflows restrict the accumulation of sediment.

Given reasonably stable environmental conditions, it can be assumed that almost any area of land which can support life will develop a climax vegetation. If the process starts from bare rock or sand, it is called **primary succession**. More usually, it starts after clearance has occurred, as on farmland left to waste. This is called **secondary succession**.

Most of the present day vegetation in Britain is the result of secondary succession. Forest was the natural vegetation of nearly all of Britain before Neolithic man. The landscape has been radically changed since then by agriculture and human settlement. The introduction of rabbits in the Middle Ages had a dramatic effect by limiting the regeneration of tree seedlings. Most of the heath and grassland typical of the countryside is artificially maintained but over many hundreds of years characteristic and stable communities have developed.

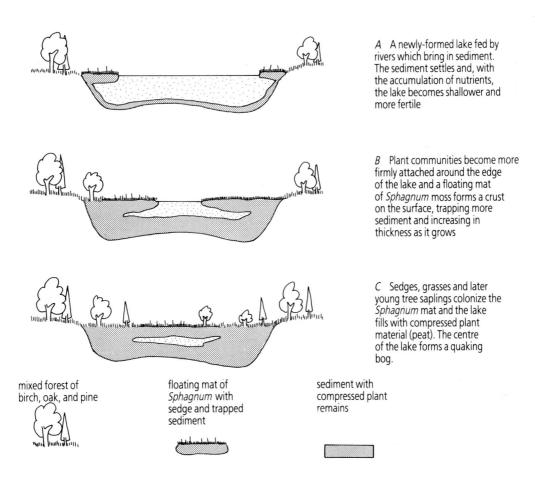

A A newly-formed lake fed by rivers which bring in sediment. The sediment settles and, with the accumulation of nutrients, the lake becomes shallower and more fertile

B Plant communities become more firmly attached around the edge of the lake and a floating mat of *Sphagnum* moss forms a crust on the surface, trapping more sediment and increasing in thickness as it grows

C Sedges, grasses and later young tree saplings colonize the *Sphagnum* mat and the lake fills with compressed plant material (peat). The centre of the lake forms a quaking bog.

mixed forest of birch, oak, and pine

floating mat of *Sphagnum* with sedge and trapped sediment

sediment with compressed plant remains

Fig 58·3 Succession in an aquatic ecosystem

59 SOIL

Objectives

After studying this Unit you should be able to:–

- Draw and label a simple soil profile and describe the major horizons

- Describe the microstructure of topsoil

- List the components of a fertile soil

- Define the following terms: clay, silt, fine sand, coarse sand, gravel

- Compare and contrast the properties of clay and sandy soils and define a loam

- Explain how humus is formed and distinguish it from litter and peat

- Comment on the importance of earthworms for soil fertility

- Explain the relationship between drainage and capillary water and define field capacity

- Define and explain flocculation and ion exchange

59·1 Soil structure ■ ■ ■ ■ ■ ■ ■ ■ ■ ■ ■ ■ ■ ■ ■

Soil is a complex mixture of rock particles and organic materials formed at the Earth's surface by weathering and biological processes. This Unit describes the physical, chemical and biological characteristics of fertile soils.

Most soils consist of a number of layers which can be seen by cutting vertically down through the soil to expose the soil **profile**. Within such a profile, it is usually possible to distinguish several different **horizons**, as illustrated in Figure 59·1. The O horizon on the soil surface consists of fresh and partly decomposed leaf litter. Below this is the A horizon or **topsoil**, which is often dark in colour because it contains a high proportion of organic matter. The fertility of an area can be judged by the depth and colour of its topsoil. Humus is incorporated into the topsoil by the activity of earthworms and microorganisms and nutrients tend to be leached out in water draining through to lower layers. Horizon B, or **subsoil**, is usually more compacted than topsoil and may be comparatively rich in inorganic nutrients. Horizon C forms a transitional zone between rock and soil known as the **weathering crust**. The unweathered rock below is sometimes called the R horizon. The thickness, colour and properties of these layers

vary considerably depending on local geology and biotic conditions. The soil profile gives an indication of potential soil fertility and a useful method of comparing soils from different areas.

The microscopic structure of a typical soil is illustrated in Figure 59·2. The most important components are as follows.

1 Inorganic particles. These originate from rock by weathering processes. Collectively they form the **mineral skeleton** of the soil.
2 Humus. This is the decayed remains of dead organisms. Humus binds inorganic particles together to form larger aggregations, called **soil crumbs**.
3 Water. Water contributes to weathering and is essential for the growth of plants and soil micro-organisms.
4 Air. Air spaces in the soil provide oxygen for the respiration of plant roots and soil organisms.
5 Dissolved inorganic salts. These can be derived from rock by the action of rainwater or from humus by the action of microorganisms. **Nitrates** in the soil are formed from humus and by nitrogen-fixing bacteria which convert atmospheric nitrogen to ammonia.

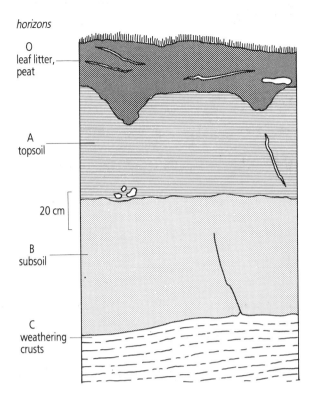

Fig 59·1 Profile of a fertile soil

6 Living organisms. The most important living organisms are decomposers including bacteria and fungi, but larger organisms such as earthworms also play a vital part in maintaining and improving soil fertility.

A fertile soil must contain all these components in appropriate amounts and must have a well-developed soil crumb structure.

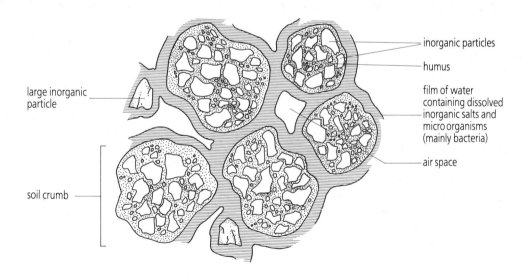

Fig 59·2 Microscopic structure of topsoil of a fertile soil

59·2 Inorganic particles ■ ■ ■ ■ ■ ■ ■ ■ ■ ■ ■ ■ ■ ■ ■

If a soil sample is shaken vigorously with water in a glass container and then allowed to stand, particles of different sizes settle at different rates and a number of layers can be observed corresponding to the variety of sizes of soil particles. The largest particles settle to the bottom quickly, while the smallest particles may remain suspended in the water for long periods. Particles

511

are described as clay, silt, fine sand, coarse sand, or gravel according to their size, as defined in Table 59·1

Table 59·1 Sizes of inorganic particles in soil

PARTICLES	AVERAGE DIAMETER
clay	less than 0.002 mm
silt	0.002–0.02 mm
fine sand	0.02–0.2 mm
coarse sand	0.2–2.0 mm
gravel	more than 2.0 mm

Inorganic particles are formed by weathering. The first stages of weathering from rock involve physical agents such as frost, ice and thermal expansion. Biological agents, such as burrowing animals or the intrusive growth of plant roots, combine with the physical processes to split rocks along cracks and lines of fissure. Additional grinding and abrasion results in the formation of small particles. Some of these small particles may be broken down still further by chemical processes involving the solvent action of water, often made acidic by dissolved gases like carbon dioxide and oxides of nitrogen. This produces extremely fine particles such as clay. Particles which resist chemical weathering or which are exposed to weathering for shorter periods remain larger, forming sand.

The properties of soils vary dramatically according to their particle content. The differences between clay-based soils and sandy soils are summarized in Table 59·2. Fertile soils possess a mixture of particle sizes and combine some of the desirable properties of both clay and sand. A well-balanced mixture of particle sizes within the mineral skeleton is called **loam**.

Table 59·2 Comparison of clay and sandy soils

PROPERTY	CLAY SOIL	SANDY SOIL
particle size	small particles	large particles
aeration	poor – small spaces often filled with water	good – large spaces between particles
drainage	poor	good
water retention	good – small spaces hold water by capillary attraction	poor
nutrient retention	good – inorganic ions held by adsorption to clay particles	poor – inorganic ions leached out by rainwater
acidity	acid if waterlogged	rarely acid
soil temperature	slow to warm up in spring, less variable	warms up quickly in spring, more variable
drying	cracks into hard lumps when dry	does not crack when dry
plant growth	plant roots do not penetrate easily	plant roots penetrate easily
microorganisms	micro-organisms active except when waterlogged	activity of microorganisms and earthworms low
cultivation	heavy – difficult to cultivate	light – easy to cultivate
improvement	can be improved by deep autumn ploughing and by addition of lime to cause flocculation (see 59.6)	can be improved by adding humus or mulching, laying wet rotting manure or peat on the soil surface

59·3 Humus ■ ■ ■ ■ ■ ■ ■ ■ ■ ■ ■ ■ ■ ■ ■

The undecomposed and partly decomposed dead remains of animals and plants form a layer of **litter** on the soil surface. In temperate zones, this varies in composition and thickness depending on the seasons. Litter is broken down by organisms of the detritus food web living in the topsoil.

The process of decomposition is also seasonal in temperate zones because the activity of bacteria and fungi is reduced by the cold winter weather. In ideal conditions of temperature and humidity, which sometimes occur in tropical forests, litter is completely broken down to inorganic compounds with the release of carbon dioxide and water. More usually however, the process is incomplete and the products of decomposition are large organic molecules called **humic acids** and collectively termed **humus**.

Humus has an important effect on the physical and chemical structure of the soil as well as providing a nutrient source for plants. Organic chemicals in humus improve water retention. In addition, humus helps to bind mineral particles together to form soil crumbs. This is important because the larger spaces between soil crumbs allow excess water to drain through easily. The sticky mucilage surrounding the cells of many soil microorganisms also contributes to the formation of crumbs.

In acid or waterlogged conditions, the formation of humus is prevented because bacterial growth is inhibited. Decomposition is minimal and large quantities of partly decomposed remains may accumulate. This material, called **peat**, can become compacted into layers many feet thick.

Peat contains no rock particles and is not regarded as a true soil. However, it can support a limited range of specialized acid-loving plants like heather, cotton-grass and *Sphagnum* moss. These plants compete for the very limited supplies of inorganic ions available in the water.

59·4 Earthworms ■ ■ ■ ■ ■ ■ ■ ■ ■ ■ ■ ■ ■ ■ ■

The soil environment supports a great diversity of living things, most of which form part of the detritus food web. The most important of the larger organisms are earthworms.

Earthworms feed by ingesting the soil as they burrow, extracting and digesting the organic fragments as they pass through the gut. An important result of this method of detritus feeding is a reduction in the size of rock particles caused by grinding forces in the alimentary tract and by the chemical action of digestive juices. The casts left behind are rich in humus and nutrient materials.

Other consequences of burrowing include the aeration of the soil and the physical mixing of topsoil components. Earthworms drag leaves from the litter down into the topsoil. Some of this material is eaten, but the remainder is decayed rapidly by bacteria and fungi which thrive in the moist conditions of the burrow.

59·5 Water ■ ■ ■ ■ ■ ■ ■ ■ ■ ■ ■ ■ ■ ■ ■

Some of the water entering the soil as rain drains through, while some is trapped between soil particles. The ratio of water drained to water retained depends on the mineral skeleton and crumb structure. Rooted plants rely to a large extent on **capillary water**, that is, the water held by surface tension between large particles and soil crumbs (see Fig 59·2).

Drainage water eventually finds a variable level called the **water table** which rises or falls according to rainfall. As the surface layers of the soil dry, more water may be drawn up from lower layers by capillary action. Some water is lost from the ecosystem through underground run-offs. The maximum amount of water which can be held by the soil after loss through drainage is known as the **field capacity**.

A relatively large amount of the water retained by fertile soils is not available to plants. This is the water which is **adsorbed** onto the surfaces of tiny clay and humus particles by electrostatic forces which are too strong to be overcome by plant roots. This process is explained in the next section.

59·6 Ionic properties of clay particles ■ ■ ■ ■ ■ ■ ■ ■ ■

Clay particles are flat alumino-silicate crystals with an extremely large surface area to volume ratio (Fig. 59·3). Their overall negative charge derives from the fact that immediately after their formation by weathering, Al^{3+} and Fe^{2+} ions from the soil solution replace some of the Si^{4+} ions in the crystal lattice, leaving a deficit of positive ions. Clay particles thus act as anions. For plants this is an extremely valuable property because it enables mineral cations from water which otherwise might drain through the soil to be held bound to clay particles by electrostatic forces. This property is called **adsorption** and allows clay to 'store' a reserve of mineral ions.

A further consequence of adsorption is that ions like Ca^{2+} and Mg^{2+} may form bridges between clay and humus particles. This binding, illustrated in Figure 59·2, is called **flocculation** and it is the basis of a good crumb structure. Flocculation can easily be observed by adding a small quantity of calcium hydroxide to a clay suspension and comparing its settling rate to that of a similar suspension without the calcium hydroxide. Bridging ions also play a vital role in holding inorganic anions such as phosphates, PO_4^{3-}, and nitrates, NO_3^- (refer to Fig. 59·2).

In the absence of bridging ions, clay particles disperse and fill the available air spaces as they settle, causing poor drainage. A striking example of this is the damage which results from the flooding of arable land by sea water. Excess Na^+ ions replace the adsorbed Ca^{2+} and Mg^{2+} ions which together with Cl^- ions are leached away with the drainage water. The clay particles separate and sink and the result is a dense mass which does not drain and which cannot easily be restored by artificial means.

The replacement of one ion adsorbed to a clay particle with another, described in this example, is called **ion exchange**. Plant roots are thought to be able to extract minerals bound to clay particles by exchanging them for H^+ ions which they secrete. Rainwater, rich in dissolved carbon dioxide forms carbonic acid which has the same effect. It supplies an excess of H^+ ions which can replace and release cations held to clay particles by adsorption.

The acidity of the soil is partly determined by

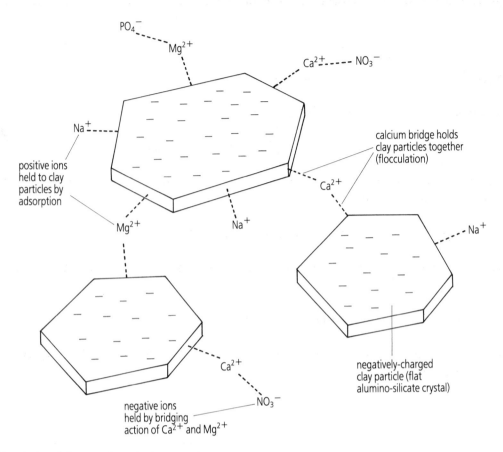

Fig 59·3 Flocculation of clay particles and adsorption of inorganic ions

the numbers of H^+ ions adsorbed by clay particles. The more there are, the more acid the soil and the more difficult the uptake of minerals by plant roots. The loss of mineral ions through leaching generally results in the clay particles acquiring more H^+ ions making the soil even more acid.

In general, the most important factor determining soil pH is the calcium carbonate content. A soil rich in calcium carbonate is alkaline and is called **calcicole**. Soils poor in calcium carbonate are acid and called **calcifuge**. Both are easily recognized and distinguished by their characteristic vegetation. The yew tree and *Clematis* for example, always indicate calcicole

conditions, while ling heather is a sure sign of a calcifuge soil.

Ecologists can infer a great deal about the physical, chemical and biological properties of a soil by identifying the plants and animals living in the region. Nevertheless all the components of an ecosystem, biotic and abiotic, interact so closely with each other that it is often easier to characterize and describe the whole system than it is to pick out and define the role of any particular factor. This will be seen again in the next Unit, which considers water as a medium for life and describes different kinds of aquatic environment.

60 AQUATIC ECOSYSTEMS

Objectives

After studying this Unit you should be able to:–

- Define the following terms: neuston, plankton, nekton, benthos

- Explain how light penetration, dissolved gases, dissolved solids, and water currents influence aquatic communities

- Compare the ecosystems of fast and slow moving rivers

- Comment on the special adaptations of estuarine organisms and give reasons for the high productivity of this ecosystem

- Describe thermal zonation in lakes and oceans

- Outline the relationship between currents, mineral nutrients and productivity in deep and shallow waters

60·1 Aquatic ecosystems and habitats ■ ■ ■ ■ ■ ■ ■ ■ ■ ■

Aquatic environments support a great diversity of life forms. Water provides a more constant and protective environment than land, and is far less prone to sudden and drastic changes in physical or chemical conditions, although some changes are brought about by climatic or seasonal variations. Aquatic organisms receive support and obtain dissolved oxygen and nutrients from the surrounding water. In addition, water currents facilitate fertilization and aid in the dispersal of organisms.

Aquatic ecosystems are classified as **salt water**, **freshwater** or **brackish water** (estuarine) systems according to the concentration of dissolved salts they contain. They can be further separated into fast-flowing streams, slow-moving rivers, lakes and oceans according to whether the water is constantly in motion or relatively static.

Organisms which live in water are categorized according to whether they are found in rivers, lakes or seas, and also by the particular habitat they occupy within these ecosystems. Four different habitats are distinguished and give rise to the following collective terms for aquatic communities.

1 Neuston. The term neuston is used to describe organisms such as pond-skaters, springtails, air-breathing diving beetles, water boatmen, and floating plants like duckweed and bladderwort, all of which live at or close to the water surface. Some of these organisms survive above the water supported by the surface film, while others cling to the surface film from beneath, or swim in the upper waters. The neuston forms a significant part of most freshwater communities but is not very important in marine ecosystems.

2 Plankton. This is the mass of small plants (**phytoplankton**) and animals (**zooplankton**) found in most aquatic ecosystems with the exception of fast flowing rivers. Their powers of locomotion are restricted to small vertical movements or to catching prey and their distribution is determined by water currents.

3 Nekton. The nekton comprises all the free-swimming organisms strong enough to swim against water currents. Some of them, like swimming insects in a pond, may be small, while others, including some bony fish, sharks and whales, are extremely large.

4 Benthos. Benthic organisms live at or close to the bottom. There are many different types, most of which contribute in one way or another to the detritus food web.

60·2 Limiting factors in aquatic ecosystems ■ ■ ■ ■ ■ ■ ■

Limiting factors restrict the distribution of living organisms, preventing colonization of otherwise favourable environments. For example, low rainfall and high temperature are limiting factors in a desert, while extreme cold in winter prevents the growth of trees in tundra habitats. The most important factors with potentially limiting effects for aquatic ecosystems are as follows:

Light penetration. Water absorbs light energy, converting it to heat. However, not all wavelengths are absorbed equally. Blue and green wavelengths penetrate to greater depths than red or violet, helping to explain the characteristic colour of a body of standing water.

The depth to which light can penetrate depends partly on the amount of suspended particulate matter in the water. Thus, the relatively sterile surface layers of deep oceans allow penetration down to 150 m, while inshore waters, with a far greater mass of living organisms and sedimentary particles in the surface layers, may have only the top 20 m illuminated.

All photosynthetic organisms are restricted to the illuminated region, which is called the **photic zone**. Although plants may be capable of photosynthesis even at the extreme limit of the photic zone, they cannot survive at this depth if they use up food faster than they produce it. The most important level therefore is not the extent of the photic zone, but the depth at which production by photosynthesis equals the rate of consumption through respiration, called the **compensation level**. This level varies according to the latitude and the season, being nearer to the surface in winter and deeper in summer. Primary production is only possible above the compensation level and virtually all aquatic organisms depend on the surface community for their energy requirements.

Dissolved gases. Gases from the atmosphere dissolve in water at the surface. However, some gases are more soluble than others, so that the proportions of gases available to aquatic organisms are quite different from those in air (Table 60·1).

Table 60·1 *Comparison between atmospheric concentration of gases and concentration of dissolved gases in water at equilibrium*

GAS	ATMOSPHERIC CONCENTRATION /cm³ dm⁻³	SATURATION IN WATER /cm³ dm⁻³
Oxygen	210	7
Nitrogen	780	14
Carbon dioxide	0.3	0.3

Oxygen is 30 times less abundant in water than in air and often limits the distribution of living organisms. In a typical lake, approximately $100 \ \text{g m}^{-2} \ \text{day}^{-1}$ of oxygen is absorbed. This is sufficient to saturate the water to depths of about 10 m and is far in excess of the amount of oxygen produced by photosynthesis in open water. In shallow aquatic habitats such as ponds, production by photosynthesis may be more significant and may cause the water to become supersaturated with oxygen. Turbulent water in fast-flowing streams is usually saturated with oxygen.

The diffusion of dissolved gases through deep bodies of water is a slow process. In some places, diffusion is aided by currents and wave action, but in still waters very little oxygen is transferred to lower levels. Consequently, the available oxygen can easily be used up by bottom-dwelling aerobic decomposers. This can have marked limiting effects on the whole of the aquatic community.

Nitrogen, though less soluble than oxygen, makes up the largest proportion of dissolved gases. It is used directly by some nitrogen-fixing bacteria and blue-green algae in the manufacture of proteins and released in gaseous form by other (denitrifying) bacteria.

The effects of carbon dioxide are more complex because of the formation of carbonic acid which dissociates to give H^+, HCO_3^- and CO_3^{2-} ions. The proportions of these ions present depends on the amounts of other dissolved substances in the water.

Dissolved solids The oceans have a nearly uniform and constant content of dissolved ions, as summarized in Table 60·2. These ions occur in

Table 60.2 *Dissolved ions present in sea water*

ANIONS	g dm⁻³
Chloride (Cl^-)	19.35
Sulphate (SO_4^{2-})	2.70
Hydrogencarbonate (HCO_3^-)	0.15
Bromide (Br^-)	0.07

CATIONS	g dm⁻³
Sodium (Na^+)	10.75
Magnesium (Mg^{2+})	1.30
Calcium (Ca^{2+})	0.42
Potassium (K^+)	0.39
Total salinity	35.13

(Ocean waters vary in salinity from 30 to 37 g dm⁻³)

such definite proportions that the salinity of water in coastal regions and estuaries is often calculated from the concentration of chloride ions determined by titrating sea water against a standard silver nitrate solution. On the other hand, freshwater ecosystems show a considerable variation in salt content, depending on the minerals present in drainage water from the surrounding countryside and also on the activity of living organisms.

The most important limiting ions in aquatic ecosystems are phosphate and nitrate, although sulphate and calcium are sometimes in short supply. Phosphate is withdrawn from the water by living organisms and its concentration may fall to virtually nil in summer. In winter the concentration rises again as organisms die and are decomposed. Some algae are adapted to survive seasonal scarcity of phosphate by means of specialized storage mechanisms.

Rain acquires small amounts of inorganic substances high in the atmosphere as droplets of water condense around dust particles or tiny salt crystals from sea spray. In addition, gases such as carbon dioxide, oxides of nitrogen and air pollutants such as sulphur dioxide dissolve in water droplets, so that rain becomes a mixture of weak acids. Water draining from the land into streams acquires organic compounds in various stages of decomposition and inorganic ions from the chemical weathering of rocks. Springs bring up dissolved minerals from underground layers in a similar way.

Water currents. Water currents become a limiting factor in fast flowing streams and on the seashore, where continual buffeting by the waves prevents colonization by delicate or weak swimming organisms.

The remainder of this Unit describes a number of ecosystems in more detail, including adaptations of the animal and plant communities typical of each.

60·3 River ecosystems ■ ■ ■ ■ ■ ■ ■ ■ ■ ■ ■ ■ ■ ■

A river cannot be considered as a single ecosystem because its characteristics and living communities change radically from source to mouth. Among the most important features from a biological point of view are the rate of flow, that is, the volume of water passing a given point per unit time measured in $m^3 s^{-1}$, the velocity of flow, measured in $m\ s^{-1}$, and whether the flow is smooth (laminar) or turbulent. These factors help to determine the sculpturing of the river bed, the type and settling rate of sedimentary particles and consequently the turbidity of the water.

A river changes gradually along its course so that a smooth environmental gradient exists with no clear boundary lines. Nevertheless, it is possible to distinguish two major types of ecosystem on the basis of differences in flow and turbulence, namely the fast-flowing hill stream and the slow-flowing river. A third and very different ecosystem exists at the mouth of the river where it merges with the sea. Estuarine communities are more closely related to those of the sea shore than the upper reaches of the river. These three ecosystems are illustrated in Figure 60·1.

Fast-flowing hill stream. (Fig. 60·1A) This part of the river has a rapid turbulent flow which sweeps away everything which is not either attached or weighted down. Consequently, the stream bed is stony with little sediment for rooted plants. The organisms present show a range of adaptations for resisting the current and are best considered in terms of the microhabitat they occupy within the stream.

Organisms exposed on top of the rocks have flattened streamlined bodies and cling tightly to the surface. Examples include the fresh water limpet, fresh water sponges and some caddis-fly larvae which are cemented to the rock encased in a weighted jacket made of tiny stones.

Typical of organisms living in the microhabitat formed by spaces between the rocks are mayfly and stonefly nymphs which have flattened bodies, but are not cemented onto the rock. They cling on by reflex behaviour, orientating themselves to face upstream. Other types of insect larvae have spikes enabling them to lodge in the cracks and crevices.

Beneath the rocks, the current is not so strong and a greater variety of organisms can be found including annelids, flatworms, snails, freshwater shrimps, and the larvae of more species of insects. All the organisms living in the fastest portion of the river are small bottom-dwellers. Except for a few algae stuck to the rocks, there is an almost total lack of plant life. Consequently, productivity is low and the main source of food is detritus washed into the stream.

Slow-flowing river. (Fig. 60·1B) In this region, flow is laminar and there is a substantial bottom sediment consisting of particles carried down from the upper reaches. A more varied community is established, consisting of neuston, plankton, nekton and benthos.

Of particular importance is the occurrence of rooted plants along the water margins and phytoplankton under the surface. Both of these constitute a productive food source. The river bed is rich in detritus feeders and a certain amount of nutrient recycling is achieved by them, although much of the biomass is washed away by the flow of the river.

The most important limiting factor in a slow moving river is the amount of dissolved oxygen. Decomposers among the bottom-dwelling benthos tend to starve the rest of the animal community of oxygen if large amounts of organic matter accumulate on the river bed. This may happen naturally, but is more commonly the result of pollution.

Estuaries. (Fig 60·1C) Estuaries are semi-enclosed regions where rivers and oceans merge. The distinctive ecosystem of an estuary is the result of a mixing of the waters. Sediment and organic discharge are gained from the river and salt and tidal currents from the sea. Superficially,

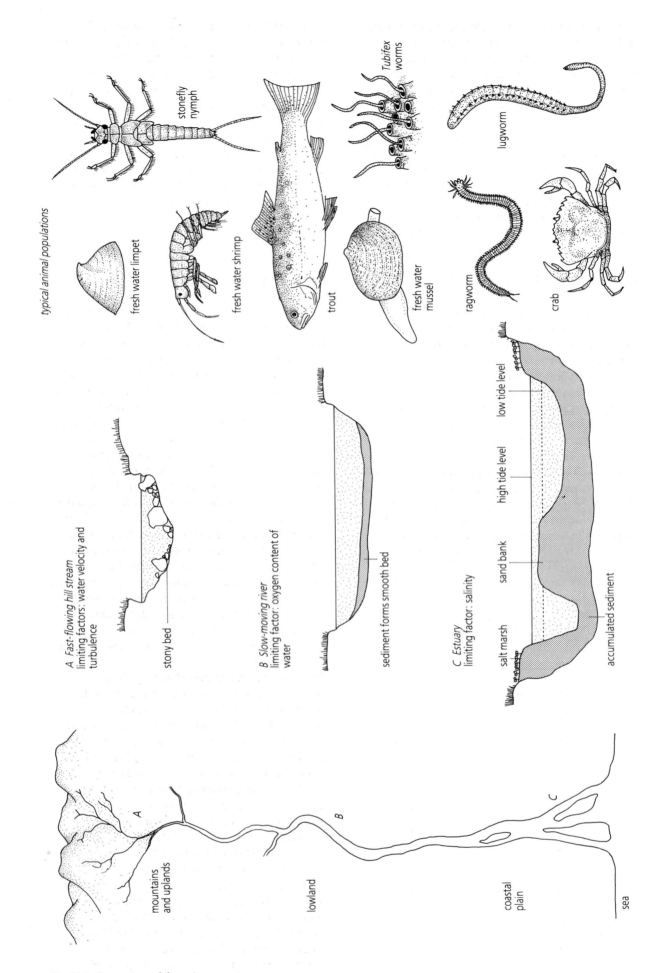

typical animal populations

stonefly nymph

fresh water limpet

fresh water shrimp

trout

fresh water mussel

Tubifex worms

lugworm

ragworm

crab

A Fast-flowing hill stream
limiting factors: water velocity and turbulence

stony bed

B Slow-moving river
limiting factor: oxygen content of water

sediment forms smooth bed

C Estuary
limiting factor: salinity

salt marsh

sand bank

high tide level

low tide level

accumulated sediment

mountains and uplands

lowland

coastal plain

sea

A

B

C

Fig 60·1 Comparison of three river ecosystems

518

estuaries often appear desolate and sterile places, but this is far from the truth, for their unseen burrowing community is second only to that of coral reefs in terms of energy flow. Furthermore, estuaries are vital spawning grounds and nursery feeding beds for a great number of marine animals.

Two features account for the high productivity: the water is shallow and turbulent and hence rich in oxygen, and the river current and ocean tide combine to make the whole area a nutrient trap. Decomposition can therefore occur at a very high rate and a great number of organisms are supported by the detritus food web. Detritus feeders include lobsters, crabs, annelids and a variety of bivalve molluscs, all of which form a rich food source for carnivorous fish and birds.

The major limiting factor of the ecosystem is salinity, which varies greatly even at the same spot between high and low tides. Apart from marine algae, only a small number of plants can tolerate the high and variable salinity of the mud flats. Such plants are called **halophytes** and make up a distinctive salt marsh vegetation, their principle specialization being the ability to prevent excessive loss of water by osmosis despite a high external salt concentration.

60·4 Lakes and oceans ■ ■ ■ ■ ■ ■ ■ ■ ■ ■ ■ ■ ■ ■

Although there are obvious individual differences between fresh and salt water organisms, the feeding relationships of lake and ocean communities are very similar. The most helpful way to classify these still-water ecosystems is on the basis of depth rather than salinity.

Deep water ecosystems. A thermal gradient is established in deep water from top to bottom, forming three broad vertical zones: the surface layer or **epilimnion**, a bottom layer or **hypolimnion** and, between the two, a transitional layer called the **thermocline**. Figure 60·2 illustrates this phenomenon in a temperate lake in summer. Water is most dense at 3.94°C so that water at this temperature invariably forms the bottom layer. The surface waters are warmed by the sun but generally do not mix with the lower levels. For this reason, the thermocline shows a sharp change rather than a smooth gradient.

An important consequence of this zonation is that there is no recycling of mineral nutrients from the detritus food web on the bottom to the planktonic producers. In deep oceans this situation may be permanent: the surface waters are very poor in nutrients while the sea bed is nutrient rich. In some areas, deep currents rise to the surface, lifting detritus and producing localized regions of high productivity such as that formed by the Humboldt current off the west coast of South America. In lakes, the thermocline disappears in autumn as the surface cools to 4°C and the entire lake contents are mixed and circulated in what is called an 'overturn'. In winter the surface may cool further forming a layer over the warmer, denser hypolimnion and creating a new thermocline in which the temperature changes in the opposite direction. In spring the process is reversed with a second overturn.

Shallow water ecosystems. It has been stressed that high productivity is only possible where mineral nutrients are in abundance. This depends on the circulation of materials between the detritus food web consisting mainly of bottom-dwelling organisms and the grazing food web consisting of organisms near the surface. In shallow waters nutrients circulate freely between these food webs, and the main limit to diversity and energy flow in living communities is usually the availability of oxygen. In this respect, shallow lakes resemble slow-flowing rivers whose oxygen is depleted by respiring decomposers.

In contrast, oceanic waters above the continental shelf are well oxygenated by constant wave action and support a great variety and number of living organisms. The richness of marine life is most evident in the coral reef ecosystem which is comparable in complexity with a tropical rain forest.

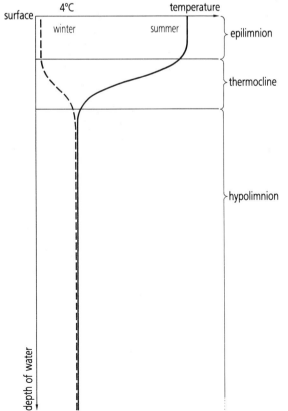

Fig. 60·2 Thermal zonation of a deep lake in summer and winter

61 MEASUREMENT IN ECOLOGY

Objectives

After studying this Unit you should be able to:-

- Comment on the importance of microclimate and outline methods for the measurement of the following factors: temperature, light, wind speed and direction, humidity

- Outline methods for the measurement of the following edaphic factors: particle content, water content, organic content, air content, drainage, water retention, pH

- Suggest suitable methods for estimating population numbers of microorganisms and small invertebrates

- Comment on the capture/recapture method of estimating total population

- Describe the use of frame and pin quadrats in the analysis of plant populations

- Comment on the analysis of environmental gradients by line transects, belt transects and isonomes

61·1 Measurement of abiotic factors ■ ■ ■ ■ ■ ■ ■ ■ ■ ■ ■

The modern study of ecosystems is a quantitative science based on measurement techniques. A wide variety of methods exist to sample, count and estimate the abundance, density and effects of different organisms and to estimate the importance of climatic and edaphic (soil) factors. This Unit outlines the range of methods available to students of ecology and is intended as an informative background to field work.

The term 'abiotic' refers to the non-living environment. Abiotic factors can be classified into two kinds, climatic and edaphic.

Climatic factors

Climatic factors have important effects in all ecosystems. Gross features of climate such as annual rainfall and seasonal variations in light levels and temperature contribute to the development of different biome types, as outlined in Unit 58. Equally important are the smaller variations which constitute the **microclimate** of different parts of the habitat. Although exposed to the same annual variations, the microclimates of an open hillside and a rocky gully may be very

different and directly affect the distribution of different species of organisms.

Some of the more important features of microclimate are described below.

1 Temperature. Soil, water and air temperatures are easily measured with a mercury thermometer and can reveal important differences, for example, between the soil temperature of a meadow on a valley floor and that of soil or peat near the tops of the surrounding hills. Similarly, the air within a forest may be several degrees cooler than air above open countryside.

Daily variations in temperature are conveniently measured using a thermistor which gives an electrical output and can be coupled to electronic recording instruments or computers. Some habitats such as the seashore show far greater daily variations in temperature than others.

2 Light. The intensity, wavelength and duration of light are all important for photosynthetic organisms. These factors can be measured with appropriate instruments. Variations in light intensity are often monitored by means of a light-dependent resistor (LDR) which can be used for continuous electronic recording of light levels.

3 Wind speed and direction. Wind speed
can be estimated using a commercially available
wind speed meter (Fig. 61·1). Wind direction can
be determined using a compass and a home-made
wind vane. Records kept over a period of weeks or
months are needed before worthwhile comparisons
between different habitats can be made.

4 Humidity. An enclosed volume of air in
contact with a water surface eventually becomes
saturated with water vapour. The amount of water
vapour required for saturation varies with
temperature and, as might be expected, warm air
retains more water vapour than cold air. The
relative humidity is expressed in terms of the
percentage saturation of air at a particular
temperature. If the relative humidity is less than
100% more water can be absorbed, while in excess
of 100% water will condense to form droplets.

The relative humidity is usually determined
using a hygrometer of the type illustrated in Figure
61·2. This consists of two thermometers mounted
in a frame which is attached to a handle and
whirled at arms length. One thermometer is
exposed directly to the air, while the other is
surrounded by a wick dampened with water.
Evaporation of water from the wick cools the bulb
of the thermometer so that a lower reading is
obtained. The difference between the temperatures

The position of the sphere gives an approximate
indication of wind speed when the meter is held
facing the wind

Fig 61·1 Wind speed meter

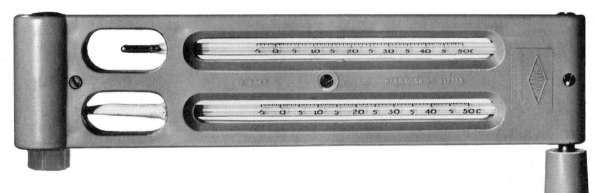

Fig 61·2 Whirling hygrometer

recorded gives an indication of the amount of
water vapour in the air and the relative humidity
can be calculated from hygrometer tables.

Relative humidity largely determines the rate of
evaporative water loss from exposed surfaces and is
a major factor restricting plants like mosses and
liverworts, and animals like woodlice to damp
habitats.

Edaphic factors

The composition and properties of soil contribute to the development of typical plant communities in most terrestrial ecosystems. Sophisticated techniques of soil analysis are essential for agriculture and research, but worthwhile comparisons between soil samples from different areas can be made using the simple methods outlined in Figure 61·3.

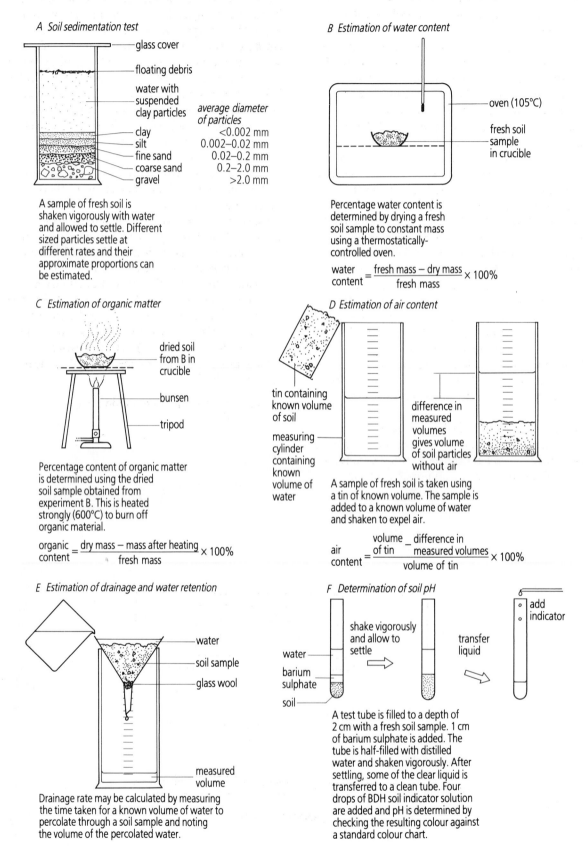

A Soil sedimentation test

- glass cover
- floating debris
- water with suspended clay particles

	average diameter of particles
clay	<0.002 mm
silt	0.002–0.02 mm
fine sand	0.02–0.2 mm
coarse sand	0.2–2.0 mm
gravel	>2.0 mm

A sample of fresh soil is shaken vigorously with water and allowed to settle. Different sized particles settle at different rates and their approximate proportions can be estimated.

B Estimation of water content

- oven (105°C)
- fresh soil sample in crucible

Percentage water content is determined by drying a fresh soil sample to constant mass using a thermostatically-controlled oven.

$$\frac{\text{water}}{\text{content}} = \frac{\text{fresh mass} - \text{dry mass}}{\text{fresh mass}} \times 100\%$$

C Estimation of organic matter

- dried soil from B in crucible
- bunsen
- tripod

Percentage content of organic matter is determined using the dried soil sample obtained from experiment B. This is heated strongly (600°C) to burn off organic material.

$$\frac{\text{organic}}{\text{content}} = \frac{\text{dry mass} - \text{mass after heating}}{\text{fresh mass}} \times 100\%$$

D Estimation of air content

- tin containing known volume of soil
- measuring cylinder containing known volume of water
- difference in measured volumes gives volume of soil particles without air

A sample of fresh soil is taken using a tin of known volume. The sample is added to a known volume of water and shaken to expel air.

$$\frac{\text{air}}{\text{content}} = \frac{\begin{matrix}\text{volume} \\ \text{of tin}\end{matrix} - \begin{matrix}\text{difference in} \\ \text{measured volumes}\end{matrix}}{\text{volume of tin}} \times 100\%$$

E Estimation of drainage and water retention

- water
- soil sample
- glass wool
- measured volume

Drainage rate may be calculated by measuring the time taken for a known volume of water to percolate through a soil sample and noting the volume of the percolated water.

F Determination of soil pH

- water
- barium sulphate
- soil

shake vigorously and allow to settle

transfer liquid

add indicator

A test tube is filled to a depth of 2 cm with a fresh soil sample. 1 cm of barium sulphate is added. The tube is half-filled with distilled water and shaken vigorously. After settling, some of the clear liquid is transferred to a clean tube. Four drops of BDH soil indicator solution are added and pH is determined by checking the resulting colour against a standard colour chart.

Fig 61·3 Simple methods used for investigating the composition and properties of soil

If valid conclusions are to be drawn about the roles and behaviour of plant and animal populations within an ecosystem, then quantitative information is needed. In other words, it is necessary to count the numbers of organisms which make up the community.

It is comparatively easy to count the number of trees in a park or the numbers of large animals like elephants and giraffes in a given area of the African savannah, but how can the numbers of mackerel in the North Sea, or the numbers of bacteria in a soil sample be determined? Obviously, representative parts of the population must be sampled and total numbers estimated accordingly. The sampling procedure used depends on the sort of organism being investigated. The following methods outline some of the possibilities.

Microorganisms.
Microorganisms suspended in water can be counted under a microscope using a special type of slide, developed for counting blood cells, called a **haemocytometer** (see Fig. 61.4). This has a grid marked with 0.05 mm squares at the bottom of a shallow trough. The distance between the coverslip and the surface of the slide is 0.1 mm. Usually, the number of microorganisms in 50 squares is counted, allowing the number per mm^3 to be estimated with reasonable accuracy. In this way, the number of organisms present in the original sample can be calculated. This method is suitable for algae, protozoa, single-celled fungi and bacteria.

An alternative method often used for estimating populations of bacteria involves adding samples to nutrient agar plates and counting the colonies produced.

Small invertebrates.
Soil organisms which are capable of active movement may be counted after extraction from a weighed sample by means of a Tullgren or Baermann funnel. These funnels are illustrated in Figure 61·5 and operate in different ways. The **Tullgren funnel** is used to extract small insects and mites. As heat from the bulb warms and dries the sample, small organisms move down and eventually fall into the preserving alcohol below. The **Baermann funnel** is used to extract nematode worms which migrate downwards away from the warm water near the light bulb, collecting at the bottom of the funnel where they can be removed and counted.

Capture-recapture method.
This is a useful method of estimating the population numbers, particularly of fast-moving or flying terrestrial organisms, such as insects. The approximate number of ladybirds in a field, for example, could be calculated by first collecting about a hundred (number collected $= f_1$) using a sweep net, marking them with a tiny spot of quick-drying cellulose paint and then releasing them. A few days later a new collection is made from the same area ($= f_2$). The number of marked individuals in this new sample is noted ($= f_3$). The following equations are used to estimate the total number (N) of ladybirds:

Assuming the samples are random and representative:

$$\frac{f_1}{N} = \frac{f_3}{f_2}$$

Therefore: $\quad N = \frac{f_1 f_2}{f_3}$

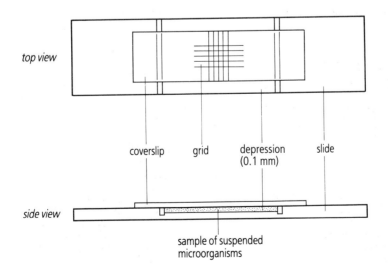

top view

coverslip   grid   depression (0.1 mm)   slide

side view

sample of suspended microorganisms

Fig 61·4 Haemocytometer slide

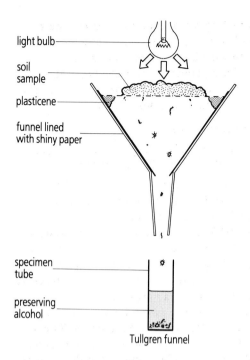

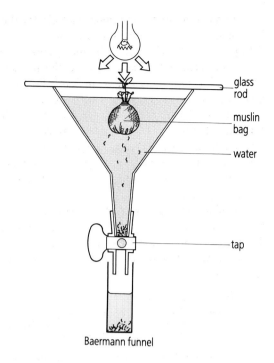

Fig 61·5 *Apparatus for collecting soil organisms*

The capture-recapture method gives accurate results only if a number of important assumptions are met:

(1) The population must be confined to an area with clear geographical boundaries, such as a pond or woodland surrounded by open country.

(2) Organisms must be distributed evenly throughout the selected area.

(3) Marking must not interfere with the organism's survival or affect the probability of its recapture.

(4) The marked individuals must mix randomly with the rest of the population. The time taken for this to happen will depend on the mobility of the organisms concerned.

Populations of slow-moving and sedentary animals are often sampled and counted by methods designed to analyse vegetation, as described in the next section.

Analysis of vegetation

It is difficult to make a total count of plants of a particular species growing in an area unless the individual plants are large or comparatively rare. The analysis of vegetation involves a statistical approach whereby a sample area is studied in some detail and inferences are made about the community as a whole.

Decisions about the position and size of sample areas are problematic. How, for example, can one be sure that the sample is sufficiently large to allow statistically valid conclusions to be made concerning the relative abundance of species present? One way is to construct a **minimal area curve**. To do this, a small area is measured out

and a list is made of all the different species which are found in it. A new larger sample area is marked out and the species present recorded. The area is increased again and the procedure is repeated until no new species are encountered. If the region is fairly uniform in its composition, then a smooth curve will be obtained like that illustrated in Figure 61·6, from which the smallest area likely to contain all the species present can be estimated. Further samples taken from areas of this size allow accurate comparisons between different parts of the habitat except where there is a high degree of association between different species or where individuals are non-randomly distributed.

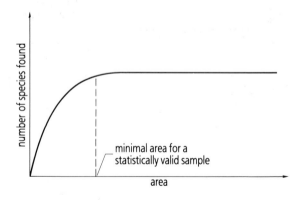

Fig 61·6 *Minimal area curve*

The most important methods available for the investigation of plant populations are as follows.

1 Quadrat methods.

A **frame quadrat** is a square or rectangular frame which delineates an area, or plot, often of 100 cm² or 1 m². Small quadrats made of strong metal wire are designed to be dropped at random over the shoulder of the experimenter who walks around the area under investigation. Where it lands, the quadrat encloses a known area which can be analysed in a number of ways to reveal information about the abundance of plant species.

The simplest analysis involves recording the presence or absence of particular species. Having thrown a minimum of 100 quadrats, the **percentage frequency** of the species is calculated as follows:

% frequency of species A
$$= \frac{\text{number of quadrats containing A}}{\text{total number of quadrats thrown}} \times 100\%$$

This method has the advantage of being very quick and is useful as a preliminary investigation into the relative abundance of different species. However, it reveals very little information about the **density** (numbers of individuals per m²) of the species. A more complete analysis would involve counting the number of individuals of each species enclosed by the quadrats and taking an average from which the density per m² could be calculated.

An alternative measure of abundance can be obtained using **pin quadrats** to estimate the **percentage cover** of a particular species. The area to be sampled is reduced to the size of a pin point by making the quadrat into a sort of dart. This consists of a length of wire sharpened at both ends, so that it sticks into the ground when thrown. The experimenter records 'hits' and 'misses' on a particular species having thrown the pin quadrat at random at least 100 times.

% cover of species A
$$= \frac{\text{number of 'hits' on A}}{\text{total number of pin quadrats thrown}} \times 100\%$$

Frame and pin quadrats are ideal for the analysis of low-lying vegetation in areas such as fields, sand-dunes, salt-marshes or bogs.

2 Transects.

A transect is used to illustrate and measure changes in vegetation due to zonation or succession along an environmental gradient. The simplest form of transect, called a **line transect**, is obtained by stretching a length of string horizontally above the area and noting the species vertically below the line at fixed intervals. This is often combined with topographical measurements so that the results can be expressed in the form of a profile, as illustrated by the example in Figure 61·7.

More quantitative data can be obtained by combining a simple line transect with pin or frame quadrat analysis at fixed points along its length. Alternatively, a **belt transect** can be carried out by adding another line parallel to it and sampling the vegetation enclosed by the two lines. If the area to be studied is small enough, a complete grid system can be laid out for detailed analysis. This is called the **isonome** method and it is the best way to show zonation in microhabitats such as rock pools.

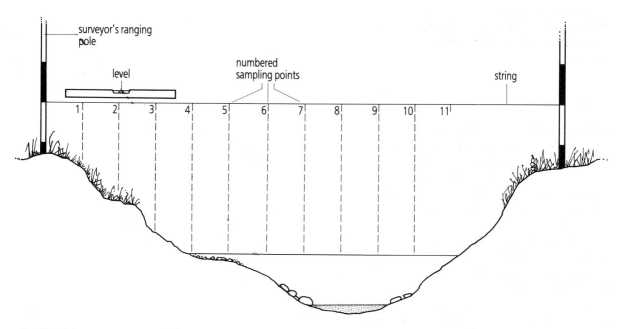

Fig 61·7 Line transect across a hill stream

62 NUTRIENT CYCLES

Objectives

After studying this Unit you should be able to:—

- Distinguish between the biotic and abiotic phases of biogeochemical cycles

- Draw a flow diagram of the carbon cycle and explain its important stages

- Describe the oxygen cycle and comment on its relationships with the carbon cycle

- Draw a flow diagram to illustrate the nitrogen cycle and give an account of the following processes: nitrogen fixation, nitrification, denitrification

- Outline the major phases of the sulphur and phosphorus cycles

- Draw a diagram to illustrate the hydrological cycle

62·1 Biogeochemical cycles ■ ■ ■ ■ ■ ■ ■ ■ ■ ■ ■ ■

The flow of energy through ecosystems was described in Unit 55. Light energy from the Sun is converted into stored chemical energy, which passes through various trophic levels and is finally lost into the atmosphere as heat. Energy transfer always occurs in the same direction so that stability of an ecosystem depends on continued energy input. In contrast, the bioelements which constitute living matter are limited in quantity and are not continually resupplied from an external source. Typically, such substances are **recycled**, that is, essential chemicals are extracted from the bodies of dead organisms and become available again for use by the living.

All bioelements show a natural cycle called a **biogeochemical cycle**, consisting of a **biotic phase**, during which the particular element is 'fixed' in living organisms in the form of organic compounds, and an **abiotic phase**, during which it forms part of the non-living environment. The rate of recycling varies for each element according to its abundance in the biosphere and the ways in which it is used by living organisms.

The remainder of this Unit discusses the carbon, oxygen, nitrogen, sulphur and phosphorus cycles, and also outlines the cycling of water through the hydrological cycle.

62·2 Carbon cycle ■ ■ ■ ■ ■ ■ ■ ■ ■ ■ ■ ■ ■ ■

Carbon forms 18% of living matter but is less abundant in the abiotic environment. The major source of carbon for living organisms is in the form of carbon dioxide obtained either from the atmosphere or dissolved in water.

Through the process of photosynthesis in green plants, CO_2 is converted into organic compounds including carbohydrates, fats, proteins and nucleic acids. In this form, carbon becomes available to organisms at higher trophic levels. As outlined in Figure 62·1, the biotic phase of the carbon cycle resembles the energy pathway, passing from producers to consumers and finally to decomposers.

During recent years, large quantities of fossil fuels, representing the stored carbon from previous ecosystems, have been extracted from the earth and burned to provide energy. Consequently, there has been a measurable increase in the amounts of CO_2 in the atmosphere. This has several predictable effects. Firstly, it increases primary productivity by increasing the rate of photosynthesis. Secondly, it increases the

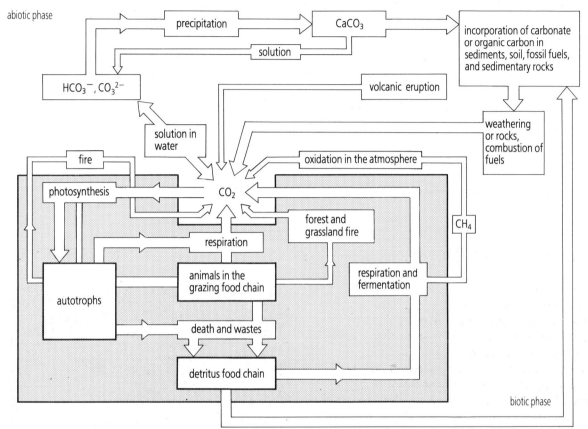

Fig 62·1 Flow diagram of the carbon cycle

Carbon dioxide is returned to the atmosphere by all respiring organisms. The decomposers are particularly important because they extract the fixed carbon from dead and decaying matter and thereby make it available once more to living organisms. However, not all the carbon fixed by living organisms is returned in this way. Wherever oxygen is in short supply, as in waterlogged soils or at the bottom of still water, decomposition is incomplete and organic matter is likely to accumulate. This leads to the formation of organic sediments and peat deposits which eventually give rise to **fossil fuels** including coal, natural gas and oil.

concentration of dissolved CO_2 in the oceans and hence the deposition of carbonate in the form of corals and carbonaceous rocks. Neither of these phenomena is alarming and neither absorbs all the excess CO_2 which continues to accumulate in the atmosphere.

Of more concern is the controversial 'greenhouse effect' whereby an increased CO_2 content in the upper atmosphere is thought to cause more of the Sun's rays to be reflected back towards the Earth, potentially disrupting the Earth's weather system and leading to melting of the polar ice-caps. The likelihood of this outcome is, however, very difficult to assess.

62·3 Oxygen cycle ■ ■ ■ ■ ■ ■ ■ ■ ■ ■ ■ ■ ■ ■ ■

The oxygen cycle (Fig. 62·2) is linked with the carbon cycle in a reciprocal way. Oxygen is released into the atmosphere by photosynthesis and is used up by the oxidation of carbon in respiration.

During photosynthesis, the amount of oxygen produced is directly proportional to the amount of carbon fixed into organic compounds. Therefore, as the quantity of organic material in living

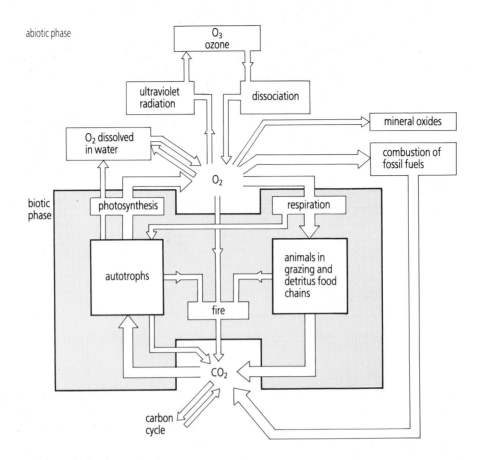

abiotic phase

O₃ ozone

ultraviolet radiation

dissociation

mineral oxides

O₂ dissolved in water

combustion of fossil fuels

O₂

biotic phase

photosynthesis

respiration

autotrophs

animals in grazing and detritus food chains

fire

CO₂

carbon cycle

Fig 62·2 Flow diagram of the oxygen cycle

organisms and in fossil fuel deposits increases, so too does the concentration of oxygen in the atmosphere. Over the millions of years that life has existed on the Earth, this has been a one-way process and the amount of oxygen in the atmosphere has steadily increased. As discussed earlier in Unit 43, the increasing availability of oxygen had an important and continuing influence on the evolution of early life forms.

In recent years, the concentration of oxygen in the atmosphere has started to fall for the first time because of the combustion of fossil fuels. A transatlantic jet, for example, uses up four tons of oxygen on each trip. Nevertheless, the percentage change in the amount of oxygen is slight, and so far there have been no significant effects on living organisms.

62·4 Nitrogen cycle ■ ■ ■ ■ ■ ■ ■ ■ ■ ■ ■ ■ ■ ■ ■ ■

Nitrogen gas constitutes almost 80% of the atmosphere. Nevertheless, nitrogen is often in short supply to living organisms. Molecules of nitrogen gas consist of pairs of atoms covalently bonded together in a very stable arrangement. Only a few living organisms possess the enzymes needed to split these molecules and are able to use nitrogen gas directly. All other living organisms must obtain nitrogen in a fixed form, that is, incorporated into other compounds. In the case of plants, nitrate ions (NO_3^-) represent the most useful form of nitrogen although sometimes they are able to use ammonium ions (NH_4^+) or urea ($CO(NH_2)_2$) instead. Animals and other heterotrophic organisms often require nitrogen in organic form, for example, as amino acids.

The nitrogen cycle, shown in Figure 62·3, can be considered in three main sections:

1 Nitrogen fixation. In this process nitrogen gas from the atmosphere is incorporated into nitrogen compounds such as nitrites, nitrates or ammonia.

Because nitrogen molecules are so stable, a great deal of energy is required to split them so that their atoms can combine with other elements. **Lightning** provides one energy source for nitrogen fixation, causing it to combine with oxygen in the air. The nitrogen oxide formed (equation (1)) may be oxidized further to nitrogen dioxide (equation(2)) which can combine with rainwater to form nitrate ions (equation(3)) as follows:

$$\text{lightning}$$
$$(1)\ N_2 + O_2 \longrightarrow 2NO$$
$$(2)\ 2NO + O_2 \longrightarrow 2NO_2$$
$$(3)\ 3NO_2 + H_2O \longrightarrow 2H^+ + 2NO_3^- + NO$$

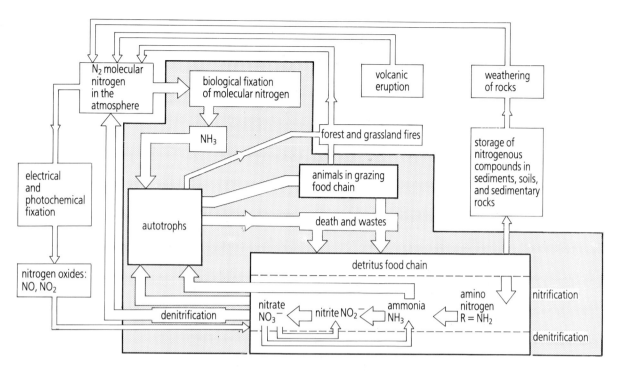

Fig 62·3 Flow diagram of the nitrogen cycle

About 6% of the nitrogen withdrawn from the atmosphere each year is fixed in this way.

A much more productive natural source of fixed nitrogen is provided by certain types of prokaryotic organisms with biochemical pathways which enable them to use nitrogen gas directly. This process requires large amounts of ATP and the enzyme **nitrogenase** and results in the production of ammonia. Examples of **nitrogen-fixing** organisms include *Rhizobium* bacteria in root nodules (Unit 57), and the free-living bacteria, *Azotobacter* and *Clostridium*. Some species of blue-green algae (Phylum Cyanophyta) are able to fix nitrogen and represent an important source of nitrogenous compounds in aquatic and tundra habitats. The ammonia which these organisms produce is quickly incorporated into protein, nucleic acids and other nitrogenous compounds including vitamins and pigments. Biological nitrogen fixation amounts to more than 60% of the annual total.

Nitrogen fixation is achieved industrially by the **Haber-Bosch process** in which hydrogen and nitrogen are reacted together at high temperature and pressure in the presence of a catalyst to form ammonia. This method is used in the manufacture of fertilizers such as ammonium nitrate and accounts for around 30% of the total nitrogen fixed from the atmosphere.

2 Nitrification. This process involves the oxidation of nitrogen compounds to produce nitrates in the soil. Nitrates taken up by plants from the soil become incorporated into nitrogenous compounds such as amino acids and are passed through the various trophic levels in organic form. Such compounds return to the soil in faeces or contained in dead material and are then acted upon by decomposers. The decomposers carry out

a process called **deamination** which results in the release of ammonia. More direct recycling occurs through the excretion by organisms of urea or uric acid.

The 'pool' of ammonia in the soil can be used as an energy source for the fixation of carbon dioxide by specialized **nitrifying bacteria**. These are described as **chemoautotrophic** because they derive energy for food manufacture from chemical reactions rather than from the Sun. One group of nitrifying bacteria, including *Nitrosomonas* bacteria, oxidize ammonia to nitrites.

$$2NH_3 + 3O_2 \longrightarrow 2NO_2^- + 2H_2O + 2H^+ \\ + \text{energy}$$

A second group, which includes *Nitrobacter*, completes the oxidation to nitrates.

$$2NO_2^- + O_2 \longrightarrow 2NO_3^- + \text{energy}$$

Not all of the ammonium compounds and nitrates produced in this way are reused by living organisms. Some are permanently incorporated into sediments. Ammonium ions tend to be retained by the soil but nitrates are quickly lost in drainage water, sometimes reappearing as a pollutant in lakes and rivers.

3 Denitrification. Some of the nitrogen contained in nitrates is returned to the atmosphere through the biochemical activity of yet another kind of bacteria called **denitrifying bacteria**. These bacteria live in soil or water where oxygen is in short supply, and use NO_3^- as a source of oxygen for aerobic respiration. Some energy has to be expended in breaking down the nitrate ion, but there is a significant gain in overall energy yield compared with anaerobic respiration. Nitrogen gas is released as a waste product from this process.

The sulphur cycle (Fig. 62·4) differs from the carbon and nitrogen cycles in that its inorganic phase is mostly sedimentary, rather than atmospheric. A small amount of sulphur exists as sulphur dioxide gas (SO_2) in the atmosphere, as a result of the combustion of sulphur-containing substances, but most of the earth's reservoir is in the form of sulphur-bearing rocks such as iron pyrites.

Sulphur is made available to plants in the form of sulphate (SO_4^{2-}) ions which are produced by the oxidation of exposed and eroded rock surfaces. Most of this oxidation is biological and is carried out by specialized bacteria which derive energy from the process.

In plants, absorbed SO_4^{2-} ions are incorporated into the thiol (-SH) groups of amino acids and proteins. In this form, the sulphur passes through the various trophic levels, being released from living organisms only as a constituent of faeces. Decomposing bacteria break down the protein of dead organisms, reducing the -SH groups to H_2S (hydrogen sulphide). It is the presence of this gas which sometimes gives the characteristic odour of rotten eggs to decomposing matter.

The H_2S produced may be oxidised to SO_4^{2-} by certain bacteria but this is only possible under aerobic conditions. Specialized photosynthetic bacteria living in sulphur springs use H_2S instead of H_2O as a raw material in the manufacture of carbohydrate as shown in the equation:

$$6CO_2 + 12H_2S \xrightarrow[\text{light}]{\text{infra-red}} C_6H_{12}O_6 + 6H_2O + 12S$$

The sulphur produced by this reaction returns to the Earth's sediments.

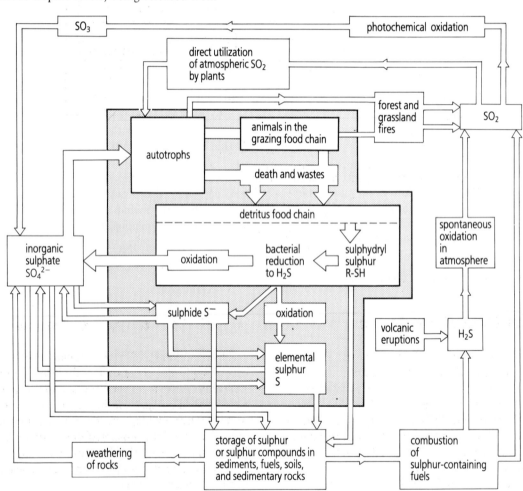

Fig 62·4 Flow diagram of the sulphur cycle

62·6 Phosphorus cycle ▪ ▪ ▪ ▪ ▪ ▪ ▪ ▪ ▪ ▪ ▪ ▪ ▪ ▪ ▪

Phosphorus enters food webs in the form of inorganic phosphates, PO_4^{3-}, HPO_4^{2-} or $H_2PO_4^{-}$ which are then built up into organic molecules such as nucleic acids, phospholipids, and ATP (See Fig. 62·5). When plants and animals die or excrete waste products, **phosphatizing bacteria** complete a simple nutrient cycle by releasing inorganic phosphate back into the soil. An unusually rich source of phosphorus comes from the accumulated droppings of sea birds on small islands, notably off the west coast of Peru. This material is called **guano** and in the past was extensively exploited as a fertilizer.

The completion of the biogeochemical cycle of phosphorus is a very slow process because of abiotic storage in the form of crystalline rocks which yield their contents only when brought to the surface by geological movements and subsequently eroded. The natural supply of phosphorus is supplemented by mining phosphorus-containing rocks for the manufacture of fertilizers. Leaching of phosphates from agricultural land increases the amount of phosphorus available in aquatic habitats and may stimulate the growth of plants, causing dramatic increases in algal populations, often with potentially damaging consequences for the rest of the ecosystem.

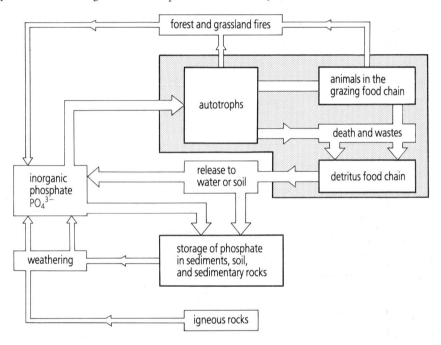

Fig 62·5 Flow diagram of the phosphorus cycle

62·7 Hydrological cycle ▪ ▪ ▪ ▪ ▪ ▪ ▪ ▪ ▪ ▪ ▪ ▪ ▪ ▪

The hydrological cycle or water cycle differs from the biogeochemical cycles described above because the cycling of a chemical compound is followed rather than that of a bioelement. Water is an essential component of all living organisms both as a solvent and as a medium for chemical reactions. Although it may undergo chemical change in living organisms, particularly in the process of photosynthesis where it is split into H^+ and OH^- ions, it is recycled in the abiotic environment almost entirely in the form of water molecules.

Figure 62·6 summarizes the major processes involved in the hydrological cycle. The oceans form the principal abiotic reservoir, containing about 97% of the Earth's water. The quantity of water available to terrestrial organisms depends on rainfall and also on the time taken for water to return to the atmosphere by evaporation, or to the oceans in streams and rivers. Human activities can have a significant effect on this drainage period. Irrigation methods increase the amount of time water spends on land and hence promote fertility in dry areas. City life, however, has the opposite effect: 'land water', consumed in vast quantities for industrial and domestic purposes, is usually channelled directly into pipes and sewers leading to the rivers and oceans.

All the nutrient cycles discussed in this Unit involve a balance between biotic and abiotic phases. Although vast quantities of substances are involved, slight changes in nutrient availability may have dramatic effects on ecosystems. Human activities can upset the natural balance of nutrient cycles, giving rise to a range of ecological problems some of which are considered in the final Unit.

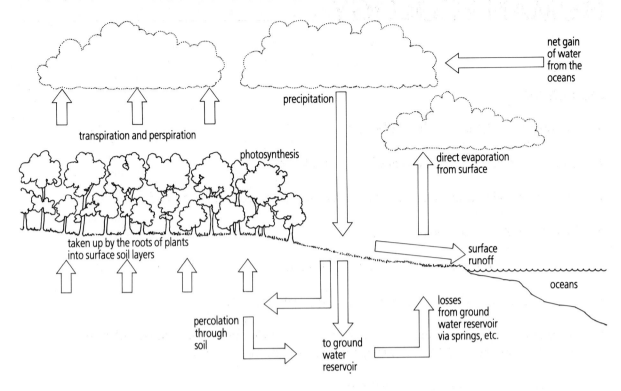

net gain
of water
from the
oceans

transpiration and perspiration

precipitation

direct evaporation
from surface

photosynthesis

surface
runoff

taken up by the roots of plants
into surface soil layers

oceans

percolation
through
soil

to ground
water
reservoir

losses
from ground
water reservoir
via springs, etc.

Fig 62·6 Summary of the hydrological cycle

63 HUMAN ECOLOGY

Objectives

After studying this Unit you should be able to:–

- Comment briefly on the causes of overpopulation

- List and explain some of the possible methods of increasing food production

- Give examples of soil erosion

- List and explain some of the agricultural techniques used to reduce soil erosion

- Give an account of the problems associated with irrigation

- Comment on the advantages and disadvantages of the following methods of pest control: pesticides, cultural control, biological control

- Outline some of the causes of water pollution

- List the most important sources of air pollution and comment on possible methods of pollution control

63·1 Introduction ■ ■ ■ ■ ■ ■ ■ ■ ■ ■ ■ ■ ■ ■ ■ ■ ■

The earliest human populations lived as hunter gatherers in ecological balance with their surroundings. They ate fruit and berries, plant roots and small animals and cooperated in family groups for hunting large prey. Predators, disease and limited supplies of food kept population numbers in check. In most parts of the world, the hunter gatherer life style gradually gave way to semi-agricultural societies with herding of domestic animals and deliberate cultivation of crops. Often, small areas of forest were cleared, used for growing crops for a few seasons and later abandoned. Provided cleared areas are given sufficient time to regenerate to their natural state, this form of agriculture causes no permanent damage to the environment.

Primitive agricultural societies consume all the food they produce: this is called **subsistence agriculture**. An important stage in the development of human societies was reached when agricultural techniques improved, providing a surplus of food which could be exchanged for weapons and tools. Permanent settlements arose and larger areas of land were claimed for agriculture. Colonization of new parts of the world

and gradual increases in human population continued for thousands of years, but, even in the most advanced societies, most people were occupied on the land. The industrial revolution triggered a massive increase in urbanization. Together with advances in medical science, this stimulated an explosive growth in human populations which continues today.

The threat to the environment posed by the increasing human population was first recognized by **Thomas Malthus** in his famous 'Essay on Population' published in 1798. Malthus was a professor of Economics and History at Cambridge University. He took as his example the United States which then offered seemingly unlimited scope for agricultural development and settlement. Estimating that the population would double every 25 years, in hindsight a fairly accurate assessment, he predicted that food supplies would become increasingly limited and that famine, war and disease would inevitably follow.

'Too many people for not enough food' is a familiar, but oversimplified view of what is termed **Malthusian overpopulation** today. Despite the fact that 40 000 people die each day as a direct

The Sheffield College

Hillsborough LRC
Telephone: 0114 260 2254

533

result of starvation or malnutrition, the global population shows a net increase of 2,000,000 per day and 90% of this increase is in the poorer countries of Asia, Africa and Latin America where food supplies are most limited. It must be remembered of course, that starvation in poor countries is more than partly the product of global economics, particularly trade tariffs and multinational exploitation, and should not be viewed as a simple relationship between the distribution of arable land and the number of hungry mouths.

In the developed nations the equation is even more complex. The populations of Western Europe and the United States show a relatively small net increase but have a huge impact on the environment. Food production technology with its attendant problems of urbanization, energy use and pollution require a revised view of Malthus's doom theory. The term **Neomalthusiasm overpopulation** has been coined to describe the condition of the technologically advanced nations. Clearly, there is a significant difference between the maximum number of people the Earth can support and what might be considered as the optimum population size. Any estimation of the gap between optimum and maximum numbers entails value judgements about the quality, as opposed to the quantity of life. What follows in this unit is a discussion of the criteria upon which such value judgements can be made.

63·2 Food production ■ ■ ■ ■ ■ ■ ■ ■ ■ ■ ■ ■ ■ ■ ■

Urban development almost invariably takes place at the expense of good agricultural land so that the remaining agricultural land must be made even more productive. Brazil for example loses an area of potential crop land the size of Wales every year through the growth of cities and towns. Intensive agricultural methods like those listed below are necessary if there is to be any hope of keeping pace with increased demand for food.

1 Fertilizers. The use of fertilizers dramatically increases crop yields and is necessary to maintain soil fertility. The most frequently required nutrients are nitrates, phosphates and potassium. After analysis of the soil, these are applied in proportions matched to the needs of the crop species.

While nitrogen can be fixed from the atmosphere in large amounts, the natural supply of phosphates and potassium, both obtained from rocks, is more limited. In addition, the manufacture and distribution of fertilizer is energy-intensive and relies heavily on the use of fossil fuels. Excessive application of fertilizers leads to pollution of lakes and rivers.

2 Pest control. It is estimated that as much as 45% of the world's food production capacity is lost through the activity of pest species, either as a result of disease, or because crops are eaten before harvesting or destroyed in storage. Infestations are usually combatted with synthetic pesticides derived from oil. Such chemicals are expensive to manufacture and are an important cause of environmental damage, as will be explained shortly.

3 Plant breeding. Many thousands of plant species are potential sources of food, but only about 80 species have been cultivated for human consumption. By far the most important are the cereal crops, wheat, rice and maize. Plant breeding has contributed to a dramatic increase in crop yields known as the **green revolution**. Benefits include resistance to disease and greater tolerance of temperature and rainfall variations. Growing seasons have been shortened to allow multiple harvests and the food value of some species has been increased by the development of strains with enhanced protein content and increased amounts of essential amino acids like lysine and tryptophan.

In spite of these successes, the green revolution has had a limited impact in many poorer countries because of the scarcity of good arable land, exposure to frequent droughts or floods and the expensive input costs of irrigation, fertilizers, pesticides and energy.

4 Farming of new species. In tropical countries, wild game are much more efficient at exploiting available food resources than cattle and sheep bred from European stocks. In addition, they are more resistant to disease and better adapted to withstand climatic extremes. Some efforts are now being made to farm these animals in wild game ranches, but long-term success depends on cultural changes in long-established tribal society. Similar possibilities exist for exploiting marginal land in developed countries, as for example in the Scottish highlands, where farming red deer may prove to be an economic proposition.

Significant long-term benefits in food production might be achieved by applying farming techniques to improving the yields from coastal fisheries. Commercial oyster beds and freshwater trout farms indicate some of the potential in this area.

5 Development of new arable land. Less than 50% of the land surface with agricultural potential has so far been cultivated. Each year new regions of tropical rain forest, desert and wetland are transformed into productive arable land. However, many large scale clearance, irrigation, drainage and reclamation schemes are doomed to failure and result in a massive loss of natural resources with only short-term rewards. The reasons for such failure will be discussed later in this Unit.

6 Factory farming. Battery farming of chickens and the rearing of pigs in pens make an important contribution to food production in developed countries and can be more efficient in energy terms than traditional farming methods. Such methods may be seen as unacceptable by many concerned with animal welfare and factory farming can pose the additional hazard of disease which spreads rapidly amongst closely confined animals.

A less contentious form of factory farming is the culture of organisms such as yeast, bacteria and algae to produce food with a high protein content. Significantly, a large production can be concentrated into a small space and cheap raw materials like paper pulp or cannery waste can be exploited. Food produced in this way is usually too expensive for poorer countries, but extracted substances are sometimes added to basic local diets. For example, the amino acid lysine which is deficient in cereal foods has been used to fortify bread in India and other Asian countries. Advances in biotechnology seem likely to give rise to new applications of this kind.

63·3 Soil erosion ■ ■ ■ ■ ■ ■ ■ ■ ■ ■ ■ ■ ■ ■ ■ ■

Soil particles are normally held in position by plant roots and humus. When land is cleared of its covering vegetation, wind and rain combine to carry away the loosened topsoil and its valuable nutrients. Dust particles enter the air or are washed into rivers, forming silt. These effects are particularly severe on hillsides and may result in **gully erosion** leaving deep scars where soil has been lost.

Tropical rain forests are vulnerable when clearance occurs because the topsoil is very shallow. Bacterial decomposition and recycling occur so quickly and so completely that very little humus is added to the soil. Instead, the vegetation absorbs nutrients from the shallow layer of litter which carpets the forest floor. When such a forest is cleared, the litter is liable to be carried off in streams and its nutrients are lost. This change is irreversible and prevents natural regeneration of the forest.

Drought and erosion from hill farms were the most important causes of famine in Ethiopia and Sudan in 1985–88 (Fig. 63·1). About 6 mm of topsoil is lost each year from the deforested central highlands and, as firewood becomes more scarce, animal dung is used for fuel so that even this supply of nutrients is lost.

A similar situation exists in Northern China where extensive hill farming over hundreds of years has resulted in severe gully erosion. Large quantities of silt are carried away in the Hwang Ho or Yellow River. In low-lying areas the sediment settles and may cause the river bed to rise by as much as 5 cm per year, so that monsoon rains cause extensive flooding of the fertile and densely populated plains. Dyke building hardly keeps pace with the rising river level and the major dam built in 1957 is already so clogged with silt that it is unable to supply electricity or to control the floods. It is easy to understand why the Hwang Ho is popularly known as 'China's sorrow'.

Preventing soil erosion

1 Contour farming. This involves ploughing and cultivation across the slope rather than up and down the hillside, a simple change which helps to reduce water loss. In **strip cropping**, the different crops are planted in alternating bands and the crops are rotated from year to year to make the most efficient use of nutrients. In peasant-farmed hill areas where the slope is steep, erosion control is improved by **terracing** (Fig. 63·2).

2 Reduced ploughing. The degree to which soil is exposed to erosion may be minimized by shallow ploughing or by planting crops through the existing vegetation using seed drills. This is possible for soya bean, citrus and certain cereal crops. Erosion can be reduced by a factor of 20 with a 20% saving in field preparation costs. However, an increased input of pesticides and weekillers is needed once the crop is established.

3 Wind-breaks. English hedgerows are a characteristic and picturesque feature of the landscape and are also vital for the conservation of

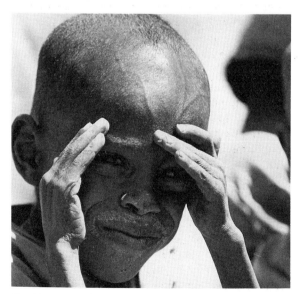

Fig 63·1 Famine in North-East Africa

Fig 63·2 Terracing to reduce soil erosion

soil, acting as wind-breaks to protect the fields. In the wake of the dust bowl disaster in the U.S.A, a programme of planting wind-breaks was begun in the Great Plains, but has now been discontinued. On a smaller scale, thousands of fast-growing eucalyptus trees are being planted in lines throughout the cotton-growing Pacific plains of Nicaragua. These serve a dual purpose, preventing excessive wind erosion and providing firewood for cooking, an important consideration since up to 70% of domestic timber is used for this purpose.

4 Gully reclamation. It is now recognized by the Chinese government that silting and flooding of the Hwang Ho river is best tackled at its source in the hills, rather than by expensive dams and dykes downstream. Gully erosion is checked by building small dams, as illustrated in Figure 63·3. Much of the silt is trapped behind these dams, eventually creating flat productive soil.

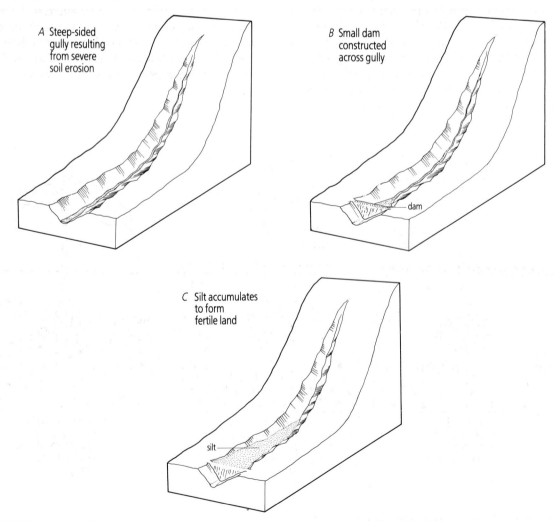

A Steep-sided gully resulting from severe soil erosion

B Small dam constructed across gully

dam

C Silt accumulates to form fertile land

silt

Fig 63·3 Gully reclamation

63·4 Irrigation ■ ■ ■ ■ ■ ■ ■ ■ ■ ■ ■ ■ ■ ■ ■ ■ ■

The Egyptian, Chinese, and Babylonian civilizations owed their existence to irrigation of river basins. The history of Babylonia, now modern Iraq, illustrates a fundamental limitation of virtually all irrigation schemes. What was once a fertile plain between the Tigris and Euphrates rivers is now mostly desert and swampland. Some of the reasons for this transformation are explained below.

When rivers are diverted to flow over arable land, water evaporates depositing a crust of salt at the soil surface (Fig. 63·4A). Water draining through carries some of this salt into the subsoil (Fig. 63·4B). After prolonged irrigation, the salinity of the soil is significantly increased and may cause damage to crop plants. Sometimes, the surface crust of salt is removed by temporary flooding of the land, but this tends to raise the water table of the soil and can expose plant roots to the even more saline conditions of the subsoil (Fig. 63·4C). Unless the soil can be successfully drained by underground pipes, only salt-tolerant plants can survive, and economic losses become so severe that the land is abandoned to its former state.

The Colorado river has nine major dams and new irrigation schemes are being opened. The Central Arizona Project was completed in 1985 at a cost of well over 1000 million dollars. Drains were laid to remove salty groundwater, dumping it back into the Colorado river near the Mexican border. Mexico understandably objected to the increased salinity of the river water and a further 200 million dollars was spent on a desalination

plant. Farmers are given substantial government subsidies to reduce the cost of irrigation water and are encouraged to conserve supplies, but it has become less and less profitable to grow crops on increasingly salty soil. Consequently, many farms have been sold to property developers. Phoenix has become one of the fastest growing cities in the West and water intended for irrigation is now used to supply gardens, swimming pools and golf courses.

Increased salinity is not the only problem in soil irrigation: some dissolved substances have toxic effects even in relatively low concentrations. In the highly productive San Joaquin valley of central California, irrigation water has washed out quantities of the trace element selenium from rocks at the base of the mountains. This substance causes abnormal development and death of bird embryos, a discovery only made when shortage of funds caused construction work to be halted on a new channel intended to carry drainage water into San Francisco bay. Instead, the water was emptied onto a wetland nature reserve, poisoning the area to such an extent that it is now classified as a toxic dump. It is interesting to note that unfavourable geological reports predicting this outcome were ignored by powerful commercial interests. Farmers must pay the consequences as drainage channels are closed and selenium floods back onto their lands. By 1990, half the irrigated land will be lost and the whole region may eventually revert to desert.

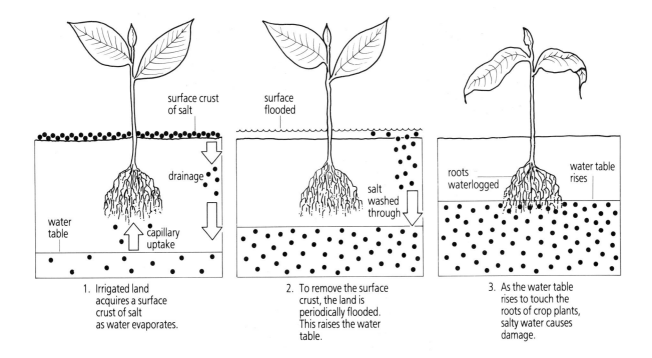

1. Irrigated land acquires a surface crust of salt as water evaporates.

2. To remove the surface crust, the land is periodically flooded. This raises the water table.

3. As the water table rises to touch the roots of crop plants, salty water causes damage.

Fig 63·4 Problem of increasing salinity in irrigated soil

Cultivation of a single crop species in a large area greatly simplifies food webs and reduces the natural diversity of plants and animals. Such **monocultures** are particularly liable to disease and to attack by pest species, which may multiply freely without competition and in the absence of natural predators. New pest species are often accidentally introduced from other areas as communication and transport systems develop. The need for efficient pest control increases rather than diminishes with improved technology.

Pesticides

Before 1940, most pesticides were simple poisons like lead arsenate, suffocants such as kerosene, or natural substances like nicotine or pyrethrum extracted from plants. After the war, new pesticides were artificially produced from petrochemicals (Table 63·1).

One of the most widely used pesticides was DDT (**d**ichloro**d**iph**y**en**ylt**richloroethane). Initially, DDT had a dramatic effect on insect pests and raised hopes for the elimination of insect-transmitted diseases such as malaria. However, continued use of DDT and other persistent insecticides revealed a number of major problems. DDT is an extremely stable compound which can remain in the soil for up to 25 years. It is almost insoluble and difficult to excrete so that it tends to be deposited in fatty tissues. Consequently, it accumulates in the bodies of consumer organisms.

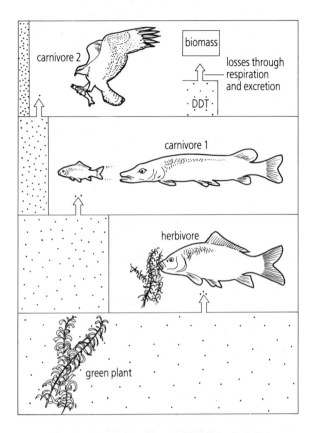

Fig 63·5 Increasing concentrations of DDT in food chains

Pesticides acquired from the environment are passed on to organisms at higher trophic levels and are often most concentrated in the tissues of top consumers (Fig 63·5). Such massive amounts of

Table 63·1 Chemical pesticides

GROUP	EXAMPLES	EFFECTS	COMMENTS
Insecticides	inorganic compounds	enzyme inhibitors	persistent
	lead arsenate		
	plant extracts	contact poisons which penetrate insect cuticle	easily degraded to harmless substances
	pyrethrum, nicotine		
	chlorinated hydrocarbons	toxic action on nervous system	very persistent, accumulate in fatty tissues, concentrated in food chains
	DDT, BHC, dieldrin, heptachlor, endrin		
	organophosphates	inhibit action of cholinesterase enzymes at nerve synapses	very toxic but easily degraded to harmless substances
	parathion, malathion		
Herbicides	2,4-D	synthetic auxin, rapid growth leads to death	selective against broad-leaved plants
	paraquat	metabolic poison	non-selective
Rodenticides	warfarin	anticoagulant, causes internal haemorrhage	relatively safe to use
	sodium fluoroacetate	hyperstimulates nervous system, causes heart failure	persistent, concentrated in food chains

DDT were used that the substance is present in all parts of the environment and can even be detected in the body fat of Antarctic penguins and seals. Its concentration in human fatty tissues may be as high as 90 parts per million according to locality.

Persistent pesticides like DDT tend to be indiscriminate or **broad-spectrum** in their effects, killing both pests and their natural predators. Long term use can actually increase the population numbers of the target organism. Soil structure is also damaged as a number of important decomposers and aerators are killed.

Ironically, many of the pest species against which DDT was used are now resistant to the poison so that newer and more powerful pesticides must be used. Modern pesticides are slowly broken down or degraded to form harmless substances in the soil. Inevitably, they are more expensive and need to be applied more often. No pesticides are selective enough to kill the pest species without damage to beneficial species such as natural predators and insect pollinators. Unfortunately, world food production would fall by at least 30% if the use of pesticides was discontinued. Damage to the environment will continue until realistic alternatives can be found.

Long term effects of substances like DDT on the human body are not fully understood but it is likely that they include an increase in the incidence of cancers. By contrast, it is well known that weedkiller byproducts can cause genetic disorders, particularly a substance called TCDD, a **dioxin** which is released in the manufacture and use of 2,4,5-T. In 1976, an explosion in a chemical factory in Seveso, Italy, released about 1800 grams of pure TCDD into a densely populated area causing the destruction of thousands of domestic animals as well as human birth defects. 2,4,5-T was used in combination with 2,4-D in a defoliant called 'agent orange' in the Vietnam war. In 1967 a clear link was established between the occurrence of abnormal births and TCDD from agent orange. As a result, the defoliation programme was discontinued, but not until 1970, following incalculable damage, not least to the ecology of South East Asia.

Cultural control

Cultural control implies a change in agricultural techniques rather than a direct attack on pest species. For example, crop rotation reduces the numbers of insects specialized to feed from a single plant species. Sometimes, pest numbers can be reduced by **intercropping**, in which two different crops are grown in alternating strips. It was found that damage to maize crops caused by the larvae of the corn borer moth, *Pyrausta nubialis* could be reduced by up to 80% by intercropping with peanuts, possibly because peanut vegetation provides shelter for the insect's natural predators. Alternatively, parts of a field may be planted with a **trap crop**. Alfalfa, for instance, is especially attractive to some species of plant bugs which otherwise destroy cotton. When the cotton is harvested, the patches of alfalfa are either burnt or treated with a chemical insecticide.

Biological control

When pest species have been accidentally introduced from other countries, a possible method of control may be to introduce the pest's natural predators or diseases. For example, adult and larval ladybirds feed on greenfly and are often introduced into greenhouses. The wasp, *Apanteles rubecula*, lays its eggs inside the bodies of cabbage worm larvae which are later eaten alive by the larvae of the wasp. The viral disease **myxomatosis** is a natural disease of some South American species of rabbits and was introduced with great effect to control rabbit populations in Australia. However, resistant strains of rabbits have evolved and populations have shown a recovery so that alternative control methods must be found.

Female screwworm flies lay their eggs under the skin of cattle, causing festering wounds as the larvae hatch out to feed on the flesh. Population numbers can be controlled by releasing males sterilized by exposure to a radioactive source. These mate with normal females which cannot reproduce because they mate only once. If large numbers of sterile males are released over several seasons, screwworm flies become locally extinct.

Another form of biological control depends on the synthesis of insect hormones or sex pheromones which can be used to attack pest species selectively. For example, a synthetic pheromone 'gyplure' has been used to attract male Gypsy moths to traps, thereby reducing the numbers of moth larvae infesting forest and fruit trees.

Biological control methods have the significant advantages of selective action against pests and minimal damage to the environment. However, complete elimination of a pest species is rarely possible so that some level of damage to crops must be tolerated.

63·6 Water pollution ■ ■ ■ ■ ■ ■ ■ ■ ■ ■ ■ ■ ■ ■

A substance becomes a pollutant when it is present in concentrations damaging to living organisms. Some pollutants, like pesticides, are poisons but others are natural substances which are sometimes present in excessive amounts.

Toxic waste

Chemical works, oil refineries, steel plants and paper mills are among the major sources of pollution in British rivers. Much of this industrial activity has been transferred to estuaries or to the coast, partly for easier access to shipping and partly to avoid the provisions of the 1964 *Pollution Act* which applies only to inland waters. The estuaries of the Mersey, Humber and Tees reveal the effects of inadequate pollution control. They are almost devoid of living organisms because of the discharge of toxic materials such as ammonia and cyanides together with organic waste from sewers. Factories manufacturing the paint pigment titanium dioxide dump 60 000 m^3 of acid iron-containing waste into the Humber estuary each day, staining coastal waters reddish-brown as far as Cleethorpes and poisoning crabs, barnacles and fish.

It is more difficult to assess the extent to which toxic waste from underground dumps finds its way into water supplies. When waste is dumped in shallow landfills without adequate geological research, containers may corrode allowing toxic chemicals to be carried off in drainage water. This is a particular problem in the United States where around 50% of domestic drinking water is supplied from undergound wells.

Thermal pollution

Coal, oil-fired and nuclear power stations generate electricity by heating water to form steam which is used to drive steam turbines. Most power stations are located near lakes or rivers or on the seashore because large amounts of cooling water are needed to recondense the steam. For a 1000 megawatt generating station, the minimum flow is about 50 m^3 per second. When cooling water is discharged into lakes and rivers, the subsequent rise in temperature stimulates bacterial growth and tends to reduce the amount of oxygen dissolved in the water. At the same time, the oxygen requirements of living organisms are increased as a result of a higher metabolic rate. These effects may be fatal for fast-swimming freshwater fish and interfere with normal spawning and migration behaviour. Increased flow rates of cooling water reduce but do not eliminate thermal pollution.

Sediment

Sediment from soil erosion can destroy the bottom-living community of a river and limits the penetration of light so that photosynthesis and primary production are reduced. In turn, this affects the whole of the aquatic ecosystem.

Fertilizers and sewage

Most newly-formed lakes are nutrient-poor or **oligotrophic**. However, over thousands of years, the availabilty of nutrients increases as minerals and organic materials are progressively carried in by streams and rivers and by drainage from the surrounding countryside. Mature lakes are typically nutrient-rich, or **eutrophic**, and support a diverse community including aerobic bacteria and other decomposers which help to recycle organic matter.

Small increases in organic material accelerate the natural process of eutrophication, but large

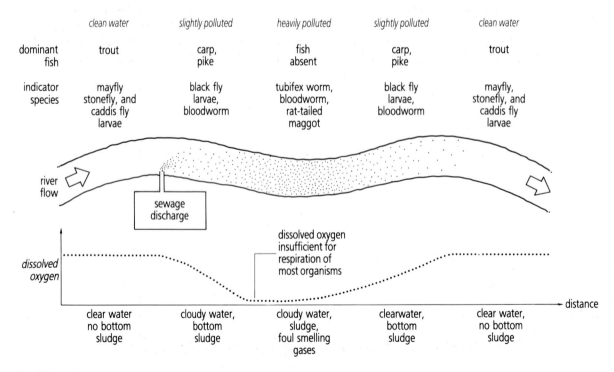

Fig 63·6 Indicator species and river pollution

increases tend to stimulate the growth of decomposing bacteria which use up most or all of the available oxygen, causing the deaths of other heterotrophic organisms. This can happen through the discharge of raw sewage but is also the result of increases in the availability of inorganic nutrients such as nitrates and phosphates. The increased supply of nutrients stimulates the growth of aquatic plants and photosynthetic algae which become so numerous that the water is coloured green. Sewage treatment works remove only about 50% of the nitrogen and 30% of the phosphorus in domestic waste. Most of the phosphorus is derived from biodegradable household detergents. These have eliminated the old problem of foaming rivers but it will be necessary in future to develop methods of removing phosphorus from sewage outlets. A better alternative would be increased availability of phosphorus-free detergents.

In a river, the first indication of pollution is often a change in animal populations, as illustrated in Figure 63·6. 'Clean water' species such as stonefly and mayfly nymphs and caddis fly larvae disappear and are replaced by low oxygen users such as *Tubifex* worms. It is possible to determine the level of pollution more exactly by estimating the **biological oxygen demand** (B.O.D.) of a water sample. This is the amount of oxygen used up by decomposers in a given volume of water incubated in the dark at 20°C for 5 days. Oxygen concentrations are measured before and after incubation in parts per million. If the pollution is not too severe only a comparatively short stretch of the river may be affected. However, spawning fish may need to pass through the polluted zone. It is an indication of the tremendous success of anti-pollution measures taken by the Thames Water Authority that salmon have recently been recorded in its upper reaches.

63·7 Air pollution ■ ■ ■ ■ ■ ■ ■ ■ ■ ■ ■ ■ ■ ■ ■

The most important atmosperic pollutants are listed in Table 63·2. Some of the reasons for heavy pollution and possible control methods are outlined below.

Smog

Smog is a mixture of smoke and gas emissions dispersed in fog. It is most often formed as a result of a **temperature inversion** in the lower atmosphere when a layer of cool air is trapped beneath a layer of still warm air. Some areas are more liable to temperature inversion than others. If combustion engines were 100% efficient, car exhausts would produce pure carbon dioxide and water as waste products. In reality, up to 200 other compounds are released. **Benzopyrene**, a cancer-producing substance is formed as petrol evaporates, and waste hydrocarbons react with nitrogen dioxide in the presence of oxygen and sunlight to produce **photochemical smog**. This phenomenon was first recognised and named in Los Angeles where unique climatic and geographical conditions often combine to hold a yellow cloud over the city. Apart from aesthetic effects it causes an increase in the amount of ozone in the lower atmosphere which irritates the eyes of pedestrians on bad days. Ozone enters the leaves of plants turning them brown and it rots rubber tyres so seriously that they are now specially impregnated to resist ozone damage. More seriously, it increases the acidity of rain. The amount of ozone close to the ground has roughly

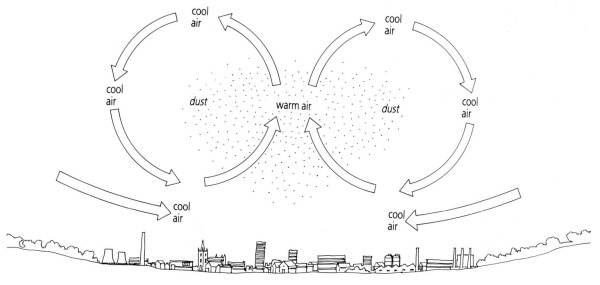

Fig 63·7 Formation of a dust dome

Table 63·2 Major atmospheric pollutants

POLLUTANT	MAJOR SOURCE	EFFECTS
carbon monoxide	natural methane oxidation, car exhausts, burning of fossil fuels	combines with blood haemoglobin to produce asphyxia
carbon dioxide	combustion of fossil fuels	possible 'greenhouse effect' (see Unit 62)
hydrocarbons	combustion of oil and petrol	carcinogenic
sulphur dioxide	coal-fired power stations	stinging eyes, lung damage, acid rain
nitrogen oxides	car exhausts	asphyxia, pneumonia
dust particles	industrial chimneys, car exhausts, volcanic eruptions	toxic effects, lung damage
radioactive isotopes	small quantities from nuclear waste, nuclear accidents	carcinogenic
chlorofluorocarbons	propellant in aerosols	destruction of ozone layer

doubled over Europe in the past 30 years due mainly to motor exhausts.

Dust particles

Sometimes pollutants are concentrated over large cities in a **dust dome** (Fig 63·7). As warm air from the city rises, cool air is drawn in from the surrounding countryside and circulates in local convection currents. In this way, dust particles are held above the city.

The effect of dust domes on local climate can be quite dramatic. For instance, it is estimated that Washington DC receives 10% less sunlight than at the beginning of the century due to the shielding effect of atmosperic pollution. In Paris, it was found that 31% more rain falls during working days than at weekends or on public holidays. This is thought to be the result of increased discharge of water vapour, soot and dust on working days: these act as condensation nuclei to make rain.

Dust particles generally distributed in the atmosphere prevent the penetration of light and absorb heat from their surroundings, causing the air to cool. Volcanic eruptions can send millions of tons of dust into the atmosphere and can produce significant changes in the weather. The cooling effect of the eruption of Krakatoa in 1883 was so severe that snow fell in Boston in June. The eruption of Mount St Helens in 1980 produced detectable if less dramatic effects.

Potentially the most serious effect of a nuclear war would be to lift massive amounts of dust into the atmosphere. It has been calculated that even a 'moderate' nuclear exchange would almost completely obscure sunlight from the Northern Hemisphere so that temperatures would fall to between −15°C and −20°C. The land surface would cool quickly and bodies of freshwater would freeze, but the oceans would remain warm for some time, giving rise to temperature differences which would cause storms of unbelievable violence.

Animal and plant species would become extinct, although bacteria and related forms might survive.

Acid rain

Sulphur dioxide from coal and oil-fired power stations and nitrogen oxides from car exhausts react with water and oxygen in the atmosphere to form acids. Unpolluted rain is slightly acid (pH 5.6) due to dissolved carbon dioxide, but polluted rain may be very acidic. For example, the average pH of rain in Pennsylvania and Ontario in 1979 was 4.2, while in Los Angeles, rain from smog heavily laden with ozone may reach pH 3·0.

The effect of acid rain on terrestrial and aquatic ecosystems can be severe. In Scandinavia, which receives a high proportion of its air pollution from Britain, once productive lakes are now completely devoid of fish. In addition to its direct effect on the pH of the water, acid rain allows metal ions such as aluminium, which is highly toxic to fish, to be leached from the soil. Acid rain inhibits tree growth, especially of spruce and pine, and reduces the activity of nitrogen-fixing bacteria.

The upper atmosphere

Combustion of fossil fuels produces increasing amounts of carbon dioxide and other gases which diffuse into the upper atmosphere. This gas layer, which has been increased by about 20% in the last 150 years causes a **greenhouse effect** allowing sunlight to pass through but reflecting heat radiation back to the earth. There is much controversy about the warming effect this has. An extreme view predicts a rise in average world temperatures of up to 3°C over the next 50 years, with a 1.65 m rise in sea levels as the polar ice caps melt. It is likely that the cooling effect of dust particles will counteract this effect to some degree and meteorological research is far from complete on this issue.

Another contemporary concern is the ozone layer. Whilst ozone near the ground can cause damaging effects, its presence in the upper atmosphere is essential to life on earth because, as described in Unit 43, it blocks harmful ultra-violet radiation. Ozone is constantly being formed in the stratosphere due to a reaction between U.V. radiation and oxygen but it is being broken down at a faster rate by human activities. Nitrogen oxide emitted from high-flying aircraft reacts with ozone to produce oxygen, but present concern centres more on chlorine. Chlorine acts as a catalyst in the conversion of ozone to oxygen.

It is released into the atmosphere from compounds known as **chlorofluorocarbons**, or CFCs, which are used extensively in cooling systems, in the manufacture of foam plastic and, to a declining extent, in aerosol sprays. The main cause for alarm is the length of time, up to 100 years, that chlorine can survive in the atmosphere and the fact that holes in the ozone layer seem to be formed in a non regular way. This has been highlighted recently by the discovery of a large hole in the layer over Antarctica.

Radiation

So far the overall level of 'background radiation' has been little affected by human activities. Local increases in radiation may be associated with disposal of nuclear waste but it is doubtful whether these have a significant effect. On the other hand, nuclear war and nuclear accidents pose a real threat to the environment.

In the event of a nuclear accident, the main risks arise from inhalation of contaminated particles, fallout of radioactive isotopes on soil with subsequent incorporation into food, and contamination of water supplies. The likely consequences of serious accidents have been studied using computer modelling programs since the 1970s. The effects of fallout can be predicted for distances of up to 1000 km, given estimates of the height of dust plumes, the duration of release, and meteorological and population data.

A case study is provided by the nuclear accident which occurred on 26th April 1986 at Chernobyl in the USSR. As a result of a high plume and the absence of a temperature inversion, the nearby town of Pripyat was said to have suffered fewer than 100 acute radiation casualties, in spite of delays in warning and evacuation. (The speed of evacuation is critical because the effects of inhalation are cumulative.) Fortunately, by the time the plume reached Kiev four days after the initial explosion, most of the highly toxic caesium isotope ^{137}Cs had been lost. Without this delay, computer estimates suggest that up to 100 000 excess cancers would have resulted over a fifty year period. As it is, radiation doses to people in the populated areas of the Ukraine may have been

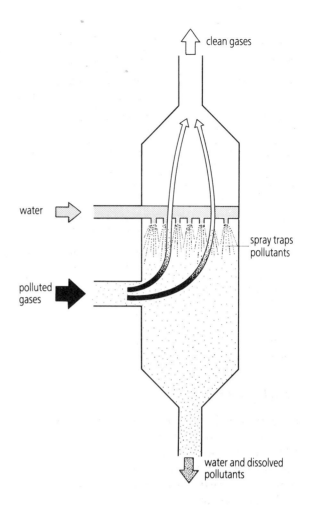

Fig 63·8 *Removal of sulphur dioxide by a wet scrubber*

considerable and predictions of about 10 000 genetic or cancer casualties are probably low.

Effects outside the USSR will be determined by the amount of radioactive material deposited and its transfer through food chains.

Control of air pollution

The technology is largely available for effective control of air pollution. Emissions of dust can and should be controlled by filters which remove solid particles before gases are discharged. Sulphur emissions from coal-fired power stations can be reduced by using low sulphur fuels or by extracting sulphur-containing impurities before use. Alternatively, chimneys can be fitted with 'wet scrubbers' which remove 80–95% of sulphur dioxide gas (Fig 63·8). Once the initial investment is made, these provide a commercially worthwhile source of sulphuric acid. Scrubbers should also be fitted to incinerators which burn plastic waste to prevent the release of toxic substances which may form if the operating temperature is too low.

The control of emissions from car exhausts is another area in which improvements can be made. The use of lead-free petrol and catalytic convertors would reduce toxic emissions to a fraction of their present level and prevent damage to the environment.

In the diagram:
clean gases

water

spray traps pollutants

polluted gases

water and dissolved pollutants

Section E Questions ▪ ▪ ▪ ▪ ▪ ▪ ▪ ▪ ▪ ▪ ▪ ▪ ▪

Short-answers and interpretation

1 Here are three pyramids of numbers from different food chains.

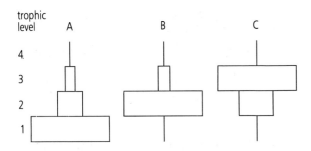

a i Name the four trophic levels in pyramid *A* and give a typical example of an organism found at each level (the organisms you suggest should all come from the same food chain). [2]
ii Account for the shapes of pyramids *B* and *C*. [2]
b the following pyramids were estimated for a single food chain in a terrestrial ecosystem.

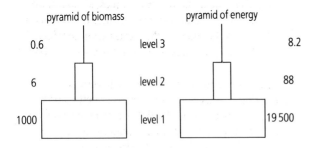

i In what units might the figures for each pyramid be expressed? [2]
ii Calculate the energy content per unit biomass of organisms in levels 1 and 3. [2]
iii Calculate the percentage energy flow from level 2 to level 3 in this ecosystem. Give one reason why this value is not 100% [2]
c The table below relates the number of fish per unit area to the annual rate of population growth. An annual rate of population growth of 1.0 signifies no net change in population.

Number of fish per unit area in year 1	Annual rate of population growth
10	0.5
15	0.75
20	1.0
25	4.3
30	5.0
35	3.9
40	1.5
45	0.1

i In each case, calculate the number of fish per unit area expected at the start of year 2. [1]
ii Consider the line in the table with the highest number of fish per unit area in year 2. How many fish could be harvested in year 2 in order to maintain the same annual rate of population growth in year 3? Explain your answer. [2]
iii Suppose there are 40 fish per unit area, of which 25 are harvested prior to reproduction. What would be the annual rate of population growth as a result of this cropping activity? [1]
iv What would be the likely effect of continuing this level of cropping activity on the future of the population? [1]

Joint Matriculation Board

2 A suspension of *Saccharomyces* (yeast) cells was added to a dilute sucrose solution at 25°C. The mixture was gently agitated for 16 days and 10 cm³ samples were withdrawn each day and the number of organisms counted. On the fifth day a small quantity of a culture of the ciliate *Paramecium* was added. The results of the experiment are shown below.

Days	Number of organisms in 10 cm³ samples		Days	Number of organisms in 10 cm³ samples	
	Yeast	*Paramecium*		Yeast	*Paramecium*
1	20	–	9	222	72
2	84	–	10	218	138
3	224	–	11	120	162
4	264	–	12	84	96
5	266	30	13	180	54
6	224	150	14	178	90
7	114	168	15	120	144
8	154	76	16	60	120

a Plot a graph to show the number of organisms of *Saccharomyces* (yeast) and *Paramecium* in the sample during the period of the experiment. [6]
b State two factors that might be responsible for the difference in population growth of the *Saccharomyces* (yeast) between days 2 and 3, and days 4 and 5. [4]
c Outline the likely causes of the changes in the *Paramecium* population between **i** days 5 and 6, and **ii** days 7 and 8. [4]
d Outline a likely cause of the change in the *Saccharomyces* (yeast) population between days 7 and 8. [2]

University of London School Examinations Board

3 The data below refer to the growth of a maize crop from 0.4 hectare of land.

Total dry mass of crop at harvest	5600 kg
Total ash content of these plants	290 kg

a What is meant by the term 'ash content'? [1]
b Name three substances likely to be present in the ash. [3]

c i Calculate the total mass of organic compounds produced by the crop per hectare. [1]

ii The equivalent mass of these organic compounds in terms of glucose was calculated to be 8070 kg. If the energy required for the synthesis of 1 kg of glucose is 15 960 kJ, how much energy was required to synthesize the 8070 kg glucose equivalent? [1]

d i If the total energy available per hectare was 8568×10^6 kJ, calculate the percentage utilization of the available energy by the maize plants. [3]

ii State three reasons why the utilization of available energy was so low. [3]

iii Suggest three ways by which greater efficiency of utilization might be achieved. [3]

Oxford and Cambridge Schools Examination Board

4 The table shows two soils *A* and *B*, which form permanent pasture (grassland) in areas of the United Kingdom similar in climate. For some years they have been spread at the same times with the same amounts of cattle dung and inorganic fertilizer.

	soil *A* %	soil *B* %
Particle diameters greater than 0.02 mm (as percentage of soil-particle volume)	83	13
Particle diameters less than 0.02 mm (as percentage of soil-particle volume)	2	47
Pore spaces (as percentage of total soil volume)	32	51

a i What type of soil is *A*? [1]

ii What type of soil is *B*? [1]

b How would you expect soils *A* and *B* to compare in their mean annual contents of **i** inorganic fertilizer and **ii** humus? Give one reason for each answer. [2]

c Soil *B* may be described as 'cold and heavy'. Suggest an explanation for each of these characteristics. [2]

Associated Examining Board

5 The following diagram shows the relationship between pH and the availability to plants of inorganic nutrients in soils.

a Which nutrients become less freely available in alkaline soils? [1]

b At which pH are the majority of nutrients most freely available? [1]

c The presence of calcium ions in soils makes potassium and phosphorus ions less soluble. Ling (*Caluna vulgaris*) has high demands for both potassium and phosphorus ions. In which soil conditions would you expect to find this plant? [2]

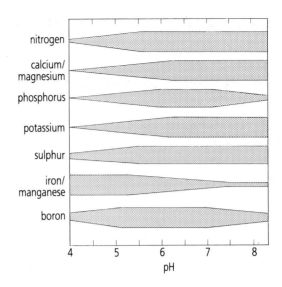

d Why do some plants growing on limestone or chalk show chlorosis? [2]

e Why are acid peat moorland soils considered poor agriculturally? [3]

f Why do many soils in conditions of high rainfall become more acid? [2]

g i How does clay contribute to the fertility of a soil? [2]

ii Why does pure sand have poor fertility? [3]

h Sulphur is said to be seldom deficient in soils in industrial areas. In some places, although not added as a fertilizer, it is increasing in quantity in the soil. What may be the explanation for this? [2]

University of London School Examinations Board

6 The histograms on the next page represent catches of various species of ground beetle along a transect in British woodland during the period from May to October. This transect passed successively through open woodland, a clearing and dense woodland.

a Compare the distribution of beetle species *A*, *B* and *C*. [6]

b Suggest reasons for the distribution of beetle species *D*. [2]

c Briefly describe a suitable procedure that could be used in a laboratory-based experiment to determine whether light intensity is an important factor in deciding the distribution of any one of these species of beetle. [3]

d Name two environmental factors, other than light, which might be operating along the transect to affect the distribution of the beetles. [2]

e An experiment was carried out to test the hypothesis that the use of a pesticide did not affect the number of beetles. The number of beetles captured in a test area treated with pesticide was compared with the number of beetles captured in a similar but untreated area where no pesticide was used. Sampling was carried out over a period of several months. The results shown in the table were obtained.

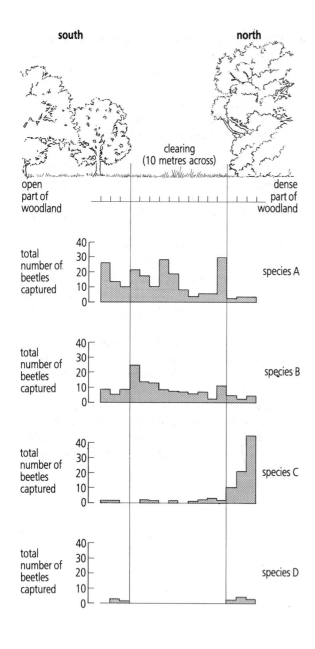

south   north

clearing
(10 metres across)

open
part of
woodland

dense
part of
woodland

total number of beetles captured — species A

total number of beetles captured — species B

total number of beetles captured — species C

total number of beetles captured — species D

Month	Number of beetles captured		Differences
	Treated area (O)	Untreated area (E)	(O–E)
1	410	400	
2	390	420	
3	325	350	
4	295	320	
5	310	330	

i Copy this table and complete the differences column by calculating the difference values (O–E) [1]

ii Using the following formula, calculate the value of Chi-squared (χ^2). Show your working.

$$\chi^2 = \Sigma \frac{(O-E)^2}{E}$$

[3]

iii A value for χ^2 of 9.49 would indicate a 5% chance, and a value for χ^2 of 5.99 a 20% chance that the original hypothesis is true. Using your calculated value for χ^2 discuss whether the pesticide has a significant effect on the number of beetles. [3]

University of London School Examinations Board

7 A factory emitting smoke containing sulphur dioxide was sited in a rural district. The table below gives distances and directions of **i** number of lichen species and **ii** sulphur dioxide concentration in the atmosphere at different directions from the factory chimney.

Distance from factory in a south, south west direction/ miles	1	2	4	8	12	16	28
Number of lichen species	0	1	2	3	7	9	14
Sulphur dioxide concentration/ parts per million	28	27	26	23	19.5	16	2
Distance from factory in a north, north east direction/ miles	1	2	4	8	12	16	28
Number of lichen species	1	2	3	4	4	5	5
Sulphur dioxide concentration/ parts per million	27	26.5	25	24	23	22	19

a Plot the information to show the relationships between the lichen distribution and the sulphur dioxide concentration using the same X-axis and two Y-axes, the one on the right for sulphur dioxide concentration and the one on the left hand for number of species. [12]

b Explain the difference in results between those obtained for the south, south west direction and those obtained for the north, north east direction. [2]

c Fully describe one example of evolution in action, e.g. industrial melanism, and its significance in the study of natural selection. [14]

University of Oxford Delegacy of Local Examinations

Essays

1 Write an essay on 'The flow of energy through an ecosystem'. [20]

University of London School Examinations Board

2 a Explain what is meant by the terms parasitism, mutualism and predation, indicating, with the help of suitable examples, how they differ from one another. [12]

b To what extent is competition important in regulating the size of a population? [6]

Cambridge University Local Examinations Syndicate

3 Discuss the ways in which the size of a population of either animals or plants may be regulated. [20]

Oxford and Cambridge Schools Examination Board

4 An ecologist walking along a valley in Britain in summer suspected that there was a difference between the vegetation growing on its north-facing and south-facing sides.

    a i Describe how it could be established that a significant difference did exist.

    ii List four factors that might account for the difference and explain how each would influence the vegetation. [14]

    b If the ecologist returned to the same valley in winter, what changes in flora and fauna might he or she expect to observe? Account for such changes. [6]

Joint Matriculation Board

5 In a field of grass grazed by cattle the distribution of daisies (*Bellis perennis*) appeared to be restricted to a public footpath. When the daisies were in flower the path stood out as white against the surrounding grass.

Describe how you would investigate two possible factors accounting for this distribution of daisies. [20]

Associated Examining Board

6 Write an essay on the role of saprophytes in the recycling of matter. [20]

Associated Examining Board

7 a What is an ecosystem? [3]

    b Describe the flow of energy and the cycling of carbon and nitrogen in any named ecosystem. [12]

    c Suggest reasons why the felling and removal of forest trees result in changes in the levels of nutrients in the soil. [3]

Cambridge University Local Examinations Syndicate

8 Describe the sequence of events in which a molecule of carbon dioxide expired by a herbivore might become incorporated into a molecule of glycogen in another animal of the same species. [20]

University of London School Examinations Board

9 Write an essay entitled, 'The biological consequences of pollution'. [20]

Welsh Joint Education Committee

10 Discuss the statement that: 'Technology enables man to produce a total world food surplus and yet there is starvation in many countries'. [20]

Associated Examining Board

INDEX-GLOSSARY

Simple definitions of basic terms are in italics. See page 2 for guidance on the use of the index-glossary.

A

Abduction, 223

ABO antigens, 149

Abomasum, 101–102

Abscess, 146

Abscisic acid, 291, 293, 294

Absorption [*Uptake of substances into cells or tissues*]
large intestine, 97
plant roots, 280–282
plasma membrane, 35–36
small intestine, 88, 94–96
see Gas exchange

Absorption spectrum, of chlorophyll, 72–73

Accommodation [*Process of changing the shape of the lens of the eye to focus objects at different distances*], 191

Acetylcholine (ACh), 176, 186, 210

Acetyl CoA, 64–65, 106

Achene, 308

Achondroplasia, 334

Acid [*Chemical compound which releases hydrogen ions in solution*], 8

Acid rain, 542

Acoelomate [*Animal which lacks the type of body cavity called a coelom*], 423

Acraniata [*Chordates with no skull or vertebral column*], 446–447

Acromegaly, 203

Acrosome, 242

Actin [*Contractile protein found in muscle fibres and associated with cell movements*], 44, 209, 324, 406

Actinia, 419

Action potential [*Electrical event associated with nerve impulses and muscle contraction*], 210

Action spectrum, photosynthesis, 72–73

Activation energy [*Energy required to trigger some type of chemical reaction*], 24, 62

Active centre, of enzyme, 23–24

Active immunity, 148

Active transport [*Transfer of substances across cell membranes involving energy release by the cell*], 36, 95, 97, 155, 175, 284–285

Adaptation [*Process or feature making an organism better suited to its environment*], 347
aquatic habitats, 515
colonization of land, 414, 449, 467

chordates, 449
insects, 440
parasites, 421
see Evolution

Adaptive radiation, 356, 361

Adduction, 223

Adenine, 47

Adenohypophysis [*Anterior lobe of pituitary gland*], 199–200

Adenosine [*Compound of adenine with the 5-carbon sugar ribose*], 47

Adenosine monophosphate (AMP), 47

Adenosine triphosphate *see* ATP

Adenyl cyclase, 204

Adipose cells, 96, 105, 163, 164

Adrenal gland, 200, 202

Adrenalin, 106, 131, 166, 186, 200–201

Adrenocorticotrophic hormone (ACTH), 200, 202, 253

Adsorption [*Attachment of substances to external surfaces*]
ions to clay particles, 513

Adventitious roots, 270, 299, 473

Aerobic respiration [*Respiration using oxygen*], 61–64, 101

Aesculus, 483

Afferent [*adj 'leading towards'*]

Afferent branchial artery (towards gills), 112, 123

Afferent neurons (towards CNS) *see* Sensory neurons

Agaricus, 463

Agnatha, 446

Agonistic behaviour, 234

Agranulocyte, 133–134, 145

Agriculture, 492, 533

AIDS (Acquired Immune Deficiency Syndrome), 144, 401

Air pollution, 541–542

Alarm response, 201–202

Alarm signal, 234

Albinism, in mice, 345

Albumin, 106, 131, 134–135

Aldose sugar, 12

Aldosterone, 158, 200, 202

Algae [*Photosynthetic protoctists*], 451–458
economic importance, 458
life cycles, 454–455

Alimentary canal, 88
see Digestion and digestive systems

Allantois, 251, 252

Allele [*One of two or more alternative forms of a gene which may occupy the same locus on a chromosome*]

dominant/recessive, 329
frequencies, 350
multiple, 333
mutant, 333

Allopatric speciation, 362

Allosteric site, 26

Alternation of generations [*Plant life cycle with separate sporophyte (diploid) and gametophyte (haploid) generations*], 454

Altruistic behaviour, 373

Alveolus (*pl* alveoli), 116

Amacrine cell, of eye, 193

Ameloblast, 90

Amensalism, 500, 503

Amino acid [*Chemical subunit of all proteins*]
essential/non-essential, 86
general formula, 18
peptide bond, 19
pH buffer, 19
regulation in blood, 106
– R group classification, 18

Amino acid activation, 58

Aminopeptidase, 95

Ammonia, 102, 151, 529

Amnion, 251–252

Amoeba, 42, 151, 215, 240, 405

Amoeboid movement, 406

AMP, 47

Amphibia, 446

Ampulla, of semi-circular canal, 197

Amylase, enzyme, 15, 28, 91

Amylopectin, 14

Amyloplast, 291

Amylose, 14

Anabaena, 398

Anaemia, 87, 136

Anaerobic respiration [*Respiration without oxygen*], 62, 101

Analogous structures, 388

Anaphase, 322–323

Androecium [*Male parts of a flower*], 300

Androgen, 242

Aneurysm, 142

Angiospermophyta [*Flowering plants*]
classification, 481–483
structure and physiology, 255–309

Angiotensin, 158

Animals (Kingdom Animalia) [*Multicellular heterotrophic eukaryotes*]
evolution of phyla, 432
Kingdom Animalia, 390
see individual phyla

– most bases release hydroxyl ions], 8

Base-pairing, in DNA and RNA, 49

Basement membrane, 153

Basidiomycota, 460

Basilar membrane, of ear, 196

Basophil [*White blood cell containing granules which react with methylene blue (a basic dye) – basophils differentiate to become connective tissue mast cells*], 134

Behaviour, 228–238
  insect, 443
  social, 235, 373
  thermoregulation, 161

Behaviourist, 229

Benedict's solution, test for reducing sugars, 12

Benthos, 515

Beri-beri, 87

Bicuspid valve, of heart, 125

Bile [*Greenish liquid produced by liver cells, collected and stored in the gall bladder ready for release into the duodenum*]
  composition, 94

Bile canaliculi, 104

Bile duct, 104, 105

Bile pigments, 107, 138

Bile salts, 95, 105

Bilirubin, 107, 138, 151

Binary fission, 42, 239, 397, 408

Binomial system, naming organisms, 392

Bioelement [*Chemical element forming an essential part of living organisms*], 5

Biogeochemical cycle [*Process whereby chemical compounds of a particular bioelement are transferred between living organisms (biotic phase) and the non-living environment (abiotic phase)*], 526

Biological control, 402, 444

Biomass [*Total organic matter present in an ecosystem – measured in grams dry mass per square metre, g m²*]
  pyramid, 490

Biomes, 506–508

Bipolar cells, of retina, 193

Bipolar neurons, 170

Birds
  agonistic behaviour, 234
  Archaeopteryx, 369
  classification, 446
  courtship, 236, 366
  Darwin's finches, 356–359
  flight, 226–227, 347
  peppered moth predation, 349
  poultry comb shapes, 345
  Birth, 253

Biston betularia, 349

Biuret test, for protein, 22

Bivalent [*Pair of homologous chromosomes joined by synapsis during prophase I of meiosis*], 324

Bladder, 152

Bladderwort, 274

Blastocyst, 250

Blind spot, 194

Blood [*Transport medium in circulatory systems of animals*], 133–140
  clotting, 140
  components, 133–134
  defence, 145–147
  plasma, 135
  platelets, 140
  red blood cells, 136–138
  white blood cells, 134
  see Circulatory system

Blood group
  ABO, 149, 333
  Rh, 149

Blood-brain barrier, 187

Blue-green bacteria (Cyanobacteria), 398–399

B lymphocyte (B cell) [*White blood cell capable of producing and releasing antibody molecules – B lymphocytes differentiate to become plasma cells*], 146

Bohr effect, 137

Bolus, 91

Bone, 216
  see Skeleton

Bony fish (Osteichthyes)
  circulatory system, 123
  classification, 446
  gas exchange, 110
  locomotion, 225
  osmoregulation, 158–159

Bony labyrinth, of ear, 196

Bordered pit, 259, 479

Bowman's capsule, 152–153

Brachiostoma, 446, 449

Brain
  human, 179, 182–184
  see Nervous system

Broad bean (*Vicia faba*)
  flower/pollination, 303
  geotropism, 291
  germination, 266
  life cycle, 255–256
  phototropism, 290
  seed, 257, 307

Bronchiole [*Tube less than 1 mm in diameter without supporting cartilage leading to the alveoli of the lungs*], 114, 116, 186, 202

Bronchus (*pl* bronchi) [*One of many tubes supported by cartilage and forming the airway of the lungs – the left*

and right principal bronchi lead from the trachea and branch repeatedly], 113

Bryophyta (mosses and liverworts), 467, 468–471

Buccal pressure pump, 112

Budding, of *Hydra*, 418

Bundle of His, 127

Bundle sheath cells, 75

Bursa (*pl* bursae) [*Sac filled with synovial fluid which reduces friction between bones and other body tissues*], 222

Buttercup (*Ranunculus acris*) flower/pollination, 300

C

C₃ photosynthesis, 75

C₄ photosynthesis, 75

Cactus, 160

Caecum, 101

Calcitonin, *see* Thyrocalcitonin

Calcium
  bioelement, 5
  blood clotting, 140
  bone, 216
  flocculation of clay particles, 513
  functions, 86–87
  muscle contraction, 208–210
  plant nutrient, 285
  regulation of blood concentration, 224
  synaptic transmission, 176

Calorie [*Unit of heat energy defined as the amount of heat required to raise the temperature of 1 cm³ of water by 1°C – the joule is the SI unit for energy measurement (1 calorie = 4.2 joules)*], 85

Calorimeter, 85, 119

Calvin cycle, in photosynthesis, 74

Calyx [*Outer whorl of a flower formed by the sepals*], 300

cAMP *see* cyclic AMP

Cambium [*Meristematic tissue from which secondary growth arises in stems and roots*], 260, 265, 268, 269, 273

Camel, 159

Canine teeth, 90, 98, 99

Capacitation, of sperm, 249

Capillary [*Smallest and most numerous blood vessels with an average diameter of 8 μm and a wall thickness of 0.2 μm*]
  blood-brain barrier, 186
  gas exchange, 112, 116
  inflammation, 146
  loops in skin, 163, 166
  pressure filtration, kidney, 152
  pressure filtration, tissue fluid, 130–131
  remodelling of bone, 217

Fallopian tube (oviduct), 242

False fruit, 308

Fasciola, 421

Fascicle, muscle, 207

Fat [*Triglyceride substance which is a solid at room temperature – the fatty acid side chains of a fat molecule are usually saturated (single bonds only)*], 31–32
chemical tests, 32
digestion, 96
energy content, 84
respiratory quotient (RQ), 119

Fatty acid [*Chemical subunit of fats, oils and phospholipids*], 31, 101

Feather, 226

Feedback, 151, 228

Femoral artery/vein, 124

Fermentation [*Anaerobic respiration which uses organic molecules as hydrogen acceptors – lactic fermentation produces lactic acid, alcoholic fermentation produces ethanol*], 64, 402

Ferns (Filicinophyta), 467, 472

Ferredoxin, 73

Ferritin, 138

Fertility drugs, 247

Fertilization (syngamy) [*Fusion of gametes*]
Dryopteris, 472
flowering plants, 306–307, 481
Fucus, 456
human, 239, 249–250
Hydra, 418
life cycles of algae, 454
liverworts, 468
mosses, 471
Selaginella, 475

Festuca, 482

Fibrin/Fibrinogen, 106, 134, 140

Fibroblast, 215

Fibrocartilage, 216

Fibroin, 20

Fibrous joint, 221

Fibrous proteins, 17

Filicinophyta (ferns), 467, 472

First convoluted tubule, kidney, 153, 154

Fish *see* Bony fish, Cartilaginous fish

Fixed action pattern, 230

Flagellum (*pl* flagella), 44, 242, 383, 395, 411, 452, 456, 457, 469, 470, 473, 475

Flagellates, 411

Flame cell, 421

Flavin adenine dinucleotide (FAD), 48

Flexion, of limbs, 223, 438

Flight
birds, 226

insects, 442

Flocculation, 513

'Florigen', 295

Flowering plants (Angiospermophyta)
classification, 481–483
structure and physiology, 255–309

Flower [*Sexual reproductive structure of flowering plants*], 300–306

Fluid mosaic model, of plasma membrane structure, 34

Fetus [*Mammalian embryo after implantation, any vertebrate embryo in egg or uterus*]
development, 250
circulation at birth, 254
exchange across placenta, 251
haemoglobin, 139, 254

Follicle, of ovary, 243–244

Follicle stimulating hormone (FSH), 200, 242, 245

Food chain, 488

Food production, 534

Food web, 489

Foramen ovale, 254

Forebrain, 182

Formed elements, of blood, 133

Fossils, 369

Fovea, of eye, 191

Freeze etch technique, 34, 42

Fructose, 12, 84

Fructose 1, 6 bisphosphate, 63

Fruits, 307–309

Fucus, 456, 508

Funaria, 470

Fungi [*Heterotrophic eukaryotic organisms with cell walls containing chitin, usually with a mycelium consisting of thread-like hyphae – many fungi are saprotrophic*], 390, 459–465, 489, 500

G

Galactosaemia, 142, 334

Galactose, 12

Galactose-1-phosphate uridyl transferase, enzyme, 142

Galapagos archipelago, 356

Gall bladder, 89, 104

Gallstones, 107

GALP *see* glyceride 3-phosphate

Gametangium (*pl* gametangia) of algae and plants [*Cell inside which gametes are produced*], 455, 457

Gamete [*Haploid cells which fuse to form a zygote in sexual reproduction*]
flowering plants, 300–302

isogamy/anisogamy/oogamy, 454
meiosis, 324
oogenesis, human, 243–244
spermatogenesis, human, 240–242
see Fertilization

Gametophyte [*Haploid stage which produces gametes by mitosis in alternation of generations*], 454, 456, 469, 470, 473, 475, 480, 481

Gamophyta (conjugating algae), 452

Ganglion (*pl* ganglia), 178

Ganglion cells, of eye, 193

Gas exchange and gas exchange systems, 109–120
Amoeba, 406
annelids, 425
bony fish, 110
colonization of land, 449
flowering plants, 265, 271, 274
fetal, 251
insects, 441
Paramecium, 408
platyhelminths, 420

Gastric juice, 92, 94

Gastric pits, of stomach, 92

Gastrin, 41, 92, 200

Gastropoda, 428

Gel, 7

Gene [*Sequence of DNA nucleotides which codes for a particular RNA molecule – using mRNA, many genes code for particular proteins*]
activation, 204
chromosomes, 330
allele frequencies/Hardy-Weinberg method, 350
genetic engineering, 402
linkage, 343
mutation, 333, 349
protein synthesis, 55–60
self-incompatibility, 306
sex-linked, 338
see Allele, Chromosome

Gene pool, 355

Generator potential, 188, 192

Genetic code, 54, 58, 371

Genetic drift, 362

Genetic engineering, 402

Genetic material *see* DNA

Genetic recombination, 398

Geological eras, 385

Geotropism, 291

Germ cell *see* Gamete

Germination, 257, 292

Gestation, 251

Giant neurons, 172, 426

Gibberellic acid/gibberellins, 292

Gigantism, 203

Glaucoma, 190

Globular proteins, 17, 23

Globulin, 106, 134–135

Glomerulus/glomerular filtrate, 152–153

Glucagon, 106, 108, 200

Glucocorticoids, 200, 202

Gluconeogenesis [*Manufacture of glucose from non-carbohydrate sources, e.g. amino acids*], 106

Glucose, 12–13, 61, 68, 84, 95, 105, 119, 134, 154, 202, 292

Glucuronic acid, 107

Glyceraldehyde 3-phosphate (GALP), 63, 74

Glycerol [*Chemical subunit of fats, oils and phospholipids*], 31

Glycine, 107

Glycogen, 15, 84, 105, 144

Glycogenolysis [*Hydrolysis of glycogen to liberate glucose*], 106

Glycolysis, 63–64, 383, 396

Glycoproteins, 41

Glycoside linkage [*Chemical bond formed between monosaccharide subunits of complex carbohydrates by removing a molecule of water (condensation reaction)*], 13

Gnathostomata [*Craniates with jaws*], 446

Goblet cells, 116, 144

Goitre, 203

Golgi body (dictyosome), 41, 45, 171, 259

Gonorrhoea, 144

Graafian follicle, 244

Grafting, vegetative propagation, 299

Gram staining, 395

Granular cells, 243

Granulocyte [*White blood cells with irregular lobed nucleus and cytoplasm containing granules*] *see* Basophil, Eosinophil, Neutrophil

Granuloma, 146

Granum (*pl* grana), of chloroplast, 43, 71

Grey matter, 180

Ground meristem, 258, 267

Ground substance, connective tissue, 215

Growth
arthropod, 439
human, 202–203
plant, 258, 264
populations, 494

Guanine, 47

Guard cells, 272, 282

Gynaecium [*Female parts of a flower*], 300

H

Habitat, 506

Habituation, 237

Haem group, 18, 66, 136

Haemerythrin, pigment, 110

Haemocoel, 122, 442

Haemocyanin, pigment, 110

Haemocytometer, 523

Haemoglobin, pigment
α-chains/β-chains, 21
γ-chains, 139
breakdown, 138
evolution, 371
fetal, 139, 254
gas exchange, 110, 112
haemoglobin-S/sickle cell anaemia, 139, 352
transport of carbon dioxide, 138
transport of oxygen, 136–137

Haemolysis, 137

Haemolytic disease of the newborn, 252

Haemophilia, 140, 142, 339, 340, 341

Haemostasis [*Prevention of blood loss*], 139

Hair follicle, 164

Halophyte, 160, 519

Haploid [*Having the number of chromosomes found in the gametes of an organsim*], 324

Haptonasty, 296

Hardy-Weinberg method/Law, 350–351

Haustorial roots, 270

Haversian system, bone, 217

Hearing, 196

Heart, human, 125–128

Hemimetabola (Exopterygota), 436

Hepaticae (liverworts), 468

Hepatic [*adj pertaining to the liver*]

Hepatic artery/hepatic vein, 103, 104, 124

Hepatic portal vein, 93, 103, 104, 124

Hepatocytes, liver cells, 104

Herbivore, 83, 98, 426, 488

Hermaphrodite [*adj describing animals with male and female sexual organs as part of the same individual*], 420, 427

Heterogametic sex, 337

Heterostyly, 305

Heterotrophic nutrition ('other feeding') [*Nutrition in which organic compounds provide the main carbon source*], 82
bacteria, 396
see Digestion and digestive systems, Consumer, Parasitic nutrition, Saprotrophic nutrition

Heterozygous [*(of cell or organism)*

Having different alleles of a given gene (e.g. Bb)*], 330

Hexose, 6-carbon sugar, 12

Hibernation, 168

Hindbrain, 182

Hirundinea (leeches), 427

Histamine, 146

Histone, 42, 50, 321

HIV *see* AIDS

Holometabola (Endopterygota), 436

Holophytic nutrition *see* Photosynthesis

Holozoic nutrition, 83

Homeostasis [*Maintenance of a constant internal environment, e.g. the composition of tissue fluid and body temperature are kept constant: all body systems contribute to homeostasis*]
colonization of land, 450
excretion, 150–156
liver, 105
osmoregulation, 157–160
temperature control, 161–168

Homogametic sex, 337

Homoiotherm (endotherm), 162

Homologous pairs, chromosomes, 324

Homologous structures, 388

Homozygous [*(of cell or organism) Having two of the same alleles of a given gene (e.g. BB, bb)*], 330

Honeybee, 235, 237

Horizontal cells, of eye, 193

Horizons, soil, 510

Hormone [*Chemical messenger substance released by cells in one part of an organism and transported to other parts where 'target' cells respond*]
animal hormones, summary table, 200
see Endocrine system
plant hormones, 287–297

Horse, 101, 369

Horse chestnut, 483

Horsetail (*Equisetum*), 474

Human chorionic gonadotrophin (HCG), 200, 246

Human growth hormone (HGH), 200, 202–203

Human immunodeficiency virus (HIV), 144, 401

Humus, 512

Huntington's chorea, 334

Hyaline cartilage, 216

Hydathodes, 160

Hydra, 151, 178, 417

Hydrochloric acid, 92, 94, 101

Hydrocortisone, 200, 202

Hydrogen bond [*Chemical bond formed when hydrogen acts as a* bridge between oppositely charged polar groups, as between water molecules (weak linkage)], 6, 14–15, 20–21, 25, 49

Hydrogencarbonate ion, 8, 138, 155

Hydrogen carrier, 48, 62, 70, see FAD, NAD, NADP

Hydrogen transport, 63, 65–66

Hydrolase, enzyme, 29

Hydrological cycle, 531

Hydrolysis reaction [Chemical reaction in which covalent bonds linking the subunits of large molecules are split by inserting a molecule of water], 8
ester linkage, 31
glycoside linkage, 13–15
peptide bond, 19

Hydrophilic ('water-liking'), 8, 19, 21, 32

Hydrophobic ('water-hating'), 8, 21, 32

Hydrostatic skeleton, 215
Amoeba, 406
earthworm, 427
turgor pressure in plants, 277–279

Hydroxonium ion, 8

Hydroxyl ion, 8

Hydrozoa, 416

Hymen, 242

Hyperglycaemia, 108

Hypermetropia, 191

Hyperthyroidism, 120, 203

Hypocotyl, 257

Hypogeal germination, 257

Hypothalamus, 157, 165, 184, 199

Hypothyroidism, 120, 203

H-zone, muscle fibre, 208

I

IAA (indoleacetic acid), 289, 293

Ice, 6

I-bands, muscle fibre, 208

Ileum, 93

Immune responses
non-specific, 145
specific, 147

Immunity
active/passive, 148

Imprinting, 237

Incisor teeth, 90, 98, 99

Incomplete dominance see codominance

Independent assortment, 335

Indoleacetic acid (IAA), 289, 293

Induced fit, action of enzymes, 24

Industrial melanism, 349

Inflammation, 146

Inflorescence, 305

Inheritance of acquired characteristics, theory of evolution, 348

Initiation codon, 57

Inorganic elements/salts
biogeochemical cycles, 526
human diet, 87
plant nutrients, 285
storage in liver, 106

Insects, 440–444

Insect pollination, 303

Insight learning, 238

Inspiration, 111

Inspiratory centre, 117

Instinct, 229

Insulin, 41, 105, 108, 200

Integuments, 302

Intercalated discs, cardiac muscle, 126

Intercostal muscles, 115

Interferon, 147

Internal respiration, 109

Interneuron, 170, 180

Interoceptor, 188

Interphase, 321

Interstitial cell stimulating hormone (ICSH), 242

Intervertebral disc, 219

Intestinal juice, 93, 95

Intramembranous ossification, 218

Intrinsic factor, 94, 96

Inversion, chromosome mutation, 341–342

Iodine, 16, 87

Ionic bond see Electrovalent bond

Ions [Charged particles formed when atoms gain or lose electrons – cations are positively-charged, anions are negatively-charged]

Ion exchange, 284

Iris, of eye, 191

Iron, 66, 87, 106, 107, 135, 136, 146, 380, 395

Irrigation, 537

Islets of Langerhans, pancreas, 108, 200

Isogamy, 454

Isolating mechanisms, 364

Isomer [Structural isomers are molecules which contain the same number and types of atoms, but arranged differently to make substances with different chemical structures], 12

Isomerase, enzyme, 29

Isotopes [Atoms of an element with the same number of protons and electrons but with different numbers of neutrons in the atomic nucleus – chemical reactions in living systems

are investigated by using isotopes to 'label' particular substances], 70, 74, 286, 384

J

Jaundice, 107, 139

Jaw muscles,
carnivore, 99
herbivore, 101
human, 90–91

Joints
cartilaginous/fibrous/synovial, 221

Joule, SI unit of energy, 85

K

Keratin, 20, 164

Kettlewell, H.B.D., 349

Keto acids, 106

Ketose, 12

Kidney, 151–159
see Excretion and excretory systems

Kinesis, 231

Kinetochores, 323

Kinins, 146

Klinefelter's syndrome, 142, 340

Klinostat, 291

Knee jerk reflex, 181

Krebs Cycle, 64, 65, 67

K-strategists, 497

Kuppfer cells, 105

Kwashiorkor, 86

L

Lactase, enzyme, 95

Lactation, 254

Lachrymal gland (tear gland), 189

Lactate (Lactic acid/lactic fermentation) 62, 64, 212, 402

Lacteal, 93, 145

Lactose, 13, 14, 59, 84

Lamarck, Jean-Baptiste, 348

Lamellibranchiata, 428

Lamina, of leaf, 271

Lamium, 304

Large intestine, 97

Larva, insect, 444

Larynx, 113

Lateral pterygoid muscles, 91

Lateral root, 258

Law of independent assortment, 335

Law of segregation, 329

Leaf
abscission, 294
development, 271

Miller, Stanley, 381

Millon's reagent, test for protein, 22

Mimosa, 296

Mineralocorticoids, 200, 202

Mineral salts *see* Inorganic elements/salts

Minimal air, 115

Minimal area curve, 524

Mitochondrion (*pl* mitochondria), 43, 45, 65, 155, 207, 212, 383

Mitosis, 202, 320, 322, 325
see Asexual reproduction, Growth

Molar teeth, 90, 98, 99, 100

Mole [*A mole of one substance contains the same number of atoms or molecules as a mole of any other substance, the mass of a mole in grams is numerically equal to the molecular weight of the substance*]

Molecular mass (relative molecular mass) [*Total of the atomic masses of all the atoms which are combined to form the molecule*]

Molecule [*Smallest particle of a substance which retains the chemical properties of the substance*]

Mollusca (soft-bodied animals), 428

Molybdenum, 106, 284, 285

Monera (Kingdom) *see* Prokaryotae

Monocotyledoneae (monocotyledons), 482

Monocyte [*Large white blood cell with bean-shaped nucleus and cytoplasm without granules – monocytes migrate into body tissues where they become macrophages*], 134

Monoecious [*Having male and female structures on the same plant, often as part of the same flower*], 306

Monoglyceride, 31

Monohybrid inheritance, 331

Monosaccharide, 12

Monosynaptic reflex, 181

Monozygotic twins, 243

Morgan, T.H., 337

Mosses (Musci), 468

Motivation, 232

Motor neuron (effector neuron), 170, 180, 207, 210, 229

Movement *see* Locomotion, Plant responses

Multicellular organisms, 414

Multiple alleles, 333

Multipolar neuron, 170

Musci (mosses), 468

Muscle

breathing, 114
cardiac, 126, 213
jaw movement, 90, 98, 100
skeletal, 181, 206–212, 223
visceral (non-striated), 93, 128, 164, 167, 189, 213
see Locomotion, Skeleton

Muscle spindle, 181, 210

Mutation
chromosome, 341, 349
gene, 333, 349

Mutualism, 102, 500–502

Mycelium, 459

Mycophycophyta (lichens), 460

Mycorrhiza, 270, 464

Myelin sheath, 171

Myocytes [*Muscle cells*], 126

Myofibrils, 207

Myogenic [*Produced by spontaneous activity of muscle cells*], 127

Myoglobin, 21, 139, 212

Myometrium, 242, 252

Myopia, 191

Myosin [*Protein which interacts with actin to produce muscle contraction and cell movement*], 208, 324

Myotome, 225

Myxoedema, 203

N

Na^+–K^+ exchange pump, 175

NAD (nicotinamide adenine dinucleotide), 48, 62–67, 87

NADH dehydrogenase, enzyme, 66

NADP (nicotinamide adenine dinucleotide phosphate), 48, 70–74

NADP reductase, enzyme, 73

Nastic movements, 296

Natural selection, 347

Navigation, 233

Neck cells, 92

Nectary, 300

Nekton, 515

Nematoda (round worms), 430

Neo-Darwinian theory, 347

Nephridium (*pl* nephridia), 426

Nephron, 152

Nereis, 424

Nerve impulse, 172–175

Nervous system, 169–177, 178–187, 228
annelid, 426
Hydra, 418
insect, 441
platyhelminth, 420

Neurofibrils, 171

Neurohypophysis [*Posterior lobe of pituitary gland*], 199–200

Neuromuscular junction, 172, 210

Neuron, 169
see Interneuron, Motor neuron, Sensory neuron

Neuston, 515

Neutrophil [*White blood cell containing granules which do not react strongly either with acid or basic dyes – neutrophils migrate to sites of injury and engulf bacteria*], 134, 146

Niacin (vitamin B_3), 48, 87

Niche differentiation, 365

Nicotinic acid *see* Niacin

Nissl granule, 171

Nitrate reductase, enzyme, 284

Nitrification, 396, 529

Nitrogenase, enzyme, 399, 502, 529

Nitrogen cycle, 528

Nitrogen fixation, 399, 502, 529

Node of Ranvier, 171

Non-competitive inhibition, of enzymes, 26

Non-cyclic phosphorylation, photosynthesis, 73

Non-disjunction, chromosomes, 340

Non-reducing sugar, 13

Non-specific immune responses, 145

Noradrenalin, 186, 200, 202

Normal distribution, 345

Nostoc, 398

Notochord, 445

Nuclear envelope, 41

Nuclear material *see* DNA

Nuclear pores, 41, 42

Nuclease, enzyme, 94

Nucleic acid *see* DNA, RNA

Nucleolus (*pl* nucleoli), 42, 45, 321

Nucleoside, 47

Nucleosome, 50, 321

Nucleotidase, enzyme, 95

Nucleotide, 46

Nucleus, 37, 42

Nutrition *see* Autotrophic nutrition, Heterotrophic nutrition, Saprotrophic nutrition

Nyctinasty, 296

O

Obelia, 178, 419

Obligate aerobe [*Organism which requires oxygen for respiration*], 396

Obligate anaerobe [*Organism which respires anaerobically and cannot tolerate oxygen in the environment*], 396

Odontoblast, 90

Oesophagus, 91

Oestrogen, 200, 242, 244

Oestrous cycle, 243

Oil [*Triglyceride substance which is liquid at room temperature – the fatty acid side chains of an oil molecule are usually unsaturated (some double bonds)*], 31–32
  chemical tests, 32
  digestion, 96
  energy content, 84

Oligochaeta, 426

Omasum, 101

Ommatidium (*pl* ommatidia), 443

Omnivore, 83

Onychophora, 434

Oogamy, 454

Oogenesis, 243

Oosphere *see* Egg cell

Oparin, Alexander, 381

Operant conditioning, 238

Operator gene, 60

Operon, 59

Opsin, 192

Opsonization, 147

Optic lobes, 184

Optic nerve/optic chiasma/optic tract, 194

Orbit, of eye, 189

Organ [*Structure consisting of several tissues which together perform a complex function*]
  level of organization, 420

Organ of Corti, of ear, 196

Organ of perennation, 299

Organelle [*Self-contained structure which forms part of a cell*], 42–45

Organic base [*Chemical subunit of DNA and RNA, bases used are adenine, cytosine, guanine, thymine and uracil*], 47

Orgasm, 249

Origin of life, 380–385

Ornithine cycle, 107, 151

Osmoregulation [*Control of water and salt balance so that the concentration of dissolved substances in the body fluids remains constant*]
  Amoeba, 406
  annelids, 426
  desert mammals, 159
  fish, 158
  human, 157–158
  plants, 159, 282

Osmoreceptors, 157

Osmosis [*Net movement of water molecules from a less concentrated solution to a more concentrated solution across a partially permeable membrane*], 36, 131, 137, 155, 156, 277, 278, 406

Osmotic potential *see* solute potential

Ossicles, of ear, 195

Ossification, 218

Osteichthyes *see* Bony fish

Osteoblast/osteoclast/osteocyte, 216

Osteoid, 216

Osteomalacia, 224

Osteonectin, 216

Otolith, 197

Oval window, of ear, 195

Ovary
  human, 200, 242
  plant, 300

Oviduct, 242

Ovulation, 243

Ovule, 302

Ovum *see* Egg cell

Oxaloacetate (anion of oxalic acid), 64

Oxidation [*Chemical reaction involving the addition of oxygen, the removal of hydrogen, or the loss of electrons*], 62

Oxidoreductase, enzyme, 62

Oxygen
  bioelement, 6
  control of respiratory gases, 117
  gas exchange, 109, 272
  photosynthesis, 70
  transport in blood, 136–137, 139

Oxygen cycle, 527

Oxygen debt, 212

Oxyhaemoglobin, 112, 136

Oxyntic cells, 92

Oxytocin, 200, 253–254

P

Pacemaker, of heart, 127

Pacinian corpuscle, 163, 188

Paedomorphosis, 448

Palisade mesophyll, 272

Pelecypoda (Lamellibranchiata), 428

Pancreas, 93, 103, 107, 108, 199, 200

Pancreatic amylase, enzyme, 94

Pancreatic juice, 94, 108

Pancreatic lipase, enzyme, 95

Panda, 372

Paneth cells, 93

Papillary muscles, of heart, 126

Parallel flow, gas exchange, 112–113

Paramecium, 408, 503

Parapatric speciation, 363

Parasitic nutrition [*Type of heterotrophic nutrition in which one organism (the parasite) obtains food materials from the living body of another organism (the host)*], 83, 142, 270, 399, 412, 421, 431, 462, 489, 505

Parasympathetic nervous system, 185

Parathyroid hormone (PTH), 200, 224

Parazoa, 432

Parenchyma, 260, 268

Parental care, 235, 254

Parturition (birth), 253

Pascal [*SI unit of pressure*], 277

Pasteurization, 398

Pathogen, 142
  see Parasitic nutrition

Pectoral girdle, 220

Pedicel, 300

Pellagra, 87

Pellicle, 408

Pelvis (pelvic girdle), 221

Penicillin, 144, 397, 465

Penis, 240

Pentadactyl limb, adaptations, 372

Pentose, 5-carbon sugar, 12

Pepsin, enzyme, 25, 28, 92, 94, 101

Peptidase enzymes, 92, 95

Peptide bond [*Chemical bond formed between amino acid subunits of a protein by removing a molecule of water (condensation reaction)*], 19

Perianth [*Outer whorl of a flower when sepals and petals cannot be clearly distinguished, collective name for the sepals and petals*], 300

Pericardium, 125

Pericarp, 307–309

Perichondrium, 216

Pericycle, 269

Perilymph, of ear, 196

Periodontal membrane, 90

Periosteum, 218

Peripatus, 434

Peripheral nervous system, 178

Peristalsis, 91

Pernicious anaemia, 87, 136

Pesticide, 538

Petals, 300, 303, 304

Petiole, 271

pH [*pH is a measure of the hydrogen ion concentration of a solution where pH = $-log_{10}[H^+]$, neutral solutions have pH = 7, acid solutions have pH values less than 7, while basic (alkaline) solutions have pH*

Pulmonary circulation, 124

Pulmonary valves, 126

Pulmonary surfactant, 116

Pulp cavity, 90

Punnett square, 330

Pupa, insect, 444

Pupillary reflex, 191

Purine base, 47

Purkyne cell, cerebellum of brain, 170, 184

Purkyne fibres, of heart, 127

Pyloric sphincter, of stomach, 92

Pyramids, biomass/energy/ numbers, 490–491

Pyrimidine base, 47

Pyrogen, 147

Pyruvate (anion of pyruvic acid), 63

Q

Quadrat, 525

Quadriceps muscle, 181

Quaternary structure, of proteins, 21

R

Radial cleavage, 433

Radiata, 432

Radiation, 542

Radicle, 257

Reabsorption, kidney tubules, 154–156

Receptacle, 300

Recessive allele, 329

Recombination, 326, 335, 343

Rectum, 97

Red blood cells, 107, 136–137

Reducing sugar, 12

Reflex, 181, 230

Refractory period, 127, 173, 211

Releaser, 231

Releasing factors, 199

Renal   [adj pertaining to the kidney]

Renal artery/vein, 124, 152, 153

Renaturation   [Spontaneous process whereby an enzyme or other protein regains its shape in appropriate conditions of temperature or pH], 22

Renin, enzyme, 158

Rennin, enzyme, 102

Replication, of DNA, 50, 321

Repolarization, 173

Repressor gene/repressor protein, 59

Reproduction see Asexual reproduction, Life cycles, Sexual reproduction

Reptilia, 446

Residual volume, 115

Resistance responses, 201–202

Respiration
biochemistry, 61–67, 396
control of respiratory gases, 117
respiratory pigments, 110
respiratory quotient (RQ), 119
respirometer/respiratory rate, 118
see Gas exchange and gas exchange systems

Resting potential, 173

Restriction point, cell cycle, 321

Reticular formation, brain, 184

Reticulo-endothelial system, 107, 138

Retina, of eye, 190

Retinal, 192

Retrovirus, 401

Reverse transcriptase, enzyme, 403

Rhesus (Rh) factor, 149

Rhizome, 299

Rhizopoda, 405

Rhodophyta (red algae), 452

Rhodopsin, 192

Ribonucleic acid see RNA

Ribose, 47

Ribosomal RNA (rRNA), 53, 56

Ribosome, 38, 42, 45, 56

Ribulose bisphosphate (RuBP), 74

Ribulose bisphosphate carboxylase, enzyme, 74, 453

Rickets, 87, 224

River ecosystems, 517

RNA (ribonucleic acid)
messenger RNA (mRNA), 52, 55, 204, 403
ribosomal RNA (rRNA), 53, 56
transfer RNA (tRNA), 52, 54, 56
protein synthesis, 54–60
viral, 400, 401

Rod cells, of eye, 191

Roots
development, 266
functions, 270, 280
structure, 268

Root hairs, 258, 269

Root nodule bacteria, 270, 502, 529

Root pressure, 284

Rotation, 223

Roughage (dietary fibre), 84

Rough ER, 38

Round window, of ear, 195

r-strategists, 497

Rumen, 101

Ruminant, 101

S

Saccule, of ear, 195

Saccus endolymphaticus, of ear, 195

Saliva, 90, 91, 94, 102

Salivary amylase, enzyme, 91, 94, 95

Saprotrophic nutrition   [Type of heterotrophic nutrition in which organisms release enzymes into the environment and absorb the simple soluble products of digestion], 83
bacteria, 396
decomposers/detritus food web, 489
fungi, 459

Sarcolemma, 207

Sarcomere, 208

Sarcoplasm/sarcoplasmic reticulum, 207

Schistocerca, 440

Schwann cell, 171

Schwartz, K.V., 390

Sclera, of eye, 190

Sclerenchyma, 261

Scurvy, 87

Scyphozoa, 416

Sebaceous gland, 144, 163, 164

Second convoluted tubule, 153, 154

Secondary cell wall, plant cell, 259

Secondary growth, 264

Secondary nature, of proteins
α-helix, 20
β-pleated sheet, 20

Secretin, 105, 108, 200

Seed   [Reproductive structure of conifers and flowering plants formed from a fertilized ovule, comprising an embryo, and a food store inside a protective coat]
development, 307
dispersal, 308
evolution of seed plants, 476
germination, 257

Segregation, 331

Selaginella, 475

Semen, 240, 249

Semi-circular canals, of ear, 195, 196–197

Semi-conservative replication, of DNA, 50, 321

Semi-lunar valves, of heart, 125–126

Seminal vesicle, 240

Seminiferous tubules, 240

Sense organs
ear, 195–197
eye, 189–195
insect, 443
muscle spindles, 210
Pacinian corpuscle, 189
receptor types, 188, 288
skin, 164

Sensory neuron, 170, 180, 189, 193, 210, 229

Serosa, 94

Serotonin, 184

Sertoli cell, of testis, 241

Serum, 148

Sex chromosomes, 324, 337

Sex-linked genes, 338

Sexual reproduction [*Reproduction involving the fusion of haploid gametes to form a diploid zygote, offspring produced by sexual reproduction are genetically varied*]
algae, 451–457
bacteria, 398
flowering plants, 300–309
fungi, 462
human, 239–254
Hydra, 418
liverworts, 468
mosses, 471
Paramecium, 408

Shivering, 166

Short-day plants, 294

Shortsightedness, 191

SI units *see end*

Sickle cell anaemia, 139, 142, 352

Sieve tube elements, of phloem, 262

Sign stimuli, 231

Simple pit, 259

Sinoatrial node (SAN), 127

Sinusoids, of liver, 104

Skeletal muscle, 181, 206–212, 223

Skeleton [*Internal or external method of providing support for body tissues and locomotion*]
Amoeba, 406
arthropod, 437–439
cytoskeleton, 44
earthworm, 427
endoskeleton, 215
exoskeleton, 215
humans, 218–224
hydrostatic skeleton, 215
turgor pressure in plants, 277–279

Skin
structure, 163–164
temperature control, 165–167

Skull, 91, 98, 99, 219

Sliding filament theory, of muscle contraction, 208

Small intestine, 93

Smooth ER, 41

Social behaviour, 235, 373

Sodium/sodium ions (Na^+), 5, 7, 87, 97, 134, 155, 158, 175, 188, 196, 285

'Sodium pump', 175

Soil, 510–514, 522, 535

Sol, 7

Solute potential, 277

Solute/solvent, 7

Sorus (*pl* sori), 472

Special creation, 348

Speciation, 355–363

Species, 355

Specific immune responses, 147

Spermatogenesis, 241

Sperm (spermatozoa), 239, 242

Sphenophyta (horsetails), 467, 474

Spinal cord, 179–180, 182

Spinal reflexes, 181

Spindle, cell division, 44, 323

Spiracle, insect, 442

Spiral cleavage, 433

Spirogyra, 452

Spirometer, 115

Sponges (Porifera), 416

Spongy bone, 218

Spongy mesophyll, 272

Sporangium [*Cell inside which spores are formed*], 455, 456, 457, 462, 468, 472, 475

Spore [*Very small reproductive structure consisting of a single cell or several cells formed by some bacteria, fungi, protoctists, and plants*]
see Endospore, Sporangium

Sporophyte [*Diploid stage which produces spores by meiosis in alternation of generations*], 454, 456, 457, 468, 471, 473, 474

Stabilizing selection, 353

Stamen, 300

Starch, 14, 43, 84, 257

Starling's Law, heart action, 131

Statistical tests *see end*

Statocyte, 291

Stearic acid, 31

Stele, 269

Stem
development, 260
functions, 264
secondary growth, 264
structure, 260
transport of water/transpiration, 281
transport of sugars, 285

Stentor, 410

Stercobilinogen, 107

Sternum, 220

Steroid, 30, 33, 41, 200

Sticta, 500

Stigma, of flower, 300

Stolon, 298

Stoma (*pl* stomata), 282–283

Stomach, 92

Stress reactions, 201

Stretch reflexes, 181

Structural gene, 60

Structural isomer, 12

Style, of flower, 300

Substrate [*Substance upon which an enzyme acts*], 23, 26

Succession, 509

Succinic acid, 26

Succinic dehydrogenase, enzyme, 26

Sucrase, enzyme, 28, 95

Sucrose, 13, 14, 84, 285

Sulphur/sulphate, 18, 87, 285

Sulphur cycle, 530

Summation, 212

Supernormal stimulus, 231

Supination, 223

Survival curves, 497

Suspensory ligaments, of eye, 190

Sweat gland/sweating, 163, 164, 167

Swim bladder, 225

Symbiotic theory, evolution of eukaryotic cells, 383

Sympathetic nervous system, 127, 131, 185

Sympatric speciation, 363

Symplast pathway, water transport, 280

Synapse, 170, 177

Synapsis, chromosomes, 324

Synecology [*Ecology of a number of species forming an ecosystem*], 488

Synovial joint, 221

Syphilis, 144

Systemic circulation, 124

Systole, 127

T

T cell *see* T lymphocyte

Taenia, 422

Taiga, biome, 506

Tapetum, of eye, 190

Tapeworm, 422

Taste buds, 91

Taxes, 231

Taxonomy, 386

Tectorial membrane, of ear, 196

Teeth, 90, 98–100

Telophase, 322, 324

Temperate deciduous forest, biome, 506

Temperate evergreen woodland, biome, 507

Temperate grassland, 507

Temperature regulation, 165–167

Temporalis muscle, 90, 98, 100

Tendon, 207

Termites, 375

Territorial behaviour, 234

Tertiary structure, of proteins, 21

Examination Boards

Syllabuses and past examination papers can be obtained from:

Associated Examining Board (AEB)
Stag Hill House
GUILDFORD
Surrey GU2 5XJ

Joint Matriculation Board (JMB)
78 Park Road
ALTRINCHAM
Cheshire WA14 5QQ

Northern Examinations and Assessment Board (NEAB)
Devas Street
Manchester M15 6EX

Northern Ireland Schools Examination Council (NISEC)
Examinations Office
Beechill House
Beechill Road
BELFAST BT8 4RS

Oxford and Cambridge Schools Examination Board (O & C)
10 Trumpington Street
CAMBRIDGE CB2 1QB

Scottish Examination Board (SEB)
Robert Gibson & Sons (Glasgow) Ltd
17 Fitzroy Place
GLASGOW G3 7SF

Southern Universities Joint Board for School Examinations (SUJB)
Cotham Road
BRISTOL BS6 6DD

University of Cambridge Local Examinations Syndicate (UCLES)
Syndicate Buildings
1 Hills Road
CAMBRIDGE CB1 2EU

University of London Examinations and Assessment Council (ULEAC)
Stewart House
32 Russell Square
LONDON WC1B 5DN

University of London Schools Examination Board (ULSEB)
Stewart House
32 Russell Square
LONDON WC1B 5DN

University of Oxford
Delegacy of Local Examinations
Ewert Place
Summertown
OXFORD OX2 7BZ

Welsh Joint Education Committee (WJEC)
245 Western Avenue
CARDIFF CF5 2YX

SI (*Systeme International*) Units

An internationally-agreed system of units based on definitions of units for fundamental quantities such as mass, length, and time, from which units for other quantities can be derived.

Fundamental quantities

quantity	symbol	base unit	symbol
mass	m	kilogram	kg
length	l	metre	m
time	t	second	s
temperature	T	kelvin	K
amount of substance	n	mol	mol

Derived quantities

quantity	symbol	base unit	symbol
energy	W	joule	J
force	F	newton	N
frequency	f	hertz	Hz
pressure	p	pascal	Pa

Common prefixes for SI units

prefix	symbol	meaning	prefix	symbol	meaning
deca-	da	$\times 10$	deci-	d	$\times 10^{-1}$
hecto-	h	$\times 10^2$	centi-	c	$\times 10^{-2}$
kilo-	k	$\times 10^3$	milli-	m	$\times 10^{-3}$
mega-	M	$\times 10^6$	micro-	μ	$\times 10^{-6}$
			nano-	n	$\times 10^{-9}$
			pico-	p	$\times 10^{-12}$

Exceptions Non-SI units used to describe biological systems include

temperature: measured in degrees celsius (°C), where 0°C = 273 K, 100°C = 373 K

pressure: measured in millimetres of mercury (mmHg), where 1 mmHg = 133.3 Pa

Statistical terms

Statistical methods are needed for the analysis and presentation of experimental results. The meanings of the most important statistical terms are listed below.

Probability (p)   For an event which is impossible $p = 0$, while for an event which is certain $p = 1$. Between these extremes are events which are more or less likely, e.g. the probability of throwing a head with an unbiased coin is $p = 0.5$. Probabilities are often expressed as percentages e.g. the probability of throwing a head is 50%.

Mean   The mean of a series of results is their average value. If the number of eggs observed in the nests of several pairs of robins was 3, 4, 4, 5, 5, 6, 6, 6, 7, the mean number of eggs would be the total number of eggs divided by the number of nests: 46/9 = 5.1.

Median   If the results are arranged in ascending or descending order, the median is the central value.

The median number of eggs in the example above is 5. If there are two central values, the median is the value halfway between them.

Mode   This is the most frequently occurring value. When the frequency of results is plotted in the form of a bar chart or histogram, the mode is the tallest bar. In the example above the modal number of eggs is 6. There may be more than one mode.

Normal distribution   Plotting the numbers of individuals in a population against continuous variables such as height or weight usually gives a characteristic bell-shaped curve called a normal distribution curve. Most human adults have heights close to the average value, while relatively few people are very short or very tall. For a normal distribution the mean, median, and mode are equal.

Standard deviation   This value indicates the spread of heights, for example, around the mean. Any normal distribution can be divided into intervals of standard deviation. 68% of individuals have heights within plus or minus one standard deviation of the mean, 95% have heights within plus or minus two standard deviations, and 99% within plus or minus three standard deviations.

Statistical tests   These are best illustrated by an example. Suppose a survey of limpets on a seashore gives the following results

quadrat	number of limpets
1	66
2	63
3	54
4	41
5	22
6	12

The question is whether these results represent real differences in the numbers of limpets at the various sampling sites, for example, because of their position relative to low and high tide, or because of differences in rocks or availability of food.

A question like this can be answered using the chi-squared (χ^2) test. The formula for χ^2 is

$$\chi^2 = \sum \frac{(O - E)^2}{E}$$

where Σ means 'the sum of', O means 'observed value', and E means 'expected value'.

If there was no difference between the quadrats, the number of limpets found in each would be expected to be approximately the same. The 'expected value' is the average value, that is, 258/6 = 43. χ^2 is calculated using a table

quadrat	O	E	O − E	$(O - E)^2/E$
1	66	43	23	12.3
2	63	43	20	9.3
3	54	43	11	2.8
4	41	43	−2	0.1
5	22	43	−21	10.3
6	12	43	−31	22.3

$$\chi^2 = 57.1$$

Table of χ^2 distribution

df	$p = 0.05$	$p = 0.01$	$p = 0.001$
1	3.84	6.63	10.83
2	5.99	9.21	13.82
3	7.81	11.34	16.27
4	9.49	13.28	18.46
5	11.07	15.09	20.52
10	18.31	23.21	29.59
15	25.00	30.58	37.70
20	31.41	37.57	45.32
25	37.65	44.31	52.62
30	43.77	50.89	59.70

The probability of obtaining this χ^2 value as a consequence of random sample differences can be determined from a table of χ^2 values. High χ^2 values suggest that the results are unlikely to have been obtained by chance alone. The 'degrees of freedom' (df) column allows the number of samples to be taken into account. $df = N - 1$, where N is the number of samples. Here, $df = 5$. The three probability values given at the top of the table correspond to different levels of statistical 'significance'. If the χ^2 value is greater than the value for $p = 0.05$, the result is said to be significant at the 0.05 level. This means that the results of the experiment could have been obtained by chance alone on fewer than one in twenty occasions. If the χ^2 value is greater than the value for $p = 0.01$, the results would be obtained by chance on fewer than one in one hundred occasions, while with $p = 0.001$, the chance is less than one in one thousand.

The χ^2 value required for $p = 0.001$ with $df = 5$ is 20.52. Since the calculated value for χ^2 in this instance is $\chi^2 = 57.1$, the differences between the numbers of limpets recorded in each quadrat are significant at the 0.001 level. In other words, it is almost certain that the distribution of limpets in the areas sampled is affected by some environmental factor.

Many other statistical tests are available for analysing data and helping to establish the significance (or otherwise) of experimental results.

ACKNOWLEDGEMENTS

The Publisher would like to thank the following for permission to reproduce photographs:

Ardea: p. 319;

Andes Press/Carlos Reyes: p. 536;

Angel, Heather: p. 304;

Arthur Mellows Village School, Ms E J Hollamby: p. 286;

BBC Natural History Unit/John Sparkes, BBC: p. 232;

Biophoto Associates: pp. 34 (left). 38 (left), 40, 52, 126, 146, 208, 242, 244, 258, 262, 263, 264, 267 (right), 268, 270, 296 (left), 296 (right), 302, 324, 409, 415, 452, 461, 479 (left), 479 (right), 501;

Bruce Coleman Ltd/CB & DW Frith: p. 373, /**Kim Taylor** p. 350 (left and right), /**Christian Zuber** p. 308;

CNRI: pp. 394 (top left), 394 (bottom right), 152;

Cremona, Julian: pp. 537, 522 (left), 366;

GeoScience Features: p. 38 (right);

Guinness Book of Records/Franklin Berger: p. 204;

John Radcliffe Hospital: p. 219 (left and right);

Dr T Kuwabara: p. 192;

NASA: p. 527;

National Vegetable Research Station: pp. 301 (right), 400 (top left);

Oxford Scientific Films: p. 320, **OSF/G I Bernard,** pp. 109, 159, 160 (right), 267 (left), 487, 511, **OSF/Cayless:** p. 375, **OSF/Cooke:** pp. 160 (left), 275, 292 (right), 292 (left), **OSF/Devries,** p. 435, **OSF/Earth Sciences/Kent,** p. 274 (bottom), **OSF/Earth Sciences/Leszczynski,** p. 480, **OSF/ Gardener,** pp. 163, 217, **OSF/Heathcote,** p. 477, **OSF/Jeffree,** pp. 273, 431, **OSF/Shepherd,** p. 303, **OSF/Watts,** pp. 429, 459;

Phillip Harris Ltd: p. 527 (right);

Science Photo Library: p. 97, **SPL/Amos,** p. 410, **SPL/Dr Burgess,** pp. 4, 15 (left), 71, 283, **SPL/Brain,** pp. 379, 394 (top right), 397, **SPL/Caro,** p. 398, **SPL/Dohrn,** p. 274 (top), **SPL/Fawcett,** pp. 18, 34 (right), 148, **SPL/Grave,** p. 410 (top), **SPL/Hinsch,** p. 458, **SPL/Kage,** pp. 108, 411, **SPL/Lynch,** p. 105, **SPL/London School of Hygiene and Tropical Medicine,** p. 394 (bottom left), **SPL/OMIKRON,** p. 352, **SPL/Porter,** p. 135, **SPL/Science Source,** p. 400 (top right), **SPL/Smith,** p. 470, **SPL/Stammers,** pp. 368, 370, 384, 417, **SPL/University of Basel,** pp. 400 (bottom right), 401, 405;

Southampton General Hospital/Electron Microscope Unit: pp. 81, 154;

Professor N Tinbergen, p. 235 (left and right);

University of Oxford, Dept. of Plant Sciences/Dr Hawes: pp. 15 (right), 42, 291; /**J Sheldon:** p. 301 (left);

Additional photographs by Chris Honeywell, Oxford University Press ©
Cover artwork: Stuart Robertson/New Scientist
Diagrams by W.D. Phillips and Nick Hawken and Associates